Springer Series in Astrophysics and Cosmology

The series covers all areas of astrophysics and cosmology, including theory, observations, and instrumentation. It publishes monographs and edited volumes. All books are authored or edited by leading experts in the field and are primarily intended for researchers and graduate students.

Eleonora Di Valentino · Dillon Brout
Editors

The Hubble Constant Tension

Springer

Editors
Eleonora Di Valentino
School of Mathematics and Statistics
University of Sheffield
Sheffield, UK

Dillon Brout
Boston University
Boston, USA

ISSN 2731-734X ISSN 2731-7358 (electronic)
Springer Series in Astrophysics and Cosmology
ISBN 978-981-99-0176-0 ISBN 978-981-99-0177-7 (eBook)
https://doi.org/10.1007/978-981-99-0177-7

This Springer imprint is published by the registered company Springer Nature Singapore Pte Ltd.
The registered company address is: 152 Beach Road, #21-01/04 Gateway East, Singapore 189721, Singapore

Contents

Part I
Intro Measurements

Chapter 1
The Hubble Tension, Book Prologue: A Perspective Along Several Axes

Adam G. Riess

> *It is, I think, particularly in periods of acknowledged crisis that scientists have turned to philosophical analysis as a device for unlocking the riddles of their field. Scientists have not generally needed or wanted to be philosophers.*
>
> —*Thomas Kuhn*

In the Kuhnian stage of scientific progress called "normal science", a model organizes our present understanding and provides the means to predict the outcomes of new experiments. New results which fail to conform to the model are initially considered errors of the experiments. As anomalous results pile up or become too big to overlook, a crisis is reached and the model must be replaced by one which can explain the old and new data, and the cycle repeats. Cosmological models have been susceptible to such revisions largely because they have shallow roots in fundamental physics. The present cosmological model, LambdaCDM model looks ripe for audit in this context. It has been good at describing what we see but it lacks fundamental comprehension of the physics needed to explain most of the Universe (i.e., Dark Energy and Dark Matter) and lacks the ability to answer some of our biggest questions ("How does gravity operate on quantum scales?" "Did inflation occur and why?" "What is the nature of neutrinos?", the "Why now?" problem, etc.) The chapters in this book describe why the "Hubble Tension", now a decade old, provides a strong challenge to LambdaCDM. The Universe appears to be expanding faster than the prediction from LambdaCDM, calibrated by the CMB, with more than 5 sigma (>99.99998%) confidence, the standard threshold of discovery (see attached figure). Yet there is no consensus as to the cause of the Tension, nor on the nature of a model revision, leaving us in a Kuhnian purgatory.

The name "Hubble Tension" was first used in the early 2010's to describe the significance of the discrepancy between the measured and expected value of the Hubble constant (Physicists tend to call 2–3 sigma differences or 95% to 99.7%

A. G. Riess (✉)
Johns Hopkins University and Space Telescope Science Institute, Baltimore, MD 21218, USA
e-mail: ariess@stsci.edu

E. Di Valentino and D. Brout (eds.), *The Hubble Constant Tension*, Springer Series in Astrophysics and Cosmology, https://doi.org/10.1007/978-981-99-0177-7_1

confidence, a "Tension"). While the increased significance in the last decade should have elevated this tension to a "Hubble Crisis" in the physicist's vernacular, "Hubble Tension" is an even better descriptor as the level of angst this discrepancy elicits within the field of cosmology. That is, the interpretation of the Hubble Tension appears to depend, to a surprising degree, on the (possibly subconscious) priors of the beholder, or perhaps what they most value.

Let's explore several axes along which some different perspectives on the Tension are apparent. One such axis we might call "Modelers versus Empiricists". For most Empiricists, measuring the Hubble constant provides an important test of LambdaCDM which they view as requiring further scrutiny due to its unsatisfactory, phenomenological underpinnings. If the quality of the measurements is high (i.e., well designed and gathered, the data is publicly available, requires no significant model dependence for interpretation, presents no significant oddities and has survived years of scrutiny) the Tension is generally accepted as a reality with hopes for an eventual explanation. Empiricists by definition can accept the veracity of an experiment without an explanation based on the caliber of the measurements. For Modelers, the inability to explain the Tension within the context of a cosmological model is held as a strong prior against acceptance of the experimental results. This group is rarely focused on the specific details of the measurements because what really bothers them is the lack of an explanation. Therefore, the span of different, experimental results more often nurtures doubt rather than confirmation of the Tension and the standard of accepting the Tension soberly proffered as "extraordinary claims require extraordinary evidence". "Extraordinary" cannot be defined a priori but is weighed against the challenge of revising LambdaCDM. Modelers may refer to "unknown errors", an evergreen, escape hatch, though from an Empiricist's point of view the speculation itself lacks enough rigor to be falsified. Empiricists and Modelers are inclined to see the Hubble Tension as the other's problem which produces another kind of tension.

Empiricists themselves may be divided along "Averagers" and "Analyzers". The former prefer to average everything in the belief that this will reduce uncertainties, equating systematic or methodological uncertainties of different observables with random errors. While there is some wisdom in crowd-sourcing, a shortcoming of this approach for the Hubble Tension is that it can delay progress by decades. A newly emerging Tension supported by new data, space missions (e.g., Gaia, Planck, HST, JWST) or understanding will always be washed out by averaging. I would be skeptical of analyses that use older data when an objectively better source is available. (A close cousin of the Averagers are "Nostalgics" who rely on historic data, perhaps data they themselves collected, rather than state-of-the-art data.) Analyzers try to understand the differences in the hopes of progress, but must be wary of "over analyzing" which materializes as the particle Physicists's "Look Elsewhere Effect". That is, if you correlate everything against anything, you are bound to find something, and thorough accounting is required to avoid insignificant findings, or, the bane of astronomy, selection bias.

Another axis of Tension is the "Bettors vs the Seekers". Bettors use the history and their memories of past Tensions to decide what to conclude at present. Many in this group see the present Hubble constant discrepancy of 67 vs 73 km/sec/Mpc as a repeat of the 50 vs 100 km/sec/Mpc difference of the 1970's-1990's, betting that the extremes of measurements with small errors are wrong. Some may bet on an answer "in between" following the example of the past. Bettors are inclined to offer a literal wager on the final answer. The Bettors case can and often is taken to an extreme, i.e., that all unexpected experimental results are wrong which is a good, statistical bet for any individual example. They may advise to just double all the uncertainties or assume "unknown unknowns" at play to cover their bets. It is a somewhat jaded perspective; a confirmation bias against discovery. Seekers, in contrast, are more impressed by a difference in the present Tension than the past because the conflicting inferences of the Hubble constant are not mutually exclusive, their comparison depends on an incomplete cosmological model. The seekers are looking for new physics in LambdaCDM that can explain the discrepancy. Its hard but noble work.

Finally, there is the axis dividing those whose measurements seem most at odds, Early vs Late Universe experimentalists. It is common for those in one of these groups to doubt the measurements in the other as they are less familiar with them and judge the other's tools to be inferior or complex. To a large degree ignorance of the methods is the source of skepticism rather than any specific concern. Members of these groups say they would like to see another, independent confirmation of the other side, but are often slow to recognize those that already exist. I have experienced this axis first hand as I have attended conferences of each with polls undertaken pointing to the high expectation of an error from "the other side" but none as likely in their own.

A number of these axes overlap or may be seen simply as Optimists (i.e., Seekers, Analyzers, Empiricists) versus Pessimists (i.e., Bettors, Averagers, Modelers) in the way they view the Tension.

Finally, it is interesting to consider these perspectives in the context of historical examples to "back-test" their value. There are famous results which were correct but the explanation required decades such as the precession of Mercury, the Solar neutrino problem, galaxy velocities in Coma, and even currently accepted results like flat rotation curves in galaxies and the accelerating Universe have only an incomplete explanation. We may also consider the circumstances of experimental results which did not hold up such as Weber's discovery of gravitational waves in the 1960's, Arp's quantized quasar redshifts, or Van Manaan's rotating galaxies of the 1920's, or more recent claims of faster-than-light neutrinos and finding B modes from inflation in the CMB. Past examples show history is not a substitute for doing the science work. For every example there is a counter example showing history is not predictive, or as the Danish proverb says, "It is difficult to make predictions, especially about the future." Still, having considered a wide range of examples the biggest discriminator of outcome is probably the time interval over which the discrepancy lasts (provided the data has been available over that time). The scientific community has shown itself to be very adept at finding errors rather quickly, and if the data is public, errors (of more than a singular author) rarely last a year let alone a decade. In contrast, anomalies which survived scrutiny over many years are the ones where something interesting was usually going on. Judged by these standards, the Hubble Tension is one worth paying attention to.

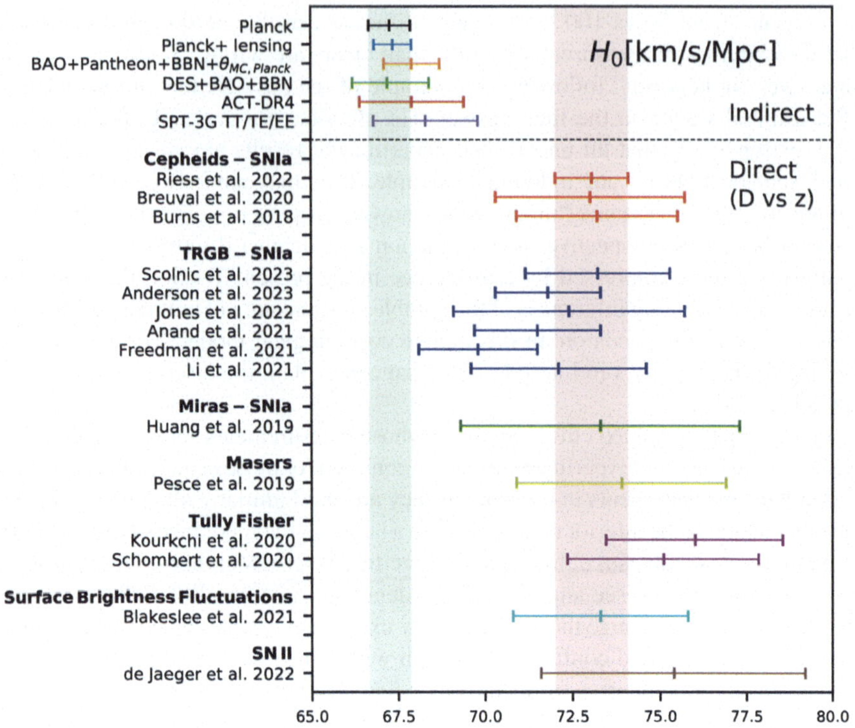

Fig. 1.1 Recent measurements of H_0 from two approaches. Direct (bottom) from distance versus redshift and Indirect (top) from an early Universe calibration (via the CMB or BBN) of LambdaCDM. Adapted from [1]

So with which perspective should you read the following chapters on the Hubble Tension? I suggest curiosity (Fig. 1.1).

Reference

1. Di Valentino, E. Mon. Not. Roy. Astron. Soc. **502**(2), 2065–2073 (2021)

Chapter 2
The Hubble Constant: A Historical Review

R. Brent Tully

2.1 In the Beginning

In the first decades of the Twentieth century, evidence was accumulating that spiral nebulae lay beyond the Milky Way as "island universes", with discussion culminating in the Great Debate [1, 2]. History books usually cite the convincing evidence for the extragalactic nature of spiral nebulae to the identification of cepheid variable stars in M31 and M33 by Hubble [3] in 1925. The relationship between pulsation period and luminosity had been established by Leavitt [4] in 1912. Hubble determined the distances to each of M31 and M33 to be 285 kpc. However, three years earlier, a better distance to M31 of 450 kpc had been calculated by Öpik [5] assuming the virial theorem relation between velocity dispersion, dimension, and mass and assuming the solar relation between luminosity and mass. The modern value for the distance to M31 is 740 kpc and to M33 is 850 kpc.

In the meantime, spectra of the spiral nebulae were being accumulated, mostly by Slipher [6], and it was being remarked that almost all spectral features were displaced to longer wavelengths from their laboratory positions. In a seminal paper [7], Hubble demonstrated the correlation between velocity displacement, subsequently called redshift, and galaxy distance, with the relationship between these parameters that now bears his name. In fact, the expansion of the Universe had been anticipated and hinted at by Lemaître [8] on dynamical grounds. The static Universe posited by Einstein [9] with his constant Λ to offset gravity was unsatisfactorily unstable. The initial estimate of Hubble's constant [7] was $H_0 = 500$ km s^{-1} Mpc^{-1}, later increased slightly by Hubble to 530 km s^{-1} Mpc^{-1} [10]. The uncertainty was large, as pointed out by Jan Oort who favored a lower value [11] and Arthur Eddington who favored a higher value [12]. Eddington provided a theoretical perspective, a prelude to an interplay that continues to be a bane of studies of the scale of the Universe.

R. B. Tully (✉)
University of Hawaii, 2680 Woodlawn Dr., Honolulu, HI 96822, USA
e-mail: tully@ifa.hawaii.edu

E. Di Valentino and D. Brout (eds.), *The Hubble Constant Tension*, Springer Series in Astrophysics and Cosmology, https://doi.org/10.1007/978-981-99-0177-7_2

Eddington claimed that the combination of relativity theory and quantum theory gave a value of H_0 with the wonderful quote: *"I have now been able to reach a solution which I regard as definitive. The result is 865 km. per sec. per megaparsec."*

2.2 Evident Problems

If the Universe has been expanding at a uniform rate since its inception then it began at a time ago that is the inverse of the Hubble constant. With Hubbles's initial value of 500 km s^{-1} Mpc^{-1} the implied age is 2 billion years. The discrete moment of the beginning of expansion has come to be called the "Big Bang." Already by 1931, radioactive dating of geological rocks was giving ages of the Earth of 2–3 Gyr [13]. Surely, the Universe is older than the Earth. A plausible resolution was offered by the steady state theory of Bondi and Gold [14] and Hoyle [15], whereby matter is continuously created as space expands so as to keep constant conditions on average in perpetuity.

However, evidence began to emerge that the early estimates of H_0 were too high. Walter Baade made a breakthrough with the discovery of distinct luminosity differences between the brightest of young (Population I) stars and old (Population II) stars when he could resolve red giant branch stars in M31, M32, and NGC 205 from Mt. Wilson during Los Angeles blackouts during WWII [16]. The stepping stone to the calibration of the cepheid period-luminosity relation had been through Galactic globular clusters which have cepheid-like variables. It was recognized that the brightest resolved stars in the central region of M31 and in the elliptical companions were similar to the brightest stars in globular clusters. It became apparent that there are two kinds of cepheid variables: Population II cepheids as found in environments of old stars and Population I cepheids found amongst young stars that are brighter at a given pulsation period by $\sim$1.5 mag [17], implying a factor 2 increase in cosmological distances. Baade provided a very nice history of this development in 1956 [18].

2.3 Grasping at Straws

Work on the extragalactic distance scale picked up in the 1950s with the inauguration of the Hale 200-inch telescope on Mt. Palomar. There was the important contribution by Humason et al. [19]. Imaging of the Virgo cluster galaxy NGC 4321 revealed that features taken earlier to be individual stars were, in fact, HI regions. The brightest individual stars are some 1.8 mag fainter. Humason, Mayall, and Sandage concluded $H_0 \sim 180$ km s^{-1} Mpc^{-1}. Soon after, Sandage [20] reviewed the situation. In addition to the confusion between old and young populations with cepheids, he argued that the apparent width of the cepheid period-luminosity relation at fixed period caused a significant under-estimation of distances. He placed H_0 between 50 and 100 km s^{-1} Mpc^{-1}.

Over the next two decades, there was increased interest in the distance scale but there were only marginal improvements. The problem to be addressed was clear but the tools were blunt. Eric Holmberg tried using a correlation between the absolute magnitude and surface brightness of galaxies [21]. Sydney van den Bergh produced a luminosity classification for mid- to late-type spirals that could be given an absolute calibration [22, 23]. Jose Sersic promoted the use of the diameter of HII regions [24]. René Racine compared the globular cluster luminosity function between M31 and a galaxy in the Virgo cluster [25]. In 1970, van den Bergh summarized results from nine methods: the diameter of HII regions, luminosity classes, brightest globular clusters, assumptions regarding mass-to-light ratios, magnitude of third brightest galaxy cluster member, supernovae, a relation between the surface brightness and diameter of galaxies, and the magnitude of the brightest star in a galaxy [26]. None of these methods survives as useful today. The use of supernovae in 1970 was rudimentary. The mass-to-light method was a variation of what was used by Öpik [5]. It was assumed that the total mass of a galaxy, M_t, is given by the virial theorem, with the dependency $M_t \propto \Delta V^2 R$, where ΔV is a measure of the galaxy kinematics and R is an estimate of the galaxy radius. A relationship is assumed between mass and luminosity. The kinematic data for this so-called indicative mass procedure were provided by Morton Roberts' observations of the linewidths of galaxies in the neutral hydrogen 21 cm line [27] or subsequent follow up.

2.4 H_0 50 or 100?

The discovery in 1965 of the permeation of space with the cosmic microwave background provided compelling evidence of a hot Big Bang beginning of the Universe [28, 29], and by implication a finite age. Necessarily, there should be compatibility between the expansion age inferred from the inverse of the Hubble constant, the ages of objects within the Universe, and a model that describes the evolution of the Universe since its inception. Two simple models to consider are a completely empty universe or a topologically flat universe filled with matter. With $H_0 = 50$ km s^{-1} Mpc^{-1} ages would be 19.3 and 13.0 Gyr, respectively, with these models. With $H_0 = 100$ km s^{-1} Mpc^{-1} the respective ages would be 9.7 and 6.5 Gyr.

Ages of stars within our galaxy could be estimated through radioactive dating, white dwarf cooling or, particularly, the main sequence turnoff magnitude within globular clusters. Into and through the 1980s, the best estimate of the oldest ages found in the Galaxy were $\sim$16 Gyr [30, 31], with the Universe presumably 1 Gyr or so older. Evidently, the Universe is not empty. Occam's razor favored models that did not invoke a cosmological constant. An open universe with the density of matter 25% of the critical density to achieve closure ($\Omega_m = 0.25$) and $H_0 = 50$ km s^{-1} Mpc^{-1} would be consistent with an age of 16 Gyr. The age of a closed universe ($\Omega_m = 1$) and $H_0 = 50$, at 13 Gyr, would be marginally consistent with stellar ages.

Such simple models were inconsistent with those stellar ages if the Hubble constant was much above 50 km s^{-1} Mpc^{-1}.

Sandage and Gustav Tammann, in a series of papers between 1974 and 1982, presented arguments that favored low values of H_0 [32–36]. The methods they used to acquire distances included sizes of HII regions, brightest stars, luminosity classifications, and bootstraps through statistical properties of the Virgo cluster to applications in the field. Estimates of H_0 in these papers ranged from 50 to 57 with typical errors ±7 or less. Gérard de Vaucouleurs, in a series of rebuttals between 1978 and 1986, favored high values of H_0 [37–41]. He and colleagues built a ladder founded on primary novae, cepheids, RR Lyrae, and horizontal branch stars, secondary open and globular clusters, brightest blue and red supergiant stars, brightest HII loops or rings, and the velocity dispersions of HII regions, then a tertiary luminosity index calibration. They found values of H_0 between 90 and 110 km s^{-1} Mpc^{-1}. So was born the H_0 50 or 100 controversy.

2.4.1 Better Methods

Already by the 1970s, there was a reasonable foundation for the distance scale through parallaxes and the distances they gave to cepheids and RR Lyrae stars on the horizontal branch, albeit subject to improvements. But there had been only crude ways to estimate distances beyond the Local Group and into the domain unaffected by random motions. It was in this time period that methodologies began to emerge that are still retained today for the measurement of galaxy distances at significant redshifts.

In 1977, this author and Richard Fisher published two papers finding $H_0 = 75 - 80$ km s^{-1} Mpc^{-1} based on the correlation between the absolute luminosities of spiral galaxies and their rotation rates [42, 43], subsequently to be referred to as the TF relation. This procedure recalls the indicative mass method but deviates from the parameterization of the virial theorem. It invokes only two observables, a distance dependent luminosity and a distance independent measure of the rotation rate (taking account of the galaxy inclination and internal obscuration). A power law was assumed between luminosity and rotation rate, with a free index subject to empirical definition. There is the obvious but simple explanation for the correlation that more massive spiral galaxies tend to be more luminous and spin faster. More massive galaxies tend to contain fractional more old stars, hence are redder. Therefore, it was not a surprise that the free index of the TF relation increases toward longer wavelengths [44], The methodology was more solidly grounded in 1979 by shifting from photographic blue magnitudes to photoelectric near-infrared magnitudes, as demonstrated by Aaronson et al. [45]. These authors demonstrated that the free index would approximate four, $L \sim \Delta V^4$, if spiral galaxies share dimensionless scale-length mass profiles/rotation curves, have disks with a common central surface brightness, and have the same mean mass-to-light ratio. This latter condition is more easily met in the infrared. The true situation is more complicated. Indeed, it is mainstream to assume that dark matter dominates the mass, so it could be considered a puzzle that the correlation between star light and kinematics is as tight as observed. The interest of most theorists has been on the implications for galaxy formation and evolution [46–51]. Treating the relation as an empirical phenomenon, Aaronson and colleagues determined values of H_0 between 90 and 95 over the years 1980–1986 [52–54]. An early detailed mapping of the velocity field of the Local Supercluster was carried out by this collaboration [55].

A contemporaneous discovery with the TF relation, was an analogous correlation between the luminosities and central velocity dispersions of early-type galaxies, ellipticals and lenticulars, by Faber and Jackson [56]. The two parameter formulation for ellipticals had higher dispersion than the TF relation for spirals but in 1987 it was found that the addition of distance-independent surface brightness as a third parameter provided a significant improvement [57, 58]. The procedure, with parameters gathered into distance dependent and independent parts, resulted in what became called the fundamental plane. Initially there was no attempt at an absolute calibration but the method was used to map *relative* distances with redshifts. A group soon known as the Seven Samurai (David Burstein, Roger Davies, Alan Dressler, Sandra Faber, Donald Lynden-Bell, Roberto Terlevich, and Gary Wegner) demonstrated the existence of a large scale flow toward a mysterious "Great Attractor" [59–62]. Even as late as 2001 when substantial samples had been accumulated, the fundamental plane was still only being used to derived peculiar velocities and velocity flows based on relative distance measurements [63–65]. However, it has found an important place today within absolute measurements of the extragalactic distance scale.

A next contribution to the pantheon of important methodologies for the measurement of galaxy distances came in 1988 with recognition that distance information is encoded in the surface brightness fluctuation amplitudes of dominantly old stellar populations [66]. The brightest stars in regions devoid of recent star formation lie on the upper red giant branch of stellar evolution. The relative number of such stars within an observer's pixel increases as the square of distance, causing a predictable damping with distance. Between 1989 and 2000, John Tonry and colleagues exploited this phenomenon to derive estimates of the Hubble constant that fell from an initial $H_0 = 88$ to a later $H_0 = 77$ km s^{-1} Mpc^{-1} [67–70]. The study culminated in the publication of 300 distances to galaxies within 4000 km s^{-1} and a study of the flow patterns discerned from this sample [71].

The foundation of the surface brightness fluctuation technique is the constancy of luminosities of stars at the "tip" of the red giant branch, where degenerate helium cores reach a critical mass for the onset of burning to carbon and subsequently a star finds itself on the horizontal branch. The phenomenon harkens back to Baade's identification of Population II stars in M31 and companions [16]. Jeremy Mould and others were using observations of such stars in the 1980s to estimate distances to galaxies within the Local Group [72]. As charge coupled device (CCD) detectors became common-place, the so-called TRGB method became well established [73, 74]. Observations from ground-based telescopes began to push slightly beyond the Local Group [75]. With space facilities the TRGB method would evolve to be an important pedestal for extragalactic distances.

Other methods of possible merit explored in those years have not found a place as important players today. There was evidence that the planetary nebulae luminosity function (PNLF) had small scatter [76, 77] and was giving values of $H_0 = 86$ km s^{-1} Mpc^{-1}. However, the PNLF has shown to have limited usefulness. It required an empirical calibration to a statistically significant number of galaxies, cross-referenced with fundamental methods, but once calibrated had limited range of application. A similar criticism can be made of the use of the Balmer spectral features in very luminous B and A spectral class supergiants [78, 79]. Potentially, this Flux weighted Gravity-Luminosity Relationship (FGLR) could be very accurate but its range is limited and requires considerable effort per target. There was once hope for the utility of the globular cluster luminosity function which gave values of H_0 in the range 75–80 [80–82]. However it was difficult to have confidence in results when galaxy globular cluster populations are so variable and the formation processes are in dispute. Novae were found to give $H_0 = 70$ km s^{-1} Mpc^{-1} with limited usage [83, 84].

2.4.2 Model-Based Methods

All of the methods discussed so far have been linked to the distance ladder that starts with stellar parallaxes and builds through properties of stars in various parts of the Hertzprung-Russell diagram like the main sequence turnoff, red giant and horizontal branches, and variable RR Lyrae and cepheid stars. Further steps on the ladder can proceed by association of such features in external galaxies, providing absolute scales to the sorts of methods that have been discussed. With exceptions yet to be discussed, these methods access limited redshifts. There are other ways of estimating galaxy distances, though, that avoid the calibration ladder and can provide information on the expansion scale with a long reach.

One of these involves gravitational lens time delays. Sjur Refsdal explored the concept in 1964, well before there was observational evidence of the phenomenon [85]. A gravitational lens caused by a mass concentration could cause light from a background source to follow multiple paths to reach an observer, with the time followed on each path slightly different. If the emission from the background source is fluctuating, say because it is a supernova or active galactic nucleus, then the arrival to the observer of that signal will be slightly different along each of the paths. Roughly, the time delay between two paths, Δt, is described by the formula [86]

$$\Delta t \sim 0.5 \left[\frac{1+z_l}{c}\right] \left[\frac{d_l d_s}{d_{ls}}\right] (\Theta_A^2 - \Theta_B^2) \tag{2.1}$$

where d are angular diameter distances, d_l from observer to the lens (at redshift z_l), d_s from observer to the source, and d_{ls} from lens to source, and Θ_A, Θ_B are angular separations from the lens of images A and B. Each of the distances depends inversely on H_0, permitting a reformulation of the time delay to give a value of the expansion parameter within an assumed world model. Early estimates of H_0 were in the range 50 to 75 and even greater than 100, with recognized large uncertainties related to the distribution of mass around the lens [87, 88].

Another model dependent methodology developed out of a phenomenon first discussed by Rashid Sunyaev and Yakov Zel'dovich [89]. The possible exploitation to obtain distances was discussed by Silk and White [90] and applied by Mark Birkinshaw and colleagues and others [91–94]. The so-called Sunyaev-Zel'dovich (SZ) effect pertains to a dimming of the cosmic microwave background radiation in the direction of rich clusters of galaxies due to inverse-Compton scattering of the radiation by electrons in the hot cluster halo. Thermal bremsstrahlung flux can also be observed from that same hot gas at X-ray wavelengths. The SZ temperature decrement depends on the electron temperature and density along the line-of-sight in the cluster. The X-ray flux depends on the square root of the electron temperature, the product of the electron and proton densities (effectively the square of electron density), the volume of the hot gas cloud, and because it is an apparent flux, inversely on the square of distance. If all of the microwave decrement, the X-ray flux, and the gas temperature can be measured then an estimate of distance can be derived from the observed size and shape of the cluster and a model of the gas distribution. It is to be assumed that the gas is in hydrostatic equilibrium, which can clearly not be the case if a cluster is experiencing a merger event but might be a reasonable approximation in some cases. There is the complication that the X-ray flux, depending on electron density squared, will particularly arise in high density regions of the gas, while the SZ effect arises from lower electron density, longer path length regions. Complex modelling is required. The early analyses gave values of the Hubble constant in the range 45–70 km s^{-1} Mpc^{-1}.

Today supernovae play an important role in measurements of the scale of the Universe but in early years their contributions were ambiguous. Walter Baade made clear the distinction between novae that can be recurring events in stars and much brighter supernovae, and with Fritz Zwicky, attempted to estimate their absolute magnitudes [95, 96]. The spectra of supernovae vary considerably over time within an event and between events. There are those without, or almost without, the spectral features of hydrogen (called Type I) and those exhibiting strong hydrogen features (called Type II). Various sub-classes can be identified by spectral or light curve characteristics. There are two fundamentally different paths to supernova cataclysms. There are core collapse events in massive stars that have exhausted their nuclear fuels, usually identified as Type II although also as Type Ib or Type Ic. Then there are Type Ia merger or accretion events that take white dwarfs over the Chandrasekhar limit of 1.44 $M_\odot$ for electron degeneracy pressure support (or at least close enough to that limit to cause ignition). The Type Ia supernovae are brighter, hence seen to greater distances and predominant in surveys because of access to greater volumes. Charles Kowal made an early attempt to define the absolute luminosities of the various supernova classes [97].

Attempts were made to specify absolute magnitudes using models. In principal, a link could be made between the distance independent expansion velocity of the supernova photosphere and the distant dependent angular expansion as determined by measurements of temperature and apparent magnitude [98–103]. The book from a meeting in 1984 provided a good summary of the status of Hubble constant estimates at that date from supernova considerations [104]. With 13 separate model-driven contributions published up to 1995, the average value of H_0 was 58 km s^{-1} Mpc^{-1}. A leader in the field who can remain anonymous once said "Either H_0 equals 50 or we don't understand core-collapse supernovae". Right, one of those.

2.5 Satellites, Supernovae, and the CMB

Changes were coming in the 1990s. At the foundational level, the Hipparcos satellite that operated from 1989 to 1993 greatly increased the number and improved the quality of stellar parallaxes. Distances to globular clusters were increased and consequently their ages decreased. By the end of the decade, it was being argued that the oldest stars in the Galaxy had ages 12–13 Gyr [105–107].

In 1990, Hubble Space Telescope (HST) was launched. After a false start, a solution was implemented with a service mission and the addition of Wide Field/Planetary Camera 2 (WFPC2) in 1993. The full HST bounty only began to be harvested, though, after the third service mission in 2002 with the activation of the much more powerful Advanced Camera for Surveys (ACS). Only hints of the considerable role that HST was to play were revealed in the last decade of the 20th century. Even before the correction of the optics, two teams began using HST to observe cepheid variables in nearby galaxies, switching to the use of WFPC2 as it became available. One team involved Abhijit Saha, Alan Sandage and collaborators [108–113]. Their results were summarized by Tamman, Sandage, and Saha in 2000 who concluded that the distances to SNIa calibrated by cepheids gave $H_0 = 58 \pm 6$ km s^{-1} Mpc^{-1} [114]. The other team lead by Wendy Freedman was formed to carry out what became known as the HST Key Project [115–119]. The Key Project study, cumulatively grounded in cepheid, tip of the red giant branch, globular cluster, planetary nebulae, and surface brightness fluctuation observations [120], arrived at a value of the Hubble constant of 72 ± 8 km s^{-1} Mpc^{-1} [121].[1]

The effort to calibrate the SNIa scale was two-pronged. Paralleling the work acquiring cepheid distances to hosts of SNIa were vast improvements in SNIa acquisitions. The utility of the SNIa tool was much improved by the adjustments that could be made to minimize scatter through the recognition by Mark Phillips that the peak brightness depends on the decay rate of luminosity after peak brightness [122]. Diverse teams other than the those of the Key Project or Saha-Sandage that looked at what could be done with SNIa in the 1990s found values of H_0 between 63 and 70 [123–127].

Throughout this period, other methodologies continued to be used, generally contributing more noise than signal. Sandage and co-workers published on studies using

[1] Up to this point in the discussion, values of H_0 have been given without errors. For the most part, errors given in the last century (and often the current one) have been wildly optimistic.

a spiral luminosity function, spiral lookalikes, the globular cluster luminosity function, and their own analysis of the TF relation, finding in each case $H_0 = 57$ within ± 5 km s^{-1} Mpc^{-1}[128–131].

The most exciting developments with supernovae at this time did not directly involve the Hubble constant. Observations of high redshift SNIa by teams lead by Schmidt and Perlmutter [132, 133] provided evidence contrary to the cold dark matter (CDM) model with sufficient density in matter for a closed universe ($\Omega_m = 1$). The most distant supernovae were too faint. Assuming that they had intrinsic luminosities similar to nearby SNIa, the observations could best be fit with a model with cosmic acceleration, such as anticipated with the inclusion of the cosmological constant, Λ, in Einstein's equation.

In quite another domain, a satellite to study the cosmic microwave background was to have a profound impact on our understanding of the scale of the Universe. The Cosmic Background Explorer (COBE) that operated between 1989 and 1993 demonstrated unambiguously the black body spectrum of the cosmic microwave background, strong proof for its origin in a hot Big Bang beginning of the Universe [134, 135]. The strong hints of a peak in the power spectrum of fluctuations was confirmed by the balloon experiments MAXIMA and BOOMERANG [136–138]. The angular scale of the first peak provided compelling evidence that the Universe has a flat topology. The results from distant SNIa indicated that cold dark matter alone could not produce a spatially flat universe and gave a preference for the existence of vacuum energy. The microwave background and SNIa observations taken together were consistent with a ΛCDM model of a universe with both dark matter with density Ω_m and dark energy with density Ω_Λ such that $\Omega_m + \Omega_\Lambda = 1$.

There was yet another ingredient in the mix. Wide angle redshift surveys were accumulating and more power was being seen on large scales than anticipated by the standard CDM model with $\Omega_m = 1$ [139–142]. The disagreements became more accute with the additional constraints imposed by the microwave background power spectrum results from COBE [143–145]. Models with $\Omega_m \sim 0.3$ were favored, whether with or without dark energy. Perversely, models with $\Omega_m = 1$ were still being offered that could be reconciled with the large scale structure and microwave background information but only if H_0 was taken to be ~ 35 km s^{-1} Mpc^{-1}, in utter contradiction with distance ladder observations [146–148].

Increasingly, wide field imaging and redshift surveys were becoming available as the 20th century came to a close. Michael Hudson was reconstructing density and velocity fields from these inputs [149, 150]. It was demonstrated that the TF relation could give distances with comparable accuracy observing at optical R, I bands as obtained in the infrared, given area photometry using CCDs [151]. Jeff Willick and colleagues made major contributions to issues of calibration and biasing [152–157]. They came to favor a value of the Hubble constant of 85 km s^{-1} Mpc^{-1} and, from the peculiar velocity field, matter density $\Omega_m \sim 0.3$ [158, 159]. Riccardo Giovanelli, Martha Haynes, and colleagues greatly enlarged the TF relation sample sizes with extensive 21cm observations with Arecibo Telescope. They derived $H_0 = 69$ km s^{-1} Mpc^{-1} [160].

A bit of a wild card was introduced with the introduction by Avishai Dekel and others of an analysis tool that came to be called POTENT [161, 162]. A smoothed three-dimensional velocity field was inferred from observed radial velocities assuming a potential flow, consistent with peculiar velocities arising from density perturbations. Initial conditions were linked with final positions using the Zel'dovich approximation assuming linear theory of the growth of structure. Dekel et al. were able to generate maps of the velocity and density field within a radius of 8,000 km s^{-1} that impressively resemble maps being built two decades later. However, it was a considerable distraction that values of matter density were being inferred that were $\Omega_m \sim 1$ or only slightly less [163]. A similar high value was found for Ω_m from a power spectrum analysis of the density fluctuations arising from the Giovanelli et al. observed peculiar velocities [164]. By contrast, an analysis based on numerical action orbit reconstructions by Ed Shaya, Jim Peebles, and this author derived a value of matter density of $\Omega_m = 0.17 \pm 0.10$ [165]. Igor Karachentsev and others were remarking on the extremely cold local flow [166]. It could be seen from orbit reconstructions that if $\Omega_m \sim 1$ then there would be the triple-value signatures of collapse [167] all over the map, a situation far from actuality. The situation was nicely summarized by contributions at a workshop organized by Stéphane Courteau in 1999 on cosmic flows in Victoria, Canada [168].

As the century came to an end, there was a remarkable paradigm shift. A foundation for it had been laid 20 years earlier by Alan Guth's ideas of the origin of the Universe in an inflation driven by vacuum energy [169]. It was thought natural that this origin would result in a universe that is close to topologically flat. The power spectrum of cosmic microwave background fluctuations with a peak on an angular scale of about a degree provided compelling evidence that this is the case. The measurements that SNIa at $z \sim 1$ are fainter than expected if $\Omega_m = 1$ favored the requirement of a positive Λ contribution. There was general agreement that redshift surveys were indicating that there was too much structure on large scale and that biasing between the distribution of galaxies and the distribution of matter was insufficient to rescue $\Omega_m = 1$ models. Even though it was demonstrated that there are large scale flows [59], flows on local scales are quiet, direct evidence for modest fluctuations in matter. Then there was abundant evidence from measurements of the Hubble parameter that, assuredly, the CDM universe with closure density in matter was untenable and even an open universe with $\Omega_\Lambda = 0$ was contradicted.

The one model that accommodated the observations posited that the Universe is spatially flat due to the combination of a currently dominant "dark energy" and subdominant matter, most of it "dark matter". This ΛCDM model relieved the Hubble constant tension at the time. Any value of H_0 in the range being proposed could be accommodated. The value from the TF relation favored by this author as of 2000 was $H_0 = 77$ km s^{-1} Mpc^{-1}, lower than earlier estimates [151] primarily because of the availability of 24 zero-point calibrators, whereas there were only 3 in 1988 [170]. The HST Key Project derived the value $H_0 = 71$ from their calibration of the TF relation [171]. Overall, the Key Project best value was $H_0 = 72$ km s^{-1} Mpc^{-1} [121]. For a brief period, there was no Hubble constant tension. That situation would not last.

2.6 H_0: 67 or 75?

Increasingly detailed observations were made of the cosmic microwave background with the Wilkinson Microwave Anisotropy Probe (WMAP) that operated between 2001 and 2010 and with the Planck satellite between 2009 and 2013. The spatial power spectrum of fluctuations could be fit in impressive detail with a ΛCDM model with $H_0 = 67.4 \pm 0.5$ km s^{-1} Mpc^{-1} and $\Omega_m = 0.315 \pm 0.007$ [172–174]. Acoustic oscillations stimulated around dark matter fluctuations in the plasma of the hot Universe would have an imprint in baryon structures in the cold Universe on the scale of the sound horizon at recombination. These baryon acoustic oscillations (BAO) give rise to a feature in two-point correlation functions over a range of redshifts, obtained in studies of large galaxy redshift surveys like the SDSS-III Baryon Oscillation Spectroscopic Survey (BOSS) and the Dark Energy Survey (DES) [175–177]. The angular scale of the BAO feature closely tracks the evolution of the Hubble parameter as a function of redshift, $H(z)$, as anticipated by the ΛCDM model defined by the Planck analysis. Similar consistent results are found from the power spectrum in the galaxy distributions of these redshift surveys [178]. These exceptionally stringent observational constraints require that $H_0 \simeq 67$ km s^{-1} Mpc^{-1} within $\sim 1\%$ if the popular ΛCDM model correctly describes our Universe.

Meanwhile since 2000, attention has been given to the various facets of the distance scale ladder and significantly higher values of H_0 were reported, giving rise to a new Hubble tension. Methodologies have been diverse and details regarding each will be discussed in the ensuing chapters in this book. Only highlights of major programs will be given attention here.

Begin at the foundation with HST. Though the number of targets is yet small, high quality stellar parallaxes have been obtained with HST to Galactic cepheids through a spatial scanning technique [179]. The observations are important because, earlier, all but two of the Galactic cepheids that had good parallax distances have shorter periods ($T < 10$ days) than the cepheids studied in external galaxies. There are to date HST observations of cepheid variables in 42 host galaxies of SNIa events. The coupling with SNIa observations out to $z \sim 1$ yields $H_0 = 73.3 \pm 1.0$ km s^{-1} Mpc^{-1} [180].

Taken at face value, the difference between this ladder H_0 and the early universe ΛCDM model H_0 is 5σ. The concern is unidentified systematics. For confirmation, a completely independent path is needed. Such a path is being explored, although it is not yet fully developed. It begins with the replacement of cepheids for distance measurements to nearby galaxies with tip of the red giant branch. Currently, TRGB absolute magnitudes are established through parallax measurements to horizontal branch and RR Lyrae stars, giving distances to Galactic clusters and Milky Way companion satellites, whence establishing the absolute magnitudes of the brightest stars on the red giant branch [107, 181]. The astronomy community was startled in 2020 when Wendy Freedman and team replaced the calibration of SNIa with TRGB distances to host galaxies in place of cepheid distances, and found $H_0 = 69.6 \pm 1.7$ km s^{-1} Mpc^{-1}, consistent with the early universe value [182]. A re-evaluation of the study determined $H_0 = 71.5 \pm 1.8$ km s^{-1} Mpc^{-1}, consistent but leaning toward tension [183].

To have a completely separate path, there needs to be an alternative to SNIa that can be applied at substantial redshifts. Core collapse supernovae (SNII) offer a possibility. It has been demonstrated that there is a correlation between expansion velocities and peak magnitudes, with scatter of about 15%. With still a small sample, a calibration based on cepheids is giving $H_0 = 76 \pm 5$ km s^{-1} Mpc^{-1} [184]. The surface brightness fluctuation technique probably offers an even greater avenue for progress because observations in the infrared with space telescopes give distances with accuracies of 5% and can reach to several hundred megaparsecs. With current applications, again grounded through cepheid calibrations, $H_0 = 73.3 \pm 2.4$ km s^{-1} Mpc^{-1} is being found [185].

Programs to study the velocity field of galaxies have been concurrent with efforts to define the Hubble constant. The peculiar velocity of a galaxy can be obtained independent of an absolute calibration and hence of H_0. Very large samples have been accumulated involving the TF and fundamental plane relations, extending to $0.05c - 0.1c$ and providing wide coverage of the sky. Individual distance uncertainties are 20–25% but sample sizes are in the 10's of thousands. There have been two noteworthy fundamental plane contributions arising from wide field imaging and redshift programs: the Six Degree Field Galaxy Survey for peculiar velocities (6dFGSv) targeting 9,000 E−S0 galaxies in the south celestial hemisphere out to 16,000 km s^{-1} [186, 187], and the SDSS peculiar velocity sample of 34,000 E−S0 galaxies in the north celestial, north Galactic quadrant extending to 30,000 km s^{-1} [188]. The latter sample was given a zero-point calibration ultimately based on cepheids, which resulted in the determination $H_0 = 75 \pm 2$ km s^{-1} Mpc^{-1}.

Complementary observations of spiral galaxies have received impetus from extensive 21cm neutral hydrogen survey programs [189, 190]. Major velocity field studies have been published based on TF relation distance measurements [191–195]. It has been suggested that the correlation with spiral galaxy rotation rate is tightened if HI flux is added to the optical or infrared flux, creating a measure of the total baryonic mass. The resultant correlation has been called the baryonic TF relation (BTFR) [196, 197]. The BTFR has been used with a sample of 10,000 galaxies, with zero point scaling set by a combination of cepheid and TRGB distances, to obtain $H_0 = 75.5 \pm 2.5$ km s^{-1} Mpc^{-1} [198].

It is an unfortunate feature of the distance ladder that there is no one methodology that is useful from the nearby within the range of parallax measurements to the redshift regime beyond peculiar velocity effects. By necessity, methodologies must be interlinked. If the Hubble constant is to be determined robustly from the ladder, the linkages must be robust, which is to say, they must be abundant and diverse. However, making linkages is tricky. For example, cepheid variables are only found in systems with young stars; likewise SNII. The use of the TF relation is restricted to spiral galaxies. The fundamental plane and surface brightness fluctuation methods are only used with massive early-type galaxies. The TRGB method can be used to derive distances to essentially any galaxy but only if it is nearby; within 10 Mpc observing with HST, unless a special effort is made. SNIa can arise in any kind of host but they are rare nearby. There is the probability that there are subtle variations in the luminosities of SNIa that depend on host properties, such as whether or not there are young stars, which raises issues if calibrations only involve cepheids [199–201].

By necessity, unless targets are within the reach of cepheid or TRGB calibration, linkages between methods must be through group affiliations. Groups can range from clusters of many hundreds of members to pairs. Linkages between methodologies across this full range are useful and desirable. Good group catalogs should have a physical basis, have a high level of completion among the galaxy samples with distances, and have minimal contamination from interlopers. Groups considered appropriate [202–204] were used to acquire linkages between the largest available samples of the procedures described above in the compilation of the Cosmicflows-4 compendium of 56,000 galaxy distances [205]. Samples were brought to a common absolute scale based on cepheid and TRGB calibrations with a Bayesian Markov chain Monte Carlo (MCMC) analysis (where a small cepheid-TRGB scale difference was reconciled with preliminary parallax information from Gaia satellite observations). The value of the Hubble constant from the Cosmicflows-4 compilation is arguably the best available determination from the distance ladder because it integrates all the major methodologies and surveys. $H_0 = 74.6 \pm 0.8$ km s^{-1} Mpc^{-1} is reported [205]. This statistical error is very small because the combined sample of 56,000 galaxies in 38,000 groups is so large. It is estimated that systematic errors could be as large as 3 km s^{-1} Mpc^{-1}.

2.6.1 H_0 from Models

A number of modeled observations that are independent of the distance ladder have been proposed that have dependence on the Hubble constant, several of which are discussed in this volume. In each instance, it must be asked if the model constrains the Hubble constant or if the Hubble constant constrains the model. To this reviewer's awareness, the strongest case of a model constraint on a galaxy distance is provided by the nuclear maser observations of NGC 4258 [206], because of the remarkably fortuitous edge-on geometry, locations, and strength of the maser signals. The occurrence of this event so near by (7.58 Mpc) is one of extreme luck because the few other nuclear maser cases that have been studied give distance results that are far less certain [207].

The detached eclipsing binary measurements that give a distance to the Large Magellanic Cloud might likewise pass the bar of informing the extragalactic distance scale [208]. It does depend on knowledge of a surface brightness-color relation and, more, on the locations of targets across a 5 degree field (a projection that is 9% of the LMC distance). The claimed precision of the LMC distance is 1% but what does that mean for an irregular system that is so large on the sky and presumably in depth? The LMC has played an important role in the calibration of the cepheid scale but, in the fullness of time, perhaps the two Magellanic clouds will be by-passed in distance scale studies.

There are a number of physical properties of the Universe that, to the degree that they are properly modeled, bear on the value of the Hubble constant. There are strong gravitational lensing time delays of AGN signal variations through cluster halos [209], gravitational waves arising from neutron star mergers [210], the Sunyaev-Zel'dovich effect involving comparisons of the X-ray flux and microwave background damping associated with cluster plasmas [211], cosmic chronometers giving record to the relative passage of time with massive, passively evolving galaxies [212], and the pair-wise galaxy separation mapping of baryon acoustic oscillation features at various epochs [213]. Generally, the evaluation of these interesting phenomena is done in the context of the ΛCDM paradigm, which is at the root of the H_0 tension controversy.

Some time ago, George Efstathiou initiated an attempt to reconcile the distance ladder measurements with the early universe implications for H_0 [214]. He explored possibilities of systematics that might affect ladder measurements; in other words, relieving the Hubble tension by postulating that the ladder distances are in error. Of course, systematic errors are possible. However, for this reviewer, the effort of reconciling distance scale observations with a favored world model too easily brings back memories of the end of the last century. The young turks of the time, of which I suppose I was one, were rather convinced that, within a Friedmann-Lemaître-Robertson-Walker framework, only a $\Lambda \neq 0$ cosmology model would accommodate the distance scale measurements that used quality tools. The theoretical community too easily discounted information on the distance scale, failing to recognize that the influential voices of Alan Sandage and Gustav Tammann were grounded in the same exercise of that of Efstathiou.

2.7 Future History

Readers can be forgiven for despairing of a resolution of the question of the current expansion rate of the Universe after a hundred years of effort, though the problem is well posed. Happily, the tools are at hand for an unambiguous resolution with the Global Astrometric Interferometer for Astrophysics (Gaia) and the James Webb Space Telescope (JWST). The problem can be separated into three parts: (1) the firm calibration of the methods founded within the Milky Way, specifically cepheids and TRGB, (2) the transfer of the calibration to methods that can be used at substantial redshifts, particularly SNIa and surface brightness fluctuation, and (3) the acquisition of large samples with those latter methods.

It can be anticipated that there will be serious advances on the first problem over the next year or so. The Gaia third data release (DR3) in June, 2022, provides parallax information on thousands of cepheids, tens of thousands of RR Lyrae stars, and hundreds of thousands of red giant branch stars with corollary metallicity information. Anything written today may be well out of date to the reader.

The way forward with problems (2) and (3) will involve paths that are independent yet connected. There are two obvious, well-defined paths; one involving young

Population I stars and the other involving old Population II stars. The Pop I path is already well known: cepheid variables calibrate SNIa which in turn are observed to high redshift. The Pop II path uses TRGB to calibrate surface brightness fluctuations in massive elliptical galaxies which in turn are observed to high redshift. These two paths can stand alone, but there are ample opportunities for cross-links through commonality of calibrator hosts, SNIa in ellipticals, and group associations. TRGB can directly calibrate SNIa in late as well as early galaxy types.

JWST is particularly well suited to exploit the Pop II path. The volume of space opened up for TRGB observations is an order of magnitude greater with JWST than with HST (from 10 Mpc to greater than 20 Mpc). The increase in volume for surface brightness fluctuation measurements is at least 30-fold (from 100 Mpc to $\sim$ 300 Mpc). If H_0 is to be determined to a precision of 1%, calibration samples must have sizes of at least 10^2 and samples at high redshift should be significantly larger. With JWST, it is more feasible to expect that the necessary large calibrator samples can be acquired for TRGB distances rather than for cepheid distances.

The SNIa tool can also be improved. While there is excitement that the Vera C. Rubin Observatory will make acquisitions of large numbers of SNIa at large redshifts, advancing our knowledge regarding cosmic acceleration, smaller telescopes already in operation are sufficient for the simpler problem of the Hubble constant. Surveys with names like Pan-STARRS, ATLAS, ZTF, and ASAS-SN are picking up several SNIa events within $0.1c$ every night. If only there are the resources for manpower, samples of 10^4+ could be gathered in a few years. With such large samples, complemented by multi-epoch spectroscopy and photometry at multiple optical and infrared bands, the information will be available to fine-tune the SNIa distance tool technology to account for variations in the paths to supernova events.

Two other large surveys in Australia will have major impacts on studies of the nature of large scale structure and how it formed. WALLABY is a 21cm blind survey with the Australian Square Kilometer Array Pathfinder that is expected to provide some 200,000 TF measures with a mean redshift of $z = 0.05$. The Taipan Galaxy Survey uses a multiplexed spectrograph on the UK Schmidt Telescope at Siding Springs Observatory and is anticipated to obtain some 50,000 fundamental plane distances within $z = 0.1$. These studies promise to make valuable contributions toward a determination of the Hubble constant. However, their uncertainties of 25% per measurement will leave them more vulnerable to systematics than can be expected with SNIa or surface brightness fluctuation measurements with 5% individual uncertainties.

As the 20th century came to an end, ladder measurements of the Hubble constant were at odds with the favored cosmological model of the time of cold dark matter with $\Lambda = 0$. The new favorite became the ΛCDM model with dark energy giving rise to acceleration of space in a topologically flat universe. Yet ladder measurements, continuously improving, create doubts that this currently favorite model is complete. Yes, there is a Hubble tension.

Acknowledgements Two of the leaders in the quest to measure the Hubble constant, Marc Aaronson and Jeff Willick, died in tragic accidents in their prime.There are so many others who have contributed, too many to be individually acknowledged so I won't even try.

References

1. H. Shapley, Bull. Nat. Res. Counc. **2**, 171 (1921)
2. H.D. Curtis, Bull. Nat. Res. Counc. **2**, 194 (1921)
3. E.P. Hubble, The Observatory **48**, 139 (1925)
4. H.S. Leavitt, E.C. Pickering, Harv.Coll. Obs. Circ. **173**, 1 (1912)
5. E. Opik, ApJ **55**, 406 (1922). https://doi.org/10.1086/142680
6. V.M. Slipher, Pop. Astron. **23**, 21 (1915)
7. E.P. Hubble, Proc. Nat. Acad. Sci. **15**, 168 (1929). https://doi.org/10.1073/pnas.15.3.168
8. G. Lemaître, Ann. Soc. Sci. Brux. **47**, 49 (1927)
9. A. Einstein, Sitzungsb. König. Preuss. Akad. 142 (1917)
10. E.P. Hubble, *Realm of the Nebulae* (Yale University Press). ISBN: 9780300025002
11. J.H. Oort, BAN **6**, 155 (1931)
12. A.S. Eddington, MNRAS **95**, 636 (1935). https://doi.org/10.1093/mnras/95.8.636
13. G.B. Dalrymple, *The Age of the Earth* (Stanford University Press). ISBN: 0804723311
14. H. Bondi, T. Gold, MNRAS **108**, 252 (1948). https://doi.org/10.1093/mnras/108.3.252
15. F. Hoyle, MNRAS **108**, 372 (1948). https://doi.org/10.1093/mnras/108.5.372
16. W. Baade, ApJ **100**, 137 (1944). https://doi.org/10.1086/144650
17. W. Baade, Trans. I.A.U. **8**, 397 (1954)
18. W. Baade, PASP **68**, 5 (1956). https://doi.org/10.1086/126870
19. M.L. Humason, N.U. Mayall, A.R. Sandage, AJ **61**, 97 (1956). https://doi.org/10.1086/107297
20. A.R. Sandage, ApJ **127**, 513 (1958). https://doi.org/10.1086/146483
21. E. Holmberg, Lund Medd. Astron. Obs. Ser. **II**, 136 (1958)
22. S. van den Bergh, Zf. Ap **49**, 198 (1960)
23. S. van den Bergh, JRASC **54**, 49 (1960)
24. J.L. Sersic, Zf Ap. **50**, 168 (1960)
25. R. Racine, JRASC **62**, 367 (1968)
26. S. van den Bergh, Nature **225**, 503 (1970). https://doi.org/10.1038/225503a0
27. M.S. Roberts, AJ **74**, 859 (1969). https://doi.org/10.1086/110874
28. A.A. Penzias, R.W. Wilson, ApJ **142**, 419 (1965). https://doi.org/10.1086/148307
29. R.H. Dicke et al., ApJ **142**, 414 (1965). https://doi.org/10.1086/148306
30. K. Janes, P. Demarque, ApJ **264**, 206 (1983). https://doi.org/10.1086/160587
31. A. Chieffi, O. Straniero, ApJS **71**, 47 (1989). https://doi.org/10.1086/191364
32. A.R. Sandage, G.A. Tammann, ApJ **194**, 223 (1974). https://doi.org/10.1086/153238
33. A.R. Sandage, G.A. Tammann, ApJ **194**, 559 (1974). https://doi.org/10.1086/153275
34. A.R. Sandage, G.A. Tammann, ApJ **196**, 313 (1975). https://doi.org/10.1086/153413
35. A.R. Sandage, G.A. Tammann, ApJ **197**, 265 (1975). https://doi.org/10.1086/153510
36. A.R. Sandage, G.A. Tammann, ApJ **256**, 339 (1982). https://doi.org/10.1086/159911
37. G. de Vaucouleurs, G. Bollinger, ApJ **233**, 433 (1979). https://doi.org/10.1086/157405
38. G. de Vaucouleurs, W.L. Peters, ApJ **248**, 395 (1981). https://doi.org/10.1086/159165
39. G. de Vaucouleurs, ApJ **268**, 468 (1983). https://doi.org/10.1086/160972
40. G. de Vaucouleurs, H.G. Corwin Jr., ApJ **297**, 23 (1985). https://doi.org/10.1086/163499
41. G. de Vaucouleurs, W.L. Peters, ApJ **303**, 19 (1986). https://doi.org/10.1086/164048
42. R.B. Tully, J.R. Fisher, A&A **54**, 66 (1977)
43. J.R. Fisher, R.B. Tully, ComAp **7**, 85 (1977)
44. R.B. Tully, J.R. Mould, M. Aaronson, ApJ **257**, 527 (1982). https://doi.org/10.1086/160009
45. M. Aaronson, J.P. Huchra, J.R. Mould, ApJ **229**, 1 (1979). https://doi.org/10.1086/156923
46. T. Okamoto, V.R. Eke, C.S. Frenk, A. Jenkins, MNRAS **363**, 1299 (2005). https://doi.org/10.1111/j.1365-2966.2005.09525.x
47. F. Governato et al., MNRAS **374**, 1479 (2007). https://doi.org/10.1111/j.1365-2966.2006.11266.x
48. S. Trujillo-Gomez, A. Klypin, J. Primack, A.J. Romanowsky, ApJ **742**, 16 (2011). https://doi.org/10.1088/0004-637X/742/1/16

49. J. Guedes, S. Callegari, P. Madau, L. Mayer, ApJ **742**, 76 (2011). https://doi.org/10.1088/0004-637X/742/2/76
50. F. Marinacci, R. Pakmor, V. Springel, MNRAS **437**, 1750 (2014). https://doi.org/10.1093/mnras/stt2003
51. M. Glowacki, E. Elson, R. Davé, MNRAS **498**, 3687 (2020). https://doi.org/10.1093/mnras/staa2616
52. M. Aaronson et al., ApJ **239**, 12 (1980). https://doi.org/10.1086/158084
53. G.D. Bothun et al., ApJ **278**, 475 (1984). https://doi.org/10.1086/161814
54. M. Aaronson et al., ApJ **302**, 536 (1986). https://doi.org/10.1086/164014
55. M. Aaronson et al., ApJ **258**, 64 (1982). https://doi.org/10.1086/160053
56. S.M. Faber, R.E. Jackson, ApJ **204**, 668 (1976). https://doi.org/10.1086/154215
57. S. Djorgovski, M. Davis, ApJ **313**, 59 (1987). https://doi.org/10.1086/164948
58. A. Dressler et al., ApJ **313**, 42 (1987). https://doi.org/10.1086/164947
59. A. Dressler et al., ApJ **313**, L37 (1987). https://doi.org/10.1086/184827
60. D. Lynden-Bell et al., ApJ **326**, 19 (1988). https://doi.org/10.1086/166066
61. D. Burstein, S.M. Faber, A. Dressler, ApJ **354**, 18 (1990). https://doi.org/10.1086/168664
62. A. Dressler, S.M. Faber, ApJ **354**, L45 (1990). https://doi.org/10.1086/164947
63. M. Colless et al., MNRAS **321**, 277 (2001). https://doi.org/10.1046/j.1365-8711.2001.04044.x
64. M.J. Hudson et al., MNRAS **327**, 265 (2001). https://doi.org/10.1046/j.1365-8711.2001.04786.x
65. M. Bernardi et al., AJ **123**, 2990 (2002). https://doi.org/10.1086/340463
66. J.L. Tonry, D.P. Schneider, AJ **96**, 807 (1988). https://doi.org/10.1086/114847
67. J.L. Tonry, E.A. Ajhar, G.A. Luppino, ApJ **346**, 57 (1989). https://doi.org/10.1086/185578
68. J.L. Tonry, ApJ **373**, L1 (1991). https://doi.org/10.1086/186037
69. J.L. Tonry, J.P. Blakeslee, E.A. Ajhar, A. Dressler, ApJ **475**, 399 (1997). https://doi.org/10.1086/303576
70. J.L. Tonry, J.P. Blakeslee, E.A. Ajhar, A. Dressler, ApJ **530**, 625 (2000). https://doi.org/10.1086/308409
71. J.L. Tonry et al., ApJ **546**, 681 (2001). https://doi.org/10.1086/318301
72. J.R. Mould, J. Kristian, ApJ **305**, 591 (1986). https://doi.org/10.1086/164273
73. M.G. Lee, W.L. Freedman, B.F. Madore, ApJ **417**, 553 (1993). https://doi.org/10.1086/173334
74. G.S. Da Costa, T.E. Armandroff, AJ **100**, 162 (1990). https://doi.org/10.1086/115500
75. S. Sakai, B.F. Madore, W.L. Freedman, ApJ **461**, 713 (1996). https://doi.org/10.1086/177096
76. G.H. Jacoby, R. Ciardullo, H.C. Ford, ApJ **356**, 332 (1990). https://doi.org/10.1086/168843
77. R.H. Mendez, R.P. Kudritzki, R. Ciardullo, G.H. Jacoby, A&A **275**, 534 (1993)
78. R.B. Tully, S.C. Wolff, ApJ **281**, 67 (1984). https://doi.org/10.1086/162075
79. R.P. Kudritzki et al., ApJ **788**, 56 (2014). https://doi.org/10.1088/0004-637X/788/1/56
80. D.A. Hanes, MNRAS **188**, 901 (1979). https://doi.org/10.1093/mnras/188.4.901
81. R. McMillan, R. Ciardullo, G.H. Jacoby, ApJ **416**, 62 (1993). https://doi.org/10.1086/173215
82. B.C. Whitmore et al., ApJ **454**, L73 (1995). https://doi.org/10.1086/309788
83. C.J. Pritchet, S. van den Bergh, ApJ **318**, 507 (1987). https://doi.org/10.1086/165387
84. M. della Valle, M. Livio, ApJ **452**, 704 (1995). https://doi.org/10.1086/176342
85. S. Refsdal, MNRAS **128**, 307 (1964). https://doi.org/10.1093/mnras/128.4.307
86. C.S. Kochanek, P.L. Schechter (2003), arXiv:astro-ph/0306040
87. U. Borgeest, S. Refsdal, A&A **141**, 318 (1984)
88. G. Rhee, Nature **350**, 211 (1991). https://doi.org/10.1038/350211a0
89. R.A. Sunyaev, Ya.B. Zel'dovich, Comm. Ap. Space Sci. **4**, 173 (1972)
90. J. Silk, S.D.M. White, ApJ **226**, L103 (1978). https://doi.org/10.1086/182841
91. M. Birkinshaw, J.P. Hughes, K.A. Arnaud, ApJ **379**, 466 (1991). https://doi.org/10.1086/170522
92. M. Birkinshaw, J.P. Hughes, ApJ **420**, 33 (1994). https://doi.org/10.1086/173540
93. T. Herbig, C.R. Lawrence, A.C.S. Readhead, S. Gulkis, ApJ **449**, L5 (1995). https://doi.org/10.1086/309616

94. Y. Rephaeli, ARA&A **33**, 541 (1995). https://doi.org/10.1146/annurev.aa.33.090195.002545
95. W. Baade, ApJ **88**, 285 (1938). https://doi.org/10.1086/143983
96. W. Baade, F. Zwicky, ApJ **88**, 411 (1938). https://doi.org/10.1086/143996
97. C.T. Kowal, AJ **73**, 1021 (1968). https://doi.org/10.1086/110763
98. D. Branch, B. Patchett, MNRAS **161**, 71 (1973). https://doi.org/10.1093/mnras/161.1.71
99. R.P. Kirshner, J. Kwan, ApJ **193**, 27 (1974). https://doi.org/10.1086/153123
100. D. Branch et al., ApJ **244**, 780 (1981). https://doi.org/10.1086/158755
101. D. Branch, ApJ **320**, L23 (1987). https://doi.org/10.1086/184970
102. B.P. Schmidt et al., ApJ **432**, 42 (1994). https://doi.org/10.1086/174546
103. A. Fisher, D. Branch, P. Hoflich, A. Khokhlov, ApJ **447**, L73 (1995). https://doi.org/10.1086/309563
104. N. Bartel, Lect. Not. Phys. **224**, (1985). https://doi.org/10.1007/3-540-15206-7
105. B. Chaboyer et al., ApJ **494**, 96 (1998). https://doi.org/10.1086/305201
106. F. Pont, M. Mayor, C. Turon, D.A. Vandenberg, A&A **329**, 87 (1998)
107. E. Carretta, R.G. Gratton, G. Clementini, F. Fusi Pecci, ApJ **533**, 215 (2000). https://doi.org/10.1086/308629
108. A. Saha et al., ApJ **425**, 14 (1994). https://doi.org/10.1086/173957
109. A. Saha, A.R. Sandage et al., ApJS **107**, 693 (1996). https://doi.org/10.1086/192378
110. A. Saha, A.R. Sandage et al., ApJ **486**, 1 (1997). https://doi.org/10.1086/304507
111. A. Saha, A.R. Sandage et al., ApJ **522**, 802 (1999). https://doi.org/10.1086/307693
112. A. Saha, A.R. Sandage et al., ApJ **562**, 314 (2001). https://doi.org/10.1086/323529
113. A.R. Sandage, A. Saha et al., ApJ **460**, L15 (1996). https://doi.org/10.1086/309973
114. G.A. Tammann, A.R. Sandage, A. Saha et al., Proc. HST Symp. **14**, 222 (2003). https://doi.org/10.1086/309973
115. D.D. Kelson et al., ApJ **463**, 26 (1996). https://doi.org/10.1086/177221
116. B.K. Gibson et al., ApJ **529**, 723 (2000). https://doi.org/10.1086/308306
117. W.L. Freedman et al., ApJ **427**, 628 (1994). https://doi.org/10.1086/174172
118. W.L. Freedman et al., Nature **371**, 757 (1994). https://doi.org/10.1038/371757a0
119. A. Turner et al., ApJ **505**, 207 (1998). https://doi.org/10.1086/306150
120. L. Ferrarese et al., ApJS **128**, 431 (2000). https://doi.org/10.1086/313391
121. W.L. Freedman et al., ApJ **553**, 47 (2001). https://doi.org/10.1086/320638
122. M.M. Phillips, ApJ **413**, L105 (1993). https://doi.org/10.1086/186970
123. A.G. Riess, W.P. Press, R.P. Kirshner, ApJ **438**, L17 (1995). https://doi.org/10.1086/187704
124. S. Jha et al., ApJS **125**, 73 (1999). https://doi.org/10.1086/313275
125. M.M. Phillips et al., AJ **118**, 1766 (1999). https://doi.org/10.1086/301032
126. Y. Wang, ApJ **536**, 531 (2000). https://doi.org/10.1086/308958
127. M. Hernandez et al., MNRAS **319**, 223 (2000). https://doi.org/10.1046/j.1365-8711.2000.03841.x
128. A.R. Sandage, G.A. Tammann, ApJ **464**, L51 (1996). https://doi.org/10.1086/310083
129. A.R. Sandage, AJ **111**, 1 (1996a). https://doi.org/10.1086/117755
130. A.R. Sandage, AJ **111**, 18 (1996b). https://doi.org/10.1086/117756
131. M. Federspiel, G.A. Tammann, A.R. Sandage, ApJ **495**, 115 (1998). https://doi.org/10.1086/305263
132. B.P. Schmidt et al., ApJ **507**, 46 (1998). https://doi.org/10.1086/306308
133. S. Perlmutter et al., ApJ **517**, 565 (1999). https://doi.org/10.1086/307221
134. G.F. Smoot et al., ApJ **396**, L1 (1992). https://doi.org/10.1086/186504
135. C.L. Bennett et al., ApJ **464**, L1 (1996). https://doi.org/10.1086/310075
136. S. Hanany et al., ApJ **545**, L5 (2000). https://doi.org/10.1086/317322
137. P. de Bernardis et al., Nature **404**, 955 (2000). https://doi.org/10.1038/35010035
138. C.B. Netterfield et al., ApJ **571**, 604 (2002). https://doi.org/10.1086/340118
139. W. Saunders et al., MNRAS **242**, 318 (1990). https://doi.org/10.1093/mnras/242.3.318
140. W. Saunders et al., Nature **349**, 32 (1991). https://doi.org/10.1038/349032a0
141. G. Efstathiou, W.J. Sutherland, S.J. Maddox, Nature **348**, 705 (1990). https://doi.org/10.1038/348705a0

142. J. Loveday et al., ApJ **400**, L43 (1992). https://doi.org/10.1086/186645
143. E.F. Bunn, M. White, ApJ **480**, 6 (1997). https://doi.org/10.1086/303955
144. J.R. Bond, A.H. Jaffe, R. Soc. London Phil. Trans. A **357**, 57 (1999). https://doi.org/10.1098/rsta.1999.0314
145. J.A. Peacock, MNRAS **284**, 885 (1997). https://doi.org/10.1093/mnras/284.4.885
146. J.G. Bartlett et al., Science **267**, 980 (1995). https://doi.org/10.1126/science.267.5200.980
147. M. White et al., MNRAS **276**, L69 (1995). https://doi.org/10.1093/mnras/276.1.L69
148. C.H. Lineweaver et al., A&A **322**, 365 (1997). https://doi.org/10.48550/arXiv.astro-ph/9610133
149. M.J. Hudson, MNRAS **265**, 43 (1993). https://doi.org/10.1093/mnras/265.1.43
150. M.J. Hudson, MNRAS **266**, 475 (1994). https://doi.org/10.1093/mnras/266.2.475
151. M.J. Pierce, R.B. Tully, ApJ **330**, 579 (1988). https://doi.org/10.1086/307707
152. J.A. Willick, ApJS **92**, 1 (1994). https://doi.org/10.1086/191957
153. J.A. Willick et al., ApJ **446**, 12 (1995). https://doi.org/10.1086/175762
154. J.A. Willick et al., ApJ **457**, 460 (1996). https://doi.org/10.1086/176746
155. J.A. Willick et al., ApJS **109**, 333 (1997). https://doi.org/10.1086/312983
156. J.A. Willick et al., ApJ **486**, 629 (1997). https://doi.org/10.1086/304551
157. M.A. Strauss, J.A. Willick, Phys. Rep. **261**, 271 (1995). https://doi.org/10.1016/0370-1573(95)00013-7
158. J.A. Willick, M.A. Strauss, ApJ **507**, 64 (1998). https://doi.org/10.1086/306314
159. J.A. Willick, P. Batra, ApJ **548**, 564 (2001). https://doi.org/10.1086/319005
160. R. Giovanelli et al., ApJ **477**, L1 (1997). https://doi.org/10.1086/310521
161. A. Dekel, E. Bertschinger, S.M. Faber, ApJ **364**, 349 (1990). https://doi.org/10.1086/169418
162. A. Dekel et al., ApJ **412**, 1 (1993). https://doi.org/10.1086/172896
163. A. Dekel et al., ApJ **522**, 1 (1999). https://doi.org/10.1086/307636
164. W. Freudling et al., ApJ **523**, 1 (1999). https://doi.org/10.1086/307707
165. E.J. Shaya, P.J.E. Peebles, R.B. Tully, ApJ **454**, 15 (1995). https://doi.org/10.1086/176460
166. I.D. Karachentsev et al., A&A **389**, 812 (2002). https://doi.org/10.1051/0004-6361:20020649
167. J.L. Tonry, M. Davis, ApJ **246**, 680 (1981). https://doi.org/10.1086/158965
168. S. Courteau, J.A. Willick, ASP Conf. Ser. **201** (2000)
169. A.H. Guth, PhRvD **23**, 347 (1981). https://doi.org/10.1103/PhysRevD.23.347
170. R.B. Tully, M.J. Pierce, ApJ **533**, 744 (2000). https://doi.org/10.1086/308700
171. S. Sakai et al., ApJ **529**, 698 (2000). https://doi.org/10.1086/308305
172. G. Efstathiou, S. Gratton, QJAp **4**, 8 (2021). https://doi.org/10.21105/astro.1910.00483
173. Planck Collaboration et al., A&A **641**, A6 (2020). https://doi.org/10.1051/0004-6361/201833910
174. Planck Collaboration et al., A&A **571**, A16 (2014). https://doi.org/10.1051/0004-6361/201321591
175. E. Aubourg et al., PhRvD **92**, 123516 (2015). https://doi.org/10.1103/PhysRevD.92.123516
176. E. Macaulay et al., MNRAS **486**, 2184 (2019). https://doi.org/10.1093/mnras/stz978
177. P. Lemos et al., MNRAS **483**, 4803 (2019). https://doi.org/10.1093/mnras/sty3082
178. G. d'Amico et al., JCAP **5**, 005 (2020). https://doi.org/10.1088/1475-7516/2020/05/005
179. A.G. Riess et al., ApJ **855**, 136 (2018). https://doi.org/10.3847/1538-4357/aaadb7
180. A.G. Riess et al., ApJ **934**, L7 (2022). https://doi.org/10.3847/2041-8213/ac5c5b
181. A.G. Rizzi et al., ApJ **661**, 815 (2007). https://doi.org/10.1086/516566
182. W.L. Freedman et al., ApJ **891**, 57 (2020). https://doi.org/10.3847/1538-4357/ab7339
183. G.S. Anand et al., ApJ **932**, 15 (2022). https://doi.org/10.3847/1538-4357/ac68df
184. T. de Jaeger et al., MNRAS **496**, 3402 (2020). https://doi.org/10.1093/mnras/staa1801
185. J.P. Blakeslee et al., ApJ **911**, 65 (2021). https://doi.org/10.3847/1538-4357/abe86a
186. C.M. Springob et al., MNRAS **445**, 2677 (2014). https://doi.org/10.1093/mnras/stu1743
187. F. Qin et al., MNRAS **477**, 5150 (2018). https://doi.org/10.1093/mnras/sty928
188. C. Howlett et al., MNRAS **515**, 953 (2022). https://doi.org/10.1093/mnras/stac1681
189. M.P. Haynes et al., ApJ **861**, 49 (2018). https://doi.org/10.3847/1538-4357/aac956
190. H.M. Courtois, R.B. Tully, MNRAS **447**, 1531 (2015). https://doi.org/10.1093/mnras/stu2405

191. C.M. Springob et al., ApJS **172**, 599 (2007). https://doi.org/10.1086/519527
192. T. Hong et al., MNRAS **487**, 2061 (2019). https://doi.org/10.1093/mnras/stz1413
193. R.B. Tully et al., ApJ **676**, 184 (2008). https://doi.org/10.1086/527428
194. R.B. Tully et al., AJ **146**, 86 (2013). https://doi.org/10.1088/0004-6256/146/4/86
195. R.B. Tully, H.M. Courtois, J.G. Sorce, AJ **152**, 50 (2016). https://doi.org/10.3847/0004-6256/152/2/50
196. S.S. McGaugh et al., ApJ **533**, L99 (2000). https://doi.org/10.1086/312628
197. F. Lelli, S.S. McGaugh, J.M. Schombert, ApJ **816**, L14 (2016). https://doi.org/10.3847/2041-8205/816/1/L14
198. E. Kourkchi et al., MNRAS **511**, 6160 (2022). https://doi.org/10.1093/mnras/stac303
199. B.T. Hayden et al., ApJ **764**, 191 (2013). https://doi.org/10.1088/0004-637X/764/2/191
200. M. Childress et al., ApJ **770**, 108 (2013). https://doi.org/10.1088/0004-637X/770/2/108
201. M. Roman et al., A&A **615**, A68 (2018). https://doi.org/10.1051/0004-6361/201731425
202. R.B. Tully, AJ **149**, 171 (2015). https://doi.org/10.1088/0004-6256/149/5/171
203. E. Kourkchi, R.B. Tully, ApJ **843**, 16 (2017). https://doi.org/10.3847/1538-4357/aa76db
204. E. Tempel et al., A&A **602**, A100 (2017). https://doi.org/10.1051/0004-6361/201730499
205. R.B. Tully et al., ApJ **944**, 94 (2023). https://doi.org/10.3847/1538-4357/ac94d8
206. M.J. Reid, D.W. Pesce, A.G. Riess, ApJ **886**, L27 (2019). https://doi.org/10.3847/2041-8213/ab552d
207. D.W. Pesce et al., ApJ **891**, L1 (2020). https://doi.org/10.3847/2041-8213/ab75f0
208. G. Pietrzyński et al., ApJ **891**, L1 (2019). https://doi.org/10.1038/s41586-019-0999-4
209. S. Birrer et al. (2022). arXiv:2210.10833
210. B.P. Abbott et al., ApJ **909**, 218 (2021). https://doi.org/10.3847/1538-4357/abdcb7
211. A. Kozmanyan et al., A&A **621**, A34 (2019). https://doi.org/10.1051/0004-6361/201833879
212. M. Moresco et al., ApJ **868**, 84 (2018). https://doi.org/10.3847/1538-4357/aae829
213. S. Alam et al., PhRvD **103**, 083533 (2021). https://doi.org/10.1103/PhysRevD.103.083533
214. G. Efstathiou (2020). arXiv:2007.10716v2

Part II
Measurements (and Systematics)

Chapter 3
Trigonometric Parallax

Lennart Lindegren

3.1 Introduction

Parallax, or more precisely *stellar trigonometric parallax*, refers to the determination of distances to objects beyond the solar system through accurate angular measurements, using the orbit of the Earth, or of a satellite near the Earth, as baseline for the triangulation. The term parallax is also used for a variety of other methods to determine distances in astronomy, such as diurnal, secular, statistical, dynamical, photometric, and spectroscopic parallax; these are not discussed here.

With the adoption of heliocentrism in the 17th century, the anticipated detection of stellar parallax was seen as the ultimate proof of the Copernican hypothesis. Over the following centuries, attempts to measure the effect stimulated instrument developments and lead to important discoveries [1, 2]. Stellar parallax was finally detected in the 1830s by Henderson, W. Struve, and, most convincingly, Bessel (Fig. 3.1) [3]. In the subsequent 50 years the number of stars with a measured parallax grew very slowly owing to the exceedingly tedious and difficult visual observation process (Table 3.1). The use of photography and long-focus refractors from the late 19th century gave a burst in numbers but substantial improvements in accuracy only came with further technological advances in plate measuring engines, telescopes, and detectors [4]. A new era began with the launch of *Hipparcos* in 1989 and the *Hubble Space Telescope* in 1990, and a quantum leap was taken when *Gaia* was launched in 2013. In the radio domain, very long baseline interferometry (VLBI) has seen a parallel development and is now routinely used to measure the parallaxes for Galactic maser sources with an accuracy approaching 1 μas [5].

Throughout the long history of parallax measurements, random and systematic errors have been of primary concern. What are their causes, how can they be reduced

L. Lindegren (✉)
Lund Observatory, Department of Physics, Lund University, 22100 Lund, Sweden
e-mail: lennart.lindegren@fysik.lu.se

E. Di Valentino and D. Brout (eds.), *The Hubble Constant Tension*, Springer Series in Astrophysics and Cosmology, https://doi.org/10.1007/978-981-99-0177-7_3

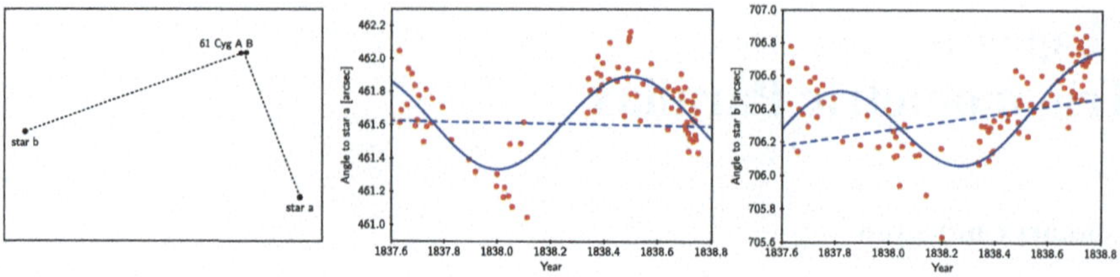

Fig. 3.1 Bessel's determination of the parallax of 61 Cyg [6]. The geometry of the double star 61 Cyg AB and the background stars 'a' (BD+37°4173) and 'b' (BD+37°4179) around 1838 is shown at left; the other panels show the individual measurements of the angular separations, together with Bessel's fit of parallax and proper motion. Bessel's result for the parallax, 316 ± 20 mas, differs by only 1.5 standard deviation from the modern value 286 mas

Table 3.1 Examples of historical and current parallax programs and catalogs. The third column is the median formal uncertainty in mas as given in the catalog, using the conversion $\sigma_\varpi = 1.5 \times$ p.e. when probable errors are given

Year	No. of stars	σ_ϖ	Reference, remark
1838	1	20	[6] 61 Cyg observed by Bessel
1889	49	30	[7] compilation by Oudemans
1924	1682	13	[8] General catalogue of stellar parallaxes
1935	3933	13	[9] General catalogue of stellar parallaxes (2nd ed.)
1952	5823	13	[10] General catalogue of trigonometric stellar parallaxes (3rd ed.)
1995	8112	10	[11] General catalogue of trigonometric stellar parallaxes (4th ed.)
1997	117955	1.1	[12] The *Hipparcos* Catalogue
2005	298	0.5	[13] USNO CCD parallax program (progress report)
2007	117955	0.9	[14] *Hipparcos*, the New Reduction
2017	105	0.2	[15] HST/FGS astrometry (review)
2018	7	0.045	[16] HST/WFC3 spatial scanning
2021	1468×10^6	0.34	[17] *Gaia* (E)DR3, all stars ($G \lesssim 21$)
	156×10^6	0.043	d:o, $G \leq 17$
	7.3×10^6	0.016	d:o, $G \leq 13$

by improved instruments and procedures, and how should they be dealt with in astrophysical applications of stellar parallax? As shown in the rest of this chapter, these questions remain highly relevant for the best modern determinations.

3.2 Astrometric Model

The reduction of astrometric observations at the microarcsec level requires that the relevant models of the object, observer, and light propagation from the object to observer are consistently formulated in a relativistic framework. To model light propagation from distant celestial objects, as well as the motion of bodies within the solar system, the Barycentric Celestial Reference System (BCRS) is the recommended framework [18]. Within the BCRS, spatial coordinates are represented as 3-vectors, here denoted $\boldsymbol{b}$, with origin at the solar system barycentre. The coordinate axes are oriented according to the standard (ICRS) astronomical reference system defining the celestial coordinates α and δ. The temporal coordinate t is the Barycentric Coordinate Time, TCB. The units of measurement are the SI meter and SI second.

In the BCRS system let $\boldsymbol{b}_{\rm s}(t)$ be the motion of an object (the 'source'), and $\boldsymbol{b}_{\rm o}(t)$ the (known) motion of the observer, for example *Gaia*. The coordinate direction towards the source at the time of observation, $t_{\rm obs}$, is given by the unit vector

$$\bar{\boldsymbol{u}}(t_{\rm obs}) = \frac{\boldsymbol{b}_{\rm s}(t) - \boldsymbol{b}_{\rm o}(t_{\rm obs})}{|\boldsymbol{b}_{\rm s}(t) - \boldsymbol{b}_{\rm o}(t_{\rm obs})|} \,. \tag{3.1}$$

Ideally, the time t should be the instant when the light is emitted from the source. For sources well beyond the solar system it is however customary to ignore most of the light travel time (which may be many years) and only consider the variable part caused by the motion of the observer around the solar system barycentre. That is also the convention adopted here, achieved by using the 'barycentric date' $t = t_{\rm obs} - \bar{\boldsymbol{u}}(t_{\rm obs})'\boldsymbol{b}_{\rm o}(t_{\rm obs})c^{-1}$ as the time argument in $\boldsymbol{b}_{\rm s}(t)$, where c is the speed of light. (The prime $'$ denotes scalar product.)

The coordinate direction $\bar{\boldsymbol{u}}(t_{\rm obs})$ varies according to the motions of the source and observer, and the astrometric observations ultimately aim to reconstruct $\boldsymbol{b}_{\rm s}(t)$ from a series of such measurements.[1] For non-single sources the function $\boldsymbol{b}_{\rm s}(t)$ can obviously be very complex. It turns out, however, that a linear model assuming constant space velocity $\boldsymbol{v}_{\rm s}$,

$$\boldsymbol{b}_{\rm s}(t) = \boldsymbol{b}_{\rm s}(t_{\rm ep}) + (t - t_{\rm ep})\boldsymbol{v}_{\rm s} \,, \tag{3.2}$$

is adequate for the astrometric reduction of the vast majority of sources, at least over an interval of several years, even though many of them are in reality binaries with more complex motions [20]. (When needed, non-linear motions can of course be

[1] The direction towards the source as measured in the instrument frame, that is relative to a set of orthogonal axes fixed in the astrometric instrument, differs from the coordinate direction because of gravitational light deflection in the solar system, stellar aberration, and the celestial orientation (attitude) of the instrument frame. The first two effects can be accurately corrected by means of the known motion of the observer and the known solar system ephemerides [19], and can be ignored for the present discussion. The attitude represents a rigid-body rotation that needs to be determined from observations; see Sect. 3.5.2.

included in the process of solving for the astrometric parameters of the source.) Here $t_{\rm ep}$ is a reference epoch, preferably chosen not far from the centre of the observed interval. Equations (3.1) and (3.2), together with the corresponding approximations, is known as the standard model of stellar motion (see Appendix A of [21]).

Following [19] we define the parallax of the source as[2]

$$\varpi = \frac{A}{|\boldsymbol{b}_{\rm s}(t_{\rm ep})|}\,, \tag{3.3}$$

where A is the astronomical unit. The parallax is the preferred way to parametrize the distance to an extrasolar object in astrometry.[3] The reason for this choice is that, in the absence of observational biases, the random errors in parallax from a series of astrometric measurements tend to be symmetric around the true value. This permits a simple interpretation of the *observed* parallax in terms of the likelihood function (Sect. 3.6). As defined in (3.3) the parallax is dimensionless, but it may be more conveniently expressed as an angle, for example in mas through multiplication by $648\,000\,000/\pi$.

The kinematic model in (3.2) has six degrees of freedom. In addition to ϖ we therefore need five more parameters to describe the stellar motion according to the standard model. Two of them are the celestial coordinates α, δ of the barycentric unit vector $\boldsymbol{r} = \boldsymbol{b}_{\rm s}|\boldsymbol{b}_{\rm s}|^{-1}$ at $t_{\rm ep}$; two more are the components $\mu_{\alpha *}$, μ_δ of the proper motion vector $\boldsymbol{\mu} = \boldsymbol{r} \times (\boldsymbol{v}_{\rm s} \times \boldsymbol{r})|\boldsymbol{b}_{\rm s}|^{-1}$ in the directions of increasing right ascension and declination, respectively; and the last one is the radial velocity $v_r = \boldsymbol{r}'\boldsymbol{v}_{\rm s}$. All six parameters refer to the adopted reference epoch $t_{\rm ep}$, although we do not explicitly write $\varpi(t_{\rm ep})$, etc. Formulas for changing $t_{\rm ep}$ are given in Appendix A of [21].

To first order in the small quantities ϖ and $(t - t_{\rm ep})\boldsymbol{\mu}$, Eqs. (3.1)–(3.3) give

$$\bar{\boldsymbol{u}}(t_{\rm obs}) = \frac{\boldsymbol{r} + \big((t - t_{\rm ep})\boldsymbol{v}_{\rm s} - \boldsymbol{b}_{\rm o}(t_{\rm obs})\big)\,\varpi/A}{\big|\boldsymbol{r} + \big((t - t_{\rm ep})\boldsymbol{v}_{\rm s} - \boldsymbol{b}_{\rm o}(t_{\rm obs})\big)\,\varpi/A\big|} \tag{3.4}$$

$$= \boldsymbol{r} + (t - t_{\rm ep})\boldsymbol{\mu} + \varpi\,\boldsymbol{f}(t_{\rm obs}) + O(\varpi^2)\,, \tag{3.5}$$

where $\boldsymbol{\mu}$ is the previously defined proper motion vector and

$$\boldsymbol{f}(t_{\rm obs}) = \boldsymbol{r} \times (\boldsymbol{r} \times \boldsymbol{b}_{\rm o}(t_{\rm obs})/A) \tag{3.6}$$

is the *parallax factor* expressed as a vector perpendicular to $\boldsymbol{r}$. It is seen that $\bar{\boldsymbol{u}}(t)$ to first order does not depend on the radial velocity. The second-order terms, known as perspective acceleration, are important for some nearby high-velocity stars [23], and for these one must include v_r in the astrometric modelling. For most Galactic

[2] The symbol ϖ is a variant of the Greek letter 'pi'. It is used for the parallax to avoid confusion with the mathematical constant π that sometimes appears in the same formulae.

[3] This definition breaks down at cosmological distances where the BCRS is no longer applicable. See for example [22] for a discussion of cosmological parallax.

stars, however, the astrometric observations are adequately modelled by just the five astrometric parameters α, δ, ϖ, $\mu_{\alpha*}$, and μ_δ.

Although the vector form of Eq. (3.4) is convenient for numerical computation, a variant using standard coordinates is often preferred. Standard coordinates (ξ, η) are obtained by gnomonic projection of the various vectors onto a tangent plane to the unit celestial sphere (e.g. [24]). With (α_c, δ_c) denoting the adopted tangent point (normally chosen close to $\boldsymbol{r}$) and

$$\boldsymbol{r}_c = \begin{bmatrix} \cos\delta_c \cos\alpha_c \\ \cos\delta_c \sin\alpha_c \\ \sin\delta_c \end{bmatrix} \quad \boldsymbol{p}_c = \begin{bmatrix} -\sin\alpha_c \\ \cos\alpha_c \\ 0 \end{bmatrix} \quad \boldsymbol{q}_c = \begin{bmatrix} -\sin\delta_c \cos\alpha_c \\ -\sin\delta_c \sin\alpha_c \\ \cos\delta_c \end{bmatrix}, \tag{3.7}$$

we have for the arbitrary direction $\boldsymbol{u}$ the standard coordinates

$$\xi = \frac{\boldsymbol{p}_c'\boldsymbol{u}}{\boldsymbol{r}_c'\boldsymbol{u}}, \quad \eta = \frac{\boldsymbol{q}_c'\boldsymbol{u}}{\boldsymbol{r}_c'\boldsymbol{u}}, \quad \text{and conversely} \quad \boldsymbol{u} = \frac{\boldsymbol{r}_c + \boldsymbol{p}_c\xi + \boldsymbol{q}_c\eta}{(1+\xi^2+\eta^2)^{1/2}}. \tag{3.8}$$

From Eq. (3.4) we find the standard coordinates of $\bar{\boldsymbol{u}}(t_{\rm obs})$,

$$\left.\begin{aligned} \bar\xi(t_{\rm obs}) &= \xi_0 + (t - t_{\rm ep})\mu_\xi + f_\xi(t_{\rm obs})\varpi \\ \bar\eta(t_{\rm obs}) &= \eta_0 + (t - t_{\rm ep})\mu_\eta + f_\eta(t_{\rm obs})\varpi \end{aligned}\right\}, \tag{3.9}$$

where (ξ_0, η_0) are the standard coordinates of $\boldsymbol{r}$, $\mu_\xi = \boldsymbol{p}_c'\boldsymbol{\mu}/\boldsymbol{r}_c'\boldsymbol{r}$, $f_\xi = \boldsymbol{p}_c'\boldsymbol{f}/\boldsymbol{r}_c'\boldsymbol{r}$, etc. Provided that the tangent point is close to $\boldsymbol{r}$, it can be seen that μ_ξ and f_ξ are approximately, but not exactly, equal to $\mu_{\alpha*}$ and $f_{\alpha*}$ (the parallax factor in right ascension), and similarly for the components in η and δ; strict equality holds only when $\boldsymbol{r}_c = \boldsymbol{r}$, in which case we also have $\xi_0 = \eta_0 = 0$. From a set of observed standard coordinates $(\bar\xi, \bar\eta)$, corrected for aberration etc., we can now obtain the linearised astrometric parameters ξ_0, η_0, μ_ξ, μ_η, and ϖ by least-squares solution of the aggregated equations (3.9). Often the equations in ξ and η are treated separately, yielding two independent estimates of the parallax, ϖ_ξ and ϖ_η. Ideally they should be the same, which can be used as a consistency check. Finally the linearised parameters ξ_0, η_0, μ_ξ, μ_η are transformed back to the normal ones α, δ, ϖ, $\mu_{\alpha*}$ using Eq. (3.8).

Equation (3.5) is sometimes written in linearised spherical coordinates, e.g. as

$$\left.\begin{aligned} \bar\alpha(t_{\rm obs}) &= \alpha + (t - t_{\rm ep})\mu_\alpha + f_\alpha(t_{\rm obs})\varpi \\ \bar\delta(t_{\rm obs}) &= \delta + (t - t_{\rm ep})\mu_\delta + f_\delta(t_{\rm obs})\varpi \end{aligned}\right\}, \tag{3.10}$$

where $\bar\alpha, \bar\delta$ are the celestial coordinates of $\bar{\boldsymbol{u}}$, f_α, f_δ the corresponding parallax factors, and $\mu_\alpha = \mu_{\alpha*}/\cos\delta$. This kind of expansion should be avoided as it introduces additional approximations proportional to $\tan\delta$, which could become catastrophic near the poles. This is avoided through the use of standard coordinates as described above, or by using the vector form in Eq. (3.4).

3.3 Differential and Global Techniques

Stellar parallax is often described as the apparent shift, caused by the Earth's orbital motion, of a nearby target star against distant background sources. This actually describes the measurement of *relative* parallax, that is of the difference between the parallax of the target star and the (mean) parallax of the background sources. Measured differential parallaxes usually need to be corrected to 'absolute' by adding an estimate of the mean parallax of the background stars (Sect. 3.4.3).

Bessel's observations from 1837–38 (Fig. 3.1) may illustrate the determination of differential and absolute parallaxes. Bessel obtained the value 316.2 ± 20.2 mas for the differential parallax if 61 Cyg with respect to the background stars 'a' and 'b'. He was fortunate in that both stars turn out to be distant red giants with parallaxes (2.2 and 2.0 mas, respectively) that are almost negligible compared with the differential parallax and its uncertainty [3]. Strictly speaking, however, his measurements thus gave an absolute parallax of about 318 mas, compared to the modern value 286.0 mas (although in 1838 the parallax was 285.0 mas, owing to the radial velocity of -65 km s^{-1}).

From Eq. (3.6) we see that the parallax factor depends only on the barycentric position of the observer at the time of observation, $\boldsymbol{b}_{\mathrm{o}}(t_{\mathrm{obs}})$, and the position of the source, $\boldsymbol{r}$. Therefore, if two sources are observed simultaneously from the same observatory within a small part of the sky, they will have practically the same parallax factor $\boldsymbol{f}$ and the difference in coordinate direction becomes

$$\bar{\boldsymbol{u}}_2(t_{\mathrm{obs}}) - \bar{\boldsymbol{u}}_1(t_{\mathrm{obs}}) = \boldsymbol{r}_2 - \boldsymbol{r}_1 + (t - t_{\mathrm{ep}})(\boldsymbol{\mu}_2 - \boldsymbol{\mu}_1) + (\varpi_2 - \varpi_1)\,\boldsymbol{f}(t_{\mathrm{obs}}) \quad (3.11)$$

From a series of measurements of $\bar{\boldsymbol{u}}_2 - \bar{\boldsymbol{u}}_1$ spread over a few years it is clearly possible to determine the differential astrometric parameters $\boldsymbol{r}_2 - \boldsymbol{r}_1$, $\boldsymbol{\mu}_2 - \boldsymbol{\mu}_1$, and $\varpi_2 - \varpi_1$, but not the individual (absolute) values. In particular, it is not possible to determine ϖ_1 and ϖ_2 separately. Parallax measurements confined to a small field ($\ll 1$ rad) therefore give differential parallaxes of the measured sources, that is up to some unknown common offset.

In a classical parallax determination, as practised from Bessel's days up to the present time with HST and large ground-based telescopes, a target source is measured with respect to a background of several reference sources, presumed to be much more distant than the target. In a second step, the measured differential parallax of the target source is corrected from relative to absolute by adding the mean parallax of the reference stars. To avoid circularity, the reference parallaxes must be estimated by some independent means. As discussed in [25] the main methods available are based on statistical, spectroscopic, and photometric distance estimates. Statistical methods attempt to determine the mean parallax of stars as a function of (primarily) magnitude and Galactic coordinates, based on proper motion surveys, star counts, and Galactic models. Although historically important, the statistical methods may be too uncertain for modern parallax determinations where it has been superseded by spectroscopic and photometric methods (e.g., [26, 27]). The spectrophotometric

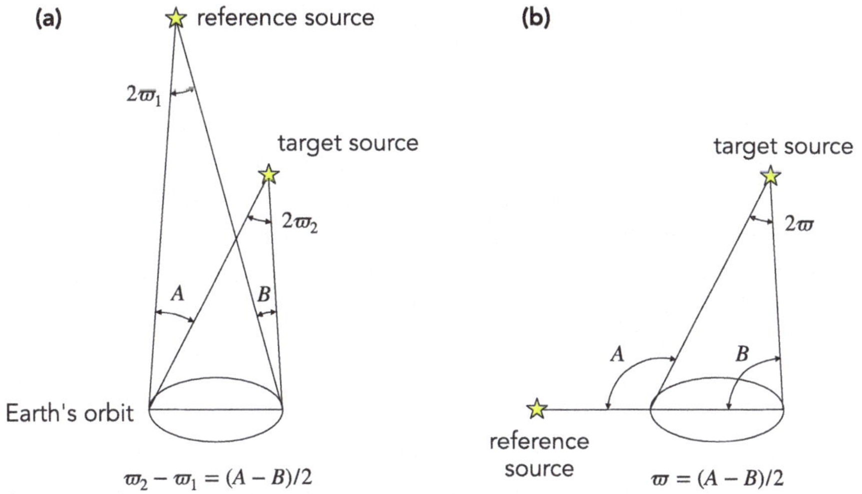

Fig. 3.2 Illustrating the principles of relative and absolute parallax determination. **a** Measuring the small angles A, B between a target source and a reference (background) source gives the relative parallax $\varpi_2 - \varpi_1$. **b** In this configuration the measurement of the large angles A, B gives the absolute parallax of the target source independent of the distance to the reference source

method is further discussed in Sect. 3.4.3 in connection with parallax measurements with the HST.

Extragalactic objects such as quasars are in principle ideal reference sources for parallax determination, thanks to their extremely small parallaxes (e.g. redshift $z \geq 0.01$ implies $\varpi \leq 0.025\,\mu$as). However, in the optical domain quasars are generally too faint and too scarce, especially in the Galactic zone, to be of much practical use. The situation is very different at radio wavelengths. Using phase-referencing VLBI [28, 29] the position difference between a target source and a radio-bright quasar used as reference can be accurately measured for angular separations up to a few degrees. Thus, absolute parallaxes for hundreds of Galactic and extragalactic sources have been measured with accuracies reaching below 10 μas [30].

As an alternative to the methods described above, the parallax degeneracy in Eq. (3.11) can be lifted by arranging the observations in such a way that the parallax factors f_1 and f_2 are significantly different in a given observation. As seen from Eq. (3.6), this requires that $\boldsymbol{r}_1$ and $\boldsymbol{r}_2$ are significantly different, in other words that the sources are separated by a large angle ($\gtrsim$ 1 rad). This principle, schematically illustrated in Fig. 3.2, was adopted for the European Space Agency missions *Hipparcos* and *Gaia*. Because optical telescopes cannot have a well-corrected field of view substantially bigger than about 1° diameter, these missions employ two telescopes or viewing directions, separated by a basic angle of 58° (*Hipparcos*) or 106.5° (*Gaia*). A major challenge of the method is to determine the precise value of the basic angle, or rather to keep it stable for a sufficiently long time that it can be accurately calibrated (Sect. 3.5.2).

3.4 Parallax Measurements Using the *Hubble Space Telescope*

The main instruments used for astrometry on-board the *Hubble Space Telescope* (HST) are the Fine Guidance Sensors (FGS), the Wide Field and Planetary Camera 2 (WFPC2, 1993–2009), the Advanced Camera for Surveys with the High-Resolution Channel (ACS/HRC, 2002–2006) and the Wide-Field Channel (ACS/WFC, since 2002), and the Wide Field Camera 3 (WFC3, since 2009). Here I focus on the determination of stellar parallaxes by means of the FGS and the spatial scanning technique using the WFC3.

3.4.1 Use of the Fine Guidance Sensors Instrument

The three Fine Guidance Sensors (FGS) of the *Hubble Space Telescope* are part of the pointing control system used for guiding the telescope, but they are also a versatile science instrument for precise relative position measurements and interferometric studies e.g. of close binaries. Each FGS comprises a two-axis, white-light shearing interferometer operating in its own 'pickle'-shaped field of view (FOV) of $\simeq 4 \times 17$ arcmin^2 located in the perimeter of the HST focal plane (Fig. 3.3a). An aperture of 5×5 arcsec^2 (the instantaneous field of view, IFOV) can be moved to any point within the FOV by means of a star selector assembly. The wavefront tilt of the light transmitted by the IFOV is interferometrically sensed in two orthogonal directions by means of Koesters prisms. For an unresolved object the resulting interferogram (know as the transfer function) is S-shaped, with an approximately linear central segment of

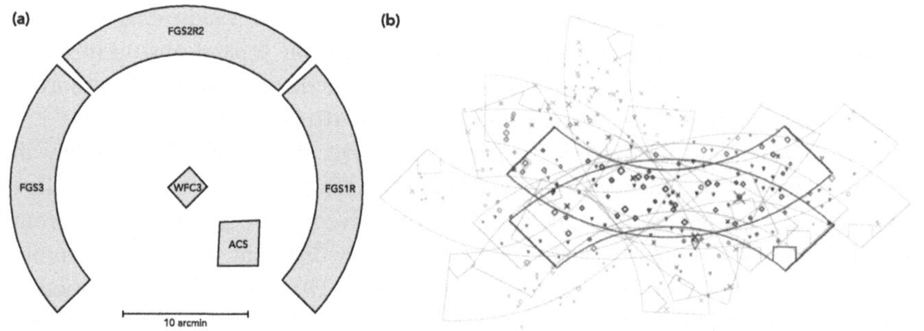

Fig. 3.3 **a** Layout of the HST focal plane since 2009, with the Wide Field Camera 3 (WFC3), Advanced Camera for Surveys (ACS), and the three Fine Guidance Sensors (FGS) as projected on the sky. FGS1R is the pickle used for astrometry. (Drawing based on [31]) **b** Calibration of the optical field angle distortion of the FGS1R pickle was achieved by measuring stars in the M35 calibration field with various pointing offsets and roll angles. (Reproduced from [31])

± 10 mas around the zero-crossing point obtained when the star is accurately centered in the IFOV.

Two of the three FGS are needed to fix the pointing and roll of the telescope, which is achieved by locking their IFOV on guide stars with known positions. The third FGS is then available for scientific observations of sources within its FOV. Initially FGS3 was the designated 'astrometer', but after the refurbishing in 1997 FGS1R is preferred. As a science instrument the FGS can observe stars in the magnitude range $V \simeq 3$ to 17.5. A detailed technical description of the FGS is in [32].

The FGS can be operated in several modes. In position mode (POS) the central zero-crossing of the fringes is tracked to sample the position of the IFOV, in FOV coordinates, at a rate of 40 Hz. By sequentially observing the selected objects in the FOV (typically the target object and several reference stars), while the pointing and roll of the telescope is held fixed, the positions of the objects in FOV coordinates are obtained after averaging and correcting for the optical field angle distortion (OFAD) and other calibrated effects (see below), and for differential stellar aberration. The POS mode is normally used for parallax work, except if the target star is a binary for which the resolved structure or orbital motion must be taken into account. In transfer mode (TRANS) the fringes are sampled over a wider range of IFOV positions, from which images of binaries and other non-point sources can be reconstructed with a resolution of about 20 mas for FGS3, and down to about 10 mas for FGS1R [33].

The OFAD includes the design distortions of the Ritchey–Chrétien telescope, which have an amplitude of several arcsec, as well as smaller contributions from alignment and figuring errors of the telescope and FGS optics and encoder errors of the star selectors. There are also small but significant temporal changes, especially to the scale and low-order distortion. For astrometric purposes the OFAD needs to be known to within a few mas. Calibration and monitoring of the long-term stability is achieved by making positional observations of stars in a calibration field centered on the open cluster M35. Combining the data collected in a short time from several different telescope pointings and roll angles (Fig. 3.3b), some 30 OFAD parameters are obtained in a robust least-squares solution simultaneously with the attitude angles of the different pointings and the celestial positions of the calibration stars [34]. For FGS1R the resulting OFAD uncertainty is about 1.5 mas in each coordinate [35], and below 1 mas in the center of the FOV [15]. Additional calibrations are needed to correct for cross-filter shifts (if a neutral filter is used, for $V \lesssim 8$) and lateral color effects.

A single POS measurement normally lasts for 1–2 min, during which the position of the object in FOV coordinates is sampled several thousand times. The adopted position is the median of the samples, which efficiently reduces photon noise and HST jitter. For parallax determination some 5–10 reference stars are used. Together with the science target they are observed multiple times during a visit (lasting up to an orbit of 90 min), which allows for intra-orbit drift correction. The complete data set typically comprises about 10 such visits spread over 1.5–2 years, most of them scheduled around maximum (positive and negative) parallax factor. The astrometric model used to analyse the data is outlined in [15]. Briefly, the measured FOV coordinates (x', y'), corrected as mentioned above, are fitted by a linear model based on Eq. (3.9),

$$\left.\begin{aligned} \xi_0 + \mu_\xi (t - t_{\text{ep}}) + f_\xi \varpi_\xi &= Ax' + By' + C \\ \eta_0 + \mu_\eta (t - t_{\text{ep}}) + f_\eta \varpi_\eta &= Dx' + Ey' + F \end{aligned}\right\} . \tag{3.12}$$

Here A through F are 'plate constants' describing the offset, scale, rotation, and shear of the FOV frame in a given visit; these are common for all the measurements in the visit, while the astrometric parameters in the left members are common for all the measurements of a given source. In view of the differential nature of the measurements (Sect. 3.3), prior information on the astrometric parameters of the reference stars are added, including uncertainties, to obtain a unique solution; in particular, spectrophotometric estimates of the reference star parallaxes (Sect. 3.4.3) are used to obtain the absolute parallax of the target object.

Astrometric results obtained with the FGS and published up to 2016 are reviewed in [15]. They include absolute parallaxes for 17 standard candles (ten Galactic Cepheids, five RR Lyr, and two dwarf Cepheids), with a median uncertainty of 0.18 mas (6.5% median relative uncertainty), and 88 absolute parallaxes for various other kinds of objects (cluster stars, white dwarfs, cataclysmic variables, M dwarf binaries, metal-poor stars, etc.). Few astrometric programs using the FGS have been carried out in the last decade following the application of the spatial scanning technique (Sect. 3.4.2) and the launch of *Gaia* (Sect. 3.5).

3.4.2 Use of the Spatial Scanning Technique on WFC3

Since its installation in 2009 the Wide Field Camera 3 (WFC3) has been the HST's primary science instrument thanks to its high-resolution imaging over a wide wavelength range [36]. The ultraviolet/visible channel (UVIS, 200–1000 nm) comprises two 2k$\times$4k CCDs with a pixel size of 40×40 mas^2; the infrared channel (IR, 800–1700 nm) features a 1k$\times$1k HgCdTe array with a pixel size of 135×121 mas^2. Although both channels have been used for astrometry, UVIS is obviously the preferred choice, except for very cool objects [37], thanks to its higher optical resolution and smaller pixels. For high-precision astrometry and photometry one must calibrate very carefully not only the geometric distortions (including telescope and camera optics, filters, and CCD irregularities), but also the effective PSF [38, 39], which is especially critical for the analysis of undersampled images [40].

The geometric distortion of UVIS is determined by means of a self-calibration procedure similar to the one illustrated in Fig. 3.3 (right), using multiple displaced and rotated images of a rich stellar field, in this case the core of ω Cen. The residual geometric distortion is about 0.01 pixel, or 0.4 mas [41], which sets the accuracy floor for the positions of well-exposed images.[4] Conventional pointed imaging with WFC3 (and its predecessor WFCP2, and ACS) has proven immensely useful in many

[4] Assuming that the geometric distortion is constant during the calibration, and given a sufficient density of sources, this procedure can determine all kinds of distortions up to a single scale factor, which can only be fixed by means of external information.

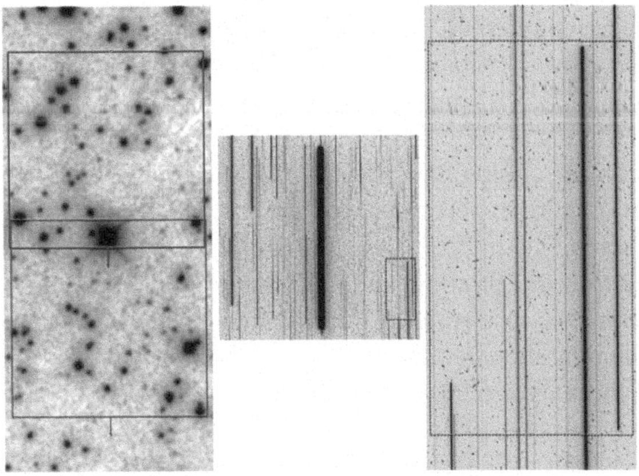

Fig. 3.4 Spatial scanning observation with WFC3. *Left:* Digital Sky Survey image of a $\simeq 3 \times 6$ arcmin area centered on the Cepheid SY Aur. The colored frames show the position of the UVIS field of view at the beginning and end of the spatial scan. *Middle:* The resulting UVIS scan image. *Right:* Part of the scan image magnified by a factor seven. (Reproduced from [42] by permission of the authors.)

investigations that benefit from the excellent cluster/field star separation that can be achieved with proper motions over just a few years; this includes population studies, the internal dynamics of clusters, and kinematic parallaxes. However, except for the IR observations already mentioned, pointed imaging with the WFC3 has not been much used for parallax determination, probably because the method would be limited to nearby, faint stars that are already well explored with ground-based techniques that can reach a similar precision (cf. Table 3.1).

Spatial scanning is an alternative observing strategy for WFC3/UVIS that overcomes many of the limitations of pointed observations for high-precision astrometric and photometric observations of bright stars [16, 42–44]. Originally motivated by the wish to obtain time-resolved, high signal-to-noise photometric observations of exo-planet transits [36], the spatial scanning spreads out the light of each star along a 'trail' on the CCD (Fig. 3.4). This is achieved by slewing the telescope in a user-defined manner, in the simplest case a fixed direction with uniform speed, as in Fig. 3.4. The maximum slew rate is just under 5 arcsec s^{-1} under FGS control, or 7.8 arcsec s^{-1} if gyros are used, which gives a smoother but less accurate motion.

For astrometry, the main advantages of spatial scanning are: (1) that many independent position measurements are obtained normal to the trail, which reduces random and (some) systematic noise by a large factor; (2) that much brighter sources can be observed without saturating the pixels; and (3) that the time-resolved measurements can be corrected for telescope jitter that would just blur the images in pointed mode. A simple calculation illustrates the potential gain in precision: At visual wavelengths the optical resolution of the HST is about $(600 \text{ nm})/(2.4 \text{ m}) \simeq 50$ mas or 1.25 pixel, so a trail of length 2000 contains $N \sim 2000/1.25 = 1600$ independent resolution elements. By averaging over all the measurements normal to the trail, the resulting positional uncertainty could be as small as (0.4 mas)/$\sqrt{N}$ = 10 μas per scan. In practice the precision obtained for well-sampled bright stars is in the range 20 to 80 μas per scan [16]. To minimize effects of charge-transfer inefficiency in the (across-trail) measurement direction, the trails are typically oriented approximately along the parallel readout direction; they should however make a small angle to the readout direction in order to average out sub-pixel phase errors. The analysis of the across-trail measurements includes corrections for the time-dependent differential aberration, frame-to-frame rotation, and roll-angle variations along the scan, as well as the known geometric distortion and other effects calibrated from pointed mode observations, including the effective PSF.

The main disadvantages of spatial scanning are (1) that precise positional measurements are only possible in one direction, that is normal to the trail; and (2) that some trails will overlap which may be a problem in crowded areas. The first disadvantage is minimal for parallax determination, if the measurement direction is chosen roughly along the major axis of the parallax ellipse (that is, in ecliptic longitude). To minimize overlap the length and precise direction of the scans need to be chosen with care. Moreover, to bridge the dynamic range gap between a bright target source and the faint fields stars used as reference objects, it may be necessary to combine measurements from both shallow and deep exposures taken at the same epoch, by a suitable choice of filters and scan speeds. For additional considerations and details on the technique and data analysis, see [16, 42, 43].

Parallaxes obtained with this technique have yielded uncertainties in the range 29 to 74 μas for bright Cepheids ($V \lesssim 10$) [16]; this includes the systematic uncertainty of the correction from relative to absolute parallax, which in this case is only about 10–20% of the total uncertainty. The resulting relative uncertainty in distance is 7–17% for the individual Cepheids.

Spatial scanning was also applied to the determination of the parallax of the globular cluster NGC 6397 [44]. The mean parallax of 39 cluster stars ($V \simeq 13$–19) was obtained with a statistical uncertainty of 13 μas, to which should be added a systematic uncertainty of 18 μas for the correction from relative to absolute. The relative uncertainty of the trigonometric distance (~2.4 kpc) is therefore about 5%.

3.4.3 Reference Stars and Correction to Absolute

As discussed in Sect. 3.3, parallax measurements made within a small field are differential, and the conversion from relative to absolute parallaxes requires knowledge of the mean parallax of the reference sources. For the FGS and WFC3 observations described above the spectrophotometric method is used for the correction to absolute. This aims to estimate the absolute magnitudes (M_V, say) and interstellar extinctions (A_V) of the individual reference sources by means of spectra and/or multi-wavelength photometry, from which their spectrophotometric parallaxes are obtained as

$$\varpi_{\rm sp} = (100\,\text{mas}) \times 10^{-0.2\mu_0}\,, \qquad (3.13)$$

where $\mu_0 = V - M_V - A_V$ is the absorption-corrected distance modulus (which should be independent of the wavelength band used, e.g. V). The combination of spectral type (or $T_{\rm eff}$), luminosity class (or $\log g$), and colors makes it possible to estimate M_V and A_V. Spectrophotometrically, it is difficult to estimate μ_0 with a precision much better than about 0.3 mag. Differentiation of Eq. (3.13) gives

$$\left|\frac{\Delta\varpi_{\rm sp}}{\varpi_{\rm sp}}\right| = 0.2(\ln 10)\,|\Delta\mu_0| \simeq 0.46\,|\Delta\mu_0|\,, \tag{3.14}$$

so an error of 0.3 mag in distance modulus corresponds to a *relative* error in parallax (and distance) of $0.46 \times 0.3 \simeq 14\%$. Using reference sources that are expected to have small parallaxes, such as distant giants, thus makes the uncertainty of their parallaxes correspondingly smaller, and the correction to absolute more accurate.

The luminosity class is generally the most problematic parameter to estimate, where a dwarf/giant misclassification can easily result in an error of 5 mag in the distance modulus, or a factor 10 in the calculated spectrophotometric parallax. To minimize the risk of such error, various strategies have been used. A classical device to discriminate between luminosity classes is the reduced proper motion [45], utilizing the fact that a nearby dwarf tends to have higher proper motion than a distant giant of the same apparent magnitude and color. Furthermore, the handle on both extinction and luminosity class is improved by using multiband photometry that extends into the infrared (e.g. from 2MASS and WISE), and by fitting synthetic colors to such data. Finally, increasing the number of reference sources per parallax field improves not only the precision of the correction to absolute but also its robustness, where for example a misclassified dwarf stands out more clearly as an outlier.

Recent programs have adopted a comprehensive modelling approach in which the astrometric parameters of the target and reference sources are estimated, together with plate constants and other local unknowns, in a global multisource, multiparameter solution. The need to correct from relative to absolute parallax is in practice eliminated by introducing the spectrophotometric parallaxes of the reference sources as prior information in the solution [16, 44]. The approach is very flexible and allows many other kinds of prior information to be incorporated in the solution, for example on the proper motions of the reference sources from a kinematic Galaxy model, and extinction maps. To avoid compromising the statistical interpretation of the resulting parallax (Sect. 3.6), it is important that no prior is assumed on the parallax of the target source.

A useful feature of the global modelling approach is that purely astrometric estimates of the parallaxes of the reference sources can be obtained by treating each star, in turn, as an additional target source. This is achieved in a series of separate solutions, one for each reference source, in which its prior information is ignored. The full solution is then cross-validated by comparing the purely spectrophotometric parallaxes against the corresponding, independent astrometric values [16]. A simulation of the global solution for the bright Cepheid SS CMa, using 1000 random fields drawn from the Besançon Galaxy Model [46], indicates that the correction to absolute in that case would be determined with a typical precision of 9 μas if ~20 reference sources are used, each having an uncertainty of 15% in its spectrophotometric distance (~0.3 mag in distance modulus) [43].

Parallax determinations using small-field telescopes in space and on ground will continue to have important applications after *Gaia*, for example in the infrared, for very bright, faint, or complex sources, and where an even higher precision is needed. However, the availability of accurate *Gaia* astrometry for virtually all potentially useful reference sources is bound to change the way such observations are analyzed. The most obvious implication is for the correction from relative to absolute, which becomes very straightforward if the *Gaia* parallaxes can be adopted for the reference sources. But this is not necessarily the best strategy. As an illustrative example let us consider the 99 reference sources used for the parallaxes of bright Cepheids observed with the spatial scanning technique [16]. The median uncertainty of their spectrophotometric parallaxes is 30 μas. All of them have parallaxes in *Gaia* DR3, with a median formal uncertainty of 28 μas. However, most of the statistical weight in the correction to absolute comes from a small number of distant giants having very small uncertainties in their spectrohotometric parallaxes. A more relevant quantity could therefore be the uncertainty corresponding to the mean statistical weight of the reference stars, or $\langle\sigma_\varpi^{-2}\rangle^{-1/2}$. This is 14 μas for the spectrohotometric parallaxes of the same stars, and 27 μas for their *Gaia* DR3 parallaxes. Thus, even the final (DR5) *Gaia* parallaxes, which are expected to be almost a factor two better than in DR3 (Sect. 3.5.4), will not make the mean weight of these particular reference sources much higher than already achieved with the spectrophotometry. In a more indirect way, *Gaia* could still contribute a lot. It should allow many more reference sources to be used, with exquisite dwarf/giant separation and screening against problematic sources, and perhaps also brighter (more nearby) sources than is possible with the spectrophotometric method. Furthermore, the geometric calibration of the images is likely to be improved through the incorporation of accurate positions and proper motions from *Gaia*, from which the target source will also benefit.

3.5 Parallax Measurements with *Gaia*

The *Gaia* satellite was launched on 19 December 2013 and arrived three weeks later to its destination near the Sun–Earth L_2 Lagrangian point, 1.5 million km from the Earth. Science operations began on 25 July 2014 and are currently (May 2023) expected to end in 2025. Three releases of science data have been made to date. The third data release was made in two stages: in December 2020 the early release EDR3 of mainly astrometric and photometric data was made, and in June 2022 the full DR3. Two more releases (DR4 and DR5) are expected in the coming years. The nominal precision of parallaxes in *Gaia* (E)DR3 is typically around 20 μas for bright stars ($G \simeq 6$ to 14), 70 μas at $G \simeq 17$, and 500 μas at $G \simeq 20$ (Fig. 3.5). See [47] for an overview of the mission, [17, 48] for summaries of the contents of EDR3 and DR3, and [49] for some of the early science results.

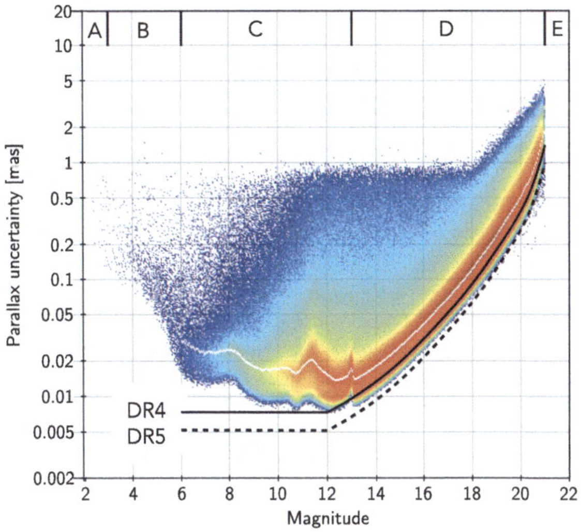

Fig. 3.5 Formal parallax uncertainty in *Gaia* (E) DR3 versus G magnitude. Colors show the density of points on a logarithmic scale, with a random subset plotted for $G > 11.5$. The white curve is the median uncertainty. The black curves show extrapolated median formal errors in DR4 and DR5. The letters at the top show the ranges for (A) sources too bright for the on-board detection; (B) partially saturated images; (C) calibration-limited performance; (D) photon-noise limited performance; (E) too faint for reliable detection

3.5.1 Measurement Principle

Gaia is optimised for accurate angular measurements in the along-scan (AL) direction, while less strict requirements apply in the across-scan (AC) direction (Fig. 3.6). This inequality is reflected in many features of the instrument and its operation. The rectangular shape of the primary mirrors, $1.45 \times 0.5\,\mathrm{m}^2$ (AL×AC), results in a point-spread function (PSF) that has three times higher angular resolution in the AL direction than AC (Fig. 3.7a). The CCD pixels are correspondingly dimensioned, $59 \times 177\,\mathrm{mas}^2$. The rotation of the satellite makes the source images move over the CCDs at a very nearly constant AL rate of $\dot{\eta} = -60\,\mathrm{arcsec\,s}^{-1}$. This is accurately compensated by reading out the CCDs in drift-scanning mode at the same rate, thus preserving the full optical resolution AL. The much smaller AC scan rate $\dot{\zeta}$ varies sinusoidally along the scanning path in the range $\pm 0.173\,\mathrm{arcsec\,s}^{-1}$, thus widening the PSF in the AC direction by up to 1 pixel. The AC motion is not compensated but analytically modelled as part of the PSF calibration (Sect. 3.5.2). Only a small pixel window centred on each detected source is sampled on-board and transmitted to the ground. To further reduce the telemetry rate and read-out noise, the full two-dimensional window is only transmitted for bright ($G \lesssim 13\,\mathrm{mag}$) sources, while for fainter it is marginalized AC, creating a one-dimensional AL image.

The accurate time t_{obs} when the image centroid passes over a fiducial 'observation line', nominally at the centre of the active CCD area, constitutes the elementary

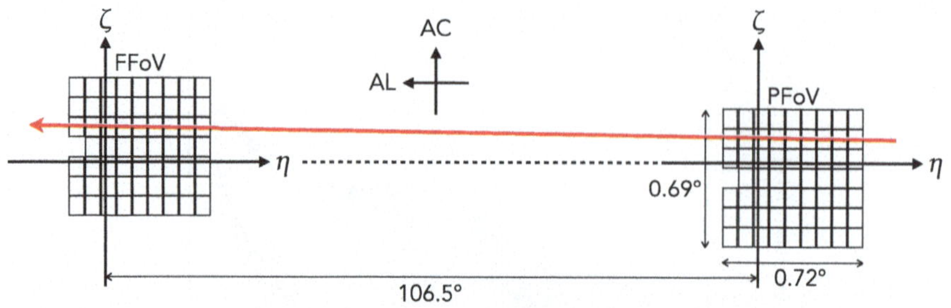

Fig. 3.6 Schematic picture of the 62 astrometric CCDs (small rectangles) projected through the *Gaia* telescopes onto the sky, creating two fields of view (preceding, PFoV, and following, FFoV) separated by the basic angle. Field angles (η, ζ) are defined separately in each field. The red curve shows the path of a source crossing first the PFoV and then the FFoV. Along- and across-scan directions (AL, AC) describe the apparent motions of source images on the CCDs, rather than the actual scanning motion of the instrument. The slow precession of the spin axis (Fig. 3.7b) generates a small positive or negative drift of the images in the AC direction. The picture shows the view from 'outside' the celestial sphere and is not drawn to scale

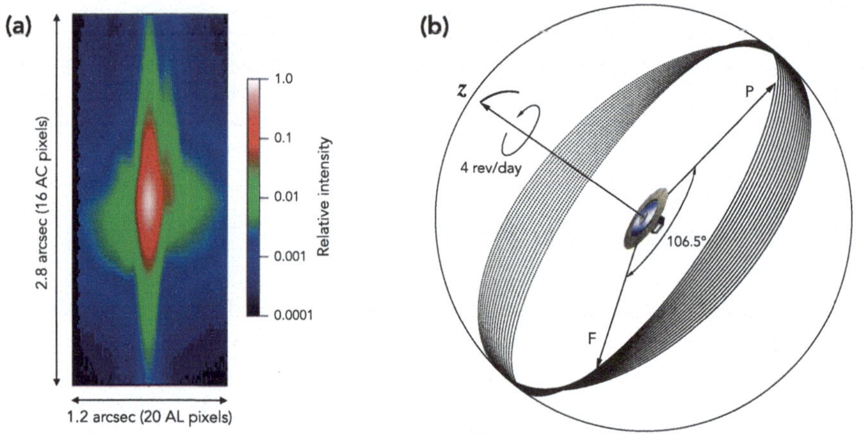

Fig. 3.7 **a** Example of the point spread function (PSF) of *Gaia*. The resolution is about 0.1 arcsec in the along-scan (AL) direction and 0.3 arcsec across-scan (AC). **b** *Gaia*'s scanning law over four days. During this time the spin axis (z) traces an arc of 16.6°, while each of the viewing directions (preceding P, following F) completes 16 revolutions. The z axis makes a constant angle of 45° to the Sun

astrometric measurement in the AL direction [50]. Up to nine such AL measurements are obtained during the 43 s it takes the image to cross the astrometric field of view (Fig. 3.6). For bright sources, less accurate measurements are also obtained in the AC direction at the same times; these are the centroid coordinates expressed in AC pixels (m) and fractions thereof. Under the most favourable conditions (which occur for $G \simeq 12$) the precision of a single measurement is about 60 μas AL and 300 μas AC. On average about 8000 AL and 40 AC observations are collected every second, but the rates vary considerably depending on the local density of sources on the sky. To date more than 2×10^{12} astrometric measurements have been collected. This does not include the skymapper (SM) observations, which are essential for the operation of the instrument but not directly used for the astrometry, or the photometric (BP, RP) and radial-velocity (RVS) observations using dedicated CCDs.

3.5.2 *Astrometric Processing*

A schematic overview of the processing chain for the main astrometric results in *Gaia* (E)DR3 is shown in Fig. 3.8. I refer to the published documentation, in particular [51] and references therein, for detailed descriptions of the various steps; here only certain aspects of relevance for the parallaxes are discussed. The separate chains for photometric and spectroscopic processing are not considered here, nor the special treatment of non-single sources.

The main processing chain is optimised for sources that can be treated as point objects (angular size $\lesssim 50$ mas; Fig. 3.7a) moving according to the standard astrometric model of Sect. 3.2. As indicated in the left part of Fig. 3.8, after an initial analysis of the down-linked data, where all non-spurious detections are matched to some particular source, the centroid positions AL ($t_{\rm obs}$) and AC (m) and other image parameters (including the flux for photometry) are estimated by fitting an image profile to the CCD samples. Normally this uses a two-dimensional PSF for bright sources ($G \lesssim 13$ mag) and a one-dimensional line-spread function (LSF) for fainter sources, where pixels are binned on-board in the AC direction. Prior to the image parameter determination the LSF and PSF were calibrated, using a subset of well-behaved observations, as functions of the field of view (P or F), position in the field, time, window class, gate, and color (effective wavenumber[5] $\nu_{\rm eff}$).

The LSF/PSF models are linear combinations of precomputed basis functions designed to represent the effective LSF or PSF for a range of colors and optical aberrations. The LSF/PSF calibration takes into account color-dependent image shifts ('chromaticity') caused by asymmetric optical aberrations and the wavelength-dependent diffraction. Chromaticity is therefore largely eliminated already in the image parameter determination stage by fitting the LSF or PSF of the appropriate color. The color information normally comes from the on-board photometric instrument (BP and RP), but for some (mainly faint) sources no reliable color was available at the time of the image parameter determination, in which case the LSF/PSF calibration for the default color $\nu_{\rm eff} = 1.43\ \mu{\rm m}^{-1}$ was used. In the subsequent astrometric processing such sources obtained a so-called six-parameter solution, in which a 'pseudocolor' was introduced as the sixth astrometric parameter (Sect. 2.3 in [51]). The pseudocolor is therefore an estimate of $\nu_{\rm eff}$ obtained by utilizing the tiny spectral dispersion of the stellar images produced by the chromaticity.

[5] A useful analytical approximation relating the effective wavenumber to the *Gaia* color index is $\nu_{\rm eff} \simeq 1.76 - (1.61/\pi){\rm atan}[0.531(G_{\rm BP} - G_{\rm RP})]\ \mu{\rm m}^{-1}$ [51].

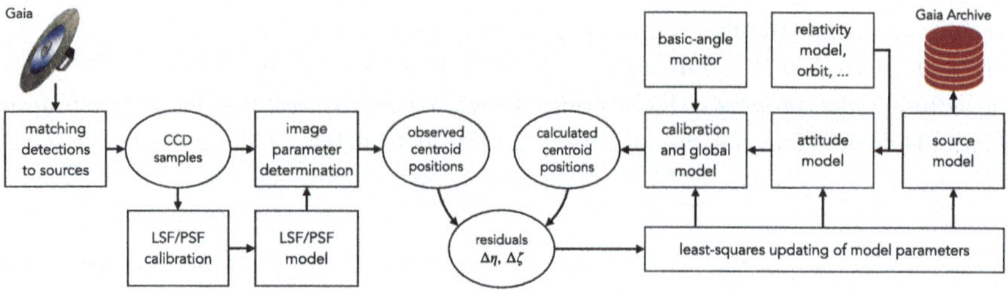

Fig. 3.8 Simplified block diagram for the processing of astrometric data from *Gaia*

It can be noted that the LSP/PSF calibration for EDR3 was restricted to the well-populated color interval $1.24 < \nu_{\rm eff} < 1.72\,\mu{\rm m}^{-1}$ ($3.0 \gtrsim G_{\rm BP} - G_{\rm RP} \gtrsim 0.14$). Beyond that range the calibration for the nearest color was used, leading to parallaxes that are more uncertain.

In binaries with a projected AL separation exceeding 2–3 pixels (0.12–0.18 arcsec) the components may produce separate peaks in the CCD samples, in which case the centroiding algorithm normally picks up the brighter peak. For smaller projected separations the centroid position is a complex function of the separation and flux ratio of the components, scanning geometry, and the LSF (see [52] for a toy model), which may create spurious signals in the astrometric results [53].

The right part of Fig. 3.8 corresponds to the processing known as the astrometric global iterative solution (AGIS). This aims to combine all the measured centroid positions into a globally consistent system of astrometric parameters for all the sources [50]. The precise AL measurements $t_{\rm obs}$ are the main input to AGIS. For a given observation $t_{\rm obs}$ depends not only on the astrometric parameters of the relevant source, but also on a number of 'nuisance parameters' that describe the behaviour of the instrument; these are the attitude, calibration, and global parameters. The attitude parameters describe the accurate celestial pointing of the instrument as a continuous function of time. The calibration parameters map the location of the fiducial observation lines of each CCD in terms of the field angles (η, ζ). The global parameters describe additional effects that are coherent on larger scales than the individual CCDs. Following the self-calibration principle, the nuisance parameters are estimated together with the astrometric parameters in a single least-squares solution, using the regular observations. For practical reasons only a subset of well-behaved, effectively single 'primary' sources is used to estimate the nuisance parameters. For *Gaia* EDR3 some 6.5 billion CCD observations of 14 million primary sources were used to determine 70 million source parameters, 11 million attitude parameters, and 3 million instrument (calibration and global) parameters [51].

A critical part of the data analysis is the definition of a good instrument model. This must be flexible and detailed enough to represent the actual instrument to an accuracy compatible with the best centroiding precision ($\gtrsim 60\ \mu$as), yet robust enough to be fully constrained by the available observations. In practice the model develops through a lengthy process of trial and error, where residual patterns in preliminary solutions may hint towards possible model improvements. A main challenge of the approach is that certain instrumental effects may be fully or partly absorbed by the astrometric parameters, in which case they do not show up in the residuals, or only partially and perhaps in a very distorted form. Simulated solutions in an AGIS-like framework are an indispensable tool for investigating possible degeneracies in various models.

In this context basic-angle variations (BAV) are of special interest. It has been known since the early phases of the *Hipparcos* project that basic-angle stability on time scales shorter than the rotation period (6 h for *Gaia*) is fundamental to the global measurement principle. In particular, a BAV proportional to $\cos\Omega$, where Ω is the rotation phase of the instrument relative to the Sun (see Fig. 1 in [54]), is tightly correlated with a global zero-point error of the measured absolute parallaxes. Very strict engineering requirements were consequently put on short-term BAV in *Gaia*. Furthermore, an internal metrology device, known as the basic-angle monitor (BAM), was introduced to measure such variations at the sub-μas level. During the commissioning of *Gaia* it was found that the basic angle has a strong dependence on Ω with an amplitude of about 1 mas, two orders of magnitude greater than pre-launch predictions [47]. The variations measured by the BAM are taken into account in the processing of the astrometric data, but this does not completely remove the BAV as sensed by the astrometric detectors. One possible reason for this could be that the separate CCDs used by the BAM are located well outside the astrometric field: because the optical field distortions also exhibit periodic variations, the BAV sensed by the BAM will be slightly different from the effective variation in the astrometric field. For EDR3, residual BAV are taken into account in the global model by means of a Fourier series in Ω, where the coefficients are slowly varying functions of time (Sect. 3.4.3 of [51]). Although this model includes a parameter describing the dependence on $\cos\Omega$, its determination is very delicate owing to the strong coupling with a global parallax offset (correlation coefficient $\simeq 0.99992$), and it was not possible to completely remove such an offset (Sect. 3.5.3).

Another instrument effect that could bias the absolute parallax measurement is a possible dependence of the AL centroid position on the AC scan rate ($\dot{\zeta}$; see Appendix-B in [51]). Such an effect could be produced by CTI in the CCDs for a PSF that is asymmetric in the AC direction. Because the CTI depends on the magnitude of the source, the overall result of the unmodelled effect could be a magnitude-dependent parallax bias. In the nominal scanning there is a strong coupling between $\dot{\zeta}$ and the AL parallax factor f_η (correlation coefficient $\simeq +0.985$), which prevented the effect to be calibrated for EDR3. During one year from 2019.5 the precession of the scanning law was however reversed, leading to an equally strong anti-correlation between $\dot{\zeta}$ and f_η, which should enable the effect to be disentangled in DR4 or DR5.

In addition to the dependencies introduced already in the LSF/PSF modelling mentioned above, the astrometric instrument model for EDR3 contains functions depending on magnitude (only for the AC calibration), saturation, subpixel phase, time since the last charge injection, and AC rate (only quadratic in $\dot{\zeta}$, to avoid degeneracy with parallax). See [51] for details.

3.5.3 Known Parallax Systematic in Gaia (E)DR3

Had the large BAV been ignored in the astrometric solution, the result would be a very large ($\simeq$ 800 µas) overall offset in the parallaxes, with considerable ($\simeq$ 100 µas) variations over the sky. Thus, corrections for the BAV, as measured by the BAM, were introduced already for *Gaia* DR1, and starting with DR2 a global model representing spin-related distortions (including BAV) was fitted in AGIS as part of the self-calibration process. Nevertheless, residual systematics of several tens of μas were clearly evident [55]. It was however decided to publish the parallaxes without applying any *ad hoc* corrections, and merely alert users to the known or suspected biases. The same policy was followed for EDR3; but an accompanying paper, here called L21 [56], examined the issue in some detail and gave a provisional recipe for its correction as a function of magnitude (G), color ($\nu_{\rm eff}$), and ecliptic latitude (β). L21 was based on an analysis of the EDR3 results for 1.1 million quasars ($G \gtrsim 14$), 2.4 million probable LMC members ($G \lesssim 19$), and 121 000 wide binaries with primary $G < 14$ and secondary $G < 18$. Importantly, the LMC sources and binaries were only used differentially, i.e. for mapping systematic differences depending on magnitude and color. The estimated parallax bias is given as a function $Z(G, \nu_{\rm eff}, \beta)$, such the corrected parallax is[6]

$$\varpi_{\rm L21} = \varpi_{\rm EDR3} - Z(G, \nu_{\rm eff}, \beta). \tag{3.15}$$

Since the publication of *Gaia* EDR3 and L21 in December 2020, a considerable number of papers have appeared in which other methods and other kinds of objects are explored for further quantification of the parallax bias. These include the use of classical Cepheids in the field and in clusters [57–59], RR Lyr variables [60, 61], asteroseismic distances [62, 63], spectrophotometric distances [64, 65], eclipsing binaries [66–68], orbital parallaxes [69], and open and globular clusters [70–72]. Similar to L21, quasars and wide binaries have also been used [73], employing a larger sample of resolved binaries constructed from EDR3 [74].

In general the cited papers agree that the L21 recipe removes most of the parallax bias in (E)DR3, at least for $G > 13$ or 14, but in many cases significant differences are found. I define the residual parallax bias in (E)DR3 relative to L21 as

$$\Delta Z = \varpi_{\rm L21} - \varpi_{\rm other}, \tag{3.16}$$

where $\varpi_{\rm other}$ is the parallax obtained by some other method; thus $Z + \Delta Z$ is the total bias in (E)DR3.[7] Fig. 3.9 is an attempt to summarize the finding of the various determinations of ΔZ to date. Several authors conclude that $\Delta Z = +10$ to $+20$ μas for $G < 11$ or 12, i.e. that L21 overcorrects the bias for the bright stars [57, 59–62,

[6] L21 provides two bias functions: Z_5 for five-parameter solutions with known $\nu_{\rm eff}$, and Z_6 for six-parameters solutions where the pseudocolor $\hat{\nu}_{\rm eff}$ was used instead. The subsequent discussion only concerns Z_5, which is applicable to most sources brighter than $G = 18$.

[7] Some authors define residual parallax offset in the opposite sense, i.e. as $-\Delta Z$ [57–59].

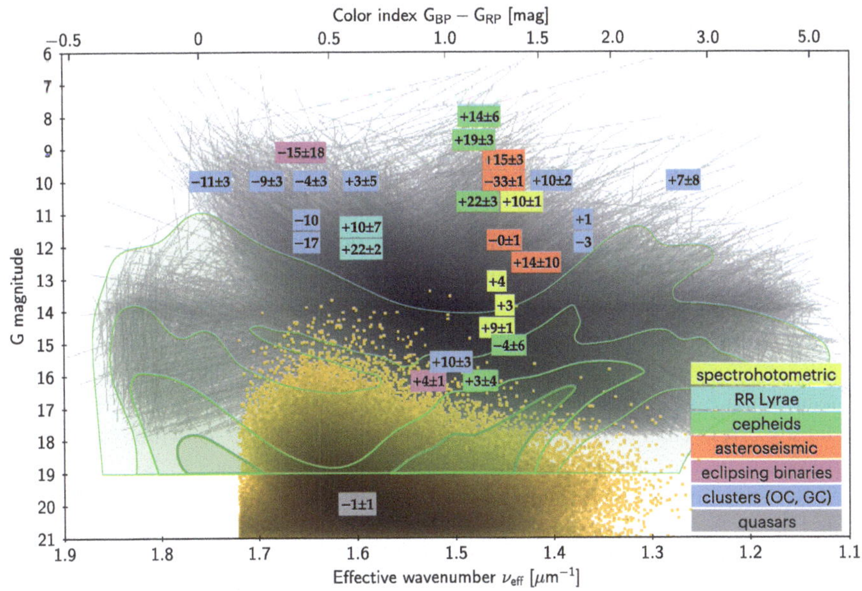

Fig. 3.9 CMD showing the residual parallax bias ΔZ (in μas relative to L21) obtained in various studies. The result of each study, with its uncertainty when available, is plotted at the approximate mean color and magnitude of the sample, color-coded according to the method used. The background shows the distribution of the three main types of sources used to derive Z in L21, namely quasars (orange dots), binaries (gray lines connecting the components), and stars in the LMC (green contours representing the smoothed density in steps of 1 dex). In L21 the binaries and LMC stars were used to map the variations of Z with magnitude, color, and position; its absolute value is anchored in the quasar parallaxes

71, 75]. However, as indicated in the figure, perhaps this only applies to stars of intermediate or red colors, while the reverse may be true for hotter stars ($G_{\mathrm{BP}} - G_{\mathrm{RP}} \lesssim 0.5$). The variation according to position is clearly more complex than just a function of ecliptic latitude; thus results may differ significantly depending on the area investigated [63, 65, 71, 73]. Overall, residual systematics are found to be at the level of about 10 µas RMS.

Another kind of systematics of the parallaxes in *Gaia* (E)DR3 is the underestimation of the uncertainty, as given by the formal standard error σ_{ϖ}, and the spatially correlated errors on angular scales of several degrees. The underestimation may require both a multiplicative factor ($k \gtrsim 1.05$) and a quadratically added contribution ($s \simeq 10$ μas) representing the overall accuracy floor. The statistical correlation of the errors implies that the uncertainty of the sample mean does not decrease as $1/\sqrt{n}$ when averaging over n sources, if they are all within a small area of the sky (as in a cluster); both effects are crudely described by

$$\sigma_{\varpi}^{\mathrm{true}} = \sqrt{(k\sigma_{\varpi})^2/n + s^2}\,, \tag{3.17}$$

where k and s may depend on factors such as the magnitude, local star density, or astrometric goodness-of-fit [70, 71, 74].

3.5.4 Expected Improvements in Gaia DR4 and DR5

The astrometric solution for *Gaia* DR3 is based on data collected during 33 months of observations. For DR4 the solution will use 63 months of data collected until early 2020, and DR5 may add another five years. The formal uncertainty of parallax determinations could therefore improve roughly by a factor 0.72 for DR4 and 0.52 for DR5, as suggested by the black curves in Fig. 3.5. The real improvement in terms of *accuracy* (i.e. including systematics) depends on the extent to which the various modelling errors can be reduced in the data processing (Fig. 3.8), which is difficult to predict.

Most modelling errors are benign in the sense that they behave almost like uncorrelated noise; their impact on the astrometry is therefore qualitatively similar to that of the photon noise and can be expected to improve by the same factors as mentioned above, even if the models are not improved. For many sources there is a potential for significantly improved calibrations, e.g. in the non-saturated bright range (C in Fig. 3.5) and for sources of extreme colors. Similarly, the non-single star processing [76], using more realistic source models than the standard one e.g. for astrometric binaries, will normally result in improved parallaxes [77]. Furthermore, epoch astrometry will be made available for all sources in DR4 and DR5, enabling users to fit their own source models to the data.

The parallax bias described in Sect. 3.5.3 is a rather different story. This bias is at least partly related to near-degeneracies with certain instrumental effects, namely the $\cos\Omega$ variation of the basic angle and centroid shifts depending on the AC rate. Although its variations with magnitude, color, and position are likely to become smaller in DR4 and DR5, thanks to the general improvement of calibration models, it may not be possible to reduce the bias to insignificant levels. Independent methods to assess the parallax bias will therefore be needed also for the future releases.

3.6 Statistical Meaning and Use of Parallax

The linearity of the standard astrometric model, Eq. (3.5), has important practical consequences for the statistical interpretation of the astrometric parameters. Let $\boldsymbol{y} = [\Delta\alpha_*, \Delta\delta, \varpi, \mu_{\alpha*}, \mu_\delta]'$ be the astrometric parameters obtained for a particular source, where the position is expressed differentially to some fixed reference point. Furthermore, let $\boldsymbol{y}_0$ be the corresponding true parameter vector—assuming that such values exist, in other words that the standard model is correct. Thanks to the linear estimation method used (essentially a weighted least-squares solution) the resulting astrometry is also unbiased, $\mathrm{E}(\boldsymbol{y}) = \boldsymbol{y}_0$, provided that the standard model holds and the individual observations are unbiased. As we have seen, both of these assumptions can be questioned, and ways to deal with possible biases are therefore needed as already discussed (concerning biases in the *Gaia* DR3 proper motions, see [78]).

Assuming unbiasedness, another consequence of the linearity is that the error distribution tends towards the multivariate normal (Gaussian), $\boldsymbol{y} \sim \mathcal{N}(\boldsymbol{y}_0, \boldsymbol{C})$, where $\boldsymbol{C}$ is the covariance matrix of $\boldsymbol{y}$. That is, the probability density function is

$$p(\boldsymbol{y}|\boldsymbol{y}_0) = (2\pi)^{-m/2} \det(\boldsymbol{C})^{-1/2} \exp\left[-\frac{1}{2}(\boldsymbol{y}-\boldsymbol{y}_0)'\boldsymbol{C}^{-1}(\boldsymbol{y}-\boldsymbol{y}_0)\right], \quad (3.18)$$

for $m = 5$ (the dimensionality of $\boldsymbol{y}$). Specifically for the parallax,

$$p(\varpi|\varpi_0) = \frac{1}{\sigma_\varpi\sqrt{2\pi}} \exp\left[-\frac{(\varpi-\varpi_0)^2}{2\sigma_\varpi^2}\right], \quad (3.19)$$

where σ_ϖ^2 is the diagonal element of $\boldsymbol{C}$ corresponding to the parallax.

In these equations $\boldsymbol{y}$ is formally regarded as a random variable and p is the limiting distribution of $\boldsymbol{y}$ when the experiment is repeated many times for fixed $\boldsymbol{y}_0$ and $\boldsymbol{C}$. In practice we only have the result from a single experiment, namely $\boldsymbol{y}$, and the true parameter vector $\boldsymbol{y}_0$ is unknown. Regarded as a function of $\boldsymbol{y}_0$, for the given $\boldsymbol{y}$, this is known as the likelihood function: $\mathcal{L}(\boldsymbol{y}_0|\boldsymbol{y}) = p(\boldsymbol{y}|\boldsymbol{y}_0)$.

Assuming that the least-squares estimation is done with the appropriate weighting of the observations, taking into account possible correlations among the data, we obtain $\boldsymbol{C}$ as part of the estimation process. In reality various approximations and simplifications are usually made. In the *Gaia* data analysis, for example, the correlations among the eight or nine CCD observations in a single transit of the astrometric field (Fig. 3.6) are currently neglected, and no marginalisation is made over the nuisance parameters. This leads to an underestimation of $\boldsymbol{C}$ and σ_ϖ, and empirical correction factors may be required, as in Eq. (3.17).

In spite of the several caveats given above, the important fact remains that the probability density in Eqs. (3.18) or (3.19) (i.e. the likelihood function) is generally known to a good approximation. This provides a solid foundation for statistical inference in most astrophysical problems where astrometric data play a major role, for example using Bayesian inference. A simple but non-trivial example, extensively discussed in the literature, is the estimation of distance $d = A/\varpi_0$ from the measured parallax value ϖ and its uncertainty σ_ϖ [79–82]. Additional considerations relevant for the interpretation of astrometric data are discussed in [83].

The published astrometric parameters $\boldsymbol{y}$, together with their covariance matrix $\boldsymbol{C}$, should thus be regarded as a concise specification of the likelihood function of the true parameters $\boldsymbol{y}_0$, using Eq. (3.18). Under the assumptions mentioned above they provide sufficient statistics of the observations with respect to the adopted source model [84], as long as that model is linear in its parameters.

3.7 Conclusion and Future Prospects of the Parallax Method

The parallax method is the most direct way to obtain information about stellar distances. Because it is based on simple geometry, with minimal assumptions concerning the nature of the sources, it is sometimes claimed to be model-independent. But as we have seen, the modelling of parallax observations is far from trivial. A basic reason is that the parallax is very small even for the nearest stars, and that the requirements in terms of measurement accuracy increase in proportion to the distance covered. In the last decades the barrier to classical optical astrometry set by the atmosphere has been overcome by the use of VLBI at radio frequencies, adaptive optics at near-infrared wavelengths, and space techniques in the visual domain. All of these techniques now produce parallax determinations at the microarcsecond level. For the future, at least two space missions dedicated to astrometric surveys in the near-infrared are currently under study, JASMINE [85] and *GaiaNIR* [86].

Thanks to *Gaia* we now have trigonometric parallaxes available for nearly two billion sources brighter than $G \simeq 21$ mag, and the next releases of *Gaia* data will bring additional improvements and new kinds of data, such as the epoch astrometry. However, results for the brightest stars ($G \lesssim 6$ mag) suffer from saturation of the detectors, and the fixed observation schedule implied by the scanning law of *Gaia* limits the integration time and hence the precision for faint objects. Bigger telescopes in space, and on ground using adaptive optics, will have important advantages over *Gaia* in crowded areas. Relative parallax determinations using HST and its successors (*James Webb Space Telecope*, *Nancy Grace Roman Space Telecope*), as well as with large ground-based facilities, will therefore remain relevant for specific targets where *Gaia* cannot deliver sufficient accuracy or resolution. For the correction from relative to absolute parallax, the statistical and spectrophotometric methods can be vastly improved by incorporating *Gaia* information on the reference sources.

Acknowledgements I wish to thank Stefano Casertano for many perceptive comments on the manuscript, and Adam Riess for important clarifications.

References

1. J.D. Fernie, JRASC **69**, 153 (1975)
2. J.D. Fernie, JRASC **69**, 222 (1975)
3. M.J. Reid, K.M. Menten, Astronomische Nachrichten **341**, 860 (2020)
4. L. Lindegren, Direct distance determination using parallax: techniques, promises and limitations, in *IAU Symposium 289, Advancing the Physics of Cosmic Distances*, ed. by R. de Grijs, p. 52 (2013)
5. M.J. Reid, M. Honma, ARA&A **52**, 339 (2014)
6. F.W. Bessel, Astronomische Nachrichten **16**, 65 (1838)
7. J.A.C. Oudemans, Astronomische Nachrichten **122**, 193 (1889)

8. F. Schlesinger, M. Palmer, A. Pond, Y.U. Observatory, in *General Catalogue of Stellar Parallaxes* (Yale University Observatory, 1924)
9. F. Schlesinger, L. Jenkins, *General Catalogue of Stellar Parallaxes: Compiled at Yale University Observatory* (Yale University Observatory, 1935)
10. L.F. Jenkins, *General Catalogue of Trigonometric Stellar Parallaxes* (Yale University Observatory, 1952)
11. W.F. van Altena, J.T. Lee, E.D. Hoffleit, *The General Catalogue of Trigonometric [Stellar] Parallaxes* (Yale University Observatory, 1995)
12. M.A.C. Perryman et al., A&A **323**, L49 (1997)
13. H.C. Harris et al., Progress in Parallaxes at USNO, in *Astrometry in the Age of the Next Generation of Large Telescopes*, ed. by P.K. Seidelmann, A.K.B. Monet. Astronomical Society of the Pacific Conference Series, vol. 338, p. 122 (2005)
14. F. van Leeuwen, A&A **474**, 653 (2007)
15. G.F. Benedict, B.E. McArthur, E.P. Nelan, T.E. Harrison, PASP **129**, 012001 (2017)
16. A.G. Riess et al., ApJ **855**, 136 (2018)
17. Gaia Collaboration, A.G.A. Brown, A. Vallenari, T. Prusti, A&A **649**, A1 (2021)
18. M. Soffel et al., AJ **126**, 2687 (2003)
19. S.A. Klioner, AJ **125**, 1580 (2003)
20. Z. Penoyre, V. Belokurov, N. Wyn Evans, A. Everall, S.E. Koposov, MNRAS **495**, 321 (2020)
21. L. Lindegren, A&A **633**, A1 (2020)
22. K. Rosquist, ApJ **331**, 648 (1988)
23. J.H.J. de Bruijne, A.C. Eilers, A&A **546**, A61 (2012)
24. C.A. Murray, *Vectorial Astrometry* (Adam Hilger, 1983)
25. W.F. van Altena, AJ **79**, 826 (1974)
26. W.C. Jao et al., AJ **129**, 1954 (2005)
27. G.F. Benedict et al., AJ **133**, 1810 (2007)
28. M.J. Rioja, R. Dodson, A&Ar **28**, 6 (2020)
29. M.J. Reid, PASP **134**, 123001 (2022)
30. M.J. Reid et al., ApJ **885**, 131 (2019)
31. E.P. Nelan, FGS Instrument Handbook v. 28.0 (2021). https://hst-docs.stsci.edu/fgsihb
32. A. Bradley, L. Abramowicz-Reed, D. Story, G. Benedict, W. Jefferys, PASP **103**, 317 (1991)
33. L. Dressel, M. Marinelli, WFC3 Instrument Handbook for Cycle 31 v. 15.0 (2023). https://hst-docs.stsci.edu/wfc3ihb
34. W. Jefferys et al., Optical field angle distortion calibration of FGS3, in *Calibrating Hubble Space Telescope*, ed. by J.C. Blades, S.J. Osmer, p. 353 (1994)
35. B.E. McArthur, G.F. Benedict, W.J. Jefferys, E. Nelan, The optical field angle distortion calibration of HST fine guidance sensors 1R and 3, in *The 2005 HST Calibration Workshop: Hubble After the Transition to Two-Gyro Mode*, ed. by A.M. Koekemoer, P. Goudfrooij, L.L. Dressel, p. 396 (2006)
36. J.W. MacKenty, Performance and calibration of the HST Wide Field Camera 3, in *Space Telescopes and Instrumentation 2012: Optical, Infrared, and Millimeter Wave*, ed. by M.C. Clampin, G.G. Fazio, H.A. MacEwen, J. Oschmann, M. Jacobus. Society of Photo-Optical Instrumentation Engineers (SPIE) Conference Series, vol. 8442, p. 84421V (2012)
37. L.R. Bedin, C. Fontanive, MNRAS **481**, 5339 (2018)
38. J. Anderson, I.R. King, PASP **112**, 1360 (2000)
39. J. Anderson, L.R. Bedin, MNRAS **470**, 948 (2017)
40. J. Anderson, Analyzing poorly sampled images: HST imaging astrometry, in *Astrometry for Astrophysics*, ed. by W.F. van Altena, p. 257 (Cambridge University Press, 2012)
41. A. Bellini, J. Anderson, L.R. Bedin, PASP **123**, 622 (2011)
42. A.G. Riess, S. Casertano, J. Anderson, J. MacKenty, A.V. Filippenko, ApJ **785**, 161 (2014)
43. S. Casertano et al., ApJ **825**, 11 (2016)
44. T.M. Brown et al., ApJL **856**, L6 (2018)
45. E.M. Jones, ApJ **173**, 671 (1972)
46. A.C. Robin, C. Reylé, S. Derrière, S. Picaud, A&A **409**, 523 (2003)

47. Gaia Collaboration, T. Prusti, J.H.J. de Bruijne, A.G.A. Brown, A&A **595**, A1 (2016)
48. Gaia Collaboration, A. Vallenari, A.G.A. Brown, T. Prusti (2022), arXiv e-prints arXiv:2208.00211
49. A.G.A. Brown, ARA&A **59**, 59 (2021)
50. L. Lindegren et al., A&A **538**, A78 (2012)
51. L. Lindegren et al., A&A **649**, A2 (2021)
52. L. Lindegren, Expected astrometric properties of binaries in (E)DR3, Technical note LL-136, Lund Observatory (2022). https://dms.cosmos.esa.int/COSMOS/doc_fetch.php?id=1566327
53. B. Holl et al. (2022), arXiv e-prints : arXiv:2212.11971
54. A.G. Butkevich, S.A. Klioner, L. Lindegren, D. Hobbs, F. van Leeuwen, A&A **603**, A45 (2017)
55. F. Arenou et al., A&A **616**, A17 (2018)
56. L. Lindegren et al., A&A **649**, A4 (2021)
57. A.G. Riess et al., ApJL **908**, L6 (2021)
58. A.G. Riess et al., ApJ **938**, 36 (2022)
59. R. Molinaro et al., MNRAS **520**, 4154 (2023)
60. A. Bhardwaj et al., ApJ **909**, 200 (2021)
61. C.K. Gilligan et al., MNRAS **503**, 4719 (2021)
62. J.C. Zinn, AJ **161**, 214 (2021)
63. S. Khan et al. (2023), arXiv e-prints: arXiv:2304.07158
64. Y. Huang, H. Yuan, T.C. Beers, H. Zhang, ApJL **910**, L5 (2021)
65. C. Wang, H. Yuan, Y. Huang, AJ **163**, 149 (2022)
66. K.G. Stassun, G. Torres, ApJL **907**, L33 (2021)
67. F. Ren et al., ApJL **911**, L20 (2021)
68. F. Ren et al., AJ **161**, 176 (2021)
69. M.A.T. Groenewegen, A&A **669**, A4 (2023)
70. E. Vasiliev, H. Baumgardt, MNRAS **505**, 5978 (2021)
71. J. Maíz Apellániz, A&A **657**, A130 (2022)
72. C. Flynn et al., MNRAS **509**, 4276 (2022)
73. M.A.T. Groenewegen, A&A **654**, A20 (2021)
74. K. El-Badry, H.W. Rix, T.M. Heintz, MNRAS **506**, 2269 (2021)
75. M. Cruz Reyes, R.I. Anderson, A&A **672**, A85 (2023)
76. J.L. Halbwachs et al. (2022), arXiv e-prints: arXiv:2206.05726
77. Gaia Collaboration, F. Arenou, C. Babusiaux, Barstow (2022), arXiv e-prints: arXiv:2206.05595
78. T. Cantat-Gaudin, T.D. Brandt, A&A **649**, A124 (2021)
79. J. Smith, Haywood, H. Eichhorn, MNRAS **281**, 211 (1996)
80. C.A.L. Bailer-Jones, PASP **127**, 994 (2015)
81. X. Luri et al., A&A **616**, A9 (2018)
82. G.M. Eadie et al. (2023), arXiv e-prints: arXiv:2302.04703
83. A.G.A. Brown, Statistical astrometry, in *Astrometry for Astrophysics*, ed. by W.F. van Altena, p. 242 (Cambridge University Press, 2012)
84. D.W. Hogg (2018), arXiv e-prints: arXiv:1804.07766
85. N. Gouda, Overview and recent progress of JASMINE near- infrared space astrometry mission (2022). https://doi.org/10.5281/zenodo.7068453
86. D. Hobbs et al., Exp. Astron. **51**, 783 (2021)

Chapter 4
Measuring H_0 with H_2O Megamasers

James A. Braatz and Dominic W. Pesce

4.1 Water Vapor Megamasers in the Nuclei of Active Galaxies

Water vapor masers emitted at a rest frequency of 22.235080 GHz (1.35 cm) are readily observed toward star forming regions and in the envelopes of evolved stars in the Galaxy. These masers have typical isotropic luminosities of $\sim 10^{-4}\ L_\odot$, though in the brightest examples such as W49 they can emit with an isotropic luminosity $\sim 1\ L_\odot$ and reach flux densities of thousands of Janskies. Extragalactic analogs to such masers, associated with stars and star formation, have been detected in nearby galaxies, where the integrated emission of the observed maser sources can reach isotropic luminosities of $\sim 10\ L_\odot$. Generally associated with enhanced star forming galaxies, these systems have been called *kilomasers*. About 20 kilomaser galaxies have so far been identified.

Another class of maser systems, the *megamasers*, are powered by active galactic nuclei (AGN). These often have apparent isotropic luminosities of hundreds of $L_\odot$, with the brightest among them being thousands of $L_\odot$. About 160 extragalactic masers in AGN have been detected to date[1]

Megamasers are highly variable over a range of timescales. Variability seen on timescales of days and longer is likely the result of changes in the maser environment; masers are highly beamed, nonlinear amplifiers that are sensitive both to the line-

[1] A catalog of extragalactic water masers is maintained by the Megamaser Cosmology Project at https://safe.nrao.edu/wiki/bin/view/Main/PublicWaterMaserList.

J. A. Braatz (✉)
NRAO, 520 Edgemont Rd., Charlottesville, VA 22903, USA
e-mail: jbraatz@nrao.edu

D. W. Pesce
Center for Astrophysics | Harvard & Smithsonian, 60 Garden St., Cambridge, MA 02138, USA
e-mail: dpesce@cfa.harvard.edu

E. Di Valentino and D. Brout (eds.), *The Hubble Constant Tension*, Springer Series in Astrophysics and Cosmology, https://doi.org/10.1007/978-981-99-0177-7_4

of-sight arrangement of molecular clouds and to the energy supply that produces the pump to create the population inversion. Several have even shown variability on timescales as short as minutes, where the rapid flux variations are thought to be caused by interstellar scintillation (e.g. [17, 42]).

Megamasers exhibit some diversity in their spectral shapes. A few have broad ($\sim$50–100 km s^{-1}) and highly variable profiles [3] and have been called "jet masers" because they appear to be affiliated with strong radio continuum jets near the nucleus, as exemplified by the maser in the radio galaxy NGC 1052 [11, 51]. Other megamaser systems have scattered narrow components, and in some cases they appear to be affiliated with outflows within a few parsecs of the AGN. Many megamasers have simple profiles with emission detected only within $\sim$ 100 km s^{-1} of the systemic recession velocity of the host galaxy, and the nature of their emission is largely unexplored. Others are rich with many spectral components at velocities that may be separated from the galaxy recession velocity by 1000 km s^{-1} or more.

In about 35 of the known water megamaser systems, the spectral profiles are characterized by three distinct groups of narrow maser components, one group near the galaxy recession velocity (systemic masers) and the other two (high-velocity masers) red- and blue-shifted in velocity by $\sim$ 250–1000 km s^{-1}. These masers arise in the thin accretion disks that surround the supermassive black holes in the galaxy nuclei, and they are called *disk masers*. The spectrum in Fig. 4.1 is representative of the characteristic spectral profile of a disk maser, where in this example the recession velocity of the host galaxy UGC 3789 is $\sim$3300 km s^{-1} and the maser disk orbits the supermassive black hole (SMBH) with velocities up to $\sim$750 km s^{-1}.

Disk megamasers originate in the warm ($\sim$ 1000 K) molecular accretion disks around low-luminosity AGN, typically $\sim$ 0.1–1 pc from the central SMBH. On these scales, the gravitational potential is dominated by the SMBH and the gas exhibits Keplerian orbital dynamics. The physical conditions of the molecular gas in these disks enable a population inversion in the rotational energy levels responsible for 22 GHz water maser activity, and the maser emission is maximally amplified along paths that (1) travel through a large volume of masing gas and (2) have small velocity gradients along the line of sight (so as not to lose resonance for stimulated emission by Doppler shifting). The exponential nature of maser amplification thus imposes a strong observational selection preference to see edge-on disk systems, such that our line of sight passes through the disk and thereby accesses the longest path lengths of masing gas. For such an edge-on disk orientation, the line-of-sight velocity gradient is minimized along sightlines parallel to or perpendicular to the orbits of the masing gas, leading to the characteristic grouping of observed spectral features into systemic and high-velocity components.

Disk masers are unique tools that have proven to be extremely productive scientific probes. They provide a direct means to map the molecular gas within a pc of the AGN, and thereby reveal the size and orientation of the thin accretion disk on these scales; they orbit the supermassive black hole within its sphere of influence and thus determine precise black hole masses; and in some cases their distances can be measured using the geometric "megamaser technique".

4.2 Measuring Geometric Distances to H_2O Megamasers

Two methods have been employed to determine geometric distances to galaxies based on observations of their accretion disk megamasers [25]. The first requires measurements of the proper motions of systemic maser components as they orbit the central supermassive black hole, achieved through a sequence of VLBI mapping observations. So far this technique has only been applied to the nearby galaxy NGC 4258. The maser disk in this galaxy has high orbital velocities approaching $\sim 1000\,\mathrm{km\,s^{-1}}$, and the observed proper motions are $\sim 30\,\mu\mathrm{as\,yr^{-1}}$, which is near the limit of detection. Measuring such proper motions becomes prohibitively difficult for more distant galaxies and for maser systems with smaller orbital velocities, and we do not discuss this method further in the present contribution.

The second method ("the megamaser technique") can be more widely applied. In essence, the linear size of the maser disk, R, can be determined through an analysis of the measured centripetal accelerations ($a = v^2/R$) of maser clouds combined with the disk's measured Keplerian rotation profile, while the angular size of the disk, θ, is determined by VLBI mapping; the distance to the galaxy ($D = R/\theta$) is therefore determined geometrically [6, 25]. In practice, a 3-D model of the maser disk is fit to the observed data to determine the galaxy distance, as described later.

The megamaser technique requires two types of observations. The first is, again, a map of the distribution on the sky for each of the detected maser components using VLBI. The second is a time sequence of spectral profiles usually spanning one to two years for the purpose of measuring secular velocity drifts (i.e., line-of-sight accelerations) of the individual, narrow maser components. Since each VLBI map includes a spectral profile, all the required data could in principle be collected through a series of VLBI observations, as in the first method. More commonly, however, spectral monitoring has been achieved with a series of single-dish observations because they are simpler to observe and process.

The maser targets of interest are typically faint, so observations to determine distances have relied on the most sensitive telescopes at K-band (18–26 GHz): the 100-m Robert C. Byrd Green Bank Telescope (GBT), the Karl G. Jansky Very Large Array (VLA), the 100-m Effelsberg Telescope, and the Very Long Baseline Array (VLBA). Other large telescopes in the 60 to 70-m class such as the Sardinia Radio Telescope, the Tianma Telescope, and the NASA DSN stations also make important contributions, especially in regard to surveys and monitoring observations.

4.2.1 VLBI Maps of Maser Disks

A VLBI array observing megamaser disks typically has maximum projected E-W and N-S baselines of $\sim$8500 km and $\sim$3500 km, respectively, making the synthesized beam ($\theta_b \approx 1.2\lambda/D$) approximately 1.0×0.4 mas at an observing frequency of 22 GHz. A representative maser disk in the Hubble flow spans an angular extent of $\sim$1–2 mas (Fig. 4.1). Because the individual maser components observed in each

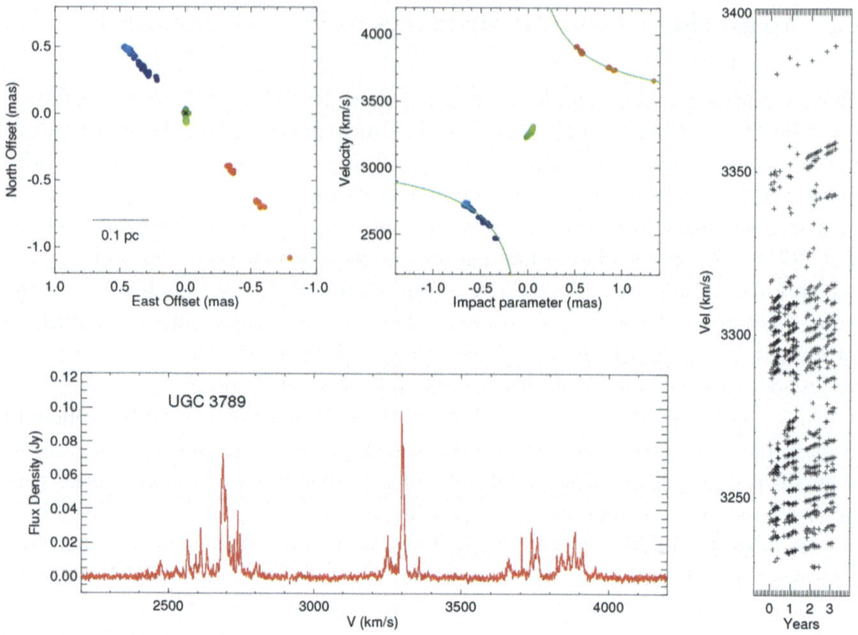

Fig. 4.1 Representative data from the megamaser galaxy UGC 3789. The bottom left panel shows a GBT spectrum with the distinctive triple-peaked spectrum of a clean disk maser. The top left panel shows the VLBI distribution of the maser spots on the sky; the colors represent line-of-sight velocities, with red-shifted and blue-shifted masers colored appropriately to identify the rotation of the edge-on maser disk. The top center panel shows a position-velocity diagram determined from the VLBI map, with a Keplerian fit to the rotation shown in green. The impact parameter is the projected angular distance from the black hole, measured along the disk diameter. The right panel shows the velocities of the maser peaks from the systemic part of the spectrum as observed over time, revealing the positive velocity drifts used to determine line-of-sight accelerations of maser components on the front side of the disk

spectral channel are spatially unresolved, it is possible to determine the relative positions of maser components within a single observation to a fraction of the beam size by measuring the centroids of the unresolved masers with 2-D Gaussian fits. The uncertainty in relative positions of maser components is then $\Delta\theta \approx \frac{\theta_b}{2S/N}$, where θ_b is the VLBI synthesized beam size in one dimension and S/N is the signal-to-noise ratio of the detected maser component in the VLBI spectral channel. Peak maser flux densities in the megamaser systems of interest typically reach $\sim$ 50–100 mJy, but the much fainter components in the rich maser spectra must also be mapped with high precision to optimize the model fit. Therefore the desired noise levels in VLBI maps is < 1 mJy beam^{-1} per (representative) $\sim 0.3\,\text{km s}^{-1}$ channel. The narrow spectral channels chosen for this noise calculation are set to have $\gtrsim 3$ spectral channels across the narrowest ($\sim 1\,\text{km s}^{-1}$) maser spectral components.

To achieve such sensitivity, VLBI mapping experiments have used the High Sensitivity Array (HSA), which comprises the VLBA, the GBT, the phased VLA, and the Effelsberg 100-m telescope. VLBI observing campaigns generally require averaging multiple "tracks" of 6–9 hours each, and it is beneficial for these tracks to be observed within a month or two of each other because the variable maser spectra may change substantially on longer timescales, and the velocity drifts of systemic maser lines will cause the lines to blur if the averaging time is long compared to the timescale of the drift. When added to VLBA observations, the GBT and VLA provide exquisitely sensitive baselines but do not affect the overall synthesized beam size significantly, while the Effelsberg telescope can increase the E-W baselines for high-declination targets. The observations are sensitive to weather and use globally dispersed telescopes under uncorrelated atmospheric conditions, so they are best done in the winter months. The sensitivity of the array enables relative positions of maser components to be measured with a precision of a few tens of μas. The VLBI map in Fig. 4.1 shows a representative case in UGC 3789 with the detected masers spanning $\sim$ 2 mas and showing orbits $\sim$ 0.1 - 0.3 pc from the supermassive black hole.

4.2.2 Spectral Monitoring of Maser Profiles

Maser profiles are observed with periodic monitoring observations to measure secular velocity drifts of the spectral components. For these maser clouds in orbit around the central black hole, such velocity drifts reveal the line-of-sight centripetal accelerations. Either sensitive single-dish observations or interferometric observations may be used to collect the spectral information. Ideally, a sequence of spectral monitoring observations would provide measured accelerations for every one of the maser spots identified in the VLBI map. The maps, however, use long integrations with the HSA and are significantly more sensitive than individual epochs from the monitoring observations, so the faintest maser lines mapped with VLBI may not have corresponding detections in the monitoring sequence to measure accelerations. To optimize the matching of VLBI-detected maser components with measured accelerations, it is beneficial to monitor the spectral profiles over a time span that straddles the VLBI observations.

Observing campaigns using GBT monitoring to measure accelerations have used a roughly monthly cadence for observations. Because maser observations at 22 GHz are sensitive to humidity and other weather conditions, the GBT does not typically achieve sensitive K-band observations during the summer months. The GBT monitoring season will therefore span about 9 months. Spectra obtained from the VLBI observations, ideally observed in the middle of the monitoring sequence, supplement the monitoring data. Summer gaps can be filled in with observations from the VLA, which has similar sensitivity to the GBT at K-band and generally much dryer observing conditions in the summer.

The line widths of individual maser components in a disk maser spectral profile are $\sim$1–2 km s^{-1}. Each of the three groups of maser features – systemic, red-, and blue-shifted – comprise many such lines spanning $\sim$50 - 250 km s^{-1} in each group (see Fig. 4.1), and the individual lines are blended and variable. Measuring individual accelerations therefore requires a global fitting procedure to model the accelerations, velocities, and amplitudes of each of the maser spectral components, decomposed from each spectrum and approximated by Gaussian profiles. Sensitive spectra are required to provide robust and precise fitting. For typical maser disk systems in the Hubble flow, spectral monitoring requires sensitivities of ~ 2 mJy per $\sim 0.3\,\mathrm{km\,s^{-1}}$ spectral channel (three channels across the narrowest maser lines), achievable with GBT observations of 2–3 hours per epoch.

Accelerations are determined for each of the narrow maser components observed throughout the spectrum. In the red- and blue-shifted components, the masers are expected to be near the midline of the maser disk, where the acceleration vectors point in the plane of the sky toward the black hole, so the measured line-of-sight accelerations are generally near zero. Small measured positive and negative deviations from zero, however, can help to place such maser clouds at small azimuthal offsets from the midline and help constrain the disk model. The systemic maser drifts, as shown in the right panel of Fig. 4.1, are always measured with positive accelerations (redward velocity drifts), indicating that the systemic masers originate on the near side of the disk, in front of the black hole. Presumably, maser clouds also exist on the back side of the disk and their emission is either attenuated by free-free absorption in the vicinity of the black hole or beamed away from our line of sight. Masers at azimuthal angles intermediate between the midline and the observed line to the black hole are fainter, if detected at all, as the line-of-sight velocity coherence of gas in the rotating disk is reduced at those intermediate angles.

4.2.3 Modeling Maser Disks to Determine Their Distance

The heterogeneous observations used to measure maser distances – VLBI mapping plus single-dish spectral monitoring – generate a (x,y,v,a) data set consisting of the positions, line-of-sight velocities, and line-of-sight accelerations for each maser spot measured from a VLBI spectral cube, along with their uncertainties. The (x,y) on-sky relative positions of each maser component are determined by fitting an elliptical Gaussian to the unresolved maser detection in each VLBI spectral channel. Uncertainties in the sky position are taken from the formal uncertainties in the Gaussian fitting procedure. Systematic uncertainties (e.g. from residual delay errors in the VLBI calibration) may be included as well. The line-of-sight velocity for each spot is determined from the center frequency of the VLBI spectral channel, and its uncertainty is taken to be the half-width of the spectral channel. The line-of-sight acceleration for each spot, and its uncertainty, are as measured from the fits to the secular velocity drifts in the monitoring observations. Each maser spot also has a measured flux density and a corresponding uncertainty, but these values are not used directly in the fitting of the disk model. The flux density of each maser component, however, determines the signal-to-noise ratio that sets the uncertainty in its measured position and contributes to the uncertainty in its measured acceleration.

The model fit to these data is a three-dimensional, thin, warped disk representing the geometry of the gas in orbit around the central supermassive black hole. While models allowing for confocal elliptical orbits have been investigated, all maser disks studied to date have been well fit by concentric circular orbits that are allowed to tilt in position angle and inclination angle to account for the warp. The warp is therefore parameterized by the position angle Ω_0 of the orbital ring at a reference radius (which may be zero) and its derivative with angular radius, $d\Omega/dr$, as well as the inclination angle at the reference radius, i_0 and its radial derivative di/dr. Second order derivatives may be applied in modeling these warp parameters as well. Each maser spot observed with VLBI is therefore placed on the disk at a modeled azimuthal angle and radius from the disk center. As described above, every VLBI maser component may not have a measured acceleration assigned to it, but spatial and velocity data alone can still help to constrain the fit, in the context of the broader model. High velocity components with unmeasured line-of-sight accelerations are assigned to fall near the disk midline (where the line-of-sight acceleration is zero) with appropriate azimuthal and radial uncertainties.

The full set of global model parameters, then, in a representative fit to a disk maser system consists of the position (x_0, y_0) of the dynamical center (i.e., the position of the supermassive black hole for relaxed Keplerian systems), the line-of-sight recession velocity v_0 of the black hole, the position angle and inclination angle warp coefficients, the mass M_{bh} of the supermassive black hole, and the angular-size distance D to the maser disk. The Hubble Constant is thus determined completely from each fit maser disk. The uncertainty in H_0 for each galaxy's measurement must account not only for the statistical uncertainties from the model fit but also an estimate of the (unknown) peculiar velocity, v_p, of the galaxy, since the Hubble recession velocity of each galaxy is $v_{rec} = v_0 + v_p$.

Early papers utilized a random walk Metropolis-Hastings Markov Chain Monte Carlo (MCMC) algorithm to sample the posterior probability distributions [48] for the global parameters of the model. The MCMC algorithm incorporates an estimate of the "error floor" on the input data parameters, tunable by the user, to account for systematic uncertainties in the measurements. Conservatively applied, these error floor estimations limit the precision of H_0 measurements. An update to the fitting procedure [44] applies a more efficient Hamiltonian Monte Carlo (HMC) sampler and fits for the error floor parameters directly, thereby reducing the systematic uncertainty and improving the overall fit uncertainties on the global parameters. Results presented in this chapter refer to fits using the HMC sampler.

4.3 The Maser Distance to NGC 4258 as an Anchor to Distance Ladders

The archetypal H_2O megamaser disk is found in the nucleus of the low-luminosity Seyfert 2 galaxy NGC 4258. Fortuitously nearby at a distance of 7.6 Mpc, the maser was first discovered [9] during a survey of 29 late-type galaxies with the OVRO 40-m telescope and using a 40 MHz bandwidth spectrometer that covered $\sim 500\,\mathrm{km\,s^{-1}}$ in velocity, centered on the galaxy's recession velocity. The discovery observation thus detected only the systemic maser emission, with an isotropic luminosity of $\sim$120 $L_\odot$, much greater than the luminosities of masers associated with star formation. Subsequent VLA observations [10] of NGC 4258 and the other newly detected megamaser in NGC 1068 (450 $L_\odot$) localized the maser emission to the central $\sim$1 pc of the AGN and clearly identified the maser emission with nuclear gas. The discovery of high-velocity maser emission in NGC 4258 [40], Doppler-shifted by $\sim \pm 1000\,\mathrm{km\,s^{-1}}$, was achieved during testing of a new 200 MHz spectrometer on the Nobeyama 45-m telescope. Early spectral monitoring of the systemic masers [16, 21] identified the secular velocity drifts of the systemic masers, now known to be affiliated with the orbital acceleration.

The first VLBI map to include the high-velocity lines [39] revealed the edge-on, warped maser disk in Keplerian rotation and secured the identification of the central object as a supermassive black hole with a mass of $\sim 4 \times 10^7\ M_\odot$. The method to determine the geometric distance to the maser [25] based on the observed orbital properties of the maser clouds was finally established and the distance measured through a series of VLBI monitoring observations. Thc initial work applied the two independent techniques to measure the maser distance, the first based on the measured proper motions, and the second using the aforementioned measurements of line-of-sight accelerations. As the proper motions of the fast-rotating systemic masers in nearby NGC 4258 are only $\sim 30\ \mu\mathrm{as\ yr^{-1}}$, just at the detection threshold for VLBI monitoring, all subsequent high-precision distance measurements of masers in this galaxy, as well as all the more distant megamasers, rely on the acceleration method.

NGC 4258 provides a critical anchor for calibrating both the Cepheid period-luminosity relationship and the tip of the red giant branch (TRGB) [49]. Because the metallicity of NGC 4258 is similar to that of many of the Cepheid-host galaxies it is used to calibrate, and the metallicity gradient in the galaxy itself is small, NGC 4258 reduces the systematic uncertainty introduced using other anchors such as the LMC. Likewise, the Cepheids in NGC 4258 have a similar mean level of "crowded backgrounds" as SN Ia host systems, further reducing systematics.

To refine the measured maser distance to NGC 4258, an exquisite set of maser monitoring data were obtained through 18 epochs over 3 years [1] and initially were fit with the warped disk model using tunable error floors for the input parameters, conservatively set, to obtain a distance 7.60 $\pm$ 0.17 (stat) $\pm$ 0.15 (sys) Mpc [26]. The same data were later re-analyzed, including the error floors as part of the formal fit, using both a modified Metropolis-Hastings MCMC approach as well as a Hamiltonian MCMC approach, to determine the distance 7.576 $\pm$ 0.082 (stat) $\pm$ 0.076 (sys) Mpc [49]. Upon applying only the NGC 4258 maser distance as the anchor to the Cepheid-to-SN Ia ladder, the Hubble Constant was determined to be $H_0 = 72.0 \pm 1.9\,\mathrm{km\,s^{-1}\,Mpc^{-1}}$. Applying the maser anchor to the TRGB method results in $H_0 = 71.1 \pm 1.9\,\mathrm{km\,s^{-1}\,Mpc^{-1}}$ [49]. Along with Gaia/HST parallaxes to MW Cepheids and measurements of Detached Eclipsing Binaries in the LMC/SMC, the maser distance to NGC 4258 marks one of the three geometric anchors to the "baseline" measurement of the SH_0ES team: $H_0 = 73.04 \pm 1.04\,\mathrm{km\,s^{-1}\,Mpc^{-1}}$ [50].

4.4 Direct Measurement of H_0 from Megamaser Distances in the Hubble Flow

NGC 4258 is too near to provide a precise, direct measurement of H_0 on its own, as its unknown peculiar velocity may be a large fraction of its observed recession velocity. At the time its maser distance was first measured, it was the only known maser disk galaxy for which the megamaser method could be applied. With peak flux densities reaching several Jy, it is by far the brightest of the "clean" disk masers in Keplerian rotation around the central supermassive black hole. The maser also extends some $\sim$15 mas in angular extent, much larger than the typical 0.4×1.0 mas VLBI beam, thus making it the largest known megamaser disk in angular size on the sky by close to an order of magnitude. Some initial speculation was that the maser was a rare system with an extremely fortunate placement in the universe at only 7.6 Mpc distance. If at 100 Mpc, the peak maser flux in NGC 4258 would be $\sim$20 mJy, detectable only by 100-m class telescopes. Given the opportunity to measure direct geometric distances in the Hubble flow, the remarkable maser system in NGC 4258 therefore inspired new surveys to identify similar systems at larger distances that could be measured using the same megamaser technique.

4.4.1 Early Surveys to Detect Megamasers in AGN

The maser in NGC 4258 was discovered as part of a series of early surveys seeking water maser systems in late-type galaxies, where accumulations of bright Orion- and W49-type masers would be expected to make them detectable in the nearest galaxies. The brightest and most luminous extragalactic masers discovered in these surveys, however, were found to be associated with the nuclear gas near AGN [10], and so search strategies later refocused on AGN in the 1990 s. Several large surveys at that time [2, 3] with the Effelsberg 100-m and Parkes 64-m telescopes, the most sensitive single dish telescopes at K-band in the northern and southern hemispheres at that time, soon identified Seyfert 2 galaxies (and some LINERs) as the most likely hosts of detectable maser emission, with the implication being that the maser amplification benefits from the edge-on orientation of the putative obscuring toroidal structures in these AGN, providing long amplification gain paths. Such early surveys revealed a diversity of maser spectra, including the prototypical jet maser in NGC 1052 [2] and the remarkable luminous gigamaser in TXFS 2226–184 [30].

Two important analogs to the maser disk in NGC 4258 were discovered as part of the early AGN surveys. Like in NGC 4258, the initial discovery of the megamaser in IC 2560 [3] revealed only maser components near the systemic recession velocity, and subsequent, more sensitive wideband observations with Nobeyama [29] identified high velocity emission, both red- and blue-shifted. Despite the relatively strong emission from this maser disk galaxy, its southern location at δ = -34° has hampered the detailed study needed to get a precise maser distance. Second, the rich maser system in Mrk 1419 [24] shows the characteristic triple-peaked spectral profile indicative of a disk maser. Despite its maser system having the signature disk profile with many bright spectral components, subsequent studies have shown it is difficult to model, possibly because of an apparent thickness in its observed maser disk as well as incompletely-traced substructure.

Progress with maser surveys focusing on AGN flourished in subsequent years, initially with programs that used the 70-m NASA Deep Space Stations [18, 31] and the Effelsberg 100-m telescope [20]. It was the significant step up in sensitivity at K-band offered by the GBT, however, that opened the field for studies of distant disk masers. The most productive GBT surveys in terms of total number of megamaser discoveries focused on narrow-line active galaxies [4, 19, 32]. However the significant discovery of the megamaser disk in the galaxy UGC 3789 resulted, surprisingly, from a nonselective GBT snapshot survey of nearby, luminous galaxies [5]. This galaxy was confirmed to host a Seyfert 2 nucleus by optical spectroscopy only after the disk maser discovery.

4.4.2 The Megamaser Cosmology Project

The Megamaser Cosmology Project [6, 47] (MCP) is an international effort formed to measure the Hubble Constant using the megamaser method. It started by measuring the distance to the maser disk in UGC 3789, and then followed up by conducting a large survey with the GBT to find and measure distances to additional megamaser disk galaxies in the Hubble Flow. The MCP and its precursor surveys with the GBT eventually surveyed > 2500 galaxies and discovered about half of the currently-known 160 megamaser galaxies. The MCP surveys focused on nearby ($z < 0.05$), narrow-line AGN selected from the Sloan Digital Sky Survey (SDSS), the 6dF galaxy survey, and the 2MASS Redshift Survey (2MRS), and included a smaller number of X-ray selected AGN and "apparently normal" galaxies. The survey also included candidates selected from infrared properties observed by the Wide-Field Infrared Survey Explorer (WISE).

Figure 4.2 shows the progress made by extragalactic maser surveys over time. The first detection of an extragalactic maser in a star-forming cloud in M33 [8] was made at Effelsberg and published in 1977. GBT surveys were particularly productive starting in 2004, and then surveys slowed in 2016 as the MCP refocused efforts on VLBI and monitoring the known disk masers to measure their distances.

The overall detection rate of $< 3\%$ for MCP surveys reflects the difficulty in identifying optimal samples of AGN candidates. The large maser surveys have mostly sampled nearby ($z < 0.05$) Seyfert 2 galaxies indiscriminately. Efforts to find improved selection criteria based on a multivariate parameter analysis of the optical and mid-infrared photometric properties show promise of increasing overall detection rates [36]. In particular, compared to a control sample of AGN, megamaser galaxies have been found to have stronger emission at mid-IR wavelengths observed with WISE, with the strongest enhancement at 22 μm. In addition, megamaser galaxies tend to

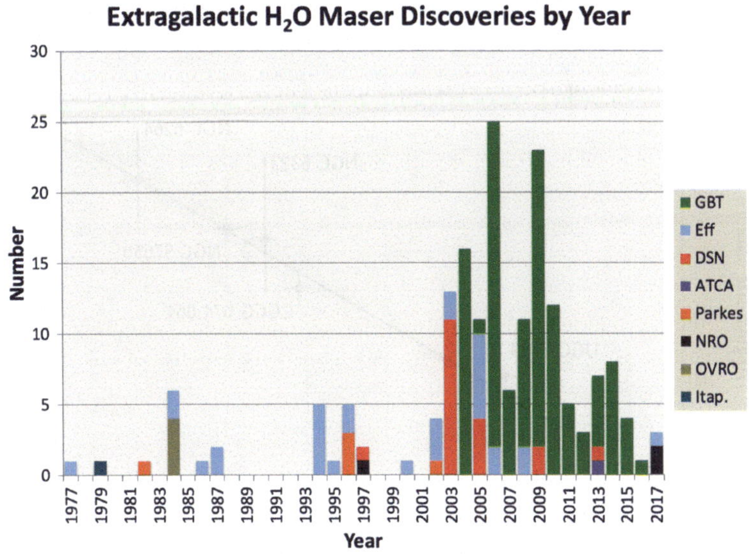

Fig. 4.2 The number of extragalactic water masers discovered each year, labeled by the telescope that made the discovery

have redder IR colors and larger integrated mid-IR luminosities [36]. The correlation of megamasers with X-ray bright galaxies and galaxies with X-ray column densities $N_H > 10^{23}$ cm^{-2} is well known [41, 54], and recent work demonstrates that the X-ray and mid-IR properties could be examined together to enhance megamaser detection rates significantly [37]. Surveys of nearby narrow-line Seyfert 1 (NLS1) galaxies have also identified an enhanced detection rate [52], but the overall sample sizes are small and the newly identified galaxies from NLS1 samples have so far not been identified with disk masers.

The combined total of extragalactic maser surveys have so far identified $\sim$ 160 megamaser galaxies in AGN. About 40 show high velocity masers consistent with a disk origin, making them candidates for precise black hole mass measurements (e.g., [33, 55]), and $\sim$ 35 of those show the triple-peaked spectra characteristic of "clean" disk masers [42, 44], making them further candidates for distance measurements. A clean disk megamaser in this context is a maser demonstrating a clear, triple-peaked spectral profile with little or no evident maser emission at intermediate velocities from non-disk components such as jets or outflows.

Because of the significant observing time requirements for VLBI and monitoring to measure each megamaser distance, only 1 or 2 maser galaxies have been observed as part of a distance campaign using the HSA in any one year. The MCP therefore selected and began measuring the best candidates from the set of clean disk masers. Those candidates are identified based on (1) having a declination $\delta >$ -20° suitable for sensitive and efficient observing with the (northern) HSA; (2) having bright and rich maser spectra in all of the systemic, red-shifted, and blue-shifted parts of the spectral

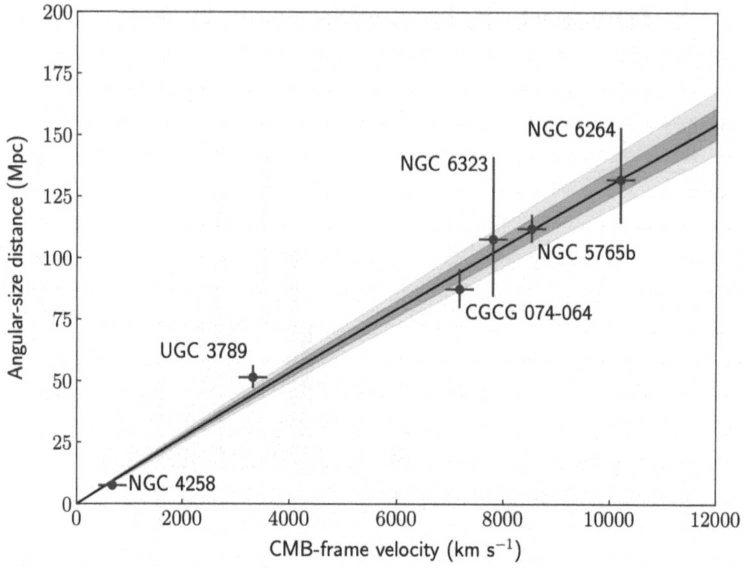

Fig. 4.3 The Hubble diagram for H_2O megamaser galaxies reproduced from [45]. Each galaxy is plotted with 1σ uncertainties in distance and 250 km s^{-1} in velocity set by the unknown peculiar velocities. The solid black line shows the best fit Hubble Constant and the shaded gray regions show 1σ and 2σ confidence intervals

profile, with many components in each to better constrain the disk model; and (3) having distinct and clearly measured velocity drifts among the systemic components in preliminary monitoring observations.

The best current megamaser measurement of H_0 by the MCP incorporates measurements for five systems in the Hubble Flow (UGC 3789 [48], NGC 6264 [34], NGC 6323 [35], NGC 5765b [13], and CGCG 074–064 [44]) as well as NGC 4258 [49]. Distances to each galaxy are calculated using a flat ΛCDM cosmology with $\Omega_m = 0.315$, but the measurement is not strongly dependent on the cosmological model. The galaxies are all within 150 Mpc and the distance calculations vary by $< 1\%$ for any $\Omega_m < 0.5$. The recession velocities are corrected to the frame of the Cosmic Microwave Background. With the number of galaxies involved in the overall measurement being small, the unknown peculiar velocities of the galaxies represents a significant source of uncertainty. When applying a conservatively-estimated uniform uncertainty of 250 km s^{-1} on the recession velocity of each galaxy, added in quadrature with the (much smaller) uncertainty on the fit velocity of the dynamical center, the MCP determines $H_0 = 73.9 \pm 3.0$ km s^{-1} Mpc^{-1} [45] ("best result"). The Hubble diagram for this determination is presented in Fig. 4.3. When applying a group recession velocity for each measured megamaser galaxy rather than the maser-determined recession velocity, the result becomes $H_0 = 73.3 \pm 2.8$ km s^{-1} Mpc^{-1}, and when applying different galaxy flow models the result varies between 71.8 and 76.9 km s^{-1} Mpc^{-1}, with similar uncertainties.

4.5 Prospects for Improving the Measurement of H_0 with Megamasers

Improvements in measuring H_0 with megamasers will progress on a number of fronts. Several known, clean disk megamaser galaxies remain to be observed and analyzed with sufficient precision to contribute to the combined MCP result. These include J0437+2456, ESO 558-G009, and Mrk 1419 [28]. In addition, the megamaser in IC 2560, which shows the promising characteristics of a high quality distance target [53] despite its southern ($\delta \approx$ -34°) location, could contribute meaningfully to the measurement. Improvements in survey strategies and the availability of new catalogs of Seyfert galaxies will lead to more efficient discovery of new megamaser systems. Meanwhile, improvements in disk modeling that relax the thin disk constraint could further reduce the statistical errors in current fits.

4.5.1 Sub-millimeter Water Megamasers and Other Molecules

Water is not the only molecule known to exhibit megamaser emission. The hydroxyl (OH) molecule is also widely observed as a megamaser, primarily associated with gas-rich major galaxy mergers where it traces densities $n(H_2) \gtrsim 10^4$ cm^{-3} [12], and other molecules have been detected as megamasers as well (e.g. [14]), tracing similar density gas. The 6_{16} - 5_{23} rotational transition of ortho-water at 22 GHz is, however, particularly useful for these cosmological studies because it is optimally excited at warm temperatures ($T_K \sim 1000$ K) and high densities $n(H_2) \sim 10^9$ cm^{-3} [15], making it a clear tracer of the Keplerian accretion disks in AGN. Indeed, water remains the only molecule detected as a maser directly in the AGN accretion disks. Moreover, the relatively high frequency of the well-studied 22 GHz water megamasers enables a smaller VLBI beam and precise mapping of the pc-scale disks.

Other rotational transitions of the water molecule at even higher frequencies have also been observed to emit as masers, with several sub-millimeter transitions detected as megamasers in AGN [27]. The development and expansion of ALMA has opened the opportunity to study these sub-mm transitions expansively. The 183 GHz maser line has particularly attracted interest because, in early ALMA studies, it has been found to accompany 22 GHz megamasers [46] widely in accretion disk systems. The 321 GHz transition has also been detected in low-z active galaxies [22, 23, 43]. The promising 380 GHz transition, which is anticipated to be more luminous even than the 22 GHz line when the gas temperature exceeds 1000 K, falls in a particularly strong atmospheric absorption band for low-z systems, but can be detected in high-z [38] QSOs, where it gets red-shifted into a frequency range where the atmosphere is more transparent. Other maser transitions from the water molecule that have been observed in Galactic sources or predicted in molecular models are yet to be explored in AGN.

Compared to the 22 GHz masers, the potential factor of 8 to 17 improvement in VLBI angular resolution from these higher frequency masers, combined with the unexplored incidence rate of sub-mm masers in AGN, has elevated the community interest in finding and exploring such masers as potential future distance indicators. Spectral line VLBI at sub-mm wavelengths is a developing field, and as VLBI facilities expand and mature at these wavelengths, newly discovered maser disk systems may generate even more precise results than the 22 GHz transition.

4.5.2 The ngVLA Era

Megamaser surveys are sensitivity limited. The advent of the sensitive GBT opened the field of direct H_0 measurement with megamasers, and the next leap forward in K-band sensitivity will come in the mid-2030s with the Next Generation Very Large Array (ngVLA). This telescope will include 244 antennas each 18-m in diameter as well as 19 more antennas 6-m in diameter and will offer better than an order of magnitude improvement in 22 GHz sensitivity over the VLA while incorporating the long baselines of the VLBA.

The sensitivity improvement will mean that, for maser systems of a given isotropic luminosity, the ngVLA will be able to sample a volume $\sim$ 30X larger than the VLA or GBT (which have similar K-band sensitivities to each other). The ngVLA will also be an efficient maser survey instrument, and every observation will include both spectral information and VLBI imaging. Combined spectral monitoring with ngVLA mapping will eliminate the heterogeneous aspect of data used in current measurements and enable an improved analysis scheme to model the disk parameters and dynamics in one global fit, rather than separate modeling of accelerations and disk structure.

By discovering and measuring new megamaser systems, and improving the measurements of known systems, the ngVLA will enable a $\sim 1\%$ geometric measurement of the Hubble Constant using the megamaser method [7]. Since individual maser distances are uncorrelated, the cumulative improvement in the uncertainty improves as $1/\sqrt{N}$ and so the telescope could attain such a 1% result by measuring, e.g., 7% distances to 50 galaxies. In practice, the uncertainty in each individual maser galaxy's distance determination will vary, being dependent on the quality of its disk maser system. If the galaxies are widely spaced on the sky, the larger number of measurements will average out the uncertainty associated with peculiar velocities and flow models.

4.6 Summary

Circumnuclear water vapor megamasers at 22 GHz in AGN provide a 1.5% anchor for extragalactic distance ladders and also enable a one-step, geometric measurement of the Hubble Constant. The best current measurement of H_0 using only the megamaser method by the MCP is $73.9 \pm 3.0\,\text{km s}^{-1}\,\text{Mpc}^{-1}$. Incremental improvements to this measurement will become available as more galaxies are measured and as megamaser methods extend into sub-mm maser transitions. The ngVLA, offering an order of magnitude better sensitivity at 22 GHz, will enable a leap forward in measurements and will permit a 1% measurement of H_0.

References

1. A.L. Argon, L.J. Greenhill, M.J. Reid, J.M. Moran, E.M.L. Humphreys, Astrophys. J. **659**, 1040 (2007). https://doi.org/10.1086/512718
2. J.A. Braatz, A.S. Wilson, C. Henkel, Astrophys. J. **437**, L99 (1994). https://doi.org/10.1086/187692
3. J.A. Braatz, A.S. Wilson, C. Henkel, Astrophys. J. Supp. Ser. **106**, 51 (1996). https://doi.org/10.1086/192328
4. J.A. Braatz, C. Henkel, L.J. Greenhill, J.M. Moran, A.S. Wilson, Astrophys. J. Lett. **617**, L29 (2004). https://doi.org/10.1086/427185. (arXiv:astro-ph/0412352 [astro-ph].)
5. J.A. Braatz, N.E. Gugliucci, Astrophys. J. **678**, 96–101 (2008). https://doi.org/10.1086/529538 (arXiv:0804.4136 [astro-ph])
6. J.A. Braatz, M.J. Reid, E.M.L. Humphreys, C. Henkel, J.J. Condon, K.Y. Lo, Astrophys. J. **718**, 657–665 (2010). https://doi.org/10.1088/0004-637X/718/2/657 (arXiv:1005.1955 [astro-ph.CO])
7. J. Braatz, D. Pesce, J. Condon, M. Reid, A.S.P. Conf. Ser. **517**, 821 (2018). (arXiv:1810.06686 [astro-ph.CO])
8. E. Churchwell, A. Witzel, W. Huchtmeier, I. Pauliny-Toth, J. Roland, W. Sieber, Astron. Astrophys. **54**, 969 (1977)
9. M.J. Claussen, G.M. Heiligman, K.Y. Lo, Nature **310**, 298 (1984). https://doi.org/10.1038/310298a0
10. M.J. Claussen, K.Y. Lo, Astrophys. J. **308**, 592 (1986). https://doi.org/10.1086/164529
11. M.J. Claussen, P.J. Diamond, J.A. Braatz, A.S. Wilson, C. Henkel, Astrophys. J. **500**, L129 (1998). https://doi.org/10.1086/311405
12. J. Darling, Astrophys. J. Lett. **669**, L9 (2007). https://doi.org/10.1086/523756. (arXiv:0710.1080 [astro-ph].)
13. F. Gao, J.A. Braatz, M.J. Reid, K.Y. Lo, J.J. Condon, C. Henkel, C.Y. Kuo, C.M.V. Impellizzeri, D.W. Pesce, W. Zhao, Astrophys. J. **817**(2), 128 (2016). https://doi.org/10.3847/0004-637X/817/2/128 (arXiv:1511.08311 [astro-ph.GA])
14. M.D. Gorski, S. Aalto, J. Mangum, E. Momjian, J.H. Black, N. Falstad, B. Gullberg, S. König, K. Onishi, M. Sato, F. Stanley, Astron. Astrophys. **654**, A110 (2021). https://doi.org/10.1051/0004-6361/202141633 ([arXiv:2108.06331])
15. M.D. Gray, A. Baudry, A.M.S. Richards, E.M.L. Humphreys, A.M. Sobolev, J.A. Yates, Mon. Not. R. Astron. Soc. **456**, 374 (2016). https://doi.org/10.1093/mnras/stv2437. ([arXiv:1510.06182])
16. L.J. Greenhill, C. Henkel, R. Becker, T.L. Wilson, J.G.A. Wouterloot, Astron. Astrophys. **304**, 21 (1995)

17. L.J. Greenhill, S.P. Ellingsen, R.P. Norris, R.G. Gough, M.W. Sinclair, J.M. Moran, R. Mushotzky, Astrophys. J. Lett. **474**, L103 (1997). https://doi.org/10.1086/310434. (arXiv:astro-ph/9611127 [astro-ph].)
18. L.J. Greenhill, P.T. Kondratko, J.E.J. Lovell, T.B.H. Kuiper, J.M. Moran, D.L. Jauncey, G.P. Baines, Astrophys. J. **582**, L11 (2003). https://doi.org/10.1086/367602
19. L.J. Greenhill, P.T. Kondratko, J.M. Moran, A. Tilak, Astrophys. J. **707**, 787–799 (2009). https://doi.org/10.1088/0004-637X/707/1/787 (arXiv:0911.0382 [astro-ph.CO])
20. Y. Hagiwara, P.J. Diamond, M. Miyoshi, E. Rovilos, W. Baan, Mon. Not. R. Astron. Soc. **344**, L53 (2003). https://doi.org/10.1046/j.1365-8711.2003.07005.x
21. A.D. Haschick, W.A. Baan, E.W. Peng, Astrophys. J. Lett. **437**, L35 (1994). https://doi.org/10.1086/187676
22. Y. Hagiwara, M. Miyoshi, A. Doi, S. Horiuchi, Astrophys. J. Lett. **768**, L38 (2013). https://doi.org/10.1088/2041-8205/768/2/L38. (arXiv:1305.0076 [astro-ph.GA].)
23. Y. Hagiwara, S. Horiuchi, A. Doi, M. Miyoshi, P.G. Edwards, Astrophys. J. **827**, 69 (2016). https://doi.org/10.3847/0004-637X/827/1/69. ([arXiv:1604.07937])
24. C. Henkel, J.A. Braatz, L.J. Greenhill, A.S. Wilson, Astron. Astrophys. **394**, L23 (2002). https://doi.org/10.1051/0004-6361:20021297. (arXiv:astro-ph/0211226 [astro-ph].)
25. J.R. Herrnstein, J.M. Moran, L.J. Greenhill, P.J. Diamond, M. Inoue, N. Nakai, M. Miyoshi, C. Henkel, A. Riess, Nature **400**, 539–541 (1999). https://doi.org/10.1038/22972. (arXiv:astro-ph/9907013 [astro-ph])
26. E.M.L. Humphreys, M.J. Reid, J.M. Moran, L.J. Greenhill, A.L. Argon, Astrophys. J. **775**, 13 (2013). https://doi.org/10.1088/0004-637X/775/1/13
27. E.M.L. Humphreys, L.J. Greenhill, M.J. Reid, H. Beuther, J.M. Moran, M. Gurwell, D.J. Wilner, P.T. Kondratko, Astrophys. J. Lett. **634**, L133–L136 (2005). https://doi.org/10.1086/498890 (arXiv:astro-ph/0511014 [astro-ph])
28. C.M.V. Impellizzeri, J.A. Braatz, C.Y. Kuo, M.J. Reid, K.Y. Lo, C. Henkel, J.J. Condon, IAU Symp. **287**, 311 (2012). https://doi.org/10.1017/S1743921312007235. (arXiv:1205.4758 [astro-ph.CO].)
29. Y. Ishihara, N. Nakai, N. Iyomoto, K. Makishima, P. Diamond, P. Hall, Publ. Astron. Soc. Jap. **53**, 215 (2001). https://doi.org/10.1093/pasj/53.2.215
30. A.M. Koekemoer, C. Henkel, L.J. Greenhill, A. Dey, W. van Breugel, C. Codella, R. Antonucci, Nature **378**, 697 (1995). https://doi.org/10.1038/378697a0
31. P.T. Kondratko, L.J. Greenhill, J.M. Moran, J.E.J. Lovell, T.B.H. Kuiper, D.L. Jauncey, L.B. Cameron, J.F. Gomez, C. Garcia-Miro, E. Moll et al., Astrophys. J. **638**, 100–105 (2006). https://doi.org/10.1086/498641. (arXiv:astro-ph/0510851 [astro-ph].)
32. P.T. Kondratko, L.J. Greenhill, J.M. Moran, Astrophys. J. **652**, 136–145 (2006). https://doi.org/10.1086/507885 (arXiv:astro-ph/0610060 [astro-ph].)
33. C.Y. Kuo, J.A. Braatz, J.J. Condon, C.M.V. Impellizzeri, K.Y. Lo, I. Zaw, M. Schenker, C. Henkel, M.J. Reid, J.E. Greene, Astrophys. J. **727**, 20 (2011). https://doi.org/10.1088/0004-637X/727/1/20. (arXiv:1008.2146 [astro-ph.CO].)
34. C. Kuo, J.A. Braatz, M.J. Reid, F.K.Y. Lo, J.J. Condon, C.M.V. Impellizzeri, C. Henkel, Astrophys. J. **767**, 155 (2013). https://doi.org/10.1088/0004-637X/767/2/155. (arXiv:1207.7273 [astro-ph.CO].)
35. C.Y. Kuo, J.A. Braatz, K.Y. Lo, M.J. Reid, S.H. Suyu, D.W. Pesce, J.J. Condon, C. Henkel, C.M.V. Impellizzeri, Astrophys. J. **800**(1), 26 (2015). https://doi.org/10.1088/0004-637X/800/1/26. (arXiv:1411.5106 [astro-ph.GA].)
36. C.Y. Kuo, A. Constantin, J.A. Braatz, H.H. Chung, C.A. Witherspoon, D. Pesce, C.M.V. Impellizzeri, F. Gao, L. Hao, J.-H. Woo, I. Zaw, Astrophys. J. **860**, 169 (2018). https://doi.org/10.3847/1538-4357/aac498
37. C.Y. Kuo, J.Y. Hsiang, H.H. Chung, A. Constantin, Y.-Y. Chang, E. da Cunha, D. Pesce, W.T. Chien, B.Y. Chen, J.A. Braatz, I. Zaw, S. Matsushita, J.C. Lin, Astrophys. J. **892**, 18 (2020). https://doi.org/10.3847/1538-4357/ab781d
38. C.Y. Kuo, S.H. Suyu, V. Impellizzeri, J.A. Braatz, https://doi.org/10.1093/pasj/psz032. (arXiv:1901.04745 [astro-ph.GA])

39. M. Miyoshi, J. Moran, J. Hernstein, L. Greenhill, N. Nakai, P. Diamond, M. Inoue, Nature **373**, 127–129 (1995). https://doi.org/10.1038/373127a0
40. N. Nakai, M. Inoue, M. Miyoshi, Nature **361**, 45 (1999)
41. F. Panessa, P. Castangia, A. Malizia, L. Bassani, A. Tarchi, A. Bazzano, P. Ubertini, Astron. Astrophys. **641**, A162 (2020). https://doi.org/10.1051/0004-6361/201937407. (arXiv:2006.08280 [astro-ph.GA].)
42. D.W. Pesce, J.A. Braatz, J.J. Condon, F. Gao, C. Henkel, E. Litzinger, K.Y. Lo, M.J. Reid, Astrophys. J. **810**(1), 65 (2015). https://doi.org/10.1088/0004-637X/810/1/65. arXiv:1507.07904 [astro-ph.GA].)
43. D.W. Pesce, J.A. Braatz, C.M.V. Impellizzeri, Astrophys. J. **827**, 68 (2016). https://doi.org/10.3847/0004-637X/827/1/68. ([arXiv:1604.03789])
44. D.W. Pesce, J.A. Braatz, M.J. Reid, J.J. Condon, F. Gao, C. Henkel, C.Y. Kuo, K.Y. Lo, W. Zhao, Astrophys. J. **890**, 118 (2020). https://doi.org/10.3847/1538-4357/ab6bcd
45. D.W. Pesce, J.A. Braatz, M.J. Reid, A.G. Riess, D. Scolnic, J.J. Condon, F. Gao, C. Henkel, C.M.V. Impellizzeri, C.Y. Kuo et al., Astrophys. J. Lett. **891**(1), L1 (2020). https://doi.org/10.3847/2041-8213/ab75f0. (arXiv:2001.09213 [astro-ph.CO])
46. D.W. Pesce, J.A. Braatz, C. Henkel, E.M.L. Humphreys, C.M.V. Impellizzeri, C.-Y. Kuo, Astrophys. J. **948**, 134 (2023). https://doi.org/10.3847/1538-4357/acc57a. ([arXiv:2302.02572])
47. M.J. Reid, J.A. Braatz, J.J. Condon, L.J. Greenhill, C. Henkel, K.Y. Lo, Astrophys. J. **695**, 287–291 (2009). https://doi.org/10.1088/0004-637X/695/1/287. (arXiv:0811.4345 [astro-ph].)
48. M.J. Reid, J.A. Braatz, J.J. Condon, K.Y. Lo, C.Y. Kuo, C.M.V. Impellizzeri, C. Henkel, Astrophys. J. **767**, 154 (2013). https://doi.org/10.1088/0004-637X/767/2/154. (arXiv:1207.7292 [astro-ph.CO].)
49. M.J. Reid, D.W. Pesce, A.G. Riess, Astrophys. J. Lett. **886**(2), L27 (2019). https://doi.org/10.3847/2041-8213/ab552d. (arXiv:1908.05625 [astro-ph.GA].)
50. A.G. Riess, W. Yuan, L.M. Macri, D. Scolnic, D. Brout, S. Casertano, D.O. Jones, Y. Murakami, L. Breuval, T.G. Brink et al., Astrophys. J. Lett. **934**(1), L7 (2022). https://doi.org/10.3847/2041-8213/ac5c5b. (arXiv:2112.04510 [astro-ph.CO].)
51. S. Sawada-Satoh, S. Kameno, K. Nakamura, D. Namikawa, K.M. Shibata, M. Inoue, Astrophys. J. **680**, 191 (2008). https://doi.org/10.1086/587886. (arXiv:0802.4149 [astro-ph].)
52. A. Tarchi, P. Castangia, A. Columbano, F. Panessa, J.A. Braatz, Astron. Astrophys. **532**, A125 (2011). https://doi.org/10.1051/0004-6361/201117213. (arXiv:1107.5155 [astro-ph.CO].)
53. A. Yamauchi, N. Nakai, Y. Ishihara, P. Diamond, N. Sato, Publ. Astron. Soc. Jap. **64**, 103 (2012). https://doi.org/10.1093/pasj/64.5.103. (arXiv:1207.6820 [astro-ph.GA].)
54. J.S. Zhang, C. Henkel, M. Kadler, L.J. Greenhill, N. Nagar, A.S. Wilson, J.A. Braatz, Astron. Astrophys. **450**, 933 (2006). https://doi.org/10.1051/0004-6361:20054138. (arXiv:astro-ph/0512459 [astro-ph].)
55. W. Zhao, J.A. Braatz, J.J. Condon, K.Y. Lo, M.J. Reid, C. Henkel, D.W. Pesce, J.E. Greene, F. Gao, C.Y. Kuo et al., Astrophys. J. **854**, 124 (2018). https://doi.org/10.3847/1538-4357/aaa95c. (arXiv:1801.06332 [astro-ph.GA].)

Chapter 5
Eclipsing Binary Stars as Precise and Accurate Distance Indicators

Grzegorz Pietrzyński and Dariusz Graczyk

5.1 Introduction

Eclipsing binary stars offer a unique opportunity to measure very precisely and accurately stellar parameters like mass, radius, etc. [21, 66]. The first precise measurements were done more than hundred years ago for famous Algol [59], see also [31] for a short historical review. Nowadays with advent of very high quality photometric data from space missions and radial velocities measurement from very stable spectrographs an amazing progress in precision of measurement of stellar parameters has been achieved. Indeed several authors reported mass and radius measurements with a precision close to 0.1% [12, 24, 38]. Eclipsing binary stars can be also used to measure geometrical distances even to nearby galaxies. This makes them a perfect tool to calibrate the best standard candles like Cepheids, TRGB, JAGB and as a consequence the SN Ia to determine H_0.

In this section we will briefly describe the progress on distance determinations with eclipsing binary stars. The detailed description of the technique and our future plans will follow in next sections.

5.1.1 Astrometric Binaries

Before proceeding to the main subject, let's discuss a bit the astrometric binary stars in context of distance measurements. In general both groups of stars overlap, and

G. Pietrzyński (✉)
Nicolaus Copernicus Astronomical Centre, Bartycka 18, Warszawa, Poland
e-mail: pietrzyn@camk.edu.pl

D. Graczyk
Nicolaus Copernicus Astronomical Centre, Rabiańska 8, Toruń, Poland
e-mail: darek@ncac.torun.pl

E. Di Valentino and D. Brout (eds.), *The Hubble Constant Tension*, Springer Series in Astrophysics and Cosmology, https://doi.org/10.1007/978-981-99-0177-7_5

techniques used to analyse them are also similar. When the spectral lines from both components can be detected the radial velocity of the orbital motion around the centre of mass of both components can be measured providing the spectroscopic orbit of the system. Together with the apparent orbit from astrometry (done with speckle or long-base interferometry) we can measure very precisely stellar masses and distance d using the following equations [66]:

$$
\begin{aligned}
M_1 &= \frac{1.036149 \times 10^{-7}(K_1 + K_2)^2 K_2 P(1-e^2)^{3/2}}{\sin^3 i} \\
M_2 &= \frac{1.036149 \times 10^{-7}(K_1 + K_2)^2 K_1 P(1-e^2)^{3/2}}{\sin^3 i} \\
a_{\mathrm{AU}} &= \frac{9.191940 \times 10^{-5}(K_1 + K_2) P(1-e^2)^{1/2}}{\sin i} \\
d &= \frac{a_{\mathrm{AU}}}{a''}
\end{aligned}
\tag{5.1}
$$

where the masses M_1, M_2 are expressed in solar units, the distance d in parsecs, semi amplitudes of radial velocities K_1 and K_2 in km s^{-1}, the period P in days, and the apparent semi-major axis a'' in arcseconds. The parameter a_{AU} is the linear semi-major axis expressed in astronomical units.

Indeed from the fit of the spectroscopic orbit we get the linear size of the semi major axis while astrometric orbit gives its projection on the sky.

This very powerful technique has been applied several times to measure geometric distances [7, 13, 40, 65, 67]. A 1% precision in distance measurement level was obtained with this method more than a decade ago. The limiting factor was the calibration of the interferometric data. Recently Gallenne et al. [12] published ultra-precise distances to ten binary systems. Observations obtained with ESO instruments GRAVITY allowed to reach 0.02% accuracy in wavelength calibration and together with spectra obtained with UVES spectrograph resulted in distances precise to 0.2% on average.

However this method is limited to the systems that can be spatially resolved. Even when using the very high resolution observations from the state-of-the-art interferometric instruments, and studying systems with periods that allow to complete data in a reasonable amount of time (e.g. 1 year or so) the range of this technique is limited to about 200 pc (Fig. 5.1).

5.1.2 *Eclipsing Binaries*

In order to reach nearby galaxies and use the full potential of binary systems to calibrate the extragalactic distance scale we have to use eclipsing binary systems. The huge potential of eclipsing binaries as precise and accurate distance indicators

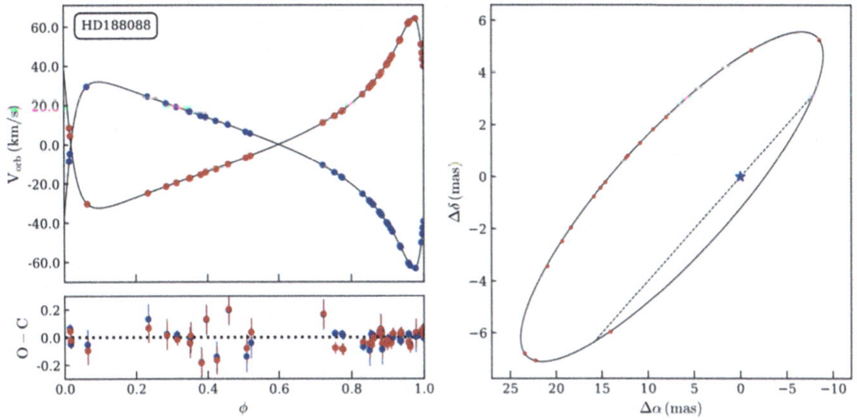

Fig. 5.1 Spectroscopic and astrometric orbits of a binary system HD 188088. Based on these data the distance of 14.1129±0.0038 pc was measured

was recognized a long time ago [25, 32, 43]. However as usual the devil is in the details and several different problems prevented astronomers to measure precise geometrical distances with detached eclipsing binaries.

First of all it is not possible to measure distances with eclipsing binaries based on solely results of analysis of the photometric light curves and spectroscopic orbit. An additional information is needed. The most common approach applied to measure extragalactic distances was to determine the intrinsic flux of each eclipsing binary component using bolometric correction and effective temperature determinations and then calculate a distance:

$$d_i \text{ [pc]} = 3.360\, R_i\, T_i^2\, 10^{0.2(BC_i + V_i)} \tag{5.2}$$

where V_i is an observed, extinction corrected V-band magnitude of i-th component and BC is the bolometric correction. This way of measuring distances makes the whole method sensitive to significant errors on effective temperature determination. One way of dealing with this difficulty is to fit spectrophotometry using stellar atmospheric models [26], however that makes the procedure strongly model-dependent.

Another complication to apply eclipsing binary method to measure distances to nearby galaxies is a limited number of good systems. The most common systems identified there are bright, massive close EBs of late O-type or early B-type. Very often they are multiple systems and they show complication in light curves due to a presence of strong stellar winds, circumstellar matter, thin accretion disc, pulsations or spot activity. In spite of this situations substantial progress was achieved and several projects reported distance determinations to nearby galaxies using early type EBs in the LMC [4, 8, 54], M33 [55, 74], and M31 [3]. All of them are subject of rather unknown systematic errors related to adopted stellar atmospheric models and strong degeneracy between effective temperature and reddening.

Another way of using eclipsing binary systems to measure distances is to use a very well established relationship between the surface brightness and color (SBCR). In this approach the angular diameters of the stars, components of the binary system comes from the surface - brightness (see more details in the next paragraph) while their linear dimensions come from modelling of the light curve and radial velocity curve.

The big complication in applying this method to measure extragalactic distances is that the SBCR is well calibrated for late type stars only (with spectral types later than middle A-type). Late typed dwarfs are intrinsically to faint to secure high-resolution spectra for such system even in the LMC. Fortunately the Magellanic Clouds have been monitored for more than 25 years by the Optical Gravitational Lensing Experiment. These unique dataset allowed to discover long period eclipsing binaries composed of helium burning giants in both LMC and SMC [45]. Such objects are perfectly suited for modeling. In the course of the Araucaria project ([49]) we started to observe these targets in order to provide high-resolution spectra, near-infrared photometry and improve the optical light curves.

The first results published for one of these systems [46] fully confirmed big prospects of this method. The first precise distance determinations with our late-typed systems to the LMC [47] and SMC [19], with precision of 2% and 3%, respectively provided the best zero points for calibration of the standard candles. These determinations were limited mostly (60% of the total error) by the systematic errors related to the SBCR. Therefore we dedicated a lot of efforts to improve on that. With new precision near-infrared photometry [33] and interferometric angular diameter determinations [14] obtained for a sample of nearby red clump stars (e.g. stars very similar to the components of our eclipsing binaries in the Magellanic Clouds) we established much better the SBCR. Based on it and studying more systems we finally provided the best so far distances to the LMC (49.50 $\pm$ 0.09 (statistical) $\pm$0.54 (systematic) kpc [48] and SMC (62.44 $\pm$ 0.47 (st.) $\pm$0.81 (sys.) kpc [22]. These results clearly show that the eclipsing binary systems can provide distances precise and accurate to about 1%. However our results are limited again by the precision of the SBCR (about 0.8% systematic contribution to the total error on distance).

5.2 Methods Details

The eclipsing binary method is combining information derived from spectroscopy and photometry. In this regard it is similar to Baade-Wesselink method of distance determination to radially pulsating stars [2, 76]. The method utillises a concept of the stellar surface brightness [77] and empirical calibrations of the surface brightness—color relations (SBCRs) in order to avoid systematic uncertainties arising from a use of bolometric corrections or absolute fluxes predicted by stellar atmospheric models [32]. Currently the most precise SBCRs are calibrated against optical-infrared colors

$(V-K)$ and $(B-K)$ and they allow to calculate angular diameters of stars with a precision of 1% or even slightly better e.g. [23, 30, 48, 58].

The simplicity of the method is actually its largest advantage: to derive a distance d to a star it is used a basic geometrical relation between a physical radius R and an angular diameter ϕ:

$$d\ [\mathrm{pc}] = 9.301\ \frac{R}{\phi} \tag{5.3}$$

where R is expressed in the solar nominal radii [52], ϕ in miilliarcseconds and the conversion factor is equal $2\mathcal{R}^N_\odot/1$ AU.

Determination of radii of eclipsing binary components is the most critical and time consuming part of the method. It consists of several steps:

1. Selection of suitable eclipsing binary candidates, usually based on light curves from large photometric surveys.
2. Spectroscopic follow up of the best candidates.
3. Securing near-infrared photometry in J and K band.
4. Combined analysis of photometry and spectroscopy using a well-tested software.

5.2.1 Candidate EBs

For the distance determination to the Magellanic Clouds we searched published catalogs of eclipsing binary stars [70, 80], based on the OGLE project photometry [69], to find good candidates. Moreover we developed our own method to identify detached eclipsing binary stars from ground-based non continuous photometry [17] and we employed it to data from the third phase of OGLE project [68] in order to find more candidates [18, 44].

In order to retain a geometrical simplicity and to avoid additional corrections an eclipsing binary should have a number of specific properties. Consequently only a small fraction of eclipsing binaries are actually suited for the method. The near-ideal system is a double-lined, well detached eclipsing binary (DEB), thus both components are seen in observed spectra and stars can be safely treated as spheres. Components should be photometrically quite avoiding this way complications introduced by stellar spots, winds, flares, etc. The orbital inclination i should be close to 90° resulting in total eclipses or deep both eclipses ($\Delta m > 0.4$ mag) making the determination of radii more unequivocal. Triple or multiple systems are disfavoured unless a light contribution of additional components to the total system's light is negligible, eventually it is small (of order of several per cents) and very well constrained. Finally, candidate DEBs with smallest interstellar extinction are preferred.

For the Magellanic Clouds even more constrains were needed. The systems should have sufficient brightness for the spectroscopic follow up with 4m-8m class telescopes. However, at the same time, they should consist of components of late spectral type in order to use high precision SBCRs. The only choice, up to now, is to use eclipsing binary stars consisting of late-type giants of F,G and K type. Such eclipsing binaries fulfilling all described above criteria are extremely rare because their orbital periods are of order of 100–1000 days. We found only 40 suitable systems in the LMC and 18 in the SMC, while total number of known eclipsing binaries in both galaxies are ∼40000 and ∼8400 in the LMC and the SMC, respectively [45].

5.2.2 *Spectroscopic Follow-Up*

The spectroscopic follow up have a number of objectives. They are:

- Obtaining radial velocity curves of both components.
- Search for eventual multiplicity of a DEB.
- Determination of spectroscopic light ratio.
- The atmospheric analysis of disentangled spectra of components.
- Determination of the interstellar extinction.

To determine precise radial velocities using spectra of binary stars a number of methods were developed. The cross correlation technique [64] based on comparison of an observed spectrum with a template spectrum is widely used, as well as its extension to two-dimensions – TODCOR [81]. Very good results, especially in cases when stellar lines are blended or S/N of spectra is small, gives another method which is based on a determination of the spectral line broadening function (BF) [56]. We used this method successfully to measure radial velocities of DEBs in the Magellanic Clouds with a dedicated software RaVeSpAn [50]. For targets as faint as 17–18 mag in V band we could extract radial velocities with a precision better than 400 m s^{-1} for spectra with S/N < 10 [20, 22].

When a third set of lines was observed in a spectrum or the systemic radial velocity showed clear and large radial velocity trend a DEB was discarded from the sample. However we found only two such cases in our full sample of systems. From observed spectra we determined the light ratio of the components in a specific photometric band, usually V-band, by measuring the strength of absorption lines of both components. Of course one needs to correct it for the temperature difference of components, and that information comes from light curve modelling. The light ratio helps constrain the ratio of components' radii in case of partial and relatively shallow eclipses eclipses ($\Delta m \sim 0.25$–0.5 mag).

The whole set of observed spectra for a particular system was separated into single spectra of individual components using the iterative method [15]. Resulting components' spectra were analysed in order to determine atmospheric parameters like the surface temperature and the metallicity. The former is important during light curve modeling to constrain the limb darkening coefficients. The latter allows for more precise comparison of components with the stellar evolution models and determination of the age of a system.

Precise determination of the interstellar extinction in direction of each eclipsing binary reduces significantly errors on distances. We employed three methods to calculate the reddening. First was based on spectra and used a calibration between reddening and equivalent width of interstellar sodium absorption lines NaI (5890.0, 5895.9Å)[42, 51]. Second method used color - effective temperature calibrations like e.g. [53, 79] to calculate intrinsic colors of components. The effective temperature are derived from atmospheric analysis of decomposed spectra. Simply comparison of observed and intrinsic colors allow for calculation of the interstellar reddening. And finally we used the reddening maps in the Magellanic Clouds [16, 27]. In order to diminish systematic uncertainties of each method we calculated an average reddening from all above methods.

5.2.3 *The Infrared Photometry*

Securing high quality near-infrared photometry is vital for the method, because most precise SBCR are calibrated for $(V-K)$ color. For every system we collected multi-epoch J,K photometry with SOFI camera installed on 3.5m NNT telescope on La Silla. Resulting precision of J and K magnitudes were about 0.01 mag or better. Finally we converted magnitudes into 2MASS system [60] in order to have photometry which is consistent with that obtained by Laney et al. [33].

5.2.4 *Combined Analysis of Photometry and Velocimetry*

Details of modelling of eclipsing binary giant stars in the Magellanic Clouds are given in [20, 22]. Here we give only a brief description. Light curves and radial velocity curves were analysed simultaneously with the Wilson-Devinney program [78] and also with JKTEBOP code [61]. An extensive modelling was performed for every system. To validate consistency of a model we compared the spectroscopic light and temperature ratios with model predictions. In most cases consistency was fully satisfactory. We put a lot of attention to constrain the third light which can origin from a blend or an unresolved physical companion star. In some cases we could strengthen the analysis by inclusion of additional B (blue) and R (red) light curves from MACHO project [1]. Modelling multi-band photometry has an advantage to better constrain the temperature ratio of components and the third light.

We used two different modelling codes in order to find the best model and to investigate eventual discrepancies in the model parameters. JKTEBOP served as auxiliary code to confirm results obtained with the WD code as well as to calculate statistical errors of model parameters with the Monte Carlo method. These errors were compared with formal errors returned by the WD code, and always the larger one was chosen as the final uncertainty on a parameter. Significant discrepancies between parameters returned by the WD and JKTEBOP were sporadic and limited to cases where components were significantly distorted due to tidal forces; then parameters and their errors were adopted after the WD code.

5.2.5 Error Analysis

We followed [9, 71] in dividing the distance error into so called the statistical and the systematic parts. We used this division in most of our papers reporting distance measurements. That was strongly motivated by a need to underline a correlation between derived individual eclipsing binary distances and also the zero-point uncertainty of the SBCR used. In other words there is a "systematic" part of the total error which does not reduce while enlarging a sample of eclipsing binary stars used to measure a distance to a particular object. This part of the error is an immanent feature of the method and only a modification/improvement of the method can change/decrease it.

This nomenclature is a bit unfortunate, because the phrase "systematic error" has in general a different meaning in physical sciences. It is a "consistent, repeatable error (bias) associated with faulty equipment or a flawed experiment design" leading to a permanent shift of a measured quantity (e.g. a distance) from the true one. Once such error is identified an instrument/method can be corrected for it and one can pretty much forget about it. However, we had something else in mind: not that the method we used is somehow "flawed", but rather that its accuracy is limited because of a number of slight but still not fully recognised effects. In our opinion a good measure of this "unknown" systematic is a standard deviation (a spread) of the SBCR fit used by us [48] because the uncertainty on the SBCR dominates the total error budget of our eclipsing binary distances.

The statistical part of the LMC distance error was found simply by the Monte Carlo simulations. We draw 10 000 times from individual distance measurements within errors assuming their Gaussian distribution. For each draw we computed the mean distance and geometrical orientation of the disc (the inclination to the line-of-sight, the longitude of nodes) and we made a histogram of the results.

In case of the SMC situation was more difficult because of the large extension of this galaxy in a direction of view and a significant distortion due to gravitational interaction with the LMC and the Milky Way. In that case statistical error was determined by analysing distance trends of individual targets in two perpendicular directions within the galaxy [22].

5.3 Toward Sub-percent Distances Within the Local Group

5.3.1 New Calibrations of SBCR

5.3.1.1 SPICA

CHARA/SPICA project [41] aims at precise measuring angular diameters of about 1000 stars with long-base optical interferometer on Mount Wilson observatory. Non-variable and non-multiple stars brighter than 10 mag in V-band were carefully selected to cover all spectral types. The main advantage of this project is a homogenous way all angular diameters will determined minimising this way systematic effects when angular diameters obtained with different instruments are combined. The angular diameters will be used to extract modern and accurate SBCRs over very wide range of colors, metallicities and the surface gravities.

The novelty of project is a use of a near-infrared fringe tracker allowing for longer exposures of the fringe signals in the visible. This way SPICA has higher sensitivity and will allow to measure angular diameters as small as 0.1–0.2 mas even for faintest stars in the sample. The resulting multi-parameter SBCRs are expected to reach sub-percent precision.

5.3.1.2 Nearby DEBs

Another, independent route to improve significantly SBCRs provide eclipsing binary stars. By reversing the eclipsing binary method one can derive angular diameters of an eclipsing binary components and calibrate the SBCR, provided a distance is known. We selected over 100 suitable, well detached eclipsing binary stars with spectral types from mid B-type to early K-type. In the analysis we use superb quality light curves from space mission *Kepler* K2 [28] and TESS [57] to find photometric solutions with the WD code. The spectroscopic follow up was performed with the HARPS spectrograph [39] well known for its high precision performance. As a result we can derive absolute radii of components usually with a precision much better 1%.

The Gaia satellite mission [10, 11] resulted in a tremendous improvement in a number of stars with precisely measured trigonometric parallaxes. Although there is still a debate on how accurate are Gaia parallaxes from latest DR3 (what is called a parallax bias) after correcting them using a general scheme proposed by Lindegren et al. [36] they have only very small zero-point parallax offset of order of several μas [6, 75]. For nearby eclipsing binary stars ($d < 200$ pc) such systematic uncertainty is practically insignificant and one can almost safely use Gaia parallaxes. We write "almost" because Gaia see photocenter movement of nearby EBs. For overwhelming majority of them Gaia DR3 provide only five- or six-parameter solutions what may, in principle, lead to some bias when an amplitude of photocenter movements is comparable or larger than a parallax. However, full 9-parameter solutions accounting for photocenter movements will likely be published by Gaia DR4.

5.3.1.3 Early Type DEBs in the LMC

There are practically no nearby well detached eclipsing binary stars containing components of O-type or early B-type. Those which are at larger distances of about 0.5–1.5 kpc lie close to the galactic plane and suffer from significant reddening. As a consequence we can't follow the same route like in case of EBs of later spectral types.

However, there is another possibility to calibrate precise SBCRs for hottest stars using EBs. The distance to the LMC was firmly established with late-type eclipsing binary stars. On the other hand there is a large number of early type EBs identified in both Magellanic Clouds [45] from which we choose about 20 best candidates. The systems are bright, well detached, little reddened and have good quality light curves. High precision spectroscopic follow-up of them with UVES spectrograph was performed in order to derive their fundamental parameters. First results for two eclipsing binaries were already published [62, 63]. We expect to obtain this way precise radii for about 40 early type stars in the LMC. Individual distances of stars follow from their positions within the disc of the LMC, the geometry of which is known [48, 72]. The resulting angular diameters would define precise SBCRs for this important type of stars.

5.3.2 Distances to More Local Group Galaxies

In order to precisely calibrate standard candles like Cepheids, TRGB etc., it is crucial to provide geometrical distances to several galaxies having a wide range of metallicities. Unfortunately, the eclipsing binaries containing late-type giant stars which were used successfully to determine distances to the Magellanic Clouds are rather too faint. The brightest such systems containing bright giants had combined absolute magnitude M_V in a range from -3.0 up to -3.5 mag. That translates into, neglecting the reddening, observed V magnitudes of 21.4–20.9 mag in M31 [35] and 21.6–21.1 mag in M33 [5]. Note that individual components will be even $\sim$0.7 mag fainter! Given their long orbital period it will be very challenging to detect such systems at a distance of M33, and then to perform follow up observations of such a faint targets.

The precise SBCRs calibrated for massive and hot stars, see Sects. 5.3.1.1 and 5.3.1.3, could be used to determine direct near-geometric distances to more distant galaxies within the Local Group like M31, M33, NGC 6822, IC 1613 etc. A number of candidate detached EBs comprising very early type components were identified in M31 and M33 by DIRECT project [29, 37], in M31 by [73] and in M31 by Pan-STARRS [34]. Three systems with eccentric orbits were selected in M31 for spectroscopic follow-up with 8-m class telescopes, based on high brightness, quality of light curves and low reddening. More effort is required to provide fully empirical distanced based on DEBs for M31 and M33. While in other galaxies dedicated surveys should be performed to select good systems first (Fig. 5.2).

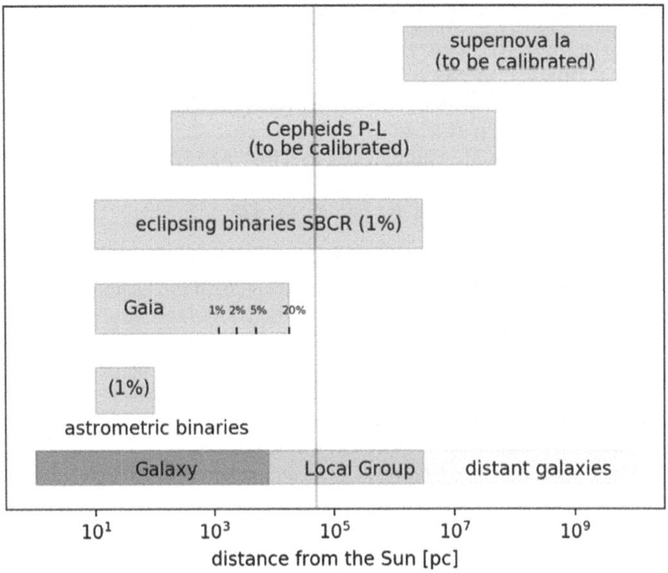

Fig. 5.2 Range and precision of the geometrical methods. The range of distances that can be measured with astrometric binaries, Gaia parallaxes, eclipsing binaries together with the surface brightness—colour relation (SBCR), the Cepheid P-L relation and type Ia supernovae are marked with the corresponding bars. Astrometric binaries, Gaia parallaxes, and eclipsing binaries with the SBCR can be used to calibrate the Cepheid P-L relation, TRGB, JAGB, and as a result the supernova Ia brightness. As can be appreciated, eclipsing binaries together with our new relation offer us the opportunity to determine distances competitive in precision to Gaia already at about 1 kpc from the Sun, while retaining their high precision for distances up to 1 Mpc. The lower bar divided into three sections shows the range of distances within our Galaxy, Local Group, and more distant galaxies. The vertical grey line marks the distance to the LMC

Acknowledgements The research leading to these results has received funding from the European Research Council (ERC) under the European Unions Horizon 2020 research and innovation program (grant agreement No. 951549). We also acknowledge support from the National Science Center, Poland grants MAESTRO UMO-2017/26/A/ ST9/00446, BEETHOVEN UMO-2018/31/G/ST9/03050, and DIR/WK/2018/09 and 2024/WK/02 grants of the Polish Ministry of Education and Science and Polish Ministry of Science and Higher Education.

References

1. C. Alcock et al., [MACHO], Astrophys. J. **486**, 697–726 (1997). https://doi.org/10.1086/304535. (arXiv:astro-ph/9606165 [astro-ph])
2. W. Baade, Astron. Nachr. **228**, 359–362 (1926)
3. A.Z. Bonanos, K.Z. Stanek, R.P. Kudritzki, L.M. Macri, D.D. Sasselov, J. Kaluzny, P.B. Stetson, D. Bersier, F. Bresolin, T. Matheson et al., Astrophys. J. **652**, 313–322 (2006). https://doi.org/10.1086/508140. (arXiv:astro-ph/0606279 [astro-ph])

4. A.Z. Bonanos, N. Castro, L.M. Macri, R.P. Kudritzki, Astrophys. J. Lett. **729**, L9–L14 (2011). https://doi.org/10.1088/2041-8205/729/1/L9. (arXiv:1101.1953 [astro-ph.SR])
5. L. Breuval, A.G. Riess, L.M. Macri, S. Li, W. Yuan, S. Casertano, T. Konchady, B. Trahin, M.J. Durbin, B.F. Williams, Astrophys. J. **951**, 118 (2023). https://doi.org/10.3847/1538-4357/acd3f4. ([arXiv:2304.00037])
6. M. Cruz Reyes, R.I. Anderson, Astron. Astrophys. **672**, A85 (2023). https://doi.org/10.1051/0004-6361/202244775. ([arXiv:2208.09403])
7. J.A. Docobo, M. Andrade, Mon. Not. R. Astron. Soc. **428**, 321–339 (2013). https://doi.org/10.1093/mnras/sts045
8. E.L. Fitzpatrick, I. Ribas, E.F. Guinan, F.P. Maloney, A. Claret, Astrophys. J. **587**, 685–700 (2003). https://doi.org/10.1086/368309. (arXiv:astro-ph/0301296 [astro-ph])
9. W.L. Freedman et al., [HST], Astrophys. J. **553**, 47–72 (2001). https://doi.org/10.1086/320638. (arXiv:astro-ph/0012376 [astro-ph])
10. Gaia Collaboration, A.G.A. Brown, A. Vallenari, T. Prusti, J.H.J. de Bruijne, F. Mignard, R. Drimmel, C. Babusiaux, C.A.L. Bailer-Jones, U. Bastian et al., Astron. Astrophys. **595**, A2 (2016). https://doi.org/10.1051/0004-6361/201629512. ([arXiv:1609.04172])
11. Gaia Collaboration, A. Vallenari, A. G. A. Brown, T. Prusti, J. H. J. de Bruijne, F. Arenou, C. Babusiaux, M. Biermann, O. L. Creevey, C. Ducourant, *et al.*, Astronomy and Astrophysics **674**, A1 (2023) https://doi.org/10.1051/0004-6361/202243940 [arXiv:2208.00211]
12. A. Gallenne, A. Mérand, P. Kervella, D. Graczyk, G. Pietrzyński, W. Gieren, B. Pilecki, Astron. Astrophys. **672**, A119 (2023). https://doi.org/10.1051/0004-6361/202245712. ([arXiv:2302.12960])
13. A. Gallenne, G. Pietrzyński, D. Graczyk, P. Konorski, P. Kervella, A. Mérand, W. Gieren, R.I. Anderson, S. Villanova, Astron. Astrophys. **586**, A35 (2016). https://doi.org/10.1051/0004-6361/201526764. ([arXiv:1511.07971])
14. A. Gallenne, G. Pietrzyński, D. Graczyk, N. Nardetto, A. Mérand, P. Kervella, W. Gieren, S. Villanova, R.E. Mennickent, B. Pilecki, Astron. Astrophys. **616**, A68 (2018). https://doi.org/10.1051/0004-6361/201833341. ([arXiv:1806.09572])
15. J.F. González, H. Levato, Astron. Astrophys. **448**, 283–292 (2006). https://doi.org/10.1051/0004-6361:20053177
16. M. Górski, B. Zgirski, G. Pietrzyński, W. Gieren, P. Wielgórski, D. Graczyk, R.-P. Kudritzki, B. Pilecki, W. Narloch, P. Karczmarek et al., Astrophys. J. **889**, 179 (2020). https://doi.org/10.3847/1538-4357/ab65ed. ([arXiv:2001.08242])
17. D. Graczyk, L. Eyer, Acta Astron. **60**, 109–119 (2010)
18. D. Graczyk, I. Soszynski, R. Poleski, G. Pietrzynski, A. Udalski, M.K. Szymanski, M. Kubiak, L. Wyrzykowski, K. Ulaczyk, Acta Astron. **61**, 103 (2011). (arXiv:1108.0446 [astro-ph.SR])
19. D. Graczyk, G. Pietrzynski, I.B. Thompson, W. Gieren, B. Pilecki, P. Konorski, A. Udalski, I. Soszynski, S. Villanova, M. Górski et al., Astrophys. J. **780**, 59 (2014). https://doi.org/10.1088/0004-637X/780/1/59. (arXiv:1311.2340 [astro-ph.CO])
20. D. Graczyk, G. Pietrzyński, I.B. Thompson, W. Gieren, B. Pilecki, P. Konorski, S. Villanova, M. Górski, K. Suchomska, P. Karczmarek et al., Astrophys. J. **860**, 1 (2018). https://doi.org/10.3847/1538-4357/aac2bf. ([arXiv:1805.04952])
21. D. Graczyk, G. Pietrzyński, W. Gieren, J. Storm, N. Nardetto, A. Gallenne, P.F.L. Maxted, P. Kervella, Z. Kołaczkowski, P. Konorski et al., Astrophys. J. **872**, 85 (2019). https://doi.org/10.3847/1538-4357/aafbed. ([arXiv:1902.00589])
22. D. Graczyk, G. Pietrzyński, I.B. Thompson, W. Gieren, B. Zgirski, S. Villanova, M. Górski, P. Wielgórski, P. Karczmarek, W. Narloch et al., Astrophys. J. **904**, 13 (2020). https://doi.org/10.3847/1538-4357/abbb2b. ([arXiv:2010.08754])
23. D. Graczyk, G. Pietrzyński, C. Galan, W. Gieren, A. Tkachenko, R.I. Anderson, A. Gallenne, M. Górski, G. Hajdu, M. Kałuszyński et al., Astron. Astrophys. **649**, A109 (2021). https://doi.org/10.1051/0004-6361/202140571. ([arXiv:2103.02077])
24. D. Graczyk, G. Pietrzyński, C. Galan, J. Southworth, W. Gieren, M. Kałuszyński, B. Zgirski, A. Gallenne, M. Górski, G. Hajdu et al., Astron. Astrophys. **666**, A128 (2022). https://doi.org/10.1051/0004-6361/202244122. ([arXiv:2208.07257])

25. E.F. Guinan, D.H. Bradstreet, L.E. Dewarf, *The Origins, Evolution, and Destinies of Binary Stars in Clusters*, vol. 90, p. 196 (1996)
26. E.F. Guinan, E.L. Fitzpatrick, L.E. DeWarf, F.P. Maloney, P.A. Maurone, I. Ribas, A. Gimenz, J.D. Pritchard, D.H. Bradstreet, Astrophys. J. Lett. **509**, L21–L24 (1998). https://doi.org/10.1086/311760. (arXiv:astro-ph/9809132 [astro-ph])
27. R. Haschke, E.K. Grebel, S. Duffau, Astron. J. **141**, 158 (2011). https://doi.org/10.1088/0004-6256/141/5/158. (arXiv:1104.2325 [astro-ph.GA])
28. S.B. Howell, C. Sobeck, M. Haas, M. Still, T. Barclay, F. Mullally, J. Troeltzsch, S. Aigrain, S.T. Bryson, D. Caldwell et al., Publ. Astron. Soc. Pacific **126**, 398 (2014). https://doi.org/10.1086/676406. ([arXiv:1402.5163])
29. J. Kaluzny, K.Z. Stanek, M. Krockenberger, D.D. Sasselov, J.L. Tonry, M. Mateo, Astron. J. **115**, 1016 (1998). https://doi.org/10.1086/300235. (arXiv:astro-ph/9703124 [astro-ph])
30. P. Kervella, F. Thevenin, E. Di Folco, D. Segransan, Astron. Astrophys. **426**, 297–307 (2004). https://doi.org/10.1051/0004-6361:20035930. (arXiv:astro-ph/0404180 [astro-ph])
31. A. Kruszewski, I. Semeniuk, Acta Astron. **49**, 561 (1999). (arXiv:astro-ph/9912270 [astro-ph])
32. C.H. Lacy, Astrophys. J. **213**, 458–463 (1977). https://doi.org/10.1086/155176
33. C.D. Laney, M.D. Joner, G. Pietrzynski, Mon. Not. R. Astron. Soc. **419**, 1637 (2012). https://doi.org/10.1111/j.1365-2966.2011.19826.x. (arXiv:1109.4800 [astro-ph.SR])
34. C.H. Lee, J. Koppenhoefer, S. Seitz, R. Bender, A. Riffeser, M. Kodric, U. Hopp, J. Snigula, C. Gossl, R.P. Kudritzki et al., Astrophys. J. **797**, 22 (2014). https://doi.org/10.1088/0004-637X/797/1/22. ([arXiv:1411.1115])
35. S. Li, A.G. Riess, M.P. Busch, S. Casertano, L.M. Macri, W. Yuan, Astrophys. J. **920**, 84 (2021). https://doi.org/10.3847/1538-4357/ac1597. ([arXiv:2107.08029])
36. L. Lindegren, U. Bastian, M. Biermann, A. Bombrun, A. de Torres, E. Gerlach, R. Geyer, J. Hernández, T. Hilger, D. Hobbs et al., Astron. Astrophys. **649**, A4 (2021). https://doi.org/10.1051/0004-6361/202039653. ([arXiv:2012.01742])
37. L.M. Macri, K.Z. Stanek, D.D. Sasselov, M. Krockenberger, J. Kaluzny, Astron. J. **121**, 870 (2001). https://doi.org/10.1086/318773. (arXiv:astro-ph/0102123 [astro-ph])
38. P.F.L. Maxted, P. Gaulme, D. Graczyk, K.G. Hełminiak, C. Johnston, J.A. Orosz, A. Prša, J. Southworth, G. Torres, G.R. Davies et al., Mon. Not. R. Astron. Soc. **498**, 332–343 (2020). https://doi.org/10.1093/mnras/staa1662. ([arXiv:2003.09295])
39. M. Mayor, F. Pepe, D. Queloz, F. Bouchy, G. Rupprecht, G. Lo Curto, G. Avila, W. Benz, J.-L. Bertaux, X. Bonfils et al., The Messenger **114**, 20–24 (2003)
40. H.A. McAlister, Publ. Astron. Soc. Pac. **88**, 957–959 (1976). https://doi.org/10.1086/130053
41. D. Mourard, P. Berio, C. Pannetier, N. Nardetto, F. Allouche, C. Bailet, J. Dejonghe, P. Geneslay, E. Jacqmart, S. Lagarde et al., *Optical and infrared interferometry and imaging VIII*, vol. 12183, 1218308 (2022). https://doi.org/10.1117/12.2628881. ([arXiv:2210.09096])
42. U. Munari, T. Zwitter, Astron. Astrophys. **318**, 269–274 (1997)
43. B. Paczynski, The Extragalactic Distance Scale 273 (1997)
44. M. Pawlak, D. Graczyk, I. Soszyński, P. Pietrukowicz, R. Poleski, A. Udalski, M.K. Szymański, M. Kubiak, G. Pietrzyński, Ł. Wyrzykowski et al., Acta Astronomica **63**, 323–338 (2013). https://doi.org/10.48550/arXiv.1310.3272. ([arXiv:1310.3272])
45. M. Pawlak, I. Soszyński, A. Udalski, M.K. Szymański, Ł. Wyrzykowski, K. Ulaczyk, R. Poleski, P. Pietrukowicz, S. Kozłowski, D.M. Skowron et al., Acta Astronomica **66**, 421–432 (2016). https://doi.org/10.48550/arXiv.1612.06394. ([arXiv:1612.06394])
46. G. Pietrzynski, I. Thompson, D. Graczyk, W. Gieren, A. Udalski, O. Szewczyk, D. Minniti, Z. Kolaczkowski, F. Bresolin, R.P. Kudritzki, Astrophys. J. **697**, 862 (2009). https://doi.org/10.1088/0004-637X/697/1/862. (arXiv:0903.0855 [astro-ph.CO])
47. G. Pietrzyński, D. Graczyk, W. Gieren, I.B. Thompson, B. Pilecki, A. Udalski, I. Soszyński, S. Kozłowski, P. Konorski, K. Suchomska et al., Nature **495**, 76–79 (2013). https://doi.org/10.1038/nature11878. (arXiv:1303.2063 [astro-ph.GA])
48. G. Pietrzyński, D. Graczyk, A. Gallenne, W. Gieren, I.B. Thompson, B. Pilecki, P. Karczmarek, M. Górski, K. Suchomska, M. Taormina et al., Nature **567**, 200–203 (2019). https://doi.org/10.1038/s41586-019-0999-4. ([arXiv:1903.08096])

49. G. Pietrzyński, W. Gieren, P. Karczmarek et al., Book **1**, 1–123 (2023). https://doi.org/10.48550/arXiv.2305.17247. ([arXiv:2305.17247])
50. B. Pilecki, P. Konorski, M. Gorski, From interacting binaries to exoplanets: essential modeling tools **282**, 301–302 (2012). https://doi.org/10.1017/S174392131102761X
51. D. Poznanski, J.X. Prochaska, J.S. Bloom, Mon. Not. Roy. Astron. Soc. **426**, 1465 (2012). https://doi.org/10.1111/j.1365-2966.2012.21796.x. (arXiv:1206.6107 [astro-ph.IM])
52. A. Prsa, P. Harmanec, G. Torres, E. Mamajek, M. Asplund, N. Capitaine, J. Christensen-Dalsgaard, É. Depagne, M. Haberreiter, S. Hekker et al., Astron. J. **152**, 41 (2016). https://doi.org/10.3847/0004-6256/152/2/41. ([arXiv:1605.09788])
53. I. Ramirez, J. Melendez, Astrophys. J. **626**, 465–485 (2005). https://doi.org/10.1086/430102. (arXiv:astro-ph/0503110 [astro-ph])
54. I. Ribas, E.L. Fitzpatrick, F.P. Maloney, E.F. Guinan, A. Udalski, Astrophys. J. **574**, 771–782 (2002). https://doi.org/10.1086/341003 (arXiv:astro-ph/0204061 [astro-ph])
55. I. Ribas, C. Jordi, F. Vilardell, E.L. Fitzpatrick, R.W. Hilditch, E.F. Guinan, Astrophys. J. Lett. **635**, L37–L40 (2005). https://doi.org/10.1086/499161. (arXiv:astro-ph/0511045 [astro-ph])
56. S.M. Rucinski, Astron. J. **104**, 1968 (1992). https://doi.org/10.1086/116372
57. G.R. Ricker, J.N. Winn, R. Vanderspek, D.W. Latham, G.Á. Bakos, J.L. Bean, Z.K. Berta-Thompson, T.M. Brown, L. Buchhave, N.R. Butler et al., J. Astron. Telesc. Instrum. Syst. **1**, 014003 (2015). https://doi.org/10.1117/1.JATIS.1.1.014003
58. A. Salsi, N. Nardetto, D. Mourard, D. Graczyk, M. Taormina, O. Creevey, V. Hocdé, F. Morand, K. Perraut, G. Pietrzynski et al., Astron. Astrophys. **652**, A26 (2021). https://doi.org/10.1051/0004-6361/202140763. ([arXiv:2106.01073])
59. F. Schlesinger, R.H. Curtiss, Publ. Allegheny Obs. Univ. Pittsburgh **1**, 25–32 (1910)
60. M.F. Skrutskie et al., 2MASS. Astron. J. **131**, 1163–1183 (2006). https://doi.org/10.1086/498708
61. J. Southworth, Mon. Not. R. Astron. Soc. **386**, 1644 (2008). https://doi.org/10.1111/j.1365-2966.2008.13145.x. (arXiv:0802.3764 [astro-ph].)
62. M. Taormina, R.-P. Kudritzki, J. Puls, B. Pilecki, E. Sextl, G. Pietrzyński, M.A. Urbaneja, W. Gieren, Astrophys. J. **890**, 137 (2020). https://doi.org/10.3847/1538-4357/ab6bd0. ([arXiv:2001.04762].)
63. M. Taormina, G. Pietrzyński, B. Pilecki, R.-P. Kudritzki, I.B. Thompson, D. Graczyk, W. Gieren, N. Nardetto, M. Górski, K. Suchomska et al., Astrophys. J. **886**, 111 (2019). https://doi.org/10.3847/1538-4357/ab4b57. ([arXiv:1910.08111])
64. J. Tonry, M. Davis, Astron. J. **84**, 1511 (1979). https://doi.org/10.1086/112569
65. G. Torres, A.S.P. Conf. Ser. **318**, 123 (2004). (arXiv:astro-ph/0312147 [astro-ph])
66. G. Torres, J. Andersen, A. Gimenez, Astron. Astrophys. Rev. **18**, 67–126 (2010). https://doi.org/10.1007/s00159-009-0025-1. (arXiv:0908.2624 [astro-ph.SR])
67. G. Torres, A. Claret, K. Pavlovski, A. Dotter, Astrophys. J. **807**, 26 (2015). https://doi.org/10.1088/0004-637X/807/1/26. ([arXiv:1505.07461].)
68. A. Udalski, Acta Astron. **53**, 291 (2003). (arXiv:astro-ph/0401123 [astro-ph])
69. A. Udalski, M. Kubiak, M. Szymanski, Acta Astron. **47**, 319–344 (1997). (arXiv:astro-ph/9710091 [astro-ph])
70. A. Udalski, I. Soszynski, M. Szymanski, M. Kubiak, G. Pietrzynski, P. Wozniak, K. Zebrun, Acta Astron. **48**, 563 (1998). (arXiv:astro-ph/9812348 [astro-ph])
71. A. Udalski, M. Szymanski, M. Kubiak, G. Pietrzynski, P. Wozniak, K. Zebrun, Acta Astron. **48**, 1 (1998). (arXiv:astro-ph/9803035 [astro-ph])
72. R.P. van der Marel, D.R. Alves, E. Hardy, N.B. Suntzeff, Astron. J. **124**, 2639–2663 (2002). https://doi.org/10.1086/343775. (arXiv:astro-ph/0205161 [astro-ph])
73. F. Vilardell, I. Ribas, C. Jordi, Astron. Astrophys. **459**(1), 321–331 (2006). https://doi.org/10.1051/0004-6361:20065667. (arXiv:astro-ph/0607236 [astro-ph])
74. F. Vilardell, I. Ribas, C. Jordi, E.L. Fitzpatrick, E.F. Guinan, Astron. Astrophys. **509**, A70 (2010). https://doi.org/10.1051/0004-6361/200913299. (arXiv:0911.3391 [astro-ph.CO])
75. C. Wang, H. Yuan, Y. Huang, Astron. J. **163**, 149 (2022). https://doi.org/10.3847/1538-3881/ac4dec. ([arXiv:2201.09438])

76. A.J. Wesselink, Bull. Astron. Inst. Netherlands **10**, 91 (1946)
77. A.J. Wesselink, Mon. Not. R. Astron. Soc. **144**, 297 (1969)
78. R.E. Wilson, E.J. Devinney, Astrophys. J. **166**, 605 (1971). https://doi.org/10.1086/150986
79. G. Worthey, H.C. Lee, Astrophys. J. Suppl. **193**, 1 (2011). https://doi.org/10.1088/0067-0049/193/1/1. (arXiv:astro-ph/0604590 [astro-ph])
80. L. Wyrzykowski, A. Udalski, M. Kubiak, M. Szymanski, K. Zebrun, I. Soszynski, P.R. Wozniak, G. Pietrzynski, O. Szewczyk, Acta Astron. **53**, 1 (2003). https://doi.org/10.1016/S0094-5765(02)00195-9. (arXiv:astro-ph/0304458 [astro-ph])
81. S. Zucker, T. Mazeh, Astrophys. J. **420**, 806 (1994). https://doi.org/10.1086/173605

Chapter 6
On Cepheid Distances in the H_0 Measurement

Richard I. Anderson

6.1 Cepheids: From Variable Stars to Cosmic Yardsticks

The variability of classical Cepheids (henceforth for simplicity: Cepheids, cf. Sect. 6.2) was first discovered by visual observations in the 18th century by John Goodricke and Edward Piggott [1]. Friedrich Argelander began systematically surveying stars on both hemispheres in the 1850s, and Charles Pickering pushed for systematic observations of variable stars towards the end of the 19th century. Pickering hired female assistants (the so-called Harvard "Computers") to process the observations collected [2]. The photographic plates were astrometrically aligned and variable stars identified by blinking observations of the same sky area observed at different epochs. Henrietta Leavitt thus conducted an unprecedented census of 1777 variable stars in the Small Magellanic Cloud [3].

In 1912, Pickering reported on Leavitt's groundbreaking discovery based on 25 classical Cepheids, that the apparent magnitude m of 25 SMC Cepheids correlated well with the logarithm of their period of variability P, both near minimum and maximum light [4]. Since all SMC stars were assumed to reside at a common distance, it was clear that the observed relation between brightness and period must be an intrinsic property of the stars. This period-luminosity relation (PLR) is nowadays referred to as Leavitt's law (LL).

Just one year later, Hertzsprung [5] made the first attempt to measure distance using Leavitt's discovery. He estimated the absolute magnitude of Cepheids at a period near 6.6 d using secular parallax and, in turn, determined the distance modulus of the SMC Cepheids by comparing their apparent magnitude at the same period to that of the MW Cepheids. Hertzsprung's work laid the foundation for the distance ladder in use today. Despite concluding on an incorrect distance (2000 lightyears), he convincingly demonstrated that the SMC is external to the MW based on a compar-

R. I. Anderson (✉)
Institute of Physics, École Polytechnique Fédérale de Lausanne (EPFL), Observatoire de Sauverny, Chemin Pegasi 51, 1290 Versoix, Switzerland
e-mail: richard.anderson@epfl.ch

E. Di Valentino and D. Brout (eds.), *The Hubble Constant Tension*, Springer Series in Astrophysics and Cosmology, https://doi.org/10.1007/978-981-99-0177-7_6

ison of the scale height of stars in the Solar Neighborhood. Thus, Cepheid distances already informed the (extragalactic) distance scale before their variability mechanism was understood [6–8], and well before the Great Debate of 1920, Hubble's works on M33 [9], as well as the subsequent discovery of the Hubble-Lemaître law [10, 11].

6.1.1 The Hubble Constant Measured Using the Distance Ladder

The relation between Hubble's constant, H_0, and D_L is given by the Friedmann equation, developed here to second order using the Robertson-Walker metric:

$$D_L = \frac{cz}{H_0}\left[1 + \frac{1}{2}(1 - q_0)z - O(z^2) + \ldots\right], \tag{6.1}$$

where $q_0 \approx -0.55$ is the *deceleration* parameter, whose observed negative value implies the Universe's accelerated expansion [12, 13]. Clearly, the wish to measure H_0 accruately is intimately tied to measuring accurate distances.

Equation 6.1 requires the measured redshifts to be purely cosmological, that is, any motions due to graviational attraction (large scale flows, peculiar motions, etc.) must be subtracted or negligible, so that H_0 is best measured in the Hubble flow at distances of 90–600 Mpc ($0.01 < z < 0.0233$) [14]. Such distances are accessible thanks to the extremely luminous type-Ia supernovae (SNeIa, cf. Vincenzi this volume) that enable ~5% relative distance estimates [15]. However, since SNeIa are very rare, no SNIa has as yet been observed in a galaxy whose distance is known by geometrical means. SNeIa thus require external calibration by other means, such as Cepheids, via the *distance ladder (DL)*.

The modern DL consists of three rungs as illustrated in Fig. 6.1. On the first rung, Cepheids are calibrated as standard candles (SCs) using distances measured by geometric methods. On the second rung, Cepheids are used to determine distances to galaxies (SN-hosts) where at least one SNIa has been observed, out about 80 Mpc [16]. On the last rung, SNeIa map cosmic expansion in the Hubble flow.

Seen the other way around, the DL connects the SNeIa distance-redshift relation to an angular scale provided by geometry using Cepheids as the intermediary. By further connecting the Hubble diagram to the early Universe's angular scale (the cosmological sound horizon) via the cosmic expansion history ($H(z)$) using baryon acoustic oscillations, the DL enables a crucial cosmological end-to-end test involving two angular scales at opposite ends of the cosmos—trigonometric parallaxes and the quantum fluctuations of the early Universe [17].

H_0 measurements using a DL have a rich history (cf. Tully, this volume), involving much controversy and disagreements. The *HST* Key Project to measure the Hubble

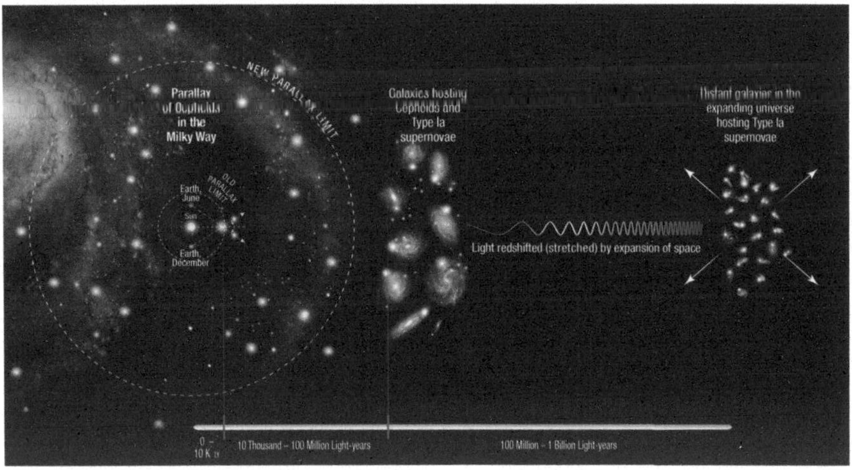

Fig. 6.1 The extragalactic distance ladder uses a two-rung approach for measuring H_0. Geometric distances calibrate stellar standard candles for determining luminosity distances, which in turn calibrate the luminosity of SNeIa that trace the Hubble flow. Credit: NASA, ESA, A. Feild (STScI), and A. Riess (STScI/JHU)

constant was designed to resolved the ongoing disputes and for the first time achieved a 10% measurement of H_0 [18].

In the late 2000s, the SH_0ES (Supernova H_0 for the Equation of State) project was launched to measure H_0 in order to determine the nature of Dark Energy using a differential DL built on the principle of data homogeneity [19, 20]. In its most recent form, the SH_0ES DL is implemented as a least squares problem that minimizes [16]:

$$\chi^2 = (y - Lq)^T C^{-1} (y - Lq) \,, \tag{6.2}$$

with the data vector y, the design matrix L that encodes the relevant equations, such as the LL, distance moduli, etc., the covariance matrix C, and the best-fit parameter vector q. Covariance among observables can be directly considered by quantifying the corresponding elements of C, and very significant amount of covariance information is already being considered, notably among background corrections of Cepheids in SN host galaxies [15, 16, 21]. Using 3445 degrees of freedom, 5 best fit parameters are determined by the SH_0ES DL (see Sect. 6.1.2 for definitions): the LL slope β (b_W), metallicity effect γ (Z_W), the fiducial absolute magnitude of a 10-day Cepheid in the *HST* NIR Wesenheit magnitude $M^W_{H,1}$, the absolute $B-$band magnitude of SNeIa, M^0_B, and Hubble's constant $5\log_{10} H_0$. Alternative symbols used in literature are listed in parenthesis.

The matrix formalism is both simple and accurate, and allows straightforward investigation of analysis variants, notably to assess the systematic effects of modeling choices or dataset selection. Nevertheless, the results from the matrix inversion are cross-checked with computationally intensive Markov chain Monte Carlo (MCMC) simulations that allow inspection of marginalized posterior distributions and correlations among fit parameters. TRGB distances have also been included in this procedure and can help to further improve precision. Including the latest results based on cluster Cepheids, this DL yields $H_0 = 73.15 \pm 0.97\,\mathrm{km\,s^{-1}\,Mpc^{-1}}$ [22].

The specific construction of the SH_0ES DL, with the simultaneous fit to all observables resembles a tightly controlled *experiment* that seeks to maximize the benefits of purely differential flux comparisons to leverage statistical precision while clearly quantifying the systematic uncertainties involved in calibrating the absolute luminosity scale. In this spirit, the SH_0ES DL emphasized the use of homogeneous and very well calibrated Hubble Space Telescope (*HST*) photometry. The greatly improved spatial resolution and enhanced sensitivity of the James Webb Space Telescope (*JWST*) is increasingly replacing the near-infrared (NIR) photometry observed using *HST* WFC3/IR [23]. Careful cross-calibration between *HST* and *JWST* NIR photometry is thus required to ensure direct compatibility. As a result of its specific construction, any modifications to the experimental setup must consider consequences across the entire DL in order to ensure apples-to-apples comparisons.

6.1.2 On Standard Candles, Calibration, and Standardization

Astronomical standard candles (SCs) are objects that fulfill at least two conditions: (a) their intrinsic brightness (absolute magnitude M) can be calibrated in a given photometric band, or combinations of several bands, and (b) they can be faithfully recognized in different observational contexts, notably in different galaxies. SCs are crucial to astronomy because they allow to determine *luminosity distances*, D_L (cf. Eq. 6.2), by exploiting the fact that the observed flux F decreases with distance squared, that is, $F \propto L/D_L^2$. In magnitude space, this becomes the distance modulus equation $\mu = m - M - A = 5\log_{10} D_L + 25$, with D_L in Mpc and extinction A. As noted by Hertzsprung, using standard candles for distance determination presumes that the absolute magnitude established in one context, e.g., in the Milky Way, M_{cal}, is representative of the absolute magnitude of sources seen in other contexts, such as SN host galaxies, M_0. Assuming this correspondence, distance is determined by comparing the observed (apparent) magnitude to the calibrated absolute magnitude, $\mu = m_{\mathrm{obs}} - M_{\mathrm{cal}} - A$.

Calibrating standard candles to establish M_{cal} requires observing them at known distance, $M_{\mathrm{cal}} = m - \mu - A$. Galaxies where this is possible are frequently referred to as *anchors*, and three anchors are of particular importance for Cepheids: The Milky Way (MW), the Large Magellanic Cloud (LMC), and the water megamaser-host galaxy NGC 4258. Despite an available geometric distance estimate of the SMC, complications related to its elongated line-of-sight orientation have thus limited its usefulness for measuring H_0 at the current levels of precision. Interestingly, the distance measurements used for all three anchors are systematically very different. In the MW, trigonometric parallaxes of SCs, notably from the European Space Agency *Gaia* mission (cf. Lindegren, this volume, and references therein) [24–26], set the current gold standard. In particular, parallaxes of open star clusters that host Cepheids have provided the most accurate Cepheid calibration to date (0.9% in distance) [27] thanks to better understood systematics of the fainter, non-variable cluster stars, and the statistical gain of averaging over many cluster members. The typical uncertainty of such cluster parallaxes is 7 μarcsec, three times better than for individual field Cepheids, so that one cluster Cepheid carries as much weight as 9 field Cepheids [22].

The geometric distance measurements of the LMC (detached eclipsing binaries) and NGC4258 (Keplerian disk model of observed megamaser features observed using VLBI) rely on very different systematics, including instrumentation used and modeling assumptions, rendering them truly independent checks of the calibration based on parallax. The close agreement among the Cepheid calibrations based on these anchors is therefore a strong attestation to the accuracy of the anchor distances as well as the consistency among Cepheid observations spanning an enormous contrast range from approximately 8th to 28th magnitude (8 dex in flux).

Standardization refers to the process of mitigating differences that disturb the equivalence between $M_{\rm cal}$ and M_0. All stellar standard candles (including Cepheids, SNeIa, and the TRGB method) require standardization to provide distances to better than a few percent.

As Leavitt showed more than a century ago, Cepheids differ very significantly in luminosity depending on their pulsation period (by about 3 mag/dex of period), which ranges from a few days to more than 100 days. Cepheid distances are thus obtained by comparing *fiducial* magnitudes. The most commonly adopted form of the Leavitt law of Cepheids is

$$M(P) = \alpha + \beta \cdot \log P/P_0 + \gamma \cdot [\mathrm{Fe/H}] \,, \tag{6.3}$$

with α the fiducial absolute magnitude of a Solar metallicity Cepheid at the pivot period P_0 (also often referred to as the zero-point), β the LL slope, and γ the metallicity term that allows to correct the impact of chemical composition using iron (also: oxygen) abundances, cf. below. Importantly, α, β, and γ generally depend on the photometric passband (or combinations thereof) used.

The fiducial absolute Cepheid magnitude in the SH_0ES DL, $M^W_{H,1}$, represents a 10 d Cepheid at the center of the instability strip in a specific combination of photometric passbands (cf. Sect. 6.2.2.2). The commonly used pivot period of 10 d is motivated by literature reports of LL non-linearity [28] and the period-amplitude diagram of Cepheids [29], and corresponds to an intermediate period among Cepheids in anchor galaxies, although Cepheids at greater distance tend to have increasingly longer periods. β and γ are determined in the global DL fit using all Cepheids in anchor and SN host galaxies simultaneously.

At present, there is no clearly demonstrated need to consider LL non-linearity when using Cepheids to measure distances, and allowing for two different slopes near the pivot period does not significantly improve the fit. Moreover, the astrophysical motivation for using non-linear LLs is limited, since stellar models of single chemical composition yield log-linear LLs across a large range of periods [e.g. [30, 31]]. However, stellar evolution models do predict a small dependence of chemical composition on LL slope [30]. Since the effect is rather small, it has not yet been confidently detected [32]. LL slope changes due to metallicity are furthermore particularly small in the infrared, rendering the SH_0ES Wesenheit formalism very insensitive to them.

It seems that observed LL non-linearity arises in a context-specific way from an uneven sampling of the period-color relation, which is caused by the finite width of the classical instability strip. Multi-phase PL relations [33] compare LLs at different pulsation periods and have demonstrated broken PL relations. However, Cepheid magnitudes used for measuring distances correspond to the average flux of Cepheids, and cannot be directly compared to PL relations observed at specific phases. The coincidence of the reported LL breaks near 10 d strongly suggest an origin in the Hertzsprung progression [34].

Two further considerations are in order concerning LL non-linearity. Firstly, H_0 bias due to LL non-linearity arises only in case of systematic differences among the distributions of Cepheid periods in anchor and SN-host galaxies, and this effect is of order $\lesssim 0.15\%$ (the author & Bastian Lengen, priv. comm.). Secondly, variants of the SH_0ES DL that include a break period of $10\,d$ are presented alongside the baseline analysis. These variants also indicate a very insignificant effect on H_0, while not being preferred according to the change in χ^2.

Aside from pulsation period, the most common parameter used for standardization considers metallicity differences [35–38]. For Cepheids, most studies assume a global offset to the LL using a linear dependence on chemical composition usually quantified via oxygen or iron abundances relative to Solar composition, that is, $\Delta M_{\mathrm{met}} = M_0 - M_{\mathrm{cal}} = \gamma \cdot \Delta[O/H]$. Determining γ is challenging because the effect is rather small (~ -0.25 mag/dex) [16, 35], and metallicity differences among most Cepheids used in the DL are typically less than 0.5–0.6 dex. Oxygen abundances estimated in SN-host galaxies fall within a rather narrow range that is fully contained within the abundance ranges of Cepheids in the Milky Way, LMC, and the SMC [16]. As a result, the uncertainty in the metallicity effect does not significantly affect H_0. However, metallicity can be a limiting factor for determining distances to individual Cepheids, e.g., to map Galactic structure, or for determining *Gaia*'s parallax offsets [39].

Standardization is best done using quantities than can be directly observed in the specific sources of interest, such as individual elemental abundances derived from Cepheid spectra. Such comparisons are presently limited to MW and LMC Cepheids, although future 30m class telescopes will significantly increase the ability to observe individual stellar spectra in other nearby galaxies. However, the majority of SN-host galaxies will remain off-limits to individual star spectroscopy. At such distances, equivalent information based on the closest proxys is typically used, such as oxygen abundances of H II regions derived by the strong line method [40, 41], or abundances of blue supergiants [42, 43]. Ensuring the compatibility of proxy information used for standardization is thus crucial to obtaining unbiased distance estimates. For the realm where this has been feasible, such agreement has been shown, for example among HII regions in the Milky Way and Cepheid abundances determined by optical spectra [16].

6.1.2.1 A Sidenote on TRGB Standardization

As all SCs, the TRGB method [44] is also subject to standardization. However, metallicity corrections are complicated by the statistical nature of the measurement (m_{TRGB} is measured based on a sample, not for each star individually) and the diverse population of giant stars composed of different ages and chemical composition. As a result, there is less agreement over how population differences should be treated, with some arguing for color-based metallicity corrections [38] and others assuming that differences among the bluest (interpreted to be the oldest) stars will be negligible [45].

Ubiquitous small amplitude pulsations observed in luminous red gaints near the tip [46, 47] provide useful addition information that may resolve this deadlock. [48] recently showed that all stars near the TRGB feature are variable, and that the variabilty of red giants provides crucial information about the composition of the diverse red giant population. Comparison between the Large and Small Magellanic Clouds (Koblischke & Anderson, in prep.) reveals that the PL sequences of small amplitude red giants in the OGLE (Optical Gravitational Lensing Experiment) catalog of variable stars (OSARGs) are systematically shifted to shorter period in the SMC. This is indicative of a metallicity effect that causes lower-metallicity stars to be more compact, increasing the average density of stars, and thus reducing their periods, since $P \propto 1/\sqrt{\bar{\rho}}$. Thus, OSARG-like variability can be used to standardize metallicity differences for the TRGB method, provided such low levels of variability (amplitudes of order 0.01–0.04 mag) can be detected.

Another form of standardization is required for TRGB distances due to methodological choices. The CATs (Comparative Analysis of TRGBs) team [49–51] demonstrated a significant tip-contrast relation (TCR) for TRGB measurements, whereby fields observed in the same (very distant) galaxy differed in TRGB magnitude according to the ratio of the number of stars that are fainter and brighter ($R = N_+/N_-$) than the TRGB feature. Standardization to a fiducial value of $R = 4$ was thus proposed to ensure consistency. Since TCR standardization does not depend on distance, the effect mainly increases the scatter of TRGB distances to SN host galaxies, rather than biasing H_0. Meanwhile, [48] showed that the TCR is a methodological artifact introduced by weighting the edge detection response (Sobel filter output) in an effort to assign statistical significance to it [52] and that unweighted edge detection responses do not exhibit a significant TCR. Moreover, simulations showed that unweighted EDRs yield less biased TRGB magnitudes than weighted EDRs [48]. Since R is defined *after* determining m_{TRGB}, TCR corrections are slightly circular, and unweighted EDRs should be the preferred choice of measuring m_{TRGB}. Additional details on the TRGB method are presented in Beaton (this volume), and [53] provides a detailed account of other methodological aspects of TRGB measurements that must be considered to obtain consistent results.

6.2 Cepheids and Where to Find Them—Observational Considerations

6.2.1 On the Usefulness of Cepheids in the DL

Classical Cepheids afford the best precision for measuring distances thanks to the fortuitous combination of (a) high luminosity, (b) highly regular variability, (c) comparatively straightforward identification thanks to large (optical) amplitudes, (d) and relatively high prevalence in star forming galaxies. Of course, Cepheids also come with their own observational challenges and are not suitable for every context. For example, Cepheids in highly inclined galaxies suffer from enhanced crowding and reddening issues. In such cases, the Tip of the Red Giant Branch (TRGB) method can offer a suitable alternative [44].

An important advantage of Cepheids is that they are observed specifically and individually. This differs notably from the detection of the TRGB feature, which is ascribed to an astrophysically extremely short-lived phenomenon and which is always measured using an inhomogeneous and contaminated (e.g., by AGB stars) sample of stars. Galaxies where multiple types of SCs can be observed thus provide cross-checks essential for understanding systematic uncertainties while allowing to cross-calibrate multiple SC types, e.g., to increase the number of SN host galaxies in the DL. The following reviews observational aspects of Cepheids and any known issues involving their calibration or use as SCs.

6.2.2 Cepheid Measurements Used in the DL

6.2.2.1 Cepheid Types, Variability, and Classification

This chapter uses the term "Cepheid" synonymous with classical (type-I) Cepheids that belong to the young population I, and whose prototype is δ Cephei. However, it is important to note that other Cepheid types exist, notably the type-II (subgroups W Vir, BL Her, and RV Tau) and anomalous Cepheids that belong to the old population II and exhibit relatively similar light variability, at least in terms of timescales and amplitudes. Distinguishing these groups is crucial to ensure distance accuracy and indeed led to a major revision of the distance scale by Walter Baade [54].

Classical Cepheids are, first and foremost, identified by their variability. Additional context, such as Galactic latitude for MW stars, or the position within a galaxy, usefully separates between young and old stars. However, misclassification between type-I and type-II Cepheids is relatively common if parallax information is not reliable or available. For stars at a common distance, PL sequences are extremely powerful at distinguishing among different types, notably in the Magellanic Clouds, where many PL sequences have been identified [55]. Since classical Cepheids are easily identified by their high luminosity, large number, and period range, there is little room for confusion among Cepheid types in SN host galaxies.

Where available, multi-band time-series data are helpful to establish temperature variations that distinguish chromatic pulations from usually achromatic eclipse-related phenomena. Spectroscopy can provide certainty by allowing to measure effective temperature variations, asymmetric line shapes (a telltale of pulsations), RV variations, as well as surface gravity and abundances.

Gaia's magnitude limited survey has provided the largest homogeneous sample of Cepheids classified solely based on (multi-band) light curves, containing 15 006 Cepheids of all sub-types, including approximately 12 554 classical Cepheids. These Cepheids belong to galaxies within the Local Group (< 1 Mpc), including the MW, the Magellanic Clouds, M31, and M33, and nearby dwarfs. In the MW, there are approximately 3500 classical Cepheids [56, 57], whereas nearly 10 000 Cepheids are nowadays known in the Magellanic System [55, 57].

The prime targets for H_0 are Cepheids that pulsate in the fundamental radial mode (FM Cepheids). These are the most numerous, most regular in terms of their variability, and the longest-period Cepheids. Their detection is further aided by saw-tooth shaped light curves that exhibit a characteristic dependence on pulsation period. This so-called Hertzsprung progression [34] provides important evidence of the similarity of nearby and distant Cepheids. FM Cepheids have their highest amplitudes between approximately 5–7 and 15–30 d, and amplitudes decrease for the longest-period Cepheids [29].

Shorter-period Cepheids that pulsate in the first overtone (FO Cepheids) feature sinusoidal light curves of lower amplitudes, and higher overtones feature progressively lower amplitudes that can become challenging to detect and distinguish from other types of variability. Approximate relations between FO and FM pulsation periods [58] are occasionally used to combine samples of FO and FM Cepheids. In these cases, FO Cepheid periods are "fundamentalized" using period ratios of FO and FM Cepheids in well observed samples, such as the LMC. However, the metallicity dependence of $P_{\mathrm{FO}}/P_{\mathrm{FM}}$ [59] complicates this process and adds systematic uncertainty that is difficult to quantify [e.g. [60]]. As a result, FO Cepheids are not usually considered in the modern DL.

Besides being more luminous, FM Cepheids also have the advantage of being more regular in their variability than FO Cepheids [61–65]. However, the very long period Cepheids ($P \gtrsim 25$ d) exhibit period fluctuations that can be stochastic or (semi-)periodic in nature [66], notably causing issues with de-phasing to mean magnitudes due to variable pulsation ephemerides [67]. However, even very fast period changes of 100 s/yr ($\dot{P}/P \sim 4 \cdot 10^{-5}$ at 30 d, $\Delta \log P \sim 6 \cdot 10^{-7}$) are irrelevant for determining distances using LLs.

6.2.2.2 Reddening-Free Wesenheit Magnitudes

Near-infrared (NIR) photometry significantly improves the use of Cepheids as standard candles due to the intrinsically reduced sensitivity to extinction, lower amplitudes, and an intrinsically reduced width of the instability strip, which results in lower dispersion. Until recently, the main drawback of NIR photometry for the DL was the reduced spatial resolution of *HST*'s WFC3/IR channel compared to WFC3/UVIS. However, NIRCAM on board the *JWST* now provides $4\times$ better spatial resolution than WFC3/IR, with a larger aperture, resulting in much improved Cepheid photometry at large distances [23]. Unfortunately, the reduced amplitudes of Cepheids in the NIR compared to the optical render Cepheid discovery using *JWST* not applicable to large samples.

Issues related to dust extinction are commonly mitigated by computing Wesenheit[1] magnitudes [68, 69], m^W, which are reddening-free by construction for a given reddening law [SH_0ES uses $R_V = 3.3$ from [70, 71]]. The SH_0ES near-IR Wesenheit formalism combines $H-$band (F160W) magnitudes with optical colors, $m_H^W=$ F160W $- R^W \cdot$ (F555W $-$ F814W), where $R^W = A_H/(A_V - A_I) = 0.386$ set by the reddening law. For Cepheids observed by *Gaia*, $R_G^W = A_G/(A_{G_{BP}} - A_{G_{RP}}) \approx 1.91$. Importantly, R^W depends on intrinsic source color when using wide-band photometry due to the shape of the spectral energy distribution incident on the photometric passband [72]. The SH_0ES approach is thus particularly robust against reddening issues thanks to the combination of intrinsically extinction-insensitive NIR photometry with a rather small color-dependent term. For example, a 10% change in reddening law (from $R_V = 3.3$ to 3.0) would result in a $0.01 - 0.05$ change in R_H (depending on reddening law), which multiplied by a typical color excess of $E(V - I) \approx 0.6$ would merely change m_H^W by $\sim 0.03 \cdot 0.6 = 0.02$ mag, that is, much less than the dispersion in most SN host galaxies, cf. Fig. 22 in [16].

Wesenheit magnitudes have the desirable side-effect of narrowing LL dispersion due to the finite width of the instability strip, since R^W happens to be close to the slope of the period-color relation. Of course, Wesenheit magnitudes require knowledge of the reddening law, which can vary among sightlines, and this can be an issue for specific sightlines, such as Cepheids in the SMC, which has been reported to have an abnormal reddening law [73]. However, given the thousands of sightlines among Cepheids in anchor galaxies (MW, LMC, NGC4258) and the thousands of sightlines to Cepheids in SN-host galaxies it is highly unlikely for reddening-law related bias to significantly affect H_0, as was also found empirically [16, 74].

6.2.2.3 Cepheids in the Milky Way

MW Cepheids are particularly important for the DL due to the ability to measure trigonometric parallax [75]. However, MW Cepheids are subject to several challenges that arise from their location inside the MW's disk, where dust extinction is high and a strong function of distance, and where the LL cannot be recovered using apparent magnitudes alone. Observationally, the brightness of Cepheids can be an asset in that high-resolution spectra can be observed [76–78]. However, the same brightness

[1] The German word "Wesenheit" relates to the abstract innate nature (or essence) of an entity, cf. https://www.dwds.de/wb/Wesenheit.

can easily saturate photometric observations of Cepheids, notably from space, using large apertures, and in the infrared. Another practical issue is the distribution of MW Cepheids across a very large sky area, requiring MW Cepheids to be targeted individually and causing large observational overheads.

Drift scanning observations with *HST* have allowed the collection of photometry in the SH_0ES Wesenheit system for a legacy sample of 75 MW FM Cepheids that feature low extinction and are sufficiently nearby ($\sim 1 - 5$ kpc) for high-quality *Gaia* parallaxes [79, 80]. *HST* drift scan observations have further provided parallaxes of 8 Cepheids, with a typical accuarcy of $40 - 50\,\mu$as [81–83].

Hundreds of high-quality parallaxes of MW Cepheids have been published by *Gaia* [24–26, 84, 85], an order of magnitude increase over ESA's predecessor *Hipparcos* [86, 87]. Clearly, *Gaia* has been a *game changer* for DL calibration and has ushered in a new era for SCs. However, *Gaia* (E)DR3 parallaxes are subject to unfortuante systematics at the level of 10–20μas (a 3–6% error at 3 kpc) that correlate strongly with sky position (ecliptic latitude), magnitude, and color, cf. Lindegren, this volume. [84] (L21) determined parallax offset corrections based on these parameters using bright physical pairs, LMC stars, and quasars. However, several follow-up studies [16, 39, 88] have shown that stars brighter than $G \lesssim 11$ mag require additional correction at levels similar to the L21 corrections [89]. Very recently, [90] showed that the L21 corrections require significant further corrections at bright ($G \lesssim 11$ mag) magnitudes and that these residual corrections depend significantly on sky position, likely because the bright physical pairs insufficiently sample the full sky. As a result, all SC calibrations in this magnitude range must simultaneously solve for a sample parallax offset, which requires careful analysis and generally comes at the cost of precision, and this notably includes Cepheids that are typically brighter than 11 mag.

The best parallax accuracy for MW Cepheids is afforded by the fainter non-variable member (main sequence) stars of open star clusters that host Cepheids [22, 27]. The numerous member stars improve statistical precision while *Gaia*'s parallax sysetmatics are best understood in the magnitude range probed by the members stars [84, 91]. This fortuitous combination yields a typical accuracy of 7 μarcsec for cluster Cepheids, affording them the same weight as ~ 9 field Cepheids for determining $M_{H,1}^{W}$. Combining cluster and field Cepheids thus has allowed to calibrate $M_{H,1}^{W}$, and an analogous magnitude based on *Gaia* photometry, to better than 0.020 mag (0.9% in distance), while simultaneously solving the sample's overall parallax offset to $-13 \pm 5\,\mu$as ($-19 \pm 3\,\mu$as for the larger sample based on *Gaia* photometry) [27].

6.2.2.4 Cepheids in the Magellanic Clouds

The Magellanic Clouds host very large Cepheid populations that have provided the most important proving grounds for testing stellar models thanks to the ability to observe the Cepheids in great detail at a virtually constant distance.

LMC Cepheids are particularly valuable due to the extremely accurate distance determined based on 20 detached eclipsing binaries (cf. Pietrzynski & Graczyk, this volume; [92]). The intrinsic depth of the SMC combined with the relatively few (15) detached eclipsing binaries that define its distance [93] has thus far limited the use of the SMC as an anchor galaxy however, see Breuval et al. [94]. Photometric observations using *HST*'s DASH mode have enabled efficient photometric observations of a significant sample of long-period Cepheids, despite the large angular separation among LMC Cepheids [95]. While ground-based photometry of LMC Cepheids abounds from several surveys [58, 96–98], *HST*'s photometry provides unmatched photometric consistency and avoids issues related to Earth's atmosphere, notably in the NIR.

Cepheids in the Magellanic Clouds usefully constrain the slope of the metallicity term, γ [99], since they are less metallic than most Cepheids in the Solar neighborhood, are not heavily extincted, and not subject to *Gaia* parallax offsets. However, the internal depth of both galaxies must be considered for best results, and correcting for the LMC plane geometry [97] reduces the scatter of the observed Cepheid LL. This effect is even more pronounced in the SMC [100], which also allows to extend the metallicity lever to the lowest values [35].

6.2.2.5 Cepheids in NGC 4258

The spiral galaxy NGC 4258 (Messier 106) is a crucial DL anchor thanks to a water megamaser that orbits its supermassive black hole [101], cf. Braatz & Pesce this volume. Modeling the observed maser emission as a warped Keplerian disk has allowed to determine the distance to this galaxy to within 1.5% [102]. Importantly for the DL, this measurement shares virtually no common systematics with *Gaia* parallaxes, or the modeling of binary star orbits (apart from Kepler's laws) that underlies the LMC distance. The largest population of 669 Cepheids in NGC 4258 has been observed in the SH_0ES Wesenheit system by [103]. However, long-period Cepheids in NGC 4258 are sufficiently close to detect them using ground-based, 8m-class telescopes [104, 105].

At a distance of 7.6 Mpc, Cepheids in NGC 4258 are observed under conditions that resemble the conditions of other galaxies much more than Cepheids in the MW, or the LMC, notably with regards to the process by which the photometry is collected (cf. Sect. 6.2.3), the typical flux range, the blending of sources, and the metallicity corrections based on HII regions. This makes NGC 4258 the only anchor galaxy where Cepheids, as well as other SCs, such as the TRGB and J-region AGB methods [106, 107], can be calibrated in the same way as they are used for determining distances. Thus, NGC 4258 allows very direct apples-to-apples comparisons for determining distances to individual galaxies, such as M51 [60]. Additionally, the consistent absolute LL calibrations determined using MW parallaxes, the LMC distance, and NGC 4258 underline the accuracy of the measurements in all three anchor galaxies.

6.2.2.6 Cepheids in SN-Host Galaxies

The SH_0ES DL uses 2150 Cepheids in 37 SN-host galaxies to calibrate the SNIa luminosity zero-point, and these galaxies are located across a wide range of distances, from 6.7 (M101) to 80 Mpc (NGC 105), with an average distance of approximately 27 Mpc. This dataset represents a major time investment, totaling approximately 1050 orbits, or the equivalent of $\sim$ 66 full days of continuous *HST* observations. In the nearest galaxies, several fields can be monitored to map Cepheids across galaxies, whereas a single frame (WFC3/NIR footprint is 2.7×2.7 arcmin^2) centered on the galactic bulge captures the majority of the Cepheids in distant galaxies.

Cepheids in SN-hosts are discovered at optical wavelengths where their amplitudes are largest, $\sim$ 1 mag. $V-$ (F555W) and $I-$band (F814W) observations are de-phased to estimate mean magnitudes, whereas $H-$band (F160W) observations can usually be used at random phase thanks to the reduced amplitudes. The combination of $H-$band magnitudes, and $V-I$ colors is then used to compute the Wesenheit magnitudes. Further information on Cepheid photometry beyond the Local Group and Wesenheit magnitudes is provided in Sect. 6.2.3.

Bona fide Cepheids used for the DL are discovered based on their light variations and subsequently vetted using several quality criteria, cf. [103]. These cuts include amplitudes, amplitude ratios, dispersion of Cepheid observations around template light curve fits, minimum period of 5 d, and average color to ensure a matching temperature range. Finally, the sample is cleaned by 3σ clipping in all three bands. Impressively, binning the light curves of many Cepheids in period ranges allows to illustrate the Hertzsprung progression, which adds further confidence concerning the equivalence of the distant and nearby Cepheids.

The constant angular resolution of the photometric system corresponds to increasingly larger physical scales, the more distant the SN host galaxy. This creates a fairly clean correlation between the degree to which Cepheids are blended with other sources and the distance at which they are observed. Variations in surface source density, e.g., due to inclination or the region of the galaxy observed, further impact the amount of blending. At large source densities, the term *crowding* refers to heavy blending in fields with many overlapping sources.

Crowding corrections are used to remove background light contributions due to unresolved sources [108]. The background due to blended sources is measured using artificial star tests that locally insert and remeasure the flux of sources in the vicinity of the target star, and with similar brightness [14, 109]. This is a significant effect in many SN host galaxies (mean of $\sim$ 0.4 mag, up to nearly 1 mag at the largest distances), although corrections as small as a few hundredths of a magnitude are also estimated in low density regions of nearby galaxies. The accuracy of crowding corrections has been extensively tested [16] and recently been validated using *JWST*'s higher spatial resolution [23].

Stellar association bias arises because wide binaries and open star clusters add flux to Cepheids in NGC 4258 and SN-host galaxies, while being resolved (and hence not contributing) to Cepheid flux measurements used to calibrate the LL in the LMC and MW [110]. Given a typical physical scale of clusters being approximately

4 pc, the artificial star tests used for crowding corrections cannot correct for this effect, leading to a small, albeit one-sided bias for the DL because the effect occurs primarily in the SN-host galaxies. Using M31 as a SN-host analog, [110] estimated an effect of ~ 0.007 mag per SN-host Cepheid, and a correction of this size has been included in the SH_0ES analysis. A similar estimate has been found in M33 [111], and additional work based on M101 and NGC 4258 is ongoing (Spetsieri et al., in prep.). While a better quantification of this bias will further improve H_0 accuracy, it is important to note that stellar association bias cannot resolve the H_0 tension, since both the requirements of Cepheid discovery (notably sufficiently large amplitudes and characteristic light curve shapes) and the quality cuts applied to Cepheid candidates (notably based on color and LL σ clipping) provide very strong protection against significant bias. In essence, Cepheids blended by very luminous clusters are either not discovered due to diminished amplitudes, or rejected due to their blue color.

Metallicity corrections in SN-host galaxies can unfortunately not (yet) be measured directly from Cepheid spectra. Instead, oxygen abundances of Cepheids in SN-host galaxies are estimated based on their galactocentric location and metallicity gradients estimated using HII regions [40]. Care must be taken to ensure the consistency of the derived oxygen abundances, notably with regards to the Solar oxygen abundance problem. In that case, a high degree of agreement between Cepheid oxygen abudances and HII region abundances has been shown [16][Appendix C].

Relativistic effects can bias H_0 measurements based on the DL because SN-host galaxies are systematically redshifted relative to the anchor galaxies due to cosmic expansion. For example, time dilation leads to systematically overestimated Cepheid periods at greater distances, and thus, to overestimated luminosities and underestimated distances. Correcting the dilated observed Cepheid periods therefore has the effect of *increasing* H_0, by approximately 0.2% [112]. This effect is already accounted for in the SH_0ES DL. Two additional relativistic effects have been considered [72]: $K-$corrections and the redshifting of the reddening law on Wesenheit magnitudes. While both effects are very small when working with NIR Wesenheit magnitudes, single-band IR observations (e.g., of TRGB stars observed with *JWST*) can exceed 1% distance bias at 100 Mpc and must therefore be considered as the use of stellar SCs is pushed to ever greater distances.

6.2.3 Cepheid Photometry in SN-Hosts and NGC 4258

The quality of the observed photometry is paramount for accurate luminosity estimates and H_0. As such, a few words on the measurement process are in order.

Point spread function (PSF) photometry remains the state-of-the-art method for measuring Cepheid light curves, mostly using DAOPHOT and DOLPHOT [113, 114]. PSF photometry treats each star as a point source that can be distinguished from neighboring sources and background, and it is used for busy and inhomogeneous scenes, where the simpler aperture photometry approach is not suitable. An important benefit of PSF photometry over aperature photometry is its ability to distinguish resolved and unresolved sources, and PSF fitting produces quantitative information that can be used to study the shape of the light's distribution over the detector. However, the key limitation of PSF photometry is the need to know how light is dispersed by the optical system either a priori, or by measuring the PSF directly from isolated sources. The details of how temporal variations in the PSF, e.g., due to telescope breathing or focus changes, affect photometry are challenges where observer experience frequently outperforms automated procedures.

In many ways, measuring photometry *well* is an *art* that requires a significant time investment to master. Given the high hurdle for participating and the prevalence of two methods commonly adopted as black boxes, there is certainly room for more independent cross-checks. A notable recent study presented new photometric measurements based on the raw *HST* frames of NGC 5584 [115]. Comparison with the photometry by the SH_0ES team revealed very good agreement. While further cross-checks would be welcome, it is worth noting that H_0 bias would primarily arise due to inconsistency among photometric measurements between near and far galaxies, rather than due to errors involving the absolute flux scale of all galaxies.

6.2.4 Baade-Wesselink Distances

The large scale radial pulsations of Cepheids enable distance determination using the ratio of the observed angular ($\Delta\Theta$) and linear radius variations (ΔR), since $D \propto \Delta R/\Delta\Theta$. Such distances are commonly referred to as Baade-Wesselink distances [8, 116], and their development was intricately linked with establishing pulsations as the cause of Cepheid pulsations [6, 117, 118]. $\Delta\Theta$ is observable for a relatively small number of MW Cepheids using very long baseline optical interferometry [119–121] as well as surface brightness color relations (SBCRs) [122], which require only photometric measurements and are thus applicable to longer distances. The SpectroPhoto-Interferometry of Pulsating Stars (SPIPS) algorithm performs simultaneous modeling of interferometry, multi-band photometry, and radial velocity observations to determine distance using the same concepts [123, 124]. $\Delta R = p \cdot \int v_r d\phi$ is the product of the integral of the line of sight (radial) velocity variations measured using optical spectra and the projection factor p, which translates the observed line of sight velocity to the pulsational velocity as measured from the center of the star. The requirement for optical high resolution spectra limits the use of BW distances to relatively nearby objects, such as Cepheids in the MW and the Magellanic Clouds [125–127]. However, the projection factor p has been a limiting factor to the precision attainable for BW distances due to its complex task of simultaneously correcting for geometric effects, limb darkening, and dynamical aspects of stellar atmospheres [61, 128–131]. For example, despite the availability of *Gaia* parallaxes, it has not been possible to determine whether or how p may depend on Cepheid properties, such as pulsation period P [124, 132, 133]. While BW distances thus provide a potentially interesting alternative for measuring individual distances to Cepheids, they do not currently afford the accuracy required to inform H_0.

6.2.5 *On the Impact of Cepheid Multiplicity on H_0*

The majority of Cepheids reside in multiple systems [134–137], with estimates of the multiplicity fraction ranging from 60% to more than 100% (many Cepheids occur in triple or even quadruple systems). Cepheid companions can affect distances either astrometrically or photometrically.

Astrometric issues falsify the parallax measurements of Cepheids and can arise either due to real or apparent orbital motion. The former is caused by the movement around the common center of gravity, whereas apparent orbital motion is a photometric effect in disguise and caused by the shifting photocenter in a binary system containing stars of similar magnitude. Timescales can be used to separate these two effects, with true orbital motion generally being much longer than the pulsation period, and apparent motion occuring in phase with the Cepheid's variability. Poor astrometry thus obviously can affect LL calibration, albeit on a star-by-star basis, increasing scatter rather than biasing LL calibration. Thankfully, radial velocity monitoring can very sensitively estimate whether orbital motion can significantly affect parallax measurements [138]. *Gaia*'s astrometric quality flags, such as RUWE, are also commonly used to identify high quality astrometric solutions.

Photometric distance bias can in principle arise if the calibrated LL does not correspond to the observed LL due to differences in the nature of the companion stars, or the prevalence of multiple star systems. However, these concerns are essentially limited to individual Cepheids [139], which may stand out significantly from observed LLs. For example, abnormally luminous and red LL outliers in the LMC have been successfully identified as high-value double-lined spectroscopic binaries (SB2s) [140, 141]. A similar population of SB2 Cepheids in the MW has not yet been detected, which prompts the interesting question of how much the companion star population of Cepheids may differ from galaxy to galaxy. Reasonable assumptions on the distribution of wide binary separations and contrasts have allowed to constrain a possible DL bias to $< 0.004\%$ [110].

6.2.6 *Cepheids in (Open) Star Clusters*

In the Milky Way, approximately 10% of Cepheids are associated with (gravitationally bound to) open clusters [27]. This estimate is based on *Gaia* (E)DR3 data and considers the 2 kpc surrounding the Sun, where a rather complete census of both Cepheids and star clusters is currently feasible. Different galaxies present different clustered Cepheid fractions, with the LMC having the highest known fraction. In M31, the fraction of Cepheids in star clusters is much less, approximately 2.5% [110, 142], and M33 has a clustered Cepheid fraction of 3.3% [111]. HST UV photometry of M101 shows that Cepheid clustering is highly variable across the two fields in which Cepheids have been observed and that it correlates strongly with star formation (Spetsieri et al. in prep.).

Dynamical simulations [143] explain the low fraction of Cepheids that occur in clusters as a consequence of cluster dissociation over time. Cluster dispersal is driven by the violent gas expulsion due to massive star winds at very early times and the dynamical interactions over longer timescales, and mostly occurs well before intermediate-mass stars reach the Cepheid stage. Despite sensitivity to initial conditions, the observed trend that younger Cepheids cluster more frequently strongly supports the paradigm that a majority, if not all, stars are born in clusters [144]. Cepheid clustering is thus both dependent on global and local properties of galaxies, as is observed.

6.2.7 Circumstellar Environments

Circumstellar environments (CSEs) of Cepheids have been detected using mid-IR photometry [145, 146] and HI imaging [147], near- and mid-IR interferometry [148–150], as IR excess among spectral energy distributions [151–153], and using *Spitzer* spectroscopy [154]. Based on the IR excesses, it appears that $H-$band is just short of the region where IR excess becomes most noticeable, and that only a fraction of Cepheids exhibit significant mid-IR excess. At present, it is unclear whether and how CSEs can bias Cepheid distances. However, the following preliminary conclusions appear reasonably secure: (a) CSEs can lead to DL bias only if the nature of CSEs differs systematically among anchor and SN-host galaxies — in this regard, CSE distance bias behaves like photometric bias due to companions; (b) CSEs may lead individual Cepheids to be outliers from LLs; (c) long-wavelength observations, such as *JWST* NIRCAM/F277W and F444W are likely to be more sensitive to CSEs than $H-$band and shorter wavelength observations; (d) if chemical composition significantly affects the presence of CSEs, then metallicity corrections in use today likely already partially correct the effect induced by CSEs.

6.2.8 Stellar Evolution Models

Cepheids pose many interesting challenges to stellar evolution theory, starting from the occurrence of blue loops during core He burning, to observed rates of period change, to predicting the correct instability strip boundaries and the right luminosity (mass) range where Cepheids occur, to the mass-luminosity relation and its dependence on metallicity, convection, and rotation, and the ability to predict Cepheid variability, among others. Stellar models thus benefit from a large number of observational constraints that allow to understand the physical nature of these important stars and to rigorously test model predictions along several axes at once. With thousands of Cepheids in dozens of galaxies, the DL offers unprecedented means for comparisons, albeit at a reduced level of detail at greater distances. However, given their systematic uncertainties, stellar models have limited predictive power for Cepheid distances.

Areas where stellar models *can* usefully inform the DL include, among others, (a) the linearity of the LL (cf. above); (b) the dispersion of the LL observed in different passbands; (c) differential comparisons of Cepheid luminosity under otherwise identical conditions as required, e.g., to determine the effects of chemical composition on the LL. For example, predictions based on Geneva stellar evolution models [30] predict a negative sign for the metallicity term, γ, both in the individual $V-$ and $H-$band PL-relations and in the optical $W_{\rm VI}$ and near-IR Wesenheit systems $W_{\rm H,VI}$ used in the SH_0ES analysis. Conversely, a widely used set of Cepheid models [31, 155–157] predicts a positive sign for PL-relations in the individual passbands ($\gamma_\lambda > 0$ in those cases) and negatively signed period-Wesenheit relations, that is, when com-

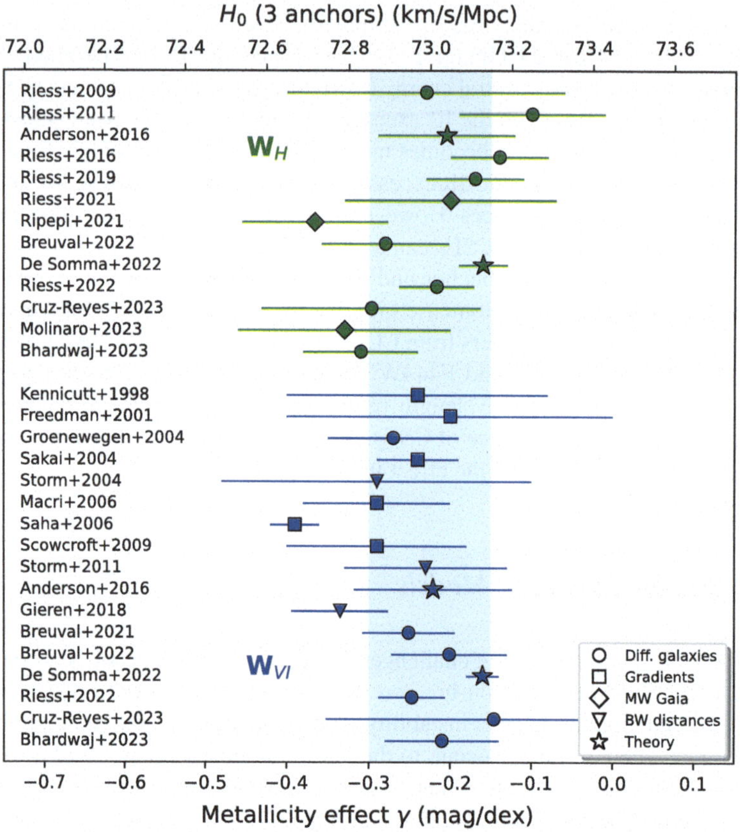

Fig. 6.2 The slope of the metallicity effect on Wesenheit magnitudes determined by different methods (bottom x-axis) and the impact of metallicity on H_0 (top x-axis). Circles show results based on different galaxies, squares results based on galactic abundance gradients, diamonds results based on MW Cepheids observed with *Gaia*, downward triangles results based on BW-type distances, and asterisks results based on stellar evolution models. Figure courtesy of Louise Breuval

bining a single photometric passband with a color ($\gamma_W < 0$). Recent empirical work by different teams has shown the metallicity effect to have negative sign across several individual photometric bands as well as in the Wesenheit formulations [35, 77, 125, 158–160], cf. Fig. 6.2. Hence, it appears that the disagreement between the two sets of models is due to different predicted locations of the instability strip boundaries, as well as their dependence on metallicity.

6.3 Conclusions

Classical Cepheids provide the base calibration for the H_0 measurement using the extragalactic DL. They benefit from more than 110 years of study for measuring distances. Recent progress in quantifying systematics has centered on sub-percent effects for H_0, which ended up sharpening the accuracy of the measurement and the significance of the Hubble tension. In the process, corrections that increase and decrease the value of H_0 have been identified. *Gaia* has revolutionized the calibration of the LL in the MW based on trigonometric parallaxes, and *JWST* stands to significantly improve H_0 precision by better sensitivity and reduced LL dispersion in SN-host galaxies due to lower background (enabled by the 4× better spatial resolution than *HST*'s WFC3/IR).

Further improvements of the distance ladder and its experimental setup will yield significant improvements in H_0 accuracy, hopefully to deliver the desired 1% direct measurement. Further improvements to *Gaia*'s parallax systematics are paramount to achieving this level of accuracy. However, despite very significant efforts to identify systematics that could bias H_0 sufficiently to explain the H_0 constant tension, no such biases have been identified. For this reason, the author is cautiously optimistic that the problem with the H_0 constant will eventually enable new interesting insights into the physics governing our Universe.

Acknowledgements RIA is funded by the SNSF through an Eccellenza Professorial Fellowship, grant number PCEFP2_194638 and acknowledges support from the European Research Council (ERC) under the European Union's Horizon 2020 research and innovation programme (grant agreement No 947660).

References

1. J. Goodricke, A series of observations on, and a discovery of, the period of the variation of the light of the star marked δ by Bayer, near the head of Cepheus, in a letter from John Goodricke, Esq. to Nevil Maskelyne, D. D. F. R. S. and astronomer royal. Philos. Trans. R. Soc. London Ser. I **76**, 48–61 (1786)
2. D. Sobel, *The Glass Universe: The Hidden History of the Women Who Took the Measure of the Stars* (HarperCollins Publishers, 2016)
3. H.S. Leavitt, 1777 variables in the magellanic clouds. Ann. Harvard College Obs. **60**, 87–108.3 (1908)

4. H.S. Leavitt, E.C. Pickering, Periods of 25 variable stars in the small magellanic cloud. Harvard College Obs. Circular **173**, 1–3 (1912)
5. E. Hertzsprung, Über die räumliche Verteilung der Veränderlichen vom δ Cephei-Typus. Astron. Nachr. **196**, 201 (1913)
6. A.S. Eddington, The pulsation theory of Cepheid variables. The Observatory **40**, 290–293 (1917)
7. H. Shapley, On the nature and cause of cepheid variation. ApJ **40**, 448 (1914)
8. A.J. Wesselink, The observations of brightness, colour and radial velocity of δ Cephei and the pulsation hypothesis (Errata: 10 258, 310). Bull. Astron. Inst. Netherlands **10**, 91 (1946)
9. E.P. Hubble, A spiral nebula as a stellar system: messier 33. ApJ **63**, 236–274 (1926)
10. E. Hubble, A relation between distance and radial velocity among extra-galactic nebulae. Proc. Natl. Acad. Sci. **15**(3), 168–173 (1929)
11. G. Lemaître, Un Univers homogène de masse constante et de rayon croissant rendant compte de la vitesse radiale des nébuleuses extra-galactiques. Annales de la Société Scientifique de Bruxelles **47**, 49–59 (1927)
12. S. Perlmutter, G. Aldering, G. Goldhaber, R.A. Knop, P. Nugent, P.G. Castro, S. Deustua, S. Fabbro, A. Goobar, D.E. Groom, I.M. Hook, A.G. Kim, M.Y. Kim, J.C. Lee, N.J. Nunes, R. Pain, C.R. Pennypacker, R. Quimby, C. Lidman, R.S. Ellis, M. Irwin, R. G. McMahon, P. Ruiz-Lapuente, N. Walton, B. Schaefer, B. J. Boyle, A.V. Filippenko, T. Matheson, A.S. Fruchter, N. Panagia, H.J.M. Newberg, W.J. Couch, T.S.C. Project, Measurements of Ω and Λ from 42 high-redshift supernovae. ApJ **517**(2), 565–586 (1999)
13. A.G. Riess, A.V. Filippenko, P. Challis, A. Clocchiatti, A. Diercks, P.M. Garnavich, R.L. Gilliland, C.J. Hogan, S. Jha, R.P. Kirshner, B. Leibundgut, M.M. Phillips, D. Reiss, B.P. Schmidt, R.A. Schommer, R.C. Smith, J. Spyromilio, C. Stubbs, N.B. Suntzeff, J. Tonry, Observational evidence from supernovae for an accelerating universe and a cosmological constant. AJ **116**(3), 1009–1038 (1998)
14. A.G. Riess, L.M. Macri, S.L. Hoffmann, D. Scolnic, S. Casertano, A.V. Filippenko, B.E. Tucker, M.J. Reid, D.O. Jones, J.M. Silverman, R. Chornock, P. Challis, W. Yuan, P.J. Brown, R.J. Foley, A 2.4% determination of the local value of the hubble constant. ApJ **826**(1), 56 (2016)
15. D. Brout, D. Scolnic, B. Popovic, A.G. Riess, A. Carr, J. Zuntz, R. Kessler, T.M. Davis, S. Hinton, D. Jones, W.D. Kenworthy, E.R. Peterson, K. Said, G. Taylor, N. Ali, P. Armstrong, P. Charvu, A. Dwomoh, C. Meldorf, A. Palmese, H. Qu, B.M. Rose, B. Sanchez, C.W. Stubbs, M. Vincenzi, C.M. Wood, P.J. Brown, R. Chen, K. Chambers, D.A. Coulter, M. Dai, G. Dimitriadis, A.V. Filippenko, R.J. Foley, S.W. Jha, L. Kelsey, R.P. Kirshner, A. Möller, J. Muir, S. Nadathur, Y.-C. Pan, A. Rest, C. Rojas-Bravo, M. Sako, M.R. Siebert, M. Smith, B.E. Stahl, P. Wiseman, The Pantheon+ analysis: cosmological constraints. ApJ **938**(2), 110 (2022). https://doi.org/10.3847/1538-4357/ac8e04
16. A.G. Riess, W. Yuan, L.M. Macri, D. Scolnic, D. Brout, S. Casertano, D.O. Jones, Y. Murakami, G.S. Anand, L. Breuval, T.G. Brink, A.V. Filippenko, S. Hoffmann, S.W. Jha, W. D'arcy Kenworthy, J. Mackenty, B.E. Stahl, W. Zheng, A comprehensive measurement of the local value of the hubble constant with 1 km s^{-1} Mpc^{-1} uncertainty from the hubble space telescope and the SH_0ES team. ApJ **934**(1), L7 (2022)
17. A.G. Riess, The expansion of the universe is faster than expected. Nat. Rev. Phys. **2**(1), 10–12 (2020)
18. W.L. Freedman, B.F. Madore, B.K. Gibson, L. Ferrarese, D.D. Kelson, S. Sakai, J.R. Mould, J. Kennicutt, C. Robert, H.C. Ford, J.A. Graham, J.P. Huchra, S.M.G. Hughes, G.D. Illingworth, L.M. Macri, P.B. Stetson, Final results from the hubble space telescope key project to measure the hubble constant. ApJ **553**(1), 47–72 (2001)
19. A.G. Riess, L. Macri, S. Casertano, M. Sosey, H. Lampeitl, H.C. Ferguson, A.V. Filippenko, S.W. Jha, W. Li, R. Chornock, D. Sarkar, A redetermination of the hubble constant with the hubble space telescope from a differential distance ladder. ApJ **699**(1), 539–563 (2009)
20. A.G. Riess, L. Macri, W. Li, H. Lampeitl, S. Casertano, H.C. Ferguson, A.V. Filippenko, S.W. Jha, R. Chornock, L. Greenhill, M. Mutchler, M. Ganeshalingham, M. Hicken, Cepheid

calibrations of modern type Ia supernovae: implications for the hubble constant. ApJS **183**(1), 109–141 (2009)
21. D. Scolnic, D. Brout, A. Carr, A.G. Riess, T.M. Davis, A. Dwomoh, D.O. Jones, N. Ali, P. Charvu, R. Chen, E.R. Peterson, B. Popovic, B.M. Rose, C.M. Wood, P.J. Brown, K. Chambers, D.A. Coulter, K.G. Dettman, G. Dimitriadis, A.V. Filippenko, R.J. Foley, S.W. Jha, C.D. Kilpatrick, R.P. Kirshner, Y.-C. Pan, A. Rest, C. Rojas-Bravo, M.R. Siebert, B.E. Stahl, W. Zheng, The Pantheon+ analysis: the full data set and light-curve release. ApJ **938**(2), 113 (2022)
22. A.G. Riess, L. Breuval, W. Yuan, S. Casertano, L.M. Macri, J.B. Bowers, D. Scolnic, T. Cantat-Gaudin, R.I. Anderson, M. Cruz Reyes, Cluster cepheids with high precision gaia parallaxes, low zero-point uncertainties, and hubble space telescope photometry. ApJ **938**(1), 36 (2022)
23. A.G. Riess, G.S. Anand, W. Yuan, S. Casertano, A. Dolphin, L.M. Macri, L. Breuval, D. Scolnic, M. Perrin, R.I. Anderson, Crowded no more: the accuracy of the hubble constant tested with high resolution observations of cepheids by JWST (2023). arXiv e-prints: arXiv:2307.15806
24. Gaia Collaboration, A.G.A. Brown, A. Vallenari, T. Prusti, J.H.J. de Bruijne, C. Babusiaux, M. Biermann, O.L. Creevey, D.W. Evans, L. Eyer, A. Hutton, F. Jansen, C. Jordi, S.A. Klioner, U. Lammers, L. Lindegren, X. Luri, F. Mignard, C. Panem, D. Pourbaix, S. Randich, P. Sartoretti, C. Soubiran, N.A. Walton, F. Arenou, C.A.L. Bailer-Jones, U. Bastian, M. Cropper, R. Drimmel, D. Katz, M.G. Lattanzi, F. van Leeuwen, J. Bakker, C. Cacciari, J. Castañeda, F. De Angeli, C. Ducourant, C. Fabricius, M. Fouesneau, Y. Frémat, R. Guerra, A. Guerrier, J. Guiraud, A. Jean-Antoine Piccolo, E. Masana, R. Messineo, N. Mowlavi, C. Nicolas, K. Nienartowicz, F. Pailler, P. Panuzzo, F. Riclet, W. Roux, G. M. Seabroke, R. Sordo, P. Tanga, F. Thévenin, G. Gracia-Abril, J. Portell, D. Teyssier, M. Altmann, R. Andrae, I. Bellas-Velidis, K. Benson, J. Berthier, R. Blomme, E. Brugaletta, P.W. Burgess, G. Busso, B. Carry, A. Cellino, N. Cheek, G. Clementini, Y. Damerdji, M. Davidson, L. Delchambre, A. Dell'Oro, J. Fernández-Hernández, L. Galluccio, P. García-Lario, M. Garcia-Reinaldos, J. González-Núñez, E. Gosset, R. Haigron, J.L. Halbwachs, N. C. Hambly, D. L. Harrison, D. Hatzidimitriou, U. Heiter, J. Hernández, D. Hestroffer, S.T. Hodgkin, B. Holl, K. Janßen, G. Jevardat de Fombelle, S. Jordan, A. Krone-Martins, A. C. Lanzafame, W. Löffler, A. Lorca, M. Manteiga, O. Marchal, P.M. Marrese, A. Moitinho, A. Mora, K. Muinonen, P. Osborne, E. Pancino, T. Pauwels, J. M. Petit, A. Recio-Blanco, P.J. Richards, M. Riello, L. Rimoldini, A. C. Robin, T. Roegiers, J. Rybizki, L. M. Sarro, C. Siopis, M. Smith, A. Sozzetti, A. Ulla, E. Utrilla, M. van Leeuwen, W. van Reeven, U. Abbas, A. Abreu Aramburu, S. Accart, C. Aerts, J. J. Aguado, M. Ajaj, G. Altavilla, M.A. Álvarez, J. Álvarez Cid-Fuentes, J. Alves, R. I. Anderson, E. Anglada Varela, T. Antoja, M. Audard, D. Baines, S.G. Baker, L. Balaguer-Núñez, E. Balbinot, Z. Balog, C. Barache, D. Barbato, M. Barros, M.A. Barstow, S. Bartolomé, J. L. Bassilana, N. Bauchet, A. Baudesson-Stella, U. Becciani, M. Bellazzini, M. Bernet, S. Bertone, L. Bianchi, S. Blanco-Cuaresma, T. Boch, A. Bombrun, D. Bossini, S. Bouquillon, A. Bragaglia, L. Bramante, E. Breedt, A. Bressan, N. Brouillet, B. Bucciarelli, A. Burlacu, D. Busonero, A. G. Butkevich, R. Buzzi, E. Caffau, R. Cancelliere, H. Cánovas, T. Cantat-Gaudin, R. Carballo, T. Carlucci, M. I. Carnerero, J. M. Carrasco, L. Casamiquela, M. Castellani, A. Castro-Ginard, P. Castro Sampol, L. Chaoul, P. Charlot, L. Chemin, A. Chiavassa, M. R. L. Cioni, G. Comoretto, W. J. Cooper, T. Cornez, S. Cowell, F. Crifo, M. Crosta, C. Crowley, C. Dafonte, A. Dapergolas, M. David, P. David, P. de Laverny, F. De Luise, R. De March, J. De Ridder, R. de Souza, P. de Teodoro, A. de Torres, E. F. del Peloso, E. del Pozo, M. Delbo, A. Delgado, H. E. Delgado, J. B. Delisle, P. Di Matteo, S. Diakite, C. Diener, E. Distefano, C. Dolding, D. Eappachen, B. Edvardsson, H. Enke, P. Esquej, C. Fabre, M. Fabrizio, S. Faigler, G. Fedorets, P. Fernique, A. Fienga, F. Figueras, C. Fouron, F. Fragkoudi, E. Fraile, F. Franke, M. Gai, D. Garabato, A. Garcia-Gutierrez, M. García-Torres, A. Garofalo, P. Gavras, E. Gerlach, R. Geyer, P. Giacobbe, G. Gilmore, S. Girona, G. Giuffrida, R. Gomel, A. Gomez, I. Gonzalez-Santamaria, J. J. González-Vidal, M. Granvik, R. Gutiérrez-Sánchez, L. P. Guy, M. Hauser, M. Haywood, A. Helmi, S. L. Hidalgo, T. Hilger, N. Hładczuk, D. Hobbs, G. Holland, H. E. Huckle, G. Jasniewicz, P. G. Jonker, J. Juaristi Campillo, F. Julbe, L. Karbevska,

P. Kervella, S. Khanna, A. Kochoska, M. Kontizas, G. Kordopatis, A.J. Korn, Z. Kostrzewa-Rutkowska, K. Kruszyńska, S. Lambert, A.F. Lanza, Y. Lasne, J.F. Le Campion, Y. Le Fustec, Y. Lebreton, T. Lebzelter, S. Leccia, N. Leclerc, I. Lecoeur-Taibi, S. Liao, E. Licata, E. P. Lindstrøm, T. A. Lister, E. Livanou, A. Lobel, P. Madrero Pardo, S. Managau, R. G. Mann, J. M. Marchant, M. Marconi, M. M. S. Marcos Santos, S. Marinoni, F. Marocco, D. J. Marshall, L. Martin Polo, J. M. Martín-Fleitas, A. Masip, D. Massari, A. Mastrobuono-Battisti, T. Mazeh, P. J. McMillan, S. Messina, D. Michalik, N. R. Millar, A. Mints, D. Molina, R. Molinaro, L. Molnár, P. Montegriffo, R. Mor, R. Morbidelli, T. Morel, D. Morris, A. F. Mulone, D. Munoz, T. Muraveva, C.P. Murphy, I. Musella, L. Noval, C. Ordénovic, G. Orrù, J. Osinde, C. Pagani, I. Pagano, L. Palaversa, P.A. Palicio, A. Panahi, M. Pawlak, X. Peñalosa Esteller, A. Penttilä, A. M. Piersimoni, F.X. Pineau, E. Plachy, G. Plum, E. Poggio, E. Poretti, E. Poujoulet, A. Prša, L. Pulone, E. Racero, S. Ragaini, M. Rainer, C.M. Raiteri, N. Rambaux, P. Ramos, M. Ramos-Lerate, P. Re Fiorentin, S. Regibo, C. Reylé, V. Ripepi, A. Riva, G. Rixon, N. Robichon, C. Robin, M. Roelens, L. Rohrbasser, M. Romero-Gómez, N. Rowell, F. Royer, K. A. Rybicki, G. Sadowski, A. Sagristà Sellés, J. Sahlmann, J. Salgado, E. Salguero, N. Samaras, V. Sanchez Gimenez, N. Sanna, R. Santoveña, M. Sarasso, M. Schultheis, E. Sciacca, M. Segol, J.C. Segovia, D. Ségransan, D. Semeux, S. Shahaf, H. I. Siddiqui, A. Siebert, L. Siltala, E. Slezak, R.L. Smart, E. Solano, F. Solitro, D. Souami, J. Souchay, A. Spagna, F. Spoto, I.A. Steele, H. Steidelmüller, C.A. Stephenson, M. Süveges, L. Szabados, E. Szegedi-Elek, F. Taris, G. Tauran, M.B. Taylor, R. Teixeira, W. Thuillot, N. Tonello, F. Torra, J. Torra, C. Turon, N. Unger, M. Vaillant, E. van Dillen, O. Vanel, A. Vecchiato, Y. Viala, D. Vicente, S. Voutsinas, M. Weiler, T. Wevers, Ł. Wyrzykowski, A. Yoldas, P. Yvard, H. Zhao, J. Zorec, S. Zucker, C. Zurbach, T. Zwitter, Gaia Early Data Release 3 summary of the contents and survey properties. A&A **649**, A1 (2021)
25. Gaia collaboration, T. Prusti, J. De Bruijne, A.G. Brown, A. Vallenari, C. Babusiaux, C. Bailer-Jones, U. Bastian, M. Biermann, D.W. Evans, L. Eyer et al., The Gaia mission. A&A **595**, A1 (2016)
26. Gaia Collaboration, A. Vallenari, A.G.A. Brown, T. Prusti, J.H.J. de Bruijne, F. Arenou, C. Babusiaux, M. Biermann, O.L. Creevey, C. Ducourant, D.W. Evans, L. Eyer, R. Guerra, A. Hutton, C. Jordi, S.A. Klioner, U.L. Lammers, L. Lindegren, X. Luri, F. Mignard, C. Panem, D. Pourbaix, S. Randich, P. Sartoretti, C. Soubiran, P. Tanga, N.A. Walton, C.A.L. Bailer-Jones, U. Bastian, R. Drimmel, F. Jansen, D. Katz, M.G. Lattanzi, F. van Leeuwen, J. Bakker, C. Cacciari, J. Castañeda, F. De Angeli, C. Fabricius, M. Fouesneau, Y. Frémat, L. Galluccio, A. Guerrier, U. Heiter, E. Masana, R. Messineo, N. Mowlavi, C. Nicolas, K. Nienartowicz, F. Pailler, P. Panuzzo, F. Riclet, W. Roux, G.M. Seabroke, R. Sordo, F. Thévenin, G. Gracia-Abril, J. Portell, D. Teyssier, M. Altmann, R. Andrae, M. Audard, I. Bellas-Velidis, K. Benson, J. Berthier, R. Blomme, P. W. Burgess, D. Busonero, G. Busso, H. Cánovas, B. Carry, A. Cellino, N. Cheek, G. Clementini, Y. Damerdji, M. Davidson, P. de Teodoro, M. Nuñez Campos, L. Delchambre, A. Dell'Oro, P. Esquej, J. Fernández-Hernández, E. Fraile, D. Garabato, P. García-Lario, E. Gosset, R. Haigron, J. L. Halbwachs, N. C. Hambly, D. L. Harrison, J. Hernández, D. Hestroffer, S. T. Hodgkin, B. Holl, K. Janßen, G. Jevardat de Fombelle, S. Jordan, A. Krone-Martins, A. C. Lanzafame, W. Löffler, O. Marchal, P. M. Marrese, A. Moitinho, K. Muinonen, P. Osborne, E. Pancino, T. Pauwels, A. Recio-Blanco, C. Reylé, M. Riello, L. Rimoldini, T. Roegiers, J. Rybizki, L. M. Sarro, C. Siopis, M. Smith, A. Sozzetti, E. Utrilla, M. van Leeuwen, U. Abbas, P. Ábrahám, A. Abreu Aramburu, C. Aerts, J. J. Aguado, M. Ajaj, F. Aldea-Montero, G. Altavilla, M. A. Álvarez, J. Alves, F. Anders, R. I. Anderson, E. Anglada Varela, T. Antoja, D. Baines, S. G. Baker, L. Balaguer-Núñez, E. Balbinot, Z. Balog, C. Barache, D. Barbato, M. Barros, M.A. Barstow, S. Bartolomé, J. L. Bassilana, N. Bauchet, U. Becciani, M. Bellazzini, A. Berihuete, M. Bernet, S. Bertone, L. Bianchi, A. Binnenfeld, S. Blanco-Cuaresma, A. Blazere, T. Boch, A. Bombrun, D. Bossini, S. Bouquillon, A. Bragaglia, L. Bramante, E. Breedt, A. Bressan, N. Brouillet, E. Brugaletta, B. Bucciarelli, A. Burlacu, A. G. Butkevich, R. Buzzi, E. Caffau, R. Cancelliere, T. Cantat-Gaudin, R. Carballo, T. Carlucci, M. I. Carnerero, J. M. Carrasco, L. Casamiquela, M. Castellani, A. Castro-Ginard, L. Chaoul, P. Charlot, L. Chemin, V. Chiaramida, A. Chi-

avassa, N. Chornay, G. Comoretto, G. Contursi, W. J. Cooper, T. Cornez, S. Cowell, F. Crifo, M. Cropper, M. Crosta, C. Crowley, C. Dafonte, A. Dapergolas, M. David, P. David, P. de Laverny, F. De Luise, R. De March, J. De Ridder, R. de Souza, A. de Torres, E. F. del Peloso, E. del Pozo, M. Delbo, A. Delgado, J.B. Delisle, C. Demouchy, T.E. Dharmawardena, P. Di Matteo, S. Diakite, C. Diener, E. Distefano, C. Dolding, B. Edvardsson, H. Enke, C. Fabre, M. Fabrizio, S. Faigler, G. Fedorets, P. Fernique, A. Fienga, F. Figueras, Y. Fournier, C. Fouron, F. Fragkoudi, M. Gai, A. Garcia-Gutierrez, M. Garcia-Reinaldos, M. García-Torres, A. Garofalo, A. Gavel, P. Gavras, E. Gerlach, R. Geyer, P. Giacobbe, G. Gilmore, S. Girona, G. Giuffrida, R. Gomel, A. Gomez, J. González-Núñez, I. González-Santamaría, J. J. González-Vidal, M. Granvik, P. Guillout, J. Guiraud, R. Gutiérrez-Sánchez, L. P. Guy, D. Hatzidimitriou, M. Hauser, M. Haywood, A. Helmer, A. Helmi, M. H. Sarmiento, S. L. Hidalgo, T. Hilger, N. Hładczuk, D. Hobbs, G. Holland, H. E. Huckle, K. Jardine, G. Jasniewicz, A. Jean-Antoine Piccolo, Ó. Jiménez-Arranz, A. Jorissen, J. Juaristi Campillo, F. Julbe, L. Karbevska, P. Kervella, S. Khanna, M. Kontizas, G. Kordopatis, A. J. Korn, Á. Kóspál, Z. Kostrzewa-Rutkowska, K. Kruszyńska, M. Kun, P. Laizeau, S. Lambert, A.F. Lanza, Y. Lasne, J.F. Le Campion, Y. Lebreton, T. Lebzelter, S. Leccia, N. Leclerc, I. Lecoeur-Taibi, S. Liao, E.L. Licata, H.E.P. Lindstrøm, T. A. Lister, E. Livanou, A. Lobel, A. Lorca, C. Loup, P. Madrero Pardo, A. Magdaleno Romeo, S. Managau, R. G. Mann, M. Manteiga, J.M. Marchant, M. Marconi, J. Marcos, M.M.S. Marcos Santos, D. Marín Pina, S. Marinoni, F. Marocco, D. J. Marshall, L. Martin Polo, J.M. Martín-Fleitas, G. Marton, N. Mary, A. Masip, D. Massari, A. Mastrobuono-Battisti, T. Mazeh, P. J. McMillan, S. Messina, D. Michalik, N. R. Millar, A. Mints, D. Molina, R. Molinaro, L. Molnár, G. Monari, M. Monguió, P. Montegriffo, A. Montero, R. Mor, A. Mora, R. Morbidelli, T. Morel, D. Morris, T. Muraveva, C. P. Murphy, I. Musella, Z. Nagy, L. Noval, F. Ocaña, A. Ogden, C. Ordenovic, J. O. Osinde, C. Pagani, I. Pagano, L. Palaversa, P. A. Palicio, L. Pallas-Quintela, A. Panahi, S. Payne-Wardenaar, X. Peñalosa Esteller, A. Penttilä, B. Pichon, A. M. Piersimoni, F. X. Pineau, E. Plachy, G. Plum, E. Poggio, A. Prša, L. Pulone, E. Racero, S. Ragaini, M. Rainer, C.M. Raiteri, N. Rambaux, P. Ramos, M. Ramos-Lerate, P. Re Fiorentin, S. Regibo, P.J. Richards, C. Rios Diaz, V. Ripepi, A. Riva, H.W. Rix, G. Rixon, N. Robichon, A.C. Robin, C. Robin, M. Roelens, H.R.O. Rogues, L. Rohrbasser, M. Romero-Gómez, N. Rowell, F. Royer, D. Ruz Mieres, K.A. Rybicki, G. Sadowski, A. Sáez Núñez, A. Sagristà Sellés, J. Sahlmann, E. Salguero, N. Samaras, V. Sanchez Gimenez, N. Sanna, R. Santoveña, M. Sarasso, M. Schultheis, E. Sciacca, M. Segol, J.C. Segovia, D.Ségransan, D. Semeux, S. Shahaf, H.I. Siddiqui, A. Siebert, L. Siltala, A. Silvelo, E. Slezak, I. Slezak, R.L. Smart, O. N. Snaith, E. Solano, F. Solitro, D. Souami, J. Souchay, A. Spagna, L. Spina, F. Spoto, I.A. Steele, H. Steidelmüller, C. A. Stephenson, M. Süveges, J. Surdej, L. Szabados, E. Szegedi-Elek, F. Taris, M.B. Taylor, R. Teixeira, L. Tolomei, N. Tonello, F. Torra, J. Torra, G. Torralba Elipe, M. Trabucchi, A. T. Tsounis, C. Turon, A. Ulla, N. Unger, M.V. Vaillant, E. van Dillen, W. van Reeven, O. Vanel, A. Vecchiato, Y. Viala, D. Vicente, S. Voutsinas, M. Weiler, T. Wevers, Ł. Wyrzykowski, A. Yoldas, P. Yvard, H. Zhao, J. Zorec, S. Zucker, T. Zwitter, Gaia Data Release 3 summary of the content and survey properties. A&A **674**, A1 (2023)

27. M. Cruz Reyes, R.I. Anderson, A 0.9% calibration of the Galactic Cepheid luminosity scale based on Gaia DR3 data of open clusters and Cepheids. A&A **672**, A85 (2023)
28. A. Bhardwaj, S.M. Kanbur, L.M. Macri, H.P. Singh, C.C. Ngeow, E.E.O. Ishida, Large magellanic cloud near-infrared synoptic survey—III a statistical study of non-linearity in the Leavitt Laws. MNRAS **457**(2), 1644–1665 (2016)
29. P. Klagyivik, L. Szabados, Observational studies of cepheid amplitudes I period-amplitude relationships for galactic cepheids and interrelation of amplitudes. A&A **504**(3), 959–972 (2009)
30. R.I. Anderson, H. Saio, S. Ekström, C. Georgy, G. Meynet, On the effect of rotation on populations of classical cepheids II pulsation analysis for metallicities 0.014, 0.006, and 0.002. A&A **591**, A8 (2016)
31. G. De Somma, M. Marconi, S. Cassisi, V. Ripepi, A. Pietrinferni, R. Molinaro, S. Leccia, I. Musella, Period-age-metallicity and period-age-colour-metallicity relations for classical

Cepheids: an application to the Gaia EDR3 sample. MNRAS **508**(1), 1473–1488 (2021)
32. V. Ripepi, G. Catanzaro, R. Molinaro, M. Marconi, G. Clementini, F. Cusano, G. De Somma, S. Leccia, I. Musella, V. Testa, Period-luminosity-metallicity relation of classical cepheids. A&A **642**, A230 (2020)
33. C.-C. Ngeow, S.M. Kanbur, E.P. Bellinger, M. Marconi, I. Musella, M. Cignoni, Y.-H. Lin, Period-luminosity relations for cepheid variables: from mid-infrared to multi-phase. Ap& SS **341**(1), 105–113 (2012)
34. E. Hertzsprung, On the relation between period and form of the light-curve of variable stars of the δ Cephei type. Bull. Astron. Inst. Netherlands **3**, 115 (1926)
35. L. Breuval, A.G. Riess, P. Kervella, R.I. Anderson, M. Romaniello, An improved calibration of the wavelength dependence of metallicity on the cepheid Leavitt law. ApJ **939**(2), 89 (2022)
36. W.L. Freedman, B.F. Madore, Two new tests of the metallicity sensitivity of the cepheid period-luminosity relation (the Leavitt Law). ApJ **734**(1), 46 (2011)
37. J. Kennicutt, C. Robert., P.B. Stetson, A. Saha, D. Kelson, D.M. Rawson, S. Sakai, B.F. Madore, J.R. Mould, W.L. Freedman, F. Bresolin, L. Ferrarese, H. Ford, B.K. Gibson, J.A. Graham, M. Han, P. Harding, J.G. Hoessel, J.P. Huchra, S.M.G. Hughes, G.D. Illingworth, L.M. Macri, R.L. Phelps, N.A. Silbermann, A.M. Turner, P.R. Wood, The hubble space telescope key project on the extragalactic distance scale XIII the metallicity dependence of the cepheid distance scale. ApJ **498**(1), 181–194 (1998)
38. L. Rizzi, R.B. Tully, D. Makarov, L. Makarova, A.E. Dolphin, S. Sakai, E.J. Shaya, Tip of the red giant branch distances II zero-point calibration. ApJ **661**(2), 815–829 (2007)
39. R. Molinaro, V. Ripepi, M. Marconi, M. Romaniello, G. Catanzaro, F. Cusano, G. De Somma, I. Musella, J. Storm, E. Trentin, Cepheid metallicity in the Leavitt law (C-MetaLL) survey—III simultaneous derivation of the Gaia parallax offset and period-luminosity-metallicity coefficients. MNRAS **520**(3), 4154–4166 (2023)
40. F. Bresolin, Revisiting the abundance gradient in the maser host galaxy NGC 4258. ApJ **729**(1), 56 (2011)
41. D. Zaritsky, J. Kennicutt, C. Robert, J.P. Huchra, HII regions and the abundance properties of spiral galaxies. ApJ **420**, 87 (1994)
42. F. Bresolin, R.-P. Kudritzki, M.A. Urbaneja, The metallicity and distance of NGC 2403 from blue supergiants. ApJ **940**(1), 32 (2022)
43. R.-P. Kudritzki, M.A. Urbaneja, F. Bresolin, N. Przybilla, W. Gieren, G. Pietrzyński, Quantitative spectroscopy of 24 a supergiants in the sculptor galaxy NGC 300: flux-weighted gravity-luminosity relationship, metallicity, and metallicity gradient. ApJ **681**(1), 269–289 (2008)
44. M.G. Lee, W.L. Freedman, B.F. Madore, The tip of the red giant branch as a distance indicator for resolved galaxies. ApJ **417**, 553 (1993)
45. W.L. Freedman, Measurements of the hubble constant: tensions in perspective. ApJ **919**(1), 16 (2021)
46. L. Eyer, Search for QSO candidates in OGLE-II data. Acta Astron. **52**, 241–262 (2002)
47. Y. Ita, T. Tanabé, N. Matsunaga, Y. Nakajima, C. Nagashima, T. Nagayama, D. Kato, M. Kurita, T. Nagata, S. Sato, M. Tamura, H. Nakaya, Y. Nakada, Pulsation at the tip of the first giant branch? MNRAS **337**(3), L31–L34 (2002)
48. R.I. Anderson, N.W. Koblischke, L. Eyer, Reconciling astronomical distance scales with variable red giant stars (2023). arXiv e-prints: arXiv:2303.04790
49. S. Li, A.G. Riess, D. Scolnic, G.S. Anand, J. Wu, S. Casertano, W. Yuan, R. Beaton, R.I. Anderson, Standardized luminosity of the tip of the red giant branch utilizing multiple fields in NGC 4258 and the CATs algorithm (2023). arXiv e-prints: arXiv:2306.10103
50. D. Scolnic, A.G. Riess, J. Wu, S. Li, G.S. Anand, R. Beaton, S. Casertano, R. Anderson, S. Dhawan, X. Ke, CATS: the hubble constant from standardized TRGB and type Ia supernova measurements (2023). arXiv e-prints: arXiv:2304.06693
51. J. Wu, D. Scolnic, A.G. Riess, G.S. Anand, R. Beaton, S. Casertano, X. Ke, S. Li, Comparative analysis of TRGBs (CATs) from unsupervised multi-halo-field measurements: contrast is key. ApJ **954**(1), 87 (2023)

52. B.F. Madore, V. Mager, W.L. Freedman, Sharpening the tip of the red giant branch. ApJ **690**(1), 389–393 (2009)
53. R.L. Beaton, G. Bono, V.F. Braga, M. Dall'Ora, G. Fiorentino, I.S. Jang, C.E. Martínez-Vázquez, N. Matsunaga, M. Monelli, J.R. Neeley, M. Salaris, Old-aged primary distance indicators. Space Sci. Rev. **214**(8), 113 (2018)
54. W. Baade, The period-luminosity relation of the cepheids. PASP **68**(400), 5 (1956)
55. I. Soszyński, A. Udalski, M. K. Szymański, D. Skowron, G. Pietrzyński, R. Poleski, P. Pietrukowicz, J. Skowron, P. Mróz, S. Kozłowski, Ł. Wyrzykowski, K. Ulaczyk, M. Pawlak, The OGLE collection of variable stars: classical cepheids in the magellanic system. Acta Astron. **65**(4), 297–312 (2015)
56. P. Pietrukowicz, I. Soszyński, A. Udalski, Classical cepheids in the milky way. Acta Astron. **71**(3), 205–222 (2021)
57. V. Ripepi, G. Clementini, R. Molinaro, S. Leccia, E. Plachy, L. Molnár, L. Rimoldini, I. Musella, M. Marconi, A. Garofalo, M. Audard, B. Holl, D.W. Evans, G. Jevardat de Fombelle, I. Lecoeur-Taibi, O. Marchal, N. Mowlavi, T. Muraveva, K. Nienartowicz, P. Sartoretti, L. Szabados, L. Eyer, Gaia DR3: specific processing and validation of all-sky RR Lyrae and Cepheid stars—The Cepheid sample (2022). arXiv e-prints: arXiv:2206.06212
58. C. Alcock, R.A. Allsman, D.R. Alves, T.S. Axelrod, A.C. Becker, D.P. Bennett, D.F. Bersier, K.H. Cook, K.C. Freeman, K. Griest, J.A. Guern, M. Lehner, S.L. Marshall, D. Minniti, B.A. Peterson, M.R. Pratt, P.J. Quinn, A.W. Rodgers, C.W. Stubbs, W. Sutherland, A. Tomaney, T. Vandehei, D.L. Welch, The MACHO project LMC variable star inventory VIII the recent star formation history of the large magellanic cloud from the cepheid period distribution. AJ **117**(2), 920–926 (1999)
59. V. Kovtyukh, B. Lemasle, F. Chekhonadskikh, G. Bono, N. Matsunaga, A. Yushchenko, R.I. Anderson, S. Belik, R. da Silva, L. Inno, The chemical composition of galactic beat cepheids. MNRAS **460**(2), 2077–2086 (2016)
60. G. Csörnyei, R.I. Anderson, C. Vogl, S. Taubenberger, S. Blondin, B. Leibundgut, W. Hillebrandt, Reeling in the Whirlpool: the distance to M 51 clarified by cepheids and the type IIP SN 2005cs (2023). arXiv e-prints: arXiv:2305.13943
61. R.I. Anderson, S. Ekström, C. Georgy, G. Meynet, N. Mowlavi, L. Eyer, On the effect of rotation on populations of classical cepheids I predictions at solar metallicity. A&A **564**, A100 (2014)
62. N.R. Evans, R. Szabó, A. Derekas, L. Szabados, C. Cameron, J.M. Matthews, D. Sasselov, R. Kuschnig, J.F. Rowe, D.B. Guenther, A.F.J. Moffat, S.M. Rucinski, W.W. Weiss, Observations of Cepheids with the MOST satellite: contrast between pulsation modes. MNRAS **446**(4), 4008–4018 (2015)
63. R. Smolec, M. Śniegowska, Non-radial pulsation in first overtone cepheids of the small magellanic cloud. MNRAS **458**(4), 3561–3577 (2016)
64. M. Süveges, R.I. Anderson, Investigating light-curve modulation via kernel smoothing—II new additional modes in single-mode OGLE classical Cepheids. MNRAS **478**(2), 1425–1441 (2018)
65. M. Süveges, R.I. Anderson, Investigating light curve modulation via kernel smoothing I application to 53 fundamental mode and first-overtone cepheids in the LMC. A&A **610**, A86 (2018)
66. D. Turner, M. Abdel-Sabour Abdel-Latif, L.N. Berdnikov, Using Cepheid period changes to test stellar evolutionary models. JRASC **99**(4), 144 (2005)
67. G. Csörnyei, L. Szabados, L. Molnár, B. Cseh, N. Egei, C. Kalup, V. Kecskeméthy, R. Könyves-Tóth, K. Sárneczky, R. Szakáts, Study of changes in the pulsation period of 148 galactic cepheid variables. MNRAS **511**(2), 2125–2146 (2022)
68. B.F. Madore, The period-luminosity relation IV intrinsic relations and reddenings for the large magellanic cloud cepheids. ApJ **253**, 575–579 (1982)
69. S. van den Bergh, The extragalactic distance scale, in *Galaxies and the Universe*, ed. by A. Sandage, M. Sandage, J. Kristian, p. 509 (1975)

70. E.L. Fitzpatrick, Correcting for the effects of interstellar extinction. PASP **111**(755), 63–75 (1999)
71. E.F. Schlafly, D.P. Finkbeiner, Measuring reddening with sloan digital sky survey stellar spectra and recalibrating SFD. ApJ **737**(2), 103 (2011)
72. R.I. Anderson, Relativistic corrections for measuring Hubble's constant to 1% using stellar standard candles. A&A **658**, A148 (2022)
73. K.D. Gordon, G.C. Clayton, K.A. Misselt, A.U. Landolt, M.J. Wolff, A quantitative comparison of the small magellanic cloud, large magellanic cloud, and milky way ultraviolet to near-infrared extinction curves. ApJ **594**(1), 279–293 (2003)
74. E. Mörtsell, A. Goobar, J. Johansson, S. Dhawan, The hubble tension revisited: additional local distance ladder uncertainties. ApJ **935**(1), 58 (2022)
75. M.W. Feast, R.M. Catchpole, The Cepheid period-luminosity zero-point from HIPPARCOS trigonometrical parallaxes. MNRAS **286**(1), L1–L5 (1997)
76. R. da Silva, J. Crestani, G. Bono, V.F. Braga, V. D'Orazi, B. Lemasle, M. Bergemann, M. Dall'Ora, G. Fiorentino, P. François, M.A.T. Groenewegen, L. Inno, V. Kovtyukh, R.P. Kudritzki, N. Matsunaga, M. Monelli, A. Pietrinferni, L. Porcelli, J. Storm, M. Tantalo, F. Thévénin, A new and homogeneous metallicity scale for galactic classical cepheids II abundance of iron and α elements. A&A **661**, A104 (2022)
77. V. Ripepi, G. Catanzaro, G. Clementini, G. De Somma, R. Drimmel, S. Leccia, M. Marconi, R. Molinaro, I. Musella, E. Poggio, Classical Cepheid period-Wesenheit-metallicity relation in the Gaia bands. A&A **659**, A167 (2022)
78. E. Trentin, V. Ripepi, G. Catanzaro, J. Storm, M. Marconi, G. De Somma, V. Testa, I. Musella, Cepheid metallicity in the leavitt law (C-MetaLL) survey-II high-resolution spectroscopy of the most metal poor galactic cepheids. MNRAS **519**(2), 2331–2348 (2023)
79. A.G. Riess, S. Casertano, W. Yuan, J.B. Bowers, L. Macri, J.C. Zinn, D. Scolnic, Cosmic distances calibrated to 1% precision with Gaia EDR3 parallaxes and hubble space telescope photometry of 75 milky way cepheids confirm tension with ΛCDM. ApJ **908**(1), L6 (2021)
80. A.G. Riess, S. Casertano, W. Yuan, L. Macri, B. Bucciarelli, M.G. Lattanzi, J.W. MacKenty, J.B. Bowers, W. Zheng, A.V. Filippenko, C. Huang, R.I. Anderson, Milky way cepheid standards for measuring cosmic distances and application to Gaia DR2: implications for the hubble constant. ApJ **861**(2), 126 (2018)
81. S. Casertano, A.G. Riess, J. Anderson, R.I. Anderson, J.B. Bowers, K.I. Clubb, A.R. Cukierman, A.V. Filippenko, M.L. Graham, J.W. MacKenty, C. Melis, B.E. Tucker, G. Upadhya, Parallax of galactic cepheids from spatially scanning the wide field camera 3 on the hubble space telescope: the case of SS Canis Majoris. ApJ **825**(1), 11 (2016)
82. A.G. Riess, S. Casertano, J. Anderson, J. MacKenty, A.V. Filippenko, Parallax beyond a kiloparsec from spatially scanning the wide field camera 3 on the hubble space telescope. ApJ **785**(2), 161 (2014)
83. A.G. Riess, S. Casertano, W. Yuan, L. Macri, J. Anderson, J.W. MacKenty, J.B. Bowers, K.I. Clubb, A.V. Filippenko, D.O. Jones et al., New parallaxes of galactic cepheids from spatially scanning the hubble space telescope: implications for the hubble constant. ApJ **855**(2), 136 (2018)
84. L. Lindegren, U. Bastian, M. Biermann, A. Bombrun, A. de Torres, E. Gerlach, R. Geyer, J. Hernández, T. Hilger, D. Hobbs, S.A. Klioner, U. Lammers, P.J. McMillan, M. Ramos-Lerate, H. Steidelmüller, C.A. Stephenson, F. van Leeuwen, Gaia Early Data Release 3 parallax bias versus magnitude, colour, and position. A&A **649**, A4 (2021)
85. L. Lindegren, S.A. Klioner, J. Hernández, A. Bombrun, M. Ramos-Lerate, H. Steidelmüller, U. Bastian, M. Biermann, A. de Torres, E. Gerlach, R. Geyer, T. Hilger, D. Hobbs, U. Lammers, P.J. McMillan, C.A. Stephenson, J. Castañeda, M. Davidson, C. Fabricius, G. Gracia-Abril, J. Portell, N. Rowell, D. Teyssier, F. Torra, S. Bartolomé, M. Clotet, N. Garralda, J.J. González-Vidal, J. Torra, U. Abbas, M. Altmann, E. Anglada Varela, L. Balaguer-Núñez, Z. Balog, C. Barache, U. Becciani, M. Bernet, S. Bertone, L. Bianchi, S. Bouquillon, A.G.A. Brown, B. Bucciarelli, D. Busonero, A.G. Butkevich, R. Buzzi, R. Cancelliere, T. Carlucci, P. Charlot, M.R.L. Cioni, M. Crosta, C. Crowley, E. F. del Peloso, E. del Pozo, R. Drimmel, P. Esquej,

A. Fienga, E. Fraile, M. Gai, M. Garcia-Reinaldos, R. Guerra, N.C. Hambly, M. Hauser, K. Janßen, S. Jordan, Z. Kostrzewa-Rutkowska, M.G. Lattanzi, S. Liao, E. Licata, T.A. Lister, W. Löffler, J.M. Marchant, A. Masip, F. Mignard, A. Mints, D. Molina, A. Mora, R. Morbidelli, C.P. Murphy, C. Pagani, P. Panuzzo, X. Peñalosa Esteller, E. Poggio, P. Re Fiorentin, A. Riva, A. Sagristà Sellés, V. Sanchez Gimenez, M. Sarasso, E. Sciacca, H.I. Siddiqui, R.L. Smart, D. Souami, A. Spagna, I. A. Steele, F. Taris, E. Utrilla, W. van Reeven, A. Vecchiato, Gaia Early Data Release 3 the astrometric solution. A&A **649**, A2 (2021)

86. *The HIPPARCOS and TYCHO catalogues. Astrometric and photometric star catalogues derived from the ESA HIPPARCOS Space Astrometry Mission*, vol. 1200. (ESA Special Publication, 1997)
87. F. van Leeuwen, M.W. Feast, P.A. Whitelock, C.D. Laney, Cepheid parallaxes and the hubble constant. MNRAS **379**(2), 723–737 (2007)
88. J.C. Zinn, Validation of the Gaia Early Data Release 3 parallax zero-point model with asteroseismology. AJ **161**(5), 214 (2021)
89. M. Groenewegen, Primary period-luminosity-relation calibrators in the milky way: cepheids and RR Lyrae physical basis, calibration, and applications (2023). arXiv e-prints: arXiv:2307.03033
90. S. Khan, R.I. Anderson, A. Miglio, B. Mosser, Y.P. Elsworth, Investigating gaia edr3 parallax systematics using asteroseismology of cool giant stars observed by kepler, k2, and tess ii. deciphering gaia parallax systematics using red clump stars (2023). arXiv e-prints: arXiv:2310.03654
91. J. Maíz Apellániz, An estimation of the Gaia EDR3 parallax bias from stellar clusters and magellanic clouds data. A&A **657**, A130 (2022)
92. G. Pietrzyński, D. Graczyk, A. Gallenne, W. Gieren, I.B. Thompson, B. Pilecki, P. Karczmarek, M. Górski, K. Suchomska, M. Taormina, B. Zgirski, P. Wielgórski, Z. Kołaczkowski, P. Konorski, S. Villanova, N. Nardetto, P. Kervella, F. Bresolin, R.P. Kudritzki, J. Storm, R. Smolec, W. Narloch, A distance to the large magellanic cloud that is precise to one per cent. Nature **567**(7747), 200–203 (2019)
93. D. Graczyk, G. Pietrzyński, I.B. Thompson, W. Gieren, B. Zgirski, S. Villanova, M. Górski, P. Wielgórski, P. Karczmarek, W. Narloch, B. Pilecki, M. Taormina, R. Smolec, K. Suchomska, A. Gallenne, N. Nardetto, J. Storm, R.-P. Kudritzki, M. Kałuszyński, W. Pych, A distance determination to the small magellanic cloud with an accuracy of better than two percent based on late-type eclipsing binary stars. ApJ **904**(1), 13 (2020)
94. L. Breuval, A.G. Riess, S. Casertano, W. Yuan, L.M. Macri, M. Romaniello, Y.S. Murakami, D. Scolnic, G.S. Anand, I. Soszyński, *Small Magellanic Cloud Cepheids Observed with the Hubble Space Telescope Provide a New Anchor for the SH0ES Distance Ladder* (2024). https://doi.org/10.48550/arXiv.2404.08038. ArXiv: 2404.08038
95. A.G. Riess, S. Casertano, W. Yuan, L.M. Macri, D. Scolnic, Large magellanic cloud cepheid standards provide a 1% foundation for the determination of the hubble constant and stronger evidence for physics beyond ΛCDM. ApJ **876**(1), 85 (2019)
96. L.M. Macri, C.C. Ngeow, S.M. Kanbur, S. Mahzooni, M.T. Smitka, Large magellanic cloud near-infrared synoptic survey I cepheid variables and the calibration of the Leavitt Law. AJ **149**(4), 117 (2015)
97. V. Ripepi, L. Chemin, R. Molinaro, M.R.L. Cioni, K. Bekki, G. Clementini, R. de Grijs, G. De Somma, D. El Youssoufi, L. Girardi, M.A.T. Groenewegen, V. Ivanov, M. Marconi, P.J. McMillan, J.T. van Loon, The VMC survey—XLVIII classical cepheids unveil the 3D geometry of the LMC. MNRAS **512**(1), 563–582 (2022)
98. I. Soszyński, A. Udalski, M.K. Szymański, P. Pietrukowicz, J. Skowron, D.M. Skowron, R. Poleski, S. Kozłowski, P. Mróz, K. Ulaczyk, K. Rybicki, P. Iwanek, M. Wrona, Final release of the OGLE collection of cepheids and RR Lyrae stars in the magellanic system the outer regions. Acta Astron. **69**(2), 87–99 (2019)
99. M. Romaniello, A. Riess, S. Mancino, R.I. Anderson, W. Freudling, R.P. Kudritzki, L. Macrì, A. Mucciarelli, W. Yuan, The iron and oxygen content of LMC classical cepheids and its implications for the extragalactic distance scale and hubble constant: equivalent width analysis with Kurucz stellar atmosphere models. A&A **658**, A29 (2022)

100. V. Scowcroft, W.L. Freedman, B.F. Madore, A. Monson, S.E. Persson, J. Rich, M. Seibert, J.R. Rigby, The carnegie hubble program: the distance and structure of the SMC as revealed by mid-infrared observations of cepheids. ApJ **816**(2), 49 (2016)
101. E.M.L. Humphreys, M.J. Reid, J.M. Moran, L.J. Greenhill, A.L. Argon, Toward a new geometric distance to the active galaxy NGC 4258 III final results and the hubble constant. ApJ **775**(1), 13 (2013)
102. M.J. Reid, D.W. Pesce, A.G. Riess, An improved distance to NGC 4258 and its implications for the hubble constant. ApJ **886**(2), L27 (2019)
103. W. Yuan, L.M. Macri, A.G. Riess, T.G. Brink, S. Casertano, A.V. Filippenko, S.L. Hoffmann, C.D. Huang, D. Scolnic, Absolute calibration of cepheid period-luminosity relations in NGC 4258. ApJ **940**(1), 64 (2022)
104. M.M. Fausnaugh, C.S. Kochanek, J.R. Gerke, L.M. Macri, A.G. Riess, K.Z. Stanek, The Cepheid distance to the maser-host galaxy NGC 4258: studying systematics with the large binocular telescope. MNRAS **450**(4), 3597–3619 (2015)
105. B.J. Shappee, K.Z. Stanek, A new cepheid distance to the giant spiral M101 based on image subtraction of hubble space telescope/advanced camera for surveys observations. ApJ **733**(2), 124 (2011)
106. B.F. Madore, W.L. Freedman, A.J. Lee, Astrophysical distance scale IV preliminary zero-point calibration of the JAGB method in the HST/WFC3-IR broad J-band (F110W) filter. ApJ **926**(2), 153 (2022)
107. B. Zgirski, G. Pietrzyński, W. Gieren, M. Górski, P. Wielgórski, P. Karczmarek, F. Bresolin, P. Kervella, R.P. Kudritzki, J. Storm, D. Graczyk, G. Hajdu, W. Narloch, B. Pilecki, K. Suchomska, M. Taormina, The Araucaria project: distances to nine galaxies based on a statistical analysis of their carbon stars (JAGB Method). ApJ **916**(1), 19 (2021)
108. L. Ferrarese, N.A. Silbermann, J.R. Mould, P.B. Stetson, A. Saha, W.L. Freedman, J. Kennicutt, C. Robert, Photometric recovery of crowded stellar fields observed with HST/WFPC2 and the effects of confusion noise on the extragalactic distance scale. PASP **112**(768), 177–201 (2000)
109. A.G. Riess, L. Macri, S. Casertano, H. Lampeitl, H.C. Ferguson, A.V. Filippenko, S.W. Jha, W. Li, R. Chornock, A 3% solution: determination of the hubble constant with the hubble space telescope and wide field camera 3. ApJ **730**(2), 119 (2011)
110. R.I. Anderson, A.G. Riess, On cepheid distance scale bias due to stellar companions and cluster populations. ApJ **861**(1), 36 (2018)
111. L. Breuval, A.G. Riess, L.M. Macri, S. Li, W. Yuan, S. Casertano, T. Konchady, B. Trahin, M.J. Durbin, B.F. Williams, A 1.3% distance to M33 from hubble space telescope cepheid photometry. ApJ **951**(2), 118 (2023)
112. R.I. Anderson, Towards a 1% measurement of the Hubble constant: accounting for time dilation in variable-star light curves. A&A **631**, A165 (2019)
113. A. Dolphin, DOLPHOT: stellar photometry. Astrophysics Source Code Library, record ascl:1608.013 (2016)
114. P.B. Stetson, DAOPHOT: crowded-field stellar photometry package. Astrophysics Source Code Library, record ascl:1104.011 (2011)
115. B. Javanmardi, A. Mérand, P. Kervella, L. Breuval, A. Gallenne, N. Nardetto, W. Gieren, G. Pietrzyński, V. Hocdé, S. Borgniet, Inspecting the cepheid distance ladder: the hubble space telescope distance to the SN Ia host galaxy NGC 5584. ApJ **911**(1), 12 (2021)
116. W. Baade, Über eine Möglichkeit, die Pulsationstheorie der δ Cephei-Veränderlichen zu prüfen. Astron. Nachr. **228**(20), 359 (1926)
117. F.A. Lindemann, Note on the pulsation theory of cepheid variables. MNRAS **78**, 639 (1918)
118. G. Tiercy, G. Abetti, Recherches sur SU Cassiopeiae et les Cepheides a courtes periodes. Osservazioni e memorie dell'Osservatorio astrofisico di Arcetri **44**, 1 (1927)
119. P. Kervella, V. Coudé du Foresto, W.A. Traub, M.G. Lacasse, Interferometric observations of the Cepheid ζ Geminorum with FLUOR/IOTA, in *Working on the Fringe: Optical and IR Interferometry from Ground and Space*, ed. by S. Unwin, R. Stachnik. Astronomical Society of the Pacific Conference Series, vol. 194, p. 22 (1999)

120. B.F. Lane, M.J. Kuchner, A.F. Boden, M. Creech-Eakman, S.R. Kulkarni, Direct detection of pulsations of the cepheid star ζ Gem and an independent calibration of the period-luminosity relation. Nature **407**(6803), 485–487 (2000)
121. D. Mourard, D. Bonneau, L. Koechlin, A. Labeyrie, F. Morand, P. Stee, I. Tallon-Bosc, F. Vakili, The mean angular diameter of δ cephei measured by optical long-baseline interferometry. A&A **317**, 789–792 (1997)
122. P. Kervella, D. Bersier, D. Mourard, N. Nardetto, P. Fouqué, V. Coudé du Foresto, Cepheid distances from infrared long-baseline interferometry. III. Calibration of the surface brightness-color relations. A&A **428**, 587–593 (2004)
123. A. Mérand, P. Kervella, J. Breitfelder, A. Gallenne, V. Coudé du Foresto, T.A. ten Brummelaar, H.A. McAlister, S. Ridgway, L. Sturmann, J. Sturmann, N.H. Turner, Cepheid distances from the spectrophoto-interferometry of pulsating stars (SPIPS): application to the prototypes δ cephei and η aquilae. A&A **584**, A80 (2015)
124. B. Trahin, L. Breuval, P. Kervella, A. Mérand, N. Nardetto, A. Gallenne, V. Hocdé, W. Gieren, Inspecting the cepheid parallax of pulsation using Gaia EDR3 parallaxes: projection factor and period-luminosity and period-radius relations. A&A **656**, A102 (2021)
125. W. Gieren, J. Storm, P. Konorski, M. Górski, B. Pilecki, I. Thompson, G. Pietrzyński, D. Graczyk, T.G. Barnes, P. Fouqué, N. Nardetto, A. Gallenne, P. Karczmarek, K. Suchomska, P. Wielgórski, M. Taormina, B. Zgirski, The effect of metallicity on cepheid period-luminosity relations from a Baade-Wesselink analysis of cepheids in the milky way and magellanic clouds $\star$. A&A **620**, A99 (2018)
126. J. Storm, W. Gieren, P. Fouqué, T. G. Barnes, G. Pietrzyński, N. Nardetto, M. Weber, T. Granzer, K.G. Strassmeier, Calibrating the Cepheid period-luminosity relation from the infrared surface brightness technique I the p-factor, the Milky Way relations, and a universal K-band relation. A&A **534**, A94 (2011)
127. J. Storm, W. Gieren, P. Fouqué, T. G. Barnes, I. Soszyński, G. Pietrzyński, N. Nardetto, D. Queloz, Calibrating the Cepheid period-luminosity relation from the infrared surface brightness technique II the effect of metallicity and the distance to the LMC. A&A **534**, A95 (2011)
128. R.I. Anderson, Discovery of cycle-to-cycle modulated spectral line variability and velocity gradients in long-period cepheids. MNRAS **463**(2), 1707–1723 (2016)
129. R.I. Anderson, A. Mérand, P. Kervella, J. Breitfelder, J.B. LeBouquin, L. Eyer, A. Gallenne, L. Palaversa, T. Semaan, S. Saesen, N. Mowlavi, Investigating cepheid ℓ carinae's cycle-to-cycle variations via contemporaneous velocimetry and interferometry. MNRAS **455**(4), 4231–4248 (2016)
130. N. Nardetto, A. Fokin, D. Mourard, P. Mathias, P. Kervella, D. Bersier, Self consistent modelling of the projection factor for interferometric distance determination. A&A **428**, 131–137 (2004)
131. N. Nardetto, D. Mourard, P. Mathias, A. Fokin, D. Gillet, High-resolution spectroscopy for Cepheids distance determination II: a period-projection factor relation. A&A **471**(2), 661–669 (2007)
132. J. Breitfelder, A. Mérand, P. Kervella, A. Gallenne, L. Szabados, R.I. Anderson, J.B. Le Bouquin, Observational calibration of the projection factor of cepheids II application to nine cepheids with HST/FGS parallax measurements. A&A **587**, A117 (2016)
133. R.I. Anderson, G. Viviani, S. Shreeya Shetye, N. Mowlavi, L. Eyer, L. Palaversa, B. Holl, S. Blanco-Cuaresma, K. Kravchenko, M. Pawlak, M. Cruz Reyes, S. Khan, H.E. Netzel, L. Löbling, P.I. Pápics, A. Postel, M. Roelens, Z.T. Spetsieri, A. Thoul, J. Zák, V. Bonvin, D.V. Martin, M. Millon, S. Saesen, A. Wyttenbach, P. Figueira, M. Marmier, S. Prins, G. Raskin, H. van Winckel, *VELOcities of CEpheids (VELOCE) I. High-precision Radial Velocities of Cepheids* (2024). https://doi.org/10.48550/arXiv.2404.12280. ArXiv: 2404.12280
134. N.R. Evans, L. Berdnikov, J. Lauer, D. Morgan, J. Nichols, H.M. Günther, N. Gorynya, A. Rastorguev, P. Moskalik, Binary properties from cepheid radial velocities (CRaV). AJ **150**(1), 13 (2015)

135. P. Kervella, A. Gallenne, N.R. Evans, L. Szabados, F. Arenou, A. Mérand, N. Nardetto, W. Gieren, G. Pietrzynski, Multiplicity of galactic cepheids and RR Lyrae stars from Gaia DR2 II resolved common proper motion pairs. A&A **623**, A117 (2019)
136. P. Kervella, A. Gallenne, N. Remage Evans, L. Szabados, F. Arenou, A. Mérand, Y. Proto, P. Karczmarek, N. Nardetto, W. Gieren, G. Pietrzynski, Multiplicity of galactic cepheids and RR Lyrae stars from Gaia DR2 I binarity from proper motion anomaly. A&A **623**, A116 (2019)
137. M. Moe, R. Di Stefano, Mind your Ps and Qs: the interrelation between period (P) and mass-ratio (Q) distributions of binary stars. ApJS **230**(2), 15 (2017)
138. R.I. Anderson, S. Casertano, A.G. Riess, C. Melis, B. Holl, T. Semaan, P.I. Papics, S. Blanco-Cuaresma, L. Eyer, N. Mowlavi, L. Palaversa, M. Roelens, Vetting galactic leavitt law calibrators using radial velocities: on the variability, binarity, and possible parallax error of 19 long-period cepheids. ApJS **226**(2), 18 (2016)
139. A. Gallenne, P. Kervella, N.R. Evans, C.R. Proffitt, J.D. Monnier, A. Mérand, E. Nelan, E. Winston, G. Pietrzyński, G. Schaefer, W. Gieren, R.I. Anderson, S. Borgniet, S. Kraus, R.M. Roettenbacher, F. Baron, B. Pilecki, M. Taormina, D. Graczyk, N. Mowlavi, L. Eyer, A geometrical 1% distance to the short-period binary cepheid V1334 Cygni. ApJ **867**(2), 121 (2018)
140. B. Pilecki, G. Pietrzyński, R.I. Anderson, W. Gieren, M. Taormina, W. Narloch, N.R. Evans, J. Storm, Cepheids with giant companions I revealing a numerous population of double-lined binary cepheids. ApJ **910**(2), 118 (2021)
141. B. Pilecki, I.B. Thompson, F. Espinoza-Arancibia, R.I. Anderson, W. Gieren, W. Narloch, J. Minniti, G. Pietrzyński, M. Taormina, G. Bono, G. Hajdu, Discovery of a binary-origin classical cepheid in a binary system with a 59 day orbital period. ApJ **940**(2), L48 (2022)
142. P. Senchyna, L.C. Johnson, J.J. Dalcanton, L.C. Beerman, M. Fouesneau, A. Dolphin, B.F. Williams, P. Rosenfield, S.S. Larsen, Panchromatic hubble andromeda treasury XIV the period-age relationship of cepheid variables in M31 star clusters. ApJ **813**(1), 31 (2015)
143. F. Dinnbier, R.I. Anderson, P. Kroupa, On the dynamical evolution of Cepheids in star clusters. A&A **659**, A169 (2022)
144. F. Dinnbier, P. Kroupa, R.I. Anderson, Do the majority of stars form as gravitationally unbound? A&A **660**, A61 (2022)
145. P. Barmby, M. Marengo, N.R. Evans, G. Bono, D. Huelsman, K.Y.L. Su, D.L. Welch, G.G. Fazio, Galactic cepheids with spitzer II search for extended infrared emission. AJ **141**(2), 42 (2011)
146. M. Marengo, N.R. Evans, P. Barmby, L.D. Matthews, G. Bono, D.L. Welch, M. Romaniello, D. Huelsman, K.Y.L. Su, G.G. Fazio, An infrared nebula associated with δ cephei: evidence of mass loss? ApJ **725**(2), 2392–2400 (2010)
147. L.D. Matthews, M. Marengo, N.R. Evans, G. Bono, New evidence for mass loss from δ Cephei from HI 21 cm line observations. ApJ **744**(1), 53 (2012)
148. A. Gallenne, A. Mérand, P. Kervella, O. Chesneau, J. Breitfelder, W. Gieren, Extended envelopes around galactic cepheids IV T Monocerotis and X Sagittarii from mid-infrared interferometry with VLTI/MIDI. A&A **558**, A140 (2013)
149. P. Kervella, A. Mérand, G. Perrin, V. Coudé du Foresto, Extended envelopes around galactic cepheids I. ℓ carinae from near and mid-infrared interferometry with the VLTI. A&A **448**(2), 623–631 (2006)
150. A. Mérand, P. Kervella, V. Coudé du Foresto, G. Perrin, S.T. Ridgway, J.P. Aufdenberg, T.A. Ten Brummelaar, H.A. McAlister, L. Sturmann, J. Sturmann, N.H. Turner, D.H. Berger, Extended envelopes around galactic cepheids II polaris and δ Cephei from near-infrared interferometry with CHARA/FLUOR. A&A **453**(1), 155–162 (2006)
151. M.A.T. Groenewegen, Analysing the spectral energy distributions of galactic classical cepheids. A&A **635**, A33 (2020)
152. M.A.T. Groenewegen, J. Lub, Spectral energy distributions of classical cepheids in the magellanic clouds. A&A **676**, A136 (2023)

153. E.G. Schmidt, Excess mid-infrared flux: an indicator of mass loss in cepheids? ApJ **813**(1), 29 (2015)
154. V. Hocdé, N. Nardetto, E. Lagadec, G. Niccolini, A. Domiciano de Souza, A. Mérand, P. Kervella, A. Gallenne, M. Marengo, B. Trahin, W. Gieren, G. Pietrzyński, S. Borgniet, L. Breuval, B. Javanmardi, A thin shell of ionized gas as the explanation for infrared excess among classical Cepheids. A&A **633**, A47 (2020)
155. G. De Somma, M. Marconi, R. Molinaro, V. Ripepi, S. Leccia, I. Musella, An updated metal-dependent theoretical scenario for classical cepheids. ApJS **262**(1), 25 (2022)
156. G. Fiorentino, I. Musella, M. Marconi, Cepheid theoretical models and observations in HST/WFC3 filters: the effect on the Hubble constant H_0. MNRAS **434**(4), 2866–2876 (2013)
157. M. Marconi, I. Musella, G. Fiorentino, Cepheid pulsation models at varying metallicity and $\Delta Y/\Delta Z$. ApJ **632**(1), 590–610 (2005)
158. A. Bhardwaj, A.G. Riess, G. Catanzaro, E. Trentin, V. Ripepi, M. Rejkuba, M. Marconi, C.C. Ngeow, L.M. Macri, M. Romaniello, R. Molinaro, H.P. Singh, S.M. Kanbur, High-resolution spectroscopic metallicities of milky way cepheid standards and their impact on the Leavitt Law and the hubble constant (2023). arXiv e-prints: arXiv:2309.03263
159. E. Trentin, V. Ripepi, G. Catanzaro, J. Storm, M. Marconi, G. De Somma, V. Testa, I. Musella, Cepheid metallicity in the leavitt law (C-MetaLL) survey: II High-resolution spectroscopy of the most metal poor galactic cepheids. arXiv e-prints: arXiv:2209.03792 (2022)
160. P. Wielgórski, G. Pietrzyński, W. Gieren, M. Górski, R.-P. Kudritzki, B. Zgirski, F. Bresolin, J. Storm, N. Matsunaga, D. Graczyk, I. Soszyński, A precision determination of the effect of metallicity on cepheid absolute magnitudes in VIJHK bands from magellanic cloud cepheids. ApJ **842**(2), 116 (2017)

Chapter 7
The Role of Type Ia Supernovae in Constraining the Hubble Constant

Dan Scolnic and Maria Vincenzi

7.1 Introduction

When Edwin Hubble first measured the expansion of the universe, no supernovae (SNe) were used in this measurement [1]. Instead, Hubble measured distances to galaxies by observing the Cepheids in the galaxy and Cepheids' properties discovered by Henrietta Leavitt [2]. Today, both Cepheids and SNe are often used together to infer what is now called the Hubble constant H_0, which parameterizes the relation between expansion velocity and distance such that $v = H_0 d$. The Cepheids allow for a projection of a geometric scale, like that from parallax or megamasers, out onto the Megaparsec scale. However, the range that Cepheids can be found and measured with the Hubble Space Telescope is limited to $\sim$ 40 Mpc, where the measurements of motions of galaxies are still strongly affected by gravitational pulls of nearby galaxies (motions called peculiar velocities). Therefore, SNe are needed as an extra rung along the distance ladder to reach out further into what is now deemed the 'Hubble flow' ($\sim$100 to 600 Mpc), and precisely measure the expansion rate of the universe.

In the typical three rung distance ladder, SNe are used in the second and third rung. In the second rung, their luminosities are calibrated with stellar distance indicators in the same galaxy. In the third rung, their brightnesses calibrate the Hubble relation. Ideally, one could remove the intermediate step, like Cepheids or Tip of the Red Giant Branch (TRGB), and go straight from geometric calibration to SNe. Unfortunately, the rate of SNe in the local universe is not nearly frequent enough to provide multiple SNe in the few galaxies used for geometric anchors (the Milky Way, the Large Magellanic Cloud, NGC 4258). Even with Cepheids or TRGB as a go-between, the

D. Scolnic (✉) · M. Vincenzi (✉)
Department of Physics, Duke University, Durham, NC 27708, USA
e-mail: daniel.scolnic@duke.edu

M. Vincenzi
e-mail: maria.vincenzi@duke.edu

E. Di Valentino and D. Brout (eds.), *The Hubble Constant Tension*, Springer Series in Astrophysics and Cosmology, https://doi.org/10.1007/978-981-99-0177-7_7

low rate of SNe in the nearby universe (roughly one per galaxy per 100 years) is the limiting component of the precision of H_0 measurements [3]. At this point, the SH_0ES team [3] for instance, utilizes every SN Ia that pass cosmological quality requirements and Cepheid suitability within 40 Mpc.

SNe Ia are used in the most cited local distance ladder analyses due to their high precision, large luminosity, and number of discovered sources. However, as we will discuss in this review, studies have attempted to replace SNe Ia with hydrogen-rich core collapse SNe [4], which while being less-precise standardizable candles, still provide an effective crosscheck. Along the same lines, the role of SNe can be replaced with measurements of Surface Brightness Fluctuations [5] or the Tully Fisher relation [6], a different type of standard candle that allows reach into the Hubble Flow. With a similar purpose, there have been a tremendous number of re-analyses and studies of potential systematics of analyses of SNe Ia. These SNe have been used in cosmological measurements for decades, so there is no shortage of checks on their accuracy.

A related benefit of SNe Ia in particular is their usage for measurements of the expansion history, beyond just the current expansion rate. There are multiple large surveys that have been charged over the last two decades to discover and measure SNe Ia in order to constrain properties of dark energy and the dark matter content of the Universe [7–11]. Leverage in understanding these properties comes from modelling the change of brightnesses of SNe Ia over a range of redshift (like within $0 < z < 2$). As such, low redshift SNe are critically important, and the SNe that make up the third rung of the distance ladder are the same that provide the anchor for the relative distance constraints for measurements of expansion history [8, 9, 11]. Thereby, progress made in growing and understanding samples of low redshift SNe has been directly beneficial to both H_0 measurements of dark energy / dark matter measurements.

For this review, we first discuss in Sect. 7.2 the common three-rung path to constraining H_0 using SNe Ia. We then discuss in Sect. 7.3 the top systematic uncertainties in this measurement on the SN side, and how they have been quantified. In Sects. 7.4 and 7.5, we present an accounting of the different crosschecks, using replacements or substitutions of the SNe sample. In Sect. 7.6, we discuss the implications of the inverse distance ladder approach. Finally, we present our discussions and conclusions in Sects. 7.7 and 7.8.

7.2 The Canonical Path to H_0 with Type Ia Supernovae

We review here the formalism for deriving the Hubble constant with SNe Ia in the local distance ladder, as measured in [11, 12] and used in [3]. We follow the typical three-rung ladder, as shown on the left hand side of Fig. 7.1. As also shown in Fig. 7.1, the third rung of the distance ladder is the low-redshift ($z < 0.15$) part of the Hubble diagram, used to measure cosmological parameters like the equation-of-state of dark energy w.

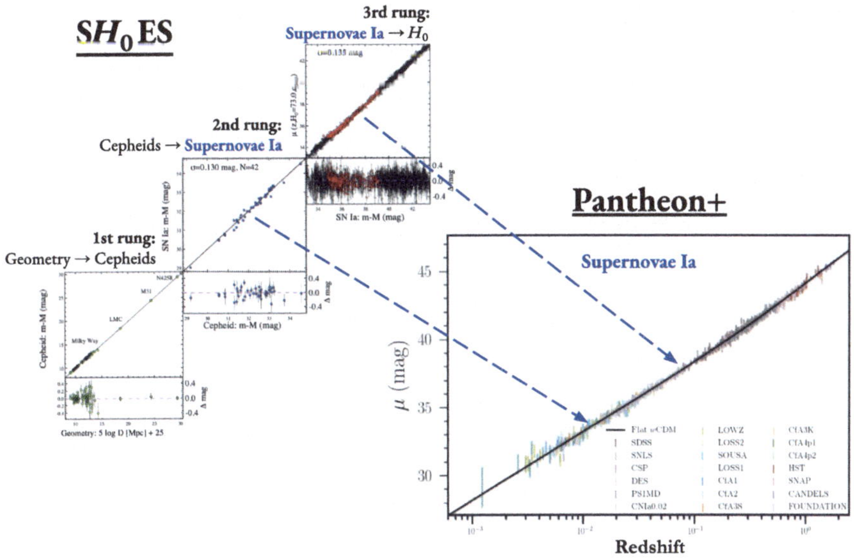

Fig. 7.1 **Left panel**: The 'distance ladder' from [3]. **Right panel**: The Pantheon+ "Hubble diagram" from [11] showing the distance modulus μ versus redshift z. The SNe from the second rung make up the majority of the $z < 0.01$ SNe in the Pantheon+ Hubble diagram, and the third rung makes up the rest of the $z < 0.15$ SNe. The uncertainties for $z < 0.01$ shown on the Hubble diagram are much larger than those in the second rung, due to propagation of redshift uncertainties

We assume that a set of Cepheid or TRGB distances are calibrated with geometric measurements like parallax or megamasers. This set of Cepheid or TRGB distances, expressed as μ_0 here, can then be compared to SNe Ia brightnesses m_X in order to find a single offset M_B, which describes the absolute magnitude of a SN Ia.

In most recent cosmological analyses with SNe Ia, the standardized brightness is measured with the Tripp formula such that

$$m_X = m_B + \alpha x_1 \beta c - \delta_{Bias} + \delta_{Host}, \tag{7.1}$$

where m_B, x_1 and c are all independent properties of each light curve, α and β are correlation coefficients that help standardize the brightness, δ_{Bias} is a correction due to selection effects and other biases as predicted by simulations, and δ_{Host} is a final correction due to residual correlations with host galaxy properties. We call the standardized brightness m_X instead of m_B as in [3] to be clear that the brightness is standardized. We note that typically SN Ia analyses like that in [11] will subtract off an absolute luminosity M_B of SNe Ia to derive a distance modulus, but this must assume a fiducial H_0 value, which is what we are trying to derive.

For a SN Ia in the i-th Cepheid host,

$$m_{X,i} = \mu_{0,i} + M_B, \tag{7.2}$$

where M_B is the fiducial SN Ia luminosity, and $\mu_{0,i}$ is the distance derived from Cepheid measurements for each galaxy.

The ladder is completed with a set of SNe Ia that measure the expansion rate quantified as the intercept, a_B, of the distance (or magnitude)–redshift relation. For an arbitrary expansion history and for $z > 0$ as

$$a_B = \log cz \left\{ 1 + \frac{1}{2}[1 - q_0]z - \frac{1}{6}\left[1 - q_0 - 3q_0^2 + j_0\right]z^2 + O(z^3) \right\} - 0.2m_X^0, \tag{7.3}$$

measured from a set of SNe Ia (z, m_X^0), where z is the redshift due to expansion, q_0 is the deceleration parameter, and j_0 is the jerk parameter. Typically, for ΛCDM, j_0 is set to 1. The determination of H_0 follows from

$$\log H_0 = 0.2M_B^0 + a_B + 5. \tag{7.4}$$

The approach of [13, 14] is now often used to account for covariance between rungs, as measurements of SNe Ia used in the second and third rung are correlated. Importantly, q_0 (and j_0) can only be constrained from SN Ia data, without the requirement of any additional information.

7.3 Top Systematics on the Path to H_0 with SN Ia

Cosmological measurements using SNe Ia are affected by various source of systematic uncertainties. However, measurements of w using SNe are significantly more sensitive than measurements of H_0 with SNe. A simple explanation of the sensitivity is that H_0 is constrained by comparing SNe at $z \sim 0.005$ to SNe in the Hubble flow at $z \sim 0.05$, not too different in redshift and therefore similar in terms of SN properties and survey methods and telescopes used for SN observations (see Fig. 7.2). On the other hand, w is constrained by comparing SNe at $z \sim 0.05$ to $z \sim 0.5$, in which evolution of SN properties is possible, and the surveys and telescopes used to find and measure the SNe will be quite different. While new surveys like Foundation [10, 15], ZTF [16] and DEBASS are attempting to measure more low-redshift SNe with the same telescopes as high-redshift SNe have been measured, it will take on the order of 10–20 years for these surveys to measure an equivalent amount of second-rung SNe (~ 40) as those in the current second rung. One route to overcome this hurdle is by extending the local volume for which host galaxy distances can be obtained, e.g., [17].

There have been a number of papers that have focused on systematics impact on H_0 and we list these in the top of Table 7.1. [11] give a comprehensive overview of

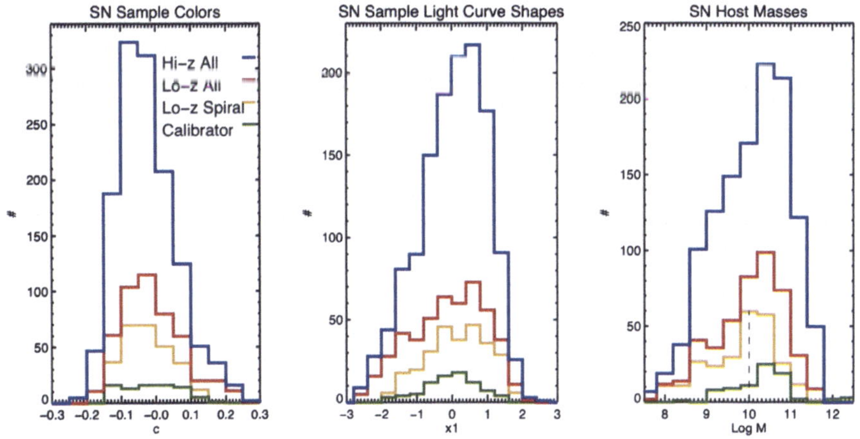

Fig. 7.2 Adapted from [3], a comparison of SN Ia light-curve shape (x_1), colour (c) and the log host galaxy stellar mass (log $M_{\odot}$) for calibrator and Hubble flow samples (with different host galaxy type and redshift cuts)

many of these systematics and how they may affect measurements of w or H_0. In Fig. 7.3, from [11], we show the impact in units of H_0 for applying 1σ shift of each systematic. We group the systematics into various categories: Redshifts, SN Physics, Selection, Calibration, and Milky Way Dust.

Before discussing in more details each category of systematics, it is important to understand which are the systematics H_0 is more sensitive to and, on the contrary, which are generally mitigated and have little impact on H_0. The distance ladder is constructed such that the same probe is used in two of the three rungs (e.g. Cepheids in 1 and 2, SN Ia in 2 and 3). If a systematic is disproportionately affecting SN Ia in the second and third rung, or if it is introducing some substantial differences in the second rung and third rung SN Ia population, this will have a large impact on H_0. On the contrary, if a source of systematic is introducing some consistent offset that affects all SN Ia coherently, the impact of that systematic will be null due to the formalism described above. In other words, the offset will cancel and the effect of the systematic will be mitigated.

Calibration: While calibration dominates the systematic error budget for w, its impact is small for H_0 because of the systematic mitigation discussed above. An illustration of this is given in [20, 30]. In particular, [20] check the possibility of 'gray' calibration offsets per survey and finds a total uncertainty due to this effect no larger than 0.2 km/s/Mpc. They show that since many of the same surveys are used to measure SNe in the second and third rung of the distance ladder, the impact of potential biases in photometric calibration of the surveys will cancel. This is shown to not be the case when different surveys are used for the second and third rung. This approach is done in [22, 31], which while emphasizing the role of the CSP SNe, increases the sensitivity to calibration errors.

Table 7.1 A summary of the various crosschecks and systematics on the supernova component of the distance ladder. If two uncertainties are given, the first one is the statistical uncertainty and the second one is the systematic uncertainty

References	Notes	Result (km/s/Mpc)
Specific systematic check for optical SNIa:		
Murakami et al. [18]	Uses spectral feature twinning process to improve standardization; checks dust modelling, intrinsic scatter modelling;	73.01 ± 0.92
Peterson et al. [19]	Checks different models of peculiar velocities/bulk flows	$\sigma_{H_0} < 0.2$
Brownsberger et al. [20]	Checks SNIa Calibration by allowing individual SN survey offsets	$\sigma_{H_0} < 0.2$
Jones et al. [21]	Checks impact of mass step, global vs local correlations	$\sigma_{H_0} < 0.15$
Burns et al. [22]	Checks light-curve fitting method; also does NIR fits	73 ± 2
Garnavich et al. [23]	Uses 4 rung distance ladder, checks SNIa host demographic systematic	$74.6 \pm 0.9 \pm 2.7$
Dhawan et al. [16]	Uses ZTF data alone, check on SNIa calibration	76.94 ± 6.4
NIR SN Ia:		
Dhawan et al. [24]	Uses literature NIR SN (restframe J) and peak fitting	72.8 ± 2.8
Galbany et al. [25]	Uses literature NIR SN (restframe J and H band)	$72.3 \pm 1.4 \pm 1.4$
Jones et al. [26]	Uses RAISIN\+literature NIR SN (restframe Y band) and SNooPy fitting, check on dust	75.9 ± 2.2
Dhawan et al. [27]	Uses literature NIR SN and BayesN fitting	$74.82 \pm 0.97 \pm 0.84$
Removing SNIa:		
Kenworthy et al. [28]	Eliminates SNIa rung entirely with 2 rung distance ladder	$72.9^{+2.4}_{-2.2}$
De Jaeger et al. [4]	Uses SN II instead of SNIa	$77.6^{+5.2}_{-4.8}$
Blakeslee et al. [5]	Uses SBF standard candles instead of SNIa	$73.3 \pm 0.7 \pm 2.4$
Kourkchi et al. [29]	Uses TF relation instead of SNIa	$76.0 \pm 2.3 \pm 1.5$
Schombert et al. [6]	Uses TF relation	$75.1 \pm 1.1 \pm 2.3$

Redshifts: H_0 measurements are more sensitive to redshift related systematic because redshift information is only used in one rung of the distance ladder (the third), so there is no systematic mitigation. [19] explore a large number of models of bulk flows in the nearby universe, given various models that best improve the Hubble residual scatter from supernova measurements, and find that there could be changes in H_0 up to 0.2 km/s/Mpc. They also find that the impact of including peculiar velocity corrections of the nominal method versus not including them is $\Delta H_0 \sim 0.5$. Peculiar velocities are discussed further in chapters of this book. [32] check biases due in redshift measurements and find uncertainties on the level of 0.1 km/s/Mpc. A predecessor study was done by [33], which similarly found small changes to H_0, though saw possible discrepancies in $\Omega_M H_0^2$ plane between subsamples of SNe for measured redshifts with different levels of precision. No evidence for subsample differences was found by [32].

SN Physics and Selection: SN intrinsic astrophysics and the role of SN dust remain one of the most poorly understood aspects of SN Ia cosmology. Recent analyses have investigated how systematics due to uncertainties in the physics of the supernova explosions and extragalactic dust extinction impact H_0 measurements [18, 34]. Most

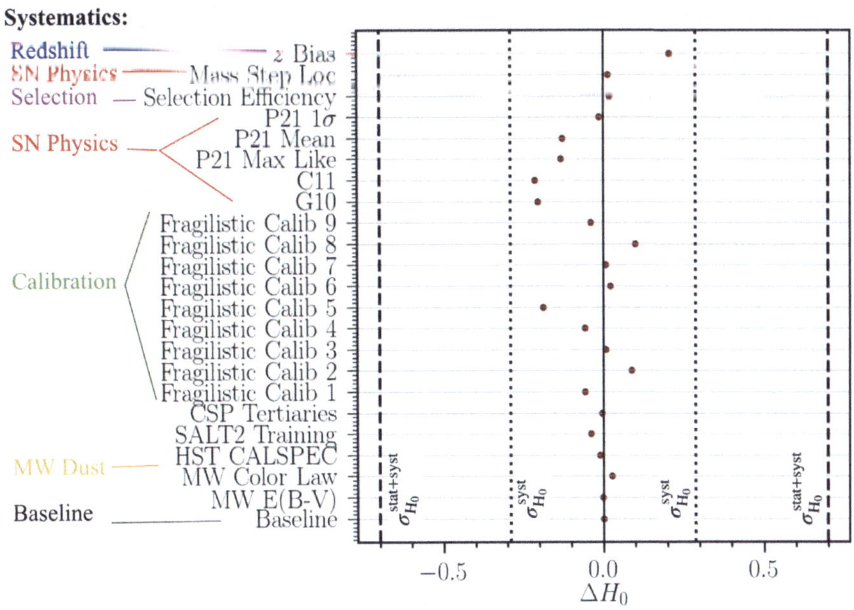

Fig. 7.3 Adapted from [11], the impact on recovery of H_0 of the various systematic uncertainties tabulated. The units of these measurements are km/s/Mpc. The dashed lines are given at ΔH_0 of 0.7, which is the entire contribution of the uncertainty in [3] from SN measurements. We add on labels explaining categories of systematic uncertainties

analyses have found that these effects have a small impact on H_0 because the SN subpopulations selected in the second and third rung are not expected to be significantly different.

A good example of a potential systematic related to SN physics was discussed in [35] following earlier SH_0ES analyses (e.g., [36]). [35] showed evidence for a correlation between standardized brightness and the age of the host galaxy (quantified estimating the specific star-formation at the SN location). In earlier measurements like [36], the third rung of SNe had no galaxy-based selection applied, but the second rung, by tying to Cepheid discovery, favored star-forming galaxies. This would potentially lead to a bias in the recovery of H_0. The size of the bias would depend on the relative differential fraction of host-galaxy demographics between the second and third rung multiplied by the size of the correlation. Subsequent analyses [21] showed that this effect would likely be insufficient to explain the Hubble tension.

Still, in the most recent SH_0ES analysis [3], the selection of SNe in the third rung of the distance ladder was done to be as similar as possible as the second rung. The SH_0ES team only selected SNe found in star-forming galaxies which thereby removed the sensitivity to this systematic; the impact from this change was less than the statistical uncertainty from the supernova component of the distance ladder. Similarly, significant differences in dust extinction and/or color-related effects between SNe in the second and third rung could potentially bias H_0 measurements [37]. SN dust extinction and color-dependent corrections are encapsulated in the nuisance parameter β (see Eq. 7.1). As a crosscheck, the β parameter has been fitted separately in SN used in the second and third rung, and it has been found to be consistent [11]. This test, together with various NIR SN H_0 measurements [24–27], see Table 7.1, suggests that dust or color-dependent effects are not expected to significantly bias H_0, or to be a significantly underestimated systematic in current H_0 measurements.

In general, it is relatively simple to modify the SN sample selection in the second and third rung to ensure better consistency in terms of SN physics/dust and sample selection between rungs, and this approach has been implemented in the latest SH_0ES analysis.

Other systematics: Finally, there have been various analyses (e.g., [18, 22]) that implemented additional systematic tests, e.g., changing the light-curve fitter or adding spectroscopic information, but overall these appear to give very consistent (within ~ 0.3 km/s/Mpc) results. Similarly, isolating only to a single survey, like done in [27], gives consistent results, though with much larger uncertainties due to the small number of calibrator SNe.

7.4 Variants on the Path to H_0 with Supernovae Ia

Some of the main crosschecks on the SNe Ia used for these analyses is varying the wavelength regime in which light curves are measured (i.e., optical to NIR) or the dataset used (i.e., the survey used to measure light curves). As H_0 constraints are limited by the number of SNe found within 40 Mpc, there is considerable overlap in the data between these various studies. The most popular path to check/improve the distance ladder with SN Ia is with measuring brightnesses from NIR light curves. We list these papers in Table 7.1. Overall, even though the rest-frame band in which light curves are measured varies between these analyses, and the fitting method varies between these methods, there is generally very good agreement in recovered values of H_0. One challenge multiple of these studies have found (e.g. [24, 26]) is larger calibration offsets between samples than those found for optical studies. A benefit of this type of study is the possibility of improved precision of distance measurements from NIR data, but the quality of older light curves has not typically been good enough to evaluate this possibility [38].

An additional path is creating a "4 rung distance ladder", as done in [23] and shown in Fig. 7.4. SNe Ia used in the SH_0ES distance ladder are those found in late-type galaxies. To avoid this specific subsample, one can add another rung in the distance ladder between TRGB/Cepheids and SNe - that from Surface Brightness Fluctuations (SBF). The analysis of [23] improve on that of [39], which follow a similar method, but uses an inhomogeneous set of SBF measurements, which significantly increases the scatter of the tie between SBF and supernova measurements, and appears to

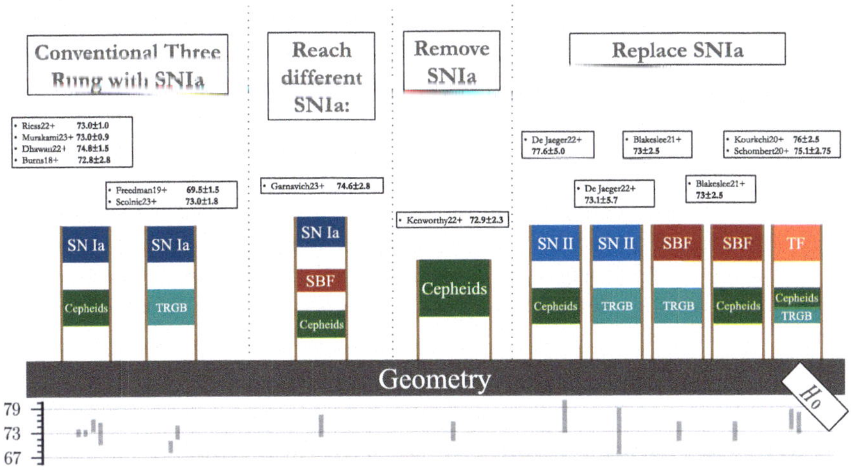

Fig. 7.4 A summary of papers and H_0 results given different uses and non-uses of SNe Ia. All H_0 measurements are in km/s/Mpc

bias H_0 to lower values. [23] find a value of $H_0 = 74.6 \pm 2.8$ km/s/Mpc, in good agreement with the SH_0ES value.

7.5 Removing Type Ia Supernovae on the Path to H_0

An alternative path is to remove SNe Ia altogether. First, one can attempt to measure H_0 with a SH_0ES-like distance ladder, but without the third rung; this is called a "two-rung" distance ladder (see Fig. 7.4). This relies on knowing the redshifts to the nearby galaxies in which Cepheids are discovered. As these galaxies are mostly $z < 0.01$, uncertainties in redshifts on the order of 300 km/s can contribute more than 10% of the uncertainty per galaxy. Given the low signal-to-noise, any systematic biases from selection effects or peculiar velocity corrections may have a large effect on the measured value of H_0. [28] attempt this with well-calibrated HST measurements of Cepheids from the SH_0ES team, and measures $H_0 = 73.1 \pm 2.5$ km/s/Mpc. A key component of this type of analysis is accounting for the inherent volumetric bias that there are more galaxies further away than nearby.

A separate path, instead of removing the third rung entirely, is to replace SN Ia with another type of standard candle. [4] attempt to replace SN Ia with SN II; while SN II are not typically considered standardizable candles, there has been good progress to improve their precision as distance indicators. The scatter in the Hubble flow is 0.27 mag, compared to the ~ 0.17 mag found for SNe Ia. The precision is enabled by leveraging a correlation between the luminosity and photospheric velocity and a color correction [40, 41]. The study finds weaker constraints on H_0, but good agreement with the SH_0ES analysis: $H_0 = 77.6^{+5.2}_{-4.8}$ km/s/Mpc.

Another is using Surface Brightness Fluctuations, but in the same role as SNe Ia [5]. With those, they find $H_0 = 73.3 \pm 0.7 \pm 2.4$ km/s/Mpc. This combines first-rung calibration from Cepheids and TRGB, though they yield similar results ($H_0 =$ 73.44 for Cepheids and $H_0 = 73.20$ km/s/Mpc for TRGB, each with $\sim 5\%$ total uncertainty). The number of calibrators is just 7, compared to the 42 that SH_0ES uses, thereby limiting the overall precision. Additionally, the Tully-Fisher (TF) relation can be done in this third-rung role. References [6, 29] calibrate the TF relation with Cepheids and TRGB; after doing so, [6] recovers $75.1 \pm 2.3 \pm 1.5$ km/s/Mpc and [29] finds $76.0 \pm 1.1 \pm 2.3$ km/s/Mpc.

7.6 Inverse Distance Ladder to H_0 with Supernova

The same set of SNe Ia that are calibrated to a physical distance scale in the SH_0ES distance ladder, can also be used as un-calibrated relative distance indicators to constrain H_0 in combination with other probes. Here we describe the process of combining the un-calibrated SNe Ia with constraints from Baryon Acoustic Oscillation (BAO). BAO constraints on the expansion history $H(z)$ and that extrapolate H_0 have been shown to achieve low values of H_0 [42, 43, 45, 46]. We note that such constraints (1) assume the sound horizon from CMB constraints and (2) are model dependent because they assume ΛCDM to infer $H(z = 0)$ from BAO datasets at $z \sim 0.5$ and do not allow for late-time physical solutions. However, the SNe Ia can be used to solve the latter of these two.

Instead of calibrating the SN Ia intrinsic magnitude to the distance ladder, they can be calibrated at the typical BAO redshift to the distance scale set by BAO and for which there is substantial overlap in z of many SNe Ia in current datasets. Having calibrated the SNe Ia to the BAO scale at moderate redshift, one can use the SNe Ia to constrain the expansion history without assuming ΛCDM to infer H_0 (see smoothed blue curve of Fig. 7.5). In this case, one finds a very similar result ($H_0 = 68.57 \pm 0.9$) to that of the CMB under the assumption of ΛCDM (grey curve of Fig. 7.5). We note here that using SNe Ia calibrated to the BAO, which itself obtains the physical scale from the sound speed in the early universe, to recover a low value of H_0 has spurred many discussion to understand the impact of the value of the sound horizon.

7.7 Discussion

Improving measurements of H_0 in the future:
While constraints on w should easily improve with upcoming large SN samples, the road to improving constraints on H_0 is more challenging. There are a limited number of SNe Ia that will explode in the near future within a ~ 40 Mpc radius, a

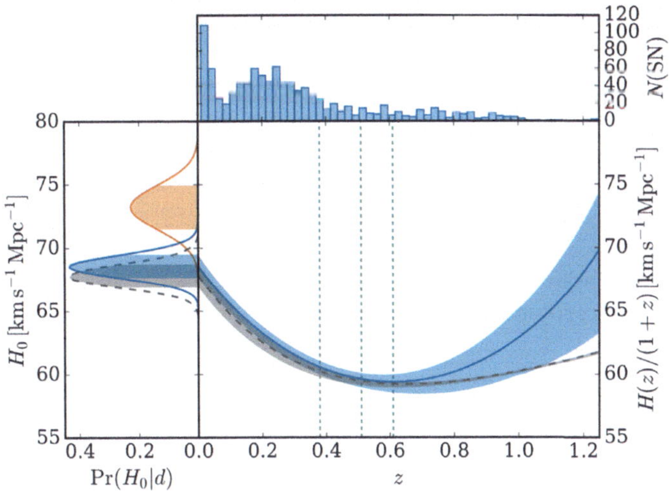

Fig. 7.5 From [43], history for BOSS BAO, Pantheon SNe and *Planck* r_d assuming smooth expansion and early-time physics only (blue), and for Planck assuming ΛCDM (grey). BAO redshifts are shown as short-dashed lines. Left panel: corresponding H_0 posteriors and Cepheid distance ladder measurement (orange). Top panel: redshift distribution of Pantheon SNe

constraint due to *HST* discovery limits of Cepheids. At roughly one to three SN Ia per year, it will likely take more than one decade to double the current sample of 42 SNe calibrated by SH_0ES Cepheid hosts. An exciting possibility is that the distance range for precise measurements of Cepheids or TRGB can be extended. As the volume of discovered SNe roughly increases with distance cubed, then even a 25% increase in the range of Cepheids or TRGB measurements would allow a doubling in the number of usable SN Ia in the second rung of the distance ladder. This may be a possibility with new telescopes such as the James Webb Space Telescope, or further into the future, the Nancy Grace Roman Space Telescope, or ground-based Extremely Large Telescope (ELT).

The other path towards improving the constraint from SN Ia is to improve the precision of the measurements. This is the path followed by papers like [18] which tried using spectral features to improve the standardization, or the large number of papers that measure SNe Ia in the NIR. [18] are able to improve the scatter of SN Ia from 0.14 to 0.12 mag, which is similar to increasing the number of SN Ia by $\sim 35\%$. The main challenge of this approach is that the benefit of the new idea depends on how much data is available for its application in past literature measurements. New types of measurements can be made for future nearby supernovae, but we can not re-measure past SN Ia.

The role of SNe Ia when comparing local probes:
Within the larger discussion of tension between early and late universe probes, there has been various discussions of smaller tensions between late-universe probes. One of note has been potential discrepancies when using Cepheids or TRGB in the distance

Table 7.2 Sources of differences in H_0 Between TRGB analysis by CATs [44] and CCHP [31] (in units of H_0)

Term	ΔCCHP (km/s/Mpc)
SN related	
1. Include SN 2021pit, 2021rhu, 2007on	0.6
2. No TRGB detected in N5584, N3021, N1309, N3370	0.0
3. Peculiar flows (Pantheon+)	0.4
4. Hubble flow surveys (Pantheon+)	1.1
SN subtotal	**2.0**
TRGB related	
5. Fiducial TRGB calibration/Tip-contrast relation	1.4
Total	**3.4**

Comments: Adapted from [44], this shows how differences that appear to be from anchor measurements of Cepheids versus TRGB are more due to differences in the supernova analysis. Here, ΔCCHP= differences between [31]. Descriptions of individual entries: (1) CCHP did not include the two SNe from 2021 and excluded SN 2007on, (2) CATs did not detect the TRGB in these four most-distant SN host galaxies; (3) Pantheon+ accounts for peculiar motions which produce a highly significant improvement in the dispersion of the Hubble diagram (see [19]) (4) CCHP measures the Hubble flow from a single SN survey which has an offset with respect to the mean of many surveys in Pantheon+ as shown in [20]; (5) This term is difference in the calibration of TRGB from NGC 4258 as applied to SN hosts and is discussed by [44] (see Sect. 7.4)

ladder. While a lot of this focus has been on Cepheids versus TRGB, we note that often the SNe that are used to project out the distance ladder can cause discrepancies. An interesting example of that is shown from [44], which breaks down changes in H_0 from the CATS and CCHP teams [31]. As shown in Table 7.2, the CATS team explains that SN choices account for more than 60% of the difference with other analyses, whereas the TRGB side is less than 40%. A number of these choices have been discussed in this review, including the application of peculiar velocity corrections and consistency amongst surveys used for the different rungs of the distance ladder.

7.8 Conclusion

We discuss the various systematic uncertainties of as well as the crosschecks on the SN Ia component of the distance ladder, which occupies two of the three rungs. We find generally that systematics in the measurement of H_0 from the SNe are at the scale of 0.3 km/s/Mpc. The range of H_0 values recovered from crosschecks is from $72 < H_0 < 78$ km/s/Mpc. Therefore, it appears unlikely that SN Ia are artificially causing a high value of H_0 to be found with the local distance ladder. Still, additional studies of SN Ia are important in order to compare other parts of the distance ladder, like TRGB versus Cepheids, as the sample overlap does not always yield high significance from a direct comparison.

Acknowledgements D. S. thanks the John Templeton Foundation for their support of grant #62314. D. S. is supported by DOE grant DE-SC0010007, the David and Lucile Packard Foundation. M. V. is supported by NASA through the NASA Hubble Fellowship grant HST-HF2-51546.001-A awarded by the Space Telescope Science Institute, which is operated by the Association of Universities for Research in Astronomy, Incorporated, under NASA contract NAS5-26555.

References

1. E. Hubble, Proc. Natl. Acad. Sci. **15**, 168 (1929). https://doi.org/10.1073/pnas.15.3.168
2. H.S. Leavitt, E.C. Pickering, *Harvard College Observatory Circular*, vol. 173 (1912), pp. 1–3, https://ui.adsabs.harvard.edu/abs/1912HarCi.173....1L
3. A.G. Riess, W. Yuan, L.M. Macri et al., ApJL **934**, L7 (2022). https://doi.org/10.3847/2041-8213/ac5c5b
4. T. de Jaeger, L. Galbany, A.G. Riess et al., MNRAS **514**(3), 4620–4628 (2022). https://doi.org/10.1093/mnras/stac1661
5. J.P. Blakeslee, J.B. Jensen, C.-P. Ma et al., ApJ **911**(1), 65 (2021). https://doi.org/10.3847/1538-4357/abe86a
6. J. Schombert, S. McGaugh, F. Lelli, AJ **160**, 71 (2020). https://doi.org/10.3847/1538-3881/ab9d88
7. M. Sullivan, J. Guy, A. Conley et al., ApJ **737**(2), 102 (2011). https://doi.org/10.1088/0004-637X/737/2/102
8. M. Betoule, R. Kessler, J. Guy et al., A&A **568**, A22 (2014). https://doi.org/10.1051/0004-6361/201423413
9. D.M. Scolnic, D.O. Jones, A. Rest et al., ApJ **859**, 101 (2018). https://doi.org/10.3847/1538-4357/aab9bb
10. D.O. Jones, D.M. Scolnic, R.J. Foley et al., ApJ **881**, 19 (2019). https://doi.org/10.3847/1538-4357/ab2bec
11. D. Brout, D. Scolnic, B. Popovic et al., ApJ **938**, 110 (2022). https://doi.org/10.3847/1538-4357/ac8e04
12. D. Scolnic, D. Brout, A. Carr et al., ApJ **938**, 113 (2022). https://doi.org/10.3847/1538-4357/ac8b7a
13. A. Conley, J. Guy, M. Sullivan et al., ApJS **192**, 1 (2011). https://doi.org/10.1088/0067-0049/192/1/1
14. S. Dhawan, D. Brout, D. Scolnic et al., ApJ **894**, 54 (2020). https://doi.org/10.3847/1538-4357/ab7fb0
15. R.J. Foley, D. Scolnic, A. Rest et al., MNRAS **475**, 193 (2018). https://doi.org/10.1093/mnras/stx3136
16. S. Dhawan, A. Goobar, J. Johansson et al., ApJ **934**(2), 185 (2022). https://doi.org/10.3847/1538-4357/ac7ceb
17. D. Jones, S. Casertano, A.V. Filippenko et al., HST Proposal Cycle **28** ID #16269 (2020). https://ui.adsabs.harvard.edu/abs/2020hst..prop16269J
18. Y.S. Murakami, A.G. Riess, B.E. Stahl et al. (2023). arXiv e-prints: arXiv:2306.00070 [astro-ph.CO]
19. E.R. Peterson, W.D. Kenworthy, D. Scolnic et al., ApJ **938**(2), 112 (2022). https://doi.org/10.3847/1538-4357/ac4698
20. S. Brownsberger, D. Brout, D. Scolnic et al. (2021). arXiv e-prints arXiv:2110.03486, [astro-ph.CO]

21. D.O. Jones, A.G. Riess, D.M. Scolnic et al., ApJ **867**(2), 108 (2018). https://doi.org/10.3847/1538-4357/aae2b9
22. C.R. Burns, E. Parent, M. Phillips et al., ApJ **869**(1), 56 (2018). https://doi.org/10.3847/1538-4357/aae51c
23. P. Garnavich, C.M. Wood, P. Milne et al. ApJ **953**(1), 35 (2023). https://doi.org/10.3847/1538-4357/ace04b
24. S. Dhawan, S.W. Jha, B. Leibundgut, A&A **609**, A72 (2018). https://doi.org/10.1051/0004-6361/201731501
25. L. Galbany, T. de Jaeger, A.G. Riess et al. (2022). arXiv e-prints: arXiv:2209.02546, [astro-ph.CO]
26. D.O. Jones, K.S. Mandel, R.P. Kirshner et al., ApJ **933**, 172 (2022). https://doi.org/10.3847/1538-4357/ac755b
27. S. Dhawan, S. Thorp, K.S. Mandel et al., MNRAS **524**(1), 235–244 (2023). https://doi.org/10.1093/mnras/stad1590
28. W.D. Kenworthy, A.G. Riess, D. Scolnic et al., ApJ **935**(2), 83 (2022). https://doi.org/10.3847/1538-4357/ac80bd
29. E. Kourkchi, R.B. Tully, G.S. Anand et al., ApJ **896**(1), 3 (2020). https://doi.org/10.3847/1538-4357/ab901c
30. D. Brout, G. Taylor, D. Scolnic et al., ApJ **938**, 111 (2022). https://doi.org/10.3847/1538-4357/ac8bcc
31. W.L. Freedman, B.F. Madore, D. Hatt et al., ApJ **882**, 34 (2019). https://doi.org/10.3847/1538-4357/ab2f73
32. A. Carr, T.M. Davis, D. Scolnic et al., PASA **39**, e046 (2022). https://doi.org/10.1017/pasa.2022.41
33. C.L. Steinhardt, A. Sneppen, B. Sen, VizieR Online Data Catalog. J. ApJ. **902** (14) (2022). DOIurlhttps://doi.org/10.26093/cds/vizier.19020014
34. B. Popovic, D. Brout, R. Kessler, D. Scolnic (2021) arXiv e-prints: arxiv:2112.04456
35. M. Rigault, G. Aldering, M. Kowalski et al., ApJ **802**, 20 (2015). https://doi.org/10.1088/0004-637X/802/1/20
36. A.G. Riess, L. Macri, S. Casertano et al., ApJ **730**, 119 (2011). https://doi.org/10.1088/0004-637X/730/2/119
37. R. Wojtak, J. Hjorth, MNRAS **515**(2), 2790–2799 (2022). https://doi.org/10.1093/mnras/stac1878
38. E.R. Peterson, D.O. Jones, D. Scolnic et al., MNRAS **522**, 2478 (2023). https://doi.org/10.1093/mnras/stad1077
39. N. Khetan, L. Izzo et al., A&A **647**, A72 (2021). https://doi.org/10.1051/0004-6361/202039196
40. M. Hamuy, P.A. Pinto, ApJL **566**, L63 (2002). https://doi.org/10.1086/339676
41. T. de Jaeger, B.E. Stahl, W. Zheng et al., MNRAS **496**, 3402 (2020). https://doi.org/10.1093/mnras/staa1801
42. E. Macaulay, R.C. Nichol, D. Bacon et al., MNRAS **486**, 2184–2196 (2019). https://doi.org/10.1093/mnras/stz978
43. S.M. Feeney, H.V. Peiris, A.R. Williamson et al., Phys. Rev. Lett. **122**(6), 061105 (2019). https://doi.org/10.1103/PhysRevLett.122.061105
44. D. Scolnic, A.G. Riess, J. Wu et al., ApJL **954**, L31 (2023). https://doi.org/10.3847/2041-8213/ace978
45. D. Camarena, V. Marra, MNRAS **495**, 2630 (2020). https://doi.org/10.1093/mnras/staa770
46. S. Taubenberger, S.H. Suyu, E. Komatsu et al., A&A **628**, L7 (2019). https://doi.org/10.1051/0004-6361/201935980

Chapter 8
Tip of the Red Giant Branch

Siyang Li and Rachael L. Beaton

8.1 Introduction

8.1.1 Physical Basis of the TRGB

The tip of the red giant branch (TRGB) marks the evolutionary transition of stars from the red giant branch (RGB) to the horizontal branch and can be used as a standardizable candle to construct an extragalactic distance ladder to measure H_0. After a low mass red giant star ($\sim \leq 2\ M_{\odot}$) finishes burning through the hydrogen in its core, it leaves a degenerate Helium core. The star then continues to burn hydrogen in a shell surrounding the Helium core, which increases the core temperature and stellar luminosity. Once the core temperature reaches a critical temperature, helium burning begins. This results in runaway reaction called the Helium Flash [1]. Here, core degeneracy is lifted, which prevents the star from further increasing in luminosity. After the Helium Flash, the star decreases in luminosity and transitions onto the horizontal branch. The maximum luminosity reached during this evolutionary sequence is a function of initial Helium core mass, which varies very little (~ 0.001 $M_{\odot}$) for stars from roughly 1.5 to 3 Gyr to $\sim$13 Gyr [2]. This small range of core masses in turn results in a small range of maximum luminosities that allows the TRGB to be used as a standardizable candle.

S. Li (✉)
Department of Physics and Astronomy, Johns Hopkins University, Baltimore, MD 21218, USA
e-mail: sli185@jh.edu

R. L. Beaton
Space Telescope Science Institute, Baltimore, MD 21218, USA
e-mail: rbeaton@stsci.edu

E. Di Valentino and D. Brout (eds.), *The Hubble Constant Tension*, Springer Series in Astrophysics and Cosmology, https://doi.org/10.1007/978-981-99-0177-7_8

8.1.2 Observational Basis of the TRGB

The increase, then decrease, of a star's luminosity before and after the Helium flash creates a distinct 'tip' of the red giant branch that can be visualized in a color magnitude diagram (CMD). To show this, we plot in Fig. 8.1 a CMD and luminosity function of stars simulated with the Python Artpop package [3] and the noise model from Li et al. [4] in the *Hubble Space Telescope (HST)* $F814W$ and $F606W$ filters. We mark the location of the TRGB with a dashed red line, which corresponds to a discontinuity in the corresponding $F814W$ luminosity function and separates the RGB population below the TRGB from the asymptotic giant branch (AGB) population above the TRGB.

The TRGB is measured statistically with a population of stars, as the TRGB feature can only be reliably identified and measured when using stars at multiple locations along their evolution leading up to the TRGB. A variety of approaches have been developed to measure the TRGB; we refer the reader to Beaton et al. [2] for a detailed historical review. Generally, these approaches first apply selection cuts to remove possible contaminants and isolate the red giant branch. These can involve color cuts, which are often based on color calibrations such as from Rizzi et al. [5] and Jang et al. [6] to isolate metal-poor stars, and spatial cuts involving young star clipping [7], B mag 25th isophotal radius [8] or other elliptical cuts [9]. We describe algorithms used to measure the TRGB after such cuts in the context of H_0 in Sects. 8.3.1 and 8.3.1.3.

Observationally, the TRGB is typically measured in the relatively metal-poor and old halos of galaxies. This is to primarily reduce the effects of contamination from younger stars [9] and to avoid internal extinction and crowding. In addition, the TRGB exhibits a metallicity dependence, which can be traced with color due to the

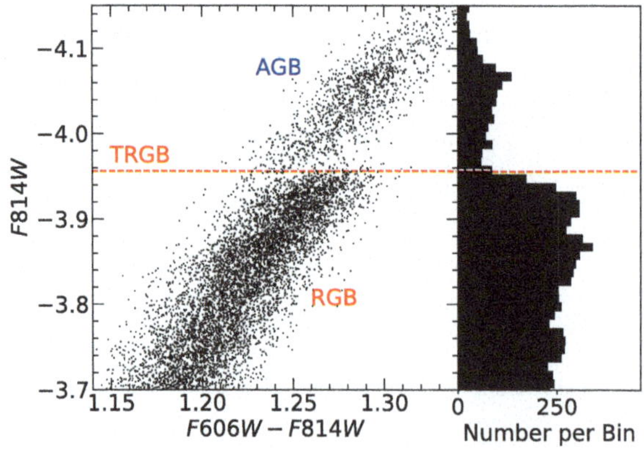

Fig. 8.1 CMD and LF simulated using Artpop [3] in the *HST* $F814W$ and $F606W$ filters and 10 Gyr stars with [Fe/H] = −1 dex. We add noise using the noise model from Li et al. [4]. The location of the TRGB discontinuity is marked with the red dashed line. The RGB and AGB lie below and above the TRGB, respectively

line-blanketing effect. The TRGB becomes fainter and brighter at redder colors in the optical and infrared, respectively. Empirically, the TRGB shows the least color dependency in the I or HST $F814W$ filters [5, 6]; for this reason, these filters are most commonly used to calibrate and measure H_0. There is, however, an increasing interest in measuring the TRGB in redder colors to extend TRGB measurements in the Hubble flow and take advantage of the IR capabilities of the *James Webb Space Telescope* ($JWST$). We discuss this further in Sect. 8.4.1.

8.2 The TRGB Path to H_0

In this section, we briefly describe how the TRGB can be used to construct a distance ladder to measure H_0. We emphasize that this description is not comprehensive and encourage the reader to delve into the literature to learn more about the various methodologies and considerations needed to measure a robust TRGB-based H_0.

8.2.1 *Absolute Calibrations*

Measuring extragalactic distances with the TRGB to measure H_0 requires knowledge of its apparent magnitude, absolute magnitude, and extinction:

$$\mu_0 = m_{host} - M - A \tag{8.1}$$

where μ_0 is the distance modulus to the host galaxy, m_{host} is the apparent magnitude of the TRGB in the host galaxy, M is the absolute magnitude of the TRGB, and A is the extinction (where we combine foreground and internal extinctions). m_{host} can be directly measured using the methods described later in this chapter. The extinction, A, can be obtained from dust maps, for instance from Schlafly & Finkbeiner 2011 [10], among others, the discussion of which is outside the scope of this chapter.

With measurements of m_{host} and A, we are left with two unknowns, μ_0 and M. To obtain M, one can invert Eq. 8.1 such that

$$M = m - \mu_0 - A \tag{8.2}$$

M is intrinsic to the TRGB and varies very little galaxy to galaxy, with variations that can be standardized via measurements of the color and contrast ratio, for instance. This means Eq. 8.2 is general and we can, in principle, exchange m_{host} with a general m and use any galaxy to obtain M provided we also have a direct measurement of its distance, μ_0, that is obtained independently of the TRGB. Examples of galaxies with distances measured geometrically and independently from the TRGB that are most commonly used to calibrate, or 'anchor', the TRGB are the Milky Way, Large Magellanic Cloud (LMC), Small Magellanic Cloud (SMC), and NGC 4258. Distances to

Milky Way red giants can be determined via parallaxes, the most precise of which have been measured by the *Gaia* mission [11]. The LMC, SMC, and NGC 4258 have geometric distances measured via eclipsing binaries [12, 13] and water masers [14], respectively. For derivations of the LMC, SMC, and NGC 4258 distances using these systems, we refer the reader to their respective references.

With measurements of the geometric distance to and apparent TRGB magnitude in the anchor galaxy (as well as an estimate of the extinction, A), one can solve for the absolute magnitude, M, of the TRGB using Eq. 8.2. With M as well as m_{host} and A for the host galaxy, one can then return to Eq. 8.1 and solve for μ_0.

We briefly pause to highlight differences between two distinct approaches of calibrating the TRGB and how they affect H_0 measurements. The first approach is to minimize uncertainties in the TRGB by applying particular selections. For instance, Anderson et al. [15] utilize the variability of Optical Gravitational Lensing Experiment (OGLE) Small Amplitude Red Giants (OSARGS) [16–19] in the LMC to achieve a 1.39% calibration of the TRGB. Another instance can be seen in Hoyt [20], who selects 5 out of 20 LMC fields that have more symmetric Sobel responses and narrower bootstrap widths to achieve a 1.5% calibration. While both these methods are effective at lowering uncertainties in thc TRGB calibration to the 1% level, it is important to note that these calibrations are distinct from calibrations that can be used (at the moment) to measure H_0, as these selections are not currently feasible in SN Ia host galaxies with the current resolution of space telescopes such as *HST* and *JWST*. For instance, the amplitudes of red giants are the same order magnitude of the photometric uncertainties in many SN Ia host galaxies, which makes it difficult to differentiate intrinsic and extrinsic variability. Li et al. 2023 demonstrate in their Appendix D that the same bootstrap width criterion of 0.025 mag used to reject 15 out of 20 LMC fields used in Hoyt [20] would eliminate all TRGBs measured in SN Ia host galaxies, precluding its use to measure H_0 [4]. The key concept here is that similar selections should be applied to both the anchor and host galaxies. Inconsistent selections can introduce biases, such as mismatch of the calibration due to different stellar populations.

The second approach applies the same or similar measurement and selection criterion to both the anchor and host galaxies, even if this results in a higher uncertainty in the calibration. Here, the calibration and selection of the TRGB remain consistent across the distance ladder such that the calibration more closely reflects the measured luminosity of the TRGB in the host galaxy such as in Scolnic et al. [21]. We emphasize that both these approaches are very useful for advancing understanding of the TRGB. One should be careful, however, to be aware of these two conceptually different approaches when choosing a calibration to anchor distance and H_0 measurements. We look forward to future instrumentation that can provide the resolution that erases this distinction.

8.2.2 Calibrating SN Ia

Now with a calibration of the TRGB, one can measure distances to galaxies using Eq. 8.1. To measure H_0, one in particular would want to measure TRGB distances to galaxies that contain both a well-observed TRGB as well as a luminous standard candle that can reach further into the Hubble flow, such as SN Ia, to obtain a calibration of the more luminous standard candle. The Tully-Fisher relation is another way one can reach far into the Hubble flow; we refer the reader to the corresponding section in this book for more information. For simplicity, we will refer to SN Ia when describing the distance ladder for the remainder of this chapter.

A TRGB distance measurement to a galaxy, combined with a measurement of the apparent magnitude of SN Ia (such as from Pantheon+ [22] or the Carnegie Supernovae Project [23, 24]), allows one to calibrate the SN Ia luminosity using the same principles described in Sect. 8.2.1. A luminosity calibration of SN Ia can then be combined with apparent magnitude SN Ia measurements to measure distances to galaxies further in the Hubble flow where the TRGB is too faint to be observed.

One illustrative example of how to calibrate SN Ia is described in Scolnic et al. [21]. They first find the difference in magnitude between the apparent TRGB and mean SN Ia apparent magnitude in a given galaxy:

$$\Delta S = m_{I,TRGB} - m_B^0 \tag{8.3}$$

where ΔS is the difference between the apparent TRGB and SN Ia magnitudes, $m_{I,TRGB}$ is the apparent TRGB magnitude, and m_b^0 is the apparent SN Ia magnitude. They then obtain the difference between the absolute magnitudes of the TRGB and SN Ia by taking the mean ΔS from all SN Ia host galaxies and subtracting it from the TRGB luminosity:

$$M_B^0 = M_{I,TRGB} - \Delta\bar{S} \tag{8.4}$$

to obtain the luminosity of SN Ia, M_B^0. We note that Eq. 8.4 is essentially an algebraic rearrangement of Eq. 8.1 for two standard candles to the same galaxy. $\mu_0 = m - M$ with TRGB should yield the same distance when using $\mu_0 = m - M$ with SN Ia when both are in the same galaxy. By plugging in ΔS from Eqs. 8.3 into 8.4, one can rearrange to find $(m - M)_{TRGB} = (m - M)_{SN\ Ia}$.

8.2.3 Measuring H_0

Measuring H_0 requires knowledge of both redshifts and distances to galaxies. The derivations of the equations used to measure H_0 are outside the scope of this chapter; we instead briefly mention that one can conveniently obtain H_0 using the Hubble diagram intercepts from previous studies. With a calibration of SN Ia obtained with the method from the previous section, one can use the equation:

$$log H_0 = 0.2M_B^0 + a_B + 5 \tag{8.5}$$

where M_B^0 is the SN Ia luminosity, and a_B is the intercept of the Hubble diagram at low redshift and can be found from various studies, such as Riess et al. [25]. Equation 8.5 can also be used to measure H_0 without a_B by comparing distances relative to another H_0 measurement.

8.2.4 Measurements of H_0 with the TRGB

In this section, we highlight a few H_0 measurements from the past couple of years. We note that this discussion is not exhaustive and encourage the reader to read the references within these studies for more examples of TRGB-based H_0 measurements.

1. The most recent H_0 measurement from the Carnegie-Chicago Hubble Program (CCHP) results in $H_0 = 69.8 \pm 0.6$ (stat) $\pm$ 1.6 (sys) km s^{-1} Mpc^{-1} [26]. This work updates the Freedman et al. [27] measurement, which used a Monte Carlo Markov Chain approach and 18 SN Ia calibrators to measure H_0. In this update, Freedman 2021 [26] replaces the Freedman et al. [27] LMC only calibration with a combination of calibrations from NGC 4258 [9], Milky Way globular clusters [28], LMC [20], and SMC [20]. They use a Sobel-filter based TRGB measurement methodology which will be described in Sect. 8.3.1.
2. Anand et al. [8] independently reduced the data used by CCHP to find $H_0 = 71.5 \pm 1.8$ km s^{-1} Mpc^{-1}. They incorporate *HST* NGC 4258 observations that were not used in Freedman 2021 to anchor their distance ladder and notably found that they could not reliable identify the TRGB in 4 out of the 15 hosts used by Freedman et al. 2019. Anand et al. [8] also used a model based least-squares fit instead of a Sobel method, a different photometry pipeline, and apply color corrections from Rizzi et al. [5]. They find a −0.028 mag difference (Extragalactic Distance Database - CCHP) distance scales and did not find a significant difference in their H_0 measurement when using SN Ia from the Carnegie Supernovae Project [23, 24] or Pantheon [29].
3. Scolnic et al. [22] recently measured $H_0 = 73.22 \pm 2.06$ km s^{-1} Mpc^{-1}. In their measurement, Scolnic et al. 2023 [22] correct their measured TRGBs to a fiducial contrast ratio to improve standardization across rungs in the distance ladder. They calibrate their TRGB to NGC 4258 [4] and include three new SN Ia in their analysis that were not previously used in Freedman 2021 or Anand et al. 2022 [8, 26], in addition to applying peculiar flow corrections and the Pantheon+ sample. They find that applying contrast ratio corrections to their TRGB measurements increases the Freedman [26] and Anand et al. [8] H_0 by

1.4 and $-0.3\,\mathrm{km\,s^{-1}\,Mpc^{-1}}$, respectively. The largest differences, however, came from different treatments of SN Ia, which would change the Freeman [26] and Anand et al. [8] H_0 by 2 and $1.3\,\mathrm{km\,s^{-1}\,Mpc^{-1}}$, respectively. Their algorithm and data are publicly available via the link provided in Sect. 8.3.2.

4. Dhawan et al. [30] used a hierarchical Bayesian SED model, BayeSN, to infer SN Ia distances calibrated with Cepheids and TRGB to find $H_0 = 70.92 \pm 1.14$ (stat) $\pm$ 1.49 (sys) $\mathrm{km\,s^{-1}\,Mpc^{-1}}$ with the TRGB calibration. Their hierarchical Bayesian SED model allowed them to simultaneously model optical and NIR SN Ia light curves, decreasing H_0 uncertainty by $\sim 15\%$ compared to their optical only case.
5. Dhawan et al. [31] measured $H_0 = 76.94 \pm 6.4\,\mathrm{km\,s^{-1}\,Mpc^{-1}}$ using only SN Ia from the from the Zwicky Transient Facility [32] to minimize systematics from host-galaxy bias and different treatments of photometry. They use a single host galaxy NGC 7814 containing SN Ia ZTF SN Ia SN 2021rhu and measure the TRGB with the same pipeline from CCHP. The large uncertainty is primarily due to their single SN Ia calibrator; they note that future Zwicky Transient Facility observations of SN Ia combined with $JWST$ observations of TRGB will significantly improve this measurement.
6. Blakeslee et al. [33] tied Cepheid and TRGB calibrations to surface brightness fluctuation measurements of 64 galaxies ranging from 19 to 99 Mpc to find $H_0 = 73.3 \pm 0.7 \pm 2.4\,\mathrm{km\,s^{-1}\,Mpc^{-1}}$. They obtain their sample from a variety of programs using the HST WFC3/IR $F110W$ filter and four different treatments of galaxy velocities.

8.3 Challenges for H_0 Measurements

8.3.1 *Different Measurement Algorithms*

The TRGB can be measured using a variety of different methodologies, each of which can introduce challenges for ensuring consistency along the distance ladder to measure H_0. Here, we focus on describing the Sobel filter and least-squares fit methods, the two of which are most commonly used to measure TRGB H_0.

8.3.1.1 Sobel Filter

The Sobel filter was first used to measure the TRGB by Lee et al. [34] and has since been adopted in several different forms. For a detailed review of these variations, we refer the reader to Beaton et al. [2]. Here, we focus on describing the primary method adopted by the Carnegie-Chicago Hubble Program (CCHP) [35], which has been used extensively to measure H_0. An unsupervised variant of this method has been developed by the Comparative Analysis of TRGBs (CATs) team [4, 7, 21]; we describe this method separately in Sect. 8.3.1.3.

The Sobel filter concept originates from the field of computer vision and is used to identify edges by evaluating the first derivative of intensities. In computer vision, this may involve finding the edges of an object in a 2D image to isolate the object from its surroundings. For a 1D luminosity function around the TRGB, the Sobel filter can be used to locate the TRGB discontinuity in the luminosity function with the maximum first derivative. To use this method, one first finely bins the luminosity function around the TRGB in bin widths of order 0.005 mag. Then, the binned luminosity function is smoothed using Gaussian-windowed, Locally Weighted Scatterplot Smoothing (GLOESS) [35–37] to suppress false detections caused by noise. Poisson weights are also applied to de-weight sparsely populated bins. Finally, a Sobel kernel of the form [-1, 0, 1] is evaluated across the luminosity function to compute the first derivatives of the smoothed and weighted luminosity function. The TRGB is taken to be at the location of the maximum Sobel response.

This method has its advantages and disadvantages. GLOESS is effective at reducing false TRGB detections caused by noise and can help suppress smaller local maxima in the Sobel response to isolate the strongest discontinuities in a luminosity function. In addition, the method is computationally inexpensive compared to maximum likelihood estimation, for instance, and can be used to detect multiple TRGBs in a sample containing mixed stellar populations. However, the smoothing filter, by nature, reduces the TRGB discontinuity by moving red giant branch (RGB) stars fainter than the TRGB discontinuity into the region brighter than the TRGB discontinuity containing asymptotic giant branch (AGB) stars, as RGB stars are more abundant than AGB stars. This can cause a smoothing bias at certain levels of smoothing, where changing the strength of the smoothing changes the location of the measured TRGB. The amount of smoothing bias can also depend on the heights of the TRGB discontinuities. This effect was noted by Hatt et al. [35] and investigated and documented by Anderson et al. [15] and Li et al. [4]. Hatt et al. [35] describes an artificial star luminosity function test, which is used to choose a smoothing level by minimizing the quadrature sum of statistical and systematic errors. As the smoothing bias can be a function of the contrast ratio [4, 15], it is essential that the artificial star luminosity function accurately reflects the observed sample for which is being used to estimate uncertainties. One should keep in mind that using different smoothing levels for different TRGB measurements in different galaxies could potentially lead to inconsistencies along the distance ladder. In addition, there can often be several local maximum in the Sobel response that are close together in height and magnitude even after applying GLOESS and Poisson weighting. Some studies suggest to increase the smoothing interval in these cases until there is a single, well defined peak (for instance, in Hatt et al. [35]). However, doing so runs the risk of combining Sobel peaks corresponding to multiple true TRGBs from different stellar populations or merging a local maximum corresponding to the true TRGB with another local maximum caused by noise.

8.3.1.2 Least Squares Fitting

The least-squares fitting method is frequently used by the Extragalactic Distance Database team and has been used to measure H_0 [8]. This method assumes that the RGB/AGB luminosity function around the TRGB is well described by a broken power law model of the form:

$$\psi(m) = \begin{cases} 10^{a(m-m_{TRGB})+b} & m \geq m_{TRGB} \\ 10^{c(m-m_{TRGB})} & m < m_{TRGB}. \end{cases} \tag{8.6}$$

where m_{TRGB} is the magnitude of the TRGB, m is the magnitude of a star, a and c are the logarithmic slopes of the luminosity function corresponding to the RGB and AGB, respectively, and b represents the strength or height of the TRGB break. This form of the RGB/AGB luminosity function was first described by Zoccali and Piotto [38] and subsequently used for TRGB measurements by Mendez et al. 2002, Makarov et al. 2006, Wu et al. 2014, Li et al. 2022, and Li et al. 2023, and references therein [39–43]. We note that Eq. 8.6 is also used for the maximum likelihood approach to measuring the TRGB, which is also used by the EDD team [27] and has been used to measure the TRGB using Milky Way field giants [29, 30]. Although both the least-squares and maximum likelihood methods are model-based, we focus here on the the least-squares fitting method here due to its use to measure H_0.

The TRGB least-squares fitting procedure as described in Wu et al. [41] is as follows: The luminosity function model described in Eq. 8.6 is first convolved with the completeness, uncertainty, and bias of the observations. Then, an initial guess for the location of the TRGB is obtained by taking first derivatives of the luminosity function and locating their maximum. The luminosity function is binned in 0.05 mag intervals across $\pm$ 1 mag from the anticipated TRGB magnitude. Finally, a least-squares fit using the Levenberg-Marquardt algorithm is applied to the luminosity function to determine m_{TRGB}, a, b, and c. Uncertainties of the fitted parameters are estimated with the square root of the variance.

This method has the advantage of also being computationally inexpensive and is easily implemented in Python, with least-squares fitting routines such as from `scipy curve_fit` [44] readily available. However, this method assumes a single, true TRGB in its model and cannot separate out two TRGB breaks that may arise from the mixture of stellar populations. In principle, one could modify the model to include multiple TRGB discontinuities, however, this would still lack the flexibility of the Sobel filter at detecting multiple Sobel peaks. In addition, one must select an appropriate binning width; for well-populated stellar populations, the measured TRGB may be insensitive to binning, however, the measurements using sparsely populated populations may be susceptible to noise and sensitive to the choice of binning. The least-squares fit can also potentially converge onto unrealistic model parameters depending on the initial guess. For instance, in some cases the fit may identify the faint magnitude cutoff as the primary discontinuity if the TRGB discontinuity itself is not well defined or identify the tip of the AGB as the tip of the RGB (for examples, see Li et al. [42] and Anand et al. [45]).

FAQ: Which method should I use to measure the TRGB?
Response: Each TRGB measurement method has its advantages and disadvantages and should be selected based on characteristic of the RGB/AGB sample and end goal. For instance, a Sobel filter paired with GLOESS can be more effective at reducing noise and false detections in noisy data compared to the binned power law least-squares fit, but the latter can also provide information about the RGB and AGB slopes, which may be useful for comparing stellar populations in other fields or galaxies. In many cases when the sample is well defined and populated, different methodologies agree well. However, we emphasize that regardless of the method used, it is important to closely examine potential sources of measurement biases to ensure that the methods are consistently across a distance ladder such that any biases do not propagate into the final distance or H_0 measurement.

8.3.1.3 Combining Algorithms and Stellar Populations

The TRGB measurement methods described in Sect. 8.3.1 have been used extensively to measure extragalactic distances and H_0. However, significant variations in calibrated TRGB luminosities (see compilations in Blakeslee et al. 2021, Freedman 2021, and Li et al. 2022 [26, 33, 42]) and H_0 (for instance, see Freedman 2021 and Anand et al. 2021 [8, 26]) suggest there may be additional considerations that need to be taken into account to improve the consistency of TRGB measurements. In particular, one must ensure that the TRGB calibrated in an anchor galaxy is standarized such that it accurately reflects the true luminosity of the TRGB in a host galaxy. Mismatches between the TRGBs in anchor and host galaxies can occur with inconsistent measurement or selection choices in the two galaxies. The stellar population used to calibrate the TRGB in a host galaxy should have similar characteristics or be corrected to a fiducial value. One must also be careful that the measurement method also does not introduce biases, or that these biases cancel out along the distance ladder (Fig. 8.2).

Some of these issues can manifest when using Sobel filter and least-squares fit TRGB measurement methods. Model based methods, such as the least-squares fit described in Sect. 8.3.1.2, have the disadvantage of assuming only a single TRGB break. While such model based methods are effective at measuring populations that show a single, well defined TRGB discontinuity, a population consisting of stars of several characteristics may exhibit multiple TRGB discontinuities [4, 7, 21]. Multiple TRGB discontinuities cannot be separated with a broken power law model that assumes a single TRGB discontinuity. On the other hand, it is difficult to differentiate

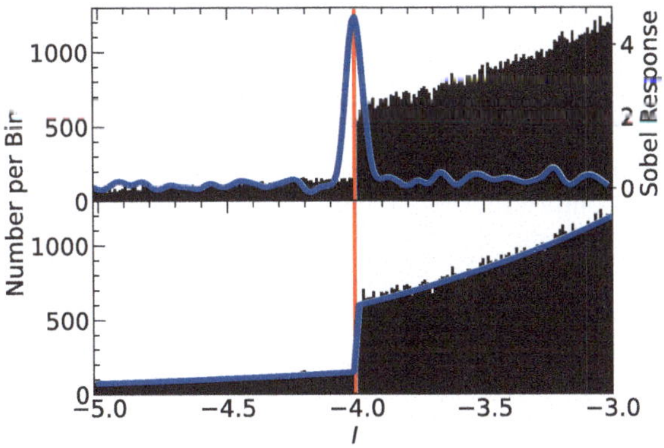

Fig. 8.2 Examples of a Sobel filter response (top; blue) and broken power law fit (bottom, blue) TRGB measurements on a simulated luminosity function (black). The red vertical line marks the location of the simulated TRGB

between local maxima in a Sobel response that correspond to noise and true TRGBs. An approach adopted by Hatt et al. [35] is to increase the smoothing until a single local maximum in the Sobel response remains. However, this can potentially combine multiple true TRGBs or a true TRGB with nearby local maxima in the Sobel response corresponding to noise, and different choices of smoothing in the calibration and host can lead to inconsistencies along the distance ladder. The selection of different smoothing parameters across the distance ladder, or one local maximum in the Sobel response over another of similar height, introduces subjective choices in the measurement method that can easily lead to confirmation bias, if one is not careful.

To address these issues, the CATs team developed an unsupervised Sobel-filter based algorithm that can be used to measure TRGBs using objectively determined measurement parameters and improve the consistency of TRGB measurements across the distance ladder to measure H_0 [7]. The algorithm first applies a step called spatial clipping, which removes regions in an observed field that contain high ratios of blue to red stars, as defined on a CMD, that might contaminate the red giant sample. Then, the algorithm applies a color band with a fixed width that maximizes the number of stars in the sample and applies GLOESS and Poisson weights to a luminosity function. A Sobel filter of the form [-1, 0, 1] is then evaluated across the luminosity function to identify the TRGB. The key differences in the CATs algorithm compared to previous Sobel filter algorithms are that the measurement parameters, such as the width of the color bands, choice of smoothing, and quality cuts, are first optimized to minimize the scatter in the observed field-to-field variation in the measured TRGBs in multiple fields from the Galaxy Halos, Outer disks, Substructure, Thick disks, and Star clusters (GHOSTS) galaxies [46]. These measurement parameters are subsequently fixed when measuring extragalactic distances and H_0. The algorithm also defines a contrast ratio, taken to be the ratio between number of stars in the 0.5 mag interval fainter and brighter than the measured TRGB break. In addition to being a useful for estimating ages [7, 47], the contrast ratio provides a metric for differentiating high and low signal TRGB discontinuities. A high contrast ratio corresponds to a well defined break, while a low contrast ratio corresponds to an ill-defined break that may be sensitive to noise. The CATs algorithm filters out contrast ratios less than 3 to avoid false detections [7]. This algorithm is publicly available via the GitHub repository link provided in Sect. 8.3.2.

In addition to developing an unsupervised Sobel-based TRGB measurement algorithm, the CATs team discovered a 5 σ relationship between the measured TRGB and contrast ratio (tip-contrast ratio, or TCR). This finding suggests that the TRGB can be further standardized via the contrast ratio so that it is used more consistently across the distance ladder. The CATs team used their unsupervised algorithm and TCR to calibrate the TRGB luminosity using NGC 4258 [4] and measure H_0 [21] and find a H_0 of 73.22 ± 2.06 km s^{-1} Mpc^{-1}. There are four key differences in the CATs H_0 compared to previous Sobel-filter based H_0 measurements. First, the CATs algorithm uses measurement parameters such as smoothing and spatial clipping consistently across the distance ladder to facilitate first order cancellation of potential biases. Second, the CATs algorithm allows for the possibility of multiple true TRGB breaks in a luminosity function due to mixtures of stellar populations. Third, the CATs team corrects measured TRGBs to a fiducial contrast ratio to facilitate closer alignment between the anchor and host TRGB and to account for variations in the measured TRGB as seen in the TCR. Fourth, the CATs measurement follows a different treatment of SN Ia, such as applying peculiar velocity corrections (see Scolnic et al. [21] for more details).

The CATs team hypothesize that the TCR is caused by a combination of astrophysical properties as well as the measurement process such as smoothing [4, 7, 21]. However, the exact nature of the TCR is inconsequential to the validity of the trend; it is important to remember that the TCR is an *empirical* relationship, such that, at the fundamental level, there is a variation in the measured TRGB that correlates with another parameter. The TRGB is corrected to account for this variation irrespective of the exact origin of the variation, whether due to astrophysical differences, measurement artifacts, or both. There are still improvements that can be made to this approach, and more studies can be conducted to better understand the nature of the TCR and whether the TCR exists when using other TRGB measurement methods. While there may be other avenues to further standardize the TRGB, these efforts show the beginning of a paradigm shift in TRGB measurements with the recognition that there are still small variations in measured TRGBs using the Sobel-filter method that were not previously accounted for, and that additional standardization should be applied to the measured TRGB as used to measure H_0.

FAQ: How much does correcting for the contrast ratio change H_0?
Response: Scolnic et al. [21] compared the CATs H_0 result with previous H_0 measurements in their Table 5 and found that applying contrast ratio standardization increased H_0 from Freedman et al. [26] by 1.4 km s^{-1} Mpc^{-1} and decreased H_0 from Anand et al. [8] by 0.3 km s^{-1} Mpc^{-1} . It's worth noting that some of the largest differences between the different H_0 values also come

from different treatments of SNe Ia. We refer the reader to Table 5 and its corresponding text in Scolnic et al. [21] for more details.

8.3.2 *Open Datasets*

Open datasets and algorithms help improve transparency and community-wide discussions in the field. We list here currently public TRGB datasets and algorithms available through websites that we encourage the reader to explore in addition to datasets made directly available via journal publications. We encourage future studies to make their datasets and algorithms available to the public.

- Photometry from the Extragalactic Distance Database [48, 49]: https://edd.ifa.hawaii.edu/.
- Unsupervised Sobel-filter algorithm and results from the CATs team: [4, 7, 21]: https://github.com/JiaxiWu1018/Unsupervised-TRGB, https://github.com/JiaxiWu1018/CATS-\inlinealttextupperH0H_0.
- 2D Maximum Likelihood algorithm used to calibrate TRGB using field stars and *Gaia*: [43]: https://github.com/siyangliastro/Gaia-DR3-MilkyWay-TRGB.

8.3.3 *TRGB at the Crossroads*

We pause here to highlight two distinct ideologies in constructing a TRGB distance ladder to measure H_0 that arise as a natural consequence of developments in observational abilities. The first we refer to as artisanal measurements, which involve adjusting measurement and selection parameters to optimize TRGB measurements. For instance, one may increase the smoothing parameter when using a Sobel filter method until there is a single and well defined TRGB detection, as suggested by Hatt et al. [35], with the optimal smoothing parameter being different for each galaxy. Advocates for this approach may argue that using different measurement and selection choices is necessary when using heterogeneous datasets having different levels of photometric uncertainties and sample size, for instance, which requires different treatments to isolate the TRGB. We note that when doing so, it is especially important to understand and minimize potential biases of the measurement method when working with data of varying qualities.

On the other hand, applying different measurement and selection criteria runs the risk of introducing inconsistencies along the distance ladder. For instance, Anderson et al. 2023 and Li et al. 2023 [4, 15] demonstrate that different smoothing, as applied to the same luminosity function, introduces different biases. Advocates for this approach argue that measurements must be applied consistently across the distance ladder for all galaxies to ensure first order cancellation of any biases and minimize inconsistencies between calibrator and host measurements. This approach is more feasible when working with datasets with taken with the same telescope, such as *HST* observations of NGC 4258 and GHOSTS galaxies.

These two ideologies can be seen in the context of a natural shift that occurs as more homogeneous observations are made possible with improved telescopes such as *HST* and *JWST*. We leave the question of which approach is best open to the reader. Both have their advantages and disadvantages. While there is not a single, accepted answer, it is important to be aware of this distinction, especially when interpreting the results and methodologies of different H_0 results.

8.4 Future Developments

8.4.1 New Prospects with JWST

The successful launch of *JWST* opened up the possibly of probing the TRGB further in the IR in greater resolutions and distances than *HST*. In addition, the configuration of detector modules on the *JWST* Near Infrared Camera (NIRCam) opens up the possibility for observational programs to simultaneously observe multiple standard candles at once. NIRCam contains two cameras each consisting of four chips that creates a rectangular array. This setup allows *JWST* to simultaneously image both the halos and disks of galaxies to observe red giants for TRGB, Cepheid variables, Mira variables, and the Carbon asymptotic giant branch stars for the J-region Asymptotic Giant Branch (JAGB), potentially enabling four H_0 measurements with one set of anchor and host observations once the JAGB is better understood and standardized. This setup also will allow for a powerful cross-calibration of these standard candles. In addition, NIRCam is a two-channel instrument which allows observations to be taken simultaneously in the short and long wavelength IR, further increasing the breadth of observations possible within a single observing program. However, as the TRGB exhibits a greater metallicity dependence in the IR compared to the I or $F814W$ bands [50], further calibration and investigation will be needed before the IR TRGB can be used to measure robust extragalactic distances. We provide an example of the metallicity dependence of the NIRCam $F150W$ TRGB in Fig. 8.3, where we use the Python ArtPop package [3] to generate 10 Gyr isochrones for the red giant branch with metallicities ranging from [Fe/H] of -2 to -1 in 0.1 increments starting from the left most isochrone. *JWST* programs designed with these features in mind have already been completed; we highlight some of these in Sect. 8.4.1.1.

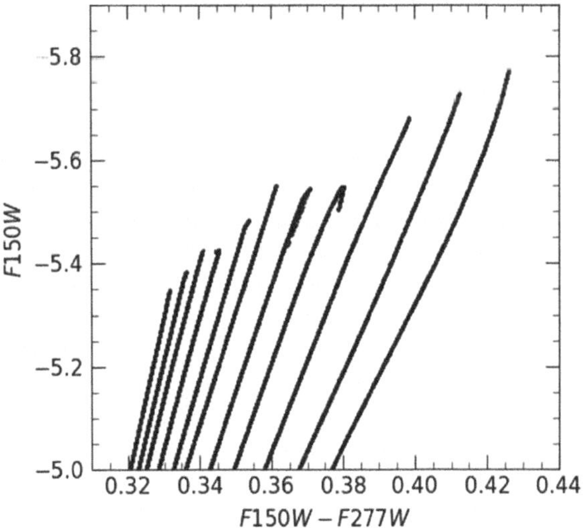

Fig. 8.3 Isochrones generated using the Python ArtPop package [3] in *JWST* NIRCam *F150W* and *F277W* using stars of 10 Gyr and metallicities ranging from [Fe/H] of −2 to −1 dex

8.4.1.1 Early *JWST* Observational Programs

As of writing this chapter, *JWST* has undergone two general observer (GO) proposal cycles in addition to the Director's Discretionary (DD) Early Release Science Programs (ERS), all of which contain accepted proposals containing observations that can be used to measure the TRGB. We briefly highlight some of these programs that contain observations that can be used for TRGB measurements and encourage the reader to explore images from these proposals on MAST.[1] Proposal abstracts can be obtained from STScI.[2,3] We anticipate this list to grow as *JWST* undergoes more proposal cycles in the future.

- **ERS DD-1334 (PI: Daniel Weisz)**: The Resolved Stellar Populations Early Release Science Program
- **Cycle 1 GO-1638 (PI: Kristen McQuinn)**: Securing the TRGB Distance Indicator: A Pre-Requisite for a JWST Measurement of H_0
- **Cycle 1 GO-1685 (PI: Adam Riess)**: Uncrowding the Cepheids for an Improved Determination of the Hubble Constant
- **Cycle 1 GO-1995 (PI: Wendy Freedman & Barry Madore)**: Answering the Most Important Problem in Cosmology Today: Is the Tension in the Hubble Constant Real?
- **Cycle 2 GO-3055 (PI: Richard Brent Tully)**: A TRGB calibration of Surface Brightness Fluctuations

---

[1] https://mast.stsci.edu/portal/Mashup/Clients/Mast/Portal.html.

[2] https://www.stsci.edu/jwst/science-execution/approved-ers-programs.

[3] https://www.stsci.edu/jwst/science-execution/approved-programs.

8.4.2 Future Gaia Data Releases

The *Gaia* mission [11] has observed over a billion stars in the Milky Way and significantly augmented the understanding of our home galaxy. Among the measurements of Milky Way field stars made by *Gaia* include low-resolution spectra that can be used to construct synthetic photometry and parallaxes. These measurements can be retrieved via the *Gaia* archive[4] and *Gaia* Synthetic Photometry Catalogue [51] and have been used to calibrate the TRGB. Dixon et al. [52] used *Gaia* Early Data Release (EDR) 3 parallaxes and SkyMapper Data Release (DR) 3, American Association of Variable Star Observers (AAVSO) Photometric All-Sky Survey (APASS) DR9, Asteroid Terrestrial-impact Last Alert System (ATLAS) Refcat2, and *Gaia* DR3 synthetic photometry to calibrate the TRGB using Milky Way field stars. They directly convert apparent magnitudes into absolute magnitudes using distances constructed from *Gaia* parallaxes and apply a Sobel filter to find a weighted average calibration of $M_I^{TRGB} = -4.042 \pm 0.041$ (stat) $\pm$ 0.031 (sys). Caution should be used, however, when directly converting apparent magnitudes to absolute magnitudes via parallaxes due to asymmetric probability distributions (see Bailer-Jones et al. [53]). Li et al. developed a two-dimensional maximum likelihood estimation algorithm to ensure simultaneous treatment of uncertainties in parallaxes and magnitudes. The algorithm was first introduced in Li et al. [42] and optimizes model parameters for a broken power law luminosity function and exponential density distribution to find the set of parameters that maximizes the likelihood of observing the sample of field giants. This first study used *Gaia* EDR3 parallaxes and APASS DR9 photometry and was subsequently improved using *Gaia* DR3 synthetic photometry in Li et al. 2023 [43] to find a calibration of $M_I^{TRGB} = -3.970\ ^{+0.042}_{-0.024}$ (sys) $\pm$ 0.062 (stat). The algorithm used for these two studies by Li et al. is publicly available via the GitHub repository link provided in Sect. 8.3.2. The differences between the Dixon et al. [52] and Li et al. [43] field star results can be due to several factors, such as a different treatment of the *Gaia* parallax zero-point offset (differences of 10 μas), which is still not well understood, and differences in measurement methodologies, with one being in magnitude space and the other in parallax space.

Gaia parallaxes can also be used to calibrate the TRGB in globular clusters such as ω Centauri. Soltis et al. [54] used *Gaia* EDR3 parallaxes to calibrate the TRGB measured by Bellazzini et al. [55] and find $M_{I,TRGB} = -3.97 \pm 0.06$ mag. Li et al. [43] investigated the possibility of using *Gaia* DR3 synthetic photometry to measure the TRGB in ω Centauri but found evidence of blending that prevents it from being used to obtain an accurate TRGB calibration. They instead update the TRGB measurement in ω Centauri using the database from Stetson et al. [56] to find $M_{I,TRGB} = -3.97 \pm 0.04$ (stat) $\pm$ 0.10 (sys) mag.

Higher precision parallaxes and synthetic photometry in future *Gaia* data releases can further improve these TRGB calibrations. We note that a better understanding of the *Gaia* parallax zero-point offset will be essential before *Gaia* data can be used to obtain a TRGB calibration that is competitive in precision with calibrations

[4] https://gea.esac.esa.int/archive/.

using other anchors such as NGC 4258. At the moment, the parallax zero-point offset is not well understood; the standard offset is described in Lindegren et al. [57], but other estimates can differ [58–62]. Regardless, *Gaia* has the potential to help increase leverage in TRGB H_0 calibrations through observations independent from those used for other anchors.

Acknowledgements The authors would like to thank Dr. Gagandeep Anand for helpful comments. S. L. is supported by the National Science Foundation Graduate Research Fellowship Program under grant number DGE2139757. R. L. B. is supported by the National Science Foundation through grant number AST-2108616.

References

1. I. Iben, A. Renzini, Ann. Rev. Astron. Astrophys. **21**, 271–342 (1983). https://doi.org/10.1146/annurev.aa.21.090183.001415
2. R.L. Beaton, G. Bono, V.F. Braga, M. Dall'Ora, G. Fiorentino, I.S. Jang, C.E. Martínez-Vázquez, N. Matsunaga, M. Monelli, J.R. Neeley et al. https://doi.org/10.1007/s11214-018-0542-1. ([arXiv:1808.09191 [astro-ph.GA]].)
3. J.P. Greco, S. Danieli, Astrophys. J. **941**(1), 26 (2022). https://doi.org/10.3847/1538-4357/ac75b7. ([arXiv:2109.13943 [astro-ph.GA]].)
4. S. Li, A.G. Riess, D. Scolnic, G.S. Anand, J. Wu, S. Casertano, W. Yuan, R. Beaton, R.I. Anderson, Astrophys. J. **956**(1), 32 (2023). https://doi.org/10.3847/1538-4357/acf4fb. ([arXiv:2306.10103 [astro-ph.GA]].)
5. L. Rizzi, R.B. Tully, D. Makarov, L. Makarova, A.E. Dolphin, S. Sakai, E.J. Shaya, Astrophys. J. **661**, 815–829 (2007). https://doi.org/10.1086/516566. ([arXiv:astro-ph/0701518 [astro-ph]].)
6. I.S Jang, M.G. Lee, Astrophys. J. **835**(1), 28 (2017). https://doi.org/10.3847/1538-4357/acdd7b. ([arXiv:2211.06354 [astro-ph.CO]].)
7. J. Wu, D. Scolnic, A.G. Riess, G.S. Anand, R. Beaton, S. Casertano, X. Ke, S. Li, Astrophys. J. **954**(1), 87 (2023). https://doi.org/10.3847/1538-4357/acdd7b. ([arXiv:2211.06354 [astro-ph.CO]].)
8. G.S. Anand, R.B. Tully, L. Rizzi, A.G. Riess, W. Yuan, Astrophys. J. **932**(1), 15 (2022). https://doi.org/10.3847/1538-4357/ac68df. ([arXiv:2108.00007 [astro-ph.CO]].)
9. I.S Jang, T.J. Hoyt, R.L. Beaton, W.L. Freedman, B.F. Madore, M.G. Lee, J.R. Neeley, A.J. Monson, J.A. Rich, M. Seibert, Astrophys. J. **906**(2), 125 (2017). https://doi.org/10.3847/1538-4357/abc8e9. ([arXiv:2008.04181 [astro-ph.GA]].)
10. E.F. Schlafly, D.P. Finkbeiner, Astrophys. J. **737**, 103 (2011). https://doi.org/10.1088/0004-637X/737/2/103. ([arXiv:1012.4804 [astro-ph.GA]].)
11. T. Prusti et al., [Gaia], Astron. Astrophys. **595**(1), A1 (2016). https://doi.org/10.1051/0004-6361/201629272. ([arXiv:1609.04153 [astro-ph.IM]].) (Gaia Data Release)
12. G. Pietrzyński, D. Graczyk, W. Gieren, I.B. Thompson, B. Pilecki, A. Udalski, I. Soszyński, S. Kozłowski, P. Konorski, K. Suchomska et al., Nature **495**, 76–79 (2013). https://doi.org/10.1038/nature11878. ([arXiv:1303.2063 [astro-ph.GA]].)
13. D. Gracyzk, G. Pietrzyński, I.B. Thompson, W. Gieren, B. Zgirski, S. Villanova, M. Górski, P. Wielgórski, P. Karczmarek, W. Narloch et al., Astrophys. J. **904**(1), 13 (2020). https://doi.org/10.3847/1538-4357/abbb2b. ([arXiv:2010.08754 [astro-ph.GA]].)
14. M.J. Reid, D.W. Pesce, A.G. Riess, Astrophys. J. Lett. **886**(2), L27 (2019). https://doi.org/10.3847/2041-8213/ab552d. ([arXiv:1908.05625 [astro-ph.GA]].)
15. R.I. Anderson, N.W. Koblischke, L. Eyer, [arXiv:2303.04790 [astro-ph.CO]]

16. L.L. Kiss, T. Bedding, Mon. Not. R. Astron. Soc. **343**, L79 (2003). https://doi.org/10.1046/j.1365-8711.2003.06931.x. ([arXiv:astro-ph/0306426 [astro-ph]].)
17. J.J. Wray, L. Eyer, B. Paczynski, Mon. Not. R. Astron. Soc. **349**, 1059 (2004). https://doi.org/10.1111/j.1365-2966.2004.07587.x. ([arXiv:astro-ph/0310578 [astro-ph]].)
18. I. Soszynski, A. Udalski, M. Kubiak, M. Szymanski, G. Pietrzynski, K. Zebrun, O. Szewczyk, L. Wyrzykowski, Acta Astron. **54**, 129 (2004). ([arXiv:astro-ph/0407057 [astro-ph]])
19. I. Soszynski, A. Udalski, M.K. Szymanski, M. Kubiak, G. Pietrzynski, L. Wyrzykowski, O. Szewczyk, K. Ulaczyk, R. Poleski, Acta Astron. **59**, 239 (2009). ([arXiv:0910.1354 [astro-ph.SR]])
20. T.J. Hoyt, [arXiv:2106.13337 [astro-ph.GA]]
21. D. Scolnic, A.G. Riess, J. Wu, S. Li, G.S. Anand, R. Beaton, S. Casertano, R.I. Anderson, S. Dhawan, X. Ke, Astrophys. J. Lett. **954**(1), L31 (2023). https://doi.org/10.3847/2041-8213/ace978. ([arXiv:2304.06693 [astro-ph.CO]].)
22. D. Scolnic, D. Brout, A. Carr, A.G. Riess, T.M. Davis, A. Dwomoh, D.O. Jones, N. Ali, P. Charvu, R. Chen et al., Astrophys. J. **938**(2), 113 (2022). https://doi.org/10.3847/1538-4357/ac8b7a. ([arXiv:2112.03863 [astro-ph.CO]].)
23. K. Krisciunas, C. Contreras, C.R. Burns, M.M. Phillips, M.D. Stritzinger, N. Morrell, M. Hamuy, J. Anais, L. Boldt, L. Busta et al., Astron. J. **154**(5), 211 (2017). https://doi.org/10.3847/1538-3881/aa8df0. ([arXiv:1709.05146 [astro-ph.IM]].)
24. M. Hamuy, G. Folatelli, N.I. Morrell, M.M. Phillips, N.B. Suntzeff, S.E. Persson, M. Roth, S. Gonzalez, W. Krzeminski, C. Contreras et al., Publ. Astron. Soc. Pac. **118**, 2–20 (2006). https://doi.org/10.1086/500228. ([arXiv:astro-ph/0512039 [astro-ph]].)
25. A.G. Riess, W. Yuan, L.M. Macri, D. Scolnic, D. Brout, S. Casertano, D.O. Jones, Y. Murakami, L. Breuval, T.G. Brink et al., Astrophys. J. Lett. **934**(1), L7 (2022).https://doi.org/10.3847/2041-8213/ac5c5b. ([arXiv:2112.04510 [astro-ph.CO]].)
26. W.L. Freedman, Astrophys. J. **919**(1), 16 (2021). https://doi.org/10.3847/1538-4357/ac0e95. ([arXiv:2106.15656 [astro-ph.CO]].)
27. W.L. Freedman, B.F. Madore, D. Hatt, T.J. Hoyt, I.S. Jang, R.L. Beaton, C.R. Burns, M.G. Lee, A.J. Monson, J.R. Neeley et al., Astrophys. J. **882**, 34 (2019). https://doi.org/10.3847/1538-4357/ab2f73. ([arXiv:1907.05922 [astro-ph.CO]].)
28. W. Cerny, W.L. Freedman, B.F. Madore, F. Ashmead, T. Hoyt, E. Oakes, N.Q.H. Tran, B. Moss (arXiv:2012.09701 [astro-ph.GA]].)
29. D.M. Scolnic et al., [Pan-STARRS1], Astrophys. J. **859**(2), 101 (2018). https://doi.org/10.3847/1538-4357/aab9bb. ([arXiv:1710.00845 [astro-ph.CO]].)
30. S. Dhawan, S. Thorp, K.S. Mandel, S.M. Ward, G. Narayan, S.W. Jha, T. Chant, Mon. Not. R. Astron. Soc. **524**(1), 235–244 (2023). https://doi.org/10.1093/mnras/stad1590. ([arXiv:2211.07657 [astro-ph.CO]].)
31. S. Dhawan, A. Goobar, J. Johansson, I.S. Jang, M. Rigault, L. Harvey, K. Maguire, W.L. Freedman, B.F. Madore, M. Smith et al., Astrophys. J. **934**(2), 185 (2022). https://doi.org/10.3847/1538-4357/ac7ceb. ([arXiv:2203.04241 [astro-ph.CO]].)
32. M.J. Graham, S.R. Kulkarni, E.C. Bellm, S.M. Adams, C. Barbarino, N. Blagorodnova, D. Bodewits, B. Bolin, P.R. Brady, S.B. Cenko et al., Publ. Astron. Soc. Pac. **131**(1001), 078001 (2019). https://doi.org/10.1088/1538-3873/ab006c. ([arXiv:1902.01945 [astro-ph.IM]].)
33. J.P. Blakeslee, J.B. Jensen, C.P. Ma, P.A. Milne, J.E. Greene, Astrophys. J. **911**(1), 65 (2021). https://doi.org/10.3847/1538-4357/abe86a. ([arXiv:2101.02221 [astro-ph.CO]].)
34. M.G. Lee, W.L. Freedman, B.F. Madore, Astrophys. J. **417**, 553 (1993). https://doi.org/10.1086/173334
35. D. Hatt, R.L. Beaton, W.L. Freedman, B.F. Madore, I.S. Jang, T.J. Hoyt, M.G. Lee, A.J. Monson, J.A. Rich, V. Scowcroft et al., Astrophys. J. **845**(2), 146 (2017). https://doi.org/10.3847/1538-4357/aa7f73. ([arXiv:1703.06468 [astro-ph.CO]].)
36. S.E. Persson, B.F. Madore, W. Krzemiński, W.L. Freedman, M. Roth, D.C. Murphy, Astron. J. **128**, 5 (2004). https://doi.org/10.1086/424934
37. A.J. Monson, R.L. Beaton, V. Skowcroft, W.L. Freedman, B.F. Madore, J.A. Rich, M. Seibert, J.A. Kollmeier, G. Clementini, Astron. J. **153**, 3 (2017). https://doi.org/10.1086/424934. ([arXiv:1703.01520 [astro-ph.SR]].)

38. M. Zoccali, G. Piotto, Astron. Astrophys. **358**, 943 (2000). ([arXiv:astro-ph/0005034 [astro-ph]])
39. B. Mendez, M. Davis, J. Moustakas, J. Newman, B.F. Madore, W.L. Freedman, Astron. J. **124**, 213 (2002). https://doi.org/10.1086/341168. ([arXiv:astro-ph/0204192 [astro-ph]].)
40. D. Makarov, L. Makarova, L. Rizzi, R.B. Tully, A.E. Dolphin, S. Sakai, E.J. Shaya, Astron. J. **132**, 2729–2742 (2006). https://doi.org/10.1086/508925. ([arXiv:astro-ph/0603073 [astro-ph]].)
41. P.F. Wu, R.B. Tully, L. Rizzi, A.E. Dolphin, B.A. Jacobs, I.D. Karachentsev, Astron. J. **148**, 7 (2014). https://doi.org/10.1088/0004-6256/148/1/7. ([arXiv:1404.2987 [astro-ph.GA]].)
42. S. Li, S. Casertano, A.G. Riess, Astrophys. J. **939**(2), 96 (2022). https://doi.org/10.3847/1538-4357/ac7559. ([arXiv:2202.11110 [astro-ph.GA]].)
43. S. Li, S. Casertano, A.G. Riess, Astrophys. J. **950**(2), 83 (2023. https://doi.org/10.3847/1538-4357/accd69. ([arXiv:2304.06695 [astro-ph.GA]].)
44. P. Virtanen, R. Gommers, T.E. Oliphant, M. Haberland, T. Reddy, D. Cournapeau, E. Burovski, P. Peterson, W. Weckesser, J. Bright et al., Nature Meth. **17**, 261 (2020). https://doi.org/10.1038/s41592-019-0686-2. ([arXiv:1907.10121 [cs.MS]].)
45. G.S. Anand, R.B. Tully, L. Rizzi, I.D. Karachentsev, Astrophys. J. Lett. **872**(1), L4 (2019). https://doi.org/10.3847/2041-8213/aafee6. ([arXiv:1901.05981 [astro-ph.GA]].)
46. D.J. Radburn-Smith, R.S. de Jong, A.C. Seth, J. Bailin, E.F. Bell, T.M. Brown, J.S Bullock, S. Courteau, J.J. Dalcanton, H.C. Ferguson et al., Astrophys. J. Suppl. **195**, 2 (2011). https://doi.org/10.1088/0067-0049/195/2/18
47. B. Harmsen, E.F. Bell, R. D'Souza, A. Monachesi, R.S. de Jong, A. Smercina, I. Jang, B.W. Holwerda, Mon. Not. R. Astron. Soc. **525**(3), 4497–4514 (2023). https://doi.org/10.1093/mnras/stad2480. ([arXiv:2308.11499 [astro-ph.GA]].)
48. R.B. Tully, L. Rizzi, E.J. Shaya, H.M. Courtois, D.I. Makarov, B.A. Jacobs, Astron. J. **138**, 323–331 (2009). https://doi.org/10.1088/0004-6256/138/2/323. ([arXiv:0902.3668 [astro-ph.CO]].)
49. G.S. Anand, L. Rizzi, R.B. Tully, E.J. Shaya, I.D. Karachentsev, D.I. Makarov, L. Makarova, P. Wu, A.E. Dolphin, E. Kourkchi, Astron. J. **162**(2), 80 (2021). https://doi.org/10.3847/1538-3881/ac0440. ([arXiv:2104.02649 [astro-ph.GA]].)
50. K.B.W. McQuinn, M. Boyer, E.D. Skillman, A.E. Dolphin, Astrophys. J. **880**(1), 63 (2019). https://doi.org/10.3847/1538-4357/ab2627. ([arXiv:1904.01571 [astro-ph.GA]].)
51. P. Montegriffo et al., [Gaia], Astron. Astrophys. **674**(3), A33 (2023). https://doi.org/10.1051/0004-6361/202243709 (Gaia Data Release 3)
52. M. Dixon, J. Mould, C. Flynn, E.N. Taylor, C. Lidman, A.R. Duffy, Mon. Not. R. Astron. Soc. **523**(2), 2283–2295 (2023). https://doi.org/10.1093/mnras/stad1500. ([arXiv:2305.09215 [astro-ph.GA]].)
53. C.A.L. Bailer-Jones, J. Rybizki, M. Fouesneau, M. Demleitner, R. Andrae, Astron. J. **161**(3), 147 (2021). https://doi.org/10.3847/1538-3881/abd806. ([arXiv:2012.05220 [astro-ph.SR]].)
54. J. Soltis, S. Casertano, A.G. Riess, Astrophys. J. Lett. **908**(1), L5 (2021). https://doi.org/10.3847/2041-8213/abdbad. ([arXiv:2012.09196 [astro-ph.GA]].)
55. M. Bellazzini, F.R. Ferraro, E. Pancino, Astrophys. J. **556**, 635–640 (2001). https://doi.org/10.1086/321613. ([arXiv:astro-ph/0104114 [astro-ph]].)
56. P.B. Stetson, E. Pancino, A. Zocchi, N. Sanna, M. Monelli, Mon. Not. R. Astron. Soc. **485**(3), 3042–3063 (2019). https://doi.org/10.1093/mnras/stz585. ([arXiv:1902.09925 [astro-ph.SR]].)
57. L. Lindegren, S.A. Klioner, J. Hernández, A. Bombrun, M. Ramos-Lerate, H. Steidelmüller, U. Bastian, M. Biermann, A. de Torres, E. Gerlach et al., Astron. Astrophys. **649** Gaia Data Release 3, A2 (2021). https://doi.org/10.1051/0004-6361/202039709. ([arXiv:2012.03380 [astro-ph.IM]].)
58. Y. Huang, H. Yuan, T.C. Beers, H. Zhang, Astrophys. J. Lett. **910**(1), L5 (2021). https://doi.org/10.3847/2041-8213/abe69a.Huang:2021. ([arXiv:2101.09691 [astro-ph.GA]].)
59. M.A.T. Groenewegen, Astron. Astrophys. **654**, A20 (2021). https://doi.org/10.1051/0004-6361/202140862. ([arXiv:2106.08128 [astro-ph.GA]].)

60. F. Ren, X. Chen, H. Zhang, R. de Grijs, L. Deng, Y. Huang, Astrophys. J. Lett. **911**(2), L20 (2021). https://doi.org/10.3847/2041-8213/abf359. ([arXiv:2103.16096 [astro-ph.SR]].)
61. J.C. Zinn, Astron. J. **161**(5), 214 (2021). https://doi.org/10.3847/1538-3881/abe936. ([arXiv:2101.07252 [astro-ph.GA]].)
62. J. Maíz Apellániz, Astron. Astrophys. **657**, A130 (2022). https://doi.org/10.1051/0004-6361/202142365. ([arXiv:2110.01475 [astro-ph.IM]].)
63. I. Ribas, C. Jordi, F. Vilardell, E.L. Fitzpatrick, R.W. Hilditch, E.F. Guinan, Astrophys. J. Lett. **635**, L37–L40 (2005). https://doi.org/10.1086/499161. ([arXiv:astro-ph/0511045 [astro-ph]].)
64. F. Vilardell, I. Ribas, C. Jordi, E.L. Fitzpatrick, E.F. Guinan, Astron. Astrophys. **509**, A70 (2010). https://doi.org/10.1051/0004-6361/200913299. ([arXiv:0911.3391 [astro-ph.CO]].)
65. L. Lindegren, J. Hernandez, A. Bombrun, S. Klioner, U. Bastian, M. Ramos-Lerate, A. de Torres, H. Steidelmuller, C. Stephenson, D. Hobbs et al., Astron. Astrophys. **616**, A2 (2018). https://doi.org/10.1051/0004-6361/201832727. ([arXiv:1804.09366 [astro-ph.IM]].)

Chapter 9
Surface Brightness Fluctuations

Michele Cantiello and John P. Blakeslee

9.1 Introduction

Stellar counts have long been used to study unresolved or partially-resolved stellar populations [1]. However, it was not until the development of CCD astronomy that the statistical characteristics of the discrete nature of stellar counts allowed for the formulation of the surface brightness fluctuations (SBF) method [50, 51], one of the most reliable and stable extragalactic distance indicators used within the range of a few to about 150 Mpc.

As the distance to a stellar population increases, individual stars blend together and the brightness profile becomes smoother, but there remain statistical fluctuations on small scales because of the discrete nature of the stars. The SBF method leverages the random nature of stellar counts and luminosities to measure a quantity that corresponds to the luminosity-weighted mean brightness of the stars in the stellar population. For evolved stellar systems, this is approximately equal to the mean brightness of the stars on the red giant branch (RGB).

The SBF method was initially developed to determine distances of relatively nearby elliptical galaxies observed from the ground. Over the past few decades, the technique has undergone improvements in the analysis tools and benefited from increasingly advanced astronomical facilities. Consequently, SBF distances are now derived for a wider class of objects beyond ellipticals: spiral bulges, dwarf galaxies, irregulars, and others. Moreover, the distance limit has increased by a factor of $\sim$7 from the original predictions, and will increase further in the coming years.

M. Cantiello (✉)
INAF-Astronomical Observatory of Abruzzo, Via Maggini Snc, 64020 Teramo, Italy
e-mail: michele.cantiello@inaf.it

J. P. Blakeslee (✉)
NSF's NOIRLab, Tucson, AZ 85719, USA
e-mail: john.blakeslee@noirlab.edu

E. Di Valentino and D. Brout (eds.), *The Hubble Constant Tension*, Springer Series in Astrophysics and Cosmology, https://doi.org/10.1007/978-981-99-0177-7_9

Quantitatively, what is measured in the SBF analysis is the ratio between the intrinsic brightness variance (in the absence of blurring) and the average surface brightness over a specific region of a galaxy. This ratio has the units of flux and it is equivalent to the ratio between the second and first moments of the stellar luminosity function within the region analyzed. Although the definition is simple, there are a number of intricacies in practice. These involve how to measure the SBF amplitudes in astronomical images, the optimal targets for the measurement, and how to calibrate SBF magnitudes, either empirically or using stellar population synthesis.

In the following sections, we provide a detailed description of the method and summarize the recent results, especially regarding the "Hubble tension." We conclude with a discussion of the future prospects for the method.

9.2 Characterization of the SBF Signal

The main conceptual difficulty related to SBF does not pertain to the measurement of fluctuations or their calibration, which we discuss below. Rather, a cause of confusion, for some, concerns the nature of the indicator itself. Unlike other commonly used distance indicators like parallaxes, Cepheid variables, and Type Ia supernovae, the SBF signal is not easily "visualized", and it is not immediately apparent how fluctuations in star counts can be converted into a reliable estimate of distance.

To grasp how the SBF signal is produced and used for distance estimation, we can consider a simplified model of a CCD image of a stellar population unblurred by a point spread function (PSF). Let all the stars have the same (unknown) luminosity L_*, with n_* being the average number of stars per pixel and d the distance. Then, $f_* = L_*/(4\pi d^2)$ is the flux contributed per star. We cannot directly measure f_*, but we can measure the surface brightness F_* as the mean flux per pixel, $F_* = n_* \times f_*$. Since $n_* \propto d^2$ and $f_* \propto 1/d^2$, F_* is independent of the distance of the stellar system.

We can also measure the standard deviation in the flux per pixel. Because of Poisson fluctuations in the numbers of stars, the standard deviation is $\sigma_F = \sqrt{n_*} f_*$. Now, we define the SBF flux as the ratio of the variance σ_F^2 to the mean:

$$\bar{f} = \sigma_F^2 / F_* = n_* f_*^2 / (n_* f_*) = f_* = L_*/(4\pi d^2). \tag{9.1}$$

Thus, in this trivial case where the stars are all identical, the SBF flux reduces to the mean flux per star, which depends inversely on the square of the distance.

In reality, stars follow a luminosity function, and a more realistic treatment [39] shows that the ratio of the variance to the mean surface brightness reduces to:

$$\bar{f} = \frac{\sum_i n_i f_i^2}{\sum_i n_i f_i} \equiv \frac{\bar{L}}{4\pi d^2}, \tag{9.2}$$

where n_i is the number of stars of apparent brightness f_i, and $\bar{L}$ is the ratio of the second and first moments of the stellar luminosity function, corresponding to the luminosity-weighted mean stellar luminosity of the population. For a more detailed derivation of $\bar{f}$, see [18] or Sect. 3.9.1 of [39]. $\bar{L}$ is easily calculated from stellar population models; because of the luminosity weighting, it is sensitive to the brightest stars in the population. Thus, the SBF flux $\bar{f}$, or the equivalent SBF magnitude $\overline{m}$, is solely determined by the stellar luminosity function and the galaxy distance.

9.3 From SBF Magnitudes to Distances

Similar to other standard candle methods, deriving SBF distances requires measuring an apparent magnitude, $\overline{m}$, and then adopting an absolute $\overline{M}$ based on a calibration relation to obtain the distance modulus, $\overline{m}-\overline{M}$. The following sections describe the process for measuring $\overline{m}$ and estimating $\overline{M}$, as well as the characteristics of suitable target galaxies and the preferred passbands for reliable SBF measurements.

9.3.1 SBF Measurements Step by Step

Accurate SBF measurements require correcting and accounting for the different sources of noise or contamination that afflict the signal. In a sense, the fluctuation signal itself is a sort of noise in the star counts per pixel; in the absence of detector noise, photon shot noise, contamination from background sources, and blurring by the PSF, the SBF would simply be the Poisson variance among pixels due to the varying number and luminosity of stars in each pixel, normalized to the local mean flux of the galaxy. However, all of these effects are present and must be dealt with.

Since the SBF signal is convolved with the PSF, it can be distinguished from photon and detector noise, which have white-noise power spectra, at least prior to distortion correction. For this reason, the analysis is done in the Fourier domain, where the SBF signal is measured by determining the amplitude of the variance component in the image power spectrum on the scale of the PSF. Numerous papers have described the steps involved in SBF measurements [6, 7, 14, 17, 28, 30, 38, 39, 51]. In general, the procedure can be summarized as follows.

After combining individual dithered exposures to make a cleaned, stacked image free from cosmic rays, satellite trails, and detector defects, we estimate and subtract the sky background, mask any bright stars and neighboring galaxies, then derive a smooth isophotal model of the galaxy surface brightness. This model is then subtracted from the image, the remaining point-like and extended sources are masked, and a low-order fit to the background is derived and subtracted to eliminate large-scale residuals from the galaxy model subtraction. This masking, modeling and subtraction procedure is typically iterative. We then detect and mask all compact objects (foreground stars, globular clusters in the galaxy, faint background galaxies) down to a signal-to-noise threshold of ~ 4.5 using an automatic photometry program. These objects, along with any other sources of non-SBF variance (visible dust, brighter satellite galaxies, tidal features, regions of poor model residuals, etc.) are masked after the smooth model and large-scale residuals are removed.

The residual masked frame is then normalized by the square root of the model and contains both the variance from the stellar fluctuations of interest as well as fluctuations from unexcised sources fainter than the detection limit, along with the unconvolved photon and detector noise. The next step is to analyze the power spectrum of the normalized residual masked frame. The fluctuations from the stellar counts and unexcised sources are convolved with the instrumental PSF in the spatial domain; hence, they are multiplied by the Fourier transform of the PSF in the Fourier domain. For accurate SBF measurements, it is crucial to have an accurate characterization of the local PSF. This can be achieved using individual point sources in the field or by utilizing PSF modeling and reconstruction methods.

After azimuthally averaging the power spectrum,[1] the amplitude P_0 of the astrophysical component of the normalized and masked residual frame's power spectrum $P(k)$ is obtained as the constant term that multiples the "expectation power spectrum" $E(k)$, which is calculated as: $E(k) = \widetilde{\mathrm{PSF}}(k) \otimes \widetilde{\mathrm{M}}(k)$, where $\widetilde{\mathrm{M}}(k)$ is the power spectrum of the normalized PSF, and $\widetilde{\mathrm{PSF}}(k)$ is the power spectrum of the mask window function. There is also the white noise (approximately independent of wavenumber k) component P_1; thus the power spectrum is modeled as: $P(k) = P_0 \times E(k) + P_1$.

If the large-scale light distribution is well-subtracted, the power spectrum will be well represented by these two components over most the k range. Additional power may exist at low k (largest spatial scales) due to imperfect galaxy subtraction. The power may also be suppressed at some values of k due to correlation of the pixel noise resulting from geometric distortion correction; this is especially problematic when using linear interpolation of the pixel values. However, for sinc-like interpolation kernels [14, 38], the "damage" to the power spectrum is confined to high k (the smallest scales). In this case, the highest and lowest wavenumbers can simply be omitted from the fit range to ensure an accurate measurement.

It is crucial to note that the fitted P_0 value includes contamination from unexcised sources fainter than the detection limit. Any astrophysical source of fluctuation that is not generated from the stellar counts must be estimated and removed by subtracting the contribution from P_0. The presence of dust in some galaxies may also affect the power spectrum, and hence the fitted P_0, but in ways that depend on the distribution and are not easy to quantify. Optical images and color maps can be used to identify and mask any dusty regions. Typically the SBF measurement is confined to "clean regions," and we assume all dust patches are identified and masked; thus, no further contribution is considered in P_0.

In most cases, the globular clusters (GCs) within the target galaxy are the main source of contamination for P_0, with faint background galaxies being a lesser contaminant. We estimate the extra power from these sources by fitting, down to the

[1] The power spectrum is by definition the squared modulus of the Fourier transform; this explains the normalization of the residual frame by the square root of the galaxy model.

detection limit, a combined luminosity function that includes a radially-dependent Gaussian GC luminosity function (GCLF; e.g., [26]) and a power-law LF for the galaxies. We then estimate the "residual power" P_r from sources fainter than this limit (which varies with radius) by extrapolating the fitted combined LF, multiplied by the square of the source flux, from the radially-dependent detection limit down to zero flux (see [10]). This residual power is then subtracted from P_0 to obtain the intrinsic stellar fluctuations, typically denoted $P_f = P_0 - P_r$.

Because the GCs have a radial density distribution that increases toward the galaxy center, while the detection limit also gets brighter near the center, the fluctuations from GCs in the central region can overwhelm the signal from the stellar fluctuations, especially for massive galaxies with rich GC systems. Space-based observations (or high-resolution ground-based data obtained with adaptive optics [AO]) can greatly improve the distance limit for SBF by reducing the effect of crowding and pushing the detection limit fainter for the GCs. The surface density of background galaxies is typically lower and relatively uniform over the area, which means that their impact is less dramatic and is easier to robustly constrain.

Finally, the SBF magnitude is then calculated as $\overline{m} = -2.5 \log(P_0 - P_r) + m_{ZP}$, where m_{ZP} is the zero point magnitude. For a schematic illustration of the intricate SBF analysis procedure, see Fig. 9.1.

9.3.2 SBF Calibration

To convert $\overline{m}$ into a distance, knowledge of the absolute SBF magnitude $\overline{M}$ of the population is required. The most common method for estimating $\overline{M}$ is through empirical relationships. Studies of large samples of galaxies at known distances or in compact groups have shown that $\overline{M}$ can be described as a one-parameter family with integrated color (e.g., $V-I$ or $g-z$) as a useful parameter. The ground-based SBF survey of Tonry and collaborators [7, 52–54] found that a linear relation describes accurately the color dependence of $\overline{m}_I$ in nearby groups and clusters. They also showed that $\overline{M}$ in the I band is a universal function of color, with an intrinsic scatter in the range of 0.05 to 0.10 mag. Later studies confirmed a similar intrinsic scatter in the calibration relation for red galaxies in suitable passbands [9, 28], although for high-quality SBF measurements over a large color range, the $\overline{M}$-vs-color relation is more accurately described by a higher-order polynomial curve [9].

Since the introduction of SBF, many empirical calibrations of $\overline{M}$ as a function of various optical and near-IR colors have been presented [6, 11, 28, 37]. However, a major drawback of empirical approaches is the need to establish a new calibration whenever the reference passband for SBF or the reference color of the observing campaign changes. An alternative approach is to use SBF magnitude versus color relations derived from stellar population synthesis models [11, 16, 56]. The advantage of theoretical calibrations is that $\overline{M}$ can easily be computed for any passband and reference color. Furthermore, fully theoretical calibrations are independent of other distance indicators, avoiding systematic uncertainties associated with the primary

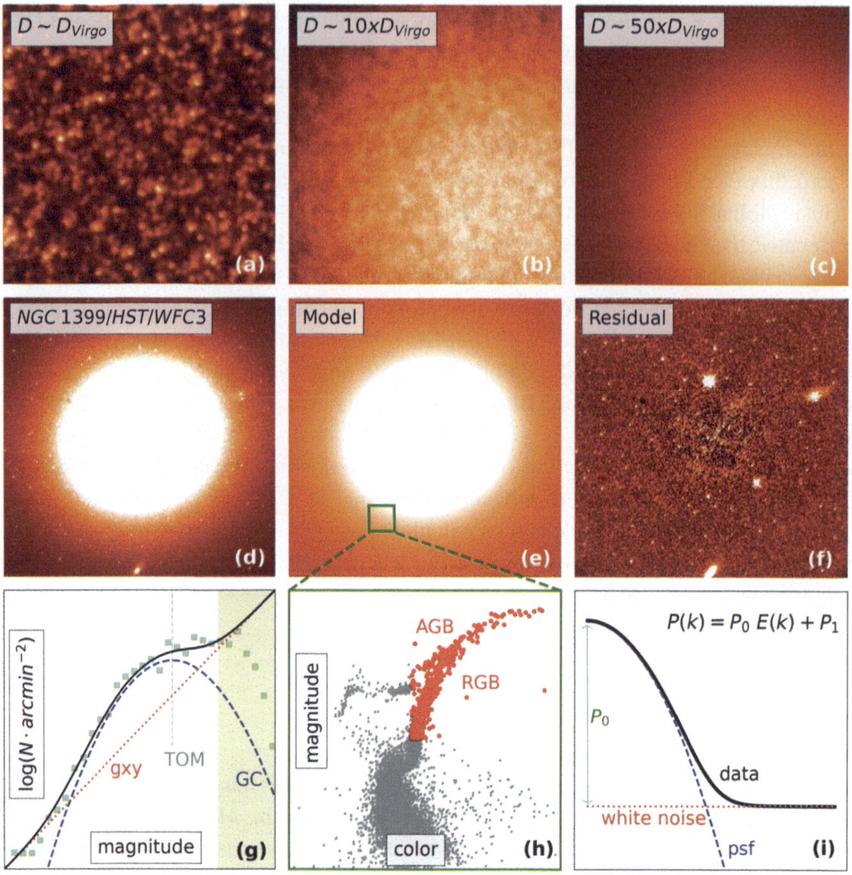

Fig. 9.1 Illustration of SBF observations and measurements. (*a*) Simulation of the stellar population in a spheroidal galaxy at the distance of the Virgo cluster ($D_{Virgo} \simeq 16.5$ Mpc [9]) as observed with the E-ELT in ~1 h. (*b*) Same as in panel (*a*), but for a galaxy ten times more distant. (*c*) Same as in panel (*a*), but for a galaxy fifty times more distant. Stars, which appear marginally resolved in panel (*a*), blend together into a smooth brightness profile at larger distances. (*d*) Near-infrared image of NGC 1399 from the HST WFC3 camera. (*e*) Model of NGC 1399s surface brightness distribution derived from the WFC3/IR image. (*f*) Residual frame, obtained from the galaxy image (*d*) minus the model (*e*). (*g*) Typical luminosity function analysis for estimating the "residual variance" P_r due to contaminating sources: green squares show the data, the blue curve and red line show the fits to the globular cluster and background galaxy luminosity functions, respectively, and the solid black line is the combined model luminosity function (data and fits are from [12]). The vertical gray dashed line indicates the GCLF turnover magnitude and the shaded area shows the magnitude interval where the detection is incomplete. (*h*) Color-magnitude diagram of an old stellar population (data for the MW globular cluster NGC 1851 [43]); the RGB/AGB population is highlighted with red dots. (*f*) A schematic illustration of the SBF power spectrum analysis. Images reproduced with permission from [39, 43]

indicator. Consequently, theoretically calibrated SBF can be regarded as a primary (though not geometric) distance indicator [3].

There is generally good agreement in SBF predictions from independent stellar population synthesis models in red optical bands. Discrepancies arise when comparing SBF magnitudes predicted by different models in bluer passbands or in the near-IR, where the models are not yet as robust [28]. These discrepancies can exceed ~0.2 mag. Such a difference is considerably larger than the empirically derived scatter, rendering stellar population model predictions unreliable as an alternative to empirical $\overline{M}$. Therefore, except for a few notable cases [2], numerically derived $\overline{M}$ vs color relations serve mainly as validation tools for empirical calibrations.

Before concluding this subsection, it is important to emphasize that accurate distance estimation using $\overline{M}$ calibrations requires precise galaxy colors. Hence, a reliable photometric calibration is crucial to ensure consistency and precision in the distances derived.

9.3.3 Reference Targets and Passbands

Owing to the definition of SBF and the measurement procedures outlined above, the ideal target galaxy for fluctuation measurements should have minimal dust, a brightness profile that is sufficiently regular, and a relatively homogeneous population of stars with a constant or at least not highly variable stellar luminosity function across its isophotes. This ensures a relatively constant SBF signal across the regions of the galaxy where the measurement is carried out, reducing the intrinsic variation of $\overline{m}$ and resulting in a more accurate distance estimate with lower errors.

In short, the ideal SBF targets are passively evolving, massive red ellipticals that are free from dust contamination and have a well-mixed population of stars with no recent star formation. Empirical calibrations have shown that such galaxies exhibit a tight correlation between absolute SBF magnitude and optical or near-IR colors, enabling SBF distances with a contribution from intrinsic scatter as low as 0.06 mag in certain passbands[2] [6, 9, 17]. Stellar population models also support the narrow scatter of $\overline{M}$ for colors typical of "red and dead" galaxies with stellar populations dominated by an old, metal-rich component [11, 16].

While the SBF method was originally developed for bright, morphologically regular early-type galaxies, it has more recently been used for a wider variety of stellar systems, including bulges of spirals and the low mass ultra-diffuse galaxies [4, 20]. In contrast to massive red ellipticals, bluer galaxies with intermediate masses have properties consistent with younger ages, lower metallicities, or both. Galaxies in this color regime have more uncertainties from the age-metallicity degeneracy,[3] which

[2] The final distance errors are larger due to measurement uncertainties, as discussed in Sec. 9.4.

[3] The age-metallicity degeneracy refers to the effect of stellar populations of a specific age and metallicity displaying similar colors, SBF magnitudes, and certain spectral indices as populations with a younger age and higher metallicity, or vice versa [57].

can result in targets with different stellar populations having the same integrated color. However, having the same integrated colors (which are mostly dominated by main sequence stars) does not always mean the same SBF absolute magnitude. This effect implies a larger scatter in the $\overline{M}$-vs-color relations as compared to red massive galaxies, leading to larger scatter associated with the derived distances.

Additionally, galaxies with intermediate colors often host an intermediate-age stellar population (3–7 Gyr), which includes many asymptotic giant branch (AGB) stars. These stars are in a bright and rapidly evolving phase, hence their stochastic appearance can make the SBF signal less stable, especially in near-IR bands [45]. Although AGBs have little impact on integrated colors they represent an additional source of scatter to the SBF in these galaxies. As a result of degeneracy effects and AGB stars, in some bands intermediate-mass targets suffer a larger scatter in SBF amplitude at fixed color, which makes it challenging to measure individual distances with an accuracy similar to the red-massive galaxies.

The accuracy of the SBF technique further drops for bluer, low-mass, low surface brightness galaxies. This is partly because of increased scatter from stellar population effects, including stochastic variations in the AGB stars. Additionally, measuring SBF in diffuse blue galaxies is challenging because low stellar densities make it harder to achieve an adequate signal-to-noise for the stellar-count fluctuations. As a result, a precision of approximately 15% should be considered the norm for such galaxies under optimal conditions [25].

Despite these challenges, the development of large-format imagers and wide-field surveys on 4m-class telescopes has increased the interest in SBF as a tool for determining distances to smaller galaxies. Recent studies have applied the SBF technique as a distance indicator for galaxies with surface brightness levels as low as $\mu_r \sim 24$ mag/arcsec^2 [4, 17, 25, 33]. In these cases, the technique is typically not exploited for its accuracy but rather because these faint and diffuse galaxies do not offer many reliable alternatives for measuring distances. Despite the larger uncertainty, the SBF technique represents a unique opportunity to confirm galaxy associations and group or cluster membership for candidate satellite galaxies.

Concerning the preferred passbands, SBF distances have been measured in various filters, ranging from visible to near-infrared wavelengths. Shorter wavelength bands are generally avoided because the SBF magnitudes are much fainter and more susceptible to contamination from GCs and dust. Nevertheless, SBF data in bluer bands can be valuable tracers for studying unresolved stellar populations due to their sensitivity to stellar population properties [11, 16, 47, 56]. Redder optical (I and z) and nearer IR (YJH) bands are instead preferred for distance measurements.

Empirical studies have shown that the intrinsic scatter of $\overline{M}$ at a fixed galaxy color is around ~ 0.06 mag [6, 9] for the most massive ellipticals in optical bands. In the near-IR bands, the intrinsic dispersion of $\overline{M}$ is closer to 0.10 mag due to the larger variance arising mainly from bright AGB stars [15, 28]. Despite the somewhat larger instrinsic scatter, near-IR bands such as J and H have several advantages. The SBF is intrinsically brighter, up to three magnitudes brighter in K than I, although the intrinsic scatter in K is significant and not well characterized. Near-IR bands also offer a much more favorable contrast with respect to GCs. Reliable $\overline{m}$ measurements need to mask GCs to at least ~ 0.5 mag fainter than the GCLF peak in the I band, while contamination can be reduced to the same level in the K band by reaching ~ 2 mag *brighter* than the GCLF peak. Finally, the impact of residual dust contamination in the near-IR is negligible in most cases [36].

The region around $\sim 1\ \mu m$ (Y band) is particularly interesting for SBF. According to independent stellar population models, and as expected from the inversion of the slope of the empirical SBF vs color relation between optical and near-IR bands for intermediate-color evolved galaxies, this wavelength should exhibit nearly complete degeneracy with respect to stellar population: $\overline{m}$ remains constant with color, making it an ideal distance indicator [16, 56]. In practice, this does not hold for the reddest, most massive ellipticals, for the which the SBF continues to get fainter with color, but these galaxies are comparatively rare. For future wide-area surveys the $\sim Y$ band may be the best choice for SBF measurements, although further study is needed.

In summary, there is not a definitive spectral "sweet spot" for measuring SBF, and any passband from ~ 0.8 to $\sim 2\ \mu m$ has its pros and cons. With the rapid development of near-IR technologies and the significant investment in future astronomical facilities in these passbands (e.g., Roman, the ELTs, and the recently launched JWST), along with the brighter intrinsic luminosity and negligible effect of dust contamination, it seems likely that the future of the SBF method, in terms of distance/depth and number of targets reached, lies primarily in the near-IR.

9.4 Uncertainties

9.4.1 Statistical Uncertainties

Statistical uncertainties affecting SBF can be grouped into three main categories: $\overline{m}$ measurement errors, uncertainty in $\overline{M}$ due to photometric errors in the galaxy color and intrinsic stellar-population scatter in the calibration relation, and uncertainties from data processing and calibration. The last category includes effects like flat fielding and uncertainty in the photometric calibration. While not unique to SBF, it is important to note their presence since they can impact various parts of the SBF measurement process in non-trivial ways. The uncertainty in the flat-fielding multiplicative factor affects the $\overline{m}$ measurement by affecting both the variance and average term in Eq. 9.2 and the color used in the calibration equation. Any sources of uncertainty in the reduction stage impact in a relatively intertwined way the final uncertainty on the distances from SBF. However, these uncertainties are typically well-controlled with standard data reduction and calibration procedures, with a typical level of $\sim 1\%$. Regarding SBF measurement uncertainties, these are directly tied to the specific steps required for the measurement. Usually, each error term introduced during any of the various analysis stages described in Sect. 9.3.1, can be independently estimated and then combined in quadrature with the others. The most relevant error terms are the following.

- *Sky background.* The process of sky subtraction has an associated error due to spatial and/or temporal variations in the sky emission (especially in the near-IR, where the sky background varies on the time scale of the observations), scattered light, and the extended nature of the galaxy, especially if the field of view is comparable to the galaxy size. The effect of the sky uncertainty increases significantly in the outer parts of the galaxy and for galaxies of lower surface brightness. To determine the final error associated to the sky uncertainty, the entire SBF measurement may be repeated by offsetting the background by the uncertainty level; this alters the final $\overline{m}$, and the size of the change is then taken into account for the final uncertainty. The sky uncertainty typically contributes ~ 0.02 mag of the total error on $\overline{m}$ in the near-IR, and less than that in optical bands.
- *Point Spread Function.* Accurate characterization of the PSF is essential for a reliable SBF measurement, and this depends on the number of bright, unsaturated stars in the field. Thus, for a given galaxy, the quality of the PSF template depends on the size of the field, stability of the PSF across the field, signal-to-noise, saturation level, etc. Once a reliable set of reference stars is identified, the SBF measurement is repeated using each available PSF and possibly a composite PSF model. The standard deviation of these measurements, typically less than ~ 0.03 mag, is then used as the PSF-fitting uncertainty.
- *Power Spectrum Fitting.* The accuracy of the power spectrum fit is affected by several factors, such as the number of good pixels in the annulus, the spatial structure of the mask, and the presence of patterns in the residual image caused by bars, shells, tidal interaction features or spiral arms. Low- and high-wavenumber power excesses or deficits can be filtered out during the power spectrum fit, yet the range of useful k-numbers needs to be large enough for reliable fitting. The fitting of the PSF power spectrum to the galaxy-subtracted data involves a statistical fit uncertainty of $\sim$0.02 mag.
- *Residual Variance P_r.* Various approaches have been used to determine the uncertainty in correcting for variance from GCs and background galaxies that are too faint for direct detection. One approach is to examine the change in P_r for different LF parameters and source detection thresholds. Another involves using different, independent detection and photometry tools (SExtractor, DAOphot, DoPHOT, etc.). Based on a variety of tests, a typical uncertainty of about 25% is adopted for the P_r estimate; thus, to ensure the contribution to the total error on $\overline{m}$ from the P_r correction is < 0.05 mag requires that $P_r/P_f \sim 0.2$. To keep this error term at an acceptable level, it is essential for the images to have sufficient depth; this is vastly easier to achieve with the superior resolution and reduced background of space-based observations.
- *Extinction.* Another component in the error budget is the uncertainty in the extinction correction, typically about 10% of the extinction correction itself. This is added in quadrature to the other sources of statistical error.

As for uncertainty in the adopted value of $\overline{M}$ for a given galaxy, apart from the systematic zero-point error from the distance calibration discussed in the next section, the statistical error in $\overline{M}$ includes terms arising from the error in the integrated color of the galaxy (needed to estimate $\overline{M}$ from the calibration relation) and from the intrinsic, irreducible scatter in the calibration equation. The latter effect results from stellar population variations and varies with the bandpass.

Overall, SBF measurement errors for well designed observations can achieve *statistical* uncertainties of ~ 0.05 mag in $\overline{m}$. For a red galaxy observed in a filter close to 1 μm, the intrinsic scatter in $\overline{M}$ is ~ 0.06 mag. Hence, when combined, these errors result in a purely statistical uncertainty that can be as low as 0.08 mag, or 4% on the distance. Nonetheless, fainter, blue galaxies and sub-optimal observing conditions will have larger statistical uncertainties.

9.4.2 Systematic Uncertainties

The main systematic uncertainty in SBF distances is in the $\overline{M}$ zero point[4]. Empirical calibrations of $\overline{M}$ are typically based on the zero point from the SBF ground-based survey by Tonry and colleagues, which obtained SBF for bulges of spiral galaxies with distance estimates from Cepheids [54]. A recent reassessment of this calibration reports a total systematic uncertainty of 0.09 mag in the Cepheid-based SBF distance zero point [8]. This includes contributions from the tie between near-IR and optical SBF distances, the tie between SBF and Cepheids, and the current uncertainty in the Cepheid zero point, taking into account the improved distance to the Large Magellanic Cloud from detached eclipsing binaries [42].

The largest source of systematic uncertainty in this calibration is the tie between SBF and Cepheids. This can be improved to a limited degree with further refinements in Cepheid luminosities with Gaia parallaxes (e.g., [19]) and by using more modern Cepheid distances, including a recently revised distance to M31 (one of the SBF calibrators) with a precision better than 2% [35]. New space-based SBF measurements for all the calibrator galaxies would also help. However, it is important to note that the ideal galaxies for measuring SBF (red ellipticals, Sect. 9.3.3), will never have Cepheid distances because they have no recent star formation.

An alternative to calibrating SBF from Cepheids is to use the tip of the red giant branch (TRGB) method. The TRGB method is ideal for measuring distances to early-type galaxies and can be calibrated using geometric distances from Gaia (e.g., [21, 34]). Unlike the case for Cepheids, the stellar populations underlying both the SBF and TRGB methods are the same, namely old low-mass stars. Given this, TRGB distances are the natural choice for calibrating SBF. In an early effort to calibrate SBF using TRGB, [40] examined a set of 16 galaxies within 10 Mpc. These were predominantly blue, low-mass early-type galaxies, and thus the scatter was large, although the zero point was consistent with the Cepheid calibration.

---

[4] As future facilities become capable of reaching distances well beyond 100 Mpc, SBF k-corrections that take the galaxy spectrum redshift properly into account may be necessary to avoid introducing further systematic effects [29]. As of now, the 100 Mpc limit is usually not surpassed, so the correction does not pose any problems when addressing systematic uncertainties.

A high-quality TRGB-based SBF calibration must be based on massive red early-type galaxies. Recently, [8] compared SBF and TRGB distances to the few such galaxies for which both types of measurment exist, including the bright ellipticals M 60 and M 87 in Virgo, and the dusty merger remnant NGC 1316 in Fornax. The authors found that the mean offset between the TRGB and Cepheid-calibrated SBF distances of the two galaxies in Virgo was only −0.01 mag. NGC 1316, a far less ideal galaxy, had a larger offset of 0.14 mag. Although preliminary, this result shows no significant difference between Cepheid- and TRGB-based SBF calibrations.

With the launch of JWST, the prospects have never been better for improving the SBF zero-point calibration using only modest amounts of observing time. A recently approved Cycle 2 program will measure TRGB distances and SBF magnitudes for 14 luminous, red early-type galaxies reaching out to 20 Mpc. The sample is designed to produce an absolute calibration for the SBF method of better than 2%. Further reduction in the zero-point uncertainty can be achieved by combining the TRGB- and Cepheid-based calibrations.

In summary, the current systematic uncertainty on SBF distances using the Cepheid-based calibration is 0.09 mag or about 4.2%. This value could be improved by obtaining new SBF measurements for the bulges of spiral galaxies that host Cepheids, but this approach will never be ideal because the best targets for the SBF method are massive red ellipticals. The preferred route for substantially reducing the systematic uncertainty in $\overline{M}$ is via the the TRGB method, which relies on the same underlying stellar population as the SBF method. A new JWST program, measuring both the TRGB and SBF for a well-selected set of early-type calibrators, should soon decrease the systematic uncertainty in SBF distances to below 2%. Further improvements can be achieved with additional TRGB calibrators and/or combining the TRGB and Cepheid calibrations.

9.5 Hubble-Lemaitre Constant

9.5.1 Current Constraints from SBF

Since its introduction for measuring extragalactic distances, the SBF method has been applied to about 400 galaxies. The random uncertainty associated with these distances varies greatly, from ∼4% for space-based observations of massive, red galaxies, to ∼20% for dwarf galaxies observed from the ground. The distances obtained have been used for a variety of scientific cases. For example, SBF has been used to measure distances to:

- NGC 4993, the lenticular galaxy host of the gravitational wave event GW 170817 which had the first observed electromagnetic counterpart [15];
- M 87, the cD host galaxy of the first supermassive black hole to have its "shadow" imaged using the EHT [23];

- hundreds of other member galaxies in the Virgo and Fornax clusters to explore the substructure in these environments [9, 13, 22, 37];
- numerous ultra-diffuse galaxies [20], including NGC 1052-DF2, the galaxy proposed to be devoid of dark matter [4, 55].

In addition, as the method reaches into the Hubble flow, many authors have used this technique for the estimation of the Hubble constant.

The first study to estimate H_0 directly from SBF distances decisively in the Hubble flow used a sample of 15 galaxies from 40 to 130 Mpc observed with NICMOS on HST and obtained $H_0 \approx 76$ km s^{-1} Mpc^{-1} (dropping to ~ 72 when restricted to the six most distant targets).[5] Another early direct measurement of H_0 from SBF used archival HST/ACS observations combined with a theoretical calibration based on population models and likewise found $H_0 = 76$ km s^{-1} Mpc^{-1} [2].

Among recent results, we mention three studies that are closely related to each other. The first study, by [32] (hereafter K21), presented a re-calibration of the peak magnitude for a sample of 24 local Type Ia supernovae (SNe Ia) using SBF distances and a hierarchical Bayesian approach. The authors then extended the calibration to a sample of 96 SNeIa at redshifts $0.02 < z < 0.08$ and obtained $H_0 = 71.2 \pm 2.4$ (stat) $\pm$ 3.4 (sys) km s^{-1} Mpc^{-1}, where we have updated the value to use the improved LMC distance from [42] for consistency with the other recent H_0 values discussed here. K21 used a diverse sample of SBF distances, which were collected from observations over a span of twenty years, along with an equally diverse sample of SNeIa. While normalizing the SBF distance sample to a common reference calibration was difficult, the heterogeneity of the sample provided some benefits, as it should in principle be less prone to systematic effects, such as those related to specific instruments or data analysis procedures. However, this also resulted in large final statistical uncertainties. The value of H_0 reported by K21 was roughly midway between those found directly from Cepheid-calibrated SNe Ia [46] and predicted from analysis of the cosmic microwave background (CMB) [44].

In another recent study, Blakeslee et al. [8] (hereafter B21) used the catalog of SBF distances from Jensen et al. [29] to estimate H_0. This sample comprises 63 bright early-type galaxies observed with the WFC3/IR instrument on the HST, reaching distances up to 100 Mpc. The resulting value of $H_0 = 73.3 \pm 0.7$ (stat) $\pm$ 2.4 (sys) km s^{-1} Mpc^{-1} agrees well with most other direct measurements in the local universe, but deviates significantly (by about 2.5 σ, including the systematic uncertainty) from the CMB prediction, assuming the standard ΛCDM model. It is noteworthy that the random uncertainty is about one third of that reported in K21.

To ensure the reliability of their findings, B21 conducted an analysis of H_0 involving four different treatments of galaxy velocities, including group-averaged CMB frame velocities, individual CMB velocities, and flow-corrected velocities from two different models. Additionally, their final H_0 combines the mutually consistent SBF calibrations from Cepheids and the TRGB in order to reduce the final systematic

---

[5] These numbers should be updated to the most recent Cepheids calibration, which has a minor impact on the final H_0 value.

error down to the $\sim 3.3\%$ level. As noted previously, the number of TRGB calibrators in this analysis was very small, and future studies incorporating a larger sample of giant ellipticals with TRGB distances measured by JWST and tied to Gaia parallaxes should vastly reduce this systematic error.

The third recent work [24] related to H_0 combines an approach like that of K21 with the WFC3/IR SBF distances from [29]. The authors of [24] used SBF distances for 25 host galaxies of SNe Ia to provide an absolute calibration for the Pantheon+ compilation [49] of 1500 SNe Ia extending far out into the Hubble flow. Using the original Pantheon+ "SALT2" parameters for the dependence of peak luminosity on decline rate and color, these authors find $H_0 = 74.6 \pm 0.9\,(\text{stat}) \pm 2.7\,(\text{syst})$ km s^{-1} Mpc^{-1}, which agrees to within $\sim 1\sigma$ with the value obtained by B21 (the systematic error is common between the two studies, so the comparison must be done with respect to the statistical error only).

However, it is important to note that the SNe Ia whose luminosities can be calibrated with SBF distances are generally fast decliners in massive host galaxies. When [24] limit their analysis to fast-declining SNe Ia in massive hosts and rederive the SALT2 parameters, the SBF-calibrated H_0 from these SNe becomes $73.3 \pm 1.0 \pm 2.7$ km s^{-1} Mpc^{-1}, which is identical to the value derived by B21 from SBF distances in the Hubble flow. This $\sim 1\sigma$ change in H_0 suggests a slight difference between the best-fit SALT2 parameters for the relatively small set of fast-declining SNe Ia in massive galaxies as compared to the full Pantheon+ sample.

9.5.2 Forecasts for New and Future Facilities

As outlined in the previous section, recent SBF-based estimates of H_0 align more closely with the results found by Cepheid-calibrated SNe Ia than with the predictions from the CMB. The potential use of SBF, particularly with JWST and upcoming astronomical facilities, offers a promising route to derive a more precise and accurate value of H_0 that is fully independent of SNe Ia and Cepheids. Here, we present the expected results of the SBF method for estimating H_0 using various existing and forthcoming facilities: JWST, Rubin Observatory, ESA's Euclid mission, the Roman Space Telescope, and the Extremely Large Telescopes (ELTs).

To evaluate the potential for constraining H_0 from SBF with each new or forthcoming facility, we proceed as follows. First, we estimate the maximum distance d_{Max} that can be reached with the specific facility, as explained below. Then, we count the number of massive elliptical galaxies that can be observed by the telescope/instrument within a distance range of $d_{min} \leq D \leq d_{Max}$. Here, we assume $d_{min} = 40$ Mpc to exclude galaxies in the local universe where flow motions and peculiar velocities are large compared to distance errors. To estimate d_{Max}, we assume (as is generally the case) that contamination by GCs is the limiting factor for well-designed SBF measurements. Thus, d_{Max} is the distance to which the GCs can be detected and removed to a faint enough limit, and the residual contamination reliably estimated, so that the error in the correction drops below the intrinsic scatter in the method. For instance, in the I band, where the peak of the GC luminosity function (GCLF) is at $M_I \approx -8.0$ AB mag, accurate SBF measurements require detecting and removing sources to +0.5 mag fainter than the GCLF peak, or $M_I \approx -7.5$ AB mag. In the H-band (which has slightly larger intrinsic scatter), the same relative amount of contamination from GCs can be reached for a detection limit about 1.5 mag *brighter* than the GCLF peak; [28]. We estimate the near-IR GCLF peak magnitudes from [41] and convert to the AB system.

The expected depth is determined based on available information for the specific facilities. For instance, in the case of Rubin Observatory's LSST, we use the expected 10-year 5σ point-source depth for the i band, with a GCLF peak of ~ -8.0 AB mag and sky coordinates with Dec(J2000) $\leq 15°$.

Another important factor is the selection of an appropriate list of targets. We use the 2MASS Redshift Survey (2MRS), which comprises approximately 43,000 galaxies brighter than $K_s = 11.75$ mag (Vega) [27]. This survey is nearly complete, with the exception of regions near the Galactic plane, which account for only 9% of the sky, and an L^* galaxy is included if it is within ~135 Mpc. We first focus on "optimal" targets, i.e., bright galaxies with a morphological T-type ≤ -4 (indicating morphologically regular ellipticals) and an absolute K-band Vega magnitude $M_K \leq -25$ mag (similar to M87), where distance is estimated from the 2MRS redshift assuming $H_0 = 75$ km s^{-1} Mpc^{-1}. Under these assumptions, Table 9.1 (column 7) reports the numbers of optimal targets reachable by each of the new and upcoming facilities mentioned above.

Now, if we relax the selection somewhat on the galaxy morphology and luminosity, accepting $T \leq -1$ and $M_K \leq -22.5$ mag as "good" candidates, then the numbers of SBF targets increase dramatically, as again shown in column 7 of the table. It is important to note that these numbers represent conservative lower limits. For reference, the NGVS survey on the CFHT, with a median i-band seeing of 0.″7 and 5σ depth three magnitudes fainter than the GCLF peak, enabled SBF measurements for about 300 galaxies in the Virgo cluster [13] down to $M_K \sim -16$ Vega mag. It was also possible to measure SBF in many galaxies with positive T-type, such as bulges of spirals and galaxies with some dust contamination, thanks to the availability of u-band data. Thus, the numbers of candidates measurable with the facilities specified in Table 9.1 could increase by a factor of a few, or even orders of magnitude, when fainter and more irregular targets are considered.

For estimating the constraints on H_0, we also need to adopt values for the expected distance errors. Thus, for ground-based observations in natural seeing (Rubin/LSST), we conservatively assume a total error in the SBF distance of 8% for the optimal target galaxies, and 16% for the "good" targets. For high-resolution space-based and AO-assisted observations, we assume total errors of 4% and 8% for the optimal and "good" targets, respectively.

All the necessary parameters required for estimating the number of reachable targets with the selected facilities are listed in Table 9.1, along with the estimated maximum distances and sample sizes for each facility in selected bands. Taking Rubin as an example, it is expected that approximately 50 optimal targets within $d_{Max} \approx 70$ Mpc will be measurable in the Baseline survey area, resulting in a constraint on H_0

of $\sim 1.2\%$ (stat). For "good" targets, the estimated number increases to 1500, and the statistical error drops well below the systematic uncertainties from the distance calibration and possible large-scale flow motions.

The Roman Space Telescope is especially promising for future constraints on H_0 using SBF because of its combination of HST-like resolution, excellent PSF stability, near-IR coverage, and planned large-area surveys [5]. While awaiting the Roman telescope, however, the JWST has unsurpassed capabilities for targeting galaxies one at a time out to ~ 300 Mpc, an unprecedented distance for SBF that exceeds the limit of claimed "Hubble bubbles" (e.g., [31, 48]). In the NIRCam passbands reported in Table 9.1, we find that around a thousand candidates from the 2MRS catalog will be optimal targets for JWST. Of course, only a few dozen are needed to bring the statistical error comfortably below 1%, so there will be no shortage of targets from which to choose.

To demonstrate the feasibility of SBF measurements with JWST/NIRCam, we conducted a preliminary analysis on NGC 7317, the elliptical galaxy in Stephan's Quintet at a distance of about 100 Mpc, using the public Early Release Data.

The characteristic SBF signal is clearly observed across all passbands, including mid-infrared bands that have not previously been used in SBF studies. The left panels of Fig. 9.2 display the measured power spectra in the F150W band of the

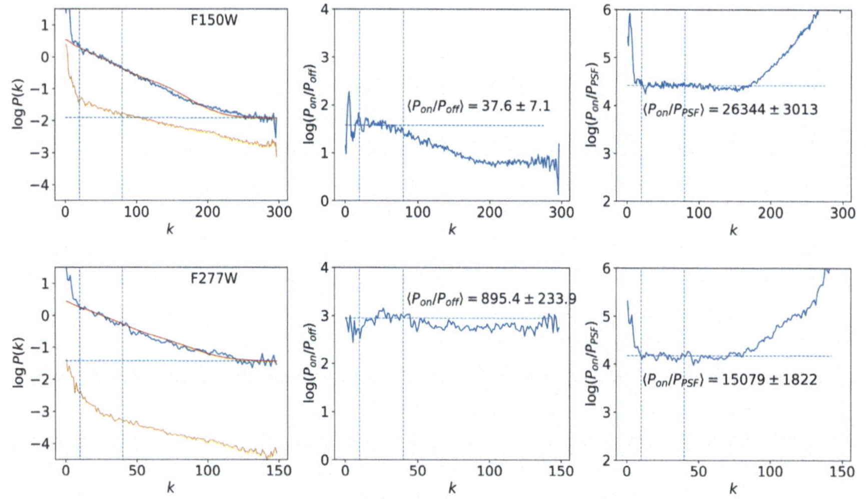

Fig. 9.2 Power spectrum analyses of JWST data on NGC 7317, an elliptical galaxy at ~ 100 Mpc, in the F150W (top row) and F277W (bottom row) passbands. Left panels: power spectra of the central masked region after galaxy subtraction (blue solid curve), the scaled PSF (red curve), and an empty region away from the galaxy (orange curve); dotted blue lines show the range of the starting waveneumbers k for the fits. Middle panels: ratio of the power spectra on- and off-galaxy; the structure in this ratio for F150W may indicate a variable PSF in the combined image. Right: Uncorrected power spectrum normalization P_0 derived from the fits in the left panels as a function of the starting k value (the fitted P_0 values here are insensitive to the starting k)

short-wavelength channel and the F277W band of the long-wavelength channel. The other panels show the ratio of the fluctuation amplitudes on- and off- the galaxy center and the P_0 coefficient of the PSF term in the power spectrum, using a star within the image as the reference PSF. Analysis of the SBF signal for this target is in progress, and further work is needed to improve the agreement with the PSF templates, but the signal is very strong, given the short exposure times for this distance.

With the forthcoming generation of wide-area imaging surveys such as Rubin/LSST, Euclid, and the Roman Community Surveys providing hundreds or even thousands of galaxies suitable for SBF measurements, it is essential that the SBF analysis should be carried out in a robust, streamlined, and automatic way. Historically, the limited number of available SBF measurement pipelines have been poorly automated and required extensive user intervention. This is especially true for the galaxy fitting and residual masking, as these steps must be optimized to ensure clean power spectra. However, it is clear that the method needs to be adapted to run in "production mode" to deliver robust results with minimal human intervention. Multiple parallel efforts are ongoing in this regard.

The prospects for SBF measurements with AO-assisted observations on 30–40m telescopes are also exciting. Table 9.1 reports the expectations for E-ELT/MICADO with the MORFEO AO module. With their extremely large apertures and diffraction-limited resolution $< 0\farcs02$, such facilities should be able to measure SBF distances out to redshift $z \sim 0.1$. However, measuring SBF with AO-assisted instruments may

Table 9.1 Maximum distances and numbers of candidates for SBF analysis

Telescope (constraint) (1)	Band (2)	Depth$^{(a)}$ (mag)	GCLF peak$^{(b)}$ (4)	GCLF offset$^{(c)}$ (5)	d_{Max} (6)	Optimal/good candidates (7)	$\delta H_0/H_0$ (%) optimal/good (8)
Rubin (Dec$\leq 15°$)	i	26.8	−8.5	+0.5	70	50/1500	1.2/<0.5
Euclid ($\lvert b\rvert \geq 23°$)	H_{Euclid}	24.0	−8.6	−1.5	55	20/600	1.0/<0.5
Roman$^{(e)}$	J_{F129}	28.0	−8.3	−1.0	230	1500/>5000	0.7/0.5$^{(g)}$
Roman$^{(e)}$	H_{F158}	28.0	−8.6	−1.5	300	1600/>5000	0.7/0.5$^{(g)}$
JWST$^{(e)}$	J_{F115W}	27.5	−8.3	−1.0	180	900/>5000	0.9/0.5$^{(g)}$
JWST$^{(e)}$	H_{F150W}	27.7	−8.6	−1.5	290	1600/>5000	0.7/0.5$^{(g)}$
JWST$^{(e)}$	K_{F200W}	27.8	−8.1	−2.0	310	1600/>5000	0.7/0.5$^{(g)}$
ELT$^{(f)}$ (Dec$\leq 0°$)	K	28.5	−8.1	−2.0	400	750/>4000	1.0/0.7$^{(g)}$

(a) Nominal 5σ point-source depth accounting for host galaxy background
(b) Peak, or turnover, magnitude (AB) of the GCLF in each band
(c) Offset of the required detection limit with respect to the GCLF peak (see text)
(d) Estimated number of optimal/suitable SBF targets based on 2MRS [27]
(e) 5σ depth in one hour integration time
(f) Depth derived from the preliminary ELT/MICADO exposure time calculator, assuming a $S/N = 5$ in one hour in multi-conjugate adaptive optics mode
(g) Assuming that a fraction of $\sim 2\%$ of possible candidates are indeed observed

be challenging because of the spatially and temporally varying PSF, significant overheads, and the requirement for accurate k-corrections at these distances [29]. Nevertheless, once the technique is optimized for 30m-class telescopes, a carefully selected set of targets could have a significant impact, potentially providing an independent direct detection of cosmic acceleration.

9.6 Conclusions

Over the last 30 years, the SBF method has become one of the most powerful methods for measuring galaxy distances from the Local Group to the Hubble flow and constraining the value of H_0. Unlike other precision distance indicators (Cepheids, TRGB, SN Ia, masers), SBF measurements require only a modest investment of telescope time, especially when using high-resolution space-based data. SBF distances to large numbers of galaxies observed with HST using WFC3/IR have already constrained H_0 to 3.4%, where the main source of uncertainty is the zero-point calibration. Now, with JWST/NIRCam, it becomes possible to directly calibrate SBF using TRGB distances to the same galaxies, opening a path to reduce the uncertainty on H_0 from SBF to below 2%. The large aperture, superior resolution, PSF stability, and near-infrared capability of JWST also make it possible to measure robust SBF distances out to $\sim$300 Mpc. This will reduce the uncertainty from possible large-scale bulk motions and vastly increase the number of optimal targets in range of the method. Other forthcoming facilities such as Euclid, Roman, Rubin, and the ELTs also hold great potential for the method. With an improved TRGB-based calibration and samples of hundreds, or even thousands, of SBF distances reaching out to hundreds of Mpc, a fully independent measurement of H_0 to a precision approaching 1% will be in reach by the end of the current decade.

References

1. W.A. Baum, M. Schwarzschild, A comparison of stellar populations in the Andromeda galaxy and its elliptical companion. AJ **60**, 247 (1955)
2. I. Biscardi, G. Raimondo, M. Cantiello, E. Brocato, Optical surface brightness fluctuations of shell galaxies toward 100 Mpc. Astrophys. J. **678**(1), 168–178 (2008)
3. J.P. Blakeslee, Surface brightness fluctuations as primary and secondary distance indicators. Ap&SS **341**, 179–186 (2012)
4. J.P. Blakeslee, M. Cantiello, Independent analysis of the distance to NGC 1052-DF2. Res. Notes Am. Astron. Soc. **2**(3), 146 (2018)
5. J.P. Blakeslee, M. Cantiello, M.J. Hudson, L. Ferrarese, N. Hazra, J.B. Jensen, E.W. Peng, G. Raimondo, Gathering galaxy distances in abundance with roman wide-area data (2023). arXiv e-prints: arXiv:2306.15170
6. J.P. Blakeslee, M. Cantiello, S. Mei, P. Côté, R. Barber DeGraaff, L. Ferrarese, A. Jordán, E.W. Peng, J.L. Tonry, G. Worthey, Surface brightness fluctuations in the hubble space telescope

ACS/WFC F814W bandpass and an update on galaxy distances. Astrophys. J. **724**(1), 657–668 (2010)
7. J.P. Blakeslee, M. Davis, J.L. Tonry, A. Dressler, E.A. Ajhar, A first comparison of the surface brightness fluctuation survey distances with the galaxy density field: implications for H_0 and Ω. Astrophys. J. **527**(2), L73–L76 (1999)
8. J.P. Blakeslee, J.B. Jensen, C.-P. Ma, P.A. Milne, J.E. Greene, The hubble constant from infrared surface brightness fluctuation distances. Astrophys. J. **911**(1), 65 (2021)
9. J.P. Blakeslee, A. Jordán, S. Mei, P. Côté, L. Ferrarese, L. Infante, E.W. Peng, J.L. Tonry, M.J. West, The ACS fornax cluster survey V measurement and recalibration of surface brightness fluctuations and a precise value of the Fornax-Virgo relative distance. Astrophys. J. **694**, 556–572 (2009)
10. J.P. Blakeslee, J.L. Tonry, Measurements of globular cluster specific frequencies and luminosity function widths in coma. Astrophys. J. **442**, 579 (1995)
11. J.P. Blakeslee, A. Vazdekis, E.A. Ajhar, Stellar populations and surface brightness fluctuations: new observations and models. MNRAS **320**, 193 (2001)
12. M. Cantiello, I. Biscardi, E. Brocato, G. Raimondo, VLT optical BVR observations of two bright supernova Ia hosts in the Virgo cluster-surface brightness fluctuation analysis. A&A **532**, A154 (2011)
13. M. Cantiello, J.P. Blakeslee, L. Ferrarese, P. Côté, J.C. Roediger, G. Raimondo, E.W. Peng, S. Gwyn, P.R. Durrell, J.C. Cuillandre, The next generation virgo cluster survey (NGVS) XVIII measurement and calibration of surface brightness fluctuation distances for bright galaxies in virgo (and beyond). Astrophys. J. **856**(2), 126 (2018)
14. M. Cantiello, J.P. Blakeslee, G. Raimondo, S. Mei, E. Brocato, M. Capaccioli, Detection of radial surface brightness fluctuations and color gradients in elliptical galaxies with the advanced camera for surveys. Astrophys. J. **634**, 239 (2005)
15. M. Cantiello, J.B. Jensen, J.P. Blakeslee, E. Berger, A.J. Levan, N.R. Tanvir, G. Raimondo, E. Brocato, K.D. Alexander, P.K. Blanchard, M. Branchesi, Z. Cano, R. Chornock, S. Covino, P.S. Cowperthwaite, P. D'Avanzo, T. Eftekhari, W. Fong, A.S. Fruchter, A. Grado, J. Hjorth, D.E. Holz, J.D. Lyman, I. Mandel, R. Margutti, M. Nicholl, V.A. Villar, P.K.G. Williams, A precise distance to the host galaxy of the binary neutron star merger GW170817 using surface brightness fluctuations. Astrophys. J. **854**(2), L31 (2018)
16. M. Cantiello, G. Raimondo, E. Brocato, M. Capaccioli, New optical and near-infrared surface brightness fluctuation models: a primary distance indicator ranging from globular clusters to distant galaxies? Astrophys. J. **125**, 2783 (2003)
17. S.G. Carlsten, R.L. Beaton, J.P. Greco, J.E. Greene, Using surface brightness fluctuations to study nearby satellite galaxy systems: calibration and methodology. Astrophys. J. **879**(1), 13 (2019)
18. M. Cerviño, V. Luridiana, L. Jamet, On surface brightness fluctuations: probabilistic and statistical bases I. Stellar population and theoretical surface brightness fluctuations. A&A **491**(3), 693–701 (2008)
19. G. Clementini, V. Ripepi, R. Molinaro, A. Garofalo, T. Muraveva, L. Rimoldini, L.P. Guy, G.J. de Fombelle, K. Nienartowicz, O. Marchal, M. Audard, B. Holl, S. Leccia, M. Marconi, I. Musella, N. Mowlavi, I. Lecoeur-Taibi, L. Eyer, J. De Ridder, S. Regibo, L.M. Sarro, L. Szabados, D.W. Evans, M. Riello, Gaia data release 2. specific characterisation and validation of all-sky Cepheids and RR Lyrae stars. A&A **622**, A60 (2019)
20. Y. Cohen, P. van Dokkum, S. Danieli, A.J. Romanowsky, R. Abraham, A. Merritt, J. Zhang, L. Mowla, J.M.D. Kruijssen, C. Conroy, A. Wasserman, The Dragonfly nearby galaxies survey V HST/ACS observations of 23 low surface brightness objects in the fields of NGC 1052, NGC 1084, M96, and NGC 4258. Astrophys. J. **868**(2), 96 (2018)
21. M. Dixon, J. Mould, C. Flynn, E.N. Taylor, C. Lidman, A.R. Duffy, A geometric calibration of the tip of the red giant branch in the Milky Way using Gaia DR3. MNRAS **523**(2), 2283–2295 (2023)
22. L.P. Dunn, H. Jerjen, First results from SAPAC: toward a three-dimensional picture of the Fornax cluster core. AJ **132**(3), 1384–1395 (2006)

23. Event Horizon Telescope Collaboration, K. Akiyama, A. Alberdi, W. Alef, K. Asada, R. Azulay, A.K. Baczko, D. Ball, M. Baloković, J. Barrett, D. Bintley, L. Blackburn, W. Boland, K.L. Bouman, G.C. Bower et al., First M87 event horizon telescope results VI the shadow and mass of the central black hole. Astrophys. J. **875**(1), L6 (2019)
24. P. Garnavich, C.M. Wood, P. Milne, J.B. Jensen, J.P. Blakeslee, P.J. Brown, D. Scolnic, B. Rose, D. Brout, Connecting infrared surface brightness fluctuation distances to type Ia supernova hosts: testing the top rung of the distance ladder. Astrophys. J. **953**(1), 35(2023). https://doi.org/10.3847/1538-4357/ace04b
25. J.P. Greco, P. van Dokkum, S. Danieli, S.G. Carlsten, C. Conroy, Measuring distances to low-luminosity galaxies using surface brightness fluctuations. Astrophys. J. **908**(1), 24 (2021)
26. W.E. Harris, Globular cluster systems, in *Saas-Fee Advanced Course 28: Star Clusters* (2001)
27. J.P. Huchra, L.M. Macri, K.L. Masters, T.H. Jarrett, P. Berlind, M. Calkins, A.C. Crook, R. Cutri, P. Erdoğdu, E. Falco, T. George, C.M. Hutcheson, O. Lahav, J. Mader, J.D. Mink, N. Martimbeau, S. Schneider, M. Skrutskie, S. Tokarz, M. Westover, The 2MASS redshift survey-description and data release. Astrophys. J. Supp. Ser. **199**(2), 26 (2012)
28. J.B. Jensen, J.P. Blakeslee, Z. Gibson, H.-C. Lee, M. Cantiello, G. Raimondo, N. Boyer, H. Cho, Measuring infrared surface brightness fluctuation distances with HST WFC3: calibration and advice. Astrophys. J. **808**, 91 (2015)
29. J.B. Jensen, J.P. Blakeslee, C.-P. Ma, P.A. Milne, P.J. Brown, M. Cantiello, P.M. Garnavich, J.E. Greene, J.R. Lucey, A. Phan, R.B. Tully, C.M. Wood, Infrared surface brightness fluctuation distances for MASSIVE and type Ia supernova host galaxies. Astrophys. J. Supp. Ser. **255**(2), 21 (2021)
30. J.B. Jensen, J.L. Tonry, G.A. Luppino, Measuring distances using infrared surface brightness fluctuations. Astrophys. J. **505**(1), 111–128 (1998)
31. R.C. Keenan, A.J. Barger, L.L. Cowie, Evidence for a $\sim$ 300 Megaparsec scale under-density in the local galaxy distribution. Astrophys. J. **775**(1), 62 (2013)
32. N. Khetan, L. Izzo, M. Branchesi, R. Wojtak, M. Cantiello, C. Murugeshan, A. Agnello, E. Cappellaro, M. Della Valle, C. Gall, J. Hjorth, S. Benetti, E. Brocato, J. Burke, D. Hiramatsu, D.A. Howell, L. Tomasella, S. Valenti, A new measurement of the Hubble constant using Type Ia supernovae calibrated with surface brightness fluctuations. A&A **647**, A72 (2021)
33. Y.J. Kim, M.G. Lee, Calibration of surface brightness fluctuations for Dwarf galaxies in the hyper suprime-cam gi filter system. Astrophys. J. **923**(2), 152 (2021)
34. S. Li, S. Casertano, A.G. Riess, A Gaia data release 3 view on the tip of the red giant branch luminosity. Astrophys. J. **950**(2), 83 (2023)
35. S. Li, A.G. Riess, M.P. Busch, S. Casertano, L.M. Macri, W. Yuan, A sub-2% distance to M31 from photometrically homogeneous near-infrared Cepheid period-luminosity relations measured with the Hubble space telescope. Astrophys. J. **920**(2), 84 (2021)
36. G.A. Luppino, J.L. Tonry, Infrared surface brightness fluctuations: K-band observations of M31, M32, and the distance to Maffei 1. Astrophys. J. **410**, 81 (1993)
37. S. Mei, J.P. Blakeslee, P. Côté, J.L. Tonry, M.J. West, L. Ferrarese, A. Jordán, E.W. Peng, A. Anthony, D. Merritt, The ACS Virgo cluster survey XIII. SBF distance catalog and the three-dimensional structure of the Virgo cluster. Astrophys. J. **655**(1), 144–162 (2007)
38. S. Mei, J.P. Blakeslee, J.L. Tonry, A. Jordán, E.W. Peng, P. Côté, L. Ferrarese, D. Merritt, M. Milosavljević, M.J. West, The ACS Virgo cluster survey IV. Data reduction procedures for surface brightness fluctuation measurements with the advanced camera for surveys. Astrophys. J. Supp. Ser. **156**, 113–125 (2005)
39. M. Moresco, L. Amati, L. Amendola, S. Birrer, J.P. Blakeslee, M. Cantiello, A. Cimatti, J. Darling, M. Della Valle, M. Fishbach, C. Grillo, N. Hamaus, D. Holz, L. Izzo, R. Jimenez, E. Lusso, M. Meneghetti, E. Piedipalumbo, A. Pisani, A. Pourtsidou, L. Pozzetti, M. Quartin, G. Risaliti, P. Rosati, L. Verde, Unveiling the universe with emerging cosmological probes. Living Rev. Relativ. **25**(1), 6 (2022)
40. J. Mould, S. Sakai, The extragalactic distance scale without Cepheids II surface brightness fluctuations. Astrophys. J. **694**(2), 1331–1334 (2009)

41. J.B. Nantais, J.P. Huchra, P. Barmby, K.A.G. Olsen, T.H. Jarrett, Nearby spiral globular cluster systems I: luminosity functions. AJ **131**(3), 1416–1425 (2006)
42. G. Pietrzyński, D. Graczyk, A. Gallenne, W. Gieren, I.B. Thompson, B. Pilecki, P. Karczmarek, M. Górski, K. Suchomska, M. Taormina, B. Zgirski, P. Wielgórski, Z. Kołaczkowski, P. Konorski, S. Villanova, N. Nardetto, P. Kervella, F. Bresolin, R.P. Kudritzki, J. Storm, R. Smolec, W. Narloch, A distance to the large Magellanic cloud that is precise to one per cent. Nature **567**(7747), 200–203 (2019)
43. G. Piotto, I.R. King, S.G. Djorgovski, C. Sosin, M. Zoccali, I. Saviane, F. De Angeli, M. Riello, A. Recio-Blanco, R.M. Rich, G. Meylan, A. Renzini, HST color-magnitude diagrams of 74 galactic globular clusters in the HST F439W and F555W bands. A&A **391**, 945–965 (2002)
44. Planck Collaboration, N. Aghanim, Y. Akrami, M. Ashdown, J. Aumont, C. Baccigalupi, M. Ballardini, A.J. Banday, R.B. Barreiro, N. Bartolo, S. Basak, R. Battye, K. Benabed, J.P. Bernard, M. Bersanelli, P. Bielewicz, J.J. Bock, J.R. Bond et al., Planck 2018 results VI cosmological parameters. A&A **641**, A6 (2020)
45. G. Raimondo, Joint analysis of near-infrared properties and surface brightness fluctuations of large Magellanic cloud star clusters. Astrophys. J. **700**(2), 1247–1261 (2009)
46. A.G. Riess, S. Casertano, W. Yuan, L.M. Macri, D. Scolnic, Large magellanic cloud Cepheid standards provide a 1% foundation for the determination of the Hubble constant and stronger evidence for physics beyond ΛCDM. Astrophys. J. **876**(1), 85 (2019)
47. P. Rodríguez-Beltrán, A. Vazdekis, M. Cerviño, M.A. Beasley, Surface brightness fluctuations to constrain secondary stellar populations: revealing very low-metallicity stars in massive galaxies. MNRAS **507**(2), 3005–3029 (2021)
48. A.E. Romano, Hubble trouble or Hubble bubble? Int. J. Mod. Phys. D **27**(9), 1850102 (2018)
49. D. Scolnic, D. Brout, A. Carr, A.G. Riess, T.M. Davis, A. Dwomoh, D.O. Jones, N. Ali, P. Charvu, R. Chen, E.R. Peterson, B. Popovic, B.M. Rose, C.M. Wood, P.J. Brown, K. Chambers, D.A. Coulter, K.G. Dettman, G. Dimitriadis, A.V. Filippenko, R.J. Foley, S.W. Jha, C.D. Kilpatrick, R.P. Kirshner, Y.C. Pan, A. Rest, C. Rojas-Bravo, M.R. Siebert, B.E. Stahl, W.K. Zheng, The Pantheon + analysis: the full data set and light-curve release. Astrophys. J. **938**(2), 113 (2022)
50. J. Tonry, D.P. Schneider, A new technique for measuring extragalactic distances. Astrophys. J. **96**, 807–815 (1988)
51. J.L. Tonry, E.A. Ajhar, G.A. Luppino, Observations of surface-brightness fluctuations in Virgo. Astrophys. J. **100**, 1416 (1990)
52. J.L. Tonry, J.P. Blakeslee, E.A. Ajhar, A. Dressler, The SBF survey of galaxy distances I sample selection, photometric calibration, and the hubble constant. Astrophys. J. **475**, 399 (1997)
53. J.L. Tonry, J.P. Blakeslee, E.A. Ajhar, A. Dressler, The surface brightness fluctuation survey of galaxy distances II local and large-scale flows. Astrophys. J. **530**(2), 625–651 (2000)
54. J.L. Tonry, A. Dressler, J.P. Blakeslee, E.A. Ajhar, A.B. Fletcher, G.A. Luppino, M.R. Metzger, C.B. Moore, The SBF survey of galaxy distances IV SBF magnitudes, colors, and distances. Astrophys. J. **546**, 681 (2001)
55. P. van Dokkum, S. Danieli, Y. Cohen, A.J. Romanowsky, C. Conroy, The distance of the dark matter deficient galaxy NGC 1052-DF2. Astrophys. J. **864**(1), L18 (2018)
56. G. Worthey, The dependence of the brightness fluctuation distance indicator on stellar population age and metallicity. Astrophys. J. **409**, 530–536 (1993)
57. G. Worthey, Comprehensive stellar population models and the disentanglement of age and metallicity effects. Astrophys. J. Supp. Ser. **95**, 107 (1994)

Chapter 10
The Pursuit of the Hubble Constant Using Type II Supernovae

Thomas de Jaeger and Lluís Galbany

10.1 Introduction

Thermonuclear explosions of a carbon-oxygen white dwarf in multi-star systems, known as Type Ia supernovae (SNe Ia), are among the best distance indicators [30, 45, 49]. For more than three decades, they have been used to measure extragalactic distances with an accuracy of 5–10% [7, 9] and have been fundamental to identify a tension between local [50] and distant [46] measurements of the Hubble constant (H_0). There are two complementary ways to solve this tension: first, increasing the level of precision of the different methods, and second, developing as many independent methods as possible with their own systematic uncertainties. In this chapter, as an independent approach, we describe the use of Type II supernovae (SNe II) to estimate H_0.

After being overshadowed by the better standardized SNe Ia, recent studies have demonstrated that SNe II can be used to measure extragalactic distances and have a role to play in the "H_0 tension." Observationally, their optical spectra are characterized by the presence of strong Balmer lines exhibiting broad P-Cygni profiles (e.g., [24, 26]), and their light curves display a plateau of varying steepness and length ([1, 3, 16, 27]) owing to hydrogen recombination. Among all the supernova types, SN II is the one for which the physics and the nature of their progenitors are the most well-understood. Unlike SNe Ia, for which no clear progenitors have

T. de Jaeger (✉)
LPNHE, CNRS/IN2P3 & Sorbonne Université, 4 place Jussieu, 75005 Paris, France
e-mail: thomas.dejaeger@lpnhe.in2p3.fr

L. Galbany
Institute of Space Sciences (ICE-CSIC), Campus UAB, Can Magrans S/N, 08193 Barcelona, Spain
e-mail: lg@csic.es

Institut d'Estudis Espacials de Catalunya (IEEC), 08034 Barcelona, Spain

E. Di Valentino and D. Brout (eds.), *The Hubble Constant Tension*, Springer Series in Astrophysics and Cosmology, https://doi.org/10.1007/978-981-99-0177-7_10

been yet identified, SNe II have dozens of direct progenitor detections [54]. It is now accepted that their progenitors have retained a significant fraction of their Hydrogen envelopes and arise from the explosion of red supergiants in late-type galaxies via core collapse. The understanding of the explosion mechanisms of SN II itself has also made remarkable progress in the past few decades [36, 60].

Although SNe II are not perfect standard candles and display a large range of peak luminosities ([1]), they have the potential to be excellent distance indicators with different systematics than SNe Ia. Their extrinsic differences can be calibrated using theoretical or empirical distance measurement methods such as the expanding photosphere method (EPM; [40]), the spectral expanding atmosphere method (SEAM; [6, 21]), the tailored EPM ([58]), the standard-candle method (SCM; [33]), the photospheric-magnitude method (PMM; [51, 52]), and the only method based solely on photometry, the photometric-color method (PCM; [12, 13, 17]).

Among this wide variety of methods to measure distances from SN II, here, we review only those used to measure H_0. For each method, we will describe the underlying theory behind them and present the most updated Hubble diagram together with a nearly exhaustive list of their H_0 measurements.

10.2 EPM: Expanding Photosphere Method

The Expanding Photosphere Method is a geometric-based method based on the approach used by [2] to measure distances for pulsating stars. The distance (D) is directly obtained by measuring the physical photosphere radius (R_{phot}) and the angular size (θ) of the SN II. Unfortunately, extragalactic SNe are unresolved and those parameters cannot be measured directly, they need to be derived by making some assumptions. First, supported by the fact that SNe II show a low polarization degree during the plateau phase ([41]), we assume a spherically symmetric photosphere. Second, the ejecta is freely expanding, so we can determine the photosphere radius from its velocity (v) and time since explosion ($t - t_0$): $R_{phot}(t) = v(t - t_0) + R_0$. Due to a fast expansion, it is safe to assume that the initial radius (R_0 at $t = t_0$) is negligible after a few days. The velocities of the photosphere are determined through the minimum flux of the absorption component of the P-Cygni line profile of weak lines such as those of Fe II (λ5018,5169). Finally, to derive the theoretical angular size θ, we assume that the photosphere radiates as a blackbody (BB), i.e., the thermalization layer from which the thermal photons originated (R_{therm}) and the photosphere have the same radius. Therefore, we have $4\pi D^2 f_\nu = 4\pi R_{therm}^2 \pi B_\nu(T_c) 10^{-0.4A_\lambda}$, where $B_\nu(T_c)$ is the Planck function at color temperature T_c of the BB radiation estimated by fitting Planck curves to observed broadband magnitudes, A_λ the total dust extinction (host galaxy and Milky Way), and f_ν is the flux received at Earth. Finally, for $z << 1$, θ can be written as,

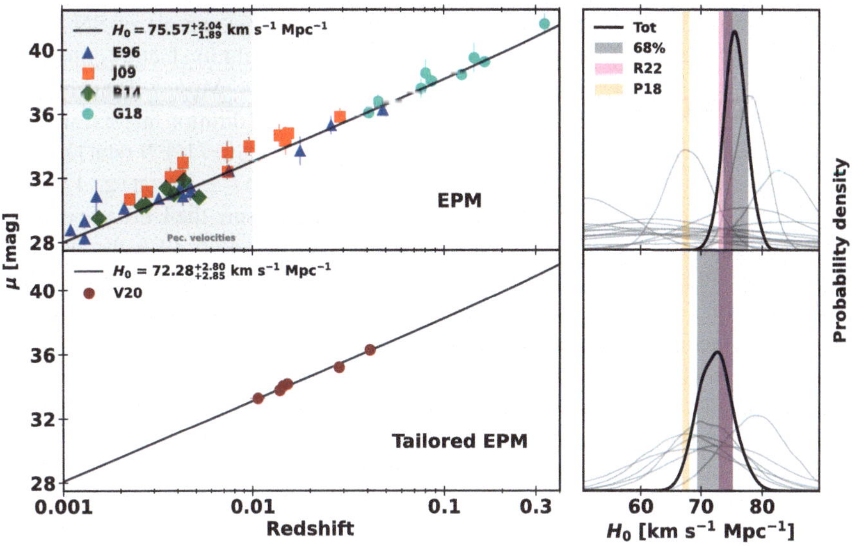

Fig. 10.1 *Left Top:* Figure updated from figure of [29] representing the Type II supernova Hubble diagram using the EPM. The EPM was applied to the data taken from [23] (blue triangles; E96), [37] (red squares; E96), [8] (green diamonds; B14), and [29] (cyan dots; G18). *Left Bottom:* Type II supernova Hubble diagram of six objects using the distance luminosities derived by [58] (brown dots; V20) when applying the tailored EPM. *Right Top* and *Right bottom* show the probability distributions of H_0 for the individual SNe in grey and the combined estimate in black. We include the value from local measurement [50] (pink filled region) and from the high-redshift results [46] (orange filled region) for comparison

$$\theta = \frac{R_{phot}}{D} \overset{\text{BB}}{=} \frac{R_{therm}}{D} = \sqrt{\frac{f_\nu}{\pi B_\nu(T_c)10^{-0.4A_\lambda}}}. \tag{10.1}$$

Combining $R(t) = v(t - t_0)$ and the above equation implies that $t = D(\theta/v) + t_0$, i.e., the distance D corresponds to the slope of a straight line made in the t versus (θ/v) plot and the explosion date to the y-intercept.

The EPM was first used by [40] to calculate distances of two SNe II (69L and 70g). They also obtained the first H_0 from SN II and found a value of $H_0 = 65 \pm 15\,\text{km}\,\text{s}^{-1}\ \text{Mpc}^{-1}$. However, a few years later, [59] noticed that the photosphere radiation deviates from that of a BB. The reason is that the thermalization layer from which the BB radiation is generated is deeper than the photosphere. The result is that the photosphere radiates less strongly than a BB of the same color temperature and therefore, appears diluted. To take into account this effect, the EPM was refined by incorporating a dilution factor (ξ) in Eq. 10.4 which parameterized the relative position of the two surfaces ($R_{phot} = R_{therm}/\xi$). In principle, the dilution factor should account for all deviations of the spectrum from BB emission, including the chemical composition or density structure of the progenitor star. However, theoretical dilution factors computed from non-LTE models with a wide range of properties have shown that they depend only on the color temperature ([20, 23]).

[53] were the first to apply this refined method to a large sample of SN II and derive a H_0. Using 16 objects and preliminary values of the dilution factors computed by [23], they obtained a value of 73 ± 6 (stat) $\pm$ 7 (sys) km s^{-1} Mpc^{-1}. Then, using 9 well-observed SNe II, with a slightly modified version of [23] dilution factors, and different filter combinations, [32] obtained an average value H_0 =71 $\pm$ 9 (stat) km s^{-1} Mpc^{-1}. Based on [32] work, [42] derived a value of H_0 =57 $\pm$ 7 (stat)$\pm$ 13 (sys) km s^{-1} Mpc^{-1}. Two other works have used the EPM to measure the Universe expansion rate. Jones et al. [37] constructed a Hubble diagram with 12 well-observed SNe II using dilutions factors from [20, 23]. They obtained a value of H_0 =100.5 $\pm$ 8.4 (stat) km s^{-1} Mpc^{-1} and H_0 =56.9 $\pm$ 4.2 (stat) km s^{-1} Mpc^{-1}, respectively. The dilution factor choice generates a large systematic difference in the H_0 values. The factors calculated by [20] are systematically larger than [23] which implies that the EPM distances are also $\sim$ 10–20% larger. The difference between the two sets of dilution factors is unclear but it is an important source of concern for using the EPM to measure distances. Gall [28] also found different values for H_0 depending on the dilution factors. For one object (13eq) in the Hubble flow ($z = 0.041$), they obtained a value of $H_0 = 83 \pm 10$ (stat) km s^{-1} Mpc^{-1} and $H_0 = 76 \pm 9$ (stat) km s^{-1} Mpc^{-1} using dilution factors from [20, 32] respectively. Finally, [29] have successfully extended the EPM to high-z and measured the distances to 10 SNe II with a $z > 0.04$. Using these distance measurements[1], together with those of seven other SNe II in the Hubble flow ($z > 0.01$) from the literature, we construct a combined probability distribution from the individual H_0 posteriors of each SN II using Gaussian kernel-density estimates (see Fig. 10.1)[2]. We obtained $H_0 = 75.53^{+2.07}_{-1.97}$ (stat) and $H_0 = 84.70^{+2.28}_{-2.21}$ (stat) using dilution factors from [20, 32], respectively.

Pros: Absolute distances, i.e., do not require any external calibration. Less sensitive to uncertainties in the dust extinction.
Cons: Dilution factors: different sets of dilution factors yield differences of $\sim$ 20% in the inferred distance.
Requirements: Optical photometry (at least BVI), at least two optical spectra, and dilution factors.
Future: New dilution factors based on independent numerical methods (e.g. [56]).

10.3 Tailored EPM

To avoid systematic biases introduced by the use of the dilution factors with the EPM, the Spectral Expanding Atmosphere Method (SEAM; [4–6] and the tailored EPM ([20, 58]) have been developed. Both methods take advantage of the full information contained in the observed spectra through detailed spectral fitting to infer the chem-

[1] We have not considered LSQ13cuw, which is an outlier in the EPM and SCM Hubble diagrams (see Fig. 7 from [29]).
[2] Using $H_0 = cz/Dl[1 + 0.5(1 - q_0)z]$ with $q_0 = -0.55$.

ical composition, the density profile, the temperature, the ejecta velocity, and other parameters of the SN. Afterwards, because the spectral energy distribution is known, one can determine the absolute magnitude in any bands from the calculated synthetic spectra. Then a comparison with to the observed apparent magnitudes allows us to obtain the distance modulus.

Mainly three radiative-transfer codes have been used to fit observed spectra: *PHOENIX* ([6, 35]), *CMFGEN* ([20], and a modified version of *Tardis* ([39, 56]). All the codes take into account the departure from the local thermodynamic equilibrium (LTE) seen in low-density, scattering-dominated SN II atmospheres via the use of a non-LTE approximation for the ionization and excitation treatment of the species. However, radiative-transfer simulations are complicated and time-consuming, ranging from minutes to days on large supercomputers depending on the number of free parameters. Additionally, it requires high signal to noise spectrum sequence but only recently such data have been made publicly accessible. For these reasons, only three well-observed SNe II in the local Universe (99em, 05cs, 06bp) have been studied using detailed radiative-transfer models ([6, 21]). The authors provided excellent fits between the observed and synthetic spectra, with a level of accuracy on the distances $\sim 10\%$. However, none of those SNe II are enough distant to derive a reliable H_0.

Recently, radiative-transfer simulations have regained popularity thanks to the development of a fitting method using a spectral emulator ([57]) to predict the SN II spectra and magnitudes. The emulator has been trained on a large grid of spectra calculated with *Tardis*, and can produce synthetic spectra in only 10^{-2}s instead of 10^5s. It has been built in two steps: (1) data dimensionality reduction by Principal Component Analysis (PCA), (2) Gaussian processes to interpolate the spectra within the PCA space and to predict preprocessed, dimensionality-reduced spectra for new sets of input parameters including: photospheric temperature and velocity, metallicity, time since explosion, and the exponent of the density profile. Finally, a synthetic spectrum is obtained by reversing the preprocessing procedure and then compare to the observations. Vogl et al. [57] demonstrated the emulator predictions match the *Tardis* simulations with a precision of better than 99%. Vogl [58] applied their new method to six low-redshift SNe II (03bn, 06it, 10id, 12ck, 13fs, and iPTF13bjx) and constructed the Hubble diagram displayed in Fig. 10.1 (bottom left). From those six distances, they were able to derive a value of $H_0 = 75.28^{+2.80}_{-2.85}$ (stat). This result is currently limited by a small sample size and low redshift range ($z < 0.04$), and it is affected by the systematic uncertainties of atmosphere models. However, this is a promising method which is independent of local calibrations. Also, it has recently been checked using sibling SNe II, i.e., SNe II in the same host galaxy. With four different host galaxies (8 SNe II), [10] found consistent distances within the errors for each of the SN sibling pairs and with a precision better than 5% for two hosts. Finally, the adH_0cc collaboration (accurate determination of H_0 with core-collapse supernovae[3]) has now gathered observations for 26 SNe II in the Hubble flow ($z > 0.03$). This new dataset will provide the basis for a highly-competitive H_0 measurement.

[3] https://adh0cc.github.io/.

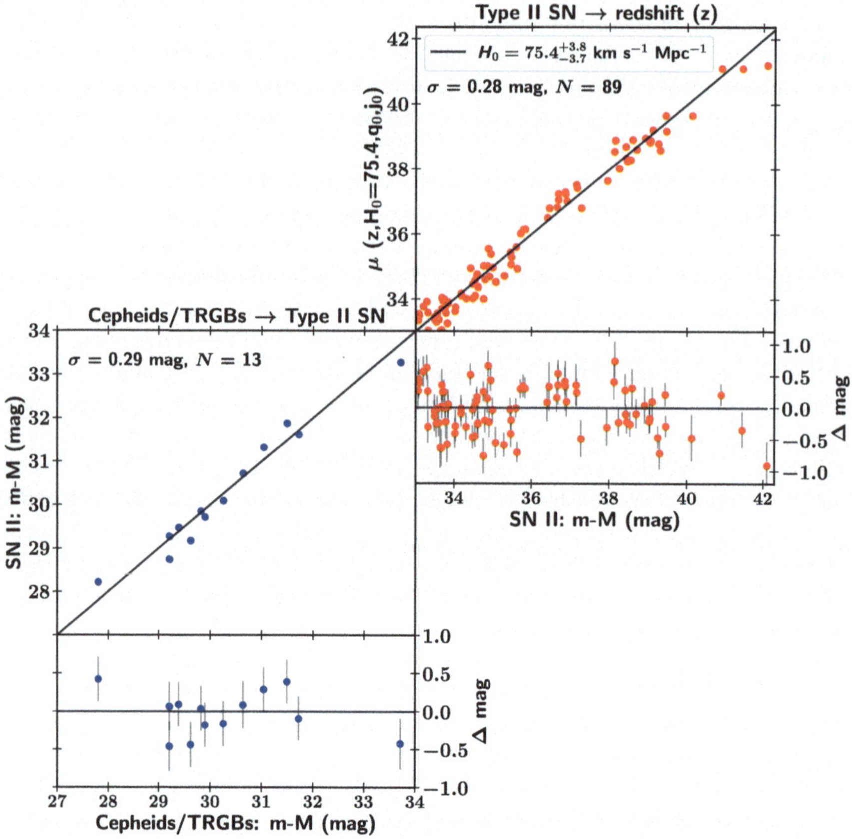

Fig. 10.2 Figure from [19] representing the last two rungs of the distance ladder with the SCM. For each rung, distances on the abscissa serve to calibrate a relative distance on the ordinate. Blue dots represent geometric, Cepheid, or TRGB distances used to calibrate SN II through the determination of the i-band absolute magnitudes (*Bottom left*). Red dots are the SNe II in the Hubble flow used to constrain H_0 (*Top right*)

Pros: Absolute distances, i.e., do not require any external calibration. Unlike the EPM, dilution factors are not needed.
Cons: Small sample size (only 6 objects) and low redshift range ($z < 0.04$). Assume fractional uncertainties for θ/v_{ph}.
Requirements: A well-defined explosion date, optical photometry (at least BVI) interpolated to the spectral epochs, multiple well-calibrated spectra in the first month after the explosion, and spectral emulator to predict synthetic spectra.
Future: Increase the sample of SN II in the Hubble flow. Better treatment of the uncertainties.

10.4 SCM: Standard Candle Method

Unlike the EPM and its tailored version, the Standard Candle Method (SCM) is an empirical method. First developed by [31], it has been later theoretically explained by [38]. SCM is based on the correlation between the SN absolute magnitude and the photospheric expansion velocity determined from the blueshift of the Fe II lines (λ5018). Such a correlation is expected when the SN is on the plateau phase, i.e., between 30–80 d after the explosion ([1]). This phase is physically well-understood and is only due to a change in opacity and density in the outermost layers of the SN. A few weeks after the explosion, the ionized hydrogen present in the outermost layers of the progenitor starts to recombine when the temperature drops and levels of the hydrogen recombination temperature ($\sim$5,000 K). During this phase, the light curve is powered by the sudden release of energy caused by the recession of the photosphere in mass that corresponds to a fixed radius in space. Therefore, as the temperature and the radius are nearly constant, the luminosity also appears constant. Consequently, in more luminous SNe II, the hydrogen recombination front will be fixed at a larger radius, and therefore, assuming a homologous expansion, the photosphere will be maintained at higher velocities.

Hamuy [34] applied this technique to 24 SNe II in the Hubble flow and reduced the Hubble diagram scatter from 0.83 mag to only 0.32 mag, i.e., 15% in distances. Distances derived using the SCM are not absolute and must be calibrated using primary distance indicators like Cepheids or Tip Red Giant Branch (TRGB). Using four SNe II (68L, 70G, 73R, 99em) with precise Cepheid distances, [34] derived a H_0 of $75 \pm 7\,\mathrm{km\ s^{-1}\ Mpc^{-1}}$. With the same technique but with only one SN (99em) calibrated via Cepheid, [42] obtained a smaller H_0 of 59 ± 3 (stat) $\pm$ 11 (sys).

Recently, many others studies have refined the SCM to simply its use and to extend it to higher redshifts. First, [43] added an extinction correction (intrinsically brighter SNe II are bluer) like for SN Ia cosmology ($\beta \times$ color) rather than relying on a method to measure the host-galaxy extinction. Second, he adjusted the velocity of the Fe II lines with a power-law decline, reducing the need of obtaining several spectra around the middle of the plateau phase. Third, [13, 43, 47, 55] have proposed to use the Hβ λ4861 absorption line which has the advantage of being stronger than Fe II lines (λ5018), and therefore easier to measure. Finally, because the measurement of the velocity can be difficult for noisy even for the strong Hβ λ4861 absorption line, [14, 47] have developed a method based on the cross-correlation between the observed spectra and a library of high S/N SN II spectra. Velocities from direct measurement or cross-correlation techniques have shown a good agreement with a dispersion of only $400\,\mathrm{km\ s^{-1}}$ ([14]). Therefore, finally, using the SCM, the corrected magnitude is written as,

$$m_{\mathrm{corr}} = \mathrm{m} + \alpha \log_{10}(v_{\mathrm{H}\beta}) - \beta color, \tag{10.2}$$

where m is the apparent magnitude in a given passband around 50 d after the explosion, $v_{\mathrm{H}\beta}$ is the velocity measured using Hβ absorption from an optical spectrum, α

and β are nuisance parameters. The fine-tuned version of the SCM have been used by several studies ([11–14, 17, 29, 44, 47, 48]) to measure SN II distances in the Hubble flow with a precision of 12–15% in distance. However, only three works have applied this updated version to measure H_0. First, [44] used two SNe II with Cepheid distances (99em, 04dj) and obtained a value of 71 $\pm$ 12 km s^{-1} Mpc^{-1}, 69 $\pm$ 16 km s^{-1} Mpc^{-1}, and 70 $\pm$ 16 km s^{-1} Mpc^{-1} depending on the passband used (B, V, and I, respectively). More recently, using seven objects (99em, 99gi, 05ay, 05cs, 09ib, 12aw, and 13ej) with Cepheid or TRGB independent host-galaxy distance measurements, [18] found a H_0 value of $75.8^{+5.2}_{-4.9}$ (stat) $\pm$ 2.8 (sys) km s^{-1} Mpc^{-1}, which differs by 1.4σ from the high-redshift result ([46]). Thanks to six additional calibrators (04et, 08bk, 14bc, 17eaw, 18aoq, 20yyz) and a better analysis of the systematic errors, [19] were able to obtain a H_0 with an uncertainty of 5%, $75.4^{+3.8\,(\mathrm{stat})}_{-3.7\,(\mathrm{stat})}$ $\pm$ 1.5 (sys) km s^{-1} Mpc^{-1}. In Fig. 10.2, the last two rungs of the distance ladder from their latest work are displayed. Their value is consistent with the local measurement [50] but differs by 2.0σ from the high-redshift results [46]. Additionally, using seven Cepheids or five TRGB, they derived values which differ by $< 1.0\sigma$ (i.e., 4.5 kms^{-1} Mpc^{-1}), suggesting that neither Cepheids nor TRGB are the source of the "H_0 tension.". It is important to note that one SN II (14bc) was discovered in NGC 4258 for which we have geometric distance. Therefore, we could use 14bc to directly calibrate the SN II in the Hubble flow, without the need for any calibrators, i.e., it will be just a two-rungs distance ladder. However, the uncertainties on its SCM distance and mostly the uncertainty on the ejecta expansion velocity are too large to constrain H_0.

Pros: Straightforward simple method and has been applied at high redshift.
Cons: Needs to be calibrated. Common reddening law for all SNe.
Requirements: Optical photometry (at least two bands) up to 50 d after the explosion, one optical spectrum during the plateau phase, a well-defined explosion date ($\sigma_{T_{exp}}$ <10 d), calibrators (Cepheids, TRGB) in the same host.
Future: Increase the number of calibrators, and find new correlations to decrease the scatter in the Hubble diagram (e.g. host galaxy environment).

10.5 PMM: Photospheric Magnitude Method

Rodríguez [51] developed another method to measure SN II distances and build Hubble diagram with a scatter of 0.15 mag. This method is a generalization of the SCM (Sect. 10.4) at any epoch during the photospheric phase. For a given bandpass X, the absolute magnitude (M_X) at an epoch t_i (since the explosion t_0) can be written as,

$$\mathrm{M}_{\mathrm{X,t_i}} = a_{X,t_i} - 5log_{10}(v_{ph}), \tag{10.3}$$

where a_{X,t_i} is a function that can be calibrated empirically and v_{ph} the photospheric velocity derived from the Fe II λ5169 line absorption minima. Finally, the SN distance modulus is obtained using,

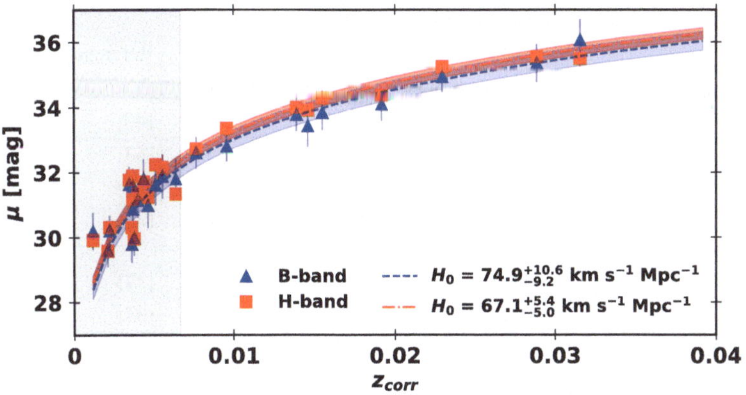

Fig. 10.3 Figure updated from figure of [52] representing SNe II distance moduli derived with the PMM for the B (blue triangles) and H (red squares) bandpass versus the redshift corrected for peculiar velocities. Red and blue lines correspond to the Hubble law fits to only the SNe II at $cz > 2000\,\mathrm{km\,s^{-1}}$ (shaded region)

$$\mu_{X,t_i} = m^{corr}_{X,t_i} - a_{X,t_i} + 5log_{10}(v_{ph}), \tag{10.4}$$

where m^{corr}_{X,t_i} is the apparent magnitude corrected for Galactic, host galaxy extinctions and K-correction. Unlike the SCM method for which a color term is added in the SN standardisation, [51] measured the host galaxy extinction using the colours and assuming that all SNe II have the same intrinsic colour. However, this assumption has recently been challenged by [15] who demonstrated that intrinsic SN II colors are most probably dominated progenitor properties like its radius.

To calibrate the PMM zero-point (encapsulated in a_{X,t_i}), [52] used four SNe II with known distances from TRGB (SN 03hn, 05cs, 12aw, and 13ej). With only 9 SNe II with $cz > 2000\,\mathrm{km\,s^{-1}}$, they were able to derive H_0 value between 67.1 and 74.9 km s^{-1} Mpc^{-1}. The large range of H_0 values is a source of concern for using the PMM as the H_0 values strongly depend on the filter chosen. For example, using the B band, they obtained $74.9^{+10.6}_{-9.2}$(stat) km s^{-1} Mpc^{-1} while using redder filter, this value decrease to $67.1^{+5.4}_{-5.0}$(stat) km s^{-1} Mpc^{-1} in H band (see Fig. 10.3).

Pros: Straightforward simple method. Smaller scatter in the Hubble diagram than the SCM.
Cons: Small sample size and affected by peculiar velocities. H_0 depends on the filters used. Host galaxy extinction estimated with colours. Needs to be calibrated. Never applied at high redshifts.
Requirements: Photometry (at least two bands) between 35–75 d since explosion, one optical spectrum during the plateau phase, a well-defined explosion date ($\sigma_{T_{exp}}$ <5 d), calibrators (Cepheids, TRGB) in the same host.
Future: Increase the number of calibrators and SN in the Hubble flow.

Table 10.1 H_0 values from Type II supernovae

H_0 km s^{-1} Mpc^{-1}	Method+ dilution factors or N_{cal}	Year	References
60 ± 15 (stat)	EPM	1974	[40]
73 ± 6 (stat) ± 7 (sys)	EPM+ [22]	1994	[53]
71 ± 9 (stat)	EPM+ [32]	2001	[32]
57 ± 7 (stat) ± 13 (sys)	EPM+ [32]	2003	[42]
100.5± 8.4 (stat)	EPM+ [23]	2009	[37]
59.6± 4.2 (stat)	EPM+ [20]	2009	[37]
83 ± 10 (stat)	EPM+ [32]	2016	[28]
76 ± 9 (stat)	EPM+ [20]	2016	[28]
$84.70^{+2.28}_{-2.21}$ (stat) ± 10.47 (sys)	EPM+ [32]	2023	This work
$75.57^{+2.04}_{-1.89}$ (stat) ± 13.46 (sys)	EPM+ [20]	2023	This work
$72.28^{+2.80}_{-2.85}$ (stat) ± 3.79 (sys)	Tailored EPM	2020	[58]
$74.9^{+10.6}_{-9.2}$(stat)	PMM+4+B	2019	[52]
$67.1^{+5.4}_{-5.0}$(stat)	PMM+4+H	2019	[52]
59 ± 3 (stat) ± 11 (sys)	SCM+1	2003	[42]
75 ± 7 (stat)	SCM +4	2003	[34]
71 ± 12 (stat)	SCM+2	2010	[44]
$75.8^{+5.2}_{-4.9}$(stat) ± 2.8 (sys)	SCM+7	2020	[18]
$75.4^{+3.8}_{-3.7}$(stat) ± 1.5 (sys)	SCM+13	2022	[19]

10.6 Conclusions

All H_0 measurements from SNe II are summarized in Table 10.1 and presented in Fig. 10.4 in comparison to late-Universe Cepheid ([50]) and TRGB ([25]) estimates, and the early-Universe ([46]) value from Planck. For the EPM values measured in this work and the tailored EPM by Vogl et al. ([58]) we considered the dispersion of all individual H_0 measurements as an estimate of the systematic uncertainty.

Although not yet as mature as those using SNe Ia, different methods have been developed to standardize SNe II to make them useful extragalactic distance indicators and, in turn, provide independent estimates of H_0. Their main advantages over SNe Ia are: (i) the extended knowledge on their progenitor systems and the physics behind their explosion mechanism; (ii) their larger volumetric rates; and (iii) their more direct connection with the local environment, which could potentially be used to

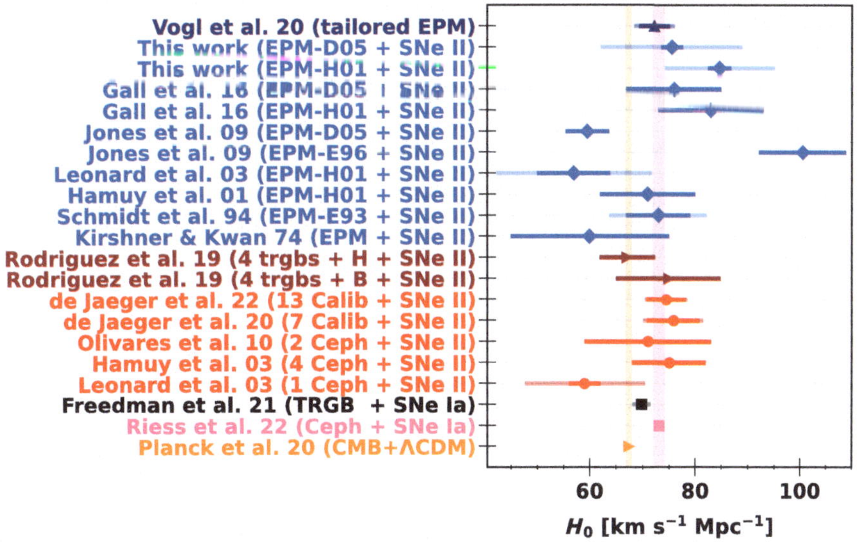

Fig. 10.4 Comparison of H_0 derived using SNe II, SNe Ia calibrated with Cepheids ([50]) and TRGB ([25]), and from the CMB anisotropies and lensing ([46]). For the Type II supernovae measurements, we represent in red those using the SCM, in brown those from the PMM method, and in blue those using the EPM, including the tailored version. Light errors bars represent statistical and sytematic uncertainties in quadrature, while solid error bars are only statistical uncertainties

improve their standardization. In addition, their constraints on H_0 are independent to those from SNe Ia since they are affected by different systematic uncertainties. Advantageously, some stand-alone methods exist that do not need calibrators as the distance ladder does. The different methods summarized here, some of which do not depend on external calibrators, are more likely to be improved in the following years thus reducing the impact of systematic uncertainties in the determination of H_0. Moreover, with more well-observed objects to be included that will reduce the statistical uncertainties, SNe II will certainly contribute to understanding the reason behind the Hubble tension.

Acknowledgements T. dJ acknowledges financial support from the French Centre National de la Recherche Scientifique (CNRS/IN2P3) and from the French Agence National de la Recherche project 21-CE31-0016. L.G. acknowledges financial support from the Spanish Ministerio de Ciencia e Innovación (MCIN), the Agencia Estatal de Investigación (AEI) 10.13039/501100011033, and the European Social Fund (ESF) "Investing in your future" under the 2019 Ramón y Cajal program RYC2019-027683-I and the PID2020-115253GA-I00 HOSTFLOWS project, from Centro Superior de Investigaciones Científicas (CSIC) under the PIE project 20215AT016, and the program Unidad de Excelencia María de Maeztu CEX2020-001058-M.

References

1. J.P. Anderson, S. González-Gaitán, M. Hamuy, C.P. Gutiérrez, M.D. Stritzinger, F.E. Olivares., M.M. Phillips, S. Schulze et al. Astrophys. J. **786**, 67 (2014). https://doi.org/10.1088/0004-637X/786/1/67[arXiv:1403.7091 [astro-ph.HE]]
2. W. Baade, Astron. Nachr. **228**, 359–362 (1926)
3. R. Barbon, F. Ciatti, L. Rosino, Astron. Astrophys. **72**, 287–292 (1979)
4. E. Baron, P.H. Hauschildt, P. Nugent, D. Branch, Mon. Not. R. Astron. Soc. **283**(1), 297–315 (1996). https://doi.org/10.1093/mnras/283.1.297
5. E. Baron, D. Branch, P.H. Hauschildt, A.V. Filippenko, R.P. Kirshner, P.M. Challis, S. Jha, R. Chevalier et al. Astrophys. J. **545**, 444–448 (2000). https://doi.org/10.1086/317795[arXiv:astro-ph/0010614 [astro-ph]]
6. E.A. Baron, P.E. Nugent, D. Branch, P.H. Hauschildt, Astrophys. J. Lett. **616**, L91–L94 (2004). https://doi.org/10.1086/426506 [arXiv:astro-ph/0410153 [astro-ph]]
7. M. Betoule et al. [SDSS] Astron. Astrophys. **568**, A22 (2014). https://doi.org/10.1051/0004-6361/201423413 ([arXiv:1401.4064 [astro-ph.CO]])
8. S. Bose, B. Kumar, Astrophys. J. **782**, 98 (2014). https://doi.org/10.1088/0004-637X/782/2/98. ([arXiv:1401.5115 [astro-ph.CO]].)
9. D. Brout, D. Scolnic, B. Popovic, A.G. Riess, J. Zuntz, R. Kessler, A. Carr, T.M. Davis et al. Astrophys. J. **938**(2), 110 (2022). https://doi.org/10.3847/1538-4357/ac8e04([arXiv:2202.04077 [astro-ph.CO]])
10. G. Csörnyei, C. Vogl, S. Taubenberger, A. Flörs, S. Blondin, M.G. Cudmani, A. Holas, S. Kressierer et al. ([arXiv:2302.03112 [astro-ph.SR]])
11. C.B. D'Andrea, M. Sako, B. Dilday, J.A. Frieman, J. Holtzman, R. Kessler, K. Konishi, D.P. Schneider et al. Astrophys. J. **708**, 661–674 (2010). https://doi.org/10.1088/0004-637X/708/1/661 ([arXiv:0910.5597 [astro-ph.CO]])
12. T. de Jaeger, S. González-Gaitán, J.P. Anderson, M. Hamuy, L. Galbany, M.M. Phillips, M.D. Stritzinger, C.P. Gutiérrez et al., Astrophys. J. **815**, 121 (2015). https://doi.org/10.1088/0004-637X/815/2/121. ([arXiv:1511.05145 [astro-ph.HE]].)
13. T. de Jaeger, S. González-Gaitán, M. Hamuy, L. Galbany, J.P. Anderson, M.M. Phillips, M.D. Stritzinger, R.G. Carlberg et al. Astrophys. J. **835**(2), 166 (2017). https://doi.org/10.3847/1538-4357/835/2/166 ([arXiv:1612.05636 [astro-ph.CO]])
14. T. de Jaeger, L. Galbany, A.V. Filippenko, S. González-Gaitán, N. Yasuda, K. Maeda, M. Tanaka, T. Morokuma et al. Mon. Not. R. Astron. Soc. **472**(4), 4233–4243 (2017). https://doi.org/10.1093/mnras/stx2300 ([arXiv:1709.01513 [astro-ph.HE]])
15. T. de Jaeger, J.P. Anderson, L. Galbany, S. González-Gaitán, M. Hamuy, M.M. Phillips, M.D. Stritzinger, C. Contreras et al. Mon. Not. R. Astron. Soc. **476**(4), 4592–4616 (2018). https://doi.org/10.1093/mnras/sty508 ([arXiv:1802.07254 [astro-ph.HE]])
16. T. de Jaeger, W. Zheng, B. E. Stahl, A. V. Filippenko, T. G. Brink, A. Bigley, K. Blanchard, P. K. Blanchard, *et al.* Mon. Not. Roy. Astron. Soc. **490**, no.2, 2799-2821 (2019) doi:https://doi.org/10.1093/mnras/stz2714[arXiv:1909.13813 [astro-ph.HE]]
17. T. de Jaeger et al. [DES], https://doi.org/10.1093/mnras/staa1402[arXiv:2005.09757 [astro-ph.HE]]
18. T. de Jaeger, B.E. Stahl, W. Zheng, A.V. Filippenko, A.G. Riess, L. Galbany, Mon. Not. R. Astron. Soc. **496**(3), 3402–3411 (2020). https://doi.org/10.1093/mnras/staa1801 ([arXiv:2006.03412 [astro-ph.CO]])
19. T. de Jaeger, L. Galbany, A.G. Riess, B.E. Stahl, B.J. Shappee, A.V. Filippenko, W. Zheng, Mon. Not. R. Astron. Soc. **514**(3), 4620–4628 (2022). https://doi.org/10.1093/mnras/stac1661 ([arXiv:2203.08974 [astro-ph.CO]])
20. L. Dessart, D.J. Hillier, Astron. Astrophys. **439**, 671 (2005). https://doi.org/10.1051/0004-6361:20053217. ([arXiv:astro-ph/0505465 [astro-ph]].)
21. L. Dessart, S. Blondin, P.J. Brown, M. Hicken, D.J. Hillier, S.T. Holland, S. Immler, R.P. Kirshner et al., Astrophys. J. **675**, 644 (2008). https://doi.org/10.1086/526451. ([arXiv:0711.1815 [astro-ph]].)

22. R.G. Eastman, P.A. Pinto, Astrophys. J. **412**, 731 (1993). https://doi.org/10.1086/172957
23. R.G. Eastman, B.P. Schmidt, R.P. Kirshner, Astrophys. J. **466**, 911 (1996). https://doi.org/10.1086/177563
24. A.V. Filippenko, Ann. Rev. Astron. Astrophys. **35**, 309–355 (1997). https://doi.org/10.1146/annurev.astro.35.1.309
25. W.L. Freedman, Astrophys. J. **919**(1), 16 (2021). https://doi.org/10.3847/1538-4357/ac0e95 ([arXiv:2106.15656 [astro-ph.CO]])
26. A. Gal-Yam, https://doi.org/10.1007/978-3-319-20794-0_35-1([arXiv:1611.09353 [astro-ph.HE]])
27. L. Galbany, M. Hamuy, M.M. Phillips, N.B. Suntzeff, J. Maza, T. de Jaeger, T. Moraga, S. González-Gaitán et al. https://doi.org/10.3847/0004-6256/151/2/33 ([arXiv:1511.08402 [astro-ph.SR]])
28. E.E.E. Gall, R. Kotak, B. Leibundgut, S. Taubenberger, W. Hillebrandt, M. Kromer, Astron. Astrophys. **592**, A129 (2016). https://doi.org/10.1051/0004-6361/201628333. ([arXiv:1603.04730 [astro-ph.CO]].)
29. E.E.E. Gall, R. Kotak, B. Leibundgut, S. Taubenberger, W. Hillebrandt, M. Kromer, W.S. Burgett, K. Chambers et al., Astron. Astrophys. **611**, A25 (2018). https://doi.org/10.1051/0004-6361/201731271. ([arXiv:1705.10806 [astro-ph.CO]].)
30. M. Hamuy, M.M. Phillips, N.B. Suntzeff, R.A. Schommer, J. Maza, R. Aviles, Astron. J. **112**, 2398 (1996). https://doi.org/10.1086/118191. ([arXiv:astro-ph/9609062 [astro-ph]].)
31. M. Hamuy, Type II supernovae as distance indicators. Ph.D. Thesis, The University of Arizona (2001)
32. M. Hamuy, P.A. Pinto, J. Maza, N.B. Suntzeff, M.M. Phillips, R.G. Eastman, R.C. Smith, C.J. Corbally et al., Astrophys. J. **558**, 615 (2001). https://doi.org/10.1086/322450. ([arXiv:astro-ph/0105006 [astro-ph]].)
33. M. Hamu, P.A. Pinto, Astrophys. J. Lett. **566**, L63–L65 (2002). https://doi.org/10.1086/339676 ([arXiv:astro-ph/0201279 [astro-ph]])
34. M. Hamuy, Springer Proc. Phys. **99**, 535–541 (2005). https://doi.org/10.1007/3-540-26633-X_71 ([arXiv:astro-ph/0309122 [astro-ph]])
35. P.H. Hauschildt, E. Baron, J. Comput. Appl. Math. **109**, 41–63 (1999). https://doi.org/10.1016/S0377-0427(99)00153-3 ([arXiv:astro-ph/9808182 [astro-ph]])
36. H.T. Janka, Ann. Rev. Nucl. Part. Sci. **62**, 407–451 (2012). https://doi.org/10.1146/annurev-nucl-102711-094901 ([arXiv:1206.2503 [astro-ph.SR]])
37. M.I. Jones, M. Hamuy, P. Lira, J. Maza, A. Clocchiatti, M. Phillips, N. Morrell, M. Roth et al. Astrophys. J. **696**, 1176–1194 (2009). https://doi.org/10.1088/0004-637X/696/2/1176 ([arXiv:0903.1460 [astro-ph.CO]])
38. D. Kasen, S.E. Woosley, Astrophys. J. **703**, 2205 (2009). https://doi.org/10.1088/0004-637X/703/2/2205. ([arXiv:0910.1590 [astro-ph.CO]].)
39. W.E. Kerzendorf, S.A. Sim, Mon. Not. R. Astron. Soc. **440**(1), 387–404 (2014). https://doi.org/10.1093/mnras/stu055 ([arXiv:1401.5469 [astro-ph.CO]])
40. R.P. Kirshner, J. Kwan, Astrophys. J. **193**, 27 (1974). https://doi.org/10.1086/153123
41. D.C. Leonard, A.V. Filippenko, Publ. Astron. Soc. Pac. **113**, 920 (2001). https://doi.org/10.1086/322151. ([arXiv:astro-ph/0105295 [astro-ph]].)
42. D.C. Leonard, S.M. Kanbur, C.C. Ngeow, N.R. Tanvir, Astrophys. J. **594**, 247–278 (2003). https://doi.org/10.1086/376831 ([arXiv:astro-ph/0305259 [astro-ph]])
43. P. Nugent et al. [SNLS], Astrophys. J. **645**, 841–850 (2006). https://doi.org/10.1086/504413 ([arXiv:astro-ph/0603535 [astro-ph]])
44. F.E. Olivares, M. Hamuy, G. Pignata, J. Maza, M. Bersten, M.M. Phillips, N.B. Suntzeff, A.V. Filippenko et al., Astrophys. J. **715**, 833–853 (2010). https://doi.org/10.1088/0004-637X/715/2/833 ([arXiv:1004.2534 [astro-ph.CO]])
45. M.M. Phillips, Astrophys. J. Lett. **413**, L105–L108 (1993). https://doi.org/10.1086/186970
46. N. Aghanim et al., [Planck], Astron. Astrophys. **641**, A6 (2020) [erratum: Astron. Astrophys. **652**, C4 (2021)]. https://doi.org/10.1051/0004-6361/201833910 ([arXiv:1807.06209 [astro-ph.CO]])

47. D. Poznanski, N. Butler, A.V. Filippenko, M. Ganeshalingam, W. Li, J.S. Bloom, R. Chornock, R.J. Foley et al., Astrophys. J. **694**, 1067–1079 (2009). https://doi.org/10.1088/0004-637X/694/2/1067 ([arXiv:0810.4923 [astro-ph]])
48. D. Poznanski, P.E. Nugent, A.V. Filippenko, Astrophys. J. **721**, 956–959 (2010). https://doi.org/10.1088/0004-637X/721/2/956 ([arXiv:1008.0877 [astro-ph.CO]])
49. A.G. Riess, W.H. Press, R.P. Kirshner, Astrophys. J. **473**, 88 (1996). https://doi.org/10.1086/178129. ([arXiv:astro-ph/9604143 [astro-ph]].)
50. A.G. Riess, W. Yuan, L.M. Macri, D. Scolnic, D. Brout, S. Casertano, D.O. Jones, Y. Murakami et al., Astrophys. J. Lett. **934**(1), L7 (2022). https://doi.org/10.3847/2041-8213/ac5c5b ([arXiv:2112.04510 [astro-ph.CO]])
51. Ó. Rodríguez, A. Clocchiatti, M. Hamuy, Astron. J. **148**, 107 (2014). https://doi.org/10.1088/0004-6256/148/6/107 ([arXiv:1409.3198 [astro-ph.CO]])
52. Ó. Rodríguez, G. Pignata, M. Hamuy, A. Clocchiatti, M.M. Phillips, K. Krisciunas, N.I. Morrell, G. Folatelli et al., Mon. Not. R. Astron. Soc. **483**(4), 5459–5479 (2019). https://doi.org/10.1093/mnras/sty3396 ([arXiv:1812.04982 [astro-ph.CO]])
53. B.P. Schmidt, R.P. Kirshner, R.G. Eastman, M.M. Phillips, N.B. Suntzeff, M. Hamuy, J. Maza, R. Aviles, Astrophys. J. **432**, 42 (1994). https://doi.org/10.1086/174546
54. S.J. Smartt, Publ. Astron. Soc. Aust. **32**, e016 (2015). https://doi.org/10.1017/pasa.2015.17. ([arXiv:1504.02635 [astro-ph.SR]].)
55. K. Takats, J. Vinko, Mon. Not. R. Astron. Soc. **419**, 2783 (2012). https://doi.org/10.1111/j.1365-2966.2011.19921.x. ([arXiv:1109.5873 [astro-ph.SR]].)
56. C. Vogl, S.A. Sim, U.M. Noebauer, W.E. Kerzendorf, W. Hillebrandt, Astron. Astrophys. **621**, A29 (2019). https://doi.org/10.1051/0004-6361/201833701. ([arXiv:1811.02543 [astro-ph.HE]].)
57. C. Vogl, W.E. Kerzendorf, S.A. Sim, U.M. Noebauer, S. Lietzau, W. Hillebrandt, Astron. Astrophys. **633**, A88 (2020). https://doi.org/10.1051/0004-6361/201936137. ([arXiv:1911.04444 [astro-ph.HE]].)
58. C. Vogl, Cosmological distances of type II supernovae from radiative transfer modeling, PhD thesis, Munich Technical University, 2020
59. R.V. Wagoner, Astrophys. J. Lett. **250**, L65–L69 (1981). https://doi.org/10.1086/183675
60. S.E. Woosley, T.A. Weaver, Astrophys. J. Suppl. **101**, 181–235 (1995). https://doi.org/10.1086/192237

Chapter 11
The Mira Distance Ladder

Caroline D. Huang

11.1 Introduction

Mira variables are fundamental-mode, radially-pulsating tip of the Asymptotic Giant Branch (AGB) stars with periods ranging from $\sim$100–1000 days or longer. They were the first class of periodic variables to be identified, starting with the discovery of Mira (*Omicron Ceti*) by David Fabricius on August 3, 1596. With their high luminosities and the largest amplitudes of any regular pulsator ($>2.5\ m_V$ [1]), they remain a popular and easy target for amateur astronomers to this day.

As highly-evolved AGB stars, they contribute to the chemical enrichment of the interstellar medium (ISM) through stellar mass loss [2] and are also extremely luminous—generally the brightest stars in an intermediate-to-old stellar population. They also also ubiquitous, as nearly all stars ($0.8M_{\odot} < M < 8M_{\odot}$) will experience a Mira phase in evolution.

Period-luminosity relations (P–L relations) for Miras have been known for several decades and have been employed to measure distances to many Local Group galaxies ([3] and references therein). However, the persistent Hubble tension has resulted in renewed interest in Mira distances as an alternative and cross-check to the more commonly-used Cepheid variables and Tip of the Red Giant Branch (TRGB) thanks to their age (intermediate-age Miras are $\sim$3 Gyr) and luminosity (up to a few $10^4 L_{\odot}$). With their low effective temperatures ($T_{\mathrm{eff}} < 3500$ K) they are particularly attractive in light of the upcoming decade's focus on NIR and IR wavelengths since the spectral intensity of Miras with thin circumstellar shells (such that those currently used in distance measurements) peaks between 1–2μm.

Miras may be classified into two spectral types—Carbon-(C-) and Oxygen-(O-) rich. Current distance measurements using Miras employ exclusively short-period ($P \lesssim 400$ d), O-rich variables, which have been shown to have a tight ($\sigma \sim 0.12$

C. D. Huang (✉)
Center for Astrophysics | Harvard & Smithsonian, 60 Garden St., Cambridge, MA 02138, USA
e-mail: caroline.huang@cfa.harvard.edu

E. Di Valentino and D. Brout (eds.), *The Hubble Constant Tension*, Springer Series in Astrophysics and Cosmology, https://doi.org/10.1007/978-981-99-0177-7_11

mag) P–L relation in the near-infrared. C-rich Miras have larger scatter at shorter wavelengths, and longer-period ($P \gtrsim 400$ days) Miras may follow a different P–L relation due to hot-bottom burning [4].

In this review, we will briefly discuss the evolutionary status of Miras in Sect. 11.2. Next, we will give an overview of the history of various Mira Period-Luminosity Relations and some local measurements and studies with Miras in Sect. 11.3. In Sect. 11.4 we will give outline the previous efforts related to measuring H_0 with Miras. In Sect. 11.5 we will briefly discuss prospects and previous observations of long-period Miras. In Sect. 11.6 we compare Miras with other stellar primary distance indicators and conclude in Sect. 11.7.

11.2 Evolutionary Status

Miras are the brightest of the AGB stars and the AGB phase is the final stage of nuclear-burning life for vast majority of stars. It is often stated that stars with masses $0.8M_\odot < M < 8M_\odot$ will experience an AGB phase of evolution, though the exact upper limit depends on how convection is handled. The AGB stage is also brief, lasting $< 1\%$ of the main sequence lifetime. On the Hertzsprung-Russell diagram, these stars are typically identified as the bright stars above the tip of the Red Giant Branch (TRGB). Figure 11.1 shows theoretical evolutionary tracks for stars from 0.4 $M_\odot$ to 16 $M_\odot$ and the location of the AGB, where the Miras reside.

AGB stars have evolved past core hydrogen and helium fusion. Instead, they have a degenerate CO core and sustain themselves against gravitational collapse primarily through hydrogen-shell burning and quasi-periodic helium-shell flashes that

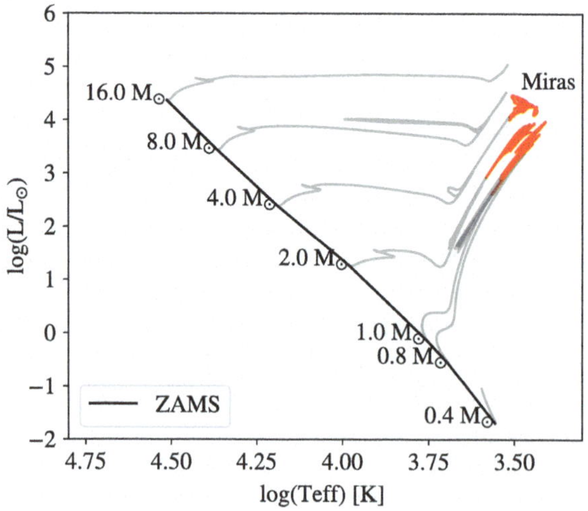

Fig. 11.1 Hertzsprung-Russell Diagram created using MIST evolutionary tracks [5–7]. The zero-age main sequence is shown in black, individual evolutionary tracks in gray, and the thermally-pulsating AGB evolutionary stage in red. Miras are located at the uppermost region of the thermally-pulsating AGB branch

ignite helium-shell burning. Stars actively undergoing this process are also known as thermally-pulsating (TP-AGB) stars.

The cycle of thermal pulses is also responsible for dredging-up carbon from the core and potentially changing the surface chemistry of a star from O-rich to C-rich. AGB stars exhibit enrichments of elements produced through the slow neutron capture process (*s*-process) and are also often experiencing rapid mass loss with rates of $10^{-8} M_{\odot} < \dot{M} < 10^{-4} M_{\odot}$ year^{-1} [8] and typical wind velocities of ~ 10 km/second.

Stars ascending the AGB will are likely to experience multiple phases of pulsational instability, first as low-amplitude, overtone pulsators, and then finally as high-amplitude, fundamentally-pulsating Miras at the tip of the AGB. During the Mira phase, stars are at their most luminous and also have the highest mass-loss rates. This stage is also quite brief, lasting $\sim 10^5$ years for low-mass stars. Depending on their mass and abundance, due to the He-shell flashes, it is also possible that during their evolution, stars may also enter and leave the Mira phase more than once.

The post-AGB evolution of a star is one of the least well-understood stages of stellar evolution. Low-to-intermediate mass stars are thought to become post-AGB stars and eventually blow off the outer layers to form planetary nebulae. The final life stage of more massive stars is even more uncertain, although they may go on to become super-AGB stars.

There have been several excellent previous reviews of the AGB evolution, including [8–14]. Here we will provide only a brief summary of the evolutionary status of Miras with a focus on the aspects most relevant to distance measurements.

11.2.1 The TP-AGB and Third Dredge up

A *dredge-up* is the process that mixes the products of nucleosynthesis from the nuclear burning zone of the star to the its upper layers via convection. The first dredge-up occurs after core hydrogen burning but before helium burning starts, as the star is ascending the red giant branch. The second dredge up occurs for stars of 4–8 $M_{\odot}$ after core helium fusion comes to an end. Lower mass stars will not experience a second dredge-up at all. Therefore, the names of the dredge up processes do not necessarily indicate the order in which they occur. We will primarily focus on the *third dredge-up*, which takes place when a star has entered the asymptotic giant branch, after a helium flash, and brings helium, carbon, and *s*-process elements to the surface.

During the AGB phase, the hydrogen-burning shell is the primary source of energy. As the star evolves up the AGB, the helium shell becomes thinner and thermally unstable, resulting in explosive, quasi-periodic ignitions of the helium-burning shell [15]. These are known as shell flashes or thermal pulses (TPs). Prior to the first TP, a star is in the early AGB phase and after the first TP, it is on the TP-AGB phase.

Each of these cycles of AGB evolution can be broken down into four phases [14, 16]:

1. **Thermal pulse:** Helium-burning is ignited explosively at the base of the intershell region. High luminosities of $\geq 8L_\odot$ are produced for a short timescale of $\sim 10^2$ years. The energy from the flash creates a convective zone that extends from the He-burning shell to the H-burning shell and mixes the abundances in the intershell region.
2. **Power Down:** During power-down, energy produced from the thermal pulse expands the star, which cools the material outwards of the He-shell and halts fusion the H-shell. The surface luminosity will dip by $\sim$ 10% during this expansion. The inner layers will cool, leading to a increase in stellar opacity.
3. **Third dredge-up:** The increase in stellar opacity results in the base of outer convective envelope moving inwards, past the previous H-burning shell and into the intershell region. This mixes the elements from He-burning to the surface, where they can be observed.
4. **Interpulse:** After the third dredge-up, the star will contract once again, reigniting the H-burning shell. The star then enters a long quiescent phase for $\sim 10^4$ years. During this time, the H-shell provides most of the surface luminosity before the cycle repeats.

Figure 11.2 shows the interior of an AGB star and the relative positions of the regions where these processes take place.

Stars enter the AGB with a photospheric ratio of C/O < 1 (the exact ratio depends on the initial oxygen abundance) and are classified as O-rich (M-) stars. On the AGB, this can change through the third dredge up process, which brings carbon synthesized through the triple-alpha reactions at the base of the helium shell to the surface via a

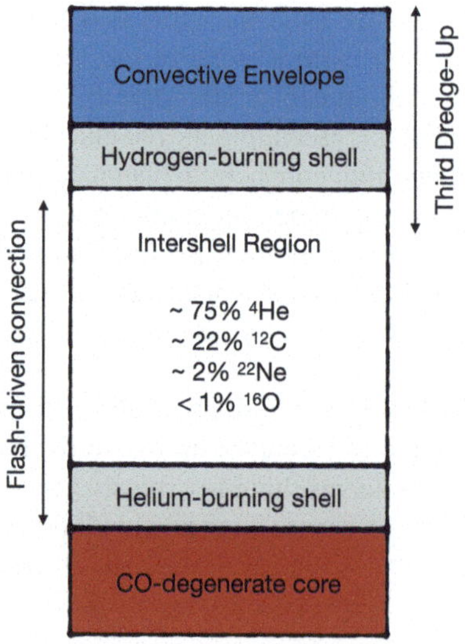

Fig. 11.2 The interior of an AGB star—not to scale—based on Fig. 1 from [17]. The degenerate CO core is surrounded by a He-burning shell. The He-burning shell is separated from the H-burning shell by an intershell region consisting mostly of helium and carbon. The H-burning shell lies directly below the deep, outer convective envelope. The ratio of radial thickness of the core compared to the envelope is $\sim 1 \times 10^{-5}$

convective zone. If the surface C/O ratio becomes >1, then the star is classified as C-rich (C-type) star. Environments with lower oxygen abundance are host to more C-rich stars, since less carbon is required to change the surface chemistry from O- to C-rich.

The two subtypes can be readily separated in the basis of the molecular features in their spectra. Carbon and oxygen will form stable CO molecules, which then leaves only the element with greater abundance for other chemical reactions.

11.2.2 Hot-Bottom Burning

Another process that affects the surface chemistry of some AGB stars is hot-bottom burning. Hot-bottom burning occurs in relatively massive AGB stars $\sim 4 - 7M_{\odot}$ and happens alongside the third dredge-up process. In stars experiencing hot-bottom burning, the surface convective envelope is deep enough to that proton-capture nucleosynthesis can occur at the base [18–21].

When the temperature at the base of the convective envelope exceeds 5×10^7K the CNO cycle is activated, which burns ^{12}C to ^{14}N. Furthermore, the entire convective envelope has a turnover time of 1 year, meaning that the entire envelope is well-mixed and will be exposed to this hot region within a year. This process prevents a C-rich atmosphere from forming and is responsible for limiting the number of C-rich variables at the longest periods and highest masses.

The exact mass at which this process activates is unknown [22]. Models indicate that AGB stars with lower metallicity experience hot-bottom burning at lower initial mass compared to stars with higher metallicity. Therefore, the masses at which hot-bottom burning occurs—and consequently, the masses of the stars that will become C-rich is not known and will almost certainly depend on the environment.

11.3 Period-Luminosity Relations

The Period-Luminosity Relation itself is typically defined as a linear relationship between mean magnitudes of variables and the logarithm of their periods. The magnitudes can be determined through averaging the flux or magnitude mean or by the mean of a (generally sinusoidal for Miras) fit to fluxes or magnitudes. Fit means are typically more consistent than an average, and therefore, preferred.

While mean magnitudes are by far the most commonly-used for P–L relations, Mira P–L relations also occasionally use bolometric magnitude or peak magnitude. Bolometric luminosities are more fundamental, easier to compare with models, and independent of circumstellar reddening, but they require more extensive observations over a longer period of time and in a large wavelength range. In addition, the results can be dependent on the color-dependent bolometric corrections used [23]. Peak magnitudes, on the other hand, are sometimes used with optical observations of

Miras [24, 25]. Due to the large amplitudes of these stars at shorter wavelengths, mean magnitudes can be difficult to measure and therefore peak magnitude can result in lower-dispersion P–L relations. In addition, the luminosity is *inversely* dependent on period in the shorter optical wavelengths, and the P–L relation becomes roughly flat around the I-band. Figure 11.3 from [26] shows the *single-epoch* P–L relations for Large Magellanic Cloud Miras at a range of wavelengths, from optical to the far infrared. With mean magnitudes the scatter in the P–L relations would be significantly smaller.

11.3.1 Basics of Mira Pulsation

Pulsating stars exist in various parts of the Hertzsprung-Russell Diagram, grouped in narrow (in color/temperature), nearly-vertical regions known as instability strips. While the relationship between luminosity and period for variable stars (specifically Cepheids) was determined empirically [27], an intuitive explanation for the existence of the P–L relation can be derived by examining the relationship between pulsational period and radius.

The boundary conditions for a pulsating star are similar to those in an organ pipe, where one end (the center) is closed, and the other (the surface) is open. In the fundamental, lowest-frequency mode, the entire star will contract and expand at the same time. Thus we expect that,

$$\Pi \sim 4\frac{R}{\bar{v}_s} \tag{11.1}$$

where Π is the period, R is the radius, and $\bar{v}_s$ is the mean sound speed. Writing sound speed in terms of density gives us,

$$v_s = \sqrt{\frac{\gamma P}{\rho}} \tag{11.2}$$

where γ is the specific heat ratio of the gas, P is the pressure, and ρ is the density. Then, taking the mean density $\bar{\rho} = M/(\frac{4}{3}\pi R^3)$ (an approximation) and plugging this back into Eq. 11.1 we quickly find

$$\Pi \propto R^{3/2} M^{-1/2} \tag{11.3}$$

which indicates that pulsational period is strongly dependent on radius, but only weakly dependent on mass.

To connect this more directly to the Period-Luminosity relation, we can rewrite R in terms of L, the luminosity. We know that luminosity of a spherical blackbody is given by $L = 4\pi R^2 \sigma T_{\rm eff}^4$. Assuming $T_{\rm eff}$ is fixed, then $L \propto R^2$. Now we can write Eq. 11.3 as

$$\Pi \propto L^{3/4} M^{-1/2} \tag{11.4}$$

Luminosity is strongly dependent on mass, however mass is relatively insensitive to luminosity. Thus, we would find $\Pi \propto L^x$, where $x \leq 3/4$. In practice, however, $T_{\rm eff}$ is not constant and thus x can be $>$3/4. Nonetheless, this indicates that there is a power law relationship between Π and L. As a result of this, P–L relations are typically shown and fit in log-log space, as a function of log period and magnitude.

In order to compare theory with observations and determine the pulsational mode of a star, pulsation is sometimes described using the pulsational constant Q, which is defined as:

$$Q = \Pi \left(\frac{M}{R^3} \right)^{1/2} \tag{11.5}$$

Here we solved for the constant term in Eq. 11.3 which we previously ignored. In practice, we typically cannot measure R directly, so we substitute R with L (luminosity) and $T_{\rm eff}$ (effective temperature) and get

$$Q = \Pi \frac{M^{1/2} T_{\rm eff}^3}{L^{3/4}} \times 5.13 \times 10^{-12} \tag{11.6}$$

for Π and Q in days and M and L in solar units. However, Q values can be difficult to calculate, because L is the bolometric luminosity. For a Mira-like star with $T_{\rm eff} \sim$ 3000 K, $Q \sim 0.1$ days for fundamental pulsators and $Q \sim 0.05$ days for overtone pulsators. In practice, some measured Q values have been more than a factor of 2 off from the theoretical values. Wood [28] argued that the differences in theoretical and measured Q could potentially be attributed to metallicity (discussed further in Sect. 11.4.4.1). However, the magnitude of metallicity dependence needed to account for this difference would be quite significant: a difference of 0.4 mag between a solar metallicity Mira one with metallicity comparable to the LMC. So far, such a strong dependence has not been observationally confirmed.

This discrepancy likely lies in the pulsational models used. Recently, [29, 30] have shown that linear calculations of overtone long-period variable sequences match well with observations. However, nonlinear pulsational models of O-rich AGB stars are required in order to reproduce the large-amplitude fundamental mode pulsations of Miras. These nonlinear models provide significantly better agreement with observations, by reproducing the earlier onset of dominant mode pulsation and shorter periods at larger radii that are present in observed P–L relations.

11.3.2 *Ages and Initial Masses of Miras*

Miras are the brightest stars in old or intermediate age populations. Similar to other fundamentally-pulsating variable stars, younger (and more massive) Miras are expected to have longer periods than older (and less massive) ones. This is known as the period-age (P-A) relation, and has been empirically motivated by stellar kinematics [31]. Theoretical investigations of the P-A relation for LPVs have generally been sparse due to the difficulty of modeling evolved red stars, but recent models have shown agreement with observations [32]. There is also evidence to suggest that C-rich Miras—which are confined to a smaller period range than O-rich Miras—may be slightly more massive than O-rich Miras at the same period and are generally younger stars.

Observational evidence comes from a range of studies in the Galaxy and Local Group. O-rich Mira variables with $\log P \sim 2.0 - 2.5$ days have been found in Galactic globular clusters [33], which suggests that they have low initial mass. Comparisons of O-rich Mira velocity dispersion with the age and velocity dispersion relations of stars in the solar neighborhood indicates that stars with $\log P \lesssim 2.3$ days have initial masses of $< 1 M_{\odot}$. Most Galactic O-rich Miras have $\log P \sim 2.5$ days and age ~ 7 Gyr. More massive Miras with $\log P \sim 2.65$ days are ~ 3 Gyr.

C-Miras on the other hand, have mean $\log P = 2.717$, age of ~ 1.8 Gyr, and initial mass of $1.8 \sim M_{\odot}$ [34]. Feast [35] provides a more detailed analysis of the relationship between ages, periods, and kinematics.

11.3.3 *Variability Classification*

Miras are a part of a broader group of variables known as long-period variables (LPVs), which have periods ranging from $\sim 100 - 1000$ days. Their P–L relations were first studied by [36], nearly a hundred years ago. He compared the periods and mean visual magnitudes of 10 LPVs with trignometric parallaxes. LPVs are very common since nearly all stars on the red giant branch or AGB will exhibit some form of variability and many of these will fall under the classification of LPV. As noted by [37], more luminous variables are also typically more variable, therefore the degree to which we can identify which stars are variable is dependent on the photometric precision of our measurements.

LPVs can typically be grouped into one of three classes: Miras, semi-regular variables (SRVs), and slow irregular variables. As the names suggest, Miras are the most regular pulsators amongst LPVs while semi-regular variables and irregular variables often exhibit variations in their periodicity. Miras are typically distinguished from SRVs by their larger amplitudes. A visual-band peak-to-trough amplitude larger than 2.5 mag within a single pulsational cycle *and* spectroscopic identification traditionally served as the cutoff between Miras and SRVs [38]. Recently, however, it has become more common to use only amplitude (often $\Delta I > 0.8$ mag, $\Delta K > 0.4$ mag) for classification since spectroscopic followup is limited and expensive [39]. While pulsational periods for Miras are generally very stable over many decades, with only an estimated 1% of them showing variations [40], it is more common to see stochastic variations in the cycle-to-cycle mean flux [41, 42]. Huang et al. [43] estimated this to be ~ 0.07 mag in H-band.

11.3.4 LMC

The Mira variables in the Large Magellanic Cloud (LMC) are the best studied and have informed our knowledge of Miras in environments where the data is sparser. The LMC has been a critical galaxy both in the history and development of Mira P–L relations and also in the extragalactic distance ladder, as an *anchor* galaxy (see Sect. 11.4.1).

With a small number of Miras from the Large Magellanic Cloud (LMC), [44] created the first P–L relation for Miras using bolometric magnitude. Subsequently, much of the history and study of Miras was a direct result of the availability of vast amounts of time-series photometry from microlensing surveys including the Optical Gravitational Lensing Experiment (OGLE, [45]) and MACHO [46]. Using MACHO data, [47] identified five parallel sequences (labeled $A - E$) on the Period-Luminosity plane, including Miras and SRVs. Miras were found to exist exclusively on sequence C and were confirmed to be radial, fundamental mode pulsators, now in good agreement with theoretical models [30].

Later, [48] determined that the K-band (2.2μm) Mira P–L relation exhibits a small scatter of $\sigma \sim 0.13$ mag. Using OGLE II and OGLE III time-series [45] and 2MASS photometry, [49] found that sequence C could further be separated into two – C and C' in the period-Wesenheit index plane. With spectroscopically-confirmed variables, they were able to attribute this separation to O- and C-rich surface chemistries. Whitelock et al. [50] examined the K-band P–L relations for the O- and C-rich subtypes separately and found a similar level of dispersion ($\sigma = 0.14$ and 0.15 mag, respectively). This dispersion is also roughly comparable to the scatter in the Cepheid K-band P–L relation ($\sigma = 0.09$ mag, [51]).

Ita and Matsunaga [26] showed that P–L relations for O-rich Miras in the LMC exist in a wide range of wavelengths, from I-band to [24] while C-rich Miras typically only exhibit a P–L relation at longer wavelengths (see Fig. 11.3). Since then, numerous other studies have examined P–L relations in the from optical ([24], using maximum light) to mid-infrared [52]. K-band is typically the bandpass of choice for ground-based observations [53] thanks to the low intrinsic scatter of the P–L and the Mira spectral intensity peaking around 1–2μm. There is also less circumstellar absorption in K than at shorter wavelengths, and less circumstellar emission than at longer wavelengths. Thus, NIR P–L relations are the most widely studied, including [3, 26, 47, 50, 54].

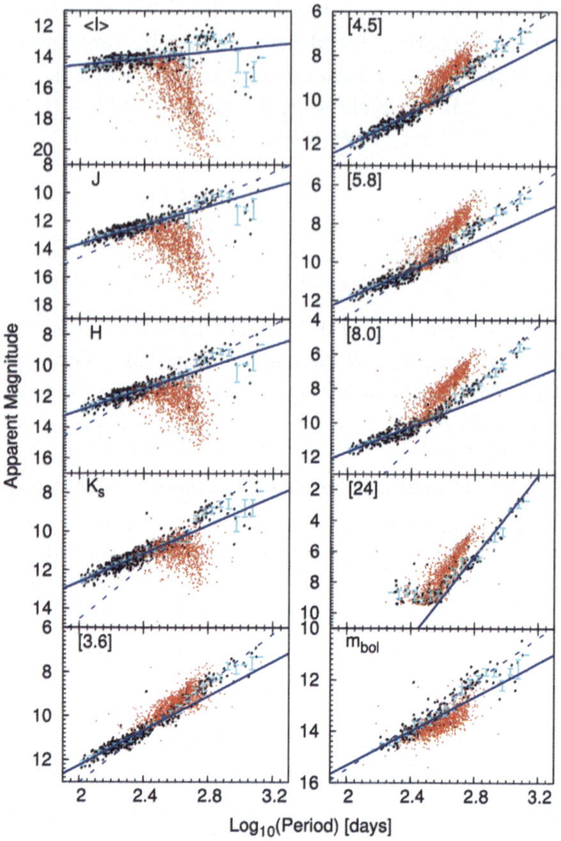

Fig. 11.3 Figure 2 from [26] showing the period-magnitude relations for OGLE III Miras in the LMC. The red points represent C-rich stars, while black points represent O-rich stars. The dashed and solid blue lines represent least-squares fits to the data. Mean magnitudes for the NIR and IR bands are derived from single-epoch observations, and thus show larger scatter

11.3.5 M33

M33 is another Local Group galaxy that has been extensively analyzed in variability studies and thus offers interesting comparisons between independent distance calibrators. There have been numerous Cepheid and TRGB distances to the galaxy, in addition to two JAGB distance determinations. Yuan et al. [55], which focused on the classification of the Miras in M33, the first to fit the Mira P–L using a quadratic relation and [56] was the first to derive a distance to M33 using Miras with both linear and quadratic P–L relations. The quadratic form of the P–L was empirically motivated by the steeper slope and excess of above-PL stars at longer periods due to hot-bottom burning. A quadratic functional form also allows Miras of all periods to be fit simultaneously. However, this form has not been physically motivated. HBB models generally still predict a linear relation at long periods and previous studies had addressed this discontinuity by using two separate P–L relations.

In [55], Miras were classified using a Random Forest classifier and novel semi-parameteric periodgram method developed and described in detail in [57]. $I-$band light curves obtained from the M33 Synoptic Stellar Survey were combined with near and mid-infrared wavelength observations from the 3.8m UK InfraRed Telescope (UKIRT), the 4-m Mayall telescope at Kitt Peak National Observatory (KPNO) and the 8-m Gemini North telescope. From this they obtained $\mu_{M33} = 24.80 \pm 0.06$ mag.

Most recently, [58] used photometry from several surveys, including PANSTARRS, PTF, and ZTF, to create well-sampled optical light curves and derive a distance modulus of 24.67 ± 0.06 mag to the galaxy using O-rich Mira variables. They fit both maximum and mean light P–L relations. Miras were classified using their separation along the $J - K_s$ period-color plane. The majority (96%) of periods were in agreement with those determined in the previous study by [55].

11.3.6 NGC 5128

NGC 5128 (Centaurus A) is the nearest giant elliptical galaxy and the first galaxy outside of the local group to have a Mira distance. It remains the only giant elliptical galaxy with a Mira distance.

Rejkuba [59] determined the distance to NGC 5128 using both Mira variables and infrared tip of the red giant branch (IR-TRGB) and found good agreement between the two methods, with $\mu_{Mira} = 27.92 \pm 0.19$, $\mu_{TRGBK} = 27.87 \pm 0.16$, and $\mu_{TRGBH} = 27.9 \pm 0.2$. Data used in the study taken using the ISAAC near-IR array on ESO VLT and absolute calibration for the Miras came from *Hipparcos* parallaxes and the K-band Mira P–L relation in the LMC. Observations consisted of 20 epochs of K_s-band observations spanning 1197 d and 1 epoch each of the J_s and H bands. Periods, amplitudes, and mean K_s magnitudes were them calculated for the variables with 10 or more measurements.

Only variables with regular periods, a sine-curve fit of $\chi^2 < 5$, and color bluer than $J_s - K_s \leq 1.4$ were used in the determination of distance. In addition, stars that were suspected to be aliased were not included in the fit. The mean log P for the remaining sample was 2.5 ($\sim$ 300 days). The color cut removed LPVs with more circumstellar reddening and also matched the colors of the Miras used for the calibration of the Mira P–L relation in the LMC ($J - K < 1.5$ [3]). Rejkuba [59] considered the possibility of hot-bottom burning in variables with $P > 400$ days. However, in practice, they also found that the distances determined with the full period range (up to just under 800 d at the longest end) and the shorter period variables only ($P < 400$ days) were nearly the same (differing by a few hundredths of a magnitude and well within the uncertainties). This could indicate that many stars, even at longer periods, are not actively undergoing hot-bottom burning.

11.3.7 Dwarf Galaxies

P–L relations also exist for some local group dwarf irregular galaxies, including NGC 6822 [60] and IC 1613 [61], which have distances determined using C-rich Miras. Perhaps most importantly for distance measurements, dwarf galaxies span a range of metallicities and star formation histories and thus are ideal environments for studying the relationship between stellar populations and galaxy evolution. While the C/O ratio is known to vary strongly with metallicity [62], studies of the AGB P–L relations in dwarf galaxies with a wide range of metallicities have been instrumental in showing that there is no measurable metallicity dependence for the P–L relation [63]. A more extensive review can be found in [64].

11.4 Measuring the Hubble Constant

Hubble's law states that:

$$v_r = H_0 \times D \tag{11.7}$$

where v_r is the recessional velocity and D is distance. Hubble's Constant can be thought of as the constant of proportionality of v_r and D. At large distances ($z \gtrsim 1$), this will relationship will begin to deviate from linearity depending on the cosmology of the universe. For nearby galaxies, local gravitational interactions can contribute significantly to or dominate the observed velocity.

Galaxies used to measure H_0 must be within the "Hubble flow". In other words, they must be in the regime where their recessional velocities are due almost entirely to the expansion of the Universe. In practice, this often means galaxies distances $D >$ 100 Mpc, or redshifts $z > 0.02$. Peculiar velocity corrections are typically applied to account for any remaining effects due to local gravitational interactions.

However, geometric methods generally cannot reach galaxies in the Hubble flow. Even using Miras as secondary distance indicators, only the most luminous and massive stars would be able to reach these distances directly and their P–L relations are less well-understood and studied (see Sect. 11.5). Thus, a typical path towards measuring H_0 consists of a three-rung ladder. The first rung uses nearby (in the case of Miras, $D < 7.5$ Mpc) geometric anchors which determine the absolute magnitude of the Mira P–L relation. Short-period Miras (typically hundreds can be discovered per galaxy) are then used to reach $<$ 40 Mpc distances, enough to measure the distances to and calibrate the peak luminosities of the nearest SN Ia, which are relatively rare (occurring at a rate of approximately 1 per galaxy per century). Thus, Miras serve as a bridge between the distances reachable by most geometric methods and the SN Ia in the Hubble flow that are capable of measuring H_0 directly.

In the low-redshift ($z \sim 0$) limit, the intercept of the Hubble diagram a_B is given by,

$$a_B = \log cz - 0.2m_B^0 \tag{11.8}$$

where m_B^0 is the standardized (corrected for variations from fiducial color, luminosity, or host dependence) maximum-light apparent magnitude of the SNe. From this equation, we can see that a_B, which is determined solely from the third rung of the distance ladder, does not depend on any Mira distance measurements.

In order to solve for H_0, we combine the absolute SN Ia peak magnitude along with the intercept for the SN Ia Hubble diagram to write,

$$\log H_0 = (M_B^0 + 5a_B + 25)/5 \tag{11.9}$$

where M_B^0 is the fiducial SN Ia absolute magnitude. We determine M_B^0 by measuring the distance to local SN Ia host galaxies using Miras (or other primary distance indicators), and using the distance to convert the observed apparent SN Ia magnitudes (m_B^0) to absolute magnitudes (M_B^0).

11.4.1 Anchors

The first rung of a cosmic distance ladder is often referred to as an *anchor* because it sets absolute scale for the later rungs using geometry. The most obvious way to obtain a geometric distance is through parallax. However, in the case of highly-extended and evolved stars like Miras the current consensus is that parallax measurements of AGB using *Gaia* have large, asymmetric, and underestimated uncertainties and should be treated with caution (see Sect. 11.4.4.2 for a more in-depth discussion). Thus, we currently do not recommend using parallax measurements for calibrating the Mira H_0 measurement.

Instead, since Miras are a ubiquitous, intermediate-age population, they can be discovered, and subsequently calibrated in almost any galaxy with a geometric distance. Mira-based H_0 measurements [65] rely primarily on geometric measurements to two anchor galaxies—the LMC and NGC 4258 [43].

11.4.1.1 NGC 4258

NGC 4258 is a relatively nearby water megamaser host galaxy which has a geometric distance of $D = 7.576 \pm 0.082$ (stat.)± 0.076 (sys.) Mpc with 1.4% precision [66]. Angular diameter distances to megamasers are obtained by combining spectral monitoring of the Keplerian motion of the masing clouds orbiting the black hole, along with VLBI mapping and a model of the maser disk geometry. Given that the galaxy is quite close, NGC 4258s peculiar motion is likely to contribute significantly to its recessional velocity. Thus, while it is unsuitable for direct H_0 measurements, it has served as an important anchor galaxy for Cepheids, TRGB, and Miras.

[43] first calibrated the distance to this galaxy using a quadratic P–L relation based on [54] and the 2.6% distance to the galaxy from [67, 68]. For P–L relation,

$$m_{\mathrm{F160W}} = a_0 - 3.59(\log P - 2.3) - 3.40(\log P - 2.3)^2 \quad (11.10)$$

where m_{F160W} is the magnitude in *F160W*, a_0 is the zeropoint, and P is period, in days. [43] obtained $a_0 = -6.15 \pm 0.09$ mag as the absolute mean magnitude of a 200-day O-rich Mira using the quadratic relation. Huang et al. [65] later updated this with the improved distance from [66] and used two linear P–L relations,

$$m_{\mathrm{F160W}} = a_0 - 3.64(\log P - 2.3) \quad (11.11)$$

$$m_{\mathrm{F160W}} = a_0 - 3.35(\log P - 2.3) \quad (11.12)$$

Eq. 11.11 used the linear slope derived for the $H-$band from [54] while Eq. 11.12 used a color transformation from H to *F160W* derived using O-rich Mira spectra to solve for a new slope. This resulted in $a_0 = -6.21 \pm 0.042$ mag for Eq. 11.11 and $a_0 = -6.25 \pm 0.042$ mag for Eq. 11.12.

The data for the Mira observations were taken using the HST WFC3/IR channel in two filters *F125W* (HST wide J) and *F160W* (HST wide H) in GO-13445 (PI: Bloom). Twelve epochs, spaced approximately monthly, were obtained of a single field over the course of a year. While the colors were originally intended to aid in separating the Miras into their O- and C-rich spectral subtypes, the color separation was found to be poor. Therefore, additional measures were taken to identify the two classes without using color information. Given that C-rich variables are rare at the shortest periods and also tend to have larger amplitudes, the sample was limited to $P < 300$ days and $0.4 \text{ mag} < \Delta F160W < 0.8$ mag to reduce C-rich contamination.

Of the 438 Mira candidates, 139 remained in the most stringent "gold" sample. For the gold sample, optical data from a previous Cepheid observing campaign [69] was used to verify that the objects were variable in the *F814W* (HST wide I) time-series. In addition, detection in *F814W* was also taken as a proxy for color since the redder C-rich Miras were expected to be difficult to detect in the Cepheid data. However, despite these additional requirements, the zeropoints obtained for the full sample and gold sample were within 0.02 magnitudes, well within the uncertainties.

Finally, the relative distance between NGC 4258 and the LMC was calculated after applying a Mira-specific color correction to ground-based H band P–L relation. This was found to be in agreement with the geoemtric distances for the two galaxies.

11.4.1.2 LMC (Again)

Huang et al. [65] used the 1.2% [70] distance to the LMC of $\mu_{\mathrm{LMC}} = 18.477 \pm 0.004(\text{stat}) \pm (\text{sys})$ mag to calculate an absolute calibration of $a_0 = -6.27 \pm 0.03$ mag for Eq. 11.12, consistent to within the uncertainties with the zeropoint of the Mira PLR from NGC 4258.

While LMC Miras were not directly observed with HST WFC3/IR, using the OGLE-III sample of O rich Mira variables [39, 43] determined the relative distance modulus between the LMC and NGC 4258,

$$\mu_{N4258} - \mu_{LMC} = 10.95 \pm 0.01(\text{stat}) \pm 0.06(\text{sys}) \text{ mag.} \quad (11.13)$$

which is consistent with the relative Cepheid distance between these two galaxies.

In order to directly compare the two samples without observations in the same bandpasses, [43] derived an H to $F160W$ transformation of

$$F160W = H + 0.39(J - H) \quad (11.14)$$

using O-rich Mira spectra from [71] as input into PySynPhot [72].

In addition, the sample of Mira variables from OGLE-III also included variables with a wider range of amplitudes and periods than those used in the NGC 4258 sample. Thus, the OGLE-III sample was restricted to only O-rich Miras with $240 < P < 400$ days and $0.4 < \Delta H < 0.8$ mag to provide a closer match to the HST sample for the calibration.

11.4.2 SN Ia Calibrators

The second rung and largest contributer to the overall H_0 error budget is the calibration of the SN Ia peak luminosity with Miras in local SN Ia host galaxies. To date, only two SN Ia have had luminosities calibrated with Mira distances.

11.4.2.1 NGC 1559

Huang et al. [65] determined the distance modulus to NGC 1559, the host of SN 2005df to be $\mu_{N1559} = 31.41 \pm 0.050(\text{stat}) \pm 0.060(\text{sys})$ using a sample of 115 Miras. At 19.1 ± 1.1 Mpc, NGC 1559 is the most distant galaxy in which Miras have been studied. The distance for this galaxy was used by [65] to determine a fiducial SN Ia luminosity of $M_B^0 = -19.27 \pm 0.13$ mag, and in agreement with the [73] Cepheid measurement of $M_B^0 = -19.360 \pm 0.106$ mag for this galaxy. Using both the NGC 4258 and LMC P–L relations as anchors, this results in $H_0 = 73.3 \pm 4.0$ km s^{-1} Mpc $^{-1}$. The methodology laid out in this paper is the blueprint for measuring Miras in other local SN Ia hosts. However, because this H_0 measurement uses only one SN Ia calibrator, the 5% uncertainty is dominated by the statistical uncertainty in the peak magnitude of the single SN Ia and the result is within 2-σ agreement with both *Planck* and the most precise Cepheid measurement.

Data used for the NGC 1559 study consisted of 10 epochs of WFC3/IR channel *F160W* observations (GO-15145; PI: Riess) with roughly monthly spacing and a baseline slightly longer than 1 year. Unlike most Mira studies, the NGC 1559 Mira sample did not use color as a selection criterion since [43, 74] showed that HST wide-band colors are an ineffective discriminant of both O- and C-rich AGB stars in general and Miras in particular. Instead, amplitude and period cuts were used to limit the C-rich contamination. The remaining C-rich contamination (estimated to have an bias of ~ -0.07 mag in the worst-case) was estimated by using the LMC Mira sample to model a pure O-rich sample and a C-rich contaminated sample. Because C-rich Miras are expected to be fainter in *F160W* than O-rich Miras and have a different period distribution than O-rich Miras, the shape of the zeropoint-vs-period curve ($Z(P)$) is different for a pure O-rich sample and a sample contaminated with C-rich Miras. The contaminated sample is then scaled to fit the $Z(P)$ relation of NGC 1559, yielding a correction of -0.057 ± 0.024 mag and no correction when applied to NGC 4258. This method was tested on the M33 Mira sample from [56] by creating a contaminated population and then correcting the zeropoint. The correction of -0.04 ± 0.01 mag was found to be in agreement with the true value of -0.034 mag.

11.4.2.2 M101

M101 is the most recent SN Ia host galaxy to have a distance measurement using Miras (Huang+2023). 288 Ω-rich Mira candidates were discovered in the field of SN 2011fe using a long baseline of HST observations, spanning $\sim$ 2900 days. Many of the observations were originally taken to the study the late-time lightcurve of the SN. From the sample of Miras with periods between 260 and 400 d, the distance modulus to M101, using the LMC and NGC 4258 as anchors, was determined to be $\mu_{\mathrm{M101}} = 29.10 \pm 0.06$ mag. In combination with the SN Ia calibration from NGC 1559, this yielded $M_B^0 = -19.27 \pm 0.09$ mag. The resulting $\mathrm{H_0}$ measurement combining both SN Ia calibrators is 73.14 ± 2.98 km s^{-1} Mpc^{-1}. $\mathrm{H_0} = 72.37 \pm 2.97$ kms^{-1} Mpc^{-1}, and confirmed previous indications that the local universe value of H_0 is higher than the early-universe value at $\sim$ 95% confidence. This 4% measurement is still dominated by the statistical uncertainty in the peak luminosity of the two SN Ia calibrators.

As the nearest SN Ia host galaxy with modern (CCD) photometry, the distance to this galaxy has also been measured numerous times using Cepheids and TRGB, offering a opportunity for comparison with the two most commonly-used primary distance indicators. The recent Mira measurement falls within 1-σ of most other recent measurements with Cepheids and TRGB.

11.4.3 Methodology

The methodology we describe here applies primarily to the targeted Mira searches in galaxies calibrated using sparse HST data. Ground-based observations have used diverse observing strategies and typically prefer K-band for P–L relations. However, Mira H_0 measurements have necessitated the use of HST to resolve stellar populations in SN Ia host galaxies and therefore use a slightly shorter bandpass – WCF3/IR's *F160W* ($\sim 1.6\mu$m). This is the longest wavelength wide-band filter available with HST and also coincides well with the peak of a typical Mira's spectral energy distribution.

11.4.3.1 Observations

Observational baselines for Miras span around one year or slightly longer and typically consist of $\sim$ 10 epochs, with roughly monthly spacing. To reduce aliasing, it is generally preferable to space the observations such that the Miras at the target period ranges (usually $\sim 100-400$ days) will be sampled a minimum of three phase points in one cycle.

The initial steps of the variability search are similar to the process for Cepheids described in several SH_0ES papers including [75–77]. Once the data is collected, a variability search is performed to cull as many nonvariable objects as possible while retaining the variables. Crowded-field photometry is used to create lightcurves for the candidates, which are then cut down to a sample of O-rich Mira candidate variables using selection criteria.

The major difference in the analysis process between Cepheids and Miras lies in the wavelengths used for determining the positions for the variables. Miras are very red stars and therefore typically not observed in optical wavelengths. With Cepheids, the finer-resolution optical observations are generally used to classify the stars and pinpoint their locations. However, given that they are an older population, Miras are generally found in less dusty areas than the younger Cepheids, and thus this entire search is performed only in the NIR.

11.4.3.2 Selection Criteria

Selection criteria are used to separate the Miras from SRVs and other potential contaminants. Unlike Cepheids, which have a distinctive sawtooth shape in optical wavelengths and are often identified on the basis of their fit to templates, there currently are no templates for Miras and their light curve shapes of Miras in optical wavelengths. However, even in the optical, most ($\sim$ 80%) short-period Miras do not show substantial deviations from a symmetric light curve [78], and $\sim$70% of them do not significantly deviate from a purely sinusoidal shape [79].

Similarly to Cepheids, the lightcurves of Miras in NIR are even more sinusoidal (see Fig. 11.4 showing composite light curves at a range of periods). Therefore, the period, amplitude, and mean magnitude of Miras are typically fit using only a sine function. When [43] applied a Bayesian information criteria before adding additional harmonics, they found that $\sim$ 99% of Miras favored a first order fit.

Amplitude cuts are used to remove SRVs or potential C-rich Miras (which often have larger amplitudes). An initial period cut is also typically applied to remove

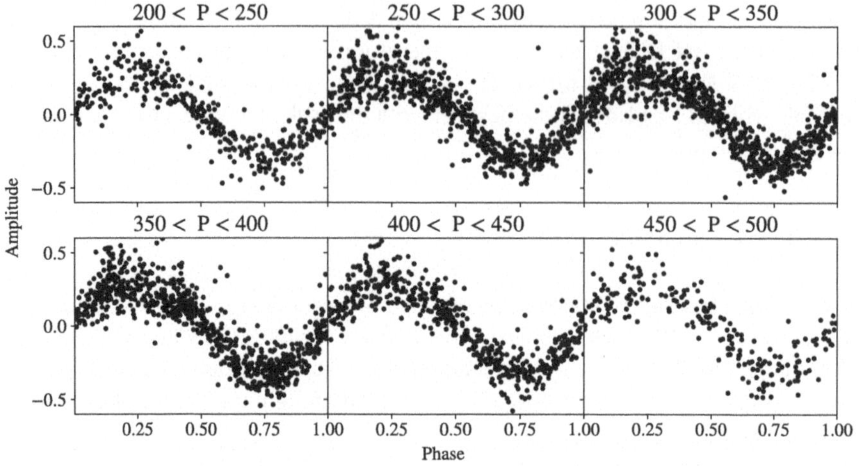

Fig. 11.4 Figure 3 from [80] showing the stacked lightcurves of ∼ 300 Miras in the SN Ia host galaxy M101. Individual observations are shown with their fit phase, mean magnitude is subtracted, but amplitudes are not scaled

Table 11.1 Mira selection criteria

	M101	NGC 1559	NGC 4258 (gold)
Period cut	200 d $< P <$ 500 d	240 d $< P <$ 400 d	$P <$ 300 d
Amplitude cut	0.4 mag $< \Delta F160W <$ 0.8 mag	0.4 mag $< \Delta F160W <$ 0.8 mag	0.4 mag $< \Delta F160W <$ 0.8 mag
Surface brightness cut	–	421 counts/second	–
Color Cut	$m_{F110W} - m_{F160W} <$ 1.3 mag	–	$m_{F125W} - m_{F160W} <$ 1.3 mag
F814W Detection	–	–	Slope-fit to *F814W* data $> 3\sigma$
F814W Amplitude	–	–	Δ *F814W* $>$ 0.3 mag
F-statistic :	$\chi_s^2/\chi_l^2 < 0.5$	$\chi_s^2/\chi_l^2 < 0.5$	–

A comparison of the criteria for the final Mira sample in M101 with those used for NGC 1559 and NGC 4258. Owing to differences in available data, we were unable to match the criteria exactly between all of the observations

objects with periods that are either too short (<100 days) or more likely to be C-rich or hot-bottom burning (>300–400 days). A summary and comparison of the cuts used in a few recent studies is shown in Table 11.1.

At this stage, the remaining objects are likely to be Miras, but further corrections may be applied to remove objects that are found to be particularly crowded (either through surface brightness cuts or cuts on the crowding corrections) or below the completeness limit and thus may bias the zeropoint of P–L relation. C-rich contamination corrections are also typically applied, with [65] estimating the level to be a few hundredths of a magnitude.

11.4.4 Other Considerations

11.4.4.1 Metallicity

As touched upon in Sect. 11.3.1, theory suggests that the Mira P–L relation may have a dependence on metallicity. Wood [28] noted that if two stars with the same mass but different metallicities begin to pulsate at the same radius (and thus, period), they will exhibit different luminosities. However, this assumes that these stars would become unstable to Mira-like pulsations at the same radius. It is also possible that Miras with the same mass but different metallicity would begin to pulsate at different radii, which may have the effect of mitigating the change in luminosity. Thus far, the agreement between theory and observations of Miras in a wide range of metallicities suggests that the dependence of the zeropoint of the P–L on metallicity is weak. For a more detailed explanation, see [81].

As discussed in Sect. 11.3.7, dwarf galaxies are the ideal testing grounds to understand the connection between galaxy evolution and stellar populations. Most recently, [63] examined the IR P–L relations of AGB stars in environments with metallicity ranging from 1.4% solar metallicity ([Fe/H] = -1.85) to 50% solar metallicity ([Fe/H] = -0.38) and found no discernible difference in the pulsational properties of stars in metal-poor and metal-rich environments. Similarly, [24] also found that the mean K_s band O-rich Mira measurement between the LMC and SMC showed no sign of a metallicity effect.

Other works have put tentative upper limits on the metallicity effect by showing that agreement in Mira P–L relations without metallicity corrections is within a few hundredths of a magnitude in the NIR. A study of the K-band P–L relation of Miras in the LMC, Galactic globular clusters, and Galaxy [82] showed agreement to within the measurement uncertainties. Whitelock et al. [50] found an upper limit of ~ 0.1 mag in the zeropoint of LMC and Galactic O-rich Miras with *Hipparcos* and VLBI parallaxes and those in globular clusters. However, when [43] used the same data, but an updated LMC distance from [70] this discrepancy dropped to 0.01 mag. For HST *F160W*, the wide-band filter used for H_0 measurements, [43] found ~ 0.03 mag agreement (within errors) in the relative distance modulus between the LMC and NGC 4258 determined using Miras that from Cepheids and from geometry. Together, this suggests that the AGB P–L relations are a reliable tool for determine distances regardless of metallicity, at least down to the ~ 0.03 mag level in the NIR.

11.4.4.2 Parallaxes

While other distance indicators such as Cepheids or RR Lyrae have been calibrated using *Gaia* parallaxes, parallax measurements are difficult to obtain for large, red, evolved stars such as Miras using an optical telescope like *Gaia*. Typical radii for Miras range from $250R_\odot - 800R_\odot$ [83]. Xu et al. [84] found that the redder the star, the larger parallax uncertainties in *Gaia* DR2. Red stars, including AGB stars, typically have more complex surface dynamics, larger physical size, and more dust [85]. Empirically, the Galactic Mira PLR using *Gaia* shows large scatter.

Gaia parallax uncertainties for Miras and AGB stars in general are both underestimated (up to a factor of 5.44 for sources with $G < 8$ mag) and asymmetrical. Andriantsaralaza et al. [86] caution against using parallaxes when uncertainties are larger than 20% and employing the *Gaia* catalogue parameter RUWE (re-normalised unit weight error) as the sole measure of quality of *Gaia* astrometric data for AGB stars. They were able to obtain distance measurements to AGB stars using maser observations with very long baseline interferometry (VLBI). Unfortunately, this method can only measure distances to Miras with substantial circumstellar envelopes, and is typically not applicable to the Miras currently used for the distance ladder. However, it is appropriate as a cross-check to the AGB distances obtained by *Gaia*.

Difficulties in parallax measurements with Miras were not unexpected and several factors—both astrophysical and instrumental, contribute. Physically, Mira radii are larger than 1 AU—in other words, their angular diameters are larger than their parallaxes. Both angular size and parallax scale as $1/d$, where d is distance to the star. Thus, the physical size of a star is a problem whether the Mira is located nearby or far away. On its face, this alone does not appear to be particularly problematic for parallax measurements since an astrometric solution can be found using only on the motion of the photocenter of the extended source. However, Miras are not uniformly-illuminated discs. Moreover, they have complex stellar surface dynamics caused by large convective cells that can result in the motion of their stellar photocenters [87]. Bright sources, including Galactic Miras, can also reach instrumental saturation which results in less accurate astrometry [88].

Hipparcos had more success obtaining parallaxes for Miras. Whitelock and Feast [89] examined subsets of the data and found that the best solution excluded short-period red group stars and also low-amplitude variables. Whitelock et al. [50] used the revised *Hipparcos* parallaxes from [90] data to obtain a calibration for Galactic, O-rich Mira variables in the K band of $\delta = -7.25 \pm 0.07$ mag.

11.5 Long-Period Miras

The long-period end of the Mira distribution is less studied than the short-period objects discussed throughout the this review for a number of reasons. In addition to more complicated physics (including the effects of hot-bottom burning, heavier mass-loss, and larger circumstellar envelopes compared to short-period Miras), they require extended observational baselines to accurately determine their periods and are formed from stars of higher initial mass and are therefore considerably rarer, younger, and more likely to be located closer to dusty regions where they were born than short-period Miras. Additionally, OH/IR stars, which are bright infrared sources with Type II OH masing emission and intermediate initial mass, fall into this category [91] and some of these extreme objects have periods as long as 1800 d. Long-period Miras also often follow a different P–L relation than their shorter-period counterparts.

Despite these difficulties, there remains great interest in studying long-period Miras both for distance ladder applications and as tracers of structure [92]. Assuming the same P–L slope as the shorter-end, their $F160W$ luminosities will *start* at -7.3 mag on the low (400 days) end. As seen in Fig. 11.3, the slope for these longer-period objects can often be steeper than at the short-period end, likely due to hot-bottom burning [4]. Thus, the brightest objects can be up to 2.5 magnitudes more luminous and reach distances >50 Mpc with Hubble Space Telescope. Long-period objects also appear to have tighter P–L relations in the mid-infrared which may be particularly advantageous with James Webb Space Telescope observations.

Very long period ($P \gtrsim 1000$ day) Miras have been discovered locally in the SMC, LMC, and Galactic Bulge by OGLE [37, 39, 93] and in the nearby galaxies Sgr dIC and NGC 3109 [94, 95] using the JHK_s bands from the InfraRed Survey Facility (IRSF). The SPIRITS [96, 97] and DUSTiNGS [2, 63, 98] collaborations have also found hundreds of long-period variables using 3.6µm and 4.5µm that are likely Miras in a wide range of host environments using *Spitzer* Space Telescope some of them with $P \sim 1000$ days.

11.6 Comparison with Other Indicators

While there are numerous excellent stellar distance indicators, we will limit this comparison to those that have currently been shown to be capable of bridging the gap between nearby, geometric measures and the local host galaxies of SN Ia. This includes Cepheids, the Tip of the Red Giant Branch (TRGB), J-Region Asymptotic Giant Branch (JAGB) stars [99], and short-period Miras and typically excludes fainter stellar indicators such as RR Lyrae and Type II Cepheids. Long-period Miras are quite different, so for brevity when we refer to Miras we refer exclusively to short-period O-rich Miras (Table 11.2).

Luminosity: Cepheids and short-period Miras are roughly comparable in luminosity the NIR. Both are $\gtrsim 10^4 L_\odot$ in H-band. JAGB (which includes C-rich Miras) are brighter than the shortest-period Miras ($P < 200$ days), but typically fainter than the brightest Miras. As a result, despite requiring time-series photometry, Miras and Cepheids can be more efficient than TRGB and JAGB larger distances (e.g. $D > 10 - 15$ Mpc). It is necessary to resolve stars $0.5 - 1.0$ mag below the tip/main clump for TRGB and JAGB measurements. On the other hand, peak-to-trough amplitudes of Cepheids and Miras generally span < 0.3 mag and $0.4 - 0.8$ mag respectively, and must also be detected in order to determine their mean magnitudes.

Age: Cepheids are young, massive stars, typically found clustered near dusty spiral arms and star-forming regions and are on average more crowded than the older indicators. They are generally found in the disk of spiral galaxies. JAGB are C-rich stars and trend younger than O-rich Miras, but are still found in many galaxy morphologies. TRGB are a ubiquitious older population and are best observed in the halos of galaxies [102, 103].

Table 11.2 A comparison of the stellar distance indicators that have been shown to be capable of reaching SN Ia host galaxy distances. For the sake of brevity, Miras refers to O-rich, short period ($P \lesssim 400$ days) Miras

Indicator	Age	Time-series	mag_H	Optical data
Miras:	$2 - 8$ Gyr	Yes^a	$< -6.25^b$	No
Cepheids:	$\lesssim 100$ Myr	Yes^c	-6.1^d	Yes
TRGB:	$\gtrsim 4 - 5$ Gyr	No	–	Yes
IR-TRGB:	$\gtrsim 4 - 5$ Gyr	No	$[-5.5 \text{ to } -6.5]^e$	No
JAGB:	300 Myr–1 Gyr	No	$-6.2[J], -6.8[H]^f$	No

A comparison of the stellar distance indicators that have either been shown (Cepheids, TRGB, Miras) or are thought to be capable of reaching SN Ia host galaxy distances (JAGB). For the sake of brevity, Miras refers to O-rich, short period ($P \lesssim 400$ days) Miras^a timescale >1 year; ^{b} for $P > 200$ d; ^{c} timescale > 60 days; ^{d}for $P > 15$ d; ^{e} IR-TRGB luminosity depends strongly on metallicity; ^{f} JAGB is most-commonly studied in the $J-$band (calibration from [100]) but a recent work by [101] provided a number of H-band absolute magnitude calibrations

Time-Series Requirements: A major advantage that TRGB and JAGB have over Cepheids or Miras is that they do not require time-series observations since both are identified using their positions on a color-magnitude diagram (CMD). This makes TRGB and JAGB more efficient for nearby galaxies $D \lesssim 10 - 15$ Mpc, since only a single epoch of observation is necessary.

11.7 Conclusions and Future Prospects

As the most luminous of the intermediate-to-old stellar distance indicators, Miras are in a unique position to expand the sample of local SN Ia with calibrated peak magnitudes and are also the only distance indicator capable of cross-checking all of the others. They will be particularly exciting in the upcoming decade with the increase focus on NIR and IR observations.

Presently, due to the small number of SN Ia calibrators, the uncertainty in the Mira-based H_0 is dominated by the statistical uncertainty in the peak magnitude of SN Ia. The best path forward to improving H_0 measurements with Miras is to increase the number of SN Ia that have been calibrated with Miras. This can be done either by revisiting the local host galaxies that already have Cepheid distances—allowing for a direct comparison of Mira and Cepheid distances—and by searching for them in the galaxies of SN Ia that have not yet been calibrated with other indicators due to their distance, or lack of star formation.

James Webb Space Telescope (JWST), while not optimally suited for obtaining time-series, is able to photometrically separate O- and C-rich Miras in SN Ia host galaxies with medium-band filters and will also be able to un-crowd the backgrounds of previously-observed variables, similar to what has currently been done for Cepheids [104]. JWST also has the sensitivity to measure a two-rung H_0 using geometry and Miras only—without the inclusion of SNe Ia. However, it suffers from large overheads that make multi-epoch visits undesirable.

The upcoming Vera C. Rubin Observatory's Legacy Survey of Space and Time, will cover nearly half of the sky in u, g, r, i, z, and y filters. Rubin has the potential to be particularly useful in developing the first rung of the Mira distance ladder, since it is expected to discover variables in galaxies out < 15 Mpc [105]. While this is too small of a volume to substantially increase the number of SN Ia hosts that can be calibrated with Miras, it is possible that some of these galaxies can eventually become additional geometric anchors. Miras will be brightest in the r, i, and z bands, and most will be discovered in the i band. As previous optical studies have found, maximum-light Period-Luminosity Relations may be more useful and have smaller scatter than mean-light relations for these shorter wavelengths.

The Nancy Grace Roman Observatory will have a similar sensitivity and resolution to Hubble, but will cover a much larger field-of-view. This will allow it to continue to study Miras at the distances of SN Ia hosts. In addition, it has a redder, K-band equivalent that may produce tighter P–L relations for Miras than the HST *F160W* (H) band currently used in the distance ladder. Given its the large field of view, many Local Group Miras are also likely to fall into survey regions, potentially providing NIR data for previously-identified stars.

Acknowledgements I would like to thank Eleonora di Valentino and Dillon Brout for inviting me to write this review chapter and for their hard work in assembling this textbook. I also thank Morgan MacLeod, John Menzies, Patricia A. Whitelock, Louise Breuval, and Andrea Sacchi for their helpful comments and discussions regarding this review. Additionally, I thank Morgan MacLeod for his help in the creation of Fig. 11.1.

References

1. N.N. Samus', E.V. Kazarovets, O.V. Durlevich, N.N. Kireeva, E.N. Pastukhova, General catalogue of variable stars: version GCVS 5.1. Astron. Rep. **61**, 80–88 (2017)
2. M.L. Boyer, K.B.W. McQuinn, P. Barmby, A.Z. Bonanos, R.D. Gehrz, K.D. Gordon, M.A.T. Groenewegen, E. Lagadec, D. Lennon, M. Marengo, M. Meixner, E. Skillman, G.C. Sloan, G. Sonneborn, J.T. van Loon, A. Zijlstra, An infrared census of dust in nearby galaxies with spitzer (dustings) I overview. ApJS **216**, 10 (2015)
3. M.W. Feast, I.S. Glass, P.A. Whitelock, R.M. Catchpole, A period-luminosity-colour relation for Mira variables. MNRAS **241**, 375–392 (1989)
4. P.A. Whitelock, M.W. Feast, J.T. van Loon, A.A. Zijlstra, Obscured asymptotic giant branch variables in the Large Magellanic Cloud and the period-luminosity relation. MNRAS **342**, 86–104 (2003)
5. A. Dotter, MESA isochrones and stellar tracks (MIST) 0: methods for the construction of stellar isochrones. ApJS **222**, 8 (2016)
6. J. Choi, A. Dotter, C. Conroy, M. Cantiello, B. Paxton, B.D. Johnson, Mesa isochrones and stellar tracks (MIST) I. Solar-scaled Models. ApJ **823**, 102 (2016)
7. B. Paxton, P. Marchant, J. Schwab, E.B. Bauer, L. Bildsten, M. Cantiello, L. Dessart, R. Farmer, H. Hu, N. Langer, R.H.D. Townsend, D.M. Townsley, F.X. Timmes, Modules for experiments in stellar astrophysics (MESA): binaries, pulsations, and explosions. ApJS **220**, 15 (2015)

8. S. Höfner, H. Olofsson, Mass loss of stars on the asymptotic giant branch: mechanisms, models and measurements. A & A Rev **26**, 1 (2018)
9. J. Iben, Asymptotic giant branch stars: thermal pulses, carbon production, and dredge up; neutron sources and s-process nucleosynthesis, in *Evolution of Stars: The Photospheric Abundance Connection*, ed. by G. Michaud, A.V. Tutukov, vol. 145, p. 257 (1991)
10. C.A. Frost, J.C. Lattanzio, On the numerical treatment and dependence of the third dredge-up phenomenon. ApJ **473**, 383 (1996)
11. P.R. Wood, Final stages of AGB evolution (Invited Review), in *Planetary Nebulae*, ed. by H.J. Habing, H.J.G.L.M. Lamers, vol. 180, p. 297 (1997)
12. M. Busso, R. Gallino, G.J. Wasserburg, Nucleosynthesis in asymptotic giant branch stars: relevance for galactic enrichment and solar system formation. ARA&A **37**, 239–309 (1999)
13. F. Herwig, Evolution of Asymptotic Giant Branch Stars. ARA&A **43**, 435–479 (2005)
14. A.I. Karakas, J.C. Lattanzio, The Dawes review 2: nucleosynthesis and stellar yields of low- and intermediate-mass single stars. PASA **31**, e030 (2014)
15. M. Schwarzschild, R. Härm, Thermal instability in non-degenerate stars. ApJ **142**, 855 (1965)
16. J. Iben, On the interior properties of red giants, in *Physical Processes in Red Giants*, ed. by J. Iben, A. Renzini. Astrophysics and Space Science Library, vol. 88 (1981), pp. 3–24
17. A.I. Karakas, J.C. Lattanzio, O.R. Pols, Parameterising the third dredge-up in asymptotic giant branch stars. PASA **19**, 515–526 (2002)
18. T. Bloecker, D. Schoenberner, A 7M AGB model sequence not complying with the core mass-luminosity relation. A&A **244**, L43–L46 (1991)
19. J.C. Lattanzio, Hot bottom burning in a 5 solar mass model. PASA **10**, 120 (1992)
20. A.I. Boothroyd, I.J. Sackmann, G.J. Wasserburg, Hot bottom burning in asymptotic giant branch stars and its effect on oxygen isotopic abundances. ApJL **442**, L21 (1995)
21. P. Marigo, A. Bressan, A. Nanni, L. Girardi, M.L. Pumo, Evolution of thermally pulsing asymptotic giant branch stars-I the COLIBRI code. MNRAS **434**, 488–526 (2013)
22. A.I. Boothroyd, I.J. Sackmann, Breakdown of the core mass-luminosity relation at high luminosities on the asymptotic giant branch. ApJL **393**, L21 (1992)
23. P.A. Whitelock, J.W. Menzies, M.W. Feast, N. Matsunaga, T. Tanabé, Y. Ita, Asymptotic giant branch stars in the Fornax dwarf spheroidal galaxy. MNRAS **394**, 795–809 (2009)
24. A. Bhardwaj, S. Kanbur, S. He, M. Rejkuba, N. Matsunaga, R. de Grijs, K. Sharma, H.P. Singh, T. Baug, C.-C. Ngeow, J.-Y. Ou, Multiwavelength period-luminosity and period-luminosity-color relations at maximum light for mira variables in the magellanic clouds. ApJ **884**, 20 (2019)
25. C.C. Ngeow, J.Y. Ou, A. Bhardwaj, J. Purdum, B. Rusholme, A. Wold, Zwicky transient facility and globular clusters: the gr-band period-luminosity relations for mira variables at maximum light and their applications to local galaxies (2023). arXiv e-prints: arXiv:2307.06749
26. Y. Ita, N. Matsunaga, Period-magnitude relation of Mira-like variables in the Large magellanic cloud as a tool to understanding circumstellar extinction. MNRAS **412**, 2345–2352 (2011)
27. H.S. Leavitt, E.C. Pickering, Periods of 25 variable stars in the small magellanic cloud. Harvard College Obs. Circ. **173**, 1–3 (1912)
28. P.R. Wood, Mira variables: pulsation, mass loss and evolution, in *Confrontation Between Stellar Pulsation and Evolution*, ed. by C. Cacciari, G. Clementini. Astronomical Society of the Pacific Conference Series, , vol. 11 (1990), pp. 355–363
29. M. Trabucchi, P.R. Wood, J. Montalbán, P. Marigo, G. Pastorelli, L. Girardi, A new interpretation of the period-luminosity sequences of long-period variables. ApJ **847**, 139 (2017)
30. M. Trabucchi, P.R. Wood, N. Mowlavi, G. Pastorelli, P. Marigo, L. Girardi, T. Lebzelter, Modelling long-period variables–II: fundamental mode pulsation in the non-linear regime. MNRAS **500**, 1575–1591 (2021)
31. M.W. Feast, The long period variables. MNRAS **125**, 367 (1963)
32. M. Trabucchi, N. Mowlavi, The period-age relation of long-period variables. A&A **658**, L1 (2022)
33. M. Feast, P. Whitelock, J. Menzies, Globular clusters and the Mira period-luminosity relation. MNRAS **329**, L7–L12 (2002)

34. M.W. Feast, P.A. Whitelock, J.W. Menzies, Carbon-rich Mira variables: kinematics and absolute magnitudes. MNRAS **369**, 791–797 (2006)
35. M.W. Feast, The ages, masses, evolution and kinematics of mira variables, in *AGB Stars and Related Phenomena*, ed. by T. Ueta, N. Matsunaga, Y. Ita (2009), p. 48
36. B.P. Gerasimovic, The absolute magnitudes of long period variable stars. Proc. Natl. Acad. Sci. **14**, 963–968 (1928)
37. I. Soszyński, A. Udalski, M.K. Szymański, M. Kubiak, G. Pietrzyński, Ł Wyrzykowski, K. Ulaczyk, R. Poleski, S. Kozłowski, P. Pietrukowicz, J. Skowron, The optical gravitational lensing experiment. The OGLE-III catalog of variable stars XV. Long-period variables in the galactic bulge. AcA **63**, 21–36 (2013)
38. P.N. Kholopov, N.N. Samus, E.V. Kazarovets, N.B. Perova, The 67th name-list of variable stars. Inf. Bull. Variable Stars **2681**, 1 (1985)
39. I. Soszyński, A. Udalski, M.K. Szymański, M. Kubiak, G. Pietrzyński, Ł Wyrzykowski, O. Szewczyk, K. Ulaczyk, R. Poleski, The optical gravitational lensing experiment. The OGLE-III catalog of variable stars IV. Long-period variables in the large magellanic cloud. AcA **59**, 239–253 (2009)
40. A.A. Zijlstra, T.R. Bedding, Period evolution in Mira variables. JAAVSO **31**, 2–10 (2002)
41. J.A. Mattei, Introducing Mira variables. JAAVSO **25**, 57–62 (1997)
42. P.A. Whitelock, F. van Leeuwen, M.W. Feast, The luminosities and diameters of Mira variables from HIPPARCOS parallaxes, in *Hipparcos–Venice 1997*, ed. by R.M. Bonnet, E. Høg, P.L. Bernacca, L. Emiliani, A. Blaauw, C. Turon, J. Kovalevsky, L. Lindegren, H. Hassan, M. Bouffard, B. Strim, D. Heger, M.A.C. Perryman, L. Woltjer. ESA Special Publication, vol. 402 (1997), pp. 213–218
43. C.D. Huang, A.G. Riess, S.L. Hoffmann, C. Klein, J. Bloom, W. Yuan, L.M. Macri, D.O. Jones, P.A. Whitelock, S. Casertano, R.I. Anderson, A near-infrared period-luminosity relation for Miras in NGC 4258, an anchor for a new distance ladder. ApJ **857**, 67 (2018)
44. I.S. Glass, T.L. Evans, A period-luminosity relation for Mira variables in the large magellanic cloud. Nature **291**, 303–304 (1981)
45. A. Udalski, M.K. Szymanski, I. Soszynski, R. Poleski, The optical gravitational lensing experiment. final reductions of the OGLE-III data. AcA **58**, 69–87 (2008)
46. C. Alcock, R.A. Allsman, T.S. Axelrod, D.P. Bennett, K.H. Cook, K.C. Freeman, K. Griest, J.A. Guern, M.J. Lehner, S.L. Marshall, H.S. Park, S. Perlmutter, B.A. Peterson, M.R. Pratt, P.J. Quinn, A.W. Rodgers, C.W. Stubbs, W. Sutherland, The MACHO project first-year large magellanic cloud results: the microlensing rate and the nature of the galactic dark halo. ApJ **461**, 84 (1996)
47. P.R. Wood, C. Alcock, R.A. Allsman, D. Alves, T.S. Axelrod, A.C. Becker, D.P. Bennett, K.H. Cook, A.J. Drake, K.C. Freeman, K. Griest, L.J. King, M.J. Lehner, S.L. Marshall, D. Minniti, B.A. Peterson, M.R. Pratt, P.J. Quinn, C.W. Stubbs, W. Sutherland, A. Tomaney, T. Vandehei, D.L. Welch, MACHO observations of LMC red giants: Mira and semi-regular pulsators, and contact and semi-detached binaries, *Asymptotic Giant Branch Stars*, ed. by T. Le Bertre, A. Lebre, C. Waelkens, vol. 191 (1999), p. 151
48. I.S. Glass, T. Lloyd Evans, The calibrating stars of the Mira P-L relation. MNRAS **343**, 67–74 (2003)
49. I. Soszynski, A. Udalski, M. Kubiak, M.K. Szymanski, G. Pietrzynski, K. Zebrun, O. Szewczyk, L. Wyrzykowski, K. Ulaczyk, The optical gravitational lensing experiment. Miras and semiregular variables in the large magellanic cloud. AcA **55**, 331–348 (2005)
50. P.A. Whitelock, M.W. Feast, F. Van Leeuwen, AGB variables and the Mira period-luminosity relation. MNRAS **386**, 313–323 (2008). (May)
51. L.M. Macri, C.C. Ngeow, S.M. Kanbur, S. Mahzooni, M.T. Smitka, Large magellanic cloud near-infrared synoptic survey. I cepheid variables and the calibration of the Leavitt law. AJ **149**, 117 (2015)
52. P. Iwanek, I. Soszyński, S. Kozłowski, Mid-infrared period-luminosity relations for Miras in the large magellanic cloud. ApJ **919**, 99 (2021)

53. P.A. Whitelock, Asymptotic giant branch variables in the galaxy and the local group. ApSS **341**, 123–129 (2012)
54. W. Yuan, L.M. Macri, S. He, J.Z. Huang, S.M. Kanbur, C.C. Ngeow, Large magellanic cloud near-infrared synoptic survey V. period-luminosity relations of Miras. AJ **154**, 149 (2017)
55. W. Yuan, S. He, L.M. Macri, J. Long, J.Z. Huang, The M33 synoptic stellar survey II Mira variables. AJ **153**, 170 (2017)
56. W. Yuan, L.M. Macri, A. Javadi, Z. Lin, J.Z. Huang, Near-infrared Mira period-luminosity relations in M33. AJ **156**, 112 (2018)
57. S. He, W. Yuan, J.Z. Huang, J. Long, L.M. Macri, Period estimation for sparsely-sampled quasi-periodic light curves applied to Miras. AJ **152**, 164 (2016)
58. J.Y. Ou, C.C. Ngeow, A. Bhardwaj, M.J. Graham, R.R. Laher, F.J. Masci, R. Riddle, A distance measurement to M33 using optical photometry of Mira variables. AJ **165**, 137 (2023)
59. M. Rejkuba, The distance to the giant elliptical galaxy NGC 5128. A&A **413**, 903–912 (2004)
60. P.A. Whitelock, J.W. Menzies, M.W. Feast, F. Nsengiyumva, N. Matsunaga, The local group galaxy NGC 6822 and its asymptotic giant branch stars. MNRAS **428**, 2216–2231 (2013)
61. J.W. Menzies, P.A. Whitelock, M.W. Feast, The local group galaxy IC 1613 and its asymptotic giant branch variables. MNRAS **452**, 910–923 (2015)
62. P. Battinelli, S. Demers, The calibration of the metallicity versus C/M relation. A&A **434**, 657–663 (2005). (May)
63. S.R. Goldman, M.L. Boyer, K.B.W. McQuinn, P.A. Whitelock, I. McDonald, J.T. van Loon, E.D. Skillman, R.D. Gehrz, A. Javadi, G.C. Sloan, O.C. Jones, M.A.T. Groenewegen, J.W. Menzies, An infrared census of DUST in nearby galaxies with spitzer (DUSTiNGS). V. The period-luminosity relation for dusty metal-poor AGB stars. ApJ **877**, 49 (2019). (May)
64. P.A. Whitelock, AGB variables in local group dwarf irregular galaxies, in *Stellar Populations and the Distance Scale*, ed. by J. Jensen, R.M. Rich, R. de Grijs. Astronomical Society of the Pacific Conference Series, , vol. 514 (2018), p. 49
65. C.D. Huang, A.G. Riess, W. Yuan, L.M. Macri, N.L. Zakamska, S. Casertano, P.A. Whitelock, S.L. Hoffmann, A.V. Filippenko, D. Scolnic, Hubble space telescope observations of Mira variables in the SN Ia host NGC 1559: an alternative candle to measure the Hubble constant. ApJ **889**, 5 (2020)
66. M.J. Reid, D.W. Pesce, A.G. Riess, An improved distance to NGC 4258 and its implications for the Hubble constant. ApJL **886**, L27 (2019)
67. E.M.L. Humphreys, M.J. Reid, J.M. Moran, L.J. Greenhill, A.L. Argon, Toward a new geometric distance to the active galaxy NGC 4258. III. Final results and the Hubble constant. ApJ **775**, 13 (2013)
68. A.G. Riess, L.M. Macri, S.L. Hoffmann, D. Scolnic, S. Casertano, A.V. Filippenko, B.E. Tucker, M.J. Reid, D.O. Jones, J.M. Silverman, R. Chornock, P. Challis, W. Yuan, P.J. Brown, R.J. Foley, A 2.4% determination of the local value of the Hubble constant. ApJ **826**, 56 (2016)
69. L.M. Macri, K.Z. Stanek, D. Bersier, L.J. Greenhill, M.J. Reid, A new Cepheid distance to the Maser-Host galaxy NGC 4258 and its implications for the Hubble constant. ApJ **652**, 1133–1149 (2006)
70. ...G. Pietrzyński, D. Graczyk, A. Gallenne, W. Gieren, I.B. Thompson, B. Pilecki, P. Karczmarek, M. Górski, K. Suchomska, M. Taormina, B. Zgirski, P. Wielgórski, Z. Kołaczkowski, P. Konorski, S. Villanova, N. Nardetto, P. Kervella, F. Bresolin, R.P. Kudritzki, J. Storm, R. Smolec, W. Narloch, A distance to the large magellanic cloud that is precise to one percent. Nature **567**, 200–203 (2019)
71. A. Lançon, P.R. Wood, A library of 0.5 to 2.5 mu m spectra of luminous cool stars. A&AS **146**, 217–249 (2000)
72. STScI Development Team, pysynphot: synthetic photometry software package. Astrophysics Source Code Library, record ascl:1303.023 (2013)
73. A.G. Riess, W. Yuan, L.M. Macri, D. Scolnic, D. Brout, S. Casertano, D.O. Jones, Y. Murakami, G.S. Anand, L. Breuval, T.G. Brink, A.V. Filippenko, S. Hoffmann, S.W. Jha, W. D'arcy Kenworthy, J. Mackenty, B.E. Stahl, W. Zheng, A comprehensive measurement

of the local value of the Hubble constant with 1 km s^{-1} Mpc^{-1} uncertainty from the Hubble space telescope and the SH_0ES team. ApJL **934**, L7 (2022)
74. J.J. Dalcanton, B.F. Williams, J.L. Melbourne, L. Girardi, A. Dolphin, P.A. Rosenfield, M.L. Boyer, R.S. de Jong, K. Gilbert, P. Marigo, K. Olsen, A.C. Seth, E. Skillman, Resolved near-infrared stellar populations in nearby galaxies. ApJS **198**, 6 (2012)
75. S.L. Hoffmann, L.M. Macri, A.G. Riess, W. Yuan, S. Casertano, R.J. Foley, A.V. Filippenko, B.E. Tucker, R. Chornock, J.M. Silverman, D.L. Welch, A. Goobar, R. Amanullah, Optical identification of Cepheids in 19 host galaxies of type Ia supernovae and NGC 4258 with the Hubble space telescope. ApJ **830**, 10 (2016)
76. W. Yuan, L.M. Macri, B.M. Peterson, A.G. Riess, M.M. Fausnaugh, S.L. Hoffmann, G.S. Anand, M.C. Bentz, E. Dalla Bontà, R.I. Davies, G. De Rosa, L. Ferrarese, C.J. Grier, E.K.S. Hicks, C.A. Onken, R.W. Pogge, T. Storchi-Bergmann, M. Vestergaard, The Cepheid distance to the narrow-line Seyfert 1 galaxy NGC 4051. ApJ **913**, 3 (2021)
77. W. Yuan, L.M. Macri, A.G. Riess, T.G. Brink, S. Casertano, A.V. Filippenko, S.L. Hoffmann, C.D. Huang, D. Scolnic, Absolute calibration of Cepheid period-luminosity relations in NGC 4258. ApJ **940**, 64 (2022)
78. M.S. Vardya, Classification of Mira variables based on visual light curve shape. A&AS **73**, 181–194 (1988)
79. T. Lebzelter, The shapes of light curves of Mira-type variables. Astronomische Nachrichten **332**, 140 (2011)
80. C.D. Huang, W. Yuan, A.G. Riess, W. Hack, P.A. Whitelock, N.L. Zakamska, S. Casertano, L.M. Macri, M. Marengo, J.W. Menzies, R.K. Smith, The Mira distance to M101 and a 4% measurement of H 0. Astrophys. J. **963**(2), 83 (2024). https://ui.adsabs.harvard.edu/abs/2024ApJ...963...83H/abstract
81. M.W. Feast, The pulsation, temperatures and metallicities of Mira and semiregular variables in different stellar systems. MNRAS **278**, 11–21 (1996)
82. P. Whitelock, J. Menzies, M. Feast, F. Marang, B. Carter, G. Roberts, R. Catchpole, J. Chapman, High-mass-loss AGB stars in the South Galactic Cap. MNRAS **267**, 711–742 (1994)
83. G.T. van Belle, R.R. Thompson, M.J. Creech-Eakman, Angular size measurements of mira variable stars at 2.2 microns II. AJ **124**, 1706–1715 (2002)
84. S. Xu, B. Zhang, M.J. Reid, X. Zheng, G. Wang, Comparison of Gaia DR2 parallaxes of stars with VLBI astrometry. ApJ **875**, 114 (2019)
85. J.L. Sanders, The period–luminosity relation for Mira variables in the Milky Way using Gaia DR3: a further distance anchor for H 0. Monthly Not. R. Astron. Soc. **523**(2), 2369–2398 (2023). https://ui.adsabs.harvard.edu/abs/2023MNRAS.523.2369S/abstract
86. M. Andriantsaralaza, S. Ramstedt, W.H.T. Vlemmings, E. De Beck, Distance estimates for AGB stars from parallax measurements. A&A **667**, A74 (2022)
87. A. Chiavassa, B. Freytag, M. Schultheis, Heading Gaia to measure atmospheric dynamics in AGB stars. A&A **617**, L1 (2018)
88. K. El-Badry, H.-W. Rix, T.M. Heintz, A million binaries from Gaia eDR3: sample selection and validation of Gaia parallax uncertainties. MNRAS **506**, 2269–2295 (2021)
89. P. Whitelock, M. Feast, Hipparcos parallaxes for Mira-like long-period variables. MNRAS **319**, 759–770 (2000)
90. F. van Leeuwen, *Hipparcos, the New Reduction of the Raw Data*, vol. 350 (2007)
91. D. Engels, E. Kreysa, G.V. Schultz, W.A. Sherwood, The nature of OH/IR stars I. Infrared Mira variables. A&A **124**, 123–138 (1983)
92. R. Urago, T. Omodaka, T. Nagayama, Y. Watabe, R. Miyanosita, N. Matsunaga, R.A. Burns, The 3D distribution of long-period Mira variables in the galactic disk. ApJ **891**, 50 (2020)
93. I. Soszyński, A. Udalski, M.K. Szymański, M. Kubiak, G. Pietrzyński, Ł. Wyrzykowski, K. Ulaczyk, R. Poleski, S. Kozłowski, P. Pietrukowicz, The optical gravitational lensing experiment. The OGLE-III catalog of variable stars XIII long-period variables in the small magellanic cloud. AcA **61**, 217–230 (2011)
94. P.A. Whitelock, J.W. Menzies, M.W. Feast, P. Marigo, A remarkable oxygen-rich asymptotic giant branch variable in the Sagittarius Dwarf Irregular Galaxy. MNRAS **473**, 173–184 (2018)

95. J.W. Menzies, P.A. Whitelock, M.W. Feast, N. Matsunaga, Luminous AGB variables in the dwarf irregular galaxy, NGC 3109. MNRAS **483**, 5150–5165 (2019)
96. M.M. Kasliwal, J. Bally, F. Masci, A.M. Cody, H.E. Bond, J.E. Jencson, S. Tinyanont, Y. Cao, C. Contreras, D.A. Dykhoff, S. Amodeo, L. Armus, M. Boyer, M. Cantiello, R.L. Carlon, A.C. Cass, D. Cook, D.T. Corgan, J. Faella, O.D. Fox, W. Green, R.D. Gehrz, G. Helou, E. Hsiao, J. Johansson, R.M. Khan, R.M. Lau, N. Langer, E. Levesque, P. Milne, S. Mohamed, N. Morrell, A. Monson, A. Moore, E.O. Ofek, D. O' Sullivan, M. Parthasarathy, A. Perez, D.A. Perley, M. Phillips, T.A. Prince, D. Shenoy, N. Smith, J. Surace, S.D. Van Dyk, P.A. Whitelock, R. Williams, SPIRITS: uncovering unusual infrared transients with spitzer. ApJ **839**, 88 (2017)
97. V.R. Karambelkar, S.M. Adams, P.A. Whitelock, M.M. Kasliwal, J.E. Jencson, M.L. Boyer, S.R. Goldman, F. Masci, A.M. Cody, J. Bally, H.E. Bond, R.D. Gehrz, M. Parthasarathy, R.M. Lau, SPIRITS Collaboration, SPIRITS catalog of infrared variables: identification of extremely luminous long period variables. ApJ **877**, 110 (2019)
98. M.L. Boyer, K.B.W. McQuinn, P. Barmby, A.Z. Bonanos, R.D. Gehrz, K.D. Gordon, M.A.T. Groenewegen, E. Lagadec, D. Lennon, M. Marengo, I. McDonald, M. Meixner, E. Skillman, G.C. Sloan, G. Sonneborn, J.T. van Loon, A. Zijlstra, An infrared census of DUST in nearby galaxies with spitzer (DUSTiNGS) II. Discovery of metal-poor dusty AGB stars. ApJ **800**, 51 (2015)
99. A.J. Lee, W.L. Freedman, B.F. Madore, K.A. Owens, I. Sung Jang, A preliminary calibration of the JAGB method using Gaia EDR3. ApJ **923**, 157 (2021)
100. B.F. Madore, W.L. Freedman, Astrophysical distance scale: the AGB J-band method. I. Calibration and a first application. Astrophys. J. **899**(1), 66 (2020). https://ui.adsabs.harvard.edu/abs/2020ApJ...899...66M/abstract
101. S. Li, A.G. Riess, S. Casertano, G.S. Anand, D.M. Scolnic, W. Yuan, L. Breuval, C.D. Huang, Reconnaissance with JWST of the J-region asymptotic giant branch in distance ladder galaxies: from irregular luminosity functions to approximation of the Hubble constant. arXiv preprint arXiv:2401.04777 (2024, Jan 9). https://ui.adsabs.harvard.edu/abs/2024arXiv240104777L/abstract
102. R.L. Beaton, G. Bono, V.F. Braga, M. Dall'Ora, G. Fiorentino, I.S. Jang, C.E. Martínez-Vázquez, N. Matsunaga, M. Monelli, J.R. Neeley, M. Salaris, Old-aged primary distance indicators. SSRv **214**, 113 (2018)
103. K.B.W. McQuinn, M. Boyer, E.D. Skillman, A.E. Dolphin, Using the tip of the red giant branch as a distance indicator in the near infrared. ApJ **880**, 63 (2019)
104. A.G. Riess, G.S. Anand, W. Yuan, S. Casertano, A. Dolphin, L.M. Macri, L. Breuval, D. Scolnic, M. Perrin, R.I. Anderson, Crowded no more: the accuracy of the Hubble constant tested with high resolution observations of Cepheids by JWST (2023). arXiv e-prints: arXiv:2307.15806
105. W. Yuan, Period-luminosity relations of Cepheid and Mira variables and their application to the extragalactic distance scale. PhD thesis, Texas A&M University System (2017)

Chapter 12
Tully–Fisher Relation

Khaled Said

12.1 Introduction

12.1.1 Description of the Tully–Fisher Relation

The Tully–Fisher (TF) relation is an empirical relation that correlates the intrinsic brightness of a spiral galaxy, measured by its total luminosity, and its dynamical properties, measured by its rotational velocity. In 1977, Brent Tully and Richard Fisher proposed the use of this relation as a distance indicator based on their observations of spiral galaxies in the Virgo cluster [1]. The main idea behind distance indicators is to use distance-independent observables to predict a distance-dependent parameter, which can subsequently be compared with the corresponding observable to derive a distance estimate. In the case of the Tully–Fisher relation, the rotational velocity of a spiral galaxy serves as the distance-independent observable that is used to predict the total absolute luminosity, which is distance-dependent. The total luminosity can then be compared with apparent magnitude to measure distance via the distance modulus. The Tully–Fisher relation has since been extensively studied and refined [2–9]. It has become a workhorse tool for measuring the distances and properties of galaxies, particularly in the context of large-scale structure surveys and cosmological studies [10–16].

12.1.2 Historical Background

The historical background of the Tully–Fisher relation can be traced back to the Great Debate in the 1920s, when Ernst Öpik used an expression between the observed rotational velocity and the absolute magnitude to measure the true distance to Andromeda

K. Said (✉)
School of Mathematics and Physics, University of Queensland, Brisbane, QLD 4072, Australia
e-mail: k.saidahmedsoliman@uq.edu.au

E. Di Valentino and D. Brout (eds.), *The Hubble Constant Tension*, Springer Series in Astrophysics and Cosmology, https://doi.org/10.1007/978-981-99-0177-7_12

[17]. This measurement helped to prove that Andromeda is an independent galaxy, not part of our own Milky Way Galaxy, as Shapley had thought.

More than 50 years later, Balkowski et al. used the Nançay radio telescope in France to measure the line widths of a sample of 13 irregular and spiral galaxies [18]. They found a correlation between the line width and the luminosity but they did not apply this correlation as a distance indicator. This laid the foundation for the ground-breaking paper by Brent Tully and Richard Fisher in 1977 [1]. In their paper, Tully and Fisher used only inclined spiral galaxies and proposed the usage of the linear relation between H I profile and absolute magnitude as a distance indicator.

The publication of the Tully–Fisher relation in 1977 and the proposal to use it as a distance indicator was significant in many different ways. Firstly, it provided a robust new tool for measuring distance at redshifts that other methods such as Cepheid variable stars cannot reach. This opened up a whole range of large scale structure and cosmological studies. Notably, similar relations like the Fundamental Plane relation [19, 20] only emerged a decade later. Secondly, Tully and Fisher measured the Hubble constant H_0 to be 80 km $s^{-1}Mpc^{-1}$ from the Virgo cluster and Ursa Major. This value was the first to deviate from the two mainstream values at that time, low value of $H_0 = 50$ km $s^{-1}Mpc^{-1}$ which was promoted by Sandage and Tammann [21] and a higher value of $H_0 = 100$ km $s^{-1}Mpc^{-1}$ claimed by de Vaucouleurs [22]. This value is much closer to the value of Hubble constant that we know today.

12.1.3 Tully–Fisher Relation in Cosmology

The Tully–Fisher relation has been heavily used in measuring the Hubble constant H_0. Additionally, it has also been used as a valuable tool in many other cosmological studies[1]:

- **Cosmography:** The Tully–Fisher relation uses the maximum rotational velocity of a galaxy to predict its luminosity. This can be used to estimate the distance to the galaxy, and to measure its peculiar velocity. The peculiar velocity is the deviation from the smooth Hubble flow and can be used to reconstruct 3D maps of the density and velocity fields of the universe. In 2014, a team led by Brent Tully used the Wiener filter reconstruction [23] to recover the 3D density and velocity maps. They used these maps to set the borders of our home supercluster which they called Laniakea [24]. More recently, Alexandra Dupuy and Hélène Courtois used the full CosmicFlows-4 [25] catalog[2] to reconstruct the 3D density and velocity fields [26] and then redefine the boundaries of the Laniakea supercluster along with other structures such as Perseus-Pisces and Shapley [27].

---

[1] I do not intend to give a comprehensive review of all cosmological studies that used TF relation, but I will discuss a few examples before focusing on the role of TF relation in the H_0 tension which is the main focus of this chapter.

[2] This catalog contains TF distances along with other distance indicators.

- **Bulk Flow:** The bulk flow is the average peculiar velocity over a sphere of a given radius R. The theoretical expected value of this measurement relies on a model of a specific set of cosmological parameters, initial conditions and a chosen gravitational law [28]. In 2009, Watkins, Feldman and Hudson used the largest TF survey available at that time, the SFI++ survey [10], to measure the bulk flow of galaxies. They used a new method, the minimal variance estimator, and found a large bulk flow amplitude of 431 ± 102 km s^{-1} within a scale of 50 Mpc h^{-1} [29]. The amplitude of this bulk flow is larger than expected from the ΛCDM model. More recently, several teams have used the state-of-the-art CosmicFlows-4 catalog [25], which contains more than 10,000 newly measured TF distances, to measure the bulk flow of galaxies. They have also found large bulk flows, arising on a much larger scale, that are higher than the expected value from the ΛCDM model [26, 30, 31]
- **Growth rate of structure:** Over the last three decades, many collaborations have started to test Einstein's general theory of relativity over a range of scales. One way to do that is by measuring the growth rate of cosmic structure, or $f\sigma_8$. This method is capable of measuring $f\sigma_8$ at low redshifts ($z < 0.1$), a range inaccessible to other methods such as the redshift space distortion. The growth rate can be constrained by comparing the density field from redshift surveys and peculiar velocity fields from TF surveys. The growth rate can then be parameterized as a function of the mass density parameter Ω_m and the growth index γ, which is determined by the theory of gravity. In 2011, Marc Davis and others used the TF survey SFI++ to find a $f\sigma_8$ value of 0.31 ± 0.06 [32]. Other teams also used the TF SFI++ catalog and found a slightly higher value of 0.401 ± 0.024 [33]. More recently, the CosmicFlows-4 catalog also was used to measure $f\sigma_8$ and found a similar value of 0.40 ± 0.07 [16].
- **Galaxy formation and evolution:** As an empirical relation between fundamental properties of galaxies, TF relation has been used to refine semi-analytical models and numerical simulations of galaxy formation and evolution [34–36].
- **Dark matter in galaxies:** The Tully–Fisher relation have also been used to probe the distribution and properties of dark matter in galaxies. Several studies put constraints on galaxy halo profiles by comparing the halo profile required by the Tully–Fisher relation to the density profile that is well-described by the Navarro-Frenk-White (NFW) profile [37–39].

12.2 The Tully–Fisher Relation

12.2.1 How Does It Work?

The Tully–Fisher relation is defined and its usages in cosmology are discussed above. But how does it work in practice?

Building the Tully–Fisher relation requires two data sets: photometry and spectroscopy. Photometry provides a measurement of the galaxy's luminosity, which is the distance-dependent parameter in the TF relation. Spectroscopy provides a measurement of the galaxy's rotational velocity, which is the distance-independent parameter in the relation. The Tully–Fisher relation establishes the connection between the luminosity and rotational velocity of spiral galaxies. The TF relation is a secondary distance indicator and as the name suggests it requires calibration from a sample of galaxies with known distances usually coming from a primary distance indicator or a sample of cluster galaxies. Figure 12.1a represents the initial step in utilizing the TF relation for distance measurements. It shows a sample of galaxies with known rotational velocities and absolute magnitudes. The TF relation of the form:

$$M = a + b \log W \tag{12.1}$$

where M is the absolute magnitude, W represents the rotation, a and b are the TF slope and intercept, respectively, can be fit to that sample. This relation, which we often call a template relation (Fig. 12.1b), can be used to predict the absolute magnitude of a galaxy with unknown distance given a measurement of its rotational velocity as shown in Fig. 12.1c. The predicted absolute magnitude can be compared to the galaxy's apparent magnitude to yield the distance via the distance modulus as:

$$\mu = m - M(W). \tag{12.2}$$

where μ is the distance modulus, m is the apparent magnitude, and $M(W)$ is the predicted absolute magnitude of a galaxy from the TF relation given its rotational velocity.

It is important to note that the derived template relation is not a universal relation. Instead, the data, parameters, and corrections used in the measurements of the TF distances should be consistent with those used in building the TF template relation. This has encouraged many teams to derive their own template relations instead of using an existing one. In the past few decades, different measurements, wavelengths, and methodologies have been used to do this. For example, shortly after the inception of the TF relation, several studies began to use near-infrared (NIR) bands because they suffer less than optical bands from dust extinction (both internal and external) and are more sensitive to stellar mass [9, 11, 40]. Other groups used isophotal magnitudes instead of total magnitudes to extend the Tully–Fisher studies into the Zone of Avoidance around the plane of the Milky Way [41, 42].

In terms of spectroscopy, many parameters have been used in an attempt to minimise the intrinsic scatter of the TF relation. Many teams have used the 20% H I line width (the width measured at 20% of the peak flux of the H I profile) including the original TF relation [1, 43, 44], while others suggested using the width measured at 50% instead [45–48]. Furthermore, other teams suggest that the flat part of the optical rotation curve is the best proxy for the rotational velocity, and they have shown that it does indeed yield the tightest Tully–Fisher relation [49, 50].

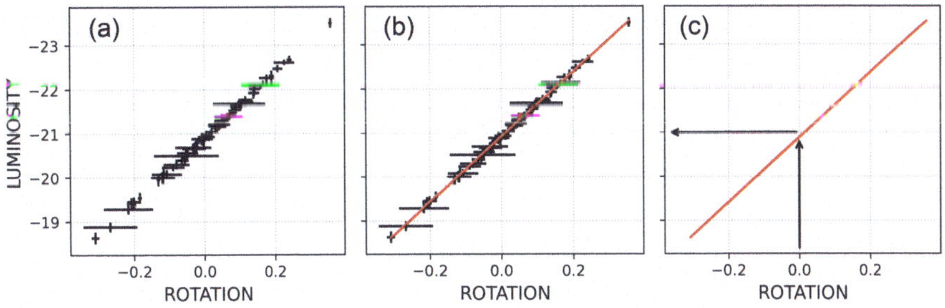

Fig. 12.1 Illustrating how the Tully–Fisher relation combines photometric and spectroscopic data to enable the estimation of galaxy distance. In order to use the Tully–Fisher relation to measure distances, we need to build something called template relation. **a** Shows a sample of galaxies with known rotation velocity, apparent magnitude, and distances which allows the determination of luminosity. **b** By fitting the TF relation to this sample, a template relation is derived. This establishes the correlation between the absolute magnitude and the rotational velocity of galaxies. **c** Applying the Tully–Fisher template to a galaxy of unknown distance, the absolute magnitude can be estimated based on its measured rotational velocity

Different methodologies have been used to extract the best-fitting parameters for the Tully–Fisher relation by applying different fitting procedures. In an attempt to overcome the Malmquist bias, Renée C. Kraan-Korteweg and others proposed the use of the inverse TF relation, which only takes into account the errors on the x-axis (rotational velocity) [4]. Since then, many teams have used both direct and inverse fitting procedures in their analyses [51, 52]. Other teams have used bivariate forms that take into account errors in both the x- and y-axes [9, 11, 45, 48]. Given the scatter in TF relation, a suggestion has been made to use a Bayesian mixture model, which is less biased by outliers [53].

In summary, the Tully–Fisher relation has been observed to hold over a wide range of applications, including different parameters, methodologies, and wavelengths. This suggests that the TF relation is one of the most robust and powerful tools for measuring galaxy distances.

12.2.2 Theoretical Basis

The physical origin of the Tully–Fisher relation is still not fully understood, but it is certainly rooted in the physics of gravity and the dynamics of galactic rotation. The TF relation links the mass of a galaxy (characterized by the luminosity) to its dynamics (characterized by the rotational velocity). Specifically, it is based on the idea that the rotation velocity of a galaxy is related to the mass contained within a certain radius from the galaxy's center, known as the circular velocity.

Michael Strauss and Jeffrey Willick provided a widely accepted theoretical explanation[3] for the Tully–Fisher relation in their 1995 review [54]. They wrote that by equating the centrifugal force of an object moving in a circle of radius r to the gravitational attraction on the same object due to the mass inside a sphere of radius r, one can write

$$V_{\text{rot}}^2 \propto \frac{M}{r} \tag{12.3}$$

where V_{rot} is the rotational velocity, M is the mass within a sphere of radius r, and r is the distance from the galaxy's center. Then, assuming a constant value for both the mass-to-light ratio (M/L) and mean surface brightness for spirals, one can obtain that

$$L \propto V_{\text{rot}}^4 \tag{12.4}$$

However, most Tully–Fisher studies have derived a power-law exponent that deviates from this theoretical explanation. This is usually attributed to the complication of the dark matter halo that surrounds a galaxy [55]. Early near-infrared Tully–Fisher analyses showed that the complications of halo mass could be avoided, and indeed derived a power-law exponent that was closer to the theoretical expectation [40]. However, their conclusion has been contradicted by many larger subsequent surveys [11, 45].

In 2000, Stacy McGaugh and his team investigated the deviation of dwarf faint galaxies from the linear Tully–Fisher relation [6]. They found that the linear TF relation could be restored if they used the sum of both stellar and gas mass components instead of luminosity. This relation takes the form

$$M_d \propto V_{\text{rot}}^4 \tag{12.5}$$

where M_d is the sum of stellar and gas masses,

$$M_d = M_* + M_{\text{gas}}. \tag{12.6}$$

Not only did they establish the baryonic TF relation as the fundamental relation, but they also found that the intrinsic scatter was smaller than the original TF relation. This result can be simply explained if we suppose that this estimate of the baryonic mass is a better proxy for the true total mass than the luminosity or the stellar mass alone.

---

[3] This was not the first such explanation, but it was particularly clear.

This result, at first glance, may seem to support the Modified Newtonian Dynamics (MOND: [56]) as an alternative to dark matter. MOND is a theory of gravity that modifies Newtonian dynamics. In MOND, the baryonic mass is actually the total mass, which means that the baryonic TF relation is the fundamental TF relation. However, it is important to note that Modified Newtonian Dynamics (MOND) has its own critics [57, 58]. We refer the reader to the literature for more information about the debate between MOND and Cold Dark Matter.

One important theoretical development in recent years has been the use of hydrodynamic simulations to study the formation and evolution of galaxies in a cosmological context [35, 59]. These simulations incorporate the effects of gravity, gas dynamics, star formation, and feedback, and have been shown to predict the Tully–Fisher relation in different environments and at different redshifts better than semi-analytical models [60].

12.2.3 Advantages and Limitations

The Tully–Fisher relation has several advantages as a tool for measuring distances of galaxies. The first and main advantage is its simplicity and robustness. The TF relation provides a linear correlation between only two easily measurable galaxy properties, luminosity and rotational velocity. Several studies have shown that, unlike other distance indicators, the scatter of the TF relation is relatively insensitive to other galaxy properties [8, 61, 62]. This makes the TF relation a fundamental, robust and reliable tool for measuring distances.

Another advantage of the Tully–Fisher relation is that it uses spiral galaxies, which are the most numerous type of galaxies at low redshifts ($z < 0.07$). This makes it a convenient tool for measuring peculiar velocities at these redshifts, where the true distances can be determined. Beyond this redshift, it is more difficult to obtain accurate peculiar velocities as the errors in measuring these velocities from distance indicators scale with redshift.

In Fig. 12.2, I show the redshift distribution of spiral and elliptical galaxies from the SDSS survey DR14 [63]. I separated spirals from ellipticals using the galaxy zoo catalog [64]. I also over plot the redshift distribution of the state-of-the-art Pantheon+ sample [65, 66]. I used only the redshift below $z < 0.15$, as this is the range used for Hubble constant (H_0) measurements [67]. It is clear that spiral galaxies (Tully–Fisher galaxies) are the most numerous type at low redshift ($z < 0.07$), with a mean redshift of $z = 0.077$. This makes TF relation one of the most precise tools for measuring H_0 at this redshift.

However, there are also some limitations to the Tully–Fisher relation as a tool for distance measurements. One of the main limitations of the TF relation is its intrinsic scatter. This scatter can lead to errors of around ~20% in distance measurements. The baryonic Tully–Fisher relation is often shown to be a tighter relation than the TF relation, with a much smaller intrinsic scatter [68, 69]. However, measuring the sum of the stellar and gas components is much more difficult than measuring just the

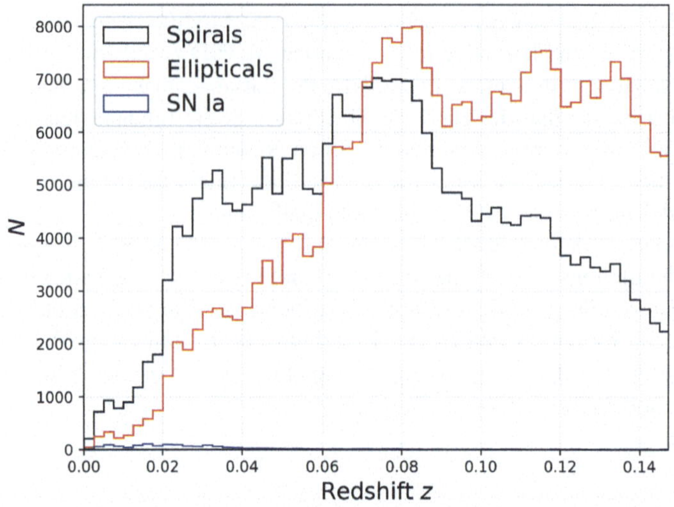

Fig. 12.2 The redshift distribution of spiral (black) and elliptical (red) galaxies in the SDSS survey, as well as the redshift distribution of the state-of-the-art SN Ia from Pantheon+ (blue), up to redshift $z < 0.15$. The number of galaxies per 0.0025 redshift bin is shown

luminosity of a galaxy. This adds a lot of observational uncertainties to the baryonic TF relation.

Another limitation of the Tully–Fisher relation and a significant source of uncertainty is the inclination of galaxies. The inclination of a spiral galaxy affects both TF parameters. It affects the derived maximum rotational velocity during the process of correcting the observed line width for projection effects (i.e. to edge-on orientation). Additionally, to correct the magnitude for the internal extinction, one needs to use the inclination to account for the path-length of the light through the galaxy. Therefore, inclination errors lead to correlated errors in the two TF parameters that have the same sense (qualitatively) as errors in distance, so an error in inclination is (at least partly) degenerate with an error in distance. Most Tully–Fisher surveys select only edge-on galaxies by applying a cut to inclinations in an attempt to minimize the line width correction. Although this seems like the right thing to do, inaccurate inclinations will lead to a selection bias in the TF distance. This is because galaxies with inaccurate inclinations may be excluded or included in the survey, depending on whether their inclinations are overestimated or underestimated. Furthermore, edge-on galaxies are subject to higher internal extinction, which means that more light is blocked by dust and gas inside the galaxy. This results in a dimmer magnitude, and thus a larger correction is needed to correct for the extinction. Some studies have even suggested that when the inclination measurements have errors larger than 10°C, it may be more prudent to omit inclinations entirely rather than assuming them to be exact [70].

Selection effects and Malmquist bias are other limitations of the TF relation. However, these limitations are not specific to the TF relation, as they affect all distance indicators.

12.3 The Role of the Tully–Fisher Relation in H_0 Measurements

The methodology for using the Tully–Fisher relation in H_0 measurements typically involves the following steps. First, a sample of galaxies with well-measured rotational velocities and luminosities (for TFR) or masses (for the BTFR) is selected (see Fig. 12.1a). These galaxies are typically chosen to be representative of the population of galaxies being studied. Next, the Tully–Fisher relation is calibrated for this sample of galaxies. This involves measuring the slope, intercept, and scatter of the relation for the sample, and correcting for any systematic biases or selection effects (see Fig. 12.1b). Once the Tully–Fisher relation has been calibrated for the sample, it can be used to infer the distances to other galaxies with similar properties (see Fig. 12.1c). The final step is measuring H_0, which can be done by plotting the derived TF distances against redshifts.

This methodology has been used in a number of studies to estimate the value of the Hubble constant (H_0). In 1977, Tully and Fisher applied their newly discovered Tully–Fisher relation to derive distances to the Virgo and Ursa Major clusters. They measured a Hubble constant of $H_0 = 84$ km s^{-1} Mpc^{-1} for Virgo and $H_0 = 75$ km s^{-1} Mpc^{-1} for Ursa Major. They conclude a preliminary value for the Hubble constant of $H_0 = 80$ km s^{-1} Mpc^{-1}. Although they called it preliminary and did not propagate their distance error to the Hubble constant value, we know today that it was the most accurate result at that time.

In 1983, Visvanathan used the Tully–Fisher relation to measure distances to 52 cluster spiral galaxies. They found that the Hubble constant derived from nearby clusters varied from 58.5 to 83.5, but the range was much smaller for distant clusters, between 76.3 and 78.9 km s^{-1} Mpc^{-1}. Visvanathan concluded that the best fit value for the Hubble constant from all clusters in their study was $H_0 = 74.4 \pm 11$ km s^{-1} Mpc^{-1} [71]. That result is very close to the value of H_0 we have today.

Shortly after, Sandage and Tammann used the Tully–Fisher relation in the infrared (IR) and blue bands to estimate distances to galaxies in the Virgo and Coma clusters, respectively. Their analysis gave a lower value of the Hubble constant $H_0 = 55 \pm 9$ km s^{-1} Mpc^{-1} [72]. This result aligned with their previous measurement of $H_0 = 50 \pm 7$ km s^{-1} Mpc^{-1} obtained through Type I supernovae [73].

Aaronson led a team in 1986 to derive distances to 10 galaxy clusters using the infrared Tully–Fisher relation. They stated that the current evidence favors a large value for H_0 of 90 km s^{-1} Mpc^{-1} [74]. In two subsequent studies, Bottinelli led a team to use the infrared Tully–Fisher relation to derive distances to galaxy clusters. They concluded that the H_0 lies between 70 and 75 km s^{-1} Mpc^{-1}. During the same time, Renée C. Kraan-Korteweg and others derived distances to a larger sample of 82 cluster galaxies using both the infrared Tully–Fisher relation and the optical Tully–Fisher relation. They found that their data is consistent with a Hubble constant of $H_0 = 56.6 \pm 0.9$ km s^{-1} Mpc^{-1} [75]. However, this low value was contradicted by three studies. One study used the TF relation to measure distances to Virgo and Ursa Major clusters, and derived a value of $H_0 = 85 \pm 10$ km s^{-1} Mpc^{-1} [43]. The other two studies derived distances to the Coma cluster, and also found a higher value of $H_0 \approx 90$ km s^{-1} Mpc^{-1} [76, 77]. Bureau, Mould, and Staveley-Smith [78] used the Tully–Fisher relation to measure distances to galaxies in the Fornax cluster, and derived a value of $H_0 = 74 \pm 11$ km s^{-1} Mpc^{-1}.

This significant discrepancy in these early measurements can be attributed to various factors, including the challenge of sample incompleteness correction and the absence of reliable distances to nearby objects for use as a calibration sample.

The Hubble Space Telescope Key Project on the Extragalactic Distance Scale used Cepheid distances to re-calibrate the Tully–Fisher relation. In a series of papers, they found results consistent with a value of H_0 of 71 ± 4 km s^{-1} Mpc^{-1} [79–81].

The Cosmicflows project combines distances measured through various methods, including the Tully–Fisher relation, surveys, and collaborations. This project has made significant contributions to measuring the Hubble constant, consistently obtaining values in the range of 74 ± 4 to 76 ± 1 km s^{-1} Mpc^{-1} over an extended period [26, 82–88].

Figure 12.3 shows a graphical representation of the published efforts to measure the Hubble constant (H_0) using the Tully–Fisher relation. The red band shows the the most recent result of Hubble constant and its associated errors measured using Type Ia supernovae by the SH_0ES team in 2022 [89]. Additionally, the blue band represent the Hubble constant value and its associated error from the CMB measured by the Planck Collaboration in 2020 [90]. The figure shows a lot of scatter in the TF results prior to 2000 with some unrealistically optimistic uncertainties. However, after 2000, the results seem consistent and in more agreement with the value derived by the SH_0ES team in 2022 [89] than the CMB value by the Planck Collaboration in 2020 [90]. Table 12.1 provides a summary of the results from these studies.

In summary, the Tully–Fisher relation has been used for decades to estimate distances to galaxies and to measure the Hubble constant. Although Tully–Fisher measurements of H_0 have shown a lot of scatter prior to 2000, they have been consistent since then with a value of H_0 that agrees with the other low-redshift probes.

12.4 Future Prospects

The largest homogeneous Tully–Fisher sample currently exists is the CosmicFlows-4 survey [87]. This survey includes 10,000 TF distances. Imminent surveys such as DESI [98] and WALLABY [15] will provide 50,000 and 200,000 TF distances, respectively, over the coming few years. With approximately 20% TF distance errors for about 250,000 galaxies covering a significant portion of the sky up to redshift

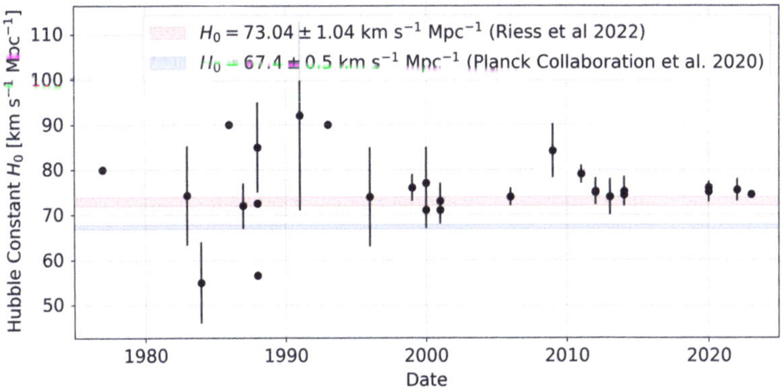

Fig. 12.3 A sample of published Hubble constant (H_0) measurements using Tully–Fisher relation spanning from 1977 to 2023. Most of these measurements are obtained by different teams using different data given in Table 12.1. The red band highlights the most recent Hubble constant value and its associated uncertainty from the baseline results of the Cepheid-SN Ia sample by the SH_0ES team [89]. The blue band represents the Hubble constant value reported by Planck Collaboration 2020 [90]. The majority of the data points align closely with the red band, particularly from 2000 onward. It is important to note that this sample of data is just a subset from the literature and additional studies may already exist beyond those represented here

Table 12.1 Evolution of published Hubble constant values from the Tully–Fisher relation (1977–2023)

Date	H_0	Refs.	Date	H_0	Refs.
1977	80.0	[1]	2001	71.0 ± 3.0	[81]
1983	74.3 ± 11.0	[71]	2001	73.0 ± 4.0	[91]
1984	55.0 ± 9.0	[72]	2006	74.0 ± 2.0	[46]
1986	90.0	[74]	2009	84.2 ± 6.0	[92]
1987	72.0 ± 5.0	[93]	2011	79.0 ± 2.0	[94]
1988	85.0 ± 10.0	[43]	2012	75.0	[82]
1988	72.5	[95]	2012	75.2 ± 3.0	[83]
1988	56.6 ± 0.9	[75]	2013	74.0 ± 4.0	[84]
1991	92.0 ± 21.0	[76]	2014	74.4 ± 1.0	[85]
1993	90.0	[77]	2014	75.2 ± 3.3	[86]
1996	74.0 ± 11.0	[78]	2020	75.1 ± 2.3	[96]
1999	76.0 ± 3.0	[79]	2020	76.0 ± 1.1	[87]
2000	77.0 ± 8.0	[97]	2022	75.5 ± 2.5	[88]
2000	71.0 ± 4.0	[80]	2023	74.5 ± 0.1	[26]

[a] This is just a subset of the published Hubble constant values using TF relation, and additional studies may already exist beyond those represented here

$z \approx 0.1$, both random and systematic errors in H_0 from large-scale structure will be considerably smaller compared to the most optimistic uncertainties associated with other low-redshift probes.

Acknowledgements KS acknowledges support from the Australian Government through the Australian Research Council's Laureate Fellowship funding scheme (project FL180100168). I would like to express my gratitude to Tamara Davis and Matthew Colless for their valuable feedback on the initial draft of this review.

References

1. R.B. Tully, J.R. Fisher, A new method of determining distances to galaxies. A&A **54**, 661–673 (1977)
2. J. Silk, Feedback, disk self-regulation, and galaxy formation. ApJ **481**, 703–709 (1997)
3. V. Avila-Reese et al., On the formation and evolution of disk galaxies: cosmological initial conditions and the gravitational collapse. ApJ **505**, 37–49 (1998)
4. R.C. Kraan-Korteweg et al., 21 CM line widths and distances of spiral galaxies, in *NATO ASIC Proceedings of the 180 Galaxy Distances and Deviations from Universal Expansion*, ed. by B.F. Madore, R.B. Tully, pp. 65–72 (1986)
5. K. O'Neil et al., Red, gas-rich low surface brightness galaxies and enigmatic deviations from the Tully–Fisher relation. AJ **119**, 136–152 (2000)
6. S.S. McGaugh et al., The Baryonic Tully–Fisher relation. ApJ **533**, L99–L102 (2000)
7. S.S. McGaugh, The Baryonic Tully–Fisher relation of galaxies with extended rotation curves and the stellar mass of rotating galaxies. ApJ **632**, 859–871 (2005)
8. S. Courteau et al., Scaling relations of spiral galaxies. ApJ **671**, 203–225 (2007)
9. K. Said et al., On how to extend the NIR Tully-Fisher relation to be truly all-sky. MNRAS **447**(2), 1618–1629 (2015)
10. C.M. Springob et al., SFI++. II a new i-band Tully–Fisher catalog, derivation of peculiar velocities, and data set properties. ApJS **172**, 599–614 (2007)
11. K.L. Masters et al., 2MTFI the Tully–Fisher relation in the two micron all sky survey J, H, and K bands. AJ **135**, 1738–1748 (2008)
12. R.B. Tully et al., The extragalactic distance database. AJ **138**(2), 323–331 (2009)
13. H.M. Courtois et al., Cosmography of the local universe. AJ **146**, 69 (2013)
14. T. Hong et al., 2MTF-VII, 2MASS Tully–Fisher survey final data release: distances for 2062 nearby spiral galaxies. MNRAS **487**(2), 2061–2069 (2019)
15. H.M. Courtois et al., WALLABY pre-pilot and pilot survey: the Tully Fisher relation in Eridanus, Hydra, Norma, and NGC4636 fields. MNRAS **519**(3), 4589–4607 (2023)
16. P. Boubel et al., Cosmic growth rate measurements from Tully–Fisher peculiar velocities (2023). arXiv e-prints: arXiv:2301.12648
17. E. Opik, An estimate of the distance of the Andromeda Nebula. ApJ **55**, 406–410 (1922)
18. C. Balkowski et al., Neutral hydrogen study of spiral and irregular dwarf galaxies. A&A **34**, 43–52 (1974)
19. S. Djorgovski, M. Davis, Fundamental properties of elliptical galaxies. ApJ **313**, 59–68 (1987)
20. A. Dressler et al., Spectroscopy and photometry of elliptical galaxies—a large-scale streaming motion in the local universe. ApJ **313**, L37–L42 (1987)
21. A. Sandage, G.A. Tammann, Steps toward the Hubble constant V the Hubble constant from nearby galaxies and the regularity of the local velocity field. ApJ **196**, 313–328 (1975)
22. G. de Vaucouleurs, G. Bollinger, The extragalactic distance scale VII the velocity-distance relations in different directions and the Hubble ratio within and without the local supercluster. ApJ **233**, 433–452 (1979)

23. S. Zaroubi et al., Wiener reconstruction of the large-scale structure. ApJ **449**, 446 (1995)
24. R.B. Tully et al., The Laniakea supercluster of galaxies. Nature **513**(7516), 71–73 (2014)
25. R.B. Tully et al., Cosmicflows-4. ApJ **944**(1), 94 (2023)
26. H.M. Courtois et al., Gravity in the local universe: density and velocity fields using CosmicFlows-4. A&A **670**, L15 (2023)
27. A. Dupuy, H.M. Courtois, Watersheds of the Universe: Laniakea and five newcomers in the neighborhood (2023). arXiv e-prints: arXiv:2305.02339
28. P. Andersen et al., Cosmology with peculiar velocities: observational effects. MNRAS **463**(4), 4083–4092 (2016)
29. R. Watkins et al., Consistently large cosmic flows on scales of $100h^{-1}$Mpc: a challenge for the standard ΛCDM cosmology. MNRAS **392**(2), 743–756 (2009)
30. R. Watkins et al., Analyzing the large-scale bulk flow using CosmicFlows4: increasing tension with the standard cosmological model (2023). arXiv e-prints: arXiv:2302.02028
31. A.M. Whitford et al., Evaluating bulk flow estimators for CosmicFlows-4 measurements (2023). arXiv e-prints: arXiv:2306.11269
32. M. Davis et al., Local gravity versus local velocity: solutions for β and non-linear bias. MNRAS **413**(4), 2906–2922 (2011)
33. J. Carrick et al., Cosmological parameters from the comparison of peculiar velocities with predictions from the 2M++ density field. MNRAS **450**(1), 317–332 (2015)
34. J.F. Navarro, M. Steinmetz, Dark halo and disk galaxy scaling laws in hierarchical universes. ApJ **538**(2), 477–488 (2000)
35. M. Vogelsberger et al., Properties of galaxies reproduced by a hydrodynamic simulation. Nature **509**(7499), 177–182 (2014)
36. A.A. Ponomareva et al., From light to baryonic mass: the effect of the stellar mass-to-light ratio on the Baryonic Tully–Fisher relation. MNRAS **474**(4), 4366–4384 (2018)
37. J.F. Navarro et al., The structure of cold dark matter halos. ApJ **462**, 563 (1996)
38. H.J. Mo, S. Mao, The Tully–Fisher relation and its implications for the halo density profile and self-interacting dark matter. MNRAS **318**(1), 163–172 (2000)
39. U. Seljak, Constraints on galaxy halo profiles from galaxy-galaxy lensing and Tully-Fisher/fundamental plane relations. MNRAS **334**(4), 797–804 (2002)
40. M. Aaronson et al., The infrared luminosity/H I velocity-width relation and its application to the distance scale. ApJ **229**, 1–13 (1979)
41. N. Bouché, S.E. Schneider, IR-TF relation in the zone of avoidance with 2MASS, in *Mapping the Hidden Universe: The Universe behind the Mily Way—The Universe in HI*, ed. by R.C. Kraan-Korteweg et al. Astronomical Society of the Pacific Conference Series, , vol. 218, pp. 111 (2000)
42. K. Said et al., NIR Tully-Fisher in the zone of avoidance—III deep NIR catalogue of the HIZOA galaxies. MNRAS **462**(3), 3386–3400 (2016)
43. M.J. Pierce, R.B. Tully, Distances to the Virgo and Ursa major clusters and a determination of H_0. ApJ **330**, 579 (1988)
44. H.M. Courtois et al., The extragalactic distance database: all digital HI profile catalog. AJ **138**(6), 1938–1956 (2009)
45. R. Giovanelli et al., The I band Tully–Fisher relation for cluster galaxies: a template relation, its scatter and bias corrections. AJ **113**, 53–79 (1997)
46. K.L. Masters et al., SFI++ I: a new I-band Tully-Fisher Template, the cluster peculiar velocity dispersion, and H_0. ApJ **653**(2), 861–880 (2006)
47. K. Said et al., NIR Tully–Fisher in the zone of avoidance—II 21 cm H I-line spectra of southern ZOA galaxies. MNRAS **457**(3), 2366–2376 (2016)
48. R. Bell et al., Calibration of the Tully–Fisher relation in the WISE W1 (3.4 μm) and W2 (4.6 μm) bands. MNRAS **519**(1), 102–120 (2023)
49. M.A.W. Verheijen, The Ursa major cluster of galaxies V HI rotation curve shapes and the Tully–Fisher relations. ApJ **563**(2), 694–715 (2001)
50. F. Lelli et al., The baryonic Tully-Fisher relation for different velocity definitions and implications for galaxy angular momentum. MNRAS **484**(3), 3267–3278 (2019)

51. J.A. Willick et al., Homogeneous velocity-distance data for peculiar velocity analysis I calibration of cluster samples. ApJ **446**, 12 (1995)
52. J.A. Willick et al., Homogeneous velocity-distance data for peculiar velocity analysis II calibration of field samples. ApJ **457**, 460 (1996)
53. D.W. Hogg et al., Data analysis recipes: fitting a model to data (2010). arXiv e-prints: arXiv:1008.4686
54. M.A. Strauss, J.A. Willick, The density and peculiar velocity fields of nearby galaxies. Phys. Rep. **261**, 271–431 (1995)
55. J. Mould, Understanding the fundamental plane and the Tully Fisher relation. Front. Astron. Space Sci. **7**, 21 (2020)
56. M. Milgrom, A modification of the Newtonian dynamics as a possible alternative to the hidden mass hypothesis. ApJ **270**, 365–370 (1983)
57. F.C. van den Bosch, J.J. Dalcanton, Semianalytical models for the formation of disk galaxies II dark matter versus modified Newtonian dynamics. ApJ **534**(1), 146–164 (2000)
58. F.C. van den Bosch, J.J. Dalcanton, Disk galaxies as cosmological benchmarks: cold dark matter versus modified Newtonian dynamics, in *Galaxy Disks and Disk Galaxies*, ed. by J.G. Funes, E.M. Corsini. Astronomical Society of the Pacific Conference Series, vol. 230, pp. 549–552 (2001)
59. J. Schaye et al., The EAGLE project: simulating the evolution and assembly of galaxies and their environments. MNRAS **446**(1), 521–554 (2015)
60. E. Papastergis et al., An accurate measurement of the baryonic Tully-Fisher relation with heavily gas-dominated ALFALFA galaxies. A&A **593**, A39 (2016)
61. S. Courteau, H.-W. Rix, Maximal disks and the Tully–Fisher relation. ApJ **513**(2), 561–571 (1999)
62. N.N.Q. Ouellette et al., The spectroscopy and H-band imaging of Virgo cluster galaxies (SHIVir) survey: scaling relations and the stellar-to-total mass relation. ApJ **843**(1), 74 (2017)
63. B. Abolfathi et al., The Fourteenth data release of the sloan digital sky survey: first spectroscopic data from the extended baryon oscillation spectroscopic survey and from the second phase of the apache point observatory galactic evolution experiment. ApJS **235**(2), 42 (2018)
64. C. Lintott et al., Galaxy Zoo 1: data release of morphological classifications for nearly 900000 galaxies. MNRAS **410**(1), 166–178 (2011)
65. A. Carr et al., The Pantheon+ analysis: improving the redshifts and peculiar velocities of Type Ia supernovae used in cosmological analyses. PASA **39**, e046 (2022)
66. D. Brout et al., The Pantheon+ analysis: cosmological constraints. ApJ **938**(2), 110 (2022)
67. A.G. Riess et al., A 2.4% determination of the local value of the Hubble constant. ApJ **826**(1), 56 (2016)
68. S.S. McGaugh, The Baryonic Tully–Fisher relation of gas-rich galaxies as a test of ΛCDM and MOND. AJ **143**(2), 40 (2012)
69. F. Lelli et al., The small scatter of the Baryonic Tully–Fisher relation. ApJ **816**(1), L14 (2016)
70. D. Obreschkow, M. Meyer, Precise Tully–Fisher relations without galaxy inclinations. ApJ **777**(2), 140 (2013)
71. N. Visvanathan, A global value of the Hubble constant. ApJ **275**, 430–444 (1983)
72. A. Sandage, G.A. Tammann, The Hubble constant as derived from 21 cm linewidths. Nature **307**(5949), 326–329 (1984)
73. A. Sandage, G.A. Tammann, Steps toward the Hubble constant VIII—the global value. ApJ **256**, 339–345 (1982)
74. M. Aaronson et al., A distance scale from the infrared magnitude/Hi velocity-width relation V distance moduli to 10 galaxy clusters, and positive detection of bulk supercluster motion toward the microwave anisotropy. ApJ **302**, 536 (1986)
75. R.C. Kraan-Korteweg et al., 21 Centimeter line width distances of cluster galaxies and the value of H_0. ApJ **331**, 620 (1988)
76. M. Fukugita et al., The distance to the Coma cluster using the B-Band Tully–Fisher relation. ApJ **376**, 8 (1991)

77. H.J. Rood, B.A. Williams, Tully–Fisher distances to M31-like galaxies in the Coma cluster. MNRAS **263**, 211–228 (1993)
78. M. Bureau et al., A new i-band Tully–Fisher relation for the Fornax Cluster. implication for the Fornax distance and local supercluster velocity field. ApJ **463**, 60 (1996)
79. B.F. Madore et al., The Hubble space telescope key project on the extragalactic distance scale. XV. a Cepheid distance to the Fornax cluster and its implications. ApJ **515**(1), 29–41 (1999)
80. S. Sakai et al., The Hubble space telescope key project on the extragalactic distance scale XXIV the calibration of Tully–Fisher relations and the value of the Hubble constant. ApJ **529**(2), 698–722 (2000)
81. W.L. Freedman et al., Final results from the Hubble Space Telescope key project to measure the Hubble constant. ApJ **553**(1), 47–72 (2001)
82. R.B. Tully, H.M. Courtois, Cosmicflows-2: i-band luminosity-HI linewidth calibration. ApJ **749**(1), 78 (2012)
83. J.G. Sorce et al., The Mid-infrared Tully–Fisher relation: calibration of the type Ia supernova scale and H_0. ApJ **758**(1), L12 (2012)
84. J.G. Sorce et al., Calibration of the mid-infrared Tully–Fisher relation. ApJ **765**(2), 94 (2013)
85. J.D. Neill et al., The calibration of the WISE W1 and W2 Tully–Fisher relation. ApJ **792**(2), 129 (2014)
86. J.G. Sorce et al., From spitzer galaxy photometry to Tully–Fisher distances. MNRAS **444**(1), 527–541 (2014)
87. E. Kourkchi et al., Cosmicflows-4: the calibration of optical and infrared Tully–Fisher relations. ApJ **896**(1), 3 (2020)
88. E. Kourkchi et al., Cosmicflows-4: the Baryonic Tully-Fisher relation providing 10000 distances. MNRAS **511**(4), 6160–6178 (2022)
89. A.G. Riess et al., A comprehensive measurement of the local value of the Hubble constant with 1 km s^{-1} Mpc^{-1} uncertainty from the Hubble space telescope and the SH_0ES team. ApJ **934**(1), L7 (2022)
90. Planck Collaboration et al., Planck 2018 results VI cosmological parameters. A&A **641**, A6 (2020)
91. M. Watanabe et al., Surface photometric calibration of the infrared Tully–Fisher relation using Cepheid-based distances of galaxies. ApJ **555**(1), 215–231 (2001)
92. D.G. Russell, The Ks-band Tully–Fisher relation—a determination of the Hubble parameter from 218 ScI galaxies and 16 galaxy clusters. J. Astrophys. Astron. **30**, 93–118 (2009)
93. L. Bottinelli et al., Cluster population incompleteness bias and the value of HO from the Tully-Fisher B relation. A&A **181**, 1–13 (1987)
94. L. Hislop et al., The extragalactic distance scale without Cepheids IV. ApJ **733**(2), 75 (2011)
95. L. Bottinelli et al., The value of HO from the infrared Tully–Fisher relation. A&A **196**, 17–25 (1988)
96. J. Schombert et al., Using the Baryonic Tully–Fisher relation to measure H_o. AJ **160**(2), 71 (2020)
97. R.B. Tully, M.J. Pierce, Distances to galaxies from the correlation between luminosities and line widths III cluster template and global measurement of H_0. ApJ **533**(2), 744–780 (2000)
98. C. Saulder et al., Target selection for the DESI peculiar velocity survey (2023). arXiv e-prints arXiv:2302.13760

Chapter 13
Determination of the Local Hubble Constant Using Giant Extragalactic H II Regions and H II Galaxies

David Fernández-Arenas and Ricardo Chávez

13.1 Introduction

The search for standard candles or rulers to measure cosmic distances is one of the fundamental efforts in the last decades in the interface of observational astronomy and cosmology. However, it has been a difficult task for a long time. In this chapter, we present a historical review of the use of H II galaxies (HIIGs) and giant H II regions (HIIRs) to carry out this task of measuring distances and its application in the determination of the Hubble constant, from its beginnings to recent years. We discuss its applicability, its potential physical origin and the associated errors which can be improved in the future.

"Peculiar" objects from Zwicky list of compact galaxies were identified by [1, 2] and further studied by [3]. Those that have spectral characteristics indistinguishable from giant extragalactic H II regions (like 30 Dor in the LMC) were designated by them as "Isolated Extragalactic H II Regions". Later on, these compact, dwarf, star-forming galaxies, were dubbed H II Galaxies [4–6, 42, 43].

D. Fernández-Arenas (✉)
Canada France Hawaii Telescope, 65-1238 Mamalahoa Hwy Kamuela, Waimea, HI 96743, USA
e-mail: arenas@cfht.hawaii.edu

R. Chávez
Instituto de Radioastronomía Y Astrofísica, UNAM, Campus Morelia, C.P., 58089 Morelia, Mexico
e-mail: r.chavez@irya.unam.mx

Consejo Nacional de Ciencia y Tecnología, Av. Insurgentes Sur 1582, 03940 Mexico City, Mexico

E. Di Valentino and D. Brout (eds.), *The Hubble Constant Tension*, Springer Series in Astrophysics and Cosmology, https://doi.org/10.1007/978-981-99-0177-7_13

In the early 1960s, Sersic discovered that the diameters of the largest HIIRs in spiral galaxies increase with the galaxy luminosity. The use of the HIIRs for the determination of distances goes back to the mid-70s, with the studies of Sandage and Tammann [7], mainly with the "classical" application, an empirical correlations between the diameters of the HIIRs and the parent galaxies luminosity. However, the correlation was not useful beyond 20 Mpc where the angular diameters become comparable with the seeing even for the largest HIIRs.

Later, Melnick in 1977 [8] studying the structures of HIIRs, noticed that their linear diameters also correlated well with the width of their integrated Hα line. This correlation between the velocity dispersion (line-width) in HIIRs and its size is analyzed and established in members of the Local Group and M81 and NGC 2403 group allowing a new determination of the distance to M101. In 1978, Melnick [9] explored the close correlation between the linear diameters of HIIRs in late-type galaxies and the width of their global Hα emission and used it to establish a close correlation between this width and the absolute magnitude of the parent galaxies; this correlation was calibrated in members of the local and NGC 2403 groups, and then used again to estimate the distance to NGC 300 and M101. The values obtained were consistent with those reported in the literature at that epoch.

In 1979, the relation between the absolute Hβ fluxes from HIIRs and the corresponding linear diameters and turbulent velocities (width of their integrated Hα line) is examined and the results presented in [10]. In this study, a strong correlation was found between the linear diameters and the absolute Hβ luminosities, no longer of the galaxies but of the HIIRs themselves. However, the large scatter precludes its application to distance determination with the properties of the dust being largely responsible for the observed scatter in the relation.

In 1981, Terlevich and Melnick [5] analyzed the correlations between Hβ luminosities, linear diameters and the widths of the global emission line profiles of 19 giant extragalactic HIIRs in 6 galaxies and 4 HII galaxies. They found that the relations: luminosity$\propto$(linewidth)^4 and size$\propto$(linewidth)^2 which are valid for pressure-supported stellar systems (elliptical galaxies, bulges of spiral galaxies and globular clusters) are also valid for giant HIIRs.

From this and other considerations, they proposed that giant extragalactic HII regions are self-gravitating systems in which the observed emission-line profile widths represent the velocity dispersion of discrete gas clouds in the gravitational potential of these massive gas-young star complexes.

The analysis also showed that the scatter observed in the Hβ luminosity-linewidth relation for their sample is largely due to a metallicity effect and after correction for this effect, the scatter in the relation is reduced with strong implications as a new method for distance calibrations and although their results are based on a very small sample, they highlight the need for more and better observations of HIIRs. They proposed that these correlations can be calibrated using HIIRs in nearby galaxies for which the distances are known to obtain distances to field spirals and irregular galaxies beyond the Virgo cluster.

From that moment on, the empirical correlations among integrated parameters of HIIRs are considered to have fundamental importance for the understanding of the nature and origin of these objects and represent an independent indicator of distance.

In 1987, with a new homogeneous sample of data for 22 HIIRs in 10 galaxies with known distances, Melnick et al. [11] studied the different correlations among integrated parameters of HIIRs. They confirmed the existence of significant correlations between core radius, emission line width, Hβluminosity, and oxygen abundance, providing a powerful distance indicator for galaxies with HIIRs. Since, the distance indicator of HIIRs relies on the correlations between Hβ luminosity, the emission line-profile widths and the oxygen abundance of giant HIIRs, they proposed a possible physical scenario where the stellar winds and gravity are the most promising mechanisms for explaining the observations, although not explaining completely the presence of supersonic motions in the ionized gas.

In 1987, Melnick et al. [12] proposed that the most useful application of HIIRs to the determination of distances, is the correlation find by [5], which, we will call the $L(\mathrm{H}\beta) - \sigma$ relation. Using this extragalactic distance indicator tied to galactic scale via cepheid variables, 41 HIIGs from the group's spectrophotometric catalogue [13] and the zero point calibrated with the giant HIIRs in nearby spiral galaxies, for the first time a value of H_0=95 $\pm$ 10 km s^{-1} Mpc^{-1}was obtained for the Hubble parameter, with the distance calibrator given by:

$$\log \mathrm{L}(\mathrm{H}\beta) = \log \frac{\sigma^5}{(\mathrm{O/H})} + 41.32 \tag{13.1}$$

A year later, Melnick et al. [14] calibrated the correlation between luminosity, line width and oxygen abundance. They reported a value for H_0 of 89.0 $\pm$ 10 km s^{-1} Mpc^{-1}, after correcting the flows for Malmquist bias and the radial velocities for the movement of the Local Group. In this study, they discuss the initial potential problems related to the application of the method: the zero point errors and the curvature of the $L(\mathrm{H}\beta) - \sigma$ relation. An underestimate of the zero point may arise if many HIIGs are composed of more than one HIIR superimposed along the line of sight. On the other hand, the curvature related to the broadening of the emission lines may be affected by the gravity of the overall galaxy or systems with rotation. If present, this effect will saturate the $L(\mathrm{H}\beta) - \sigma$ relation at high luminosities where the line broadening will be dominated by the gravity of the old stellar component. Excluding systems with a linewidth $> 60\,\mathrm{km\,s^{-1}}$ the correlation between Hβ luminosity, the width of the line and oxygen abundance, HIIGs show the same functional form and the same scatter as that for local HIIRs, suggesting that any broadening of the lines by a source not related to the young stellar population must be small. Additionally, by imposing the condition that the equivalent width of Hβ> 30Å, effectively the number of galaxies where the light of the young component is severely affected by the underlying older stellar population is reduced.

In the last two decades, with the advance of technology in new detectors, the new generation of big telescopes and given the potential use of HIIRs and HIIGs as distance indicators, a strategy to measure the Hubble constant and to constrain cosmological parameters was devised. It is a long-term program using HIIGs, as standard candles that can, in principle, be applied out to $z \sim 3.5$ [15–17] with present-day ground-based instrumentation and up to $z \sim 9$ with the Near Infrared Spectrograph (NIRSpec) component of the James Webb Space Telescope (JWST) ([18, 19] and Chavez et al. in preparation). Let's remember that the potential of HIIGs as distance indicators stems from the existence of a correlation between the luminosity of a hydrogen recombination line, e.g. L(Hβ) (proportional to the number of ionising photons), and the line width of the emission line, σ.

In what follows, we show the latest advances in the description of the $L(\mathrm{H}\beta) - \sigma$ relation, its calibrations with new data sets and the application to the determination of the local value of the Hubble constant.

13.2 The L-σ Relation

Two of the more recent works related to the $L(\mathrm{H}\beta) - \sigma$ relation for HIIGs have been presented in the works of Bordallo and Telles [20] and Chávez et al. [21] in 2011 and 2014, respectively. They presented new observational datasets of high- and low-spectral resolution. After multi-parametric fits, they found that the Hβ equivalent width is more important for determining the scatter in the $L(\mathrm{H}\beta) - \sigma$ relation than the dependence on metallicity, contrary to the suggestion by Terlevich et al. [5]. Assuming that the EW(Hβ) is an indicator of the burst's age, the interpretation is that the age of the underlying older stellar population could be contaminating the luminosity of a recent dominant young burst more than its velocity dispersion. Chávez et al. [21] show that the majority of the scatter in the $L(\mathrm{H}\beta) - \sigma$ relation is reduced using the size of the burst as a second parameter, which allows us to think of a fundamental plane defined by the luminosity, velocity dispersion and size in HIIGs and HIIRs.

Both studies conclude that, in addition, part of the scatter in the $L(\mathrm{H}\beta) - \sigma$ relation can be reduced by using the objects with the most Gaussian profiles in their emission lines. However, the cost of reducing the samples is of the order of 50% using this criterion, which can be parameterized in different ways, (see Sect. 4.2 in [20] and sec 6.1 in [21]). The main parametric relations presented in the above papers are reproduced in Table 13.1. According to different settings, constraints are placed on the slope between 3 and 5, which could be consistent with a fundamental plane similar to that followed by elliptical galaxies and old star systems supported by gravity. A study on the evolution of these young systems and their scale relationships compared to old systems can be found in Terlevich et al. [22]. However, the existence of the $L - \sigma$ relation for HIIRs and HIIGs poses an intriguing question about the origin of the supersonic velocity widths of the emission line profiles. A simple explanation would be that metal rich galaxies produce more massive starbursts, and this should

Table 13.1 Fits for the $L - \sigma$ relation in local HIIGs

$\log L = \alpha + \beta \times \log \sigma$				
Linear Regression	Intercept (α)	Slope (α)	rms	N
(BT11, Here the luminosity of Hα has been used instead of Hβ luminosity)				
OLS(Y|X)	36.21 ± 0.32	3.01 ± 0.23	0.37	81^a
OLS(X|Y)	34.52 ± 0.38	4.18 ± 0.27		
OLS Bisector	35.49 ± 0.32	3.51 ± 0.23		
OLS(Y|X)	35.29 ± 0.42	3.72 ± 0.31	0.31	53^b
OLS(X|Y)	33.73 ± 0.47	4.85 ± 0.34		
OLS Bisector	34.61 ± 0.41	4.22 ± 0.30		
OLS(Y|X)	34.80 ± 0.41	4.14 ± 0.29	0.29	37^c
OLS(X|Y)	33.45 ± 0.53	5.13 ± 0.38		
OLS Bisector	34.19 ± 0.43	4.58 ± 0.30		
(C14)				
Error in both axes	33.71 ± 0.21	4.65 ± 0.14	0.332	107^d
Error in both axes	33.22 ± 0.27	4.97 ± 0.17	0.332	69^e

[a] All galaxies with homogeneous spectrophotometry in. [b]The sample showing regular Gaussian profiles. [c]The sample showing regular Gaussian profiles and homogeneous data. [d]All galaxies with homogeneous data in. [e]The sample showing regular Gaussian profiles

be verified by observations. Another hint about the origin of velocity widths found in HIIGs is provided by the systematic difference found between the width of Balmer lines and [O III] lines, the most recent studies relating to "The $L - \sigma$ relation for HIIGs in green" can be found in Melnick et al. [23]. A possible explanation would be the existence of a well behaved ionization structure, in which more excited ions are concentrated closer to the ionizing sources and densest regions.

13.3 The Data

The use of the $L - \sigma$ relation as a distance indicator and as a tool to derive the Hubble constant requires accurate determination of both the luminosity and the FWHM or velocity dispersion of the emission lines in giant HIIRs and HIIGs. Observationally HIIGs and HIIRs represent the youngest Super Star Clusters that can be observed in any detail. The measurements that are required for the calibration of the $L(\text{H}\beta) - \sigma$ relation are the following:

1. Narrow-band imaging or low-resolution-wide slit spectroscopy to obtain the integrated Hβ flux.
2. High-resolution spectroscopy to measure the velocity dispersion from the Hβ (or [O III] for comparison) line profile.

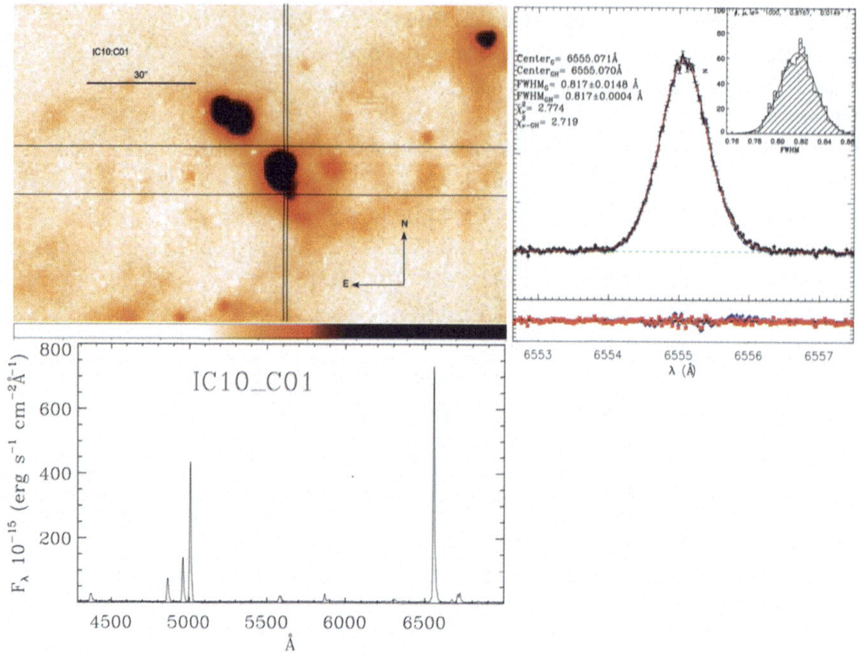

Fig. 13.1 Hα image obtained from NASA/IPAC Extragalactic Database (NED), high-resolution profile for the giant HIIR (IC10: 101) and low-resolution spectrum. Taken from [24]

In Fig. 13.1, we show as an example the data for one HIIR in a nearby galaxy used in the calibration of the zero-point of the $L - \sigma$ relation.

To guarantee the best-integrated spectrophotometry and the youngest objects, the following selection criteria are applied:

- A lower limit for the equivalent width, EW(Hβ)$>$ 50Å(HIIGs and giant HIIRs).
- An upper limit to the Balmer line widths, $\log\sigma < 1.8$ km s^{-1}(HIIGs and giant HIIRs).
- Redshift range $0.01 < z < 0.2$ (HIIGs).
- Petrosian diameter less than 6 arcsecs (HIIGs).

Figure 13.2 shows typical HIIG in the local universe and their high-resolution spectrum dominated by strong emission lines.

The above criteria allow us to exclude highly evolved regions, to diminish contamination by an underlying older stellar population and to avoid objects with a high rate of escape of ionizing photons. They also minimize the possibility of including systems supported by rotation or with multiple young ionizing clusters as it was discussed in [14]. The lower redshift limit in HIIGs is selected to avoid nearby objects that are more affected by local peculiar motions relative to the Hubble flow and the upper limit was set to minimize cosmological non-linearity effects.

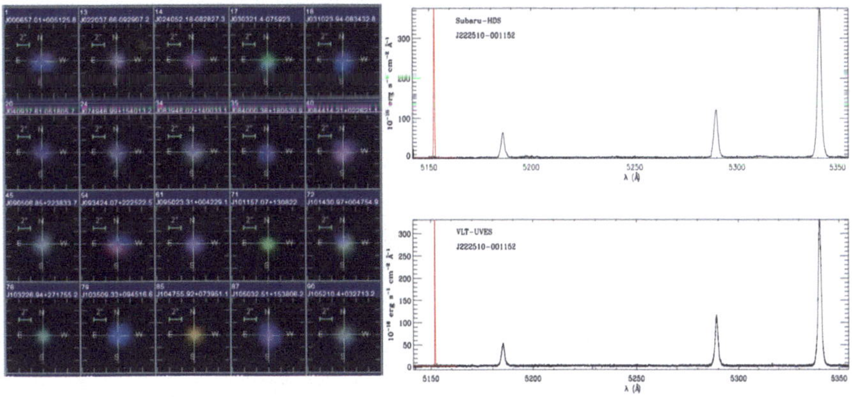

Fig. 13.2 A selection of colour images of HIIGs from the sample of [21]. The SDSS name and the index number are indicated in the stamps. The changes in colour are related to the redshift of the object. In the right panel: Examples of the high dispersion spectra obtained for the same object with Subaru HDS (top) and VLT UVES (bottom), showing the region covering Hβ and the [O III] lines at $\lambda\lambda$ 4959,5007 Å. The instrumental profile is shown in red at the left of each spectrum. Taken from [21]

In Fig. 13.3, we can see the evolution of the Hβ equivalent width for a simple instantaneous burst with metallicity $Z = 0.004$. In the right panel, we show the location of the HIIGs in the BPT diagrams [25] which are widely used to differentiate the excitation mechanism for star-forming galaxies from AGN, the EW criterion clearly can separate between other similar star-forming ionizing mechanism as Blue Compact Dwarfs and also show green peas galaxies as a class of HIIGs discovered using different selection criteria. Interesting, in recent years, making evolution studies of these objects the interpretation has been that these young massive stellar clusters, HIIGs and GHIIRs can evolve to form globular clusters and ultracompact dwarf ellipticals in about 12 Gyr so that present-day globular clusters and ultracompact dwarf ellipticals could have originated in conditions similar to those observed in today GHIIR and HIIG [22]. However, this is a topic for another discussion.

13.3.1 Giant HII Regions and HII Galaxies

HIIG are compact and massive systems experiencing luminous bursts of star formation generated by the formation of young super clusters (SSCs) with a high luminosity per unit mass and with properties similar, if not identical, to giant HIIRs. The potential of GHIIRs as distance indicators was originally realized from the existence of a correlation between the giant HIIR diameter and the luminosity [28, 29] see also [30].

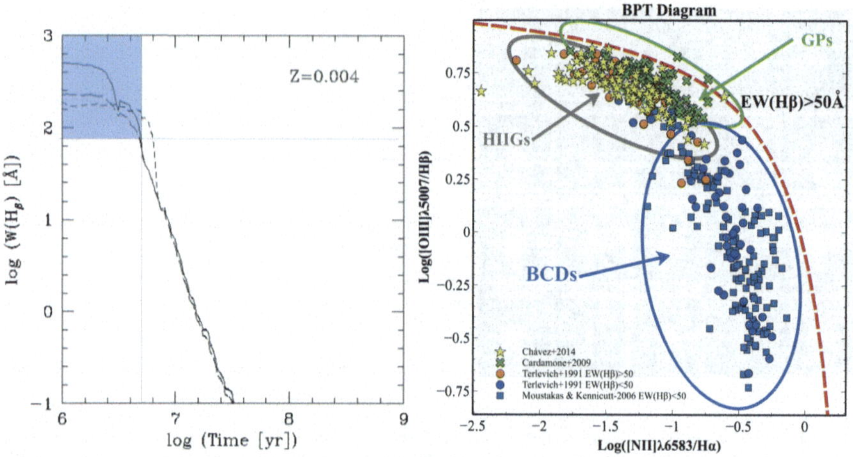

Fig. 13.3 The evolution of the Hβequivalent width for an instantaneous burst with metallicity $Z = 0.004$ and a Salpeter IMF with an upper limit of 100 M$_\odot$[26]. The blue area marks the Hβ equivalent width of 50Å, corresponding to an age of $\sim$ 5 Myrs. In the right plot: BPT diagram showing the high excitation level of a sample of HIIGs selected mainly as having high equivalent width in their Balmer emission lines. The solid line represents the upper limit for stellar photoionization, from [27]. Taken from Fernandez PhD Thesis, 2018

Most HIIG were discovered in objective-prism surveys thanks to their strong narrow emission lines. Currently, in spectroscopic surveys like Sloan Digital Sky Survey (SDSS), they are selected by very large equivalent widths in the Balmer lines. Since the luminosity of HIIG is dominated by the starburst component they can be observed even at large redshifts becoming interesting standard candles.

So far, we have analyzed a sample of HIIG and GHIIRs containing 217 objects, which can be split into: the anchor sample (36 GHIIRs in 13 local galaxies with distances from primary indicators, [24, 44]), the local sample of HIIG (107 with $z < 0.16$, [21]) and a high-z sample based on 29 KMOS, 15 MOSFIRE, 6 XShooter and 24 literature objects (see [31, 32] for more details). We used these data to constrain cosmological parameters and to determine the local value of the Hubble constant by means of the distance indicator defined by the $L - \sigma$ relation.

13.4 Methodology

The Hubble constant is determined as follows: first we fix the slope of the $L - \sigma$ relation using the velocity dispersion and luminosity of the HIIGs. The slope is independent of the actual value of H_0.

To estimate the Hubble constant we use the slope (α) of the $L - \sigma$ relation of the HIIG and new GHIIR data (anchor sample) to calibrate the zero point (Z_p) of the distance indicator (see Fig. 13.4) as follows:

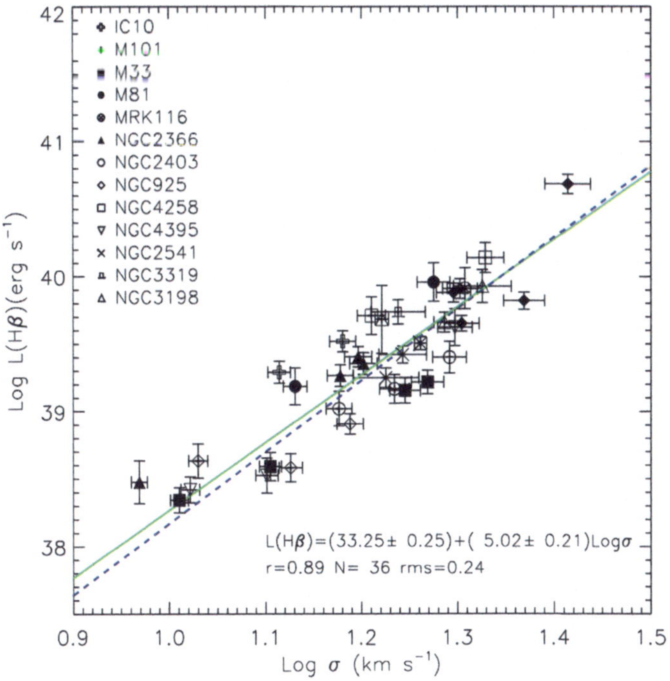

Fig. 13.4 $L - \sigma$ relation for the GHIIRs. The adopted distances to the parent galaxies come from Cepheids. The green solid line is the fit to the data given in the inset and the dashed line is the fit. Taken from [24]

$$Z_p = \frac{\sum_{i=1}^{36} W_i(\log L_{\rm GHR,i} - \alpha \times \log \sigma_{\rm GHR,i})}{\sum_{\rm i=1}^{36} W_i} \tag{13.2}$$

where $L_{GHR,i}$ is the Hβ luminosity of each GHIIR and $\sigma_{\rm GHR,i}$ is the corresponding velocity dispersion. The statistical weights W_i are calculated as:

$$W_i^{-1} = \left(0.4343\frac{\delta L_{\rm GHR,i}}{L_{\rm GHR,i}}\right)^2 + \left(0.4343\alpha\frac{\delta\sigma_{\rm GHR,i}}{\sigma_{\rm GHR,i}}\right)^2 + (\delta\alpha)^2(\sigma_{\rm GHR,i} - <\sigma_{\rm HIIG}>)^2 \tag{13.3}$$

where $<\sigma_{\rm HIIG}>$ is the average velocity dispersion of the HIIG that define the slope of the relation. Thus, the calibrated $L - \sigma$ relation or distance estimator is: $\log L(H\beta) = \alpha \log \sigma + Z_p$. To calculate the Hubble constant we minimise the function,

$$\chi^2(H_0) = \sum_{i=1}^{N} [W_i(\mu_i - \mu_{H_0,i})^2 - ln(W_i)] \tag{13.4}$$

where μ_i is the logarithmic distance modulus to each HIIG calculated using the distance indicator and the Hβ flux F(Hβ) as

$$\mu_i = 2.5[Z_p + \alpha \times \log \sigma_i - \log F_i(H\beta) - \log 4\pi] \tag{13.5}$$

and $\mu_{H_0,i}$ is the distance modulus calculated from the redshift using either the linear relation $D_L = zc/H_0$ or the full cosmological prescription with $\Omega_\Lambda = 0.71$.

The best value of H_0 is then obtained minimising χ^2 with statistical weights $W_i^{-1} = \delta\mu_i^2 + \delta\mu_{H_0,i}^2$ calculated as,

$$W_i^{-1} = 6.25[(\delta Z_p)^2 + \left(0.4343\frac{\delta F_i}{F_i}\right)^2 + \left(0.4343\alpha\frac{\delta\sigma_i}{\sigma_i}\right)^2 + (\delta\alpha)^2(\sigma_i - <\sigma>)^2] \tag{13.6}$$

13.5 Systematics

Genuine systematic errors are difficult to estimate. However, we have explored and listed a range of parameters to quantify at least part of the systematic error component. In particular, we present different systematic errors that can be taken into account in order to determine the value of H_0 using giant HIIRs and HIIGs, these effects are described in the following.

13.5.1 *Age Effects*

Even in our sample of HIIGs and giant HIIRs, chosen to be the youngest systems, it is crucial to verify that the rapid luminosity evolution of the stellar cluster does not introduce a systematic bias in the distance indicator. This would happen, for example, if the average age of the giant HIIRs is different from that of the HIIGs or if luminous and faint HIIGs have different average ages. In general, if velocity dispersion measures mass, younger clusters will be more luminous than older ones for a given velocity dispersion. The evolution effect can be scrambled by the superposition of bursts of different ages along the line of sight. Nevertheless, even small systematics can have a sizable effect on the value of H_0 so it is important to remove the evolution effect from the data in a similar fashion as for the dust extinction.

The equivalent width of Hβ [EW(Hβ)] is a useful age estimator [33–35] or at least it provides an upper limit of the age of the burst [36]. Indeed there is some empirical evidence for this as discussed by [16] and [20]. [21] explored the possibility that the age of the burst is a second parameter, using the EW(Hβ) as an age estimator in the $L(\mathrm{H}\beta) - \sigma$ relation for HIIGs, and found a rather weak dependence. Different evolution corrections have been explored in recent years, e.g. [24] and [37]. This correction must be taken into account in order to improve the estimates of the value of H_0 but it must still be explored in more detail with better observations especially to characterize the underlying older population.

13.5.2 Extinction Effects

One of the major contributions to the systematic errors is related to extinction correction, in particular the use of different extinction laws. In particular,[38] law yields larger values of H_0 than [39] typically by about 1.5~km s^{-1} Mpc^{-1}. As the [38] law was derived from a sample of eight heterogeneous starburst galaxies where only two, Tol 1924–416 and UGCS410 are bonafide HIIGs and the rest are evolved high metallicity starburst galaxies, we prefer to use the [39] extinction curve which corresponds to the LMC2 supershell near the 30 Dorado star-forming region, the prototypical giant HIIR, in the Large Magellanic Cloud.

13.5.3 Metallicity Effects

In 1981, Terlevich et al. [5] proposed that oxygen abundance is a good indicator of the long-term evolution of the system. They proposed a simple 'closed box' chemical evolution model with many successive cycles of star formation. In it, for each cycle, evolution is traced by the EW(Hβ) whereas the long-term evolution of the system, spanning two or more cycles, could be traced by the oxygen abundance, which then becomes a plausible second parameter in the $L(\mathrm{H}\beta) - \sigma$ correlation.

13.5.4 Aperture Effects

The sensitivity to Hβ photometry could be crucial for the determination of the luminosity especially when multiple bursts of star-forming regions and the nearest HIIGs are present. This effect was explored by [24] by comparing SDSS and [21] photometric measurements. The results reveal that the values of H_0 are on average systematically lower by about 4.9~km s^{-1} $Mpc^{-1}$for SDSS fluxes and no evolutionary corrections. This systematic difference is reduced to 1.7~km s^{-1} $Mpc^{-1}$when evolutionary corrections are included. The smaller value of H_0 for the SDSS photometry is related to the systematically steeper slope of the $L(\mathrm{H}\beta) - \sigma$ relation compared with that obtained when using [21] photometry, both with and without evolution correction. Since the lower luminosity HIIGs are also the closer ones, the steepening is a consequence of the smaller SDSS aperture, compared with that of [21], underestimating the line fluxes of the nearest galaxies. This effect may also introduce a bias in luminosity or velocity dispersion due to the multiplicity seen in some objects (e.g. J084220+115000) as presented by [40].

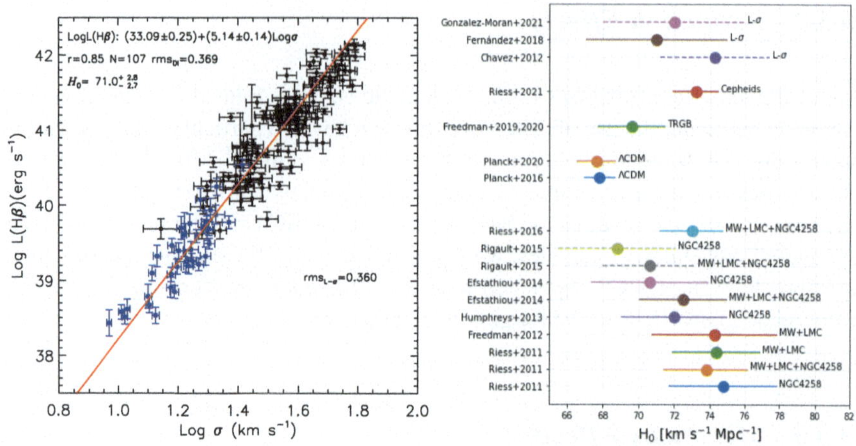

Fig. 13.5 Left: The $L-\sigma$ relation for the [21] sample using the velocity dispersions in the original paper; the fluxes have been corrected using [39] extinction law. The solid line is the fit to the HIIGs. The inset equation is the distance indicator where the slope is obtained from the fit to the HIIG and the zero point is determined following the procedure described in the text. In the right panel: Our main result incorporating the evolution correction, is H_0=71.0 $\pm$ 3.5 km s^{-1} Mpc^{-1}(random+systematic) a value that is between the most recent determination from Planck and SNIa. Taken and updated from [24]. We also plot the most recent results derived from the cosmological constrain using HIIGs from Gónzalez-Morán et al. [32]

13.5.5 *Redshift Effects*

An upper redshift cutoff of $z = 0.1$ was set (S2) instead of $z = 0.16$ (S1) for the HIIGs. Comparing S1 (in which the distances were computed using a flat cosmology with $\Omega_m = 0.29$) with sample S2, using the linear Hubble relation for the distances, we found that using $z < 0.1$ reduces the value of H_0 typically by about 2 ~km s^{-1} Mpc^{-1}with a range from 1.1 to 3.3 km s^{-1} Mpc^{-1}. The sensitivity of H_0 to the actual value of Ω_m is low, amounting in our case to an uncertainty of about 0.1% in H_0 for an uncertainty in Ω_m of 0.02 [41]. This, together with the larger uncertainty when using S2 drives us to use S1 with the distance determined using a flat cosmology with $\Omega_m = 0.29$. This reinforces the point that we are not biasing the results by using additional cosmological parameters.

13.6 Cosmological Constraints

To calculate the parameters of the $L-\sigma$ relation in a unified way including HIIGs and GHIIRs, we define the following likelihood function:

$$\mathcal{L} \propto \exp(-\frac{1}{2}\chi^2_{HII}) \tag{13.7}$$

wheie.

$$\chi^2_{HII} = \sum_n \frac{(\mu_0(\log f, \log\sigma|\alpha,\beta) - \mu_\theta(z|\theta))^2}{\epsilon^2} \tag{13.8}$$

where μ_0 is the distance modulus calculated from a set of observables as:

$$\mu_0 = 2.5(\alpha + \beta\log\sigma - \log f - 40.08) \tag{13.9}$$

α and β are the $L - \sigma$ relation's intercept and slope, respectively, $\log\sigma$ is the logarithm of the measured velocity dispersion and log f is the logarithm of the measured flux. For HIIG the theoretical distance modulus, μ_θ, is given as:

$$\mu_\theta = 5\log d_L(z,\theta) + 25 \tag{13.10}$$

where z is the redshift, d_L is the luminosity distance in Mpc and θ is a given set of cosmological parameters. For GHIIRs, the value of μ_θ is inferred from primary indicators and finally ϵ^2 are the weights in the likelihood function.

The luminosity distance d_L of the sources tracing the Hubble expansion is employed to calculate the theoretical distance moduli. We define, for convenience, an extra parameter independent of the Hubble constant as:

$$D_L(z,\theta) = (1+z)\int_0^z \frac{dz'}{E(z',\theta)} \tag{13.11}$$

i.e., $d_L = cD_L/H_0$. $E(z.\theta)$ for a flat Universe is given by:

$$E^2(z,\theta) = \Omega_r(1+z)^4 + \Omega_m(1+z)^3 + \Omega_w(1+z)^{3y}\exp\left(\frac{-3w_a z}{1+z}\right) \tag{13.12}$$

with $y = (1 + w_0 + w_a)$. The parameters w_0 and w_a refer to the dark energy equation of state (DE EoS) the general form of which is:

$$p_w = w(z)\rho_w \tag{13.13}$$

with ρ_w the pressure and ρ_w the density of the postulated Dark Energy fluid. Different DE models have been proposed and many are parametrized using a Taylor expansion around the present epoch:

$$w(a) = w_0 + w_a(1-a) \Longrightarrow w(z) = w_0 + w_a\frac{z}{1+z} \tag{13.14}$$

The cosmological constant is just a special case of DE, given for $(w_0, w_a) = (-1, 0)$, while the so called wCDM models are such that $w_a = 0$ but w_0 can take values $\neq -1$.

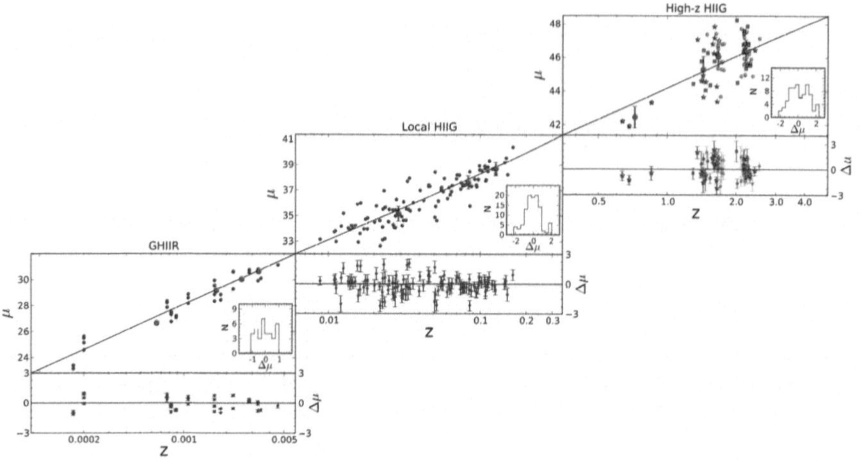

Fig. 13.6 Hubble diagram connecting our local and high redshift samples up to $z \sim 2.6$. Red circles represent averages of the distance moduli in redshift bins. The continuous line corresponds to $\Omega_m = 0.249$ and $w_0 = -1.18$ (our best cosmological model using only HIIG). The insets show the distribution of the residuals of the fit that are plotted in the bottom panels. Taken from [32]

13.7 Conclusions

The accurate determination of the Hubble constant, H_0, is considered one of the most fundamental tasks in the interface between astronomy and cosmology and the $L(\mathrm{H}\beta) - \sigma$ relation is a powerful independent distance indicator that has to be explored with more precise data in the future. Here we summarise the main results of the last year in the use of the HIIRs and HIIGs, the Hubble constant and the cosmological constraints.

This chapter presents the determination of the local value of the Hubble constant and observational constraints of the cosmological parameters making use of the GHIIRs in nearby galaxies and HIIG local and at high-z, the most recent results are summarized in the Figs. 13.4, 13.5 and 13.6, where we describe the local $L(\mathrm{H}\beta) - \sigma$ relation followed by GHIIRs, the determination of the local value of the Hubble constant, the local $L(\mathrm{H}\beta) - \sigma$ relation, the Hubble diagram tracing local GHIIRs up to high-z HIIG, the space of solutions in the cosmological parameters and the joint analysis of the other tracers of the Hubble expansion.

From the analysis of different systematic effects, we can infer that they cause changes in the $L(\mathrm{H}\beta) - \sigma$ relation that translate into r.m.s. variations in the value of H_0 of 1.5 to 2.5 km s^{-1} Mpc^{-1}.

On the other hand, combining HIIGs (the $L - \sigma$ relation as distance indicator), CMB and BAO yields: $\Omega_m = 0.298 \pm 0.012$ and $w_0 = -1.005 \pm 0.051$, fully consistent with the ΛCDM model. It is clear that the solution space of HIIG/CMB/BAO, although less constrained, is certainly compatible with the solution space of SNIa/CMB/BAO.

Acknowledgements D.F.A acknowledges the support from the National Science Foundation under grant 2109124 for SIGNALS: Unveiling Star-Forming Regions in Nearby Galaxies. R.C and D.F.A gratefully acknowledge support from the CONAHCyT of Mexico under the grant CF2022-320152. D.F.A and R.C. would like to thank professors Roberto Terlevich and Elena Terlevich for their constant support throughout these years and for their valuable suggestions, which have allowed us to improve the chapter that we present here.

References

1. W.L.W. Sargent, Astrophys. J. **159**, 765 (1970). https://doi.org/10.1086/150353
2. W. L. W. Sargent, The Astrophysical Journal, **160**, no. 405 (1970) https://doi.org/10.1086/150443
3. W.L.W. Sargent et al., Astrophys. J. **162**, L155 (1970). https://doi.org/10.1086/180644
4. A. Campbell et al., Mon. Not. R. Astron. Soc. **223**, 811–825 (1986). https://doi.org/10.1093/mnras/223.4.811
5. R. Terlevich, J. Melnick et al., Mon. Not. R. Astron. Soc. **195**, 839–851 (1981). https://doi.org/10.1093/mnras/195.4.839
6. J. Maza et al., Astron. Astrophys. Suppl. Ser. **89**, 389 (1991)
7. A. Sandage, G.A. Tammann, Astrophys. J. **190**, 525–538 (1974). https://doi.org/10.1086/152906
8. J. Melnick, Astrophys. J. **213**, 15–17 (1977). https://doi.org/10.1086/155122
9. J. Melnick, Astron. Astrophys. **70**, 157 (1978)
10. J. Melnick, Astrophys. J. **228**, 112–117 (1979). https://doi.org/10.1086/156827
11. J. Melnick et al., Mon. Not. R. Astron. Soc. **226**, 849–866 (1987). https://doi.org/10.1093/mnras/226.4.849
12. J. Melnick et al., Revista Mexicana de Astronomia y Astrofisica **14**, 158–164 (1987)
13. R. Terlevich et al., Astron. Astrophys. Suppl. Ser. **91**, 285 (1991)
14. J. Melnick et al., Mon. Not. R. Astron. Soc. **235**, 297–313 (1988). https://doi.org/10.1093/mnras/235.1.297
15. R. Terlevich et al., Mon. Not. R. Astron. Soc. **451**(3), 3001–3010 (2015). https://doi.org/10.1093/mnras/stv1128
16. J. Melnick et al., Mon. Not. R. Astron. Soc. **311**(3), 629–635 (2000). https://doi.org/10.1046/j.1365-8711.2000.03112.x
17. M. Plionis et al., Mon. Not. R. Astron. Soc. **416**(4), 2981–2996 (2011). https://doi.org/10.1111/j.1365-2966.2011.19247.x
18. de Graaff et al. (2023), arXiv e-prints: arXiv: 2308.09742, https://doi.org/10.48550/arXiv.2308.09742
19. Bunker et al. (2023), arXiv e-prints: arXiv:2306.02467, https://doi.org/10.48550/arXiv.2306.02467
20. V. Bordalo, E. Telles et al., Astrophys. J. **735**(1), 52 (2011). https://doi.org/10.1088/0004-637X/735/1/52
21. R. Chávez et al., Mon. Not. R. Astron. Soc. **442**(4), 3565–3597 (2014). https://doi.org/10.1093/mnras/stu987
22. E. Terlevich et al., Mon. Not. R. Astron. Soc. **481**, 268–276 (2018). https://doi.org/10.1093/mnras/sty2325
23. J. Melnick et al., Astron. Astrophys. **599**(A76) (2017).https://doi.org/10.1051/0004-6361/201629728
24. D. Fernández Arenas et al., Mon. Not. R. Astron. Soc. **474**(1), 1250–1276 (2018). https://doi.org/10.1093/mnras/stx2710
25. J.A. Baldwin et al., Publ. Astron. Soc. Pac. **93**, 5–19 (1981). https://doi.org/10.1086/130766

26. C. Leitherer et al., Astrophys. J. Suppl. Ser. **123**(1), 3–40 (1999). https://doi.org/10.1086/313233
27. L.J. Kewley et al., Astrophys. J. **556**(1), 121–140 (2001). https://doi.org/10.1086/321545
28. A. Sandage, Problems of extra-galactic research **15**, 359 (1962)
29. J.L. Sérsic, Zeitschrift fur Astrophysik **50**, 168 (1960)
30. R.C. Kennicutt, Astrophys. J. **228**, 394–404 (1979). https://doi.org/10.1086/156858
31. A.L. González-Morán et al., Mon. Not. R. Astron. Soc. **487**(4), 4669–4694 (2019). https://doi.org/10.1093/mnras/stz1577
32. A.L. González-Morán et al., Mon. Not. R. Astron. Soc. **505**(1), 1441–1457 (2021). https://doi.org/10.1093/mnras/stab1385
33. H.A. Dottori, E.L.D. Bica et al., Astron. Astrophys. **102**(2), 245–249 (1981)
34. G. Stasińska, C. Leitherer et al., Astrophys. J. Suppl. Ser. **107**, 661 (1996). https://doi.org/10.1086/192377
35. M.L. Martín-Manjón et al., Mon. Not. R. Astron. Soc. **385**(2), 854–866 (2008). https://doi.org/10.1111/j.1365-2966.2008.12875.x
36. R. Terlevich et al., Mon. Not. R. Astron. Soc. **348**(4), 1191–1196 (2004). https://doi.org/10.1111/j.1365-2966.2004.07432.x
37. E. Telles, J. Melnick et al., Astron. Astrophys. **615** (A55) (2018). https://doi.org/10.1051/0004-6361/201732275
38. D. Calzetti et al., Astrophys. J. **533**(2), 682–695 (2000). https://doi.org/10.1086/308692
39. K.D. Gordon et al., Astrophys. J. **594**(1), 279–293 (2003). https://doi.org/10.1086/376774
40. D. Fernández Arenas et al., Mon. Not. R. Astron. Soc. **519**(3), 4221–4240 (2023). https://doi.org/10.1093/mnras/stac3309
41. M. Betoule et al., Astron. Astrophys. **568** (A22) (2014). https://doi.org/10.1051/0004-6361/201423413
42. R.J. Terlevich, Star-forming dwarf galaxies and related objects, 395–402 (1986)
43. M.L. Martín-Manjón et al., Mon. Not. R. Astron. Soc. **385**(2), 854–866 (2008). https://doi.org/10.1111/j.1365-2966.2008.12875.x
44. R. Chávez et al., Mon. Not. R. Astron. Soc. **425**(1), L56–L60 (2012). https://doi.org/10.1111/j.1745-3933.2012.01299.x

Chapter 14
Strong Lensing and H_0

Tommaso Treu and Anowar J. Shajib

14.1 Introduction

Following Fermat's Principle, multiple images of the lensed source form at stationary points of the arrival time surface. The difference in arrival time between multiple images arises from the combination of the geometric delay, which makes some light paths longer than others, and the Shapiro delay [94], arising from the difference in gravitational potential along different paths.

Thus, if one can measure the difference in arrival time and reconstruct the gravitational potential, one obtains an absolute measurement of the difference in the length of the light paths. It is then a simple matter to convert this measurement into a combination of angular diameter distances, known as the "time-delay distance" [98], and thus obtain a direct measurement of the Hubble constant, H_0 [76].

The principle is elegant and straightforward, and Refsdal [76] recognized its potential well before strong gravitational lenses were discovered in the late 70 s [117]. Since then, breakthroughs in observations, methodology, and theory have finally made Resfdal's dream a reality. Time-delay cosmography, as it has come to be called, has been demonstrated to yield measurements of H_0 at the level of precision and accuracy of a few percent for a single strong lens system [88, 98]. As strong lens systems are being discovered and followed up at an increasing pace, the overall precision achievable by combining multiple systems is improving rapidly. Progress has also been achieved in understanding systematic errors [7, 28, 38, 39, 64, 114]

T. Treu (✉)
Physics and Astronomy Department, University of California, Los Angeles, CA 90095, USA
e-mail: tt@astro.ucla.edu

A. J. Shajib
NHFP Einstein Fellow, Department of Astronomy and Astrophysics, University of Chicago, Chicago, IL 60637, USA
e-mail: ajshajib@uchicago.edu

Kavli Institute for Cosmological Physics, University of Chicago, Chicago, IL 60637, USA

E. Di Valentino and D. Brout (eds.), *The Hubble Constant Tension*, Springer Series in Astrophysics and Cosmology, https://doi.org/10.1007/978-981-99-0177-7_14

and how to control them so as to achieve unbiased sample averages at the level of 1–2% in accuracy and precision required to help settle the Hubble tension [9].

Time-delay cosmography has several advantages as a probe of H_0. This method provides a one-step measurement of H_0 that is completely independent of other H_0 probes. The measured cosmological distances are angular diameter distances. Thus, the method is not susceptible to uncertain dust extinction laws that luminosity distance indicators could be susceptible to [31].

In this Chapter, we first provide some background in the theory and history of the method in Sect. 14.2. We then review the current state of the art in Sect. 14.3, by covering two recent case studies in some detail. First, in 14.3.1 to Sect. 14.3.4, we describe the case of RXJ1131–1231, a galaxy deflector lensing a quasar, which has been modeled based on excellent data, including ground-based spatially resolved kinematics. Second, in Sect. 14.3.5 we described the case of supernova (SN) Refsdal, multiply imaged by a foreground cluster, the first lensed SN that has been used to measure H_0. We conclude with our future outlook in Sect. 14.4. Due to space limitations, we only focus on the main points currently relevant to the measurement of H_0 and refer to previous reviews for more extensive treatments of the theory and history of the method [8, 102, 107, 110].

14.2 Background

14.2.1 Theory

This section briefly reviews the theory of time-delay cosmography. See [85] for a detailed treatment of the strong lensing formalism.

14.2.1.1 Strong Lensing Formalism

The lensing phenomenon is described by the lens equation

$$\boldsymbol{\beta} = \boldsymbol{\theta} - \boldsymbol{\alpha}(\boldsymbol{\theta}) \tag{14.1}$$

that maps a source plane coordinate vector $\boldsymbol{\beta}$ to an image plane coordinate vector $\boldsymbol{\theta}$, where $\boldsymbol{\alpha}(\boldsymbol{\theta})$ is the deflection angle vector. The lensing deflection is produced by the mass distribution between the source and the observer. In the thin lens approximation, the lensing mass distribution is described by the surface mass density $\Sigma(\boldsymbol{\theta})$ projected on the lens plane. The dimensionless lensing convergence is defined as $\kappa \equiv \Sigma/\Sigma_{\rm cr}$, where the critical density $\Sigma_{\rm cr}$ is given by

$$\Sigma_{\rm cr} \equiv \frac{c^2}{4\pi G} \frac{D_{\rm s}}{D_{\rm d} D_{\rm ds}}. \tag{14.2}$$

Here, $D_{\rm d}$ is the angular diameter distance between the observer and the deflector, $D_{\rm s}$ is the angular diameter distance between the observer and the source, and $D_{\rm ds}$ is the angular diameter distance between the deflector and the source. The angular diameter distance between two redshifts z_1 and z_2 is given by

$$D_{\rm A}(z_1, z_2) = \frac{c}{H_0(1+z_2)} f_k(z_1, z_2, \Theta), \tag{14.3}$$

where Θ is the set of cosmological parameters excluding H_0 in a given cosmology and $f_k(z_1, z_2, \Theta)$ is a function whose form depends on the sign of the curvature density Ω_k [73, 118].

The deflection angle $\boldsymbol{\alpha}$ is related to the convergence κ as

$$\kappa = \frac{1}{2}\nabla \cdot \boldsymbol{\alpha}, \tag{14.4}$$

and to the lensing potential ψ as

$$\boldsymbol{\alpha}(\boldsymbol{\theta}) = \nabla\psi(\boldsymbol{\theta}). \tag{14.5}$$

Thus, the lensing potential is connected to the surface mass distribution by the two-dimensional Poisson equation:

$$\nabla^2\psi(\boldsymbol{\theta}) = 2\kappa(\boldsymbol{\theta}). \tag{14.6}$$

We can define the Fermat potential [10, 84] as

$$\tau(\boldsymbol{\theta}, \boldsymbol{\beta}) = \frac{(\boldsymbol{\theta} - \boldsymbol{\beta})^2}{2} + \psi(\boldsymbol{\theta}). \tag{14.7}$$

The time delay between the arrival times of photons for images A and B is given by

$$\Delta t = \frac{D_{\Delta t}}{c}\left[\tau(\boldsymbol{\theta}_{\rm A}) - \tau(\boldsymbol{\theta}_{\rm B})\right], \tag{14.8}$$

where $D_{\Delta t}$ is the "time-delay distance" [76, 98] defined as

$$D_{\Delta t} \equiv (1+z_{\rm d})\frac{D_{\rm d} D_{\rm s}}{D_{\rm ds}}. \tag{14.9}$$

The time delay Δt can be measured for transient sources such as SNe or quasars. The time-delay distance is inversely proportional to the Hubble constant H_0. Therefore, if we can measure the time delay Δt and the lensing potential ψ, we can measure H_0. The two unknowns $\Delta\psi$ and $\boldsymbol{\beta}$ in Eq. 14.8 are obtained by modeling the data. The practical aspects of lens modeling with imaging data are described in Sect. 14.3.2.

The next section discusses the well-known "mass-sheet degeneracy" [MSD 32, 85]. Limiting the MSD is crucial to achieving precise and accurate H_0 measurements.

14.2.1.2 Mass-Sheet Degeneracy

The mass-sheet transformation (MST) of the convergence κ given by

$$\kappa \rightarrow \kappa' = \lambda\kappa + (1 - \lambda) \tag{14.10}$$

leaves all the imaging observables invariant with a simultaneous rescaling of the unknown source position

$$\boldsymbol{\beta} \rightarrow \boldsymbol{\beta}' = \lambda\boldsymbol{\beta}, \tag{14.11}$$

where λ is a constant. This degeneracy in the convergence κ and thus the potential ψ is called the MSD.

The time delay Δt transforms under the mass-sheet transformation as

$$\Delta t \rightarrow \Delta t' = \lambda \Delta t. \tag{14.12}$$

Thus, the time-delay distance $D_{\Delta t}$ and Hubble constant H_0 inferred from the observed time-delay Δt will change as

$$D'_{\Delta t} = D_{\Delta t}/\lambda, \tag{14.13}$$

$$H'_0 = \lambda H_0. \tag{14.14}$$

To gain an understanding of the MST, it is useful to describe the "true" physical convergence κ_{true} in terms of two components as

$$\kappa_{\text{true}}(\boldsymbol{\theta}) = \kappa_{\text{int}}(\boldsymbol{\theta}) + \kappa_{\text{ext}}, \tag{14.15}$$

where $\kappa_{\text{int}}(\boldsymbol{\theta})$ is the convergence produced by the mass distribution of the galaxy or group or cluster acting as the main deflector, while κ_{ext} is the convergence produced by the mass distribution not physically associated with the main deflector, e.g., along the line of sight (LOS). Since for a physical deflector $\lim_{\theta\rightarrow\infty} \kappa_{\text{int}}(\boldsymbol{\theta}) = 0$, we can see that $\lim_{\theta\rightarrow\infty} \kappa_{\text{true}}(\boldsymbol{\theta}) = \kappa_{\text{ext}}$. Due to the MSD, this external convergence term cannot be constrained from the imaging of the lensing system. However, it can be estimated by comparing the statistics of LOS mass distribution between cosmological simulations and that observed using photometric and spectroscopic surveys (Sect. 14.3.3). If the external convergence κ_{ext} is ignored during lens modeling, then the modeled convergence κ'_{model} will be an MST of the true convergence κ_{true} with $\lambda = 1/(1 - \kappa_{\text{ext}})$ as

$$\kappa'_{\rm model}(\boldsymbol{\theta}) = \frac{\kappa_{\rm true}(\boldsymbol{\theta})}{1-\kappa_{\rm ext}} + 1 - \frac{1}{1-\kappa_{\rm ext}} = \frac{\kappa_{\rm int}(\boldsymbol{\theta})}{1-\kappa_{\rm ext}}. \tag{14.16}$$

Here, the condition $\lim_{\theta\to\infty}\kappa'_{\rm model} = 0$ is satisfied, since $\kappa'_{\rm model}$ is only attributed to the central galaxy's or galaxies' mass distribution that ought to vanish at infinity.

In practice, lens models are often described with simply parameterized profiles, which implicitly break the MSD. Therefore, the best fit simply-parametrized model $\kappa_{\rm model}$ could be thought to be an approximate MST of $\kappa'_{\rm model}$ as

$$\kappa'_{\rm model}(\boldsymbol{\theta}) = \lambda_{\rm int}(\boldsymbol{\theta})\kappa_{\rm model}(\boldsymbol{\theta}) + 1 - \lambda_{\rm int}(\boldsymbol{\theta}). \tag{14.17}$$

Here, the internal MST parameter $\lambda_{\rm int}(\boldsymbol{\theta})$ cannot be a constant to satisfy $\lim_{\theta\to\infty}\kappa'_{\rm model} = 0$. A $\lambda_{\rm int}(\boldsymbol{\theta})$ can be designed, which acts as a constant sheet of mass near the central region of the lensing system ($\theta \lesssim \mathcal{O}(10\theta_{\rm E})$), but vanishes at $\theta \gg \theta_{\rm E}$ [7, 11], where $\theta_{\rm E}$ is the Einstein radius. As a result, the true internal mass distribution can be expressed as

$$\kappa_{\rm int}(\boldsymbol{\theta}) = (1-\kappa_{\rm ext})[\lambda_{\rm int}(\boldsymbol{\theta})\kappa_{\rm model}(\boldsymbol{\theta}) + 1 - \lambda_{\rm int}(\boldsymbol{\theta})] \tag{14.18}$$

Thus, the true Fermat potential difference relates to the modeled Fermat potential difference as

$$\Delta\tau_{\rm true} = \Delta\tau_{\rm model}\lambda_{\rm int}(1-\kappa_{\rm ext}). \tag{14.19}$$

Here, we have expressed $\lambda_{\rm int}$ as a constant since it is approximately a constant in the region where lensed images appear.

Combining Eqs. 14.8, 14.9, and 14.19, the measured Hubble constant from time-delay cosmography can be expressed as

$$H_0 = \frac{\Delta\tau_{\rm model}(1-\kappa_{\rm ext})\lambda_{\rm int}}{\Delta t}\frac{f_k(0, z_{\rm d}, \Theta)f_k(0, z_{\rm s}, \Theta)}{f_k(z_{\rm d}, z_{\rm s}, \Theta)}. \tag{14.20}$$

In this formulation, the time delay Δt is directly measured from light curves of the images (Sect. 14.3.1), $\Delta\tau_{\rm model}$ is obtained from lens modeling of high-resolution imaging data of the lens system (Sect. 14.3.2), $\kappa_{\rm ext}$ is estimated from photometric and spectroscopic surveys of the lens environment (Sect. 14.3.3), and $\lambda_{\rm int}$ is constrained from the stellar kinematics of the lens galaxy (Sect. 14.3.4).

This formulation also illustrates that H_0 measured by time-delay cosmography weakly depends on the expansion history of the Universe from redshift $z_{\rm s}$, through redshift $z_{\rm d}$, up to redshift $z = 0$. Usually, a cosmological model is assumed to compute the f_k functions, and H_0 can be slightly degenerate with other cosmological parameters, such as the dark energy's equation of state parameter w [12, 120]. However, it is also possible to infer H_0 by constraining the Universe's expansion history empirically, using relative distances of the Type Ia SNe. This inverse distance ladder method [103] allows one to obtain H_0 without assuming specific values for the other cosmological parameters.

14.2.2 A Brief History

After Refsdal's [76] suggestion, time-delay cosmography stayed dormant until the first lensed quasars were discovered some 15 years later [117]. A period of excitement followed in the 1980s when astronomers tried to measure time delays for the first time [32]. However, determining time delays proved to be a significant challenge. First, the stochastic nature of quasar light curves makes it more difficult than using the well-behaved SN light curves, as originally suggested by Refsdal. Second, light curves at optical wavelengths (corresponding usually to rest frame UV or blue) are severely affected by microlensing, which further confounds the signal. Third, the typical image separation of galaxy-scale lenses is similar to the image quality of seeing-limited ground-based optical telescopes.

Overcoming these challenges required monitoring campaigns with significantly higher precision per epoch, higher cadence, and longer duration than initially thought. By the end of the 1990s and early 2000s, such campaigns started to produce reliable time delays with few percent precision, both in the optical [57, 71, 83] and in the radio [33].

Once the ability to obtain time delays was demonstrated, attention turned to constraining the lensing potential. A two-pronged approach proved successful. On the one hand, improvements in data quality and lens modeling techniques allowed astronomers to capture the information content of multiply imaged extended sources, e.g., quasar host galaxies and radio jets and emissions. This step increases by several orders of magnitude the constraints on the mass distribution of the deflector with respect to the positions of the quasars in the image plane used in previous work.

On the other hand, it became possible to measure the stellar kinematics of the deflector, a dynamical mass tracer [112] that is crucially insensitive to the limiting factors of lensing, chiefly the MSD. Stellar kinematics also provides an independent handle on the angular diameter distance to the deflector, further enhancing the cosmological constraints [49, 50]. Conversely, lensing is insensitive to the mass-anisotropy degeneracy affecting dynamical estimates of mass. As shown in Sect. 14.3.4, the combination of lensing and stellar dynamics is crucial to the success of this methodology [87].

The final piece of the puzzle for precision time-delay cosmography at the galaxy scale was accounting for the effects of the LOS and the local environment [98]. Under or overdense LOSs, nearby perturbers, multiple-plane lensing, and group/cluster scale halos can affect the inferred distances for galaxy-scale lenses as much as 5–10% if not properly accounted for.

Fittingly, 50 years after Refsdal's paper, the first multiply imaged SN was discovered in 2014 [54]. The SN, properly named "Refsdal", was multiply imaged by a cluster of galaxies, making the geometry and arrival time sequence significantly more complex than those observed for quasars lensed by galaxy-scale potentials. The discovered "Einstein-cross" configuration is primarily due to a cluster galaxy and has little value for cosmography since the time delays are short [78]. Additional images, however, were predicted upon discovery based on the cluster mass model [54, 69, 95]. One was in the past and thus sadly lost, but the other was predicted to re-appear approximately a year later [113], and it indeed appeared when and where the model predicted [53]. Measurements of H_0 based on SN Refsdal have been reported [55, 56].

Additional multiply imaged SNe have been discovered since SN Refsdal [35], and this is an area where much growth is expected in the coming decade.

14.3 Current State of the Art

We now review the current state of the art by means of two case studies. First, in Sects. 14.3.1, 14.3.2, 14.3.3 and Sect. 14.3.4, we describe the galaxy-scale lens RXJ1131–1231. In this system, the variable source is a quasar (Fig. 14.2), and it is arguably the galaxy-scale system with the highest quality dataset to date. Then, in Sect. 14.3.5, we examine the case of the SN Refsdal, lensed by cluster MACS1149.5+2223. This is the first example of H_0 measured via a multiply imaged SN and presents a number of interesting differences with respect to the more established method of using quasars lensed by galaxies.

14.3.1 Time Delay Measurements

Time delays (i.e., Δt in Eq. 14.20) between lensed quasar images are measured using the quasar's intrinsic variability. The light curves of the individual images are measured with 1-m or 2-m class telescopes. The COSMOGRAIL collaboration [22] has monitored several lensed quasar systems, spanning up to more than a decade, providing precise measurements to a few percent [23, 63, 104]. Figure 14.1 shows light curves from COSMOGRAIL for RXJ1131–1231 based on 16 years (2003–2019) of monitoring. Courbin et al. [24] have demonstrated that robust time delays can be measured within a single season with almost a daily cadence and millimag photometric precision to detect small amplitude variations of the order of 10–50 millimag. Given the numerous discoveries of new lensed quasar systems from recent, ongoing, and future surveys, such fast turnaround for time delay measurements will be essential to rapidly produce cosmological constraints.

The main challenge in measuring the time delay is the extrinsic variability caused by the microlensing of foreground stars. This extrinsic variability pattern is unique

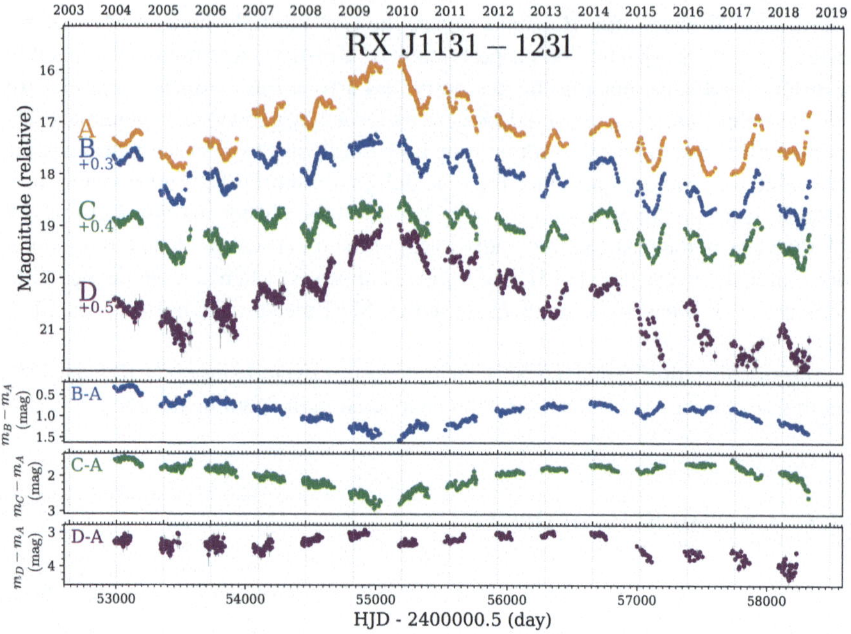

Fig. 14.1 Light curve of RXJ1131–1231 in the *R*-band from COSMOGRAIL. The bottom panels show the difference curves between pairs of images after shifting the curves corresponding to the measured delays. These difference curves illustrate the long-term extrinsic variability due to microlensing. Figure from [63]

to each lensed image. The microlensing variability can be of two types. The first one is a fast rise and fall, giving a sharp peak in the light curve. This fast variability happens when the lines of formally infinite magnification (i.e., caustics [85]) from a single foreground star cross over the quasar accretion disk. The second one is a long-term variability owing to overlapping caustics resulting from crowded stars creating a smooth change in the microlensing magnification as they move in front of the background quasar, due to internal and peculiar transverse motions.

Techniques developed to extract the time delays from the light curves while accounting for the microlensing variability include: cross-correlation that does not require explicit modeling of the microlensing variability [74]; explicit modeling of the intrinsic and microlensing variabilities with spline fitting or Gaussian processes [46, 104]. The "Time Delay Challenge" (TDC) [29, 59] validated these techniques with simulated data. The PyCS software used for the COSMOGRAIL data analysis was among the techniques achieving the target precision and accuracy in the TDC.

14.3.2 Lens Modeling

The main objective in lens modeling is to constrain the lensing potential that gives rise to the observed lensed images and distorted arcs from the background quasar and its host. This lensing potential provides $\Delta\tau_{\mathrm{model}}$ in Eq. 14.20. Usually, high-resolution imaging from the *Hubble Space Telescope* (*HST*) is used in lens modeling due to the superb stability in the point spread function (PSF) and diffraction-limited nature of the PSF to resolve lensed quasars that are separated by $\sim 1''$ (e.g., Fig. 14.2). However, imaging from large ground-based telescopes assisted with adaptive optics (AO), such as the NIRC2 imager at the Keck Observatory, has also been successfully modeled for cosmographic analysis after meticulous reconstruction of the PSF [19].

The lens model has four main components: (i) the lensing potential or the deflector mass distribution, (ii) the flux distribution in the deflector galaxy, (ii) the flux distribution in the quasar host galaxy, and (iv) the point spread function (PSF). Thus, the model for the imaging data can be reconstructed from which the likelihood function $p(\boldsymbol{d}_{\mathrm{imaging}} \mid \xi_{\mathrm{mass}}, \xi_{\mathrm{light}}, \xi_{\mathrm{source}}, \mathscr{P})$ can be computed, where $\boldsymbol{d}_{\mathrm{imaging}}$ is the imaging data, ξ_{mass} is the mass model parameters, ξ_{light} is the deflector galaxy's light model parameters ξ_{source} is the quasar host galaxy's light model parameters and $\mathscr{P}$ is the PSF model. The PSF $\mathscr{P}$ can be initially estimated from a few stars within the imaging data. However, due to color mismatch between the stellar type and the quasar, spatial variations, and undersampling, this initial PSF model needs to be improved during the lens modeling to fit the pixels around the lensed quasar images to the noise level [4, 18]. The PSF also usually needs to be reconstructed at a higher pixel resolution than the original image [91, 92]. The lens model parameters ξ_{mass}, ξ_{light}, and ξ_{source} are then constrained by sampling from the likelihood function.

The most common choices for the deflector galaxy's mass or potential model are "simply parametrized" models, where the mass distribution of the main deflector galaxy and other nearby galaxies are described with functions depending on a few free parameters. The nearby galaxies along the LOS are added to the parametric lens model when their higher order lensing effect (i.e., flexion) is non-negligible [6, 62, 89, 119]. The lensing contribution from all the other LOS structures can be accounted for by the independently estimated external convergence term (Sec. 14.3.3) and an external shear component added to the lens model's parametric description. The simplest choice that yields good residuals for the simply parametrized model of galaxy-scale lenses is the elliptical power-law mass distribution with the 3D radial density profile $\rho(r) \propto r^{-\gamma}$ [79]. The dark matter and baryonic distributions in massive galaxies are individually not power-laws. Yet, a simple power-law radial mass profile has been sufficient to describe both lensing [1, 36] and non-lensing observables such as stellar dynamics [17, 27] and X-ray intensities [47]. The total density profile is well approximated by a power-law close to the isothermal $r^{-}2$, a phenomenon known as the "bulge–halo conspiracy" [30, 109]. Suyu et al. [98] allowed departure from the power-law model using pixelated perturbations in the potential and found that large deviations ($> 2\%$) from the power-law form were not required.

An alternative choice of simply parametrized mass profile is a composite of the baryonic mass distribution that follows the observed light distribution with a spatially uniform mass-to-light ratio and the dark matter distribution described with a Navarro–Frenk–White (NFW) profile [66, 68]. The H_0 inferred using this composite mass model is reassuringly consistent with that using the simpler power-law model [64]. The composite model with the dark matter distribution described by an NFW profile has also been consistent for a sample of non-time-delay galaxy–galaxy lenses [30, 90]. These simply parametrized mass models are sufficient to fit the lens imaging data to the noise level and are internally consistent within a few percent for each individual system and within 1% for the combined H_0 values from 7 systems [64, 120]. However, the true mass distribution can potentially differ from that inferred from lensing data only, owing to the MSD [90]. A physical interpretation for such possible deviation—small and not required by the data—is shown in Fig. 14.3.

To model the deflector light profile, one, two, or three Sérsic profiles [86] are used [92, 98, 119]. The quasar host galaxy's light profile can be described on a pixellated grid with a regularization condition [99, 116]. For example, such a pixellated scheme is adopted to reconstruct the source of RXJ1131−1231 illustrated in Fig. 14.2. An alternative parametric approach is to adopt a basis set of a Sérsic profile and shapelets (i.e., 2D Gauss–Hermite polynomials [75]), whose amplitudes are determined through linear inversion of the observed image [5].

It is often necessary to make choices in the model settings, for example, the pixel resolution of the reconstructed host galaxy. To avoid any systematic bias and account for modeling uncertainty, different models are constructed by taking a combination of the plausible choices, and then this source of error is marginalized over by combining the posteriors from all these models, generally weighting by goodness of fit to avoid overcounting relatively poor models. Figure 14.4 illustrates this marginalization over multiple lens models with differing resolutions in the host galaxy reconstruction.

14.3.3 External Convergence from the LOS

The mass structures between the background source and the observer (i.e., galaxies, groups, and clusters) contribute additional lensing effect that can modify the estimated Fermat potential difference and thus the inferred H_0 by a few percent. This shift is estimated with the external convergence term κ_{ext} in Eq. 14.20. The contributing LOS structures can be grouped into two categories: (i) for lensing effects falling outside the tidal regime[1], the non-linear lensing contribution needs to be taken into account by directly including their mass distributions in the lens model (described in Sect. 14.3.2), and (ii) for lensing effect falling within the tidal regime, the combined lensing effect can be accounted for with the external convergence term κ_{ext} and an external shear term that is already included in the lens modeling. The LOS struc-

[1] Tidal regime is when the perturber's gravitational field is smaller than the gradient in the deflection angle field created by the main deflector(s).

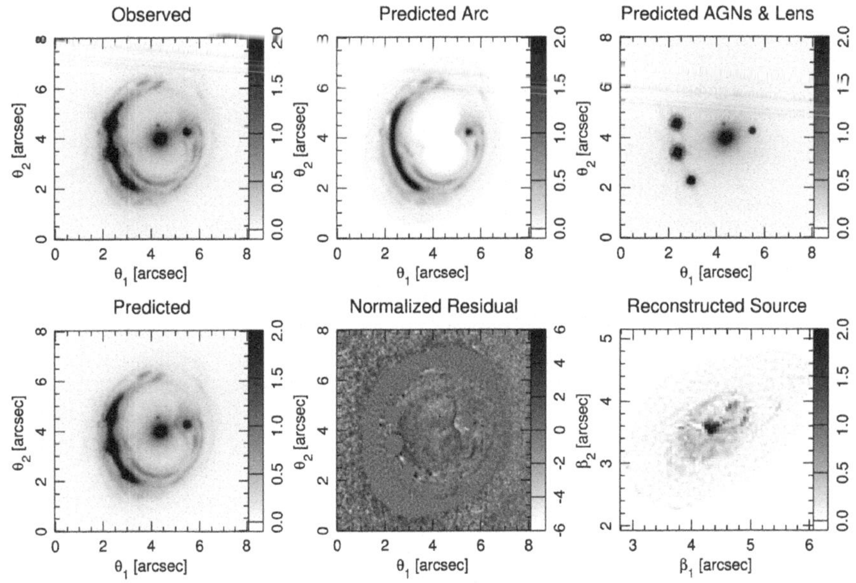

Fig. 14.2 Illustration of lens model of RXJ1131–1231 from [98]. **Top left:** Observed *HST* ACS/F814W image. **Top middle:** Reconstructed lensed arcs of the quasar host galaxy using the best-fit lens model. **Top right:** Reconstructed quasar images and the lens galaxy light. **Bottom right:** Total reconstructed image obtained by summing the top-middle and top-right panels. **Bottom middle:** Normalized residuals for the reconstructed image. **Bottom right:** The reconstructed quasar host galaxy morphology on the source plane

tures of the first category are typically within $\sim 10''$ from the central deflector. The commonly used criterion to select the perturbers in this category is a "flexion shift" threshold [62]. Estimating flexion shift requires photometric redshifts and mass measurements of the LOS galaxies. Spectroscopic redshifts are used over photometric redshifts whenever available. If the LOS galaxies form a group (or cluster), then the group-scale (or cluster-scale) halo also needs to be explicitly modeled if it satisfies the flexion-shift criterion [52, 65, 81]. The velocity dispersions of the LOS galaxies are additionally used to infer group or cluster memberships of those galaxies [14, 96].

The external convergence for the second category of perturbers can be estimated statistically from the number count of galaxies [34] around the lens system (usually within $120''$) or using weak lensing effect created by these LOS perturbers on the shapes of background galaxies. The galaxy number counts, either directly or weighted by quantities that correlate with lensing strength, such as projected distance from the central deflector, the perturber's mass, and redshift [40, 80], are used as summary statistics for the lens environment. LOS cones with matching summary statistics are then selected from cosmological simulations [44, 45]. The external convergence values computed for these simulated cones then provide a probability distribution of the external convergence of the real target (e.g., as obtained by [98] for RXJ1131–

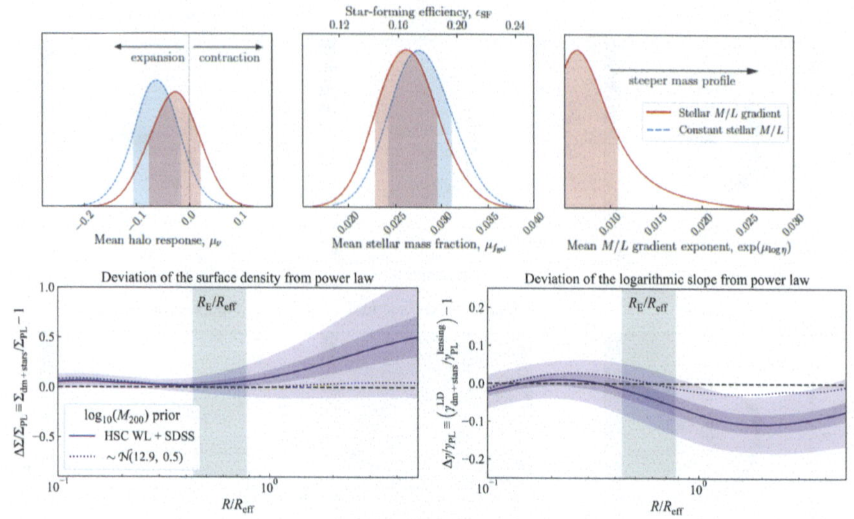

Fig. 14.3 Physical interpretation of residual uncertainty allowed by MSD on mass density profile. [89] modeled a set of non-time-delay lens galaxies with exquisite *HST* images and unresolved stellar velocity dispersion of the deflector fully accounting for the MSD and expressed the results as deviations from standard "composite" (top row) and power-law mass profiles (bottom row). The standard composite model comprising an NFW [67] dark matter halo and a stellar component with constant mass-to-light ratio is consistent with the data, although a small amount of contraction/expansion of the halo (top left panel) or a small gradient in the mass-to-light ratio (top right panel) cannot be ruled out. Similarly, a power-law mass density profile is consistent with the data, although the data cannot rule out small deviations (purple bands in the bottom panels). See [89] for more description. When available, additional information—such as spatially resolved stellar kinematics—reduces the residual freedom and thus tightens the bounds on H_0 when applied to time-delay lenses

1231). The number counts for the observed lens environment are normalized with a large number of control fields for which the photometric data is available at the same quality, and the same is done for the simulated fields. Taking these relative number counts as the summary statistics minimizes the impact of the chosen cosmological parameters in the cosmological simulation.

From high-quality and wide-field imaging, the distortion in the shapes of background galaxies can be used to constrain the weak lensing shear created by the LOS structures. In the linear regime, this shear uniquely maps to external convergence [51]. Tihhonova et al. [105, 106] applied this technique to provide alternative estimates of the external convergence κ_{ext} for two systems, finding estimates in agreement with those based on galaxy number counts.

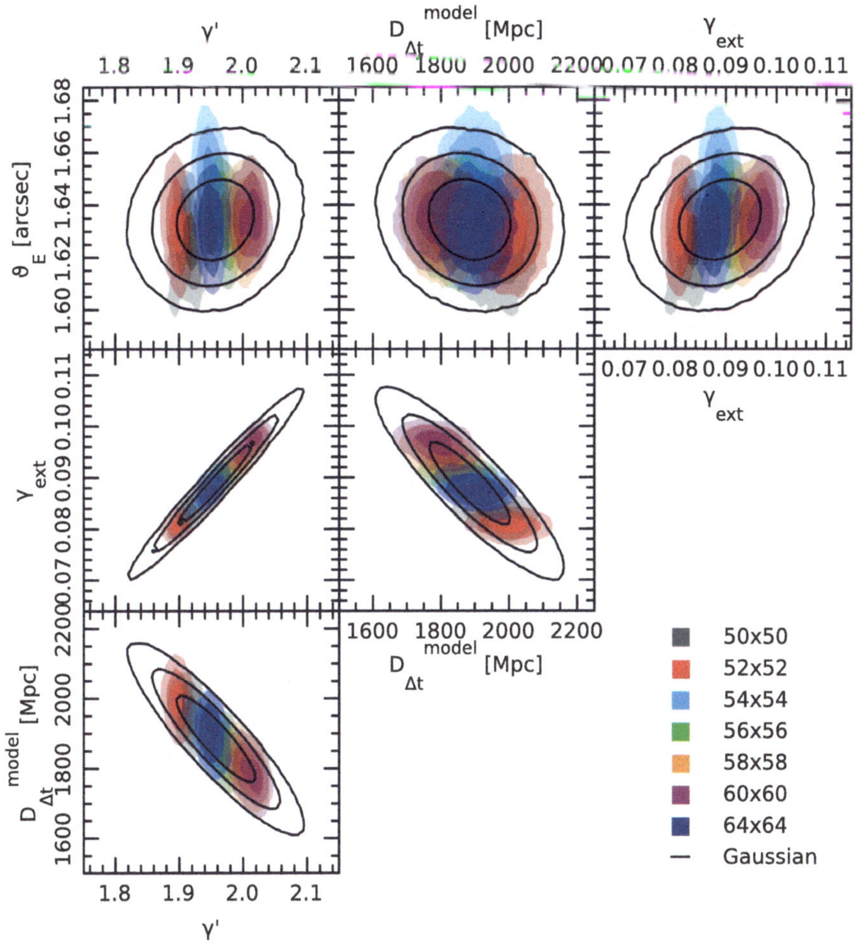

Fig. 14.4 Combined posterior of the lens model parameters power-law exponent γ, Einstein radius $\theta_{\rm E}$, and external shear magnitude $\gamma_{\rm ext}$. The illustrated model-predicted time-delay distance $D_{\Delta t}^{\rm model}$ is obtained using a fiducial cosmology from the predicted Fermat potential difference $\Delta\tau_{\rm model}$. Therefore, the illustrated $D_{\Delta t}^{\rm model}$ is simply a reformulation of $\Delta\tau_{\rm model}$, and the measured time delays are still necessary to measure H_0 using the lens model posteriors. The combined posterior marginalizes over the choice of pixel resolution in the host galaxy reconstruction, where the individual posterior for each choice is illustrated with colored distributions. Figure from [101]

14.3.4 Stellar Kinematics

Stellar kinematics traces the 3D potential of the deflector, whereas lensing traces the 2D projected potential. Thus, in combination, dynamics and lensing can break the degeneracies inherent to each method: the MSD in lensing and the mass-anisotropy degeneracy in dynamics. Therefore, stellar kinematics provide the $\lambda_{\rm int}$ term in Eq. 14.20.

Owing to the angular size and faintness of the deflectors, the most common type of measurement for stellar kinematics is an unresolved (i.e., aperture-integrated) LOS velocity dispersion from long-slit spectra. These spectra are usually taken using large ground-based telescopes, for example, the Keck Observatory and the Very Large Telescope (VLT), in seeing-limited cases in the optical.

The stellar velocity dispersion is usually modeled by solving the Jeans equation [16, 48], which is derived from the collisionless Boltzmann equation [3]. The LOS velocity dispersion can then be expressed, in the spherical case, as

$$\sigma_{\rm los}^2(R) = \frac{2G}{I(R)} \int_R^{\infty} \mathcal{K}_\beta \left(\frac{r}{R}\right) \frac{l(r)M(r)}{r} {\rm d}r, \tag{14.21}$$

where $I(R)$ is the surface brightness distribution of the kinematic tracer in the deflector, $l(r)$ is the 3D luminosity density of the same tracer, $M(r)$ is the 3D enclosed mass, $\mathcal{K}_\beta$ is a function that depends on the anisotropy profile $\beta(r)$ [61].

The anisotropy parameter $\beta(r)$ is defined as

$$\beta(r) \equiv 1 - \frac{\sigma_{\rm t}^2(r)}{\sigma_{\rm r}^2(r)}, \tag{14.22}$$

where $\sigma_{\rm t}$ is the tangential component of the velocity dispersion and $\sigma_{\rm r}$ is the radial component. In Eq. 14.21, the terms related to the light distribution, that is, $I(R)$ and $l(r)$, are well-constrained from the observed light, albeit with assumptions to deproject from the 2D $I(R)$ to the 3D $l(r)$. However, there is a degeneracy between the mass distribution giving $M(r)$ and the anisotropy profile giving $\mathcal{K}_\beta$, namely the mass–anisotropy degeneracy [25]. Although combining constraints from lensing and dynamics breaks the mass-sheet and mass–anisotropy degeneracies, the breaking power from unresolved velocity dispersion is limited. In the past, the mass-sheet degeneracy was broken chiefly through the mass profile assumption (i.e., the power-law or composite form), and then the stellar kinematics provided tighter constraint on the mass profile distribution with the mass-anisotropy degeneracy marginalized over [101]. In such cases, the internal mass-sheet transformation parameter $\lambda_{\rm int}$ (in Eq. 14.20) was effectively set to $\lambda_{\rm int} = 1$. With these assumptions, for RXJ1131–1231 H_0=$78.3^{+3.4}_{-3.3}$ km s^{-1} Mpc^{-1} based on joint lens modeling constraints from *HST*+AO imaging [19, 97].

Spatially resolved velocity dispersion constrains anisotropy more tightly than the unresolved case. Thus, spatially resolved velocity dispersion is much more effective

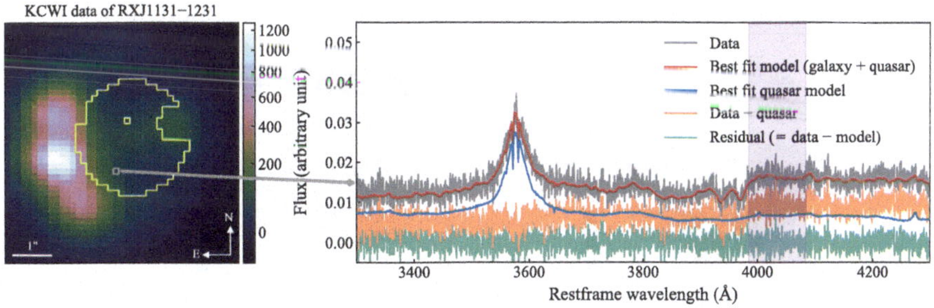

Fig. 14.5 Keck/KCWI integral field spectra of RXJ1131–1231. **Left:** 2D representation of the datacube obtained by summing across the wavelength dimension. The yellow contour shows the region within which spectra were extracted to measure the resolved velocity dispersion of the deflector. The grey box marks the pixel for which the observed spectra and model fitting are shown in the right-hand panel. **Right:** The observed spectra (grey line) and the estimated deflector spectra (orange line) after removing the modeled quasar contribution (blue line). The model for the observed spectra using X-shooter Spectral Library (XSL) and having fitted the velocity dispersion is shown with the red line. The spatially resolved velocity dispersion map is thus extracted in 41 bins within the yellow contour (Fig. 14.6). Figure from [93]

in simultaneously breaking the above-mentioned degeneracies [87]. Furthermore, an axisymmetric mass model can be constrained based on the spatial information in the resolved kinematics, allowing one to go beyond simple spherical symmetry in the mass models [15]. The oblateness or prolateness of the 3D mass shape can potentially be constrained by the misalignment between the kinematic and photometric major axes [58]. The only time-delay lens system with spatially resolved kinematics published so far is RXJ1131–1231, based on data from the Keck Cosmic Web Imager (KCWI) on the Keck Telescope (Figs. 14.5 and 14.6). Obtaining integral field spectra for kinematic measurement has been challenging from the ground since the quasar contribution will contaminate the central deflector's spectra in the seeing-limited case without AO as the typical separation between the two is $\sim 1''$. The kinematic measurement made by [93] accounted for quasar light contamination in extracting the velocity dispersion of the deflector by forward modeling it. Combining this resolved kinematics with the *HST* imaging, measured time delays, and estimated external convergence yields $H_0 = 77.3^{+7.1}_{-7.3}$ km s^{-1} Mpc^{-1} [93]. This value, obtained by combining resolved kinematics and mass modeling with relaxed assumptions, agrees very well with that obtained from simple parametric mass models, thus corroborating the standard assumptions made in time-delay cosmography [93, 120] (Fig. 14.7).

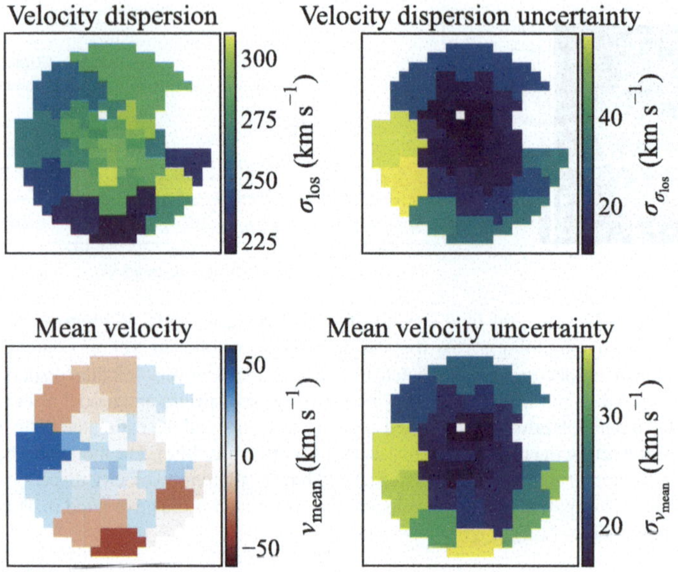

Fig. 14.6 Spatially resolved kinematic maps in 41 bins for RXJ1131–1231 from Keck/KCWI integral field spectra. The top row corresponds to the LOS velocity dispersion, and the bottom row corresponds to the LOS mean velocity. The left column shows the mean values, and the right column shows the uncertainties. Figure from [93]

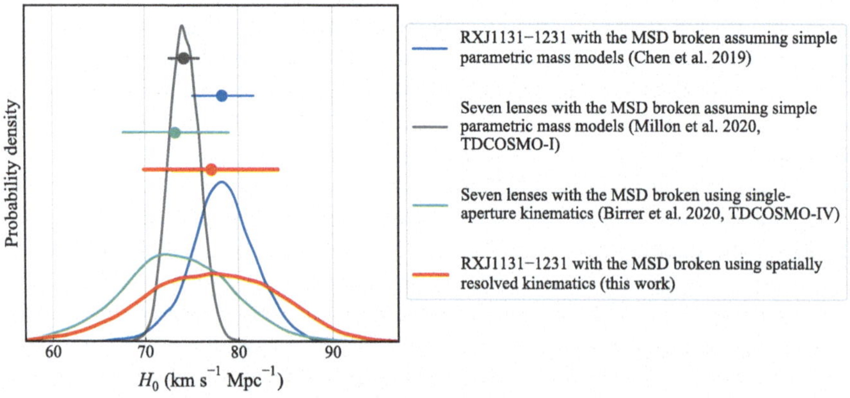

Fig. 14.7 Comparison of H_0 values with various ways to break the MSD. For the seven systems analyzed by the TDCOSMO collaboration, the MSD broken by simple parametric assumption on the mass profile (with $\lambda_{\text{int}} = 1$ fixed) gives $74.2^{+1.6}_{-1.6}$ km s^{-1} Mpc^{-1} (grey) [64], and the MSD broken by unresolved kinematics gives $73.3^{+5.8}_{-5.8}$ km s^{-1} Mpc^{-1} (emerald) [7]. For RXJ1131-1231, the MSD broken by simple parametric mass profile assumption gives $78.3^{+3.4}_{-3.3}$ km s^{-1} Mpc^{-1} (blue) [19], and the MSD broken by spatially resolved kinematics gives $77.1^{+7.3}_{-7.1}$ km s^{-1} Mpc^{-1} (red) [93]. Using spatially resolved kinematics for one system gives similar uncertainty on H_0 from seven systems with unresolved kinematics, illustrating the superior power of resolved kinematics in breaking the MSD. Figure from [93]

14.3.5 SN Time Delay Cosmography: The "Refsdal" Case Study

SN Refsdal (Fig. 14.8; [54]) is the first multiply imaged SN with measured time delays [56, 78]. It is worth studying this case in some detail as it provides some important lessons for SN time delays, which we expect to be a major contributor to cosmography in the next decade.

Before going through all the steps leading to the recent measurements of H_0 [55], it is useful to summarize the main differences between this case and the one examined in the previous sections, representing quasars lensed by galaxy-scale deflectors:

1. The intrinsic light curve of an SN (Fig. 14.8) is usually well described by a template or a low-order polynomial. This regular nature simplifies the task of obtaining a high-precision time delay with respect to the stochastic light curves of lensed quasars (Fig. 14.1). Extrinsic effects from the foreground, e.g., microlensing, can affect both cases.
2. The main deflector of SN Refsdal is a cluster of galaxies. Clusters of galaxies are not dynamically relaxed, in contrast to the inner regions of massive elliptical galaxies—the typical deflectors of galaxy-scale lenses. Thus, cluster lenses are generally significantly more complex to model, both from a lensing and dynamical point of view, with respect to galaxy-scale ones.
3. The caustics of clusters of galaxies cover a much larger solid angle on the sky than those of galaxies. Therefore, tens or even hundreds [2] of multiply imaged sources can be used to constrain the lens model in a cluster. There is generally one multiply imaged source for galaxy-scale lenses, although systems with up to a handful of them have been found [37, 88]. Note that having several families of

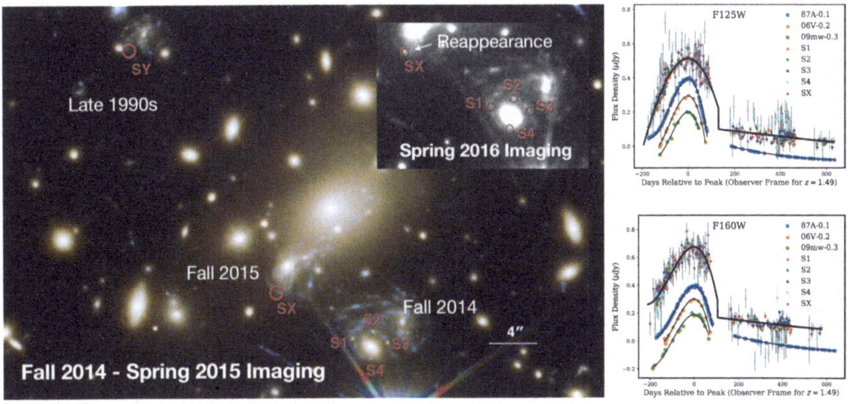

Fig. 14.8 **Left:** *HST* images of SN Refsdal, summarizing the time evolution of the phenomenon, including the original discovery of the Einstein Cross in 2014 and the appearance of SX in 2015/2016. **Right:** light curve of SN Refsdal compared to SN1987A-like light curves and polynomial fits. Figures from [56]

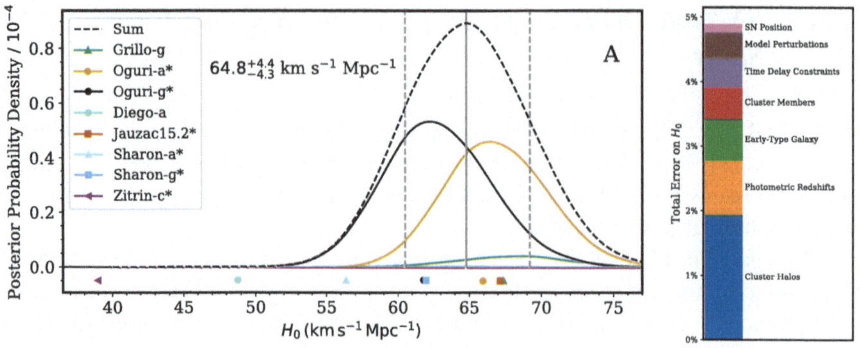

Fig. 14.9 Left: unblinded posterior distribution function of H_0 based on SN Refsdal. The lens models have been weighted according to their agreement with the H_0-independent observables, such as image positions and magnification ratios. Right: error budget on H_0 measurement. Figures from [55]

multiply imaged sources at different redshifts helps mitigate the MSD [13, 41, 42] (Fig. 14.9).

SN Refsdal was first discovered [54] as an Einstein cross in images taken as part of the Grism Lens-Amplified Survey from Space (GLASS) program [111]. Although the time delays between the cross images were expected to be of order days/week (and thus not very useful for cosmography) because of the symmetry and separation of the configuration, it was immediately recognized [54, 69, 95] that one additional and more distant image (SX) would re-appear some time in the near future, with great potential for cosmography.

Considerable effort went into predicting SX's timing, position, and brightness using updated lens models based on extensive spectroscopy [43, 113] before it appeared in the sky. Indeed, SX appeared as predicted [54], helping to build confidence in the models. The good agreement was not a foregone conclusion, considering the complexity of the cluster lens models.

The first estimate of the magnification ratio and of the long time delay (SX to the cross; the one with the most cosmographic constraining power) was used by one team to produce an estimate of H_0 [115]. That study highlighted the spread between the different cluster lens models. The spread between all the possible models is not surprising, considering that many of the lensing codes employed were not designed for precision cosmology and, therefore, did not have the necessary numerical precision and resolution. Pixelated models and models based on heavily regularized basis sets do not have by design the angular resolution needed for time-delay cosmography. The resolution requirement can be understood in terms of astrometric precision. Measuring the Hubble constant to a few percent precision usually requires knowing how the Fermat potential (and image positions) vary over tens of milliarcseconds [4]. Furthermore, many of the early models did not make full use of all the available information, e.g., cluster membership, stellar velocity dispersions of cluster members, and spectroscopic redshift of multiple images. A thorough discussion of the sources of uncertainty is given by [42].

After the re-appearance, a major effort was undertaken to obtain a blind, high-precision (1.5%) measurement of the time delay [56] and fold it in with existing lens models to obtain a blind measurement of H_0 [55]. It is crucial to stress that all the analysis choices were made blindly with respect to H_0 (the time delay was kept blind throughout the analysis, for example) and that the analysis was not modified after unblinding (with the exception of the correction of a mistake in sign convention on magnifications). This is a crucial step to prevent "experimenter bias," and we advocate that every cosmological measurement should be carried out blindly.

As highlighted by [115], the spread between lens models is clearly significant. It should be noted, however, that they include in their analysis models that were later discovered to be affected by substantial numerical errors [55], and therefore their spread was overestimated. Two options were considered by [55] to account for the spread between models. First, it was decided to consider only the two models that are based on codes designed to do high-precision cosmography, GLEE [100] and GLAFIC [70], weighted by their agreement with observables independent of H_0, yielding $H_0 = 66.6^{+4.1}_{-3.3}$ km s^{-1} Mpc^{-1}. If one had preferred to do the straight average of the two models selected prior to unblinding to be the most accurate, the result would have changed negligibly: combining the GLEE and GLAFIC models with equal weight yields $H_0 = 67.2^{+4.1}_{-3.7}$ km s^{-1} Mpc^{-1}, consistent within the errors. Alternatively, an analysis considering all models was run, again weighted by the same scheme, finding $H_0 = 64.8^{+4.3}_{-4.3}$ km s^{-1} Mpc^{-1}. The results are very similar between the two schemes because the GLEE and GLAFIC models provide, by far, the best match to the observables. The good agreement between the two options is partly due to the fact that the GLEE and GLAFIC models are by far the ones that best match the observables. This is not surprising, considering that some of the other models were not designed for precision cosmography, as discussed by [108] and [56]. Figure 14.2 in [56] shows that the models with the least weights (Zitrin and Diego) have very broad tails to significantly lower H_0, while Jauzac and Sharon yield results comparable to Grillo and Oguri.

The measurement from SN Refsdal [56] is not precise enough to help solve the "Hubble tension." Given the uncertainties [42, 56], it is consistent with both *Planck* and the local distance ladder method [77]. The latter is only 1.5σ away from the best fit, thus consistent with the SN Refsdal analysis.

The important lessons from SN Refsdal are two. First, the blind efforts succeeded to the level that only the most optimistic practitioners of cluster lens modeling would have expected. SX appeared exactly where and when it was predicted to be, and the inferred value of H_0 is perfectly consistent with those of other methods. It did not have to be this way, considering the complexity of the mass distribution in MACS1149.5+2223. SX could have appeared elsewhere or at a different time, and H_0 could have turned out to be 30 or 100 km s^{-1} Mpc^{-1}. This blind prediction helps build confidence in the models. Although it is clearly impossible to prove that a better model does not exist, the best models appear to be sufficiently precise and accurate, to the extent that we can probe them empirically. Second, 6% is a very small uncertainty for an absolute distance measurement from a single system, and compelling arguments show that it is not significantly underestimated [42, 55] when the data quality is as high as in this case. We stress that high-quality data is vital to obtain this kind of precision. The model from [42] is constrained by 89 spectroscopically confirmed multiple images arising from 28 distinct sources. Most clusters to date do not have the same level of empirical constraints [60], although the situation is rapidly improving [2]. A simple $\sqrt{N}$ scaling suggests that ten systems similar to MACS1149.5+2223 will suffice to reach $\sim 2\%$ precision and thus contribute to solving the Hubble tension if no yet-to-be-discovered systematic floor arises.

14.4 Conclusions and Future Outlook

The two recent case studies we chose to highlight represent genuine breakthroughs. We will now discuss them in the context of other measurements to give a sense of the landscape (a selection of measurements is summarized in Fig. 14.10).

RXJ1131–1231 is the first galaxy-scale lens with time delay and spatially resolved kinematics. The exquisite data for this system enabled [93] to reach 9% precision from a single lens, using what we call "conservative" assumptions about the mass distribution, i.e., allowing the MSD to be only constrained by kinematic data, and thus allowing maximum uncertainty on H_0. This is remarkable, as the precision is comparable to what was obtained by [7] with seven lenses for which only unresolved kinematics were available. With the higher angular resolution attainable with the *JWST* or with future instruments with AO, we can expect the precision per system to reach $\sim 4\%$ [121], better than that was achieved in the past for individual systems by using the more "assertive" approach of breaking the MSD by imposing a functional form for the mass density profile [89]. With this kind of precision, reaching the 2% precision, which was previously achieved under "assertive" assumptions by [64, 120], seems within reach under "conservative" assumptions with the systems known today [9]."Free form" models, that is, the ones in which the surface mass density or potential of the main deflector is rendered as pixels (see [20, 26, 72, 82]), should be able to obtain similar precision if they can be constrained by the full, high-information content from data with state-of-the-art quality. This has not been achieved yet but should be attainable. Along the way, work will need to continue in order to uncover and mitigate new sources of systematic errors, including those arising from selection effects [21].

The success of MACS1149.5+2223 paves the way for lensed SNe to become a major contributor to time-delay cosmography. A single system with excellent data quality, modeled in an "assertive" way, obtains a 6% precision on H_0, comparable to the average time-delay quasars of those analyzed by [120]. With major synoptic surveys such as *Euclid* and the Vera C. Rubin Observatory's Legacy Survey of Space and Time (LSST) about to begin, we are confident that many such systems will be discovered in the coming decade [70]. Hopefully, samples of lensed SN will soon reach comparable precision (or better!) to lensed quasars, thus providing a vital sanity

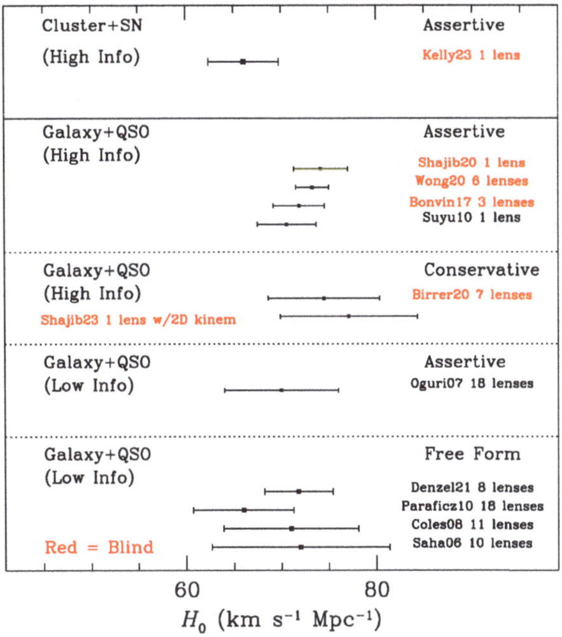

Fig. 14.10 Comparison of H_0 measurements based on time-delay cosmography, in ΛCDM cosmology. The measurements are grouped by: (i) the lensing configuration (galaxy+QSO vs. cluster+SN). (ii) Assumptions on the mass distribution of the main deflector, "assertive" and "conservative" for simply parametrized models or "free form" for pixellated models. (iii) Amount of information used per lens; in the case of a galaxy-scale lens, "low info" utilizes quasar positions and time delays, "high info" adds the extended surface brightness distribution of the multiple images of the quasar host galaxy, stellar kinematics of the main deflector, and number counts or weak lensing to estimate the LOS convergence. For the cluster+SN case, we define it as high info due to the large number of spectroscopically confirmed multiply-images and cluster members, and we define it as assertive because the GLEE and GLAFIC models used for this measurement are based on simply parametrized forms. We give the reference and the number of time delay lenses for each measurement. The measurements shown in red have been blinded to prevent experimenter bias. The figure is updated from [110]

check and increased overall precision with respect to quasar-only forecasts such as those presented by [110], thus getting us closer to solving the "Hubble Tension."

Acknowledgements TT and AJS thank their many colleagues and friends working in the field of time delay cosmography, without whom the progress described here would have been impossible. We thank Fred Courbin, Pat Kelly, Veronica Motta, and Paul Schechter for their constructive comments on an early draft of this manuscript. TT gratefully acknowledges support by the National Science Foundation, the National Aeronautics and Space Administration, the Packard Foundation, and the Moore Foundation. Support for this work was provided by NASA through the NASA Hubble Fellowship grant HST-HF2-51492 awarded to AJS by the Space Telescope Science Institute, which is operated by the Association of Universities for Research in Astronomy, Inc., for NASA, under contract NAS5-26555.

References

1. M.W. Auger et al., Astrophys. J. **724**, 511–525 (2010). https://doi.org/10.1088/0004-637X/724/1/511
2. P. Bergamini et al., Astrophys. J., in press (2023). arXiv:2303.10210. [astro-ph.GA]
3. J. Binney et al. (1987). arXiv: 1987gady . Accessed 7 Nov 2022
4. S. Birrer et al., Mon. Not. R. Astron. Soc. **489**(2), 2097–2103 (2019). https://doi.org/10.1093/mnras/stz2254. https://ui.adsabs.harvard.edu/abs/2019MNRAS.tmp.2172B
5. S. Birrer et al., Astrophys. J. **813**, 102 (2015). https://doi.org/10.1088/0004-637X/813/2/102
6. S. Birrer et al., Mon. Not. R. Astron. Soc. **484**(4), 4726–4753 (2019). https://doi.org/10.1093/mnras/stz200. http://adsabs.harvard.edu/abs/2019MNRAS.484.4726B
7. S. Birrer et al., Astron. Astrophys. **643**, A165 (2020). ISSN: 0004-6361, 1432-0746. https://doi.org/10.1051/0004-6361/202038861, https://www.aanda.org/articles/aa/abs/2020/11/aa38861-20/aa38861-20.html. Accessed 1 Apr 2021
8. S. Birrer et al. (2022). arXiv e-prints, arXiv:2210.10833 [astro-ph.CO]
9. S. Birrer, ApJ **919**(1), 38 (2021). ISSN: 0004-637X. https://doi.org/10.3847/1538-4357/ac1108. Accessed 21 Feb 2022
10. R. Blandford et al., Astrophys. J. **310**, 568 (1986). https://doi.org/10.1086/164709
11. K. Blum et al. (2020). arXiv e-prints, arXiv:2001.07182
12. V. Bonvin et al., Astron. Astrophys. **616**, A183 (2018). https://doi.org/10.1051/0004-6361/201833287
13. M. Bradač et al., Astron. Astrophys. **424**, 13–22 (2004). https://doi.org/10.1051/0004-6361:20035744. arXiv:astro-ph/0405357 [astro-ph]
14. E.J. Buckley-Geer et al., Mon. Not. R. Astron. Soc. **498**, 3241–3274 (2020). ISSN: 0035-8711. https://doi.org/10.1093/mnras/staa2563. arXiv:2020MNRAS.498.3241B. https://ui.adsabs.harvard.edu/abs/2020MNRAS.498.3241B. Accessed 21 Feb 2022
15. M. Cappellari, Mon. Not. R. Astron. Soc. **390**, 71–86 (2008). https://doi.org/10.1111/j.1365-2966.2008.13754.x
16. M. Cappellari, Mon. Not. R. Astron. Soc. **494**(4), 4819–4837 (2020). ISSN: 0035-8711. https://doi.org/10.1093/mnras/staa959. Accessed 3 June 2022
17. M. Cappellari, Ann. Rev. Astron. Astrophys. **54**, 597–665 (2016). https://doi.org/10.1146/annurev-astro-082214-122432
18. G.C.F. Chen et al., Mon. Not. R. Astron. Soc. **462**(4), 3457–3475 (2016). https://doi.org/10.1093/mnras/stw991
19. G.C. Chen et al., Mon. Not. R. Astron. Soc. **490**, 1743–1773 (2019). ISSN: 0035-8711. https://doi.org/10.1093/mnras/stz2547. arXiv: 2019MNRAS.490.1743C. https://ui.adsabs.harvard.edu/abs/2019MNRAS.490.1743C. Accessed 21 Feb 2022
20. J. Coles, Astrophys. J. **679**(1), 17–24 (2008). https://doi.org/10.1086/587635. (arXiv:0802.3219 [astro-ph])
21. T.E. Collett et al., Mon. Not. R. Astron. Soc. **462**(3), 3255–3264 (2016). https://doi.org/10.1093/mnras/stw1856
22. F. Courbin et al., **225**, 297–303 (2005). https://doi.org/10.1017/S1743921305002097. arXiv: 2005IAUS.225.297C. https://ui.adsabs.harvard.edu/abs/2005IAUS.225.297C. Accessed 20 Dec 2022
23. F. Courbin et al., Astron. Astrophys. **536**, A53 (2011). https://doi.org/10.1051/0004-6361/201015709. (arXiv:1009.1473 [astro-ph.CO])
24. F. Courbin et al., Astron. Astrophys. **609**, A71 (2018). https://doi.org/10.1051/0004-6361/201731461. https://adsabs.harvard.edu/abs/2018A%26A...609A..71C
25. S. Courteau et al., Rev. Mod. Phys. **86**, 47–119 (2014). https://doi.org/10.1103/RevModPhys.86.47
26. P. Denzel et al., Mon. Not. R. Astron. Soc. **501**(1), 784–801 (2021). https://doi.org/10.1093/mnras/staa3603. (arXiv:2007.14398 [astro-ph.CO])
27. C. Derkenne et al., Mon. Not. R. Astron. Soc. **506**(3), 3691–3716 (2021). https://doi.org/10.1093/mnras/stab1996. (arXiv:2107.05206 [astro-ph.GA])

28. X. Ding et al., Mon. Not. R. Astron. Soc. **503**, 1096–1123 (2021). ISSN: 0035-8711. https://doi.org/10.1093/mnras/stab484. arXiv: 2021MNRAS.503.1096D. https://ui.adsabs.harvard.edu/abs/2021MNRAS.503.1096D. Accessed 21 Feb 2022
29. G. Dobler et al., Astrophys. J. **799**(2), 168 (2015). https://doi.org/10.1088/0004-637X/799/2/168
30. A.A. Dutton et al., Mon. Not. R. Astron. Soc. **438**, 3594–3602 (2014). https://doi.org/10.1093/mnras/stt2489
31. G. Efstathiou, Mon. Not. R. Astron. Soc. **505**(3), 3866–3872 (2021). https://doi.org/10.1093/mnras/stab1588. (arXiv:2103.08723 [astro-ph.CO])
32. E.E. Falco et al., Astrophys. J. Lett. **289**, L1–L4 (1985). https://doi.org/10.1086/184422
33. C.D. Fassnacht et al., Astrophys. J. **581**, 823–835 (2002). https://doi.org/10.1086/344368
34. C.D. Fassnacht et al., Mon. Not. R. Astron. Soc. **410**(4), 2167–2179 (2011). https://doi.org/10.1111/j.1365-2966.2010.17591.x. arXiv: 0909.4301 [astro-ph.CO]
35. B. Frye et al., Transient Name Server AstroNote **96**, 1 (2023)
36. R. Gavazzi et al., Astrophys. J. **667**, 176–190 (2007). https://doi.org/10.1086/519237
37. R. Gavazzi et al., Astrophys. J. **677**, 1046–1059 (2008). https://doi.org/10.1086/529541
38. D. Gilman et al., Astron. Astrophys. **642**, A194 (2020). https://doi.org/10.1051/0004-6361/202038829
39. M.R. Gomer et al., A&A **667**, A86 (2022). ISSN: 004–6361, 1432–0746 (2022). https://doi.org/10.1051/0004-6361/202244324. Accessed 28 Dec 2022
40. Z.S. Greene et al., Astrophys. J. **768**, 39 (2013). https://doi.org/10.1088/0004-637X/768/1/39
41. C. Grillo et al., Astrophys. J. **860**(2), 94 (2018). https://doi.org/10.3847/1538-4357/aac2c9. (arXiv:1800.1584 [astro-ph.CO])
42. C. Grillo et al., Astrophys. J. **898**(1), 87 (2020). https://doi.org/10.3847/1538-4357/ab9a4c. (arXiv:2001.02232 [astro-ph.CO])
43. C. Grillo et al., Astrophys. J. **822**(2), 78 (2016). https://doi.org/10.3847/0004-637X/822/2/78. (arXiv:1511.04093 [astro-ph.GA])
44. S. Hilbert et al., Astron. Astrophys. **499**, 31–43 (2009). https://doi.org/10.1051/0004-6361/200811054
45. S Hilbert et al., Mon. Not. R. Astron. Soc. **382**(1), 121–132 (2007). https://doi.org/10.1111/j.1365-2966.2007.12391.x. arXiv:astro-ph/0703803 [astro-ph]
46. A. Hojjati et al., Phys. Rev. D **87**(12), 123512 (2013). https://doi.org/10.1103/PhysRevD.87.123512. (arXiv:1304.0309 [astro-ph.CO])
47. P.J. Humphrey et al., Mon. Not. R. Astron. Soc. **403**(4), 2143–2151 (2010). https://doi.org/10.1111/j.1365-2966.2010.16257.x
48. J.H. Jeans, Mon. Not. R. Astron. Soc. **82**, 122–132 (1922). ISSN: 0035-8711. https://doi.org/10.1093/mnras/82.3.122. arXiv:1922MNRAS.82.122J. https://ui.adsabs.harvard.edu/abs/1922MNRAS.82.122J. Accessed 07 Nov 2022
49. I. Jee et al., J. Cosmol. Astroparticle Phsy. **11**, 033 (2015). https://doi.org/10.1088/1475-7516/2015/11/033
50. I. Jee et al., J. Cosmol. Astroparticle Phys. **4**, 031 (2016). https://doi.org/10.1088/1475-7516/2016/04/031
51. N. Kaiser et al., Astrophys. J. **404**, 441 (1993). https://doi.org/10.1086/172297
52. C.R. Keeton et al., Astrophys. J. **612**(2), 660–678 (2004). https://doi.org/10.1086/422745. arXiv:astro-ph/0406060 [astro-ph]
53. P.L. Kelly et al., Astrophys. J. Lett. **819**(1), L8 (2016). https://doi.org/10.3847/2041-8205/819/1/L8. (arXiv:1512.04654 [astro-ph.CO])
54. P.L. Kelly et al., Science **347**, 1123–1126 (2015). https://doi.org/10.1126/science.aaa3350
55. P.L. Kelly et al., Science, in press (2023a). arXiv:2305.06367 [astro-ph.CO]
56. P.L. Kelly et al., Astrophys. J. **948**(2), 93 (2023). https://doi.org/10.3847/1538-4357/ac4ccb. (arXiv:2305.06377 [astro-ph.CO])
57. T. Kundic et al., Astrophys. J. **482**, 75 (1997). https://doi.org/10.1086/304147. (arXiv:astro-ph/9610162)

58. H. Li et al., Astrophys. J. **863**, L19 (2015). ISSN: 0004-637X. https://doi.org/10.3847/2041-8213/aad54b. arXiv:2018ApJ.863L.19L. https://ui.adsabs.harvard.edu/abs/2018ApJ.863L.19L. Accessed 17 Nov 2022
59. K. Liao et al., Astrophys. J. **800**(1), 11 (2015). https://doi.org/10.1088/0004-637X/800/1/11
60. Y. Liu et al. (2023). arXiv:2307.14833 [astro-ph.CO]
61. G.A. Mamon et al., Mon. Not. R. Astron. Soc. **363**, 705–722 (2005). https://doi.org/10.1111/j.1365-2966.2005.09400.x
62. C. McCully et al., Astrophys. J. **836**, 141 (2017). https://doi.org/10.3847/1538-4357/836/1/141
63. M. Millon et al., Astron. Astrophys. **640**, A105 (2020). https://doi.org/10.1051/0004-6361/202037740. (arXiv:2002.05736 [astro-ph.CO])
64. M. Millon et al., Astron. Astrophys. **639**, A101 (2020). https://doi.org/10.1051/0004-6361/201937351
65. I.G. Momcheva et al., Astrophys. J. Suppl. **219**(2), 29 (2015). https://doi.org/10.1088/0067-0049/219/2/29. (arXiv:1503.02074 [astro-ph.CO])
66. J.F. Navarro et al., Astrophys. J. **490**, 493–508 (1997). https://doi.org/10.1086/304888
67. J.F. Navarro et al., Astrophys. J. **490**, 493 (1997). https://doi.org/10.1086/304888. (arXiv:astro-ph/9611107)
68. J.F. Navarro et al., Astrophys. J. **462**, 563 (1996). https://doi.org/10.1086/177173
69. M. Oguri, Mon. Not. R. Astron. Soc. **449**, L86–L89 (2015). https://doi.org/10.1093/mnrasl/slv025. (arXiv:1411.6443 [astro-ph.CO])
70. M. Oguri et al., Mon. Not. R. Astron. Soc. **405**, 2579–2593 (2010). https://doi.org/10.1111/j.1365-2966.2010.16639.x
71. A. Oscoz et al., Astrophys. J. Lett. **479**(2), L89–L92 (1997). https://doi.org/10.1086/310599
72. D. Paraficz et al., Astrophys. J. **712**(2), 1378–1384 (2010). https://doi.org/10.1088/0004-637X/712/2/1378. (arXiv:1002.2570 [astro-ph.CO])
73. P.J.E. Peebles (1993). https://doi.org/10.1515/9780691206721. arXiv:1993ppc.book.P. https://ui.adsabs.harvard.edu/abs/1993ppc.book.P. Accessed 28 Apr 2023
74. J. Pelt et al., Astron. Astrophys. **305**, 97 (1996). https://doi.org/10.48550/arXiv.astro-ph/9501036. arXiv:astro-ph/9501036 [astro-ph]
75. A. Refregier, Mon. Not. R. Astron. Soc. **338**, 35–47 (2003). https://doi.org/10.1046/j.1365-8711.2003.05901.x
76. S. Refsdal, Mon. Not. R. Astron. Soc. **128**, 307 (1964). https://doi.org/10.1093/mnras/128.4.307
77. A.G. Riess et al., Astrophys. J. Lett. **934**(1), L7 (2022). https://doi.org/10.3847/2041-8213/ac5c5b. (arXiv:2112.04510 [astro-ph.CO])
78. S.A. Rodney et al., Astrophys. J. **820**(1), 50 (2016). https://doi.org/10.3847/0004-637X/820/1/50. (arXiv:1512.05734 [astro-ph.CO])
79. D. Rusin et al., Astrophys. J. **587**, 143–159 (2003). https://doi.org/10.1086/346206.eprint: astro-ph/0211229
80. C.E. Rusu et al., Mon. Not. R. Astron. Soc. **467**, 4220–4242 (2017). https://doi.org/10.1093/mnras/stx285
81. C.E. Rusu et al., Mon. Not. R. Astron. Soc. **498**, 1440–1468 (2020). https://doi.org/10.1093/mnras/stz3451
82. P. Saha et al., Astrophys. J. Lett. **650**(1), L17–L20 (2006). https://doi.org/10.1086/507583. arXiv:astro-ph/0607240 [astro-ph]
83. P.L. Schechter et al., Astrophys. J. Lett. **475**, L85+ (1997). https://doi.org/10.1086/310478. arXiv:astro-ph/9611051
84. P. Schneider, Astron. Astrophys. **143**, 413–420 (1985). ISSN: 0004-6361. arXiv:1985A and arXiv:A.143.413S. Accessed 27 Apr 2023
85. P. Schneider et al. (1992). https://doi.org/10.1007/978-3-662-03758-4
86. J.L. Sérsic (1968), http://adsabs.harvard.edu/abs/1968adga.book.S
87. A.J. Shajib et al., Mon. Not. R. Astron. Soc. **473**, 210–226 (2018). https://doi.org/10.1093/mnras/stx2302

88. A.J. Shajib et al., Mon. Not. R. Astron. Soc. **483**, 5649–5671 (2019). https://doi.org/10.1093/mnras/sty3397. http://adsabs.harvard.edu/abs/2019MNRAS.483.5649S
89. A.J. Shajib et al., Mon. Not. R. Astron. Soc. **494**(4), 6072–6102 (2020). https://doi.org/10.1093/mnras/staa828
90. A.J. Shajib et al., MNRAS **503**(2), 2380–2405 (2021). ISSN: 0035-8711. https://doi.org/10.1093/mnras/stab536. Accessed 31 Mar 2021
91. A.J. Shajib et al., Astron. Astrophys. **667**, A123 (2022a). ISSN: 0004-6361. https://doi.org/10.1051/0004-6361/202243401. https://ui.adsabs.harvard.edu/abs/2022A-abstract. Accessed 01 Feb 2023
92. A.J. Shajib et al. (2022b), arXiv: 2022arXiv220305170S. Accessed 08 Aug 2022
93. A.J. Shajib et al., Astron. Astrophys. **673**, A9 (2023). https://doi.org/10.1051/0004-6361/202345878. (arXiv:2301.02656 [astro-ph.CO])
94. I.I. Shapiro, Phys. Rev. Lett. **13**(26), 789–791 (1964). https://doi.org/10.1103/PhysRevLett.13.789
95. K. Sharon et al., Astrophys. J. Lett. **800**(2), L26 (2015). https://doi.org/10.1088/2041-8205/800/2/L26. (arXiv:1411.6933 [astro-ph.CO])
96. D. Sluse et al., Mon. Not. R. Astron. Soc. (2019). https://doi.org/10.1093/mnras/stz2483. https://ui.adsabs.harvard.edu/abs/2019MNRAS.490..613S/abstract
97. S.H. Suyu et al., Astrophys. J. Lett. **788**, L35 (2014). https://doi.org/10.1088/2041-8205/788/2/L35
98. S.H. Suyu et al., Astrophys. J. **711**, 201–221 (2010). https://doi.org/10.1088/0004-637X/711/1/201
99. S.H. Suyu et al., Astron. Astrophys. **524**, A94 (2010). https://doi.org/10.1051/0004-6361/201015481
100. S.H. Suyu et al., Astron. Astrophys. **524**, A94 (2010). https://doi.org/10.1051/0004-6361/201015481. (arXiv:1007.4815 [astro-ph.CO])
101. S.H. Suyu et al., Astrophys. J. **766**, 70 (2013). https://doi.org/10.1088/0004-637X/766/2/70
102. S.H. Suyu et al., Space Sci. Rev. **214**(5), 91 (2018). https://doi.org/10.1007/s11214-018-0524-3. (arXiv:1801.07262 [astro-ph.CO])
103. S. Taubenberger et al., Astron. Astrophys. **628**, L7 (2019). ISSN: 0004-6361, 1432-0746. https://doi.org/10.1051/0004-6361/201935980. Accessed 5 Jan 2023
104. M. Tewes et al., Astron. Astrophys. **556**, A22 (2013). https://doi.org/10.1051/0004-6361/201220352
105. O. Tihhonova et al., Mon. Not. R. Astron. Soc. **498**(1), 1406–1419. https://doi.org/10.1093/mnras/staa1436. arXiv:2005.12295 [astro-ph.GA]
106. O. Tihhonova et al., Mon. Not. R. Astron. Soc. **477**(4), 5657–5669 (2018). https://doi.org/10.1093/mnras/sty1040
107. T. Treu et al., Astrophys. J. **817**, 60 (2016). https://doi.org/10.3847/0004-637X/817/1/60
108. T. Treu et al., Astrophys. J. **817**(1), 60 (2016). https://doi.org/10.3847/0004-637X/817/1/60. (arXiv:1510.05750 [astro-ph.CO])
109. T. Treu et al., Astrophys. J. **611**, 739–760 (2004). https://doi.org/10.1086/422245. eprint:astro-ph/0401373
110. T. Treu et al. (2022). arXiv: 2022arXiv221015794T. Accessed 3 Nov 2022
111. T. Treu et al., Astrophys. J. **812**(2), 114 (2015). https://doi.org/10.1088/0004-637X/812/2/114. (arxiv:1509.00475 [astro-ph.GA])
112. T. Treu et al., Mon. Not. R. Astron. Soc. **337**, L6–L10 (2002). https://doi.org/10.1046/j.1365-8711.2002.06107.x. eprint: astro-ph/0210002
113. T. Treu et al., Astron. Astrophys. Rev. **24**(1), 11 (2016). https://doi.org/10.1007/s00159-016-0096-8. https://adsabs.harvard.edu/abs/2016A%26ARv..24...11T
114. L. Van de Vyvere et al., Astron. Astrophys. **659**, A127 (2022). ISSN: 0004-6361. https://doi.org/10.1051/0004-6361/202141551. https://ui.adsabs.harvard.edu/abs/2022A&A.659A.127V/abstract. Accessed 22 Aug 2022
115. J. Vega-Ferrero et al., Astrophys. J. Lett. **853**(2), L31 (2018). https://doi.org/10.3847/2041-8213/aaa95f. arXiv preprint: arxiv:1712.05800 [astro-ph.CO]

116. S. Vegetti et al., Mon. Not. R. Astron. Soc. **392**, 945–963 (2009). https://doi.org/10.1111/j.1365-2966.2008.14005.x
117. D. Walsh et al., Nature **279**, 381–384 (1979). https://doi.org/10.1038/279381a0
118. S. Weinberg (1972). arXiv: 1972gcpa.book.W. Accessed 28 Apr 202
119. K.C. Wong et al., Mon. Not. R, Astron. Soc. **465**, 4895–4913 (2017). https://doi.org/10.1093/mnras/stw3077
120. K.C. Wong et al., Mon. Not. R. Astron. Soc. **498**, 1420–1439 (2020). https://doi.org/10.1093/mnras/stz3094
121. A. Yıldırım et al. (2021). arXiv:2109.14615 [astro-ph]. Accessed 21 Feb 2022

Chapter 15
Addressing the Hubble Tension with Cosmic Chronometers

Michele Moresco

15.1 Introduction

The discovery of the accelerated expansion of the Universe [64, 67] has been a major revolution in modern cosmology, introducing a shift in the paradigm of the Universe considered until the end of the XX century. This discovery required scientists to introduce the presence of an unknown form of energy driving this acceleration (the "dark energy"), or to postulate that the laws of General Relativity as known would break at some large cosmological scales.

In this context, finding diverse independent ways to measure the expansion history of the Universe became one of the main goals in astrophysics, since it could provide fundamental inside into understanding the nature of dark energy, or if modified gravity models are a better explanation. The quantity that measures how the Universe is expanding as a function of time (or equivalently redshift) is the so-called Hubble parameter, $H(z)$. It is typically useful to link this quantity to the scale factor $a(t)$ (providing the relation between physical and comoving distances $r_{phys} = a(t)r_{com}$) with the equation:

$$H(z) = \frac{\dot{a}}{a} \, . \tag{15.1}$$

Several cosmological probes have been introduced during these years to address these open questions, measuring how the Universe is expanding and modeling it through cosmological parameters to characterize the behavior of dark energy and dark matter. Due to methodological, technical, and theoretical advances, several probes have reached nowadays a maturity that allowed them to be considered "standard" probes, among which e.g. the Cosmic Microwave Background (CMB), Type Ia Supernovae (SNeIa), Weak Lensing (WL), and Baryon Acoustic Oscillations (BAO)

M. Moresco (✉)
Dipartimento di Fisica e Astronomia "Augusto Righi", Università di Bologna, Viale Berti Pichat 6/2, 40127 Bologna, Italy
e-mail: michele.moresco@unibo.it

E. Di Valentino and D. Brout (eds.), *The Hubble Constant Tension*, Springer Series in Astrophysics and Cosmology, https://doi.org/10.1007/978-981-99-0177-7_15

(for a detailed review, see, e.g. [33]). Cosmological missions have been devised and carried out constantly improving the accuracy and precision of the measurements, pushing all these probes almost to their limits (e.g. Planck [2] for the CMB, Pantheon+ [8] for SNeIa, and BOSS and eBOSS [69] for BAO). It is worth mentioning among these the just launched ESA mission *Euclid* [45], which will provide the most advanced results for BAO and WL.

Recently, significant attention has been given also to the determination of the local value of the expansion rate, the Hubble constant ($H_0 = H(z = 0)$), that since first Hubble's measurement [32] in 1929 has improved its accuracy by orders of magnitude. In particular, with the observational and theoretical of recent years both direct measurements from the local Universe considering Cepheids and SNeIa (SH_0ES, [68]) and indirect measurements from the Early Universe based on the analysis of CMB (Planck, [2]) have reached the golden goal to constrain H_0 with an uncertainty of 1 km/s/Mpc and less. However, the results obtained from the two probes presented a tension increasing with the decreasing errors of the two analyses, reaching now a 4–5σ disagreement called the "Hubble tension" (for a comprehensive review, see [1, 76]). While the origin of this tension is still under investigation, as done in the past it is fundamental now to go beyond the standard probes, which are reaching their intrinsic limits, and explore new solutions to constrain H_0, providing new and independent measurements that could help to shed light in this debate between early- and late-Universe analyses.

A typical pattern of the various cosmological probes previously discussed is the search for a standard property of objects (luminosity, length) that allows us to decouple in the observable the effect due to the intrinsic evolution and to the expansion of the Universe, making it possible to measure the latter. Historically, standard candles (SNeIa) and standard rulers (BAO) have been amongst the most successful ones, but in this context it is possible to exploit also the time. Using the oldest objects in the Universe to set constraints on its age and on cosmological parameters is not new in the literature (e.g., see [14, 38, 41, 43], and for a review [52]). Moreover, given how the age of the Universe scales as a function of redshift:

$$t(z) = \int_0^z \frac{dz'}{H(z')(1+z')}. \tag{15.2}$$

finding the oldest objects at each redshift it is possible to use them to constrain the Hubble parameter inside the integral; for some applications of this method, see, e.g., Refs. [6, 74, 75]. In this way, we would be searching for standard clocks, objects whose absolute ages can be used to constrain cosmological models. This approach has, however, some downsides. First of all, robustly determining absolute ages is quite difficult, requiring an absolute calibration that increases the total error and being prone to not negligible systematic effects. Moreover, as can be evinced from Eq. 15.2, with this method we measure an integral of $H(z)$[1], and typically one has to assume a cosmological model, substitute it in Eq. 15.2, and constrain its

[1] The same holds for SNeIa, that measure the luminosity distance $d_L(z)$.

parameters. The method is, therefore, not purely cosmology-independent, since we have to assume each time a particular cosmological model.

With the cosmic chronometers method, we can improve on this approach, bypassing all these issues.

15.2 The Cosmic Chronometers Method

The idea behind the cosmic chronometer method has been introduced by Ref. [39]. Under minimal assumptions, namely just a Friedmann-Lemaître-Robertson-Walker (FLRW) metric, given the relationship between the scale factor and redshift $a(t) = 1/(1+z)$, the Hubble parameter can be derived directly from Eq. 15.1 as:

$$H(z) = -\frac{1}{1+z}\frac{dz}{dt} \tag{15.3}$$

This means that by measuring the differential age of the Universe (how much the Universe has aged between two redshifts) it is possible to obtain a direct and model-independent determination of the expansion rate $H(z)$. The main difference here is that instead of looking for some standard clocks, we will be looking for *standard chronometers*, a homogeneous population of objects with a synchronized star formation, i.e. whose clocks started "ticking" at the same time and that are therefore optimal tracers of the differential age evolution of the Universe. There are many advantages in this technique with respect to using absolute ages: (i) the measurement of differential ages has been proven to be more accurate and robust than absolute ages, since many systematic effects in age determination are minimized when estimating the differences dt; (ii) the differential approach allows us to minimize many other systematic effects that might affect this type of analysis (e.g. progenitor bias, rejuvenation, ...) since for the application of the method we are only interested in the homogeneity between the two redshift bins where the quantity dt is estimated; and (iii) it provides a direct determination of the Hubble parameter without relying on any cosmological assumptions (apart the cosmological principal and a metric). A detailed review of cosmic chronometers is provided in Ref. [52].

The pillars of the cosmic chronometers method are the following:

- **Selection of a population of optimal cosmic chronometers.** As can be evinced from Eq. 15.3, the first point to address is finding a population of objects able to trace homogeneously how much the Universe has aged between two redshifts. We also need to ensure a way to maximize the homogeneity and purity of the sample, devising criteria to minimize the potential contamination from outliers. We describe the selection process in Sect. 15.3.

- **Robust measurement of the differential age dt**. While in Eq. 15.3 there are formally two observables (dz and dt) the advances in spectroscopic surveys makes the measurement of dz remarkably accurate when a spectroscopic redshift is available (typically $\delta z/(1+z) \lesssim 10^{-3}$, see e.g. [53]). Therefore, for a proper application of the method, we need to find ways to obtain a robust and unbiased determination of dt, as we will describe in Sect. 15.4.
- **Assessment of the statistical and systematic effects.** For a proper application of the method, we need to consider carefully and assess the various sources of errors—statistical and systematic—and include those in the total error budget with a proper covariance matrix. We will discuss this at the end of Sect. 15.4.

15.3 Selecting an Optimal Sample Cosmic Chronometers

The most elementary objects in the Universe that we can date and that can be found from the local Universe up to high redshifts are, of course, galaxies. To apply the cosmic chronometer method, the idea is therefore to find at each redshift the oldest population of galaxies available. These will, of course, have an offset with respect to the age of the Universe given the fact that their redshift of formation is not infinite, but as long as our selection is homogeneous this will not impact the analysis, since the differential age dt will not be affected by this issue.

The first and more straightforward solution that can be explored is to avoid any selection, collect at each redshift all available galaxies, and determine the upper envelope of the distribution, selecting the eldest crust of objects. This approach (applied, e.g., in [39, 40, 42, 53, 71]) while having the advantage of relying on the easiest possible selection, has the problem that to be applied we need a complete sample with a very high statistic (typically $O(10^3)$ galaxies, see [40]) to robustly detect the upper envelope.

A viable and more feasible alternative suggested in the literature is instead to select the most massive ($log(M/M_{\odot})$>10.5–11) and passively evolving galaxies. There is a large literature confirming that these objects have formed at extremely high redshifts ($z > 2-3$, [11–13, 21, 22, 47]) over a very short time-scale ($\tau < 0.3$ Gyr, [12, 16, 37, 47, 73, 74]). These galaxies have been found up to $z \sim 2.5$ [9, 20, 25, 49, 61], and with the advent of JWST this boundary has been pushed up to $z \sim 5$ [10]. These systems, having formed most of their mass at very high redshifts, in a very quick episode of star formation, and having mostly exhausted their gas reservoir are expected to evolve passively as a function of cosmic time; therefore, they represent the best cosmic chronometers. This is confirmed by the fact that their stellar metallicity is found to be very stable as a function of redshift, around solar to slightly over-solar values at $0 \lesssim z \lesssim 2$ [19, 22, 26, 27, 44, 47, 62, 63]. In addition, it has also been found that more massive galaxies undergo a faster process of formation with respect to less massive ones, forming also at earlier cosmic times. This scenario, referred to as mass downsizing, suggests that most massive passive galaxies are therefore ideal cosmic chronometers, being at each redshift the oldest objects and having a much more synchronized time of formation [15, 31, 73].

Different definitions have been proposed in the literature to select passive galaxies, based on photometry, spectroscopy, physical properties, and/or morphology. However, it has been also proven that a single selection criterion is not sufficient to obtain a sample with high purity and that depending on the adopted criterion the contamination by young/star-forming outlier can be as high as 30–50% [59] (see, e.g., the lower-right panel of Fig. 15.1). This effect can be significantly mitigated by combining different selection criteria, following the workflow proposed by Ref. [57] and shown in the upper panel of Fig. 15.1.

In summary, for an optimal selection of cosmic chronometers, it has been demonstrated that the following multiple selection criteria should be combined.

- A **photometric selection**. It has been shown that the wider the photometric coverage, the better passive and star-forming galaxies can be separated. From this point of view, the near-UV and bluer optical bands are crucial, in particular to detect possible contamination by a younger population (0.1–1 Gyr). So far, the NUV-r-J color-color diagram [35] has been identified as the criterion that allows the best identification of passive galaxies (being minimally sensitive to contamination due to dust), but valid alternatives are also the U-V-J diagram [78] and the NUV-r-K [4]; as another option, it is also possible to assess the galaxy type modeling the full spectral energy distribution (as done e.g. in [34, 83]).
- A **spectroscopic selection**. Different spectroscopic features have been identified as ideal tracers of ongoing star formation, or of potential contamination by a younger population; among those, the most used are the [OII]λ3727, Hβ ($\lambda =$ 4861 Å), [OIII]λ5007, and Hα ($\lambda = 6563$ Å) emission lines. Different approaches can be used to reject objects with significant emission, either adopting a cut on the equivalent width of the line (a typical threshold is EW<5Å[7, 50, 53]) or on the signal to noise ratio [58, 77]). These cuts, while increasing the purity of the sample, are not optimized to detect the presence of potential sub-dominant contamination by a young population, which might affect the oldest galaxies e.g. via episodes of mergers. Recently, Ref. [57] proposed a novel spectroscopic indicator able to precisely assess this level of contamination. The ratio between the CaII H ($\lambda = 3969$ Å) & K (λ3934 Å) lines was found to be maximally sensitive to this ratio, since the presence of a younger population, characterized by a significant Hϵ ($\lambda = 3970$ Å) absorption lines, affects the CaII H/K inverting its typical ratio. For this reason, it is a powerful diagnostic for the presence of contamination, and it has been shown to correlate extremely well with other indicators such as star formation rate, NUV, and optical colors [7]. As an example, in the bottom-left panel of Fig. 15.1 it is shown how the CaII H/K ratio perfectly map the region of passive galaxies. Depending on the data availability, it is also advisable to combine

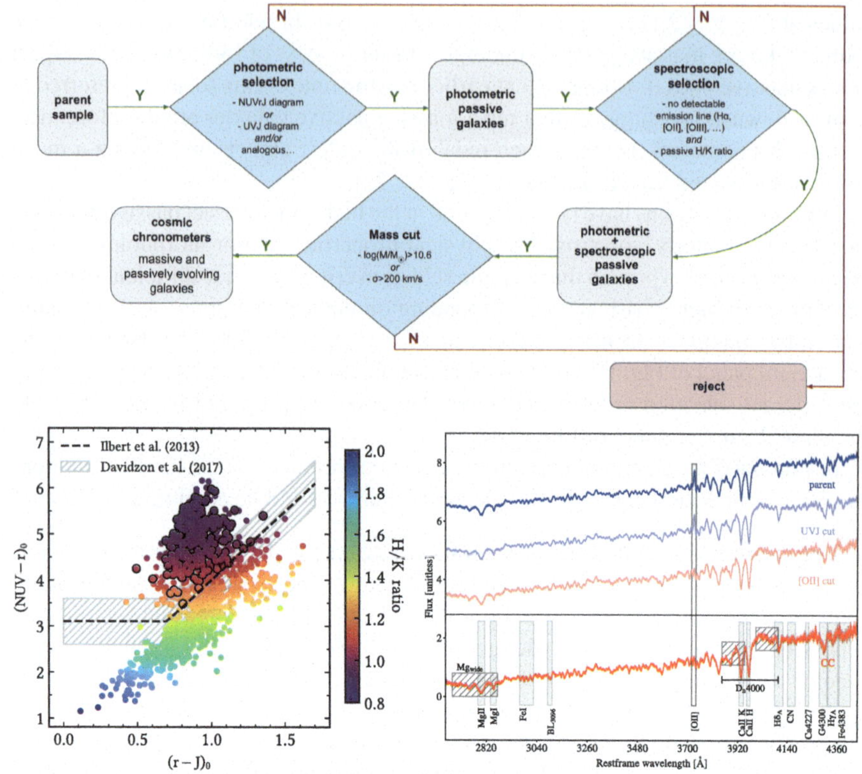

Fig. 15.1 Cosmic chronometer selection process. In the upper panel is shown the optimal workflow to select a pure sample of cosmic chronometers (reproduced from [52]). The lower left plot show a NUV-r-J color-color diagram, colored as a function of the CaII H/K ratio The dashed line displays the theoretical lines separating passive galaxies (upper part) from star-forming galaxies (lower part), showing the very good correlation with the CaII H/K ratio, demonstrating how it is an excellent diagnostic to select passive galaxies. The right panel shows the effect of different selection criteria on the spectra of passive galaxies. The sample has been obtained from Ref. [74], showing, from top to bottom, the stacked spectra of passive galaxies selecting only on photometry (parent), adding a U-V-J color-color cut (UVJ_{cut}), adding a spectroscopic cut on [OII] emission lines ($[OII]_{cut}$), and finally a cut on CaII H/K ratio (CC). It is evident how combining more and more restrictive selection criteria, the emission lines progressively disappear from the stacked spectra, displaying at the end the characteristic features of a purely passive population

several of these indicators to maximize the purity of the sample. In the end, in the spectra of the galaxies ideally should not be visible any residual emission lines, as shown by the bottom-right plot of Fig. 15.1.

- A **cut in stellar mass or stellar velocity dispersion**. As previously discussed, different studies have shown that the most massive galaxies are the ones that formed before and over the shortest timescale[15, 31, 73]; a cut in stellar mass (typically $\log(M/M_{\odot}) > 10.6$–11) ensures the synchronicity of the sample selected, and, if not available, it can equivalently be done on the stellar velocity dispersion of the galaxy, since it presents a good correlation with the stellar mass.

The combination of these criteria has been demonstrated to be effective to maximize the purity of the sample, minimizing possible residual contamination of a young and/or star-forming subdominant population. However, it might not be always feasible to combine multiple criteria, due to data quality or availability. In that case, it is crucial to try to characterize the purity of the sample and properly take into account potential issues in the total covariance matrix as discussed in Sect. 15.4.

15.4 Measuring the Differential Ages

Measuring the age of a stellar population is not a trivial task. While encoded in the integrated spectrum of the galaxy analyzed, this information presents some degeneracy with other physical properties, the most known being the age-metallicity degeneracy [24, 79] (other noticeable degeneracies are the ones with the star formation history [28] or with dust [65]). However, the information encoded in the photometry and spectrum of a galaxy allows us to break (or mitigate in some cases) these degeneracies, and different methods have been introduced and exploited to provide a robust measurement. First of all, we need to underline here that in the cosmic chronometers approach we are interested in determining the relative age dt, and not the absolute age t; this means that we will not be dependent on any offset in the determination of the age, provided that the differential measurement is unbiased, and it has been shown that the relative ages are more robust and can achieve higher accuracy than absolute ones, removing many systematic effects when taking the differences [46].

We summarize here in the following the main methods that have been used to derive dt, leaving to Ref. [52] a detailed discussion, and report in Table 15.1 and Fig. 15.2 the Hubble parameters determined with the various approaches.

Full-spectral fitting. The first approach that can be adopted is to exploit the full information available. Different codes are publicly available performing a fit to the photometric and spectroscopic data with theoretical spectra obtained combining the different physical ingredients of the stellar population of a galaxy (age, metallicity, mass, star formation history). This allows us to determine the ages of our chronometers, from which it is possible to derive the differential age by binning the data at different redshifts and estimating the quantity dt. Successful cosmic chronometers analyses have been performed with this method by several groups [66, 71, 72, 81], that have also demonstrated how with a good photometric and spectroscopic coverage the dt measurement is very robust and independent of the specific assumptions chosen to model the spectra [37, 74].

Analysis of absorption lines (Lick indices). Instead of considering the full spectral information as in the previous method, an alternative is to study instead the absorption lines that characterize the spectra of passive galaxies, since they are known

Table 15.1 Measurements of the Hubble parameter derived with the cosmic chronometers method obtained with the CC method (in units of km/s/Mpc). The error on $H(z)$ reported here accounts only for the diagonal part of the covariance matrix, while for a proper analysis also the full covariance should be considered as discussed in Sect. 15.4. The method (M) used to derive the differential age dt is also provided (full-spectrum fitting F, Lick indices L, $D4000$ D, Machine Learning ML), and the corresponding reference

z	$H(z)$	$\sigma_{H(z)}$	M	References	z	$H(z)$	$\sigma_{H(z)}$	M	References
0.07	69.0	19.6	F	[81]	0.593	104	13	D	[53]
0.09	69	12	F	[71]	0.68	92	8	D	[53]
0.12	68.6	26.2	F	[81]	0.75	98.8	33.6	L	[6]*
0.17	83	8	F	[71]	0.75	105	10.76	ML	[40]*
0.179	75	4	D	[53]	0.781	105	12	D	[53]
0.199	75	5	D	[53]	0.8	113.1	25.22	F	[37]*
0.20	72.9	29.6	F	[81]	0.875	125	17	D	[53]
0.27	77	14	F	[71]	0.88	90	40	F	[72]
0.28	88.8	36.6	F	[81]	0.9	117	23	F	[71]
0.352	83	14	D	[53]	1.037	154	20	D	[53]
0.38	83	13.5	D	[58]	1.26	135	65	F	[74]
0.4	95	17	F	[71]	1.3	168	17	F	[71]
0.4004	77	10.2	D	[58]	1.363	160	33.6	D	[51]
0.425	87.1	11.2	D	[58]	1.43	177	18	F	[71]
0.445	92.8	12.9	D	[58]	1.53	140	14	F	[71]
0.47	89.0	49.6	F	[66]	1.75	202	40	F	[71]
0.4783	80.9	9	D	[58]	1.965	186.5	50.4	D	[51]
0.48	97	62	F	[72]					

* These data have been obtained from the same sample, and should not be used together in an analysis

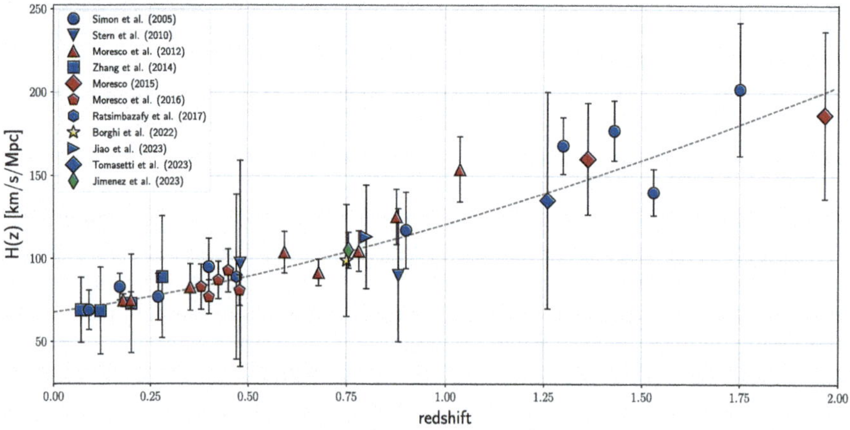

Fig. 15.2 Latest compilation of Hubble parameter measurements as a function of redshift obtained with cosmic chronometers. Different symbols indicate a measurement performed with a different method, as discussed in Sect. 15.4 and presented in Table 15.1

to be particularly sensitive to the age (e.g., Balmer lines), metal content (e.g., iron lines), and enhancement in α elements (e.g., magnesium lines) of the population. Similarly as before, it is possible to build theoretical models to interpret the strength of these features, known as Lick indices, as a function of the age, metallicity, and α/Fe of the population, and therefore extract this information by analyzing a combination of indices. Also with this method, it has been recently obtained a determination of the Hubble parameter with the cosmic chronometer approach [6].

Analysis of specific spectral features (D4000). A slightly different approach consists in still considering spectroscopic features, but focusing on a specific spectral feature known to correlate particularly well with the age of the population instead of analyzing a joint combination of those. This approach, introduced by Ref. [54], suggested using the break at 4000 Å restframe (known as $D4000$, see Fig. 15.1) as an age indicator since it has been found that for cosmic chronometers it correlates linearly with the stellar age (at fixed metallicity and star formation history). If there is a linear relation between the $D4000$ and the age modeled as $D4000 = A(Z, SFH) \times \text{age} + B$, Ref. [53] proposed to rewrite Eq. 15.3 in the form:

$$H(z) = -\frac{1}{1+z} A(SFR, Z/Z_{\odot}) \frac{dz}{dD4000}, \tag{15.4}$$

where $A(SFR, Z/Z_{\odot})$ is a calibration parameter depending on the metallicity and star formation history obtained from stellar population synthesis models. We illustratively show how this method works in Fig. 15.3. This equation improves on Eq. 15.3 since it allows to decouple statistical and systematic effects because the differential evolution will be measured on purely observational data ($dD4000$, purely including statistical errors), while the systematic errors will be all contained in the calibration parameter. To derive this properly, the metallicity of the sample has to be derived with independent measurements, or assumed in a realistic range; here we underline that given the cosmic chronometers selection described in Sect. 15.3, we expect the metal content to be rather homogeneous as a function of redshift. However, it is fundamental to keep into account this effect, as well as the dependence of the results on the model to derive the calibration parameter, in the total covariance matrix, to properly include in the error budget all systematic effects as discussed below. With this approach, the most accurate $H(z)$ determinations of $H(z)$ have been derived to date, with accuracy going from ~5% at $z \sim 0.15$, to ~10% at $z \sim 0.8$, to ~25% at $z \sim 2$ [51, 53, 58].

Other approaches. Recently, Ref. [40] demonstrated that it is possible to apply the cosmic chronometer approach also considering purely photometric data, using a Machine Learning algorithm trained on well-determined aged (obtained from Lick indices analysis). This method, while needing to rely on larger samples ($O(10^4)$) to determine the upper envelope of the age-redshift relation, is extremely promising in view of all the surveys that will provide photometry for almost the full sky in the near future (e.g. Euclid [45] and Rubin [36]).

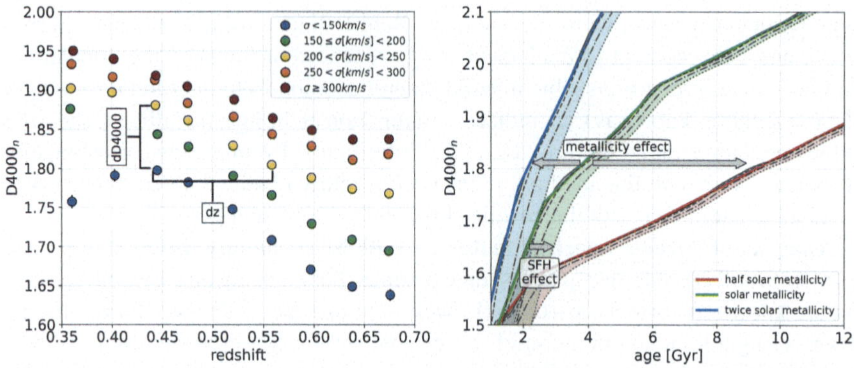

Fig. 15.3 Illustrative representation of how to apply the $D4000$ method in the cosmic chronometers approach (reproduced from [52]. The left panel shows an example of median $D4000$-z values obtained from an analysis of SDSS BOSS survey, representing how to derive the quantity $dz/dD4000$. The right panel show for a specific model the $D4000$-age relations, where it is highlighted their dependence on metallicity (different colors) and star formation history (different lines). It is also possible to see how these relations can be approximated with piecewise linear relation with good accuracy

Cosmic chronometers full covariance matrix. The full covariance matrix for cosmic chronometers has been formalized in Ref. [55], and can be divided into different contributions:

$$\begin{aligned}\mathrm{Cov}_{ij}^{\mathrm{tot}} &= \mathrm{Cov}_{ij}^{\mathrm{stat}} + \\ &+\mathrm{Cov}_{ij}^{\mathrm{met}} + \mathrm{Cov}_{ij}^{\mathrm{young}} + \\ &+\mathrm{Cov}_{ij}^{\mathrm{SFH}} + \mathrm{Cov}_{ij}^{\mathrm{IMF}} + \mathrm{Cov}_{ij}^{\mathrm{st.lib.}} + \mathrm{Cov}_{ij}^{\mathrm{SPS}} \ . \end{aligned} \tag{15.5}$$

The first row of Eq. 15.5 reports the component related to the statistical errors (e.g. the one related to the measurement of $dD4000$ if using the $D4000$ method to measure dt). The second row reports the systematic contribution to the covariance due to uncertainty in the determination of the metallicity of the sample, or to a non-perfect selection of the sample when residual contamination by a young population is not completely removed; this second effect has been quantified in Ref. [57], where a recipe to derive it directly from observable is provided. These two effects are expected to provide diagonal contributions to the covariance since they depend on uncorrelated properties of the galaxies selected. The last row describes instead the systematic component of the covariance due to the ingredients of the model assumed. It has been divided into various parts, i.e. the one related to the uncertainty on the stellar population synthesis model considered (*SPS*), the one related to the stellar library that the model adopts (*st.lib.*), the one related to the initial mass function considered in the model (*IMF*), and the one depending on the assumed star formation history (*SFH*). All these parts have been thoroughly explored in Ref. [55], where also a

Jupyter notebook is provided as an example of how to derive the total covariance matrix for cosmic chronometers.

To conclude, we note here that several other effects have been considered that might in principle affect the cosmological measurements, such as the progenitor bias or a dependence on the stellar mass of the sample. These effects, discussed in Ref. [52], are expected however not to have a significant impact on the analysis due to the strict selection process described in Sect. 15.3 and since in the differential approach relative ages are estimated in homogeneous samples really close in cosmic time, minimizing therefore by definition these effects.

15.5 Cosmic Chronometers and the Hubble Constant

As shown in Fig. 15.2, cosmic chronometers do not directly constrain the Hubble constant, but rather measure the Hubble parameter independently of any cosmological assumptions. For this reason, they represent the ideal testbed to compare different cosmological models. While not measuring H_0 directly, however, it can be derived from these data in different ways.

A possible solution is to estimate the Hubble constant as the extrapolated value of $H(z)$ at $z = 0$. To exploit at most the cosmology-independent nature of these measurements, different methods have been suggested, either based on Gaussian Process regression (e.g. [29, 30, 70], but see also [18] on the assumptions of the method and potential dependence of the measured errors on the assumed kernel) or on Machine Learning (e.g. [3, 48]). The most recent constraint obtained with these methods give $H_0 = 70.7 \pm 6.7$ km/s/Mpcwhen considering the full covariance matrix [23].

The most direct way to derive H_0 is to assume a cosmological model and fit the data presented in Table 15.1, properly considering their covariance matrix as discussed above. We assumed here two cosmological models, a flat ΛCDM model(the one with the least free parameters, H_0 and Ω_m) and the open wCDM model, where also the curvature of the Universe and the dark energy equation of state parameter are left free. This can give us a range of the expected constraints while varying the number of parameters in the fit. Moreover, we also extend our analysis by adding to the current $H(z)$ measurements the forecasts discussed by Ref. [52], where two sets of simulated data are provided as expected by the analysis of the just launched Euclid mission and from the exploitation of newly derived data (such as the latest SDSS BOSS and eBOSS surveys). In this case, two different simulations are explored, a standard one where current data are combined with these simulated data (*combined*) and one where a more optimistic impact of systematic error is considered for the simulated data, minimizing the contribution due to different models in the analysis. The constraints are shown in Fig. 15.4, and the results are reported in Table 15.2. We find that current data are able to measure $H_0$with an accuracy of 8% in a flat λCDM cosmology, and of 11% in an open wCDM cosmology, with values $H_0 = 66.7 \pm 5.3$ and $68.0^{+8.8}_{-7.1}$ km/s/Mpc, respectively. As can be seen, the constraints at the moment

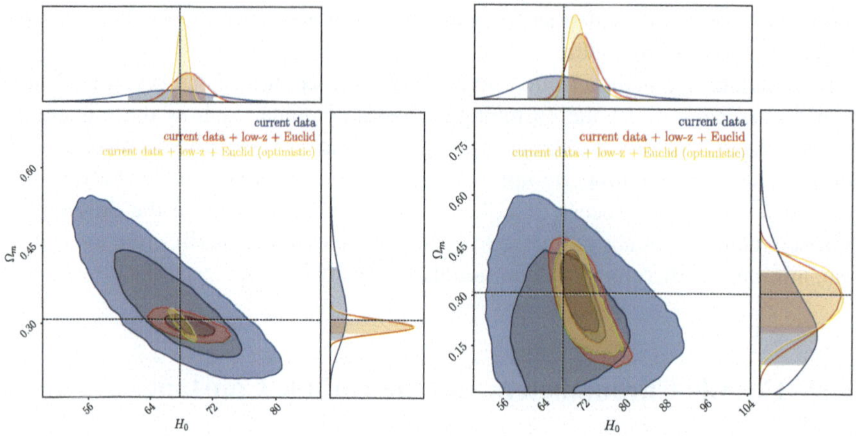

Fig. 15.4 Constraints in the H_0-Ω_m plane obtained from the analysis of cosmic chronometers for two cosmological models, a flat ΛCDM (left panel) and an open wCDM (right panel). The different colors refer to the analysis of the current dataset presented in Table 15.1 (in blue), and with the forecasts for future measurements in two configurations (a standard one in red, and an optimistic one in gold, as discussed in Sect. 15.5. Dashed lines report as a reference the cosmological parameters obtained by Planck2018 [2]

are preferring a lower value of H_0, confirming the results obtained with Gaussian Processes, but with an errorbar that does not allows us to significantly discriminate between SH_0ES and CMB values. However, it is interesting to notice how future data could significantly improve, reaching an accuracy of around 3–1% depending on the cosmological model assumed and on the simulation considered. This sheds a very promising light on this method since it suggests that future analysis and a more thorough study of the models impacting on the systematic errors might provide competitive H_0 measurements, that can help in giving important independent pieces of evidence to address the H_0 tension.

Finally, another possibility is to explore how much the constraints would improve by combining cosmic chronometers with other cosmological probes. This has been done by several authors (e.g. [5, 56, 60, 80, 82]), and the most recent measurement to date obtained combining SNeIa, BAO, cosmic chronometers and Gamma-Ray Bursts data gives $H_0 = 67.2^{+3.4}_{-3.2}$ km/s/Mpc(for a ΛCDM cosmology, $H_0 = 66.9^{+3.5}_{-3.4}$ km/s/Mpc for an open wCDM cosmology, [17]).

Table 15.2 Cosmological parameter constraints that can be obtained with present-day and future CC measurements (as presented in Sect. 15.5) in two different cosmologies, a flat ΛCDM cosmology (upper rows) and a more general open wCDM (lower rows). The Hubble constant H_0 is given in units km/s/Mpc

	H_0	% acc	Ω_m	% acc	Ω_{de}	% acc	w	% acc
	Flat ΛCDM							
Current dataset	66.7 ± 5.3	8%	$0.33^{+0.08}_{-0.06}$	20.6%	–	–	–	–
Combined	69.0 ± 2.1	3%	0.3 ± 0.01	4.4%	–	–	–	–
Optimistic	68.2 ± 0.7	1%	0.29 ± 0.01	4.2%	–	–	–	–
	Open wCDM							
Current dataset	$68.0^{+8.8}_{-7.1}$	11.7%	$0.22^{+0.15}_{-0.12}$	62%	$0.51^{+0.24}_{-0.26}$	48%	$-1.6^{+0.8}_{-0.9}$	52%
Combined	$71.6^{+3.1}_{-2.6}$	4%	$0.29^{+0.09}_{-0.09}$	31%	$0.67^{+0.18}_{-0.14}$	24%	$-1.2^{+0.3}_{-0.4}$	29%
Optimistic	$70.8^{+2.8}_{-1.9}$	3.3%	$0.3^{+0.08}_{-0.09}$	28%	$0.68^{+0.18}_{-0.11}$	22%	$-1.2^{+0.3}_{-0.4}$	29%

Acknowledgements MM acknowledges the grants ASI n.I/023/12/0 and AS I n.2018–23-HH.0., and support from MIUR, PRIN 2017 (grant 20179ZF5KS) and from grant MIUR, PRIN 2022 (grant 2022NY2ZRS_001).

References

1. E. Abdalla, G. Franco Abellán, A. Aboubrahim, A. Agnello, O. Akarsu, Y. Akrami, G. Alestas, D. Aloni, L. Amendola, L.A. Anchordoqui et al., JHEAp **34**, 49–211 (2022) https://doi.org/10.1016/j.jheap.2022.04.002 ([arXiv:2203.06142 [astro-ph.CO]].)
2. N. Aghanim et al. [Planck], Astron. Astrophys. **641**, A6 (2020) [erratum: Astron. Astrophys. **652**, C4 (2021)]. https://doi.org/10.1051/0004-6361/201833910. ([arXiv:1807.06209 [astro-ph.CO]] .)
3. R. Arjona, S. Nesseris, Phys. Rev. D **101**(12), 123525 (2020). https://doi.org/10.1103/PhysRevD.101.123525. ([arXiv:1910.01529 [astro-ph.CO]].)
4. S. Arnouts, E. Le Floc'h, J. Chevallard, B.D. Johnson, O. Ilbert, M. Treyer, H. Aussel, P. Capak, D.B. Sanders, N. Scoville et al., Astron. Astrophys. **558**, A67 (2013). https://doi.org/10.1051/0004-6361/201321768. ([arXiv:1309.0008 [astro-ph.CO]].)
5. A. Bonilla, S. Kumar, R.C. Nunes, Eur. Phys. J. C **81**(2), 127 (2021). https://doi.org/10.1140/epjc/s10052-021-08925-z. ([arXiv:2011.07140 [astro-ph.CO]])
6. N. Borghi, M. Moresco, A. Cimatti, Astrophys. J. Lett. **928**(1), L4 (2022). https://doi.org/10.3847/2041-8213/ac3fb2 ([arXiv:2110.04304 [astro-ph.CO]].)
7. N. Borghi, M. Moresco, A. Cimatti, A. Huchet, S. Quai, L. Pozzetti, Astrophys. J. **927**(2), 164 (2022). https://doi.org/10.3847/1538-4357/ac3240. ([arXiv:2106.14894 [astro-ph.GA]].)
8. D. Brout, D. Scolnic, B. Popovic, A.G. Riess, J. Zuntz, R. Kessler, A. Carr, T.M. Davis, S. Hinton, D. Jones et al., Astrophys. J. **938**(2), 110 (2022). https://doi.org/10.3847/1538-4357/ac8e04. ([arXiv:2202.04077 [astro-ph.CO]].)
9. K.I. Caputi, J.S. Dunlop, R.J. McLure, J. Huang, G.G. Fazio, M.L.N. Ashby, M. Castellano, A. Fontana, M. Cirasuolo, O. Almaini et al., Astrophys. J. Lett. **750**, L20 (2012). https://doi.org/10.1088/2041-8205/750/1/L20. ([arXiv:1202.0496 [astro-ph.CO]].)
10. A.C. Carnall, D.J. McLeod, R.J. McLure, J.S. Dunlop, R. Begley, F. Cullen et al., Mon. Not. R. Astron. Soc. **520**, 3974–3985 (2023). https://doi.org/10.1093/mnras/stad369. ([arXiv:astro-ph/2208.00986 [astro-ph]].)

11. A.C. Carnall, R.J. McLure, J.S. Dunlop, F. Cullen, D.J. McLeod et al., Mon. Not. R. Astron. Soc. **490**, 417–439 (2019). https://doi.org/10.1093/mnras/stz2544. ([arXiv:astro-ph/1903.11082 [astro-ph]].)
12. A.C. Carnall, R.J. McLure, J.S. Dunlop, R. Davé, Mon. Not. R. Astron. Soc. **480**, 4379–4401 (2018). https://doi.org/10.1093/mnras/sty2169. ([arXiv:astro-ph/1712.04452 [astro-ph]].)
13. J. Choi, C. Conroy, J. Moustakas, G.J. Graves, B.P. Holden, M. Brodwin, M.J.I. Brown, P.G. van Dokkum, Astrophys. J. **792**(2), 95 (2014). https://doi.org/10.1088/0004-637X/792/2/95. ([arXiv:1403.4932 [astro-ph.GA]].)
14. A. Cimatti and M. Moresco, [arXiv:2302.07899 [astro-ph.CO]]
15. A. Cimatti, E. Daddi, A. Renzini, P. Cassata, E. Vanzella, L. Pozzetti, S. Cristiani, A. Fontana, G. Rodighiero, M. Mignoli et al., Nature **430**, 184–187 (2004). https://doi.org/10.1038/nature02668. ([arXiv:astro-ph/0407131 [astro-ph]].)
16. A. Citro, L. Pozzetti, M. Moresco, A. Cimatti, Astron. Astrophys. **592**, A19 (2016). https://doi.org/10.1051/0004-6361/201527772. ([arXiv:1604.07826 [astro-ph.CO]].)
17. F. Cogato, M. Moresco, L. Amati and A. Cimatti, [arXiv:2309.01375 [astro-ph.CO]]
18. E.Ó. Colgáin, M.M. Sheikh-Jabbari, Eur. Phys. J. C **81**(10), 892 (2021). https://doi.org/10.1140/epjc/s10052-021-09708-2. ([arXiv:2101.08565 [astro-ph.CO]].)
19. C. Conroy, G. Graves, P. van Dokkum, Astrophys. J. **780**, 33 (2014). https://doi.org/10.1088/0004-637X/780/1/33. ([arXiv:1303.6629 [astro-ph.CO]].)
20. E. Daddi, A. Cimatti, A. Renzini, J. Vernet, C. Conselice, L. Pozzetti, M. Mignoli, P. Tozzi, T. J. Broadhurst and S. di Serego Alighieri et al., Astrophys. J. Lett. **600**, L127–L130 (2004). https://doi.org/10.1086/381020. ([arXiv:astro-ph/0308456 [astro-ph]].)
21. E. Daddi, A. Renzini, N. Pirzkal, A. Cimatti, S. Malhotra, M. Stiavelli, C. Xu, A. Pasquali, J.E. Rhoads, M. Brusa et al., Astrophys. J. **626**, 680–697 (2005). https://doi.org/10.1086/430104. [arXiv:astro-ph/0503102 [astro-ph]].)
22. V. Estrada-Carpenter, C. Papovich, I. Momcheva, G. Brammer, J. Long, R.F. Quadri, J. Bridge, M. Dickinson, H. Ferguson, S. Finkelstein et al., Astrophys. J. **870**, 133–160 (2019). https://doi.org/10.3847/1538-4357/aaf22e. ([arXiv:astro-ph/1810.02824 [astro-ph]].)
23. A. Favale, A. Gómez-Valent, M. Migliaccio, Mon. Not. R. Astron. Soc. **523**(3), 3406–3422 (2023). https://doi.org/10.1093/mnras/stad1621. ([arXiv:2301.09591 [astro-ph.CO]].)
24. I. Ferreras, S. Charlot, J. Silk, Astrophys. J. **521**, 81 (1999). https://doi.org/10.1086/307513. ([arXiv:astro-ph/9803235 [astro-ph]].)
25. A. Fontana, S. Salimbeni, A. Grazian, E. Giallongo, L. Pentericci, M. Nonino, F. Fontanot, N. Menci, P. Monaco, S. Cristiani et al., Astron. Astrophys. **459**, 745–757 (2006). https://doi.org/10.1051/0004-6361:20065475. ([arXiv:astro-ph/0609068 [astro-ph]].)
26. A. Gallazzi, E.F. Bell, S. Zibetti, J. Brinchmann, D.D. Kelson, Astrophys. J. **788**, 72 (2014). https://doi.org/10.1088/0004-637X/788/1/72. ([arXiv:1404.5624 [astro-ph.GA]].)
27. A. Gallazzi, S. Charlot, J. Brinchmann, S.D.M. White, C.A. Tremonti, Mon. Not. R. Astron. Soc. **362**, 41–58 (2005). https://doi.org/10.1111/j.1365-2966.2005.09321.x. ([arXiv:astro-ph/0506539 [astro-ph]].)
28. G. Gavazzi, C. Bonfanti, G. Sanvito, A. Boselli, M. Scodeggio, Astrophys. J. **576**, 1 (2002). https://doi.org/10.1086/341730
29. A. Gómez-Valent, L. Amendola, JCAP **04**, 051 (2018). https://doi.org/10.1088/1475-7516/2018/04/051. ([arXiv:1802.01505 [astro-ph.CO]].)
30. B.S. Haridasu, V.V. Luković, M. Moresco, N. Vittorio, JCAP **10**, 015 (2018). https://doi.org/10.1088/1475-7516/2018/10/015. ([arXiv:1805.03595 [astro-ph.CO]].)
31. A. Heavens, B. Panter, R. Jimenez, J. Dunlop, Nature **428**, 625 (2004). https://doi.org/10.1038/nature02474. ([arXiv:astro-ph/0403293 [astro-ph]].)
32. E. Hubble, Proc. Nat. Acad. Sci. **15**, 168–173 (1929). https://doi.org/10.1073/pnas.15.3.168
33. D. Huterer, D.L. Shafer, Rept. Prog. Phys. **81**(1), 016901 (2018). https://doi.org/10.1088/1361-6633/aa997e. ([arXiv:1709.01091 [astro-ph.CO]].)
34. O. Ilbert, P. Capak, M. Salvato, H. Aussel, H.J. McCracken, D.B. Sanders, N. Scoville, J. Kartaltepe, S. Arnouts, E.L. Floc'h et al., Astrophys. J. **690**, 1236–1249 (2009). https://doi.org/10.1088/0004-637X/690/2/1236. ([arXiv:0809.2101 [astro-ph]].)

35. O. Ilbert, H.J. McCracken, O.L. Fevre, P. Capak, J. Dunlop, S. Arnouts, H. Aussel, K. Caputi, J. Comparat, Q. Guo et al., Astron. Astrophys. **556**, A55 (2013). https://doi.org/10.1051/0004-6361/201321100. ([arXiv:1301.3157 [astro-ph.CO]].)
36. Ž. Ivezić et al., LSST, Astrophys. J. **873**(2), 111 (2019). https://doi.org/10.3847/1538-4357/ab042c. ([arXiv:0805.2366 [astro-ph]].)
37. K. Jiao, N. Borghi, M. Moresco, T.J. Zhang, Astrophys. J. Suppl. **265**(2), 48 (2023). https://doi.org/10.3847/1538-4365/acbc77. ([arXiv:2205.05701 [astro-ph.CO]].)
38. R. Jimenez, A. Cimatti, L. Verde, M. Moresco, B. Wandelt, JCAP **03**, 043 (2019). https://doi.org/10.1088/1475-7516/2019/03/043. ([arXiv:1902.07081 [astro-ph.CO]].)
39. R. Jimenez, A. Loeb, Astrophys. J. **573**, 37–42 (2002). https://doi.org/10.1086/340549. ([arXiv:astro-ph/0106145 [astro-ph]].)
40. R. Jimenez, M. Moresco, L. Verde, B.D. Wandelt, [arXiv:2306.11425 [astro-ph.CO]]
41. R. Jimenez, P. Thejll, U. Jorgensen, J. MacDonald, B. Pagel, Mon. Not. R. Astron. Soc. **282**, 926–942 (1996). https://doi.org/10.1093/mnras/282.3.926. ([arXiv:astro-ph/9602132 [astro-ph]].)
42. R. Jimenez, L. Verde, T. Treu, D. Stern, Astrophys. J. **593**, 622–629 (2003). https://doi.org/10.1086/376595. ([arXiv:astro-ph/0302560 [astro-ph]].)
43. L.M. Krauss, B. Chaboyer, Science **299**, 65–70 (2003). https://doi.org/10.1126/science.1075631
44. M. Kriek, S.H. Price, C. Conroy, K.A. Suess, L. Mowla, I. Pasha, R. Bezanson, P. Van Dokkum, G. Barro, Astrophys. J. Lett. **880**(2), L31 (2019). https://doi.org/10.3847/2041-8213/ab2e75. ([arXiv:astro-ph/1907.04327 [astro-ph]].)
45. R. Laureijs et al., [EUCLID], [arXiv:1110.3193 [astro-ph.CO]]
46. A. Marin-Franch, A. Aparicio, G. Piotto, A. Rosenberg, B. Chaboyer, A. Sarajedini, M. Siegel, J. Anderson, L.R. Bedin, A. Dotter et al., Astrophys. J. **694**, 1498–1516 (2009). https://doi.org/10.1088/0004-637X/694/2/1498. ([arXiv:0812.4541 [astro-ph]].)
47. R.M. McDermid, K. Alatalo, L. Blitz, F. Bournaud, M. Bureau, M. Cappellari, A.F. Crocker, R.L. Davies, T.A. Davis, P.T. de Zeeuw et al., Mon. Not. R. Astron. Soc. **448**(4), 3484–3513 (2015). https://doi.org/10.1093/mnras/stv105. ([arXiv:1501.03723 [astro-ph.GA]].)
48. A. Mehrabi, M. Rezaei, Astrophys. J. **923**(2), 274 (2021). https://doi.org/10.3847/1538-4357/ac2fff. ([arXiv:2110.14950 [astro-ph.CO]].)
49. E. Merlin, F. Fortuni, M. Torelli, P. Santini, M. Castellano, A. Fontana, A. Grazian, L. Pentericci, S. Pilo, K.B. Schmidt Mon. Not. R. Astron. Soc. **490**(3), 3309–3328 (2019). https://doi.org/10.1093/mnras/stz2615. ([arXiv:1909.07996 [astro-ph.GA]].)
50. M. Mignoli, G. Zamorani, M. Scodeggio, A. Cimatti, C. Halliday, S.J. Lilly, L. Pozzetti, D. Vergani, C.M. Carollo, T. Contini et al., Astron. Astrophys. **493**, 39 (2009). https://doi.org/10.1051/0004-6361:200810520. ([arXiv:0810.2245 [astro-ph]].)
51. M. Moresco, Mon. Not. R. Astron. Soc. **450**(1), L16–L20 (2015). https://doi.org/10.1093/mnrasl/slv037. ([arXiv:1503.01116 [astro-ph.CO]].)
52. M. Moresco, L. Amati, L. Amendola, S. Birrer, J.P. Blakeslee, M. Cantiello, A. Cimatti, J. Darling, M. Della Valle, M. Fishbach et al., Living Rev. Rel. **25**(1), 6 (2022). https://doi.org/10.1007/s41114-022-00040-z). ([arXiv:2201.07241 [astro-ph.CO]].)
53. M. Moresco, A. Cimatti, R. Jimenez, L. Pozzetti, G. Zamorani, M. Bolzonella, J. Dunlop, F. Lamareille, M. Mignoli, H. Pearce et al., JCAP **08**, 006 (2012). https://doi.org/10.1088/1475-7516/2012/08/006. ([arXiv:1201.3609 [astro-ph.CO]].)
54. M. Moresco, R. Jimenez, A. Cimatti, L. Pozzetti, JCAP **03**, 045 (2011). https://doi.org/10.1088/1475-7516/2011/03/045. ([arXiv:1010.0831 [astro-ph.CO]].)
55. M. Moresco, R. Jimenez, L. Verde, A. Cimatti, L. Pozzetti, Astrophys. J. **898**(1), 82 (2020). https://doi.org/10.3847/1538-4357/ab9eb0. ([arXiv:2003.07362 [astro-ph.GA]].)
56. M. Moresco, R. Jimenez, L. Verde, A. Cimatti, L. Pozzetti, C. Maraston, D. Thomas, JCAP **12**, 039 (2016). https://doi.org/10.1088/1475-7516/2016/12/039. ([arXiv:1604.00183 [astro-ph.CO]].)
57. M. Moresco, R. Jimenez, L. Verde, L. Pozzetti, A. Cimatti, A. Citro, Astrophys. J. **868**(2), 84 (2018). https://doi.org/10.3847/1538-4357/aae829. ([arXiv:1804.05864 [astro-ph.CO]].)

58. M. Moresco, L. Pozzetti, A. Cimatti, R. Jimenez, C. Maraston, L. Verde, D. Thomas, A. Citro, R. Tojeiro, D. Wilkinson, JCAP **05**, 014 (2016). https://doi.org/10.1088/1475-7516/2016/05/014. ([arXiv:1601.01701 [astro-ph.CO]].)
59. M. Moresco, L. Pozzetti, A. Cimatti, G. Zamorani, M. Bolzonella, F. Lamareille, M. Mignoli, E. Zucca, S.J. Lilly, C.M. Carollo et al., Astron. Astrophys. **558**, A61 (2013). https://doi.org/10.1051/0004-6361/201321797. ([arXiv:1305.1308 [astro-ph.CO]].)
60. M. Moresco, L. Verde, L. Pozzetti, R. Jimenez, A. Cimatti, JCAP **07**, 053 (2012). https://doi.org/10.1088/1475-7516/2012/07/053. ([arXiv:1201.6658 [astro-ph.CO]].)
61. A. Muzzin, D. Marchesini, M. Stefanon, M. Franx, H.J. McCracken, B. Milvang-Jensen, J.S. Dunlop, J.P.U. Fynbo, O. Le Fevre, G. Brammer et al., Astrophys. J. **777**, 18 (2013). https://doi.org/10.1088/0004-637X/777/1/18. ([arXiv:1303.4409 [astro-ph.CO]].)
62. M. Onodera, C.M. Carollo, A. Renzini, M. Cappellari, C. Mancini, N. Arimoto, E. Daddi, R. Gobat, V. Strazzullo, S. Tacchella et al., Astrophys. J. **808**(2), 161 (2015). https://doi.org/10.1088/0004-637X/808/2/161. ([arXiv:1411.5023 [astro-ph.GA]].)
63. M. Onodera, A. Renzini, M. Carollo, M. Cappellari, C. Mancini, V. Strazzullo, E. Daddi, N. Arimoto, R. Gobat, Y. Yamada et al., Astrophys. J. **755**, 26 (2012). https://doi.org/10.1088/0004-637X/755/1/26. ([arXiv:1206.1540 [astro-ph.CO]].)
64. S. Perlmutter et al., Supernova cosmology project. Astrophys. J. **517**, 565–586 (1999). https://doi.org/10.1086/307221. ([arXiv:astro-ph/9812133 [astro-ph]].)
65. L. Pozzetti, F. Mannucci, Mon. Not. R. Astron. Soc. **317**, L17 (2000). https://doi.org/10.1046/j.1365-8711.2000.03829.x. ([arXiv:astro-ph/0006430 [astro-ph]].)
66. A.L. Ratsimbazafy, S.I. Loubser, S.M. Crawford, C.M. Cress, B.A. Bassett, R.C. Nichol, P. Väisänen, Mon. Not. R. Astron. Soc. **467**(3), 3239–3254 (2017). https://doi.org/10.1093/mnras/stx301. ([arXiv:1702.00418 [astro-ph.CO]].)
67. A.G. Riess et al., Supernova Search Team. Astron. J. **116**, 1009–1038 (1998). https://doi.org/10.1086/300499. ([arXiv:astro-ph/9805201 [astro-ph]].)
68. A.G. Riess, W. Yuan, L.M. Macri, D. Scolnic, D. Brout, S. Casertano, D.O. Jones, Y. Murakami, L. Breuval, T.G. Brink et al., Astrophys. J. Lett. **934**(1), L7 (2022). https://doi.org/10.3847/2041-8213/ac5c5b. ([arXiv:2112.04510 [astro-ph.CO]].)
69. A.J. Ross, J. Bautista, R. Tojeiro, S. Alam, S. Bailey, E. Burtin, J. Comparat, K.S. Dawson, A. de Mattia, H. du Mas des Bourboux et al., Mon. Not. R. Astron. Soc. **498**(2), 2354–2371 (2020). https://doi.org/10.1093/mnras/staa2416 ([arXiv:2007.09000 [astro-ph.CO]].)
70. M. Seikel, S. Yahya, R. Maartens, R. Clarkson, Phys. Rev. D **86**, 083001 (2012). https://doi.org/10.1103/PhysRevD.86.083001. ([arXiv:astro-ph/1205.3431 [astro-ph]].)
71. J. Simon, L. Verde, R. Jimenez, Phys. Rev. D **71**, 123001 (2005). https://doi.org/10.1103/PhysRevD.71.123001. ([arXiv:astro-ph/0412269 [astro-ph]].)
72. D. Stern, R. Jimenez, L. Verde, M. Kamionkowski, S.A. Stanford, JCAP **02**, 008 (2010). https://doi.org/10.1088/1475-7516/2010/02/008. ([arXiv:0907.3149 [astro-ph.CO]].)
73. D. Thomas, C. Maraston, K. Schawinski, M. Sarzi, J. Silk, Mon. Not. R. Astron. Soc. **404**, 1775 (2010). https://doi.org/10.1111/j.1365-2966.2010.16427.x. ([arXiv:0912.0259 [astro-ph.CO]].)
74. E. Tomasetti, M. Moresco, N. Borghi, K. Jiao, A. Cimatti, L. Pozzetti, A.C. Carnall, R.J. McLure, L. Pentericci, [arXiv:2305.16387 [astro-ph.CO]]
75. S. Vagnozzi, F. Pacucci, A. Loeb, JHEAp **36**, 27–35 (2022). https://doi.org/10.1016/j.jheap.2022.07.004. ([arXiv:2105.10421 [astro-ph.CO]].)
76. L. Verde, T. Treu, A.G. Riess, Nat. Astron. **3**, 891. https://doi.org/10.1038/s41550-019-0902-0. ([arXiv:1907.10625 [astro-ph.CO]])
77. L.L. Wang, A.L. Luo, S.Y. Shen, W. Hou, X. Kong, Y.H. Song, J.N. Zhang, H. Wu, Z.H.Z. Cao, Y.H. Hou, Y.F. Wang, Y. Zhang, Y.H. Zhao, Mon. Not. R. Astron. Soc. **474**(2), 1873–1885 (2018). https://doi.org/10.1093/mnras/stx2798. ([arXiv:1710.10611 [astro-ph.CO]].)
78. R.J. Williams, R.F. Quadri, M. Franx, P. van Dokkum, I. Labbe, Astrophys. J. **691**, 1879–1895 (2009). https://doi.org/10.1088/0004-637X/691/2/1879. ([arXiv:0806.0625 [astro-ph]].)
79. G. Worthey, Astrophys. J. Suppl. **95**, 107–149 (1994). https://doi.org/10.1086/192096

80. H. Yu, B. Ratra, F.Y. Wang, Astrophys. J. **856**(1), 3 (2018). https://doi.org/10.3847/1538-4357/aab0a2. ([arXiv:1711.03437 [astro-ph.CO]].)
81. C. Zhang, H. Zhang, S. Yuan, T.J. Zhang, Y.C. Sun, Res. Astron. Astrophys. **14**(10), 1221–1233 (2014). https://doi.org/10.1088/1674-4527/14/10/002 ([arXiv:1207.4541 [astro-ph.CO]].)
82. G.B. Zhao, M. Raveri, L. Pogosian, Y. Wang, R.G. Crittenden, W.J. Handley, W.J. Percival, F. Beutler, J. Brinkmann, C.H. Chuang et al., Nat. Astron. **1**(9), 627–632 (2017). https://doi.org/10.1038/s41550-017-0216-z. ([arXiv:1701.08165 [astro-ph.CO]].)
83. E. Zucca, S. Bardelli, M. Bolzonella, G. Zamorani, O. Ilbert, L. Pozzetti, M. Mignoli, K. Kovac, S. Lilly, L. Tresse et al., Astron. Astrophys. **508**, 1217 (2009). https://doi.org/10.1051/0004-6361/200912665. ([arXiv:0909.4674 [astro-ph.CO]].)

Chapter 16
The Cosmic Microwave Background and H_0

Pablo Lemos and Paul Shah

16.1 The Cosmic Microwave Background

16.1.1 Introduction

The Cosmic Microwave Background (CMB) is a fundamental component of modern cosmology, providing a wealth of information about the early universe. It is a faint radiation that permeates the entire cosmos, and it is essentially the 'afterglow' of the Big Bang. The CMB is composed of photons that were released approximately 380, 000 years after the Big Bang when the universe had cooled sufficiently for atoms to form (at redshift $z \sim 1100$). Before that time, the universe was a hot, dense plasma where photons were constantly scattered by charged particles in a process known as Thomson scattering.

The CMB was first detected in 1965 by Arno Penzias and Robert Wilson at the Bell Telephone Laboratories [45]. They were conducting experiments using a large horn antenna designed for satellite communication. Penzias and Wilson noticed a persistent background noise in their measurements, which they initially attributed to various sources of interference. However, they encountered difficulties in eliminating the noise, and after careful consideration, they realized that they were detecting a uniform radiation coming from all directions in the sky. They had inadvertently stumbled upon the CMB. The discovery of the CMB provided strong support for

P. Lemos (✉)
Department of Physics and Astronomy, University of Sussex, Brighton BN1 9QH, UK

Mila, Montreal, QC H2S 3H1, Canada
e-mail: P.Lemos@sussex.ac.uk

P. Shah
Department of Physics and Astronomy, University College London, Gower Street, London WC1E 6BT, UK
e-mail: paul.shah.19@ucl.ac.uk

E. Di Valentino and D. Brout (eds.), *The Hubble Constant Tension*, Springer Series in Astrophysics and Cosmology, https://doi.org/10.1007/978-981-99-0177-7_16

the Big Bang theory; as opposed to the steady state theory, stating that the Universe remains unchanged [31].

The CMB radiation is highly isotropic, with a temperature of approximately 2.7 Kelvin. However, there is a lot of cosmological information on the tiny variations in the temperature of the CMB across the sky. These fluctuations were first detected by the Cosmic Background Explorer (COBE) satellite, launched in 1989 [62, 63]. In recent years, there have also been remarkable advancements in the detection of CMB polarization. The polarization of the CMB arises from the Thomson scattering of photons by electrons at the time of last scattering, and it carries independent information to the temperature fluctuations. The first detection of CMB polarization was announced by the Degree Angular Scale Interferometer (DASI, [36]) in 2002, followed by the Wilkinson Microwave Anisotropy Probe (WMAP, [30]).

However, the most significant breakthroughs in CMB temperature and polarization measurements came from the Planck satellite, launched by the European Space Agency (ESA) in 2009 [48]. The Planck mission provided unprecedented CMB data, unveiling the intricate anisotropy patterns of the CMB on both large and small angular scales. These observations have shed light on the conditions of the universe during the era of recombination and have provided insights into cosmological parameters, inflationary models, and the nature of dark matter and dark energy.

The origin of the modern Hubble constant tension can be reasonably dated to the first Planck release of cosmological parameters [48], although disagreements about the value of the Hubble constant date decades back [58, 68]. While WMAP values for H_0 were compatible with other results at the time, the Planck value of $H_0 = 67.4 \pm 0.5$km s^{-1} Mpc^{-1} is discrepant with the SH_0ES result of $H_0 = 73.04 \pm 1.04$km s^{-1} Mpc^{-1} [56] at 4.9σ (although Planck is consistent with WMAP as we discuss below). As such, CMB values for H_0 merit considerable scrutiny. Fortunately, while local measurements focus purely on H_0, the CMB delivers a "package deal" of cosmological parameters which is highly dependent on the cosmological model. The cosmological model that is most supported by present-day observations is ΛCDM, consisting of a spatially-flat cosmology with a cosmological constant and three species of neutrinos with small but non-zero masses. We will make it clear when we depart from this model. As usual $h \equiv H_0/(100\,\text{km s}^{-1}\,\text{Mpc}^{-1}$. Constraints in the parameters of this model may be compared to the values from other data sets to see if other parameters are discrepant too. Also, as we explain below, the CMB is not exclusively an early universe result, as the anisotropy maps carry the imprint of late-universe physics. If therefore the Planck value for H_0 is considered an "extraordinary claim", we discuss the evidence for it in this chapter.

16.1.2 Why Are There Anisotropies in the CMB?

The physical origin of density perturbations lies in the primordial universe, specifically during the period of cosmic inflation. Cosmic inflation is a theoretical concept that proposes a rapid exponential expansion of space in the early universe [24, 41,

42]. This expansion was driven by a hypothetical scalar field known as the inflaton. During inflation, quantum fluctuations in the inflaton field lead to variations in the energy density across space. These quantum fluctuations are inherently random and occur on extremely small scales. However, due to the exponential expansion of space during inflation, these fluctuations are stretched to cosmological scales, providing the seed for density perturbations. After inflation, the universe enters a phase known as reheating, during which the inflaton field decays into other particles and reestablishes the thermal equilibrium of the universe. As the universe cools, the density perturbations seeded by inflation begin to grow under the influence of gravity.

The theory of inflation is widely accepted in the cosmology community, despite a lack of direct observational evidence. This is because it solves some theoretical issues, such as the horizon and flatness problems; but more importantly because of its ability to explain the origin of the density perturbations that seed the formation of cosmic structures. Inflationary models make specific predictions for the spectrum of these initial fluctuations. They are expected to be approximately scale invariant (if inflation lasts long enough), but small deviations from scale invariance can occur due to the dynamics of the inflation field. The predicted density perturbations are in excellent agreement with observations, including the temperature anisotropies in the CMB and the distribution of galaxies in the universe. Inflation, therefore, provides a natural mechanism for generating the observed structure in the universe.

The density perturbations are responsible for the observed variations in the CMB temperature across the sky. During the epoch of recombination (which occurred at the same temperature everywhere) when the universe became transparent to radiation, the photons of the CMB decoupled from the baryonic matter. At this point, the temperature of the CMB was determined by the density of matter in different regions of the universe. Regions with higher matter density experienced stronger gravitational pull, causing photons to lose energy as they climbed out of the gravitational potential wells. As a result, these regions appear slightly cooler in the CMB. Conversely, regions with lower matter density experienced weaker gravitational pull, allowing photons to retain more energy and appear slightly warmer in the CMB. In addition, photons are Doppler-shifted by the velocity perturbations in the plasma.

During their long journey to our detectors, the CMB photons are perturbed by effects known as secondary anisotropies. In addition, foreground emission of radiation at the same frequencies as the CMB occurs both in our own galaxy and from extra-galactic sources. For the purposes of cosmology, these foregrounds must be subtracted, but we note they also carry interesting astrophysical information about the properties of dust, magnetic fields and the inter-galactic medium.

16.1.3 The Temperature of the CMB

The average flux intensity of the CMB observed today (that is, the monopole in the power spectrum), shown in Fig. 16.1 is a near-perfect black body spectrum (*'the most perfect black body ever measured in nature'*, [72]), demonstrating the plasma

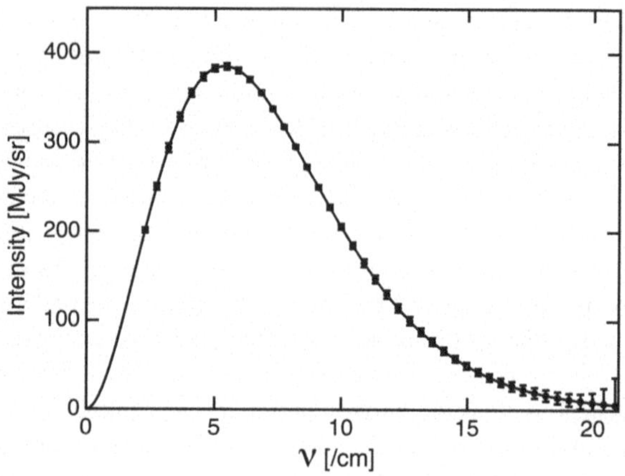

Fig. 16.1 The spectrum of the CMB measured by FIRAS. The error bars shown are $\pm 400\sigma$. The solid line is a 2.728 K blackbody. Image credit: [61]

of baryons and photons were in thermal equilibrium on the surface of last scattering (and therefore presumably at earlier times). The temperature was initially measured in balloon experiments, but was more accurately measured by the FIRAS instrument aboard the satellite COBE. The combination of data gives $T = 2.72548 \pm 0.00057$ K [23]. Because the spectrum is thermal, the energy density of the CMB is determined solely by the temperature as $U = \frac{4}{c}\sigma T^4$. This results in the physical density of photons today as $\rho_\gamma = 4.65 \times 10^{-31}$ kg m^{-3}, or in units of the critical density $\Omega_\gamma \equiv \rho_\gamma/\rho_c = 5.4 \times 10^{-5}$, where the critical density represents the average mass or energy per unit volume required to achieve a flat universe. It is worth pointing out that spectral distortions in the CMB have been proposed as a powerful cosmological probe [35], although more accurate measurements of the CMB spectrum would be required.

We use N_{eff} to refer to the effective number count of neutrinos or other relativistic particles. Assuming there is no substantial addition to the photon density from the decay of other particles or fields, knowledge of the CMB temperature today determines the energy density of radiation at earlier times as

$$\rho_r(z) = (1 + 0.227 N_{\text{eff}})(1+z)^4 (\frac{T_0}{2.72548})^4 \times 4.65 \times 10^{-31} \text{kg m}^{-3}, \quad (16.1)$$

where T_0 is the present-day CMB temperature.

In the standard model of particle physics, when $T \ll 1$ Mev, $N_{\text{eff}} = 3.043$ (it is not an integer as the spectrum of neutrinos is not precisely a black-body; the energy density of neutrinos is altered by a contribution from the annihilation of electrons and positrons after their decoupling from the primordial plasma). However, in extensions to the standard model N_{eff} is often treated as a free parameter (Fig. 16.2).

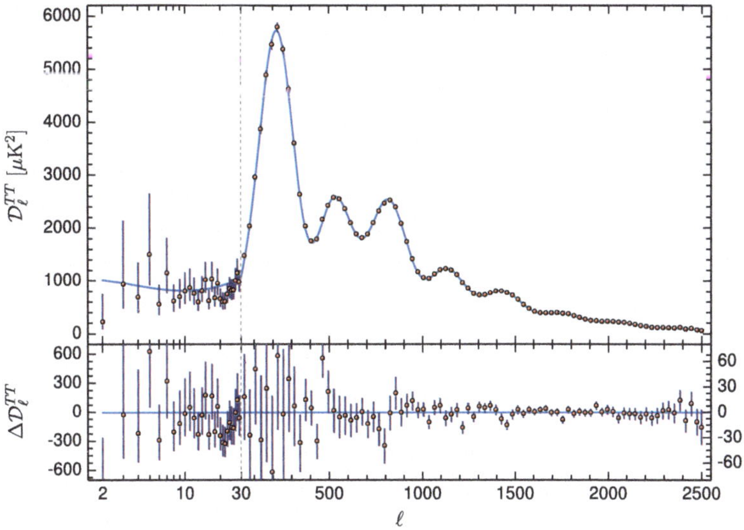

Fig. 16.2 The binned Planck power spectrum for TT and residuals to the best fit CDM model. The vertical line delineates the different methodologies used to resolve the power spectrum at $l < 30$. Figure from [51]

16.1.4 The Power Spectrum and Its Features

To analyse the anisotropies in the CMB temperature and polarization, the map of the fluctuations in the sky can be decomposed into a sum of spherical harmonics

$$a_{\ell m} = \int \Delta T(\theta, \phi)\ Y_{\ell m}(\theta, \phi) d\Omega \quad , \tag{16.2}$$

where ℓ is limited by the resolution of the instrument and m runs over $(-\ell, \ell)$. The CMB sky represents a single sampling of the random distribution of primordial fluctuations. If we want to know the "true" values of cosmological parameters, we should average over a large ensemble of realisations—but we have only one! An approximation to ensemble averaging is obtained by averaging over the $2\ell + 1$ harmonics in each mode, which is likely to give a reliable answer at high-ℓ but is of limited accuracy at low-ℓ, a floor referred to as cosmic variance.

The power spectra are then defined as

$$\mathcal{C}_\ell^{XX} = \langle |a_{\ell m}^{XX}|^2 \rangle_m \quad , \tag{16.3}$$

where X can reperesent the temperature (T), E-mode polarization (E) or B-mode polarization (B). In the case of temperature, for convenience of display, the power spectra are often presented as the dimensionless quantity

$$\mathcal{D}_\ell^{TT} = \frac{\ell(\ell+1)}{2\pi} C_\ell^{TT}. \tag{16.4}$$

This scaling factor is commonly used to highlight the characteristic angular scales of the CMB fluctuations. At present, there is no evidence of non-Gaussanity in the primordial fluctuations so the C_ℓ (or equivalently the 2-point correlation function) capture all available information on primary anisotropies (although there are claims of structure in low-ℓ $a_{\ell m}$—see below). However, gravitational lensing induces non-zero higher-order correlation function values that can be measured separately (also see below).

16.1.4.1 Primary Anisotropies and Length Scales

Primary anisotropies in the cosmic microwave background (CMB) radiation provide a wealth of information about the early universe and the processes that shaped its evolution. These anisotropies refer to the small fluctuations in temperature and polarization observed across the CMB sky. There are three main features of primary anisotropies in the CMB:

1. **The Large-Scale Plateau**: On large angular scales, typically above a few degrees, the primary anisotropies in the CMB predominantly exhibit a feature known as the large-scale plateau. This plateau corresponds to scales that were not significantly affected by pre-recombination physics and provides valuable information about the primordial Universe.
2. **Acoustic Oscillations**: On smaller angular scales, below a few degrees, an additional feature becomes prominent in the primary anisotropies: the acoustic oscillations. These oscillations arise from the competition between gravitational attraction and radiation pressure in the early universe. The density and velocity perturbations in the primordial plasma give rise to sound waves that propagate through the medium, and these waves lead to periodic oscillations in the CMB power spectrum, with peaks and troughs that correspond to the specific length scale associated with the sound horizon at the time of photon decoupling (Fig. 16.3). The position of the peaks provides a measurement of the extrema of these oscillations at last scattering, and in particular of the *sound horizon at radiation drag* r_d, which measures the distance sounds waves can travel before recombination ([71]):

$$r_d = \int_{z_\star}^{\infty} \frac{c_s(z)}{H(z)} z, \tag{16.5}$$

where $z_\star$ is the redshift at photon decoupling and $c_s(z)$ is the sound speed in the photon-baryon fluid ([32]) which by the standard thermodynamics of adiabatic perturbations is

$$c_s^2(z) = \frac{1}{3}\left[1 + \frac{3}{4}\frac{\rho_b(z)}{\rho_\gamma(z)}\right]^{-1}. \tag{16.6}$$

The sound horizon serves as a standard ruler, and the *position* of the acoustic peaks captures the history of $H(z)$ together with matter and radiation densities prior to recombination. Additionally, the *heights* of the peaks represent the competition between pressure (a function of the physical baryon density $\Omega_b h^2$) and gravity (a function of the physical matter density $\Omega_m h^2$) as we show in Fig. 16.4 below.

3. **Silk Damping Tail**: On even smaller angular scales, the anisotropies in the CMB exhibit a phenomenon known as the Silk damping tail. This effect arises from the diffusion of photons caused by scattering off free electrons just prior to recombination. The diffusion process tends to erase small-scale temperature fluctuations in the CMB, resulting in a damping of the power spectrum on scales below the *Silk scale*:

$$\lambda_D \sim \sqrt{N}\lambda_c \sim \sqrt{\eta_\star \lambda_c}, \tag{16.7}$$

where N is the number of steps in the random walk, λ_c is the mean-free path to Thomson scattering, and $\eta_\star$ is the conformal time at recombination. The Silk scale corresponds to an angular scale around $\ell \sim 1000$ and is determined by baryon density and $H(z)$ just prior to recombination (e.g. [34]).

16.1.4.2 Secondary Anisotropies

Secondary anisotropies in the CMB refer to additional fluctuations observed in the CMB sky, which arise from various astrophysical and cosmological phenomena. The main sources of secondary CMB anisotropies are:

1. **integrated Sachs-Wolfe (ISW) effect**: The ISW effect [57] is caused by the gravitational potential wells that photons traverse as they travel through evolving large-scale structures in the universe. If the potentials change photons pass through them, their energy is altered, leading to temperature fluctuations in the CMB. In particular, the Late ISW effect provides constraints on the expansion rate of the universe during the accelerated expansion era at $z < 0.6$, albeit as it primarily affects large scales this constraint is limited by cosmic variance.
2. **CMB Lensing**: The light from the CMB is deflected by foreground matter inhomogeneities, with the typical deflection angle being 3' [40]. This is in the domain of weak lensing, where just one image source is seen but the source is distorted and magnified or de-magnified. The amount of lensing depends on the degree of inhomogeneity (best captured by the combination $\sigma_8 \Omega_m^{0.25}$) and a combination of angular diameter distances between the observer, deflector and source (sometimes called the lensing efficiency). For the CMB, the lensing efficiency peaks between $1 < z < 2$ meaning it is sensitive to structure just beyond the reach of present-day galaxy weak lensing studies.

Lensing has two linked effects: firstly, it smooths the observed power spectra, most noticeably for $l > 800$ multipoles. Secondly, observations of the four-point correlation function of fluctuations can be used to reconstruct the (projected) foreground matter distribution. As a consistency check we may compare (1) the theory prediction of lensing given the amplitude of the fluctuations in the CMB and growth of structure (2) the observed smoothing of the power spectrum, and (3) the observed power spectrum of the lensing reconstruction.

3. **Reionization**: During the early stages of the universe, neutral hydrogen became ionized due to the radiation from the first stars and galaxies, a process known as reionization. This reionization process leaves characteristic imprints on the CMB through the scattering and absorption of CMB photons. At $z \sim 10$ the formation of the first stars ionizes the Universe. The presence of free electrons from this point onward means that a fraction $(1 - e^{-\tau})$ of CMB photons undergo Thomson scattering after reionization, where τ is the optical depth to reionization defined as the line-of-sight opacity of the CMB radiation with respect to Thomson scattering with free electrons:

$$\tau = \int_0^{\chi_{\rm re}} \sigma_T \, n_e \, \mathrm{d}\chi \,, \tag{16.8}$$

where $\chi_{\rm re}$ is the comoving distance to reionization, σ_T is the Thomson scattering cross-section and n_e is the number density of free electrons. As a consequence, the power spectrum on scales that entered the horizon before reionization is damped by a factor of $e^{-2\tau}$.

4. **Sunyaev-Zel'dovich (SZ) effect**: The SZ effect occurs when CMB photons interact with the hot, ionized gas in galaxy clusters through inverse Compton scattering [65, 66, 73]. There are two types of SZ effect: thermal SZ and kinematic SZ. The thermal SZ effect arises from the Compton scattering of CMB photons with the electrons in the cluster gas, boosting the energy of the photons and distorting their spectrum to higher frequencies. The kinematic SZ effect is subdominant and arises from the bulk motion of the clusters. The thermal SZ effect can be detected by its frequency dependence, while the kinematic SZ effect is frequency independent.

16.1.5 Polarization

Temperature anisotropies are not the only source of information from CMB radiation. CMB polarization provides valuable complementary information to temperature anisotropies, offering further insights into the early Universe. The polarization pattern arises due to scalar and tensor perturbations to the metric. Scalar perturbations generate a quadrupolar component of the radiation field, leading to linear polarization through Thomson scattering during recombination. On the other hand, tensor perturbations, originating from gravitational waves, also contribute to the polariza-

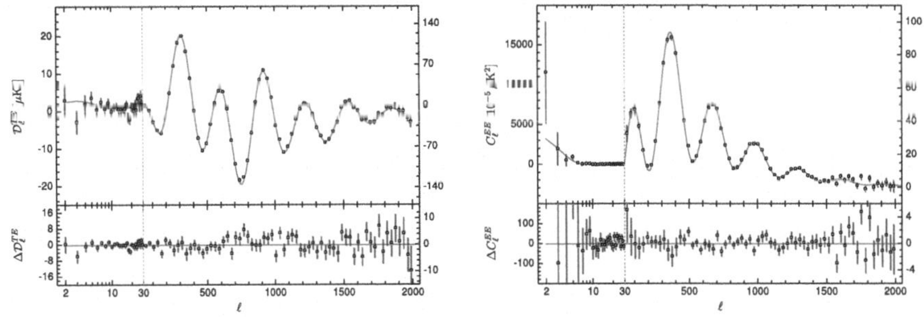

Fig. 16.3 *Planck* 2018 temperature-cross E-mode polarization power spectrum (TE, left), and E-mode autocorrelation power-spectrum (EE, right). The vertical axis on the left plot shows $\mathcal{D}_\ell \equiv \ell(\ell+1)C_\ell/(2\pi)$. The red points are the observed data and the blue line shows the best fit CDM cosmology. Scales change from logarithmic to linear at $\ell = 30$. The lower panels show residuals with respect to the best fits. Image credit: [50]

tion. The polarization can be decomposed into four Stokes parameters: intensity (I), linear polarization (Q, U), and circular polarization (V). The intensity parameter represents the total power, while the linear polarization parameters capture the orientation and amplitude of the polarization. Circular polarization is absent in the CMB due to Thomson scattering properties (for a more detailed discussion about CMB polarization see [13, 33]).

The linear polarization field can be further classified into two types: E-modes and B-modes. E-modes represent the curl-free component of the polarization field and are thus primarily generated by scalar perturbations on the surface of last scattering. B-modes arise from a combination of tensor perturbations (primordial gravitational waves) and gravitational lensing of E-modes. B-modes are the curl component of the polarization field and are particularly interesting as they provide a unique window into the presence of primordial gravitational waves. Detection of B-mode polarization, after accounting for the effects of gravitational lensing, would be a direct confirmation of the existence of primordial gravitational waves, which are predicted by some inflationary models. The *Planck* measurements of the E-mode power spectrum (C_ℓ^{EE}) and its cross-correlation with temperature (C_ℓ^{TE}) is shown in Fig. 16.3. Cosmological parameters determined from the Planck polarization and temperature power spectrums are consistent with each other (see Fig. 5 of [51]).

16.1.6 Foreground Removal

Foregrounds introduced by our own Milky Way, extragalactic sources and instrument effects are primarily identified by a comparison of multiple frequency channels. The CMB has a near-perfect blackbody spectrum, and therefore a precisely defined "colour". Foregrounds generated by non-thermal effects (such as synchrotron emission from dust) or thermal effects at different temperatures (such as the thermal

Sunyaev-Zeldovich effect) exhibit different colours, enabling their individual identification.

Foregrounds are removed by a combination of sky masks, frequency modelling and spectral templates $C_{\ell,\mathrm{fore}}$ (for unresolved foregrounds) which represent the contribution of foreground emissions at different multipole moments. These templates capture the spectral characteristics of foreground sources such as synchrotron emission from dust and the thermal Sunyaev-Zeldovich effect. The nuisance parameters associated with these templates are marginalized over during the fitting process for cosmological parameters (see [49] for a description of the process). Data from high-resolution experiments like ACT and SPT are particularly useful to constrain the nuisance parameters, and the final cleaned spectra can be compared across frequencies as a consistency check. The details are not important for this review, but the key point is: could the foreground model bias the inference of H_0?

The evidence is they do not. The foreground models provide an acceptable fit across frequency ranges as measured by their χ^2, justifying the claim that they correctly model the physics [48]. Correlations between foreground nuisance parameters and cosmological parameters are low [49]. Features such as the oscillating residuals in the region $1000 < l < 1500$ (see Fig. 16.2) and the low power for $l < 30$ which play a role in driving H_0 for Planck lower than WMAP are present in multiple frequency channels; this suggests they are real features of the spectrum and not foregrounds. The fit to foreground-cleaned temperature spectrum is consistent with the polarisation spectrum, which depends differently on the foregrounds and in particular, is not expected to be dominated by them at high-ℓ [67]. Likelihoods that use different foreground subtraction methods produce cosmological parameters consistent with Planck [20, 64, 70].

16.1.7 The Planck Cosmology

The CMB is the only cosmological probe that can simultaneously constrain all six parameters in the standard CDM model of cosmology $\left\{\Omega_b h^2, \Omega_c h^2, H_0, \tau, A_s, n_s\right\}$. Note that different combinations of these six parameters are commonly used to parameterize CDM. The values measured by the *Planck* 2018 TT+TE+EE+lowE likelihood are shown in Table 16.1, and the effect of varying them in the CMB power temperature power spectrum is shown in Fig. 16.4.

16.1.8 How Does the CMB Determine H_0?

16.1.8.1 The Angular Size of the Sound Horizon

As we discussed in the section on primary anisotropies, the sound horizon is a characteristic length imprinted on the CMB power spectrum, visible as the positions

Table 16.1 Best fit values and 68% limits for the constraints from *Planck* 2018 TT+TE+EE+lowE [50]

Parameter	Symbol	Best fit value and 68% limits
Cold dark matter density	$\Omega_c h^2$	0.1202 ± 0.0014
Baryon density	$\Omega_b h^2$	0.02236 ± 0.00015
Hubble parameter	h	0.6727 ± 0.0060
Optical depth to reionization	τ	$0.0544^{+0.0070}_{-0.0081}$
Scalar spectrum amplitude	$\log\left(10^{10} A_s\right)$	3.045 ± 0.016
Scalar spectral index	n_s	0.9649 ± 0.0044

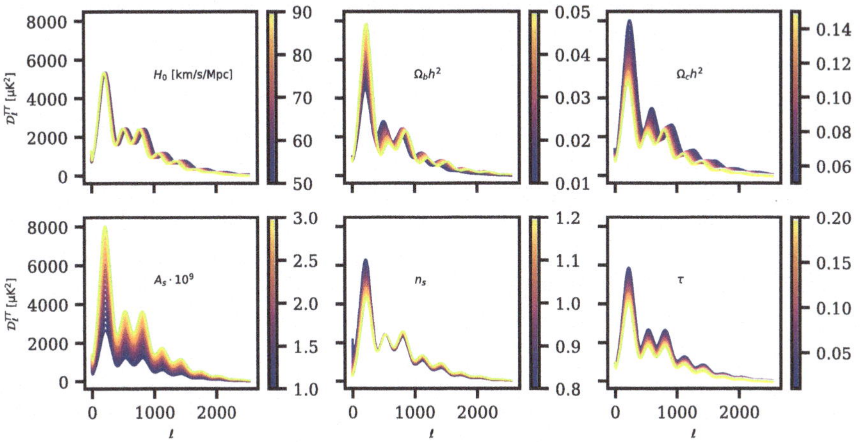

Fig. 16.4 Variations in the CMB temperature power spectrum, as different cosmological parameters are changed, one at a time, while keeping the rest of the parameters set to the Planck best-fit values shown in Table 16.1

of the multiple peaks and troughs (strictly, we should say the complete waveform of the spectrum). This is observed as a ratio $\theta_\star = r_s/d_A(z_\star)$ where $d_A(z_\star)$ is the angular diameter distance to the surface of last scattering. It is the most accurately measured parameter by Planck, and $100\theta_\star = 1.0411 \pm 0.0003$ [51].

The theory expectation can then be simply expressed as

$$\theta_\star = \frac{\int_{z_\star}^{\infty} \frac{c_s(z)}{H(z)} \mathrm{d}z}{\int_0^{z_\star} \frac{1}{H(z)} \mathrm{d}z} \ . \tag{16.9}$$

At this stage, very few assumptions about cosmology have been made. The above assumes only that the sound horizon has resulted from the propagation of adiabatic pressure waves in plasma from initial seed densities, that recombination is instantaneous and photons propagate freely to us from the surface of last scattering (these last two are straightforwardly generalized).

It is convenient to define dimensionless physical densities $\omega_i = \Omega_i h^2$. In ΛCDM cosmology, we may accurately approximate the above as

$$\theta_\star = \frac{\int_{z_\star}^{\infty} \frac{c_s(z)}{((1+z)^3 + \frac{\omega_r}{\omega_m}(1+z)^4)^{1/2}} \mathrm{d}z}{\int_0^{z_\star} \frac{1}{((1+z)^3 + \frac{\omega}{\omega_m})^{1/2}} \mathrm{d}z} . \tag{16.10}$$

The above looks rather complicated, so let us unpack it! The denominator is relatively simple, with the two terms representing the matter density and dark energy density respectively. The radiation density is mostly negligible (although it is still relevant for the first dex of expansion after recombination).

In the numerator, we have assumed the universe is comprised only of matter and radiation. Curvature and dark energy are physically negligible pre-combination. As discussed, the radiation density $\omega_r(z, T_{\mathrm{CMB}})$ is determined by the observed CMB temperature. The sound speed $c_s(z, \omega_{\mathrm{b}})$ is determined by the physical baryon density ω_b and standard thermodynamics. The redshift of the surface of last scattering has only a weak cosmology dependence: it is approximately the freeze-out redshift when the scattering rate of photons and baryons matches the Hubble expansion rate. More baryons increase the scattering rate but also make recombination happen earlier, and the factors roughly cancel (the same argument is made for the Hubble rate). It is thus mainly a function of primordial Helium abundance Y_P (which may be either measured or calculated from the baryon density, nucleosynthesis rates, and the Hubble expansion rate at that time).

As discussed above in the section on primary anisotropies, Ω_{m} and ω_{b} are determined by the relative heights of the peaks of the power spectrum, and by the Silk damping scale at high-ℓ. The only free parameter remaining to be adjusted is h.

The situation is a little more complicated if we drop the assumption of spatial flatness. In this case, we now have two free parameters, Ω_Λ and h. However, Ω_Λ is constrained by the secondary anisotropies of the CMB: the power spectrum is smoothed on a characteristic scale by gravitational lensing at $z \sim 1 - 2$, and power in $l < 30$ modes is enhanced by the late-ISW effect (which depends on the expansion history for $z < 0.6$). We show in Fig. 16.5 a schematic of the redshift range of the sensitivity of the CMB power spectra to $H(z)$: each window function effectively determines a ruler whose theoretically-determined length may be compared to the observed angular size.

16.1.9 Consistency of H_0 Derived from the CMB

Cosmological parameters influence CMB measurements in multiple ways. It is therefore useful, as a test of ΛCDM, to check the consistency of CMB-derived parameters both internally and with non-CMB data sets.

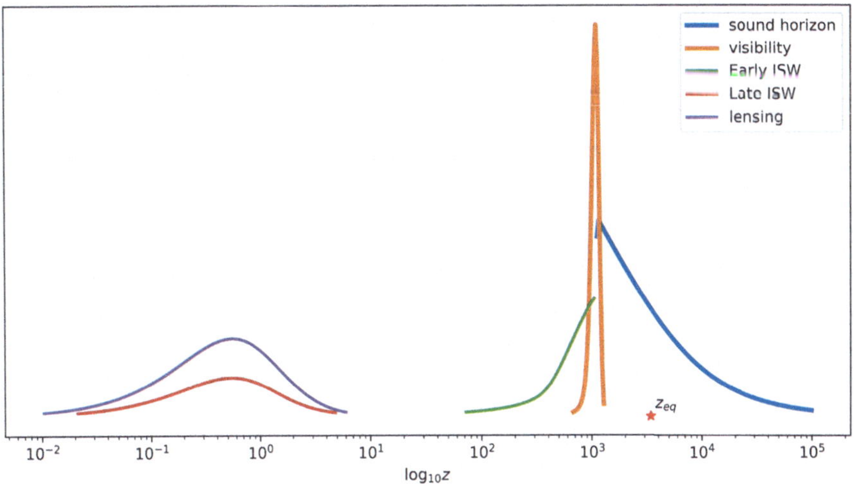

Fig. 16.5 A schematic of the $H(z)$ redshift windows to which the CMB power spectra are sensitive to. The y-axis scale for each component is arbitrary and not shown. The growth of the sound horizon (blue) takes place primarily in the last dex of expansion before recombination : solutions to the H_0-tension that change the sound horizon must therefore modify $H(z)$ at this time. The visibility function (orange) represents the depth of the surface of last scattering, manifest in the Silk damping scale and polarization spectra. The Early ISW effect (green) modifies the power spectrum near its first peak due to the effect on gravitational potentials of the residual radiation density in the first dex of expansion after z_{eq}, whereas the Late ISW effect (red) modifies the low-l power spectrum close to the onset of accelerated expansion at $z \sim 0.6$. Finally lensing (purple) peaks at a redshift corresponding to a distance that is midway (in comoving coordinates) between us and the surface of last scattering

We should however remember that we are observing one sample of a Gaussian random field, limited by cosmic variance. If we make multiple consistency tests, it would be surprising if there were *not* some statistical excursions from the underlying truth.

16.1.9.1 Consistency with Other CMB Experiments

The consistency among various cosmic microwave background (CMB) experiments provides valuable insights into the fundamental properties of the universe, particularly the Hubble constant. WMAP determined $H_0 = 70.0 \pm 2.2\text{km s}^{-1}\,\text{Mpc}^{-1}$ [10, 29] purely from the power spectrum of temperature fluctuations up to $l \sim 800$. This is consistent with the equivalent Planck measurement over the same multipole range (see Fig. 21 of [51]). Planck shifts H_0 lower due to its higher multipole range. At least part of the reason why WMAP has a higher H_0 than Planck appears to be due to low power in the $10 < l < 30$ range (as also observed by Planck—more on this below). Adding BAO information to WMAP brings the value closer to *Planck*

($H_0 = 68.76 \pm 0.84 \mathrm{km\,s^{-1}\,Mpc^{-1}}$). A remarkable result is a consistently low value of H_0 from *Planck* lensing and BAO alone ($H_0 = 68.76 \pm 0.84 \mathrm{km\,s^{-1}\,Mpc^{-1}}$ [14]).

Two notable recent experiments are the Atacama Cosmology Telescope (ACT), and the South Pole Telescope (SPT) which have played a crucial role in advancing our understanding of the CMB. ACT's combined analysis of temperature and polarization data has led to a determination of the Hubble constant, with a value of $H_0 = 67.9 \pm 1.5 \mathrm{km\,s^{-1}\,Mpc^{-1}}$ (e.g. [4, 43]). Interestingly, initial polarization analyses from the SPT yielded a value of $H_0 = 71.3 \pm 2.1 \mathrm{km\,s^{-1}\,Mpc^{-1}}$ [27], in tension with *Planck* and ACT, and closer to the value obtained using direct measurements. However, the more recent SPT-3G measurements including temperature anisotropies recovered a low value $H_0 = 68.3 \pm 1.5 \mathrm{km\,s^{-1}\,Mpc^{-1}}$ [9], consistent with other CMB experiments. Furthermore, possible issues in the analysis of [27] were raised in [50], as well as in Chap. 4 of [37].

Therefore, there is remarkable consistency observed between H_0 measurements from different CMB experiments, as shown in Fig. 16.6, despite small tensions in other parameters [26]. This convergence of results from independent experiments provides strong evidence for the low H_0 from the CMB and rules out the option of explaining the tension as a systematic effect in one of the experiments.

16.1.9.2 Internal Consistency of Planck Data

There is a moderate tension between low ($2 < l < 800$) and high ($801 < l < 2508$) multipole values for H_0 that has been noted by many authors (for example, see [2]). The overlap between the two posteriors is consistent with the independent polarization and lensing power spectra, and also by the addition of BAO to WMAP. At least part of the difference appears to be caused by a dip in power for the multipole range $10 < l < 30$ (visible in Fig. 16.2 above). This seems to be a real feature of the CMB sky.

The power spectrum of lensing reconstruction and amplitude of fluctuations are consistent with each other, both internally in Planck (see Fig. 3 of [51]) and between Planck and ACT [54]. However, the smoothing in the Planck power spectrum is not consistent with its amplitude, captured by the parameter $A_L = 1.180 \pm 0.065$ [51] (where $A_L = 1$ indicates perfect consistency). The connection with H_0 is that allowing A_L as an additional nuisance parameter in the ΛCDM solution both pulls H_0 to higher values and broadens the error bands, so the tension with SH_0ES measurements is lowered, although it remains significant at $\sim 3.8\sigma$ (see Fig. 16.7). Despite its definition, A_L is probably not due to lensing. It may be in part due to the dip in power between $10 < l < 30$: removing the $l < 30$ part of the spectrum restores consistency with $A_L = 1$ (and values of H_0 consistent with the overall Planck result). It may also be due to unknown systematics or residual foregrounds in Planck. There is no evidence for $A_L > 1$ from either ACT ($A_L = 1.01 \pm 0.11$ [4]) or SPT-3G ($A_L = 0.87 \pm 0.11$ [9]). In addition, the likelihood for Planck from the CamSpec analysis [22] including a larger area of the sky is consistent with $A_L = 1$ across the full multipole range, implying it is not a real-sky effect.

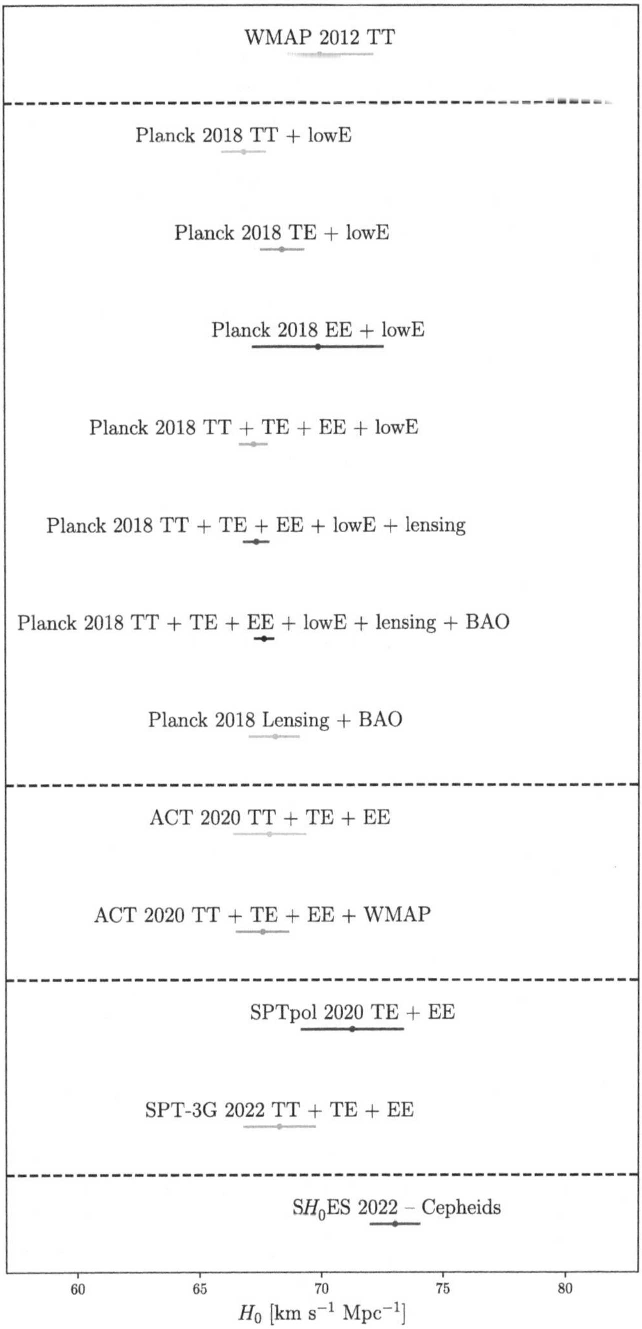

Fig. 16.6 Comparison of Hubble constant constraints between different CMB experiments, and with the latest results from SH_0ES [55]

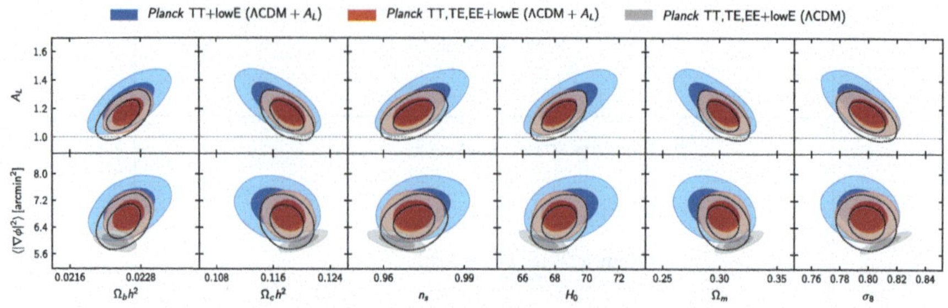

Fig. 16.7 Comparison of the *Planck* 2018 68% and 95% marginalized posterior distributions for different parameters, including and not including A_L. Note that the significance of this result decreases significantly in the CamSpec analyses [22]. Image credit: [50]

Other anomalies have been noted in Planck data, for example, a lack of large angle correlations (such as would be expected from the measured power spectrum), alignment of the quadrupole and octopole, and angular variations of the spectral power and H_0 across patches of the sky (see for example [5]). While some are simulated to have less than 0.1% chance of occurrence in random realisations of the CMB sky, the "look-elsewhere" effect[1] cautions us to avoid simple interpretations of these p-values. Nevertheless, such claims are at least at the level of curiosity; but at present there is no apparent connection with the Hubble tension or any physical process.

16.1.9.3 Consistency with Other Astrophysical Data

As we described above, the conversion of CMB spectral data into constraints on cosmological parameters is highly model dependent. Since H_0 from the CMB in ΛCDM is discrepant with local values, other discrepancies would add evidence to the argument ΛCDM is not the correct model and may provide hints about how to change it, especially where those parameters are correlated to H_0 (see Fig. 16.8 or Fig. 5 of [51]).

The relative heights and positions of the peaks of the CMB spectrum derive from the baryon density ρ_b, the matter density ρ_m and the pre-recombination expansion history. Two key events pre-combination are nucleosynthesis (BBN) at $z_\mathrm{BBN} \sim 10^{10}$ and matter-radiation equality at $z_\mathrm{eq} \sim 3400$. Additionally, $H(z)$ at the time of recombination determines the shape of the Silk damping tail.

The primordial abundance of helium is sensitive to the expansion rate $H(z_\mathrm{BBN})$ as the availability of neutrons to form it depends on the time between the decoupling of weak interactions that keep protons and neutrons in equilibrium, and the time when deuterium (D) can form. The abundance of D is very sensitive to the

[1] We refer the reader to https://en.wikipedia.org/wiki/Six_nines_in_pi and https://xkcd.com/882/ for entertaining examples.

baryon density: as D is rapidly processed into He, a higher baryon density implies a faster interaction rate and hence less D. Primordial element abundance observations are extrapolative by nature, but the Planck values for $(N_{\rm eff}, \rho_{\rm b}) = (2.89 \pm 0.37, 0.02224 \pm 0.00022)$ (Tables 4 and 5 of P18) are consistent with observations of $Y_{\rm P} = 4n_{\rm He}/n_{\rm b} = 0.2449 \pm 0.004$ [8] and $Y_{\rm D} = n_{\rm D}/n_{\rm H} \times 10^5 = 2.52 \pm 0.03$ [15]. There is also excellent agreement with calculations using experimentally determined D-burning rates [47], and Planck is therefore compatible with the standard model of particle physics which predicts $N_{\rm eff} \simeq 3.043$.

It is possible to go further and drop the CMB spectrum (but not its temperature) by combining $\rho_{\rm b}$ constraints from BBN with the standard ruler provided by baryon acoustic oscillations; these are an imprint of the sound horizon on the late universe and effectively determine $\Omega_{\rm m}$ [3]. In this case, one obtains $H_0 = 68.1 \pm 1.1 \rm km\, s^{-1}\, Mpc^{-1}$ ([17], see also [59]) and $\Omega_{\rm M} = 0.300 \pm 0.018$, fully compatible with Planck within ΛCDM.

The power spectrum $P(k)$ of matter density fluctuations has a maximum at a scale $k \sim k_{\rm eq}$ which is the scale of modes entering the horizon at $z_{\rm eq}$ (the shape near the maximum also depends weakly on $\rho_{\rm b}$ as dark matter fluctuations are suppressed by baryons). Faster expansion in the pre-recombination universe lowers the horizon and increases $k_{\rm eq}$. Therefore, the shape of the power spectrum as measured from galaxy surveys may be used to check compatibility with the Planck H_0 in ΛCDM. The evidence at present is in favour of compatibility with $H_0 = 69.6 \pm 1.2 \rm km\, s^{-1}\, Mpc^{-1}$ ([46], see also the chapter in this book).

The expansion history of the universe from $z < 1.2$ is strongly constrained by Type Ia SN (the furthest observations reach to $z \sim 2$). Type Ia SN do vary in intrinsic brightness by a factor of $\times 2$, but the majority of this fluctuation is attributable to the colour and duration of the explosion. The residual fluctuations may be as low as 0.07 mag [11] and the evidence points to them being a homogeneous population for the purposes of cosmology. While they cannot by themselves determine H_0 as this is degenerate with their fiducial absolute magnitude, the Hubble diagram constructed from the Pantheon+ sample [60] is fully compatible with Planck with $\Omega_{\rm m} = 0.338 \pm 0.018$ in ΛCDM [12]. Is it then is no surprise that calibrating the SN Ia magnitudes with BAO in turn produces a H_0 also fully compatible with Planck [44].

The values of $S_8 = \sigma_8(\Omega_{\rm m}/0.3)^{0.5}$ derived from Planck (via ΛCDM assumptions on structure growth) are high (at the level of $\sim 2.5\sigma$) compared to observations of galaxy lensing [1, 28, 39]. The relevance of this for the Hubble tension is unclear, and it has been proposed a resolution lies in correcting theory calculations of structure growth in the quasi-linear regime [7]. However, we note if one were to attempt to resolve the Hubble tension by speeding up the expansion of the universe pre-recombination, one would need to compensate for the slower structure growth by adding more matter, increasing the S_8 discrepancy and potentially creating a new discrepancy with SN Ia.

In summary, within the framework of ΛCDM and some simple one-parameter extensions, the only CMB-derived cosmological parameters that are inconsistent with other astrophysical datasets are H_0, and to a lesser extent σ_8. In particular, matter, baryon, curvature densities and number of particle species are compatible with other observations.

16.1.10 The CMB in ΛCDM Extensions

Extensions to the ΛCDM model often involve modifications to the standard physics of the early universe or the presence and nature of dark energy (either pre- or post-recombination). These modifications can leave imprints on the CMB, allowing for tests of these alternative scenarios. For instance, variations in the primordial power spectrum, generated during cosmic inflation, can lead to specific patterns in the CMB temperature and polarization anisotropies. Planck data has been instrumental in placing constraints on such modifications, providing valuable insights into the inflationary paradigm and the physics of the early universe [6], and finding strong support for a featureless primordial power spectrum.

Another area of interest is the nature of dark energy itself. While the ΛCDM model assumes a cosmological constant Λ to account for the observed accelerated expansion of the universe, alternative theories propose dynamical dark energy models or modifications to general relativity. These models can result in deviations from the standard predictions for the CMB power spectrum and other statistical properties. One common approach to parameterize dark energy deviations from ΛCDM is with a time-dependent equation of state $w(a) = w_0 + (1 - a)w_a$ (ΛCDM assumes $w_0 = -1$ and $w_a = 0$). Interestingly for the H_0 tension, Planck measurements are consistent with a value $w_0 < -1$, which also raises the value of H_0. This value falls under the regime known as "phantom" dark energy [69], as they require a second scalar field, or some other complex theoretical explanation (e.g. [16, 18]). However, the combination of SN Ia, Planck and BAO (which may be expected to give the best constraints on dark energy) does not favour phantom dark energy [12].

Another interesting extension to ΛCDM is spatial curvature ($\Omega_K \neq 0$), which raised interest due to the Planck measurement $\Omega_K = -0.044^{+0.018}_{-0.015}$, which is over 2σ away from zero [19, 25]. However, as discussed in [21], this value is strongly correlated with the A_L measurement from Planck discussed above, and is equally explained by statistical fluctuations in the temperature power spectra in the multipole range $800 < \ell < 1600$. Similarly to A_L, the CamSpec likelihood [22] strongly reduces the significance of this detection. Furthermore, a closed Universe leads to even lower values of H_0, further exacerbating the tension.

Finally, it is possible to explore the changes to neutrino physics. For example, it is possible to vary the sum of neutrino masses $\sum m_\nu$ or the effective number of neutrino species N_{eff} introduced above. None of these changes, however, is enough to alleviate the tension.

A more radical idea is to increase the size of the sound horizon by introducing "early" dark energy (EDE) to the pre-recombination universe (possibly connected with the onset of matter domination). EDE must be present at the level of $\sim 10\%$ of the energy density during the last dex of expansion prior to the surface of last

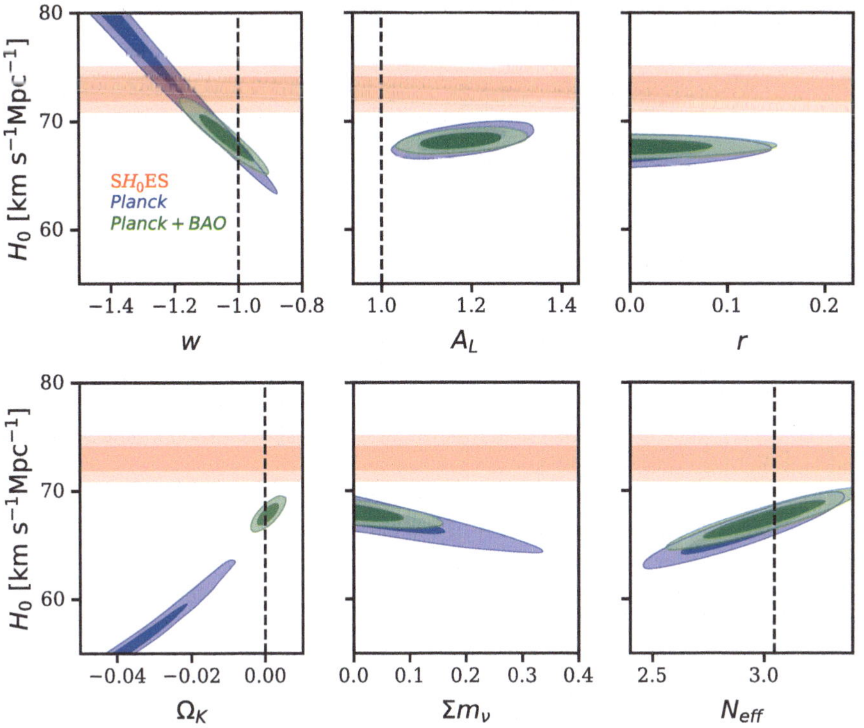

Fig. 16.8 Comparison of Hubble constant constraints between different one parameter extensions to the ΛCDM model, using Planck TTTEEE + lowE in blue, and adding BAO in green. In red, are the 68% and 95% confidence levels for H_0. The black, dotted lines show the ΛCDM value for each parameter

scattering, as that is when most of the growth in the size of the sound horizon happens (see Fig. 16.5). Silk damping and CMB polarization on small scales constrain its presence at $z = 1100$. In general, whilst EDE models cannot resolve the Hubble tension, they can alleviate it to the level of $\sim 2\sigma$ [52]. EDE models also make specific predictions for the $l > 1000$ temperature and polarization spectrums, and so future high-resolution CMB data will determine whether they are viable or not. EDE models may seem somewhat fine-tuned. They do not *predict* a high late universe H_0 but they may *explain* it (see [53] for a review). An additional problem is the boost to pre-recombination expansion requires more dark matter to reproduce the heights of the peaks of the power spectrum (as dark energy suppresses growth of structure), increasing the S_8 tension with galaxy weak lensing results.

Figure 16.8 shows the effect of the most popular one-parameter ΛCDM extensions in the Planck estimates of the Hubble parameter. As we see, there is no unambiguous evidence from CMB data alone of a departure from Flat ΛCDM.

Generally, while it is true that one possible solution to the H_0 tensions is a modification to the ΛCDM cosmological model, it is very challenging to envision new physics that solves the tension. Indeed, as described in [38] to solve the tension with new physics, it needs to lower the sound horizon by approximately 9% (to about 135 Mpc) while preserving the structure of the temperature and polarization power spectra described in this chapter; and also preserving the consistency between BBN and observed abundances of light elements.

16.1.11 Conclusions and Future Prospects

The cosmic microwave background (CMB) observations within the framework of the ΛCDM model have demonstrated remarkable consistency with a wide range of non-CMB astrophysical data at varying redshifts. The cosmological parameters derived from Planck are consistent across frequencies and likelihood methodologies, and with WMAP, ACT and SPT.

However, there are a few remaining tensions that warrant further investigation. Notably, weak lensing measurements of the parameter σ_8, which characterizes the amplitude of matter fluctuations on large scales; and, more importantly for this chapter, the local measurement of the Hubble constant (H_0) from the nearby universe appear to deviate from the predictions of the CMB. An interesting discrepancy that could be connected to the Hubble constant is the baseline Planck lensing amplitude parameter A_L. However results elsewhere are consistent with $A_L = 1$ suggesting that the observed $\sim 2\sigma$ Planck discrepancy may be a fluke. Further investigations and independent corroborating measurements are necessary to better understand these internal inconsistencies.

Proposed solutions that aim to resolve the Hubble constant discrepancy by accelerating the growth of the sound horizon during the pre-recombination era have shown the potential to exacerbate the tension in the σ_8 parameter. These proposals, which introduce variations in the early universe's expansion rate, could have broader implications for our understanding of cosmological structures and the growth of matter fluctuations. The challenge lies in reconciling these competing tensions and finding a coherent theoretical framework that can simultaneously explain both the Hubble constant and σ_8 measurements. A comprehensive analysis of late-time evolution, combining observations from Type Ia supernovae and baryon acoustic oscillations (BAOs), offers valuable insights into the cosmic expansion history and provides constraints on different cosmological models.

Future CMB experiments hold immense promise for shedding light on unresolved questions and advancing our understanding of the universe. Next-generation missions, such as the Simons Observatory, CMB-S4, and the LiteBIRD mission (planned for launch in 2028) are poised to deliver groundbreaking observations. These experiments will allow for higher precision measurements of the CMB temperature and polarization anisotropies. This will allow tests of early dark energy models and explore potential deviations from the standard ΛCDM cosmology. The combination of these next-generation CMB experiments with independent probes, such as large-scale structure surveys and supernova observations, will provide a multi-faceted approach to addressing the outstanding tensions and advancing our understanding of the universe's evolution.

Acknowledgements We thank Antony Lewis for comments on an earlier version of this chapter. PL acknowledges support from the Simons Foundation.

References

1. T.M.C. Abbott et al., Dark energy survey year 3 results: cosmological constraints from galaxy clustering and weak lensing. Phys. Rev. D **105**(2), 023520 (2022). https://doi.org/10.1103/PhysRevD.105.023520. (arXiv:2105.13549 [astro-ph.CO])
2. G.E. Addison, Y. Huang, D.J. Watts, et al.: Quantifying Discordance in the 2015 Planck CMB Spectrum. ApJ **818**(2), 132 (2016). https://doi.org/10.3847/0004-637x/818/2/132. arXiv:1511.00055
3. G.E. Addison, D.J. Watts, C.L. Bennett et al., Elucidating ΛCDM: impact of baryon acoustic oscillation measurements on the hubble constant discrepancy. ApJ **853**(2), 119 (2018). https://doi.org/10.3847/1538-4357/aaa1ed. arXiv:1707.06547
4. S. Aiola, E. Calabrese, L. Maurin et al., The Atacama cosmology telescope: DR4 maps and cosmological parameters. J. Cosmology Astropart. Phys. **12**, 047 (2020). https://doi.org/10.1088/1475-7516/2020/12/047. (arXiv:2007.07288 [astro-ph.CO])
5. Y. Akrami, Y. Fantaye, A. Shafieloo et al., Power asymmetry in WMAP and Planck temperature sky maps as measured by a local variance estimator. Astrophys. J. Lett. **784**, L42 (2014). https://doi.org/10.1088/2041-8205/784/2/L42. (arXiv:1402.0870 [astro-ph.CO])
6. Y. Akrami et al., Planck 2018 results X: constraints on inflation. Astron. Astrophys. **641**, A10 (2020). https://doi.org/10.1051/0004-6361/201833887. (arXiv:1807.06211 [astro-ph.CO])
7. A. Amon, G. Efstathiou, A non-linear solution to the S_8 tension? MNRAS **516**(4), 5355–5366 (2022). https://doi.org/10.1093/mnras/stac2429. (arXiv:2206.11794 [astro-ph.CO])
8. E. Aver, K.A. Olive, E.D. Skillman, The effects of He i λ10830 on helium abundance determinations. J. Cosmol. Astropart. Phys. **2015**(7) (2015). https://doi.org/10.1088/1475-7516/2015/07/011. (arXiv:1503.08146)
9. L. Balkenhol et al.: A measurement of the CMB temperature power spectrum and constraints on cosmology from the SPT-3G 2018 TT/TE/EE data set (2022). (arXiv:2212.05642 [astro-ph.CO])
10. C.L. Bennett, D. Larson, J.L. Weiland et al., Nine-year Wilkinson Microwave Anisotropy Probe (WMAP) observations: final maps and results. ApJs **208**(2), 20 (2013). https://doi.org/10.1088/0067-0049/208/2/20. (arXiv:1212.5225 [astro-ph.CO])
11. D. Brout, D. Scolnic, It's dust: solving the mysteries of the intrinsic scatter and host-galaxy dependence of standardized type Ia Supernova Brightnesses. ApJ **909**(1), 17 (2021). arXiv:2004.10206
12. D. Brout, D. Scolnic, B. Popovic et al., The Pantheon+ analysis: cosmological constraints. ApJ **938**(2), 110 (2022). https://doi.org/10.3847/1538-4357/ac8e04. (arXiv:2202.04077 [astro-ph.CO])
13. P. Cabella, M. Kamionkowski, Theory of cosmic microwave background polarization. Astrophysics. ArXiv e-prints astro-ph/0403392
14. J. Carron, M. Mirmelstein, A. Lewis, CMB lensing from Planck PR4 maps. JCAP **09**, 039 (2022). https://doi.org/10.1088/1475-7516/2022/09/039. (arXiv:2206.07773 [astro-ph.CO])
15. R.J. Cooke, M. Pettini, C.C. Steidel, One percent determination of the primordial deuterium abundance. ApJ **855**, 102 (2018). https://doi.org/10.3847/1538-4357/aaab53 (arXiv:1710.11129)

16. P. Creminelli, G. D'Amico, J. Noreña, F. Vernizzi, The effective theory of quintessence: the w < i-1 side unveiled. J. Cosmol. Astropart. Phys. **2**, 018 (2009). https://doi.org/10.1088/1475-7516/2009/02/018. (arXiv:0811.0827 [astro-ph])
17. A. Cuceu, J. Farr, P. Lemos, A. Font-Ribera, Baryon acoustic oscillations and the Hubble constant: past, present and future. J. Cosmol. Astropart. Phys. **10**, 044 (2019). https://doi.org/10.1088/1475-7516/2019/10/044. (arXiv:1906.11628 [astro-ph.CO])
18. C. Deffayet, O. Pujolàs, I. Sawicki, A. Vikman, Imperfect dark energy from kinetic gravity braiding. J. Cosmol. Astropart. Phys. **10**, 026 (2010). https://doi.org/10.1088/1475-7516/2010/10/026. (arXiv:1008.0048 [hep-th])
19. E. Di Valentino, A. Melchiorri, J. Silk, Planck evidence for a closed universe and a possible crisis for cosmology. Nat. Astron. **4**(2), 196–203 (2019). https://doi.org/10.1038/s41550-019-0906-9. (arXiv:1911.02087 [astro-ph.CO])
20. C. Dickinson, CMB foregrounds—a brief review, in *51st Rencontres de Moriond on Cosmology* (2016), pp. 53–62. (arXiv:1606.03606 [astro-ph.CO])
21. G. Efstathiou, S. Gratton, The evidence for a spatially flat Universe. Mon. Not. R. Astron. Soc. **496**(1), L91–L95 (2020). https://doi.org/10.1093/mnrasl/slaa093. (arXiv:2002.06892 [astro-ph.CO])
22. G.P. Efstathiou, S. Gratton, A detailed description of the CamSpec likelihood pipeline and a reanalysis of the Planck high frequency maps (2019). arXiv arXiv:1910.00483
23. D.J. Fixsen, The temperature of the cosmic microwave background. ApJ **707**(2), 916–920 (2009). https://doi.org/10.1088/0004-637X/707/2/916. (arXiv:0911.1955)
24. A.H. Guth, Inflationary universe: a possible solution to the horizon and flatness problems. Phys. Rev. D **23**, 347–356 (1981). https://doi.org/10.1103/PhysRevD.23.347
25. W. Handley, Curvature tension: evidence for a closed universe. Phys. Rev. D **103**(4), L041301 (2021). https://doi.org/10.1103/PhysRevD.103.L041301. (arXiv:1908.09139 [astro-ph.CO])
26. W. Handley, P. Lemos, Quantifying the global parameter tensions between ACT, SPT, and Planck. Phys. Rev. D **103**(6), 063529 (2021)
27. J.W. Henning, J.T. Sayre, C.L. Reichardt et al., Measurements of the temperature and E-mode polarization of the CMB from 500 square degrees of SPTpol data. ApJ **852**(2), 97 (2018). https://doi.org/10.3847/1538-4357/aa9ff4. (arXiv:1707.09353)
28. C. Heymans et al., KiDS-1000 cosmology: multi-probe weak gravitational lensing and spectroscopic galaxy clustering constraints. Astron. Astrophys. **646**, A140 (2021). https://doi.org/10.1051/0004-6361/202039063. (arXiv:2007.15632 [astro-ph.CO])
29. G. Hinshaw, D. Larson, E. Komatsu et al., Nine-year Wilkinson Microwave Anisotropy Probe (WMAP) observations: cosmological parameter results. Astrophys. J. Suppl. Ser. **208**(2), 19 (2013a)
30. G. Hinshaw, D. Larson, E. Komatsu et al., Nine-year Wilkinson Microwave Anisotropy Probe (WMAP) observations: cosmological parameter results. Astrophys. J. Suppl. Ser. **208**, 19 (2013b). https://doi.org/10.1088/0067-0049/208/2/19. (arXiv:1212.5226)
31. F. Hoyle, G. Burbidge, J.V. Narlikar, *A different approach to cosmology: from a static universe through the big bang towards reality* (Cambridge University Press, 2000)
32. W. Hu, M. White, Acoustic signatures in the cosmic microwave background. ApJ **471**, 30 (1996). https://doi.org/10.1086/177951.astro-ph/9602019
33. W. Hu, M. White, A CMB polarization primer. New **A2**(2), 323–344 (1997). https://doi.org/10.1016/S1384-1076(97)00022-5.astro-ph/9706147
34. W. Hu, M. White, The damping tail of cosmic microwave background anisotropies. ApJ **479**, 568–579 (1997). https://doi.org/10.1086/303928.astro-ph/9609079
35. A. Kogut, M. Abitbol, J. Chluba et al., CMB spectral distortions: status and prospects (2019)
36. J.M. Kovac, E.M. Leitch, C. Pryke et al., Detection of polarization in the cosmic microwave background using DASI. Nature **420**, 772–787 (2002). https://doi.org/10.1038/nature01269.astro-ph/0209478
37. P. Lemos, Tests of the Planck cosmology at high and low redshifts. PhD thesis, Cambridge University (2018). https://doi.org/10.17863/CAM.36135

38. P. Lemos, E. Lee, G. Efstathiou, S. Gratton, Model independent $H(z)$ reconstruction using the cosmic inverse distance ladder (2018). ArXiv e-prints arXiv:1806.06781
39. P. Lemos et al., Assessing tension metrics with dark energy survey and Planck data. Mon. Not. R. Astron. Soc. **505**(4), 6179–6194 (2021). https://doi.org/10.1093/mnras/stab1670 (arXiv:2012.09554 [astro-ph.CO])
40. A. Lewis, A. Challinor, Weak gravitational lensing of the CMB. Phys. Rep. **429**(1), 1–65 (2006)
41. A.D. Linde, A new inflationary universe scenario: a possible solution of the horizon, flatness, homogeneity, isotropy and primordial monopole problems. Phys. Lett. B **108**, 389–393 (1982). https://doi.org/10.1016/0370-2693(82)91219-9
42. A.D. Linde, Chaotic inflation. Phys. Lett. B **129**, 177–181 (1983). https://doi.org/10.1016/0370-2693(83)90837-7
43. T. Louis, E. Grace, M. Hasselfield et al., The Atacama cosmology telescope: two-season ACT-Pol spectra and parameters. J. Cosmol. Astropart. Phys. **6**, 031 (2017). https://doi.org/10.1088/1475-7516/2017/06/031. (arXiv:1610.02360)
44. E. MacAulay, R.C. Nichol, D. Bacon et al., First cosmological results using type Ia supernovae from the dark energy survey: measurement of the Hubble constant. MNRAS **486**(2), 2184–2196 (2019). https://doi.org/10.1093/mnras/stz978. (arXiv:1811.02376)
45. A.A. Penzias, R.W. Wilson, A measurement of excess antenna temperature at 4080 Mc/s. ApJ **142**, 419–421 (1965). https://doi.org/10.1086/148307
46. O.H.E. Philcox, M.M. Ivanov, BOSS DR12 full-shape cosmology: Λ CDM constraints from the large-scale galaxy power spectrum and bispectrum monopole. Phys. Rev. D **105**(4), 043517 (2022). https://doi.org/10.1103/PhysRevD.105.043517. (arXiv:2112.04515 [astro-ph.CO])
47. O. Pisanti, G. Mangano, G. Miele, P. Mazzella, Primordial deuterium after LUNA: concordances and error budget. J. Cosmol. Astropart. Phys. **4**, 020 (2021). https://doi.org/10.1088/1475-7516/2021/04/020. (arXiv:2011.11537 [astro-ph.CO])
48. Planck Collaboration, P.A.R. Ade, N. Aghanim et al., Planck 2013 results XXII constraints on inflation. A&A **571**, A22 (2014). https://doi.org/10.1051/0004-6361/201321569. (arXiv:1303.5082)
49. Planck Collaboration, P.A.R. Ade, N. Aghanim, et al.: Planck 2013 results XV CMB power spectra and likelihood. A&A **571**, A15 (2014b). https://doi.org/10.1051/0004-6361/201321573. (arXiv:1303.5075 [astro-ph.CO])
50. Planck Collaboration, N. Aghanim, Y. Akrami et al., Planck 2018 results VI cosmological parameters (2018). ArXiv e-prints arXiv:1807.06209
51. Planck Collaboration, N. Aghanim, Y. Akrami et al., Planck 2018 results VI: cosmological parameters. A&A **641** (2020). https://doi.org/10.1051/0004-6361/201833910. (arXiv:1807.06209)
52. V. Poulin, T.L. Smith, T. Karwal, M. Kamionkowski, Early dark energy can resolve the Hubble tension. Phys. Rev. Lett. **122**(22), 221301 (2019). https://doi.org/10.1103/PhysRevLett.122.221301. (arXiv:1811.04083)
53. V. Poulin, T.L. Smith, T. Karwal, The ups and downs of early dark energy solutions to the hubble tension: a review of models, hints and constraints circa 2023 (2023). https://doi.org/10.48550/arXiv.2302.09032. (arXiv:2302.09032 [astro-ph.CO])
54. F.J. Qu, B.D. Sherwin, M.S. Madhavacheril et al., The Atacama cosmology telescope: a measurement of the DR6 CMB lensing power spectrum and its implications for structure growth (2023). https://doi.org/10.48550/arXiv.2304.05202. (arXiv:2304.05202 [astro-ph.CO])
55. A.G. Riess, S. Casertano, W. Yuan et al., Cosmic distances calibrated to 1% precision with gaia EDR3 parallaxes and Hubble space telescope photometry of 75 milky way Cepheids confirm tension with lambda CDM. ApJ **908**(1):1–11 (2021). (arXiv:2012.08534)
56. A.G. Riess, W. Yuan, L.M. Macri et al., A comprehensive measurement of the local value of the Hubble constant with 1 km s^{-1} Mpc^{-1} uncertainty from the hubble space telescope and the SH_0ES team. ApJ **934**(1), L7 (2022). https://doi.org/10.3847/2041-8213/ac5c5b. (arXiv:2112.04510 [astro-ph.CO])
57. R.K. Sachs, A.M. Wolfe, Perturbations of a cosmological model and angular variations of the microwave background. ApJ **147**, 73 (1967). https://doi.org/10.1086/148982

58. A. Sandage, Current problems in the extragalactic distance scale. ApJ **127**, 513 (1958). https://doi.org/10.1086/146483
59. N. Schöneberg, J. Lesgourgues, D. C. Hooper, The BAO+BBN take on the Hubble tension. J. Cosmol. Astropart. Phys. **10**, 029 (2019). https://doi.org/10.1088/1475-7516/2019/10/029. (arXiv:1907.11594 [astro-ph.CO])
60. D. Scolnic, D. Brout, A. Carr et al., The Pantheon+ analysis: the full data set and light-curve release. ApJ **938**(2), 113 (2022). https://doi.org/10.3847/1538-4357/ac8b7a. (arXiv:2112.03863 [astro-ph.CO])
61. G.F. Smoot, The CMB spectrum, in *International School of Astrophysics, D. Challenge: 5th Course: Current Topics in Astro fundamental Physics* (1997), pp. 407–440. (arXiv:astro-ph/9705101)
62. G.F. Smoot, COBE observations and results, in *3K cosmology*, vol. 476, ed. by L. Maiani, F. Melchiorri, N. Vittorio (American Institute of Physics Conference Series, 1999), pp 1–10. https://doi.org/10.1063/1.59326. (astro-ph/9902027)
63. G.F. Smoot, C.L. Bennett, A. Kogut et al., Structure in the COBE differential microwave radiometer first-year maps. Astrophys. J. **396**, L1–L5 (1992). https://doi.org/10.1086/186504
64. V. Sudevan, P.K. Aluri, S.K. Yadav, R. Saha, T. Souradeep, Improved diffuse foreground subtraction with the ILC Method: CMB map and angular power spectrum using planck and WMAP observations. Astrophys. J. **842**(1), 62 (2017). https://doi.org/10.3847/1538-4357/aa7334. (arXiv:1612.03401 [astro-ph.CO])
65. R.A. Sunyaev, I.B. Zeldovich, The velocity of clusters of galaxies relative to the microwave background—the possibility of its measurement. MNRAS **190**, 413–420 (1980). https://doi.org/10.1093/mnras/190.3.413
66. R.A. Sunyaev, Y.B. Zeldovich, The observations of relic radiation as a test of the nature of X-ray radiation from the clusters of galaxies. Comments Astrophys. Space Phys. **4**, 173 (1972)
67. M. Tucci, L. Toffolatti, The impact of polarized extragalactic radio sources on the detection of CMB anisotropies in polarization. Adv. Astron. **2012**, 624987 (2012). https://doi.org/10.1155/2012/624987. (arXiv:1204.0427 [astro-ph.CO])
68. R. De Vaucouleurs, *The cosmic Distance Scale and the Hubble Constant*. Mount Stromlo and Siding Spring Observatories, Australian National University (1982). https://books.google.ca/books?id=zRSRzgEACAAJ
69. A. Vikman, Can dark energy evolve to the phantom? Phys. Rev. D **71**(2), 023515 (2005). https://doi.org/10.1103/PhysRevD.71.023515. (arXiv:astro-ph/0407107 [astro-ph])
70. S. Wagner-Carena, M. Hopkins, A. Diaz Rivero, C. Dvorkin, A novel CMB component separation method: hierarchical generalized morphological component analysis. Mon. Not. R. Astron. Soc. **494**(1), 1507–1529 (2020). https://doi.org/10.1093/mnras/staa744. (arXiv:1910.08077 [astro-ph.CO])
71. M. White, D. Scott, J. Silk, Anisotropies in the cosmic microwave background. ARA&A **32**, 319–370 (1994). https://doi.org/10.1146/annurev.astro.32.1.319
72. M.J. White, Anisotropies in the CMB, in *American Physical Society (APS) Meeting of the Division of Particles and Fields (DPF 99)* (1999). arXiv:astro-ph/9903232
73. Y.B. Zeldovich, R.A. Sunyaev, The interaction of matter and radiation in a hot-model universe. Astrophys. Space Sci. **4**, 301–316 (1969). https://doi.org/10.1007/BF00661821

Chapter 17
Measuring H_0 with Spectroscopic Surveys

Mikhail M. Ivanov and Oliver H. E. Philcox

17.1 Introduction

Spectroscopic surveys map out the large-scale structure (hereafter LSS) of the Universe through the distribution of galaxies. By measuring the positions and redshifts of a wide variety of galaxies, we can build a three-dimensional map of the galaxy overdensity, which, on comparatively large-scales, traces the underlying density of dark matter (e.g., [1]). As such, spectroscopic surveys can be used to place constraints on physics operating in the early Universe (which set the initial conditions for matter clustering), and at late times (through the growth of structure). In this chapter, we will discuss how they can be used to place competitive constraints on the Hubble parameter, H_0.

Since the late 1970s, there have been a wealth of galaxy redshift surveys, with the current state-of-the-art public data (from the Sloan Digital Sky Survey [2]) containing around one-million objects in the Northern sky. These are primarily analyzed through clustering statistics, the simplest of which are the power spectrum, $P(k)$, or two-point correlation function, $\xi(r)$, encoding the distribution of pairs of objects on some characteristic scale r or $2\pi/k$ (k denotes perturbations' wavenumber). In the coming decade the data volume will greatly increase, both due to ground-based (DESI, and, potentially, others such as MegaMapper) and space-based (SPHEREx, Euclid, Roman, and Rubin) missions [3–8]. With increasing survey volume comes tighter constraints on cosmological parameters; as such, we may expect to yield sub-percent measurements of the Hubble parameter from a variety of probes in the near future.

M. M. Ivanov (✉)
Institute for Advanced Study, Princeton, USA
e-mail: ivanov@ias.edu

O. H. E. Philcox (✉)
Columbia University, New York and Simons Foundation, New York, USA
e-mail: ohep2@cantab.ac.uk

E. Di Valentino and D. Brout (eds.), *The Hubble Constant Tension*, Springer Series in Astrophysics and Cosmology, https://doi.org/10.1007/978-981-99-0177-7_17

As for the cosmic microwave background (CMB), spectroscopic H_0 constraints arise from the combination of observed and physical scales, through the canonical 'standard ruler' mechanism. An important question, therefore, is which scales should be used. To be an accurate probe of H_0, a standard ruler must be: (a) precisely predicted form theory, and (b) well measured by experiment. In LSS, there are two candidates: the particle horizon scale at the epoch of matter-radiation equality (at $z_{\text{eq}} \approx 3600$) (e.g., [9]) and the sound horizon at recombination (strictly, at baryon drag, $z_d \approx 1060$) (e.g., [10, 11]) Each is straightforward to predict, and leaves detectable signatures in the observed galaxy clustering statistics, primarily (and fortuitously) concentrated in the linear regime. Whilst one could concoct other physical scales (such as that of Silk damping, or dark-energy-matter equality), the above are the simplest in practice in the context of galaxy surveys.

Mathematically, the directly observed angular size of a feature, $\Delta\theta$ is related to the theoretically predicted physical size, r, via

$$\Delta\theta_{\text{feature}} \equiv \frac{r_{\text{feature}}}{(1+z)D_A(z)}, \tag{17.1}$$

where $D_A(z)$ is the angular diameter distance to the galaxy survey at some redshift z. If the feature is primordial, r_{feature} does not depend on physics operating at late times, however $D_A(z)$ does, thus the measurement is not strictly 'early' in nature.

Let us use the following heuristic argument to understand where the sensitivity to H_0 comes from. Inserting the form of the angular diameter distance in Eq. (17.1) and assuming a flat Universe (for simplicity alone), we find

$$H_0 = \frac{\Delta\theta_{\text{feature}}}{r_{\text{feature}}} \int_0^z \frac{dz'}{E(z')}, \tag{17.2}$$

where $E^2(z) = \Omega_{m,0}(1+z)^3 + \Omega_{r,0}(1+z)^4 + \Omega_\Lambda \equiv H^2(z)/H_0^2$. Equation (17.2) gives the rough idea on how the combination of a measured r_{feature} and an observed $\Delta\theta_{\text{feature}}$ yields H_0 (given the cosmological expansion history); equivalently, the angular measurements translate (roughly) to constraints on $r_{\text{feature}} H_0$. In the below, we will discuss our two possibilities for the standard ruler feature, including their physical origin and non-linear evolution, and present the various associated H_0 measurements, as well as their limitations and future prospects.[1]

[1] We caution that for the minimal ΛCDM model, and for probes of cosmological fluctuations such as the CMB and LSS, Ω_Λ, Ω_m etc. implicitely depend on H_0, so that Eq. (17.2) does not actually *define* the Hubble constant.

17.2 Baryon Acoustic Oscillations

17.2.1 Theoretical Background

Prior to recombination, the Universe is filled with a strongly interacting photon-baryon fluid, pressure-supported against. As with any fluid, this is filled with sound waves, on some characteristic scale r_s, seeded by early Universe inhomogeneities. As the Universe cools and free electrons combine with protons in recombination, the fluid decouples (at the baryon drag redshift $z \approx z_d \sim 1060$) and the photon-baryon oscillations cease to propagate, effectively 'freezing' the sound waves at the sound-horizon scale, $r_s(z_d) \equiv r_d$. During the subsequent evolution of the Universe, this characteristic inhomogeneity is transferred to the dark matter distribution by gravitational interactions, yielding an excess of clustering at late-times.

Physically, these features (dubbed 'baryon acoustic oscillations', hereafter BAO) source a peak in the two-point correlation function at $r \sim r_d$, or an oscillatory feature in the galaxy power spectrum at $k \sim 2\pi/r_d$. The precise sound horizon scale can be computed explicitly:

$$r_s(z) = \int_z^\infty dz' \frac{c_s(z)}{H(z')}, \tag{17.3}$$

with the sound speed $c_s \approx c\left[3 + 9/4\rho_b/\rho_\gamma\right]^{-1/2}$, where $\rho_{b,\gamma}$ are baryon and photon densities. All quantities in the above can be estimated precisely given cosmological parameters (such as those measured by *Planck*), and are principally sensitive to the physical baryon density, ω_b, since the photon density is strongly constrained from the CMB monopole. Assuming the *Planck* ΛCDM cosmology, we find $r_d \sim 105h^{-1}$Mpc, with an approximate cosmology dependence of [12]

$$r_d \approx \frac{55.154 \exp\left[-72.3(\omega_{\nu,0} + 0.0006)^2\right]}{\omega_{m,0}^{0.25351}\omega_{b,0}^{0.12807}} \text{Mpc}, \tag{17.4}$$

where $\omega_{m,0}$, $\omega_{b,0}$, $\omega_{\nu,0}$ denote the current physical densities of matter (i.e. dark matter + baryons), baryons, and neutrinos.

Whilst (17.3) accurately predicts the BAO scale at decoupling, various physical effects could alter its manifestation in the late-time galaxy correlators. Firstly, the amplitude of oscillatory features decays with time due to bulk flows, *i.e.* the large-scale motion of matter in the post-recombination Universe. At late-times, the oscillatory part of the linear power spectrum, $P_{\rm L}^{\rm osc}$, (which contains the BAO wiggles), is suppressed according to

$$P_{\rm L}^{\rm osc}(k,z) \to e^{-k^2\Sigma^2(z)} P_{\rm L}^{\rm osc}(k,z), \tag{17.5}$$

$$\Sigma^2(z) = \tfrac{1}{6\pi^2}\int_0^\Lambda dp\, P_{\rm L}(p,z)[1 - j_0(pr_d) + 2j_2(pr_d)], \tag{17.6}$$

where $\Lambda \approx 0.1\,h\,\mathrm{Mpc}^{-1}$ is the IR cutoff scale, $j_\ell(x)$ are spherical Bessel functions and $\Sigma(z)$ is BAO damping scales extracted via a systematic theoretical procedure called IR resummation [13–17] (also see [18, 19] for early work on the subject). When extracting the BAO signal from data, it is important to include such a suppression to avoid biased results. We note that this is *not* included in simple parametrizations such as HaloFit.

Secondly, gravitational evolution modulates the clustering of matter at late times, causing various physical effects such as halo formation, which impact the galaxy power spectrum. This is important for H_0 determination, since non-linear physics could alter the position of peaks in the galaxy power spectrum, from which the absolute distance scale is extracted via (17.2). In practice, the effects of this are small, though not negligible (e.g., [10, 20]). Non-linear effects yield a (cosmology-dependent) $\sim 0.5\%$ shift in the BAO scale whose small amplitude is itself somewhat of an accident in ΛCDM [16, 17, 21]. In practice, one often marginalizes over non-linear contributions to the power spectrum, whilst assuming the BAO to be determined only by linear physics (plus damping) (e.g. [22]). This is sufficiently accurate for current surveys, but will not suffice in the future.

17.2.2 *Extracting the BAO Feature*

17.2.2.1 Template Analyses

To infer H_0 from the BAO feature, it has become commonplace to use template (or scaling) analyses, based on the measurement of Alcock-Paczynski parameters (e.g., [22, 23]). This involves a theoretical templates for the BAO feature, such that the oscillatory part of the power spectrum (or equivalently correlation function), takes the form

$$P^{\mathrm{obs}}(k_\parallel, k_\perp) \propto P^{\mathrm{template}}(k_\parallel/\alpha_\parallel, k_\perp/\alpha_\perp), \tag{17.7}$$

where we parametrize the three-dimensional space by $k_\parallel, k_\perp$ (relative to some line-of-sight direction), where the Alcock-Paczynski parameters are defined as

$$\alpha_\parallel = \frac{H^{\mathrm{fid}}(z) r_d^{\mathrm{fid}}}{H(z) r_d} \qquad \alpha_\perp = \frac{D_A(z) r_d^{\mathrm{fid}}}{D_A^{\mathrm{fid}}(z) r_d} \tag{17.8}$$

These encode both the physical BAO scale, r_d, and the distance to the galaxy sample, via $D_A(z)$ (tangential) or $H(z)$ (radial, through velocities), and are normalized such that $\alpha_{\parallel,\perp} = 1$ if the reference cosmology (used to define the templates) matches the truth. We caution that these parameters assume that the position of the BAO peak is not altered by non-linear physics.

In practice, one must additionally account for non-oscillatory effects contributing to the power spectrum, including galaxy bias, the fingers-of-God effect, and redshift-space distortions. Incorporating these effects (usually via some phenomenological framework, or with perturbation theory) gives rise to a full theoretical model for the galaxy power spectrum depending on a number of free parameters, including $\alpha_\parallel, \alpha_\perp$. Fitting this model to data then yields the best-fit parameters, themselves encoding $H(z)r_d$ and $D_A(z)/r_d$. Though these are usually said to be cosmology-independent, we caution that their estimation does require theoretical modeling to define the templates. If the templates are far from the truth, or new physics is at work that alters the BAO features, (17.7) breaks down, thus one cannot extract H_0-dependent quantities directly from the measured Alcock-Paczynski parameters. In practice, the former effect is small, assuming the background cosmology to be relatively well known.

The output of this procedure is measurements of the cosmological quantities $H(z)r_d$ and $D_A(z)/r_d$. Alone, these do not measure H_0, due to a perfect $H_0 r_d$ degeneracy; practically, they are measuring only the angular size of the BAO feature. As mentioned above, one requires also the physical size. Combining these constraints with knowledge of r_d, within ΛCDM or some other model (such as early dark energy), one can break the degeneracy and measure H_0 directly. This is usually done by combining with the CMB (e.g., [22]), or with Big Bang Nucleosynthesis probes (e.g., [24]), both of which place strong constraints on ω_b, and hence r_d. The resulting constraints on H_0 will be discussed in Sect. 17.4.

Whilst the above discussion has focused on measuring quantities from the observed galaxy power spectrum, it has become commonplace for survey analyses to include also data from the "reconstructed" power spectrum [25]. In this instance, one first performs a catalog-level transformation of the data, effectively 'undoing' the large-scale bulk flows which lead to smearing of the BAO feature. This can be performed robustly, and increases the signal-to-noise of oscillatory features by reducing the damping given in (17.5). The resulting BAO peak is well described by (weakly damped) linear theory (although we stress that the underlying physics relies on approximate cancellations between non-linear contributions), and can be used to enhance constraints on $H(z)r_d$ and $D_A(z)/r_d$. At low redshifts (where the damping is significant), this can bolster constraints by up to $\sim$40%.

17.2.2.2 Full-Shape Analyses

In the above formalism, the galaxy power spectrum is analyzed using a phenomenological model, effectively discarding any information besides the BAO feature. An alternative approach, which has found favor in recent years, is to instead model the entire power spectrum theoretically (not just the wiggles), from which a range of cosmological information can be extracted, including H_0 (e.g., [26, 27]). This is made possible via theoretical efforts such as the Effective Field Theory of Large Scale Structure (hereafter EFT, (e.g., [28–32])), which provides a first principles prediction for the galaxy power spectrum, $P_{gg}(k; \theta)$, based on perturbation theory and symmetry arguments, operating in the linear and mildly non-linear regime. This depends on a parameter vector θ, which encompasses cosmology, galaxy bias, and other nuisance parameters. By fitting $P_{gg}(k; \theta)$ to the data directly, we obtain a posterior for θ, and thus obtain constraints on the underlying cosmological parameters. Since the theory model depends on the cosmological initial conditions and subsequent evolution, it naturally picks up H_0 information through the BAO feature (and the equality scale, as discussed below), including both the physical size (from theory) and the angular size (distorted due to some choice of reference cosmology).

Full-shape analyses are naturally dependent on the cosmological model; however, this is not a limitation, since one can construct predictions in any framework, be it ΛCDM or beyond, with extension parameters including early dark energy, modified gravity, and massive neutrinos. As far as H_0 is concerned, the approach is similar to a template fitting, but the templates themselves are dynamically varied, and thus depend on cosmological parameters such as Ω_m and (through the wiggle damping), ω_b. Rather than predicting derived parameters such as $H(z)r_d$), full-shape analyses can predict the underlying H_0 parameter (or some other combination of cosmological parameters, such as the CMB temperature, or angular sound horizon), since the physical forms of $D_A(z)$, $H(z)$, and r_d are baked into the model. To extract H_0 with template-based approaches, external information on these functions must be supplied, thus BAO measurements are analyzed jointly with *Planck*, or some other external data-set, which provide estimates of the Ω_m (ω_b) parameters needed to determine D_A and H (r_d). In the full-shape case, both the parameters can be constrained from the data directly (through coordinate distortions and peak heights), though it is usually commonplace to supplement the survey with external ω_b information coming from the CMB or Big Bang Nucleosynthesis (BBN) constraints (e.g., [33]). In the limit of a well-measured BAO feature, the precision of our output constraint on H_0 is fully limited by our knowledge of ω_b, through the $r_d(\omega_b)$ relation: if this is not measured with full-shape approaches (which themselves give only comparatively weak constraints), we may not see large improvements in H_0 constraints if the external BBN information does not increase in precision.

As for template analyses, it is possible to consider the reconstructed power spectrum in the full-shape context. Once again, our logic is that reconstruction acts to sharpen the BAO oscillations by shifting information into the two-point function from the higher-point correlators, thus capturing H_0 information that would usually be found only in the bispectrum and above. Though several works have considered the full-shape modeling of the reconstructed power spectrum [34–38], it is highly non-trivial, and depends strongly on the particular flavor of reconstruction applied (quickly becoming unwieldy for multi-step reconstruction procedures). Given that the aim of reconstruction is to sharpen the BAO feature, and not to undo non-linear effects in the bulk flow (which it is known to perform poorly at, with non-linear formation being distorted by the necessary smoothing [39]), it is thus appropriate to employ a hybrid approach, performing a template analysis of the reconstructed power spectrum and a full-shape analysis of the data, pre-reconstruction (e.g., [40, 41]). This proceeds by measuring Alcock-Paczynski parameters from the former (effectively acting as a data compression step), and performing a joint cosmological analysis of these with the measured power spectrum multipoles in a ΛCDM (or beyond) context, including the covariance between the two. This has been shown to bolster H_0 information from full-shape analyses by up to $\sim$40% [40], and avoids the difficulty in modeling broadband reconstruction artifacts.

17.3 Matter-Radiation Equality

17.3.1 Theoretical Background

Around $z \approx 3600$, the dynamics of the Universe change from being dominated by radiation pressure, to being dominated by matter (both baryonic and dark). The characteristic redshift is straightforward to determine by setting equal the density of photons and matter, $\Omega_{m,0}(1+z)^3 = \Omega_{r,0}(1+z)^4$ (treating neutrinos as relativistic), where the radiation density is set by the current temperature of the CMB, as measured by FIRAS.[2] The horizon scale is set by the light horizon at this time,

$$r_{\rm eq} = \int_{z_{\rm eq}}^{\infty} dz' \frac{c}{H(z')}, \tag{17.9}$$

with the Fourier-space scale being approximately

$$k_{\rm eq} = (2\Omega_m H_0^2 z_{\rm eq})^{1/2}, \qquad z_{\rm eq} = 2.5 \times 10^4 \Omega_m h^2 \Theta_{2.7}^{-4} \tag{17.10}$$

Eisenstein and Hu [43], where $\Theta_{2.7}$ is the CMB temperature in 2.7K units. This is a precisely known scale that acts as a second standard-ruler for H_0 analyses (dependent only on the physical matter density $\Omega_m h^2$).

Physically, the equality scale corresponds to a turn-over of the matter power spectrum at wavenumbers $k \sim k_{\rm eq}$, as described by the Meszaros equation. Whilst the turnover predominantly occurs on larger scales than can be robustly probed in many spectroscopic analyses (except for quasar samples, and SPHEREx, though it will be subject to significant cosmic variance and systematic penalities), its signature can be seen also on somewhat larger scales, due to a $\ln^2(k/k_{\rm eq})$ feature in the matter power spectrum at $k \geq k_{\rm eq}$. These scales are concentrated in the linear regime, which significantly simplifies modeling, especially given that the feature is spectrally smooth, unlike the easy-to-distinguish BAO oscillations.

As before, we practically measure the ratio of the observed and theoretical equality scale: by the above arguments, this imparts a degeneracy $h \sim \Delta\theta_{\rm eq} k_{\rm eq}$, thus we in practice constrain the combination $\Omega_m h$, usually denoted Γ. The notion that galaxy surveys are sensitive to this scale arose significantly before the interest in the BAO feature, with measurements of Γ being a primary target of the spectroscopic surveys of the 1990s and early 2000s [44, 45], though interest in its use as a BAO-independent probe of H_0 is comparatively recent.

---

[2] Note that physical moment of matter-radiation equality is determined by the invariant matter-to-photon ratio and does not actually depend on the current CMB temperature [42].

17.3.2 Extracting Equality

As for the BAO feature, one can proceed to measure the equality scale via template based approaches, by constructing some phenomenological model for the low-k galaxy power spectrum, or from full-shape analyses. In this instance, most of the constraints have been derived using full-shape approaches, though there has been some work employed to connect the equality scale to effects parameters in the 'ShapeFit' formalism [46]. Though one could proceed to analyze only the galaxy power spectrum in the linear regime, it is usually preferred to utilize the whole EFTofLSS full-shape methodology to extract H_0 from the equality scale, since this (a) calibrates the galaxy bias parameters from quasi-linear scales, (b) naturally includes any non-linear contributions relevant at low-k, allowing their differentiation from the equality features in the transfer function.

In practice, applying full-shape modeling to the galaxy power spectrum (and beyond) picks up information from both the equality and sound-horizon scales, since both enter the theory model $P_{gg}(k, \theta)$. To isolate the equality feature, one employ measures to remove the BAO features; it is not sufficient to simply employ a scale cut, since there is some overlap between the smooth equality feature and the first BAO wiggle. Noting that the BAO-derived H_0 information depends strongly on sound-horizon calibration through ω_b information, one option is to simply perform the analysis without an external prior on ω_b [47]. This is sufficient for current surveys (where the external information is much more precise than the ω_b measured internally, e.g., by BAO wiggle heights), but this will cease to be the case in the future, due to higher precision survey data. In this case, one may effectively marginalize over the oscillatory features, yielding an equality-only constraint on H_0 [48]. Notably that these measurements are naturally dependent on the choice of model used to analyze them (since one requires a prediction for $P_{gg}(k)$ in terms of $\Gamma \equiv \Omega_m h$), however, this is rarely a limitation since one can straightforwardly test any theoretical model of interest within the formalism, and, furthermore, there is very different dependence on the cosmological model to BAO analyses, e.g., the measurement would be largely unaffected by changes in sound-horizon physics.

As described above, equality-based measurements are practically sensitive to $\Gamma \equiv \Omega_m h$, thus H_0 determination requires some additional constraint on Ω_m (and, to a lesser extent, n_s). This can be done within the survey itself, principally through coordinate distortions and redshift evolution, or externally. In the latter case, one can place a prior on Ω_m from uncalibrated BAO measurements or supernovae distance constraints (both of which measure the relative expansion rate, without H_0 calibration). We finally note that an analogous procedure can be used to constrain H_0 from photometric surveys or CMB lensing. The same does not hold for the BAO feature, since the projection integrals tend to wash out such features; indeed this allows straightforward baryon-independent H_0 constraints to be wrought.

17.4 Measurements

Having discussed the physics behind spectroscopic H_0 constraints, we now present a number of recent constraints from both the BAO and equality scales. For tractability, we will specialize to analyses performed on BOSS data and beyond. A representative (though not necessarily complete) set of constraints are shown in Figs. 17.1 and 17.2, which indicate the plurality of work on the subject, particularly given that most results are derived from the same data-sets (BOSS and/or eBOSS clustering). In all cases, we assume a ΛCDM cosmology, unless otherwise specified.

We first consider discuss sound-horizon-based constraints on H_0, as derived via template fitting. This is the preferred mode of analysis for the BOSS and eBOSS surveys, with the major data-product being a set of $H(z)r_d$ and $D_A(z)/r_d$ measurements at the effective redshifts of the relevant data sets. As discussed above, these can be combined with external 'shape' information from the cosmic microwave background or Big Bang Nucleosynthesis (effectively fixing ω_b), to break the r_d-degeneracy and extract H_0 directly. The official BOSS analysis (combining with *Planck*) finds (67.6 ± 0.5) km s^{-1}Mpc^{-1} [22]; this is updated to (67.6 ± 0.4) km s^{-1}Mpc^{-1} with the inclusion of eBOSS data [49] (see also [12, 50]). Removing *Planck* information leads to broader constraints (since *Planck* tightly constrains r_d and Ω_m as well as H_0 itself, through BAO features), with [24] finding (67.5 ± 1.1) km s^{-1}Mpc^{-1}, reducing to (67.6 ± 1.0) km s^{-1}Mpc^{-1} with eBOSS [51] (see also [52]).

Full-shape analyses make use of both sound-horizon and matter-radiation-equality standard rulers, which can yield tight constraints without explicit *Planck* calibration (see also the *ShapeFit* approach, which captures similar information in a template-based approach [46, 53], finding $(67.9^{+0.7}_{-0.4})$ km s^{-1}Mpc^{-1} with BOSS, eBOSS and *Planck*). A wide variety of full-shape H_0 constraints exist, predominantly differing in their choice of cosmological priors and supplementary datasets. For example, [54] find (68.7 ± 1.5) km s^{-1}Mpc^{-1} from the BOSS DR12 power spectrum with a *Planck* prior on the baryon density (see also [27, 55]). Constraining the spectral slope to that measured by *Planck* improves constraints by $\sim$40%, with consistent results found from both *Planck* and BBN ω_b priors, and three major analysis teams [26, 27, 41, 56]. Combining with the full *Planck* dataset yields $(68.0^{+0.7}_{-0.5})$ km s^{-1}Mpc^{-1} [57]).

The above results can be improved further by folding in information from the reconstructed power spectrum (which is standard practice in template analyses). As shown in [40], this improves constraints by another $\sim$ 40%, with BOSS DR12 + BBN yielding (67.8 ± 0.7) km s^{-1}Mpc^{-1} or (68.6 ± 1.1) km s^{-1}Mpc^{-1} with and without n_s priors (see also [41, 58]). A variety of extensions to these approaches have been performed, for example the inclusion of the configuration-space correlation function [41, 59], a proxy for the real-space power spectrum [60, 61] (also see [62–65]), lensing-galaxy cross correlations [66], the bispectrum monopole [67], and the bispectrum multipoles [68, 69]. The most recent result, combining the BOSS power spectrum, reconstructed power spectrum, real-space power spectrum, and bispectrum multipoles, yields (68.2 ± 0.8) km s^{-1}Mpc^{-1} [68]. This is consistent with the official BOSS results and those from *Planck*. Similar results can be obtained using eBOSS

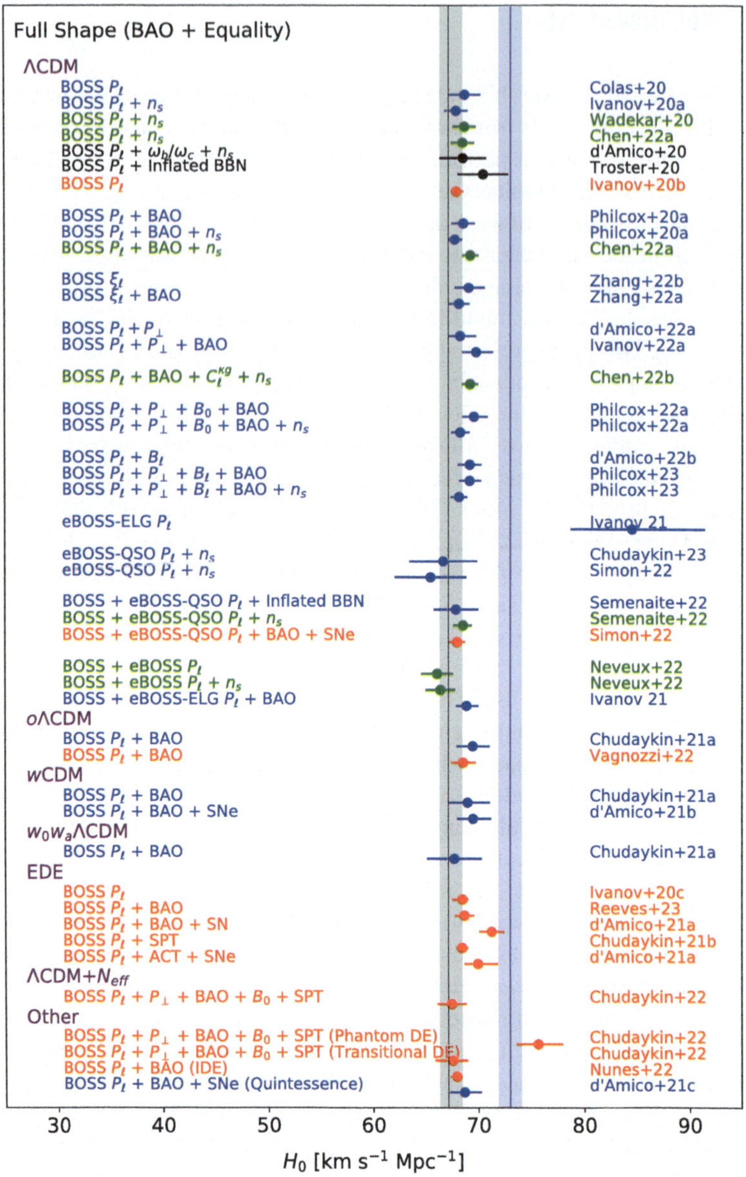

Fig. 17.1 Summary plot of recent H_0 measurements from spectroscopic data, alongisde the *Planck* CMB-only posterior (black) and SH_0ES results (blue). In each case, we plot the 1σ bound on H_0 from the given analysis. Results shown in blue include a BBN prior on ω_b, whilst those in green use an ω_b prior from *Planck*. Measurements shown in red are combined with the full *Planck* dataset. In each case we list the datasets used, with $P_\perp$ indicating real-space power spectrum proxies, B_ℓ indicating bispectrum multipoles, and $C_\ell^{\kappa g}$ indicating a galaxy-lensing cross spectrum. Entries denoted '+ SNe' add supernova Ω_m information from Pantheon or PantheonPlus, whilst '+ n_s' indicates use of the *Planck* prior on n_s. This figure shows only full-shape results (which naturally include both BAO and equality information; Fig. 17.2 shows the other measurements

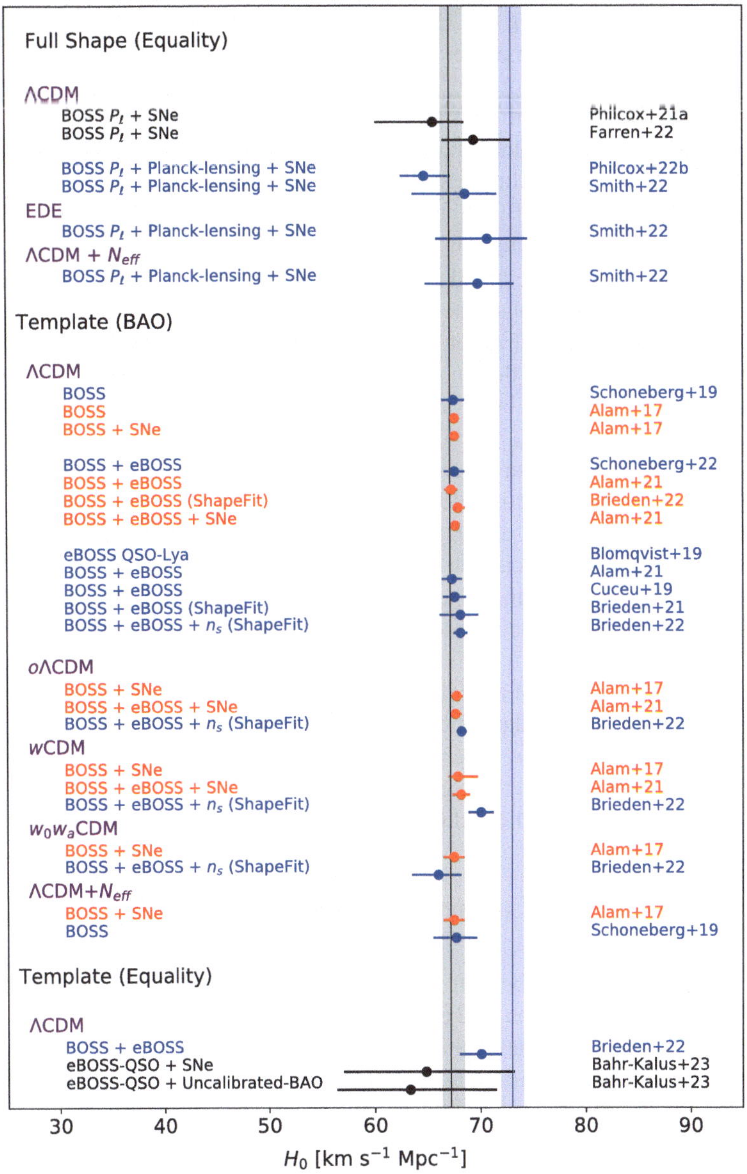

Fig. 17.2 As Fig. 17.1, but displaying results from other types of measurement; full-shape equality measurements, template BAO constraints, and template equality results. The colors are as before. Results obtained using the 'ShapeFit' compression scheme are indicated explicitly

data, both from the ELG [70], QSO [70–74], and combined samples [74, 75], though their individual constraining power is weaker than that of BOSS.

Finally, we consider H_0 constraints which derive information only from the equality scale. Since this ruler suffers from increased cosmic variance and requires Ω_m calibration, the constraints are generally weaker than for the BAO, though can still be used to facilitate interesting tests of the cosmological model [48]. From template based analyses of the BOSS and eBOSS data with BBN information, we find $(70.1^{+1.9}_{-2.1})$ km s^{-1}Mpc^{-1} using the 'ShapeFit' formalism [76], whilst results from the eBOSS quasars yield the broad constraint $(64.8^{+8.4}_{-7.8})$ km s^{-1}Mpc^{-1} [77]. In full-shape analyses, one can isolate the equality information either by omitting the ω_b prior, or by explicit marginalization; combining BOSS with CMB lensing and 'Pantheon+' supernovae, yields the final constraint $(64.8^{+2.2}_{-2.5})$ km s^{-1}Mpc^{-1} , which is sound-horizon-independent but consistent with BAO constraints and *Planck* (see also [9, 47, 48, 78].

Whilst the above results have been derived in a ΛCDM framework, similar analyses are possible in any other cosmological model of choice. For the template analyses, one reinterprets the $H(z)r_d$ and $D_A(z)/r_d$ constraints with the updated model for growth and r_d, though we caution that this requires the desired cosmology to be relatively close to *Planck*, since one assumed a ΛCDM model to generate the templates. In contrast, full-shape analyses can be extended by changing the underlying model used in computation of the linear power spectrum (and, if necessary, late-time evolution). A wide variety of alternative cosmologies have been considered, such as non-flat universes, varying dark energy, early dark energy, varying $N_{\rm eff}$, interacting dark energy, quintessence, *et cetera* [22, 49, 53], full-shape BOSS [usually with Planck]: [79–87]. In general, these give consistent constraints to those in ΛCDM but with broader errors, due to the additional degrees of freedom in the model.

17.5 Summary and Future Prospects

Spectroscopic surveys can be utilized to provide stringent constraints on the Universe's structure, composition, and expansion rate. In this chapter, we have discussed their ability to measure the latter, through standard rulers provided by baryon acoustic oscillations and the transition from radiation- to matter-domination. In each case, one observes a characteristic angular size in the galaxy distribution (either directly or indirectly) which is compared to a theoretical computation, with the ratio giving an (inverse) absolute distance measurement, at known redshift. Historically, interest has focussed on the BAO signal analyzed by way of scaling templates; more recently, theoretical and computational advances have led to H_0 being directly extracted from the spectrum (through implicit standard ruler constraints) using full-shape methodologies. To date, these measurements have allowed for tight bounds on the Universe's expansion rate, with the BOSS survey yielding CMB-independent $\sim 1.6\%$ constraints on H_0 (e.g., [68]), which is of comparable width to the *Planck* posterior.

It is interesting to discuss how such measurements will evolve in the future (see Fig. 17.3 for a selection of representative forecasts). In the coming years, both DESI

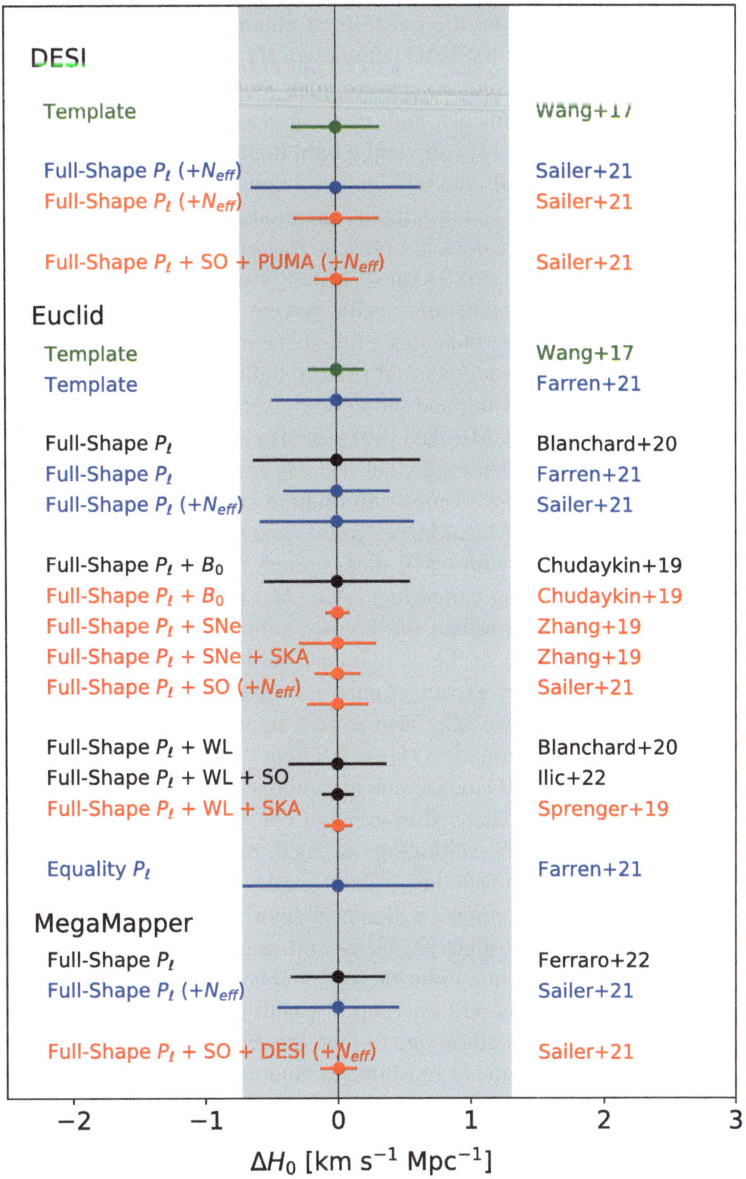

Fig. 17.3 As Fig. 17.1, but displaying a collection of H_0 forecasts for future experiments from the literature. As before, constraints marked in blue include (current) BBN priors on ω_b, those in green add *Planck* priors on ω_b, whilst those in red combine with the full *Planck* posterior. Here, we group forecasts by instrument, with the captions denoting the forecasting set-up. 'WL' indicates the inclusion of weak lensing information, whilst SO and SKA refer to the Simons Observatory and the Square Kilometer Array

[3] and Euclid [6] will measure the redshifts of hundreds of millions of galaxies, providing precise mapping of the BAO parameters $H(z)r_d$ and $D_A(z)/r_d$ from low to intermediate redshifts; moreover, surveys such as SPHEREx [5] will add significant low-redshift measurements, and, more futuristically, proposed space missions including MegaMapper [4] can yield a tight high-redshift extension. According to [52], five years of DESI data will lead to a tight BAO-derived H_0 constraint of $\sigma(H_0) = 0.33\,\mathrm{km\,s^{-1}Mpc^{-1}}$ via template methods, working at fixed ω_b (noting that the DESI white paper does not present H_0 constraints directly [3]), whilst the same paper gives $\sigma(H_0) = 0.21\,\mathrm{km\,s^{-1}Mpc^{-1}}$ from *Euclid*. As before, there are a plethora of forecasted constraints, with varying analysis choices: for example, the official Euclid forecast indicates a full-shape power spectrum constraint of $0.63\,\mathrm{km\,s^{-1}Mpc^{-1}}$ [88] (without external priors), tightening to $0.37\,\mathrm{km\,s^{-1}Mpc^{-1}}$ in conjunction with weak lensing and photometric surveys, or $0.12\,\mathrm{km\,s^{-1}Mpc^{-1}}$, additionally adding in future CMB data [89] (see also [48, 90–93]). For SPHEREx, no official H_0 forecast has been made, but one can expect similar results to DESI and Euclid spectroscopy, but with most information coming from lower redshifts. In the future, surveys such as MegaMapper could yield $\sigma(H_0) = 0.36\,\mathrm{km\,s^{-1}Mpc^{-1}}$ alone, or three times tighter with CMB data, though the precise forecasts depend on which parameters are being varied (e.g., $N_{\rm eff}$, M_ν) [93, 94]. We note that many of these results are naturally reliant on the ω_b calibration (with BBN yielding a $\approx 1.6\%$ constraint [33]), thus, if this is tightened further (for example from better BBN measurements), one may expect a tighter increase still.

From the equality scale, we may also expect significant improvements in the future. A forecast for Euclid found $\sigma(H_0) = 0.72\,\mathrm{km\,s^{-1}Mpc^{-1}}$ [48], which is only some 40% tighter than the BAO measurement (without reconstruction). We find that the two scales will be more balanced in terms of constraining power in the future, in the absence of tighter BBN calibration. As such, many of the above full-shape constraints are importantly impacted by equality-scale information. Finally, we note that a variety of other measurements will appear soon, both from additional galaxy surveys, such as Rubin and Roman [7, 8], as well as other probes including 21cm emission, line-intensity mapping, radio surveys, and beyond.

The coming measurements will open up a wealth of tests of the cosmological model. Firstly, BAO analyses allow one to map the $H(z)r_d$ and $D_A(z)/r_d$ derived parameters across a large swathe of redshifts, facilitating robust consistency tests of the low-redshift Universe; the same is true for full-shape derived constraints if we search for trends in the H_0 posteriors from different redshift samples. Secondly, the consistency of equality and BAO H_0 constraints has been previously shown to be a powerful diagnostic test of new physics in the early Universe, due to the different dependence on redshifts and recombination-scale physics. For now, the tests are weak, but with future surveys such as Euclid, one could potentially rule out many such models [48]. Furthermore, the two rulers are symbiotic; it is only by combining all sets of information available that we can obtain the strongest bounds on H_0, and thus physics new and old.

Acknowledgements We thank Eric Baxter, Giovanni Cabass, Gerrit Farren, Blake Sherwin, Marko Simonovic, and Matias Zaldarriaga, with whom many of the above H_0 constraints were derived. OHEP thanks Emirates for hosting a visit during which the majority of this draft was written. OHEP is a Junior Fellow of the Simons Society of Fellows and thanks the Simons Foundation for support.

References

1. P.J.E. Peebles, Large scale clustering in the universe, in *Large Scale Structures in the Universe*, ed. by M.S. Longair, J. Einasto (1978), vol. 79, p. 217
2. SDSS collaboration, SDSS-III: massive spectroscopic surveys of the distant universe, the milky way galaxy, and extra-solar planetary systems. Astron. J. **142**, 72 (2011) [1101.1529]. https://doi.org/10.1088/0004-6256/142/3/72
3. DESI collaboration, *The DESI Experiment Part I: Science,Targeting, and Survey Design*, 1611.00036
4. D.J. Schlegel et al., Astro2020 APC white paper: the MegaMapper: a z > 2 spectroscopic instrument for the study of inflation and dark energy. Bull. Am. Astron. Soc. **51**, 229 (2019) [1907.11171]
5. O. Doré et al., Cosmology with the SPHEREX all-sky spectral survey, 1412.4872
6. EUCLID collaboration, *Euclid Definition Study Report*, 1110.3193
7. LSST collaboration, Large synoptic survey telescope: overview. Proc. SPIE Int. Soc. Opt. Eng. **4836**, 10 (2002) [astro-ph/0302102] https://doi.org/10.1117/12.456772
8. J. Green et al., *Wide-Field InfraRed Survey Telescope (WFIRST) Final Report*, 1208.4012
9. E.J. Baxter, B.D. Sherwin, Determining the hubble constant without the sound horizon scale: measurements from CMB lensing. Mon. Not. Roy. Astron. Soc. **501**, 1823 (2021) [2007.04007]. https://doi.org/10.1093/mnras/staa3706
10. D.J. Eisenstein, H.-J. Seo, M.J. White, On the robustness of the acoustic scale in the low-redshift clustering of matter. Astrophys. J. **664**, 660 (2007) [astro-ph/0604361]. https://doi.org/10.1086/518755
11. SDSS collaboration, Detection of the Baryon acoustic peak in the large-scale correlation function of SDSS luminous red galaxies. Astrophys. J. **633**, 560 (2005) [astro-ph/0501171]. https://doi.org/10.1086/466512
12. A. Cuceu, J. Farr, P. Lemos, A. Font-Ribera, Baryon acoustic oscillations and the hubble constant: past, present and future. JCAP **10**, 044 (2019) [1906.11628]. https://doi.org/10.1088/1475-7516/2019/10/044
13. L. Senatore, M. Zaldarriaga, The IR-resummed effective field theory of large scale structures, JCAP **02**, 013 (2015) [1404.5954]. https://doi.org/10.1088/1475-7516/2015/02/013
14. T. Baldauf, M. Mirbabayi, M. Simonović, M. Zaldarriaga, Equivalence principle and the Baryon acoustic peak. Phys. Rev. D **92**, 043514 (2015) [1504.04366]. https://doi.org/10.1103/PhysRevD.92.043514
15. D. Blas, M. Garny, M.M. Ivanov, S. Sibiryakov, Time-sliced perturbation theory for large scale structure I: general formalism, JCAP **07**, 052 (2016) [1512.05807]. https://doi.org/10.1088/1475-7516/2016/07/052
16. D. Blas, M. Garny, M.M. Ivanov, S. Sibiryakov, Time-sliced perturbation theory II: Baryon acoustic oscillations and infrared resummation, JCAP **07**, 028 (2016) [1605.02149]. https://doi.org/10.1088/1475-7516/2016/07/028
17. M.M. Ivanov, S. Sibiryakov, Infrared resummation for biased tracers in redshift space. JCAP **07**, 053 (2018) [1804.05080]. https://doi.org/10.1088/1475-7516/2018/07/053
18. M. Crocce, R. Scoccimarro, Renormalized cosmological perturbation theory. Phys. Rev. D **73**, 063519 (2006) [astro-ph/0509418]. https://doi.org/10.1103/PhysRevD.73.063519

19. M. Crocce, R. Scoccimarro. Nonlinear evolution of Baryon acoustic oscillations. Phys. Rev. D **77**, 023533 (2008) [0704.2783]. https://doi.org/10.1103/PhysRevD.77.023533
20. R.E. Smith, R. Scoccimarro, R.K. Sheth, Eppur Si Muove: on the motion of the acoustic peak in the correlation function. Phys. Rev. D **77**, 043525 (2008) [astro-ph/0703620]. https://doi.org/10.1103/PhysRevD.77.043525
21. B.D. Sherwin, M. Zaldarriaga, The shift of the Baryon acoustic oscillation scale: a simple physical picture. Phys. Rev. D **85**, 103523 (2012) [1202.3998]. https://doi.org/10.1103/PhysRevD.85.103523
22. BOSS collaboration, The clustering of galaxies in the completed SDSS-III Baryon oscillation spectroscopic survey: cosmological analysis of the DR12 galaxy sample. Mon. Not. Roy. Astron. Soc. **470**, 2617 (2017) [1607.03155]. https://doi.org/10.1093/mnras/stx721
23. C. Alcock, B. Paczynski, An evolution free test for non-zero cosmological constant. Nature **281**, 358 (1979). https://doi.org/10.1038/281358a0
24. N. Schöneberg, J. Lesgourgues, D.C. Hooper, The BAO+BBN take on the hubble tension. JCAP **10**, 029 (2019) [1907.11594]. https://doi.org/10.1088/1475-7516/2019/10/029
25. D.J. Eisenstein, H.-J. Seo, E. Sirko, D. Spergel, Improving cosmological distance measurements by reconstruction of the Baryon acoustic peak. Astrophys. J. **664**, 675 (2007) [astro-ph/0604362]. https://doi.org/10.1086/518712
26. M.M. Ivanov, M. Simonović, M. Zaldarriaga, Cosmological parameters from the BOSS galaxy power spectrum. JCAP **05**, 042 (2020) [1909.05277]. https://doi.org/10.1088/1475-7516/2020/05/042
27. G. D'Amico, J. Gleyzes, N. Kokron, K. Markovic, L. Senatore, P. Zhang et al., The cosmological analysis of the SDSS/BOSS data from the effective field theory of large-scale structure. JCAP **05**, 005 (2020) [1909.05271]. https://doi.org/10.1088/1475-7516/2020/05/005
28. D. Baumann, A. Nicolis, L. Senatore, M. Zaldarriaga, Cosmological non-linearities as an effective fluid. JCAP **07**, 051 (2012) [1004.2488]. https://doi.org/10.1088/1475-7516/2012/07/051
29. J.J.M. Carrasco, M.P. Hertzberg, L. Senatore, The effective field theory of cosmological large scale structures. JHEP **09**, 082 (2012) [1206.2926]. https://doi.org/10.1007/JHEP09(2012)082
30. A. Chudaykin, M.M. Ivanov, O.H.E. Philcox, M. Simonović, Nonlinear perturbation theory extension of the Boltzmann code CLASS. Phys. Rev. D **102**, 063533 (2020) [2004.10607]. https://doi.org/10.1103/PhysRevD.102.063533
31. M.M. Ivanov, Effective field theory for large scale structure. 2212.08488
32. G. Cabass, M.M. Ivanov, M. Lewandowski, M. Mirbabayi, M. Simonović, Snowmass white paper: effective field theories in cosmology, in *2022 Snowmass Summer Study*, 3 (2022) 2203.08232
33. E. Aver, K.A. Olive, E.D. Skillman, The effects of He I λ 10830 on helium abundance determinations. JCAP **07**, 011 (2015) [1503.08146]. https://doi.org/10.1088/1475-7516/2015/07/011
34. C. Hikage, K. Koyama, A. Heavens, Perturbation theory for BAO reconstructed fields: one-loop results in the real-space matter density field. Phys. Rev. D **96**, 043513 (2017) [1703.07878]. https://doi.org/10.1103/PhysRevD.96.043513
35. S.-F. Chen, Z. Vlah, M. White, The reconstructed power spectrum in the Zeldovich approximation. JCAP **09**, 017 (2019) [1907.00043]. https://doi.org/10.1088/1475-7516/2019/09/017
36. C. Hikage, K. Koyama, R. Takahashi, Perturbation theory for the redshift-space matter power spectra after reconstruction. Phys. Rev. D **101**, 043510 (2020) [1911.06461]. https://doi.org/10.1103/PhysRevD.101.043510
37. A. Ota, H.-J. Seo, S. Saito, F. Beutler, Modeling iterative reconstruction and displacement field in the large scale structure. Phys. Rev. D **104**, 123508 (2021) [2106.46]. https://doi.org/10.1103/PhysRevD.104.123508
38. M. Schmittfull, T. Baldauf, M. Zaldarriaga, Iterative initial condition reconstruction. Phys. Rev. D **96**, 023505 (2017) [1704.06634]. https://doi.org/10.1103/PhysRevD.96.023505

39. M. Schmittfull, Y. Feng, F. Beutler, B. Sherwin, M.Y. Chu, Eulerian BAO reconstructions and N-point statistics. Phys. Rev. D **92**, 123522 (2015) [1508.06972]. https://doi.org/10.1103/PhysRevD.92.123522
40. O.H.E. Philcox, M.M. Ivanov, M. Simonović, M. Zaldarriaga, Combining full-shape and BAO analyses of galaxy power spectra: a 1.6% CMB-independent constraint on H_0. JCAP **05**, 032 (2020) [2002.04035]. https://doi.org/10.1088/1475-7516/2020/05/032
41. S.-F. Chen, Z. Vlah, M. White, A new analysis of galaxy 2-point functions in the BOSS survey, including full-shape information and post-reconstruction BAO. JCAP **02**, 008 (2022) [2110.05530]. https://doi.org/10.1088/1475-7516/2022/02/008
42. M.M. Ivanov, Y. Ali-Haïmoud, J. Lesgourgues, H_0 tension or T_0 tension?. Phys. Rev. D **102**, 063515 (2020) [2005.10656]. https://doi.org/10.1103/PhysRevD.102.063515
43. D.J. Eisenstein, W. Hu, Baryonic features in the matter transfer function, Astrophys. J. **496**, 605 (1998) [astro-ph/9709112]. https://doi.org/10.1086/305424
44. M. Tegmark, Measuring cosmological parameters with galaxy surveys, Phys. Rev. Lett. **79**, 3806 (1997) [astro-ph/9706198]. https://doi.org/10.1103/PhysRevLett.79.3806
45. 2dFGRS collaboration, The 2dF galaxy redshift survey: the power spectrum and the matter content of the universe. Mon. Not. Roy. Astron. Soc. **327**, 1297 (2001) [astro-ph/0105252]. https://doi.org/10.1046/j.1365-8711.2001.04827.x
46. S. Brieden, H. Gil-Marín, L. Verde, ShapeFit: extracting the power spectrum shape information in galaxy surveys beyond BAO and RSD. JCAP **12**, 054 (2021) [2106.07641]. https://doi.org/10.1088/1475-7516/2021/12/054
47. O.H.E. Philcox, B.D. Sherwin, G.S. Farren, E.J. Baxter, Determining the hubble constant without the sound horizon: measurements from galaxy surveys. Phys. Rev. D **103**, 023538 (2021) [2008.08084]. https://doi.org/10.1103/PhysRevD.103.023538
48. G.S. Farren, O.H.E. Philcox, B.D. Sherwin, Determining the hubble constant without the sound horizon: perspectives with future galaxy surveys. Phys. Rev. D **105**, 063503 (2022) [2112.10749]. https://doi.org/10.1103/PhysRevD.105.063503
49. eBOSS collaboration, Completed SDSS-IV extended Baryon oscillation spectroscopic survey: cosmological implications from two decades of spectroscopic surveys at the Apache point observatory. Phys. Rev. D **103**, 083533 (2021) [2007.08991]. https://doi.org/10.1103/PhysRevD.103.083533
50. M. Blomqvist et al., Baryon acoustic oscillations from the cross-correlation of Lyα absorption and quasars in eBOSS DR14. Astron. Astrophys. **629**, A86 (2019) [1904.03430]. https://doi.org/10.1051/0004-6361/201935641
51. N. Schöneberg, L. Verde, H. Gil-Marín, S. Brieden, BAO+BBN revisited—growing the hubble tension with a 0.7 km/s/Mpc constraint. JCAP **11**, 039 (2022) [2209.14330]. https://doi.org/10.1088/1475-7516/2022/11/039
52. Y. Wang, L. Xu, G.-B. Zhao, A measurement of the hubble constant using galaxy redshift surveys, Astrophys. J. **849**, 84 (2017) [1706.09149]. https://doi.org/10.3847/1538-4357/aa8f48
53. S. Brieden, H. Gil-Marín, L. Verde, Model-agnostic interpretation of 10 billion years of cosmic evolution traced by BOSS and eBOSS data, JCAP **08**, 024 (2022) [2204.11868]. https://doi.org/10.1088/1475-7516/2022/08/024
54. T. Colas, G. D'amico, L. Senatore, P. Zhang, F. Beutler, Efficient cosmological analysis of the SDSS/BOSS data from the effective field theory of large-scale structure. JCAP **06**, 001 (2020) [1909.07951]. https://doi.org/10.1088/1475-7516/2020/06/001
55. T. Tröster et al., Cosmology from large-scale structure: constraining ΛCDM with BOSS. Astron. Astrophys. **633**, L10 (2020) [1909.11006]. https://doi.org/10.1051/0004-6361/201936772
56. D. Wadekar, M.M. Ivanov, R. Scoccimarro, Cosmological constraints from BOSS with analytic covariance matrices. Phys. Rev. D **102**, 123521 (2020) [2009.00622]. https://doi.org/10.1103/PhysRevD.102.123521
57. M.M. Ivanov, M. Simonović, M. Zaldarriaga, Cosmological parameters and neutrino masses from the final planck and full-shape BOSS data. Phys. Rev. D **101**, 083504 (2020) [1912.08208]. https://doi.org/10.1103/PhysRevD.101.083504

58. G. D'Amico, L. Senatore, P. Zhang, H. Zheng, The hubble tension in light of the full-shape analysis of large-scale structure data. JCAP **05**, 072 (2021) [2006.12420]. https://doi.org/10.1088/1475-7516/2021/05/072
59. P. Zhang, G. D'Amico, L. Senatore, C. Zhao, Y. Cai, BOSS correlation function analysis from the effective field theory of large-scale structure. JCAP **02**, 036 (2022) [2110.07539]. https://doi.org/10.1088/1475-7516/2022/02/036
60. M.M. Ivanov, O.H.E. Philcox, M. Simonović, M. Zaldarriaga, T. Nischimichi, M. Takada, Cosmological constraints without nonlinear redshift-space distortions. Phys. Rev. D **105**, 043531 (2022) [2110.00006]. https://doi.org/10.1103/PhysRevD.105.043531
61. G. D'Amico, L. Senatore, P. Zhang, T. Nishimichi, *Taming redshift-space distortion effects in the EFTofLSS and its application to data* [2110.00016]
62. A.J.S. Hamilton, M. Tegmark, The real space power spectrum of the PSCz survey from 0.01 to 300 h Mpc**-1. Mon. Not. Roy. Astron. Soc. **330**, 506 (2002) astro-ph/0008392. https://doi.org/10.1046/j.1365-8711.2002.05033.x, https://arxiv.org/abs/astro-ph/0008392
63. M. Tegmark, A.J.S. Hamilton, Y.-Z. Xu, The power spectrum of galaxies in the 2dF 100k redshift survey. Mon. Not. Roy. Astron. Soc. **335**, 887 (2002) astro-ph/0111575. https://doi.org/10.1046/j.1365-8711.2002.05622.x, https://arxiv.org/abs/astro-ph/0111575
64. SDSS collaboration, The 3-D power spectrum of galaxies from the SDSS, Astrophys. J. **606**, 702 (2004) astro-ph/0310725. https://doi.org/10.1086/382125, https://arxiv.org/abs/astro-ph/0310725
65. R. Scoccimarro, Redshift-space distortions, pairwise velocities and nonlinearities, Phys. Rev. D **70**, 083007 (2004) astro-ph/0407214. https://doi.org/10.1103/PhysRevD.70.083007, https://arxiv.org/abs/astro-ph/0407214
66. S.-F. Chen, M. White, J. DeRose, N. Kokron, Cosmological analysis of three-dimensional BOSS galaxy clustering and Planck CMB lensing cross correlations via Lagrangian perturbation theory. JCAP **07**, 041 (2022) [2204.10392]. https://doi.org/10.1088/1475-7516/2022/07/041
67. O.H.E. Philcox, M.M. Ivanov, BOSS DR12 full-shape cosmology: ΛCDM constraints from the large-scale galaxy power spectrum and bispectrum monopole. Phys. Rev. D **105**, 043517 (2022) [2112.04515]. https://doi.org/10.1103/PhysRevD.105.043517
68. M.M. Ivanov, O.H.E. Philcox, G. Cabass, T. Nishimichi, M. Simonović, M. Zaldarriaga, Cosmology with the galaxy bispectrum multipoles: optimal estimation and application to BOSS data [2302.04414]
69. G. D'Amico, Y. Donath, M. Lewandowski, L. Senatore, P. Zhang, The BOSS bispectrum analysis at one loop from the effective field theory of large-scale structure [2206.08327]
70. M.M. Ivanov, Cosmological constraints from the power spectrum of eBOSS emission line galaxies. Phys. Rev. D **104**, 103514 (2021) [2106.12580]. https://doi.org/10.1103/PhysRevD.104.103514
71. A. Semenaite et al., Cosmological implications of the full shape of anisotropic clustering measurements in BOSS and eBOSS. Mon. Not. Roy. Astron. Soc. **512**, 5657 (2022) [2111.03156]. https://doi.org/10.1093/mnras/stac829
72. A. Semenaite, A.G. Sánchez, A. Pezzotta, J. Hou, A. Eggemeier, M. Crocce et al., Beyond ΛCDM constraints from the full shape clustering measurements from BOSS and eBOSS [2210.07304]
73. A. Chudaykin, M.M. Ivanov, Cosmological constraints from the power spectrum of eBOSS quasars [2210.17044]
74. T. Simon, P. Zhang, V. Poulin, Cosmological inference from the EFTofLSS: the eBOSS QSO full-shape analysis [2210.14931]
75. R. Neveux et al., Combined full shape analysis of BOSS galaxies and eBOSS quasars using an iterative emulator. Mon. Not. Roy. Astron. Soc. **516**, 1910 (2022) [2201.04679]. https://doi.org/10.1093/mnras/stac2114
76. S. Brieden, H. Gil-Marín, L. Verde, A tale of two (or more) *h's* 2212.04522
77. B. Bahr-Kalus, D. Parkinson, E.-M. Mueller, Measurement of the matter-radiation equality scale using the extended Baryon oscillation spectroscopic survey quasar sample [2302.07484]

78. T.L. Smith, V. Poulin, T. Simon, Assessing the robustness of sound horizon-free determinations of the Hubble constant [2208.12992]
79. A. Chudaykin, K. Dolgikh, M.M. Ivanov, Constraints on the curvature of the universe and dynamical dark energy from the full-shape and BAO data. Phys. Rev. D **103**, 023507 (2021) [2009.10106]. https://doi.org/10.1103/PhysRevD.103.023507
80. A. Chudaykin, D. Gorbunov, N. Nedelko, Exploring ΛCDM extensions with SPT-3G and Planck data: 4σ evidence for neutrino masses, full resolution of the hubble crisis by dark energy with phantom crossing, and all that [2203.03666]
81. S. Vagnozzi, E. Di Valentino, S. Gariazzo, A. Melchiorri, O. Mena, J. Silk, The galaxy power spectrum take on spatial curvature and cosmic concordance. Phys. Dark Univ. **33**, 100851 (2021) [2010.02230]. https://doi.org/10.1016/j.dark.2021.100851
82. S. Kumar, R.C. Nunes, P. Yadav, Updating non-standard neutrinos properties with Planck-CMB data and full-shape analysis of BOSS and eBOSS galaxies. JCAP **09**, 060 (2022) [2205.04292]. https://doi.org/10.1088/1475-7516/2022/09/060
83. A. Reeves, L. Herold, S. Vagnozzi, B.D. Sherwin, E.G.M. Ferreira, Restoring cosmological concordance with early dark energy and massive neutrinos?. Mon. Not. Roy. Astron. Soc. **520**, 3688 (2023) [2207.01501]. https://doi.org/10.1093/mnras/stad317
84. G. D'Amico, L. Senatore, P. Zhang, Limits on wCDM from the EFTofLSS with the PyBird code. JCAP **01**, 006 (2021) [2003.07956]. https://doi.org/10.1088/1475-7516/2021/01/006
85. G. D'Amico, Y. Donath, L. Senatore, P. Zhang, Limits on clustering and smooth quintessence from the EFTofLSS [2012.07554]
86. M.M. Ivanov, E. McDonough, J.C. Hill, M. Simonović, M.W. Toomey, S. Alexander et al., Constraining early dark energy with large-scale structure. Phys. Rev. D **102**, 103502 (2020) [2006.11235]. https://doi.org/10.1103/PhysRevD.102.103502
87. T. Simon, P. Zhang, V. Poulin, T.L. Smith, Updated constraints from the effective field theory analysis of BOSS power spectrum on early dark energy [2208.05930]
88. EUCLID collaboration, Euclid preparation: VII. Forecast validation for Euclid cosmological probes. Astron. Astrophys. **642**, A191 (2020) [1910.09273]. https://doi.org/10.1051/0004-6361/202038071
89. EUCLID collaboration, Euclid preparation—XV. Forecasting cosmological constraints for the Euclid and CMB joint analysis. Astron. Astrophys. **657**, A91 (2022) [2106.08346]. https://doi.org/10.1051/0004-6361/202141556
90. A. Chudaykin, M.M. Ivanov, Measuring neutrino masses with large-scale structure: Euclid forecast with controlled theoretical error. JCAP **11**, 034 (2019) [1907.06666]. https://doi.org/10.1088/1475-7516/2019/11/034
91. J.-F. Zhang, L.-Y. Gao, D.-Z. He, X. Zhang, Improving cosmological parameter estimation with the future 21 cm observation from SKA. Phys. Lett. B **799**, 135064 (2019) [1908.03732]. https://doi.org/10.1016/j.physletb.2019.135064
92. T. Sprenger, M. Archidiacono, T. Brinckmann, S. Clesse, J. Lesgourgues, Cosmology in the era of Euclid and the square Kilometre array. JCAP **02**, 047 (2019) [1801.08331]. https://doi.org/10.1088/1475-7516/2019/02/047
93. N. Sailer, E. Castorina, S. Ferraro, M. White, Cosmology at high redshift—a probe of fundamental physics. JCAP **12**, 049 (2021) [2106.09713]. https://doi.org/10.1088/1475-7516/2021/12/049
94. S. Ferraro, N. Sailer, A. Slosar, M. White, Snowmass2021 cosmic frontier white paper: cosmology and fundamental physics from the three-dimensional large scale structure [2203.07506]

Part III
Complications with Measurements

Chapter 18
Impact of Peculiar Velocities on Measurements of H_0

W. D'Arcy Kenworthy and Tamara M. Davis

18.1 Introduction

The most direct method to measure the Hubble constant, H_0, is to compare the current proper distance, D_0, to the current recession velocity, v_0, to get $H_0 = v_0/D_0$. All direct measures of H_0 use some variant of this basic idea. However, to derive H_0 accurately it is important that measurements of the recession velocity are not contaminated by (unbudgeted) local motions induced by inhomogeneities in the Universe. Motions that deviate from the homogeneous, isotropic expansion are known as *peculiar velocities*. On large scales peculiar velocities arise primarily due to gravitational acceleration. The impact of peculiar velocities on measurements of H_0 are the focus of this chapter.

The most prominent example of a direct H_0 measurement involves luminosity distances inferred from standard candles such as supernovae, calibrated using a 'distance ladder'. These are local measurements that directly infer recession velocities from observed redshifts and compare those to measured distances.

By contrast, indirect methods involve fitting an entire cosmological model to data that has some dependence on H_0. The most prominent example uses the patterns in the cosmic microwave background (CMB), which are dependent on H_0 but strongly correlated with other parameters, so H_0 only appears as one parameter in a fit to a global cosmological model.

Both direct and indirect methods assume an underlying on-average homogeneous and isotropic universe (even when they are using inhomogeneities as the signal!). However, we know fluctuations exist—after all, we are here—and these fluctuations

W. D'Arcy Kenworthy (✉)
Stockholm University, AlbaNova University Center, Fysikum, Stockholm 10691, Sweden
e-mail: darcykenworthy@me.com

T. M. Davis
School of Mathematics and Physics, The university of Queensland, Brisbane, QLD 4072, Australia
e-mail: tamarad@physics.uq.edu.au

E. Di Valentino and D. Brout (eds.), *The Hubble Constant Tension*, Springer Series in Astrophysics and Cosmology, https://doi.org/10.1007/978-981-99-0177-7_18

can cause biases in our measurements of cosmological parameters unless properly accounted for. Currently there is a tension between direct and indirect methods of measuring H_0, so it is critical to understand whether any inhomogeneities are biasing our cosmological inferences.

In this chapter we review the impact of peculiar velocities on our measurements of H_0. We begin by summarising the theoretical basis for peculiar velocities. We then look at the potential impact peculiar velocities can have on cosmological measurements, and ways to mitigate that impact to give as pure a measurement of H_0 as possible.

18.2 Theoretical Basis

The distance in the Hubble-Lemaître law is usually split into two components, the scalefactor R and the comoving coordinate χ, such that $D = R\chi$[1]. Any motion due to a change in the scalefactor, which affects all distances, is a recession velocity $v_{\rm rec} = \dot{R}\chi = HD$ (where an overdot represents differentiation with respect to time[2]), and the Hubble parameter $H \equiv \dot{R}/R$. While any motion due to movement *through* comoving coordinates is a peculiar velocity, $v_{\rm pec} = R\dot{\chi}$. Although the Hubble-Lemaître law holds at all times we often evaluate these values at the present day, which we indicate by the subscript 0.

There are two primary sources of peculiar velocities to worry about. Namely, our own motion, $v_{\rm pec}^{\rm Sun}$, and the sources' motion, $v_{\rm pec}^{\rm source}$. Both are measured with respect to the cosmic 'rest frame', which is the frame in which the cosmic microwave background (CMB) radiation appears isotropic.

18.2.1 Inferring Peculiar Velocities

Peculiar velocities can either be used as a cosmological probe in their own right, or measured in order to mitigate their impact on other cosmological probes.

Inferring peculiar velocities from observables needs two measured quantities—the observed redshift $z_{\rm obs}$ and an observed distance measure $D_{\rm obs}$. The distance measure could be any measure (e.g. luminosity or angular diameter distance) from which we can infer the true cosmological redshift.

To derive the recession velocity (the velocity that appears in the Hubble-Lemaître Law, defined above) from the cosmological redshift one needs to know the values of the cosmological parameters that accelerate and decelerate the universe such as the matter density $\Omega_{\rm m}$, dark energy density $\Omega_{\rm x}$, and dark energy equation of state, w. The basic equation for recession velocity as a function of cosmological redshift is,

[1] Throughout this work we assume a flat cosmology on large scales. In non-flat cosmologies, the analysis is only marginally more complex.

[2] The time in the Friedmann-Lemaître-Robertson-Walker metric.

$$v_{\rm rec} = c \int_0^{z_{\rm cos}} \frac{dz}{E(z)} \tag{18.1}$$

where from Friedmann's equation we use $E(z) = H(z)/H_0 = \left[\Omega_{\rm m}(1+z)^3 + \Omega_{\rm x}(1+z)^{3(1+w)} + \Omega_{\rm k}(1+z)^2\right]^{1/2}$, where $\Omega_{\rm k} = 1 - \sum_i \Omega_i$. Importantly, you can see that recession velocity as a function of redshift is *independent* of H_0. You only need H_0 if you want to know how velocity depends on distance.

A good approximation for the recession velocity that is often used is,

$$v_{\rm rec} = \frac{c z_{\rm cos}}{1+z_{\rm cos}} \left[1 + \frac{1}{2}(1-q_0)z_{\rm cos} - \frac{1}{6}(1 - q_0 - 3q_0^2 + j_0)z_{\rm cos}^2\right], \tag{18.2}$$

where in the standard cosmological model the deceleration parameter is $q_0 \sim -0.55$, and the jerk is $j_0 \sim 1.0$. This approximation is what is used in the SH_0ES papers (e.g. [1, 2]) for deriving H_0 from supernova distances and redshifts.

However, we do not have direct access to the true recession redshift, $z_{\rm cos}$. Instead we have the observed redshift, $z_{\rm obs}$, which is impacted by peculiar velocities—both our own and that of the source.

The total velocity is the sum of the peculiar and recession velocities, $v_{\rm tot} = v_{\rm rec} + v_{\rm pec}$. So in many direct measurements of peculiar velocity you see this rearranged to become $v_{\rm pec} = v_{\rm tot} - v_{\rm rec} \approx c z_{\rm obs} - H_0 D$, where $v_{\rm rec}$ was replaced with $H_0 D$ from the Hubble-Lemaître law, and the total velocity was approximated by $c z_{\rm obs}$. However, this is a terrible approximation even by redshift $z \sim 0.1$ where it will give a $\sim$700 km s^{-1} error in the derived peculiar velocity [3].[3]

A more accurate equation for expressing velocity in terms of observed redshift and observed distance is,

$$v_{\rm pec} = c\left(\frac{z_{\rm obs} - z_{\rm cos}(D_{\rm obs})}{1 + z_{\rm cos}(D_{\rm obs})}\right), \tag{18.3}$$

where $z_{\rm cos}$ is the cosmological redshift that has been inferred from the observed distance measure $D_{\rm obs}$. Calculating $z_{\rm cos}$ from a distance measure typically requires a cosmological model.[4] However, at low redshift ($z \lesssim 0.1$) the cosmological dependence is weak and any model that approximates the expansion evolution well (e.g. Eq. 18.2 is sufficient).

Direct measures of peculiar velocities are expensive (because you need to measure distances) and have large errors (because distance measurements have large uncertainties). Therefore a more precise (but modelling-dependent) way to predict the peculiar velocity of distance sources is to reconstruct a peculiar velocity field

---

[3] Using this equation is equivalent to approximating Eq. 18.5 by $z_{\rm obs} = z_{\rm cos} + z_{\rm pec}$, which neglects a factor of $z_{\rm cos} z_{\rm pec}$. It therefore gets worse linearly with redshift. Note that most modern analyses do not actually use this approximation, even if it still appears in the standard introduction in many papers.

[4] Just to describe the shape of the magnitude redshift relation, i.e. parameters such as $\Omega_{\rm m}$ and Ω_Λ but not a value of H_0.

from the density contrast, $\delta(\vec{r}) \equiv (\rho(\vec{r}) - \bar{\rho})/\bar{\rho}$, where $\bar{\rho}$ is the mean density of the Universe and ρ is the density at position $\vec{r}$. Given a density field as measured by a large redshift survey, the large-scale peculiar velocity of a point at $\vec{r}_x$ can be derived by integrating over the gravitational attraction of all other points,

$$\vec{v}(\vec{r}_x) = \frac{f}{4\pi} \int \delta(\vec{r}) \frac{\vec{r} - \vec{r}_x}{|\vec{r} - \vec{r}_x|^3} d^3\vec{r}, \tag{18.4}$$

where f is the growth rate of structure. This equation is only valid in the linear regime, that is when the density fluctuations are small perturbations on the mean density of the Universe ($\delta \ll 1$). Thus a reconstruction from the density field will only accurately reproduce large scale peculiar velocities. It will not be able to reproduce small-scale motions, such as the orbits of galaxies in a galaxy cluster (see Sect. 18.3.4).

In Sect. 18.4.2 we show how these peculiar velocity reconstructions can be used to mitigate the impact of peculiar velocities on our cosmological measurements.

18.2.2 The Impact of Peculiar Velocities on Observables

The impact of peculiar velocities on cosmological measurements arises in a number of ways. Most significantly they generate an additional peculiar redshift z_{pec} which causes the observed redshift, z_{obs}, to deviate from the true cosmological (recession) redshift z_{cos} according to [4],

$$1 + z_{\text{obs}} = (1 + z_{\text{cos}})(1 + z_{\text{pec}}). \tag{18.5}$$

The luminosity distance (defined as the distance that fulfills Flux = Luminosity/ $(4\pi D_{\text{L}}^2)$) differs from the proper distance by a factor of $(1 + z_{\text{obs}})$ and curvature terms. Ignoring curvature and concentrating only on the differences impacted by peculiar velocity, we can write,

$$D_{\text{L}} = (1 + z_{\text{obs}})D. \tag{18.6}$$

The extra factor of $(1 + z_{\text{obs}})$ arises because the luminosity drops both due to time dilation decreasing the arrival rate of photons and the redshift decreasing the energy of each photon. Note that these factors both depend on the total redshift the photons experience, both cosmological and peculiar, which is why the $(1 + z_{\text{obs}})$ factor uses the observed redshift. In the absence of peculiar velocities it reduces to $D_L = (1 + z_{\text{cos}})D$.

Peculiar velocities also induce relativistic beaming—photons are focused in the direction of travel—which means the flux of photons is higher in the direction of motion than behind it. For a source with a peculiar velocity toward us we therefore infer both a lower redshift (relative blueshift) and a lower distance (brighter) than a comoving source at the same distance, so the effects slightly cancel (see [5], Fig. 1).

However, the beaming impact is an order of magnitude lower than the impact of the redshift shift, and is therefore negligible for most realistic peculiar velocities.

Beaming due to our own motion also slightly alters the distribution of distant sources, such that there is a higher number density of sources in the direction of our motion [6, 7]. This becomes relevant for high-precision baryon acoustic oscillation (BAO) and CMB measurements of H_0, and has been seen both in the number counts of distant galaxies (e.g. [8, 9]) and in the CMB [10, 11] at a measurable level.

The impact of peculiar velocities is largest for low-redshift probes, and when few sources are available. Such is the current case for gravitational wave standard sirens. For example, the peculiar velocity correction that was applied to the neutron-star merger measurement of H_0 lowered the estimated H_0 by ~ 7 km s^{-1} Mpc^{-1} [12] relative to the uncorrected measurement, and the large value of this correction motivated several reanalyses [13–15]. As the number of sources grows the impact of peculiar velocities tends to reduce, as their random directions average out.

18.3 Power Spectrum of Peculiar Velocities

Peculiar velocities are correlated across a range of scales and patterns, from small-scale infall around clusters, through correlations between large-scale structure, to bulk flows and expansion. It is therefore convenient to discuss the impact of peculiar velocities using a power spectrum, in which we separate the signal into multipoles. Here we discuss the monopole (expansion or contraction), dipole (motion in one direction), and higher-order multipoles (correlated motion on smaller scales).

18.3.1 Monopole

The lowest multipole is the monopole, which corresponds to spherical expansion or contraction. This could occur due to local over- or under-densities. Over the years evidence has appeared and disappeared for such a local 'Hubble bubble' (e.g. [16–21]). This is usually very difficult to disentangle from the effect of the cosmological expansion, and so is one of the most insidious potential peculiar velocity errors. However, to find ourselves sitting at the centre of an abnormally large over- or under-density is disfavoured by the cosmological principle, and if we sit off-centre we would see asymmetric effects that would reveal its presence. Observationally, recent measurements of the distance-redshift relation from type Ia supernovae show no evidence for a abnormal monopole sufficient to explain the Hubble tension [20, 21]. However, other measurements continue to report potentially anomolous structure (see Sect. 18.3.3 for examples). Measurements of the local 'gravitational watershed' of our Universe, known as Laniakea [22, 23], shows that we do likely sit at the edge (not centre) of an over-density.

In general, we can quantify the expected distribution of monopole effects from different locations in the standard model of cosmology, and they would be expected to be small see also Chap. 21 [24]. However, advocates of inhomogenous cosmological solutions to the Hubble tension usually invoke a density fluctuation that is larger than expected in the standard model.

18.3.2 Dipole

The second multipole is the dipole, which corresponds to our own motion with respect to the cosmic microwave background 'rest' frame. Measurements of the dipole in the CMB reveal that the Sun is moving at 369.82 ± 0.11 km s^{-1} with respect to the cosmic rest frame [25]. Interestingly our galaxy is moving faster, at 552.2 ± 5.5 km s^{-1}, but the Sun in rotation around the galactic centre is currently going backwards relative to the galaxy's motion, so the two partially cancel out.

The dipole can be measured not only with respect to the CMB, but also with respect to shells of matter at a range of redshifts—e.g. radio galaxies at high-redshift [8, 9, 26] or supernovae at low to intermediate redshifts [27, 28]. As the radius of the shell approaches the radius of the observable Universe you expect the dipole measurement to converge on that of the CMB. However, when the radius of the shell is very small, you expect the dipole to disappear, as we are measuring with respect to galaxies that share our local motion.

It is important (and straightforward, using Eq. 18.5) to remove the dipole from the observed redshift before calculating cosmological distances, and this is standard practice. Historically this has often been done using low-redshift approximations (akin to Eq. 18.12, but given the accuracy of modern cosmological measurements it is becoming important to use the full correction as in Eq. 18.1 [29, 30].

18.3.3 Bulk Flow

The bulk flow is also a manifestation of the second multipole, but is not to be confused with the dipole. The bulk flow refers to the average motion of all the galaxies (or total matter) in a volume of space. For example, in the standard cosmological model one expects the galaxies in a sphere of 100 Mpc$/h$ to have an average speed of approximately 180 km s^{-1} Mpc^{-1} relative to the CMB rest frame.

Many theoretical calculations predicting the magnitude of the bulk flow assume it to be a spherical volume, but in most cases a bulk flow can only be measured using data sets covering incomplete patches of the sky. Moreover, the completeness of a survey as a function of redshift has a significant impact on the bulk flow prediction. Mitigating the incomplete data is a major challenge to estimating accurate bulk flows and comparing them to the expectation from theory. Recent measurements have confirmed a long-standing trend that measured bulk flows are larger than expected in the standard cosmological model [31–35].

In contrast to the dipole, you expect the bulk flow to be significant at small scales, but disappear as you increase the radius to encompass the whole observable Universe. Interestingly bulk flows stay somewhat large even out to surprisingly large distances, the expectation of a spherical bulk flow of the volume within $z \sim 1$ (about half the radius of the observable Universe) would still be about 10 km s^{-1}.

18.3.4 Higher Order Motion

Higher-order multipoles can be quantified using the peculiar velocity power spectrum, which describes the amplitude of peculiar velocity correlations as a function of scale. The auto-correlation can be supplemented with the cross-correlation between peculiar velocities and densities, which contains additional information as galaxies fall toward over-dense regions.

Perturbations on very large scales are well described by linear perturbations on a smooth background. Linear perturbation theory allows you to essentially predict analytically how large scale fluctuations will behave. This is relevant for predicting the size of the bulk flow, for example. However, on small scales perturbation theory breaks down, and we enter the non-linear regime. On scales small enough that the Universe deviates significantly from homogeneity ($\delta \gtrsim 1$) we can no longer treat density fluctuations as small perturbations, and we need numerical simulations to estimate the pattern of density and velocity fluctuations.

Scales on which you see galaxies orbiting each other (such as galaxy clusters) are *highly* non-linear, subject to the virial theorem. However, non-linear density fluctuations begin to become important even on much larger scales than clusters; often nonlinear effects are present at scales $k \gtrsim 0.1\ h\mathrm{Mpc}^{-1}$, where $k \equiv 2\pi/|\vec{r}|$, corresponding to separations of $|r| \lesssim 60\ \mathrm{Mpc}\, h^{-1}$. When nonlinear scales dominate ($k \gtrsim 1\ h\mathrm{Mpc}^{-1}$) one cannot straightforwardly perform peculiar velocity reconstruction, limiting the resolution of our peculiar velocity maps (see Sect. 18.4.2).

18.3.5 Covariances

Non-linear behavior is, in general, more complicated than can be represented by Gaussian statistics, and requires use of higher order statistics to fully model. However, given that velocities are *less* sensitive to small-scale behavior as compared to densities or number-counts by a factor of k^2, non-linear behaviors are comparatively less impactful. If necessary, the effects of nonlinear contributions can also be evaluated by reference to full N-body simulations [24]. Nevertheless, in many analyses the non-linear or virial contribution is the component that is easiest to fit to data, typically representing merely a diagonal uncorrelated scatter on the Hubble diagram. It is

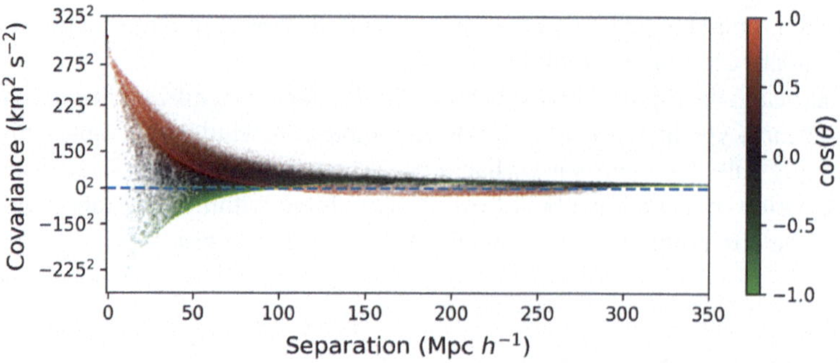

Fig. 18.1 Expected, uncorrected peculiar velocity covariance between pairs of galaxy in the low-redshift Pantheon+ sample [36]. Each point represents a pair of SN host galaxies, with their angular separation represented by color, and physical distance shown on the x-axis. Red points indicate points on the same line of sight, while green points are across the sky from one another. The power spectrum was evaluated with Planck parameters [37] and halofit corrections [38]

thus a reasonable approximation to use Gaussian statistics to quantify the effect of peculiar velocities on cosmography. The matter power spectrum is defined by the relation

$$< \tilde{\delta}(\vec{k}^*)\tilde{\delta}(\vec{k}) >= (2\pi)^3 P(|\vec{k}|)\delta^3(\vec{k} - \vec{k}^*) \tag{18.7}$$

where $\tilde{\delta}(\vec{k})$ is a Fourier transform of the density contrast $\delta = \Delta\rho/\bar{\rho}$, and δ^3 is the Dirac delta function in 3D. Using the theory of Gaussian statistics, you can then compute the covariance of the peculiar velocity between two objects at arbitrary points as

$$< v_i v_j > = \left.\frac{d\mathcal{D}}{d\tau}\right|_{\chi_i} \left.\frac{d\mathcal{D}}{d\tau}\right|_{\chi_j} \int \frac{dk}{2\pi^2} P_{\Lambda\mathrm{CDM}}(k) \times F(k\chi_i, k\chi_j, \cos(\theta)) \tag{18.8}$$

with

$$F(u, v, \cos(\theta)) = \sum_{\ell=0}^{\infty}(2\ell + 1) j_\ell'(u) j_\ell'(v) P_\ell(\cos(\theta)) \tag{18.9}$$

$$= \left.\frac{\partial^2 \mathrm{sinc}(\sqrt{x^2 - 2xy\cos(\theta) + y^2})}{\partial x \partial y}\right|_{x=u, y=v} \tag{18.10}$$

where τ is conformal time, $\mathcal{D}$ is the linear growth factor, χ is comoving distance, and θ is the angular separation between the two objects across the sky [39, 40]. The term $F(u, v, \cos(\theta))$, which has been written here as a sum over all multipoles, can alternatively be expressed in closed form using the second definition. An example of this covariance function is visualized in Fig. 18.1, where we show the covariance between pairs of galaxies found in the Pantheon+ [36] catalog. Galaxies close together have almost wholly correlated motions, while galaxies on the same line of sight but distant from one another are anti-correlated. Galaxies on the opposite side of the sky show significant anti-correlation, depending on distance. These covariances are present throughout the sample, and require mitigation through several techniques to yield an accurate measurement.

18.4 Mitigations of the Impact of Peculiar Velocities on H_0 Measurements

While the effect of peculiar velocities on leading measurements is likely insufficient to explain the Hubble tension, verification of proposed theoretical solutions will require a maximally robust, precise, and accurate measurement. Mitigating, correcting for, and budgeting peculiar velocity uncertainties will thus be an essential element of distance-scale measurements using data in the local ($z < 0.05$) universe. We discuss the strategies used by different teams below, along with suggestions for improvements in future iterations.

18.4.1 Exclusion/Weighting of Data

The simplest means of mitigating the effect of peculiar velocities on direct measurements of the Hubble constant is to exclude low-redshift data. Since the fractional error in redshifts is the quantity of concern for a direct measurement of H_0, removing data below a given redshift drastically reduces the potential impact of these uncertainties, without introducing additional dependence on modeling or external datasets. However relying entirely on high-redshift data may limit the statistical power of the sample, increase dependence on other systematic uncertainties such as evolution, and/or increase the dependence of the result on knowledge of the higher-order terms of the Hubble-Lemaître law. The SH_0ES team cut their Hubble flow sample at $z > 0.023$ for the primary measurement. As can be seen in Fig. 18.2, most of the constraining power of the supernova sample is above this value, while the variance contribution of peculiar velocities is no longer dominant. In contrast, the CCHP team [41] chose to include all 99 supernovae from the Carnegie Supernova Project [42] that passed quality cuts and were not themselves used as TRGB calibrators in their Hubble flow sample. This approach effectively represents a cut at $z > 0.004$. Similarly the Megamaser Cosmology Project [43], with only six objects to choose from, did not impose any redshift cut on their final result. It is thus evident that the choice of a redshift cut will depend on the demographics of a particular sample and the significance of peculiar velocities in the overall error budget of a given measurement.

A less extreme approach (which may be combined with redshift cuts) than entirely cutting low-redshift data is to downweight nearby data using a variance or covariance matrix [5, 39]. This often takes the form of an assumed uncorrelated velocity scatter

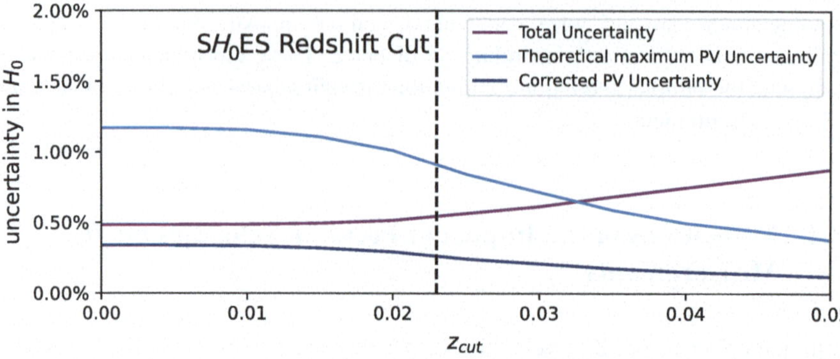

Fig. 18.2 Dependence of the uncertainty in the SH_0ES estimate of H_0 from the type Ia supernova Hubble diagram as a function of minimum redshift cut. Purple line shows total, budgeted uncertainty in the measurement from the Hubble diagram measurement (not including uncertainties from the first two rungs of the ladder), while light blue shows *potential* impact of *uncorrected* peculiar velocities based on the ΛCDM power spectrum. The power spectrum was evaluated with Planck parameters [37] and halofit corrections [38]. As the real measurement includes corrections from reconstructed peculiar velocity maps, PV uncertainty in the final measurement is significantly smaller, shown in dark blue

of $\sigma_{\mathrm{PV}} \approx 150 - 300 \mathrm{km\,s}^{-1}$, ignoring the correlations between velocities, which are small for widely separated sources. Frequently no covariant or correlated uncertainties will be included in the primary fit of the analysis, although the sensitivity of the final value to various correction methodologies (as discussed in the next section) may be investigated and ultimately included in systematic error budgets. Different studies have found quite different values for the total variance of peculiar velocities σ_{PV} [44–51] ranging from 150 to 250 $\mathrm{km\,s}^{-1}$. As was discussed earlier in Sect. 18.3, the magnitude of this quantity will include significant nonlinear and virialized contributions it is difficult to predict from theory. Accurate measurement of this term requires both an understanding of peculiar velocities and the control of other elements of the uncertainty budget. Assessment of the correlated contribution from linear theory alone is comparatively much simpler [39], as measurements of the power spectrum at scales $k < 0.1\,\mathrm{Mpc}\,h^{-1}$ are widely published and the relevant physics is thought to be well understood. Use of an appropriate covariance matrix will reduce sensitivity of the final result to peculiar velocities and improve the robustness of the result. However it can be a significant contributor to the final error budget. To further improve measurements, direct correction of peculiar velocity noise is necessary.

18.4.2 Peculiar Velocity Corrections

Previously mentioned methods work by reducing the contribution of the most contaminated data in a linear fit to the Hubble diagram. A separate category of mitigations would be the attempt to subtract the signal in the observed data by using other datasets to predict the peculiar velocities. As peculiar velocities are gravitationally sourced, measurements of the mass distribution in proximity to the Milky Way provide information on the local peculiar velocity field. One can therefore use the observed distribution of galaxies to 'reconstruct' the expected peculiar velocity field.

Given a reconstructed velocity field with line-of-sight velocity $v_{\text{pec}}^{\text{recon}}(\vec{r})$ as a function of comoving position, the velocity corrections for a given galaxy can be derived by reference to Eq. 18.5. By substitution, with observed redshift in the CMB frame z_{obs} and sky position $\hat{n}$ given as a unit vector this is evaluated as

$$(1 + z_{\text{cos}}) = \frac{1 + z_{\text{obs}}}{1 + v_{\text{pec}}^{\text{recon}}(\chi(z_{\text{cos}})\hat{n})/c}, \tag{18.11}$$

although some works use the approximated form

$$z_{\text{cos}} \approx z_{\text{obs}} - v_{\text{pec}}^{\text{recon}}(\chi(z_{\text{obs}})\hat{n})/c \tag{18.12}$$

where $\chi(z)$ is the comoving distance as a function of redshift. This approximation leaves out a term of $z_{\text{pec}}z_{\text{cos}}$. Another common approximation is the use of $\chi(z_{\text{obs}})$ rather than $\chi(z_{\text{cos}})$ to determine the position $\vec{r}$. As $v_{\text{pec}}^{\text{recon}}$ is a function of physical position, the appropriate redshift to use here is the one corresponding to the inferred cosmological distance, rather than the observed quantity. This can most easily be corrected for by resampling the reconstruction into redshift-space instead of physical space [30]. However, as discussed in Sect. 18.4.3, the complex structure of these maps near large mass concentrations, as well as scatter in redshifts caused by small-scale noise may require more sophisticated procedures.

The simplest peculiar velocity model may be the "VGAS" model used in the Hubble Key Project [52], which has seen subsequent use by the the CCHP [41, 53]. This models the local peculiar velocity field as sourced by the three largest concentrations of mass in the nearby universe, Virgo, Shapley, and the Great Attractor. This model has the advantage of simplicity and explainability, and is likely to reduce the overall systematics by some amount. However these objects combined account for less than a part per billion of the mass within 200 Mpc of the Milky Way. Further, an analysis of the associated errors/uncorrected motions is difficult to carry out due to the ad-hoc construction of the model.

Researchers have also created full reconstructions of the local volume, which provide an estimated description of all mass within a defined volume at resolutions between 1 and 10 Mpc h^{-1}. Uncertainty at scales larger than those covered by the underlying data is represented by a fitted external bulk flow parameter, constrained by both the redshift catalog as well as sometimes additional SN Ia data. The methodology underpinning these reconstructions varies widely, from filter-based approaches which convolve a defined kernel with observed data [49] to Bayesian inference of initial conditions combined with a forward model [48]. However in many cases significant portions of the underlying data are shared between these different teams, in particular the 2MASS Redshift Survey [54, 55]. The most widely used of these among local measurements of H_0 is the "2M++ model" [46], a reconstruction primarily based on the 2M++ catalog using data from the 2MASS Redshift Survey [54, 55] and the 6 Degree Field Survey [56]. The 2M++ model has been used in H_0 measurements including those made with the standard siren GW170817 [12], Type II SNe [57], by the SH_0ES project [2, 58], and the Megamaser Cosmology Project [43]. Redshift survey based approaches have provided some estimates of their uncertainties, around $150\,\mathrm{km\,s^{-1}}$ in variance. However these estimates are typically provided as either general quantities or only referred to on plots, rather than made available as data files.

Cosmicflows [59–62] is an effort to incorporate as many distance measurements as possible into a single model of the nearby universe. The most recent results include more than 55,000 distances, from the Tully-Fisher relation, the fundamental plane, surface brightness fluctuations, core-collapse supernovae, type Ia supernovae, masers, Cepheids, and the tip of the red giant branch. The methodology used to process this data has evolved from a Wiener-filter based approach to more recent efforts using Hamiltonian Monte-Carlo. However the heterogeneity of the underlying data means that associated uncertainties are highly unclear.

The Pantheon+ and SH_0ES team have analyzed the effectiveness of peculiar velocity corrections in their application to the SN Ia distance ladder. The Pantheon+ analysis [50] examined the data primarily through looking at diagonal scatter of SN Ia data on the Hubble diagram. Their results found strong evidence that the use of 2M++ [46] and 2MRS [49] reconstructions substantively improved low-redshift scatter on the Hubble diagram. At 10σ significance these two reconstructions were favored as compared to uncorrected CMB frame redshifts. However the Pantheon+ team found that there was no evidence use of corrections based on Cosmicflows-3 or the VGAS model were effective. In the final covariance matrix for SH_0ES/Pantheon+, a covariant term based on the differences between the two favored reconstructions was included. Further work by the SH_0ES team [51] attempted to empirically quantify the degree of accuracy of the 2M++ and 2MRS models in reducing correlated noise. Cepheid observations from the SH_0ES team confirmed the accuracy of these two reconstructions at $\sim 3\sigma$, and measured a factor of ~ 4 reduction in covariant uncertainties in the linear regime. These results are qualitatively consistent with the estimated uncertainties presented by the reconstruction teams. Estimated errors have been included in the SH_0ES error budget, and are subdominant in the final measurement, at the level of $\sim 0.25\%$.

18.4.3 Mathematical Treatment of Reconstructions

A direct substitution of observed redshift and reconstructed velocity along a given line of sight into Eq. 18.11 to solve for cosmological redshift is the most straightforward approach to calculating PV corrections. However, since the PV reconstructions are not perfectly precise, the problem of determining corrections is a statistical problem.

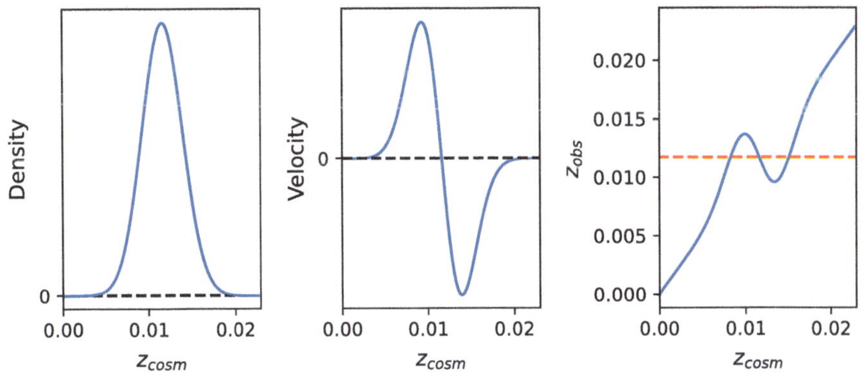

Fig. 18.3 Cartoon example of a triply-valued region. Near a dense cluster, the velocity as a function of distance takes on a characteristic Z-shaped curve which flattens the distance-redshift relation. This velocity profile is shown in the second panel. As a result, an object's observed redshift is fully consistent with any of three positions; the middle of the cluster, the near side, or the far side of it. Intuitively, this means that objects near clusters have highly uncertain true distances

The scatter in redshifts sourced by small scale motions and the uneven shape of the distance-redshift relation in the presence of inhomogeneity can bias corrected redshifts obtained from direct substitution. In extreme cases near galaxy clusters, the distance-redshift relation is bent into "triply-valued regions", where a single observed redshift is consistent with three underlying cosmological redshifts. An illustration of the phenomenon is shown in Fig. 18.3.

True triply valued regions are rare, but the principle of curvature in the Hubble diagram due to inhomogeneity is generally applicable. Due to scatter of the observed redshifts by small-scale noise, estimation of the cosmological redshift given an observed redshift is nontrivial, requiring either a Bayesian or frequentist analysis of the likelihood. A simple example of this might be based on a likelihood function for the observed redshift, given the cosmological redshift. Such a likelihood could be estimated by comparing the predicted peculiar redshift from the reconstruction, to the peculiar redshift expected from the ratio of the observed and cosmological $(1+z)$ factors,

$$P(z_{\rm obs}|z_{\rm cos}) = \frac{1}{\sqrt{2\pi\sigma_v^2}} \exp\left(-\left((1+v_{\rm pec}^{\rm recon}(\chi(z_{\rm cos})\hat{n})/c) - \frac{1+z_{\rm obs}}{1+z_{\rm cos}}\right)^2 /(2\sigma_v^2)\right). \tag{18.13}$$

where σ_v represents uncertainty due to small-scale scatter, between 150 and 250 km s^{-1} depending on sample, use of group corrections, or other variables. As $v_{\rm pec}^{\rm recon}$ is not a linear function, this likelihood function is not Gaussian as a function of $z_{\rm cos}$. A simple prior on $z_{\rm cos}$ may assume that the sample is a galaxy randomly selected from a given volume

$$P(z_{\text{cos}}) \propto \chi(z_{\text{cos}})^2 \cdot (1 + \delta(z_{\text{cos}})) \frac{d\chi(z)}{dz}|_{z=z_{\text{cos}}} \tag{18.14}$$

where δ is the matter density contrast in the redshift-space reconstruction and $\chi(z)$ is the comoving distance as a function of redshift. This prior assumes the sample is proportional to the volume element times the relative number density of galaxies, accounting for the inhomogeneous Malmquist bias [63]. However selection effects may require the use of different priors or more complicated modeling. If this simple model is appropriate, it is simple to sample from it using an MCMC, or even evaluate it directly by one-dimensional integration over the line of sight. When correctly accounted, each point will have different uncertainties based on the local environment, and the expectation of the cosmological redshift may differ from the maximum likelihood value found by direct substitution into Eq. 18.11. The Pantheon+ team determined that estimates integrated using this methodology are more accurate [50], although ultimately the simpler maximum likelihood method was used in the final analysis. Other selections of data may impose other considerations and should be considered by groups making use of peculiar velocity corrections. These line-of-sight statistical effects have been accounted for by some [14, 51, 64, 65], but not all, works in cosmography.

18.4.4 Group Assignments

Galaxies can be determined to belong to gravitationally bound groups based on the use of redshift catalogs. As the mean redshift of a group averages over virialized motion within the group, the use of group redshifts should remove noise below the resolution of full-scale reconstructions.

The Pantheon+ team [50] tested four sets of group assignments [66–69], evaluated by their effect on the Hubble scatter of the low-redshift Pantheon+ sample. All four were favored by the data at $\sim 8\sigma$ significance, although results by Tully [67] were favored compared to the others by $\sim 4\sigma$. Use of any one of these group catalogs is likely to significantly improve precision in Hubble diagram redshifts. Group assignments have been used in several more recent measurements of the Hubble constant [2, 12, 13, 43].

18.5 Conclusions and Future Outlook

Ultimately, peculiar velocity uncertainties are significantly unlikely to be responsible for the Hubble tension. However, correctly understanding these effects is critical to ensure our measurements are robust. Impact on the full SH_0ES three rung ladder is estimated at the level of $\sim 0.25\%$ from the released covariant uncertainties (see Fig. 18.1). Evaluation of the theoretical maximum contribution under ΛCDM based on the measured power spectrum yields a maximum contribution of $\sim$0.8%. In the context of a more than 10% discrepancy, these uncertainties are subdominant. When considering competing methods restricted to smaller numbers of objects at smaller distances, the role of peculiar velocities becomes much more critical. Use of redshifts based on galaxy groups, corrected for bulk motions by the use of peculiar velocity reconstructions, is significantly favored by SN Ia data.

Existing peculiar velocity reconstructions are limited in the distances which they can probe, as the completeness of most spectroscopic surveys methods limit the distances at which high-resolution models can be efficiently determined beyond $z < 0.05$. Future results by the DESI [70] and Euclid [71] instruments may greatly increase the distance over which such reconstructions can be made [72, 73]. However, the impact of peculiar velocities is typically drowned out by distance uncertainties at these redshifts regardless, and the most precise measurements of the Hubble constant are unlikely to be limited by peculiar velocities at any point in the near future. Nevertheless, improvements in modeling may significantly improve some probes, particularly the two rung Cepheid distance ladder [51], direct geometric measurements from nuclear megamasers [43], and gravitational wave standard sirens [13–15].

The utility of these reconstructions may be limited by the distribution of best-fit reconstructions, without public estimates of uncertainties at the voxel level. The methodology of the 2MRS reconstruction [49] created multiple realizations of the reconstruction, although final data products are published only as a single estimate of the density distribution. However their code is publicly available, and estimated variances as a function of radial position are available from their paper. Fully Bayesian approaches like BORG [48] offer fully quantified uncertainties with full covariance, but neither the BORG code nor data products have been publicly released. We strongly encourage the production of voxel-level and/or covariant uncertainties in future results. Scaling relations imply distribution of these data products may impose significant storage and/or computational costs, but release of multiple samples or low-ranked matrix approximations may be a viable approach.

Future work on extremely low-redshift measurements of the Hubble constant should focus on consistent modeling of peculiar velocities as an integrated component of the measurement process.

Acknowledgements W. D'Arcy Kenworthy thanks the Knut and Alice Wallenberg Foundation for their support from the research project grant 'Understanding the Dynamic Universe' under Dnr KAW 2018.0067. T. M. Davis acknowledges the support of an Australian Research Council Australian Laureate Fellowship (FL180100168) funded by the Australian Government. The authors thank Khaled Said for useful comments.

References

1. A.G. Riess, L.M. Macri, S.L. Hoffmann, D. Scolnic, S. Casertano, A.V. Filippenko, B.E. Tucker, M.J. Reid, D.O. Jones, J.M. Silverman, R. Chornock, P. Challis, W. Yuan, P.J. Brown, R.J. Foley. A 2.4% determination of the local value of the Hubble constant. APJ **826**(1), 56 (2016)
2. A.G. Riess, W. Yuan, L.M. Macri, D. Scolnic, D. Brout, S. Casertano, D.O. Jones, Y. Murakami, G.S. Anand, L. Breuval, T.G. Brink, A.V. Filippenko, S. Hoffmann, S.W. Jha, W. D'arcy Kenworthy, J. Mackenty, B.E. Stahl, W. Zheng, A comprehensive measurement of the local value of the Hubble constant with 1 km s^{-1} Mpc^{-1} uncertainty from the Hubble space telescope and the SH_0ES team. ApJL **934**(1), L7 (2022)
3. Tamara M. Davis, Morag I. Scrimgeour, Deriving accurate peculiar velocities (even at high redshift). MNRAS **442**(2), 1117–1122 (2014). (August)
4. E.R. Harrison, Interpretation of redshifts of galaxies in clusters. ApJL **191**, L51 (1974). (July)
5. Lam Hui, Patrick B. Greene, Correlated fluctuations in luminosity distance and the importance of peculiar motion in supernova surveys. PhRvD **73**(12), 123526 (2006). (June)
6. G.F.R. Ellis, J.E. Baldwin, On the expected anisotropy of radio source counts. MNRAS **206**, 377–381 (1984). (January)
7. Tobias Nadolny, Ruth Durrer, Martin Kunz, Hamsa Padmanabhan, A new way to test the cosmological principle: measuring our peculiar velocity and the large-scale anisotropy independently. JCAP **2021**(11), 009 (2021). (November)
8. M. Rubart, D.J. Schwarz, Cosmic radio dipole from NVSS and WENSS. A&A **555**, A117 (2013). (July)
9. J.D. Wagenveld, H.R. Klöckner, D.J. Schwarz, The cosmic radio dipole: Bayesian estimators on new and old radio surveys. A&A **675**, A72 (2023). (July)
10. Planck Collaboration, N. Aghanim, C. Armitage-Caplan, M. Arnaud, M. Ashdown, F. Atrio-Barandela, J. Aumont, C. Baccigalupi, A.J. Banday, R.B. Barreiro, J.G. Bartlett, K. Benabed, A. Benoit-Lévy, J.P. Bernard, M. Bersanelli, P. Bielewicz, J. Bobin, J.J. Bock, J.R. Bond, J. Borrill, F.R. Bouchet, M. Bridges, C. Burigana, R.C. Butler, J.F. Cardoso, A. Catalano, A. Challinor, A. Chamballu, H.C. Chiang, L.Y. Chiang, P.R. Christensen, D.L. Clements, L.P.L. Colombo, F. Couchot, B.P. Crill, A. Curto, F. Cuttaia, L. Danese, R.D. Davies, R.J. Davis, P. de Bernardis, A. de Rosa, G. de Zotti, J. Delabrouille, J.M. Diego, S. Donzelli, O. Doré, X. Dupac, G. Efstathiou, T.A. Enßlin, H.K. Eriksen, F. Finelli, O. Forni, M. Frailis, E. Franceschi, S. Galeotta, K. Ganga, M. Giard, G. Giardino, J. González-Nuevo, K.M. Górski, S. Gratton, A. Gregorio, A. Gruppuso, F.K. Hansen, D. Hanson, D.L. Harrison, G. Helou, S.R. Hildebrandt, E. Hivon, M. Hobson, W.A. Holmes, W. Hovest, K.M. Huffenberger, W.C. Jones, M. Juvela, E. Keihänen, R. Keskitalo, T.S. Kisner, J. Knoche, L. Knox, M. Kunz, H. Kurki-Suonio, A. Lähteenmäki, J.M. Lamarre, A. Lasenby, R.J. Laureijs, C.R. Lawrence, R. Leonardi, A. Lewis, M. Liguori, P.B. Lilje, M. Linden-Vørnle, M. López-Caniego, P.M. Lubin, J.F. Macías-Pérez, N. Mandolesi, M. Maris, D.J. Marshall, P.G. Martin, E. Martínez-González, S. Masi, M. Massardi, S. Matarrese, P. Mazzotta, P.R. Meinhold, A. Melchiorri, L. Mendes, M. Migliaccio, S. Mitra, A. Moneti, L. Montier, G. Morgante, D. Mortlock, A. Moss, D. Munshi, P. Naselsky, F. Nati, P. Natoli, H.U. Nørgaard-Nielsen, F. Noviello, D. Novikov, I. Novikov, S. Osborne, C.A. Oxborrow, L. Pagano, F. Pajot, D. Paoletti, F. Pasian, G. Patanchon, O. Perdereau, F. Perrotta, F. Piacentini, E. Pierpaoli, D. Pietrobon, S. Plaszczynski, E. Pointecouteau, G. Polenta, N. Ponthieu, L. Popa, G.W. Pratt, G. Prézeau, J.L. Puget, J.P. Rachen, W.T. Reach, M. Reinecke, S. Ricciardi, T. Riller, I. Ristorcelli, G. Rocha, C. Rosset, J.A. Rubiño-Martín, B. Rusholme, D. Santos, G. Savini, D. Scott, M.D. Seiffert, E.P.S. Shellard, L.D. Spencer, R. Sunyaev, F. Sureau, A.S. Suur-Uski, J.F. Sygnet, J.A. Tauber, D. Tavagnacco, L. Terenzi, L. Toffolatti, M. Tomasi, M. Tristram, M. Tucci, M. Türler, L. Valenziano, J. Valiviita, B. Van Tent, P. Vielva, F. Villa, N. Vittorio, L.A. Wade, B.D. Wandelt, M. White, D. Yvon, A. Zacchei, J.P. Zibin, A. Zonca. Planck 2013 results. XXVII. doppler boosting of the CMB: Eppur si muove. A&A **57**, A27 (2014)
11. Planck Collaboration, Y. Akrami, M. Ashdown, J. Aumont, C. Baccigalupi, M. Ballardini, A.J. Banday, R.B. Barreiro, N. Bartolo, S. Basak, K. Benabed, J.P. Bernard, M. Bersanelli,

P. Bielewicz, J.R. Bond, J. Borrill, F.R. Bouchet, C. Burigana, E. Calabrese, J.F. Cardoso, B. Casaponsa, H.C. Chiang, C. Combet, D. Contreras, B.P. Crill, F. Cuttaia, P. de Bernardis, A. de Rosa, G. de Zotti, J. Delabrouille, E. Di Valentino, J.M. Diego, O. Doré, M. Douspis, X. Dupac, T.A. Enßlin, H.K. Eriksen, R. Fernandez-Cobos, F. Finelli, M. Frailis, E. Franceschi, A. Frolov, S. Galeotta, S. Galli, K. Ganga, R.T. Génova-Santos, M. Gerbino, J. González-Nuevo, K.M. Górski, A. Gruppuso, J.E. Gudmundsson, W. Handley, D. Herranz, E. Hivon, Z. Huang, A.H. Jaffe, W.C. Jones, E. Keihänen, R. Keskitalo, K. Kiiveri, J. Kim, T.S. Kisner, N. Krachmalnicoff, M. Kunz, H. Kurki-Suonio, J.M. Lamarre, M. Lattanzi, C.R. Lawrence, M. Le Jeune, F. Levrier, M. Liguori, P.B. Lilje, V. Lindholm, M. López-Caniego, J.F. Macías-Pérez, D. Maino, N. Mandolesi, A. Marcos-Caballero, M. Maris, P.G. Martin, E. Martínez-González, S. Matarrese, N. Mauri, J.D. McEwen, A. Mennella, M. Migliaccio, D. Molinari, A. Moneti, L. Montier, G. Morgante, A. Moss, P. Natoli, L. Pagano, D. Paoletti, F. Perrotta, V. Pettorino, F. Piacentini, G. Polenta, J.P. Rachen, M. Reinecke, M. Remazeilles, A. Renzi, G. Roha, C. Roset, J.A. Rubiño-Martín, B. Ruiz-Granados, L. Salvati, M. Savelainen, D. Scott, C. Sirignano, G. Sirri, L.D. Spencer, R.M. Sullivan, R. Sunyaev, A.S. Suur-Uski, J.A. Tauber, D. Tavagnacco, M. Tenti, L. Toffolatti, M. Tomasi, T. Trombetti, J. Valiviita, B. Van Tent, P. Vielva, F. Villa, N. Vittorio, I.K. Wehus, A. Zacchei, A. Zonca, Planck intermediate results. LVI. Detection of the CMB dipole through modulation of the thermal Sunyaev-Zeldovich effect: Eppur si muove II. A&A **644**, A100 (2020)

12. B.P. Abbott, R. Abbott, T.D. Abbott, F. Acernese, K. Ackley, C. Adams, T. Adams, P. Addesso, R.X. Adhikari, V.B. Adya, C. Affeldt, M. Afrough, B. Agarwal, M. Agathos, K. Agatsuma, N. Aggarwal, O.D. Aguiar, L. Aiello, A. Ain, P. Ajith, B. Allen, G. Allen, A. Allocca, P.A. Altin, A. Amato, A. Ananyeva, S.B. Anderson, W.G. Anderson, S.V. Angelova, S. Antier, S. Appert, K. Arai, M.C. Araya, J.S. Areeda, N. Arnaud, K.G. Arun, S. Ascenzi, G. Ashton, M. Ast, S.M. Aston, P. Astone, D.V. Atallah, P. Aufmuth, C. Aulbert, K. Aultoneal, C. Austin, A. Avila-Alvarez, S. Babak, P. Bacon, M.K.M. Bader, S. Bae, P.T. Baker, F. Baldaccini, G. Ballardin, S.W. Ballmer, S. Banagiri, J.C. Barayoga, S.E. Barclay, B.C. Barish, D. Barker, K. Barkett, F. Barone, B. Barr, L. Barsotti, M. Barsuglia, D. Barta, J. Bartlett, I. Bartos, R. Bassiri, A. Basti, J.C. Batch, A gravitational-wave standard siren measurement of the Hubble constant. Nature **551**(7678), 85–88 (2017). (November)
13. Cullan Howlett, Tamara M. Davis, Standard siren speeds: improving velocities in gravitational-wave measurements of H_0. MNRAS **492**(3), 3803–3815 (2020). (March)
14. Suvodip Mukherjee, Guilhem Lavaux, François R. Bouchet, Jens Jasche, Benjamin D. Wandelt, Samaya Nissanke, Florent Leclercq, Kenta Hotokezaka, Velocity correction for Hubble constant measurements from standard sirens. A&A **646**, A65 (2021). (February)
15. Constantina Nicolaou, Ofer Lahav, Pablo Lemos, William Hartley, Jonathan Braden, The impact of peculiar velocities on the estimation of the Hubble constant from gravitational wave standard sirens. MNRAS **495**(1), 90–97 (2020). (June)
16. Idit Zehavi, Adam G. Riess, Robert P. Kirshner, Avishai Dekel, A local Hubble bubble from type Ia supernovae? APJ **503**(2), 483–491 (1998). (August)
17. Saurabh Jha, Adam G. Riess, Robert P. Kirshner, Improved distances to type Ia supernovae with multicolor light-curve shapes: MLCS2k2. APJ **659**(1), 122–148 (2007). (April)
18. A. Conley, R.G. Carlberg, J. Guy, D.A. Howell, S. Jha, A.G. Riess, M. Sullivan, Is there evidence for a Hubble bubble? The nature of type Ia supernova colors and dust in external galaxies. ApJL **664**(1), L13–L16 (2007). (July)
19. Benjamin Sinclair, Tamara M. Davis, Troels Haugbølle, Residual Hubble-bubble effects on supernova cosmology. APJ **718**(2), 1445–1455 (2010). (August)
20. W. D'Arcy Kenworthy, D. Scolnic, A. Riess, The local perspective on the Hubble tension: local structure does not impact measurement of the Hubble constant. APJ **875**(2), 145 (2019)
21. Sveva Castello, Marcus Högås, Edvard Mörtsell, A cosmological underdensity does not solve the Hubble tension. JCAP **2022**(7), 003 (2022). (July)
22. R. Brent Tully, Hélène Courtois, Y. Hoffman, D. Pomarède, The Laniakea supercluster of galaxies. Nature **513**(7516), 71–73 (2014)

23. A. Dupuy, H.M. Courtois, Dynamic cosmography of the local Universe: Laniakea and five more watershed superclusters (2023). arXiv e-prints arXiv:2305.02339
24. Wu. Hao-Yi, Dragan Huterer, Sample variance in the local measurements of the Hubble constant. MNRAS **471**(4), 4946–4955 (2017). (November)
25. Planck Collaboration, N. Aghanim, Y. Akrami, F. Arroja, M. Ashdown, J. Aumont, C. Baccigalupi, M. Ballardini, A.J. Banday, R.B. Barreiro, N. Bartolo, S. Basak, R. Battye, K. Benabed, J.P. Bernard, M. Bersanelli, P. Bielewicz, J.J. Bock, J.R. Bond, J. Borrill, F.R. Bouchet, F. Boulanger, M. Bucher, C. Burigana, R.C. Butler, E. Calabrese, J.F. Cardoso, J. Carron, B. Casaponsa, A. Challinor, H.C. Chiang, L.P.L. Colombo, C. Combet, D. Contreras, B.P. Crill, F. Cuttaia, P. de Bernardis, G. de Zotti, J. Delabrouille, J.M. Delouis, F.X. Désert, E. Di Valentino, C. Dickinson, J.M. Diego, S. Donzelli, O. Doré, M. Douspis, A. Ducout, X. Dupac, G. Efstathiou, F. Elsner, T.A. Enßlin, H.K. Eriksen, E. Falgarone, Y. Fantaye, J. Fergusson, R. Fernandez-Cobos, F. Finelli, F. Forastieri, M. Frailis, E. Franceschi, A. Frolov, S. Galeotta, S. Galli, K. Ganga, R.T. Génova-Santos, M. Gerbino, T. Ghosh, J. González-Nuevo, K.M. Górski, S. Gratton, A. Gruppuso, J.E. Gudmundsson, J. Hamann, W. Handley, F.K. Hansen, G. Helou, D. Herranz, S.R. Hildebrandt, E. Hivon, Z. Huang, A.H. Jaffe, W.C. Jones, A. Karakci, E. Keihänen, R. Keskitalo, K. Kiiveri, J. Kim, T.S. Kisner, L. Knox, N. Krachmalnicoff, M. Kunz, H. Kurki-Suonio, G. Lagache, J.M. Lamarre, M. Langer, A. Lasenby, M. Lattanzi, C.R. Lawrence, M. Le Jeune, J.P. Leahy, J. Lesgourgues, F. Levrier, A. Lewis, M. Liguori, P.B. Lilje, M. Lilley, V. Lindholm, M. López-Caniego, P.M. Lubin, Y.Z. Ma, J.F. Macías-Pérez, G. Maggio, D. Maino, N. Mandolesi, A. Mangilli, A. Marcos-Caballero, M. Maris, P.G. Martin, M. Martinelli, E. Martínez-González, S. Matarrese, N. Mauri, J.D. McEwen, P.D. Meerburg, P.R. Meinhold, A. Melchiorri, A. Mennella, M. Migliaccio, M. Millea, S. Mitra, M.A. Miville-Deschênes, D. Molinari, A. Moneti, L. Montier, G. Morgante, A. Moss, S. Mottet, M. Münchmeyer, P. Natoli, H.U. Nørgaard-Nielsen, C.A. Oxborrow, L. Pagano, D. Paoletti, B. Partridge, G. Patanchon, T.J. Pearson, M. Peel, H.V. Peiris, F. Perrotta, V. Pettorino, F. Piacentini, L. Polastri, G. Polenta, J.L. Puget, J.P. Rachen, M. Reinecke, M. Remazeilles, C. Renault, A. Renzi, G. Rocha, C. Rosset, G. Roudier, J.A. Rubiño-Martín, B. Ruiz-Granados, L. Salvati, M. Sandri, M. Savelainen, D. Scott, E.P.S. Shellard, M. Shiraishi, C. Sirignano, G. Sirri, L.D. Spencer, R. Sunyaev, A.S. Suur-Uski, J.A. Tauber, D. Tavagnacco, M. Tenti, L. Terenzi, L. Toffolatti, M. Tomasi, T. Trombetti, J. Valiviita, B. Van Tent, L. Vibert, P. Vielva, F. Villa, N. Vittorio, B.D. Wandelt, I.K. Wehus, M. White, S.D.M. White, A. Zacchei, A. Zonca, Planck 2018 results. I. Overview and the cosmological legacy of Planck. A&A **641**, A1 (2020)
26. Chris Blake, Jasper Wall, A velocity dipole in the distribution of radio galaxies. Nature **416**(6877), 150–152 (2002). (March)
27. U. Feindt, M. Kerschhaggl, M. Kowalski, G. Aldering, P. Antilogus, C. Aragon, S. Bailey, C. Baltay, S. Bongard, C. Buton, A. Canto, F. Cellier-Holzem, M. Childress, N. Chotard, Y. Copin, H.K. Fakhouri, E. Gangler, J. Guy, A. Kim, P. Nugent, J. Nordin, K. Paech, R. Pain, E. Pecontal, R. Pereira, S. Perlmutter, D. Rabinowitz, M. Rigault, K. Runge, C. Saunders, R. Scalzo, G. Smadja, C. Tao, R.C. Thomas, B.A. Weaver, C. Wu, Measuring cosmic bulk flows with type Ia supernovae from the nearby supernova factory. A&A **560**, A90 (2013). (December)
28. Nick Horstmann, Yannic Pietschke, Dominik J. Schwarz, Inference of the cosmic rest-frame from supernovae Ia. A&A **668**, A34 (2022). (December)
29. Tamara M. Davis, Samuel R. Hinton, Cullan Howlett, Josh Calcino, Can redshift errors bias measurements of the Hubble constant? MNRAS **490**(2), 2948–2957 (2019). (December)
30. Anthony Carr, Tamara M. Davis, Dan Scolnic, Khaled Said, Dillon Brout, Erik R. Peterson, Richard Kessler, The Pantheon+ analysis: improving the redshifts and peculiar velocities of type Ia supernovae used in cosmological analyses. PASA **39**, e046 (2022). (October)
31. Richard Watkins, Hume A. Feldman, Michael J. Hudson, Consistently large cosmic flows on scales of $100h^{-1}$Mpc: a challenge for the standard ΛCDM cosmology. MNRAS **392**(2), 743–756 (2009). (January)
32. K. Migkas, F. Pacaud, G. Schellenberger, J. Erler, N.T. Nguyen-Dang, T.H. Reiprich, M.E. Ramos-Ceja, L. Lovisari, Cosmological implications of the anisotropy of ten galaxy cluster scaling relations. A&A **649**, A151 (2021). (May)

33. H.M. Courtois, A. Dupuy, D. Guinet, G. Baulieu, F. Ruppin, P. Brenas, Gravity in the local universe: density and velocity fields using CosmicFlows-4. A&A **670**, L15 (2023). (February)
34. R. Watkins, T. Allen, C.J. Bradford, A. Ramon, A. Walker, H.A. Feldman, R. Cionitti, Y. Al-Shorman, E. Kourkchi, R.B. Tully, Analysing the large scale bulk flow using cosmicflows4: increasing tension with the standard cosmological model. MNRAS **524**(2), 1885–1892 (2023)
35. A.M. Whitford, C. Howlett, T.M. Davis, Evaluating bulk flow estimators for CosmicFlows-4 measurements (2023). arXiv e-prints arXiv:2306.11269
36. D. Brout, D. Scolnic, B. Popovic, A.G. Riess, A. Carr, J. Zuntz, R. Kessler, T.M. Davis, S. Hinton, D. Jones, W. D'Arcy Kenworthy, E.R. Peterson, K. Said, G. Taylor, N. Ali, P. Armstrong, P. Charvu, A. Dwomoh, C. Meldorf, A. Palmese, H. Qu, B.M. Rose, B. Sanchez, C.W. Stubbs, M. Vincenzi, C.M. Wood, P.J. Brown, R. Chen, K. Chambers, D.A. Coulter, M. Dai, G. Dimitriadis, A.V. Filippenko, R.J. Foley, S.W. Jha, L. Kelsey, R.P. Kirshner, A. Möller, J. Muir, S. Nadathur, Y.-C. Pan, A. Rest, C. Rojas-Bravo, M. Sako, M.R. Siebert, M. Smith, B.E. Stahl, P. Wiseman, The Pantheon+ analysis: cosmological constraints. APJ **938**(2), 110 (2022)
37. Planck Collaboration, N. Aghanim, Y. Akrami, M. Ashdown, J. Aumont, C. Baccigalupi, M. Ballardini, A.J. Banday, R.B. Barreiro, N. Bartolo, S. Basak, R. Battye, K. Benabed, J.P. Bernard, M. Bersanelli, P. Bielewicz, J.J. Bock, J.R. Bond, J. Borrill, F.R. Bouchet, F. Boulanger, M. Bucher, C. Burigana, R.C. Butler, E. Calabrese, J.F. Cardoso, J. Carron, A. Challinor, H.C. Chiang, J. Chluba, L.P.L. Colombo, C. Combet, D. Contreras, B.P. Crill, F. Cuttaia, P. de Bernardis, G. de Zotti, J. Delabrouille, J.M. Delouis, E. Di Valentino, J.M. Diego, O. Doré, M. Douspis, A. Ducout, X. Dupac, S. Dusini, G. Efstathiou, F. Elsner, T.A. Enßlin, H.K. Eriksen, Y. Fantaye, M. Farhang, J. Fergusson, R. Fernandez-Cobos, F. Finelli, F. Forastieri, M. Frailis, A.A. Fraisse, E. Franceschi, A. Frolov, S. Galeotta, S. Galli, K. Ganga, R.T. Génova-Santos, M. Gerbino, T. Ghosh, J. González-Nuevo, K.M. Górski, S. Gratton, A. Gruppuso, J.E. Gudmundsson, J. Hamann, W. Handley, F.K. Hansen, D. Herranz, S.R. Hildebrandt, E. Hivon, Z. Huang, A.H. Jaffe, W.C. Jones, A. Karakci, E. Keihänen, R. Keskitalo, K. Kiiveri, J. Kim, T.S. Kisner, L. Knox, N. Krachmalnicoff, M. Kunz, H. Kurki-Suonio, G. Lagache, J.M. Lamarre, A. Lasenby, M. Lattanzi, C.R. Lawrence, M. Le Jeune, P. Lemos, J. Lesgourgues, F. Levrier, A. Lewis, M. Liguori, P.B. Lilje, M. Lilley, V. Lindholm, M.López-Caniego, P.M. Lubin, Y.Z. Ma, J.F. Macías-Pérez, G. Maggio, D. Maino, N. Mandolesi, A. Mangilli, A. Marcos-Caballero, M. Maris, P.G. Martin, M. Martinelli, E. Martínez-González, S. Matarrese, N. Mauri, J.D. McEwen, P.R. Meinhold, A. Melchiorri, A. Mennella, M. Migliaccio, M. Millea, S. Mitra, M.A. Miville-Deschênes, D. Molinari, L. Montier, G. Morgante, A. Moss, P. Natoli, H.U. Nørgaard-Nielsen, L. Pagano, D. Paoletti, B. Partridge, G. Patanchon, H.V. Peiris, F. Perrotta, V. Pettorino, F. Piacentini, L. Polastri, G. Polenta, J.L. Puget, J.P. Rachen, M. Reinecke, M. Remazeilles, A. Renzi, G. Rocha, C. Rosset, G. Roudier, J.A. Rubiño-Martín, B. Ruiz-Granados, L. Salvati, M. Sandri, M. Savelainen, D. Scott, E.P.S. Shellard, C. Sirignano, G. Sirri, L.D. Spencer, R. Sunyaev, A.S. Suur-Uski, J.A. Tauber, D. Tavagnacco, M. Tenti, L. Toffolatti, M. Tomasi, T. Trombetti, L. Valenziano, J. Valiviita, B. Van Tent, L. Vibert, P. Vielva, F. Villa, N. Vittorio, B.D. Wandelt, I.K. Wehus, M. White, S.D.M. White, A. Zacchei, A. Zonca, lanck 2018 results. VI. Cosmological parameters. A&A **641**, A6 (2020)
38. J.A. Peacock, R.E. Smith, HALOFIT: nonlinear distribution of cosmological mass and galaxies. Astrophys. Source Code Library, record ascl:1402.032 (2014)
39. Tamara M. Davis, Lam Hui, Joshua A. Frieman, Troels Haugbølle, Richard Kessler, Benjamin Sinclair, Jesper Sollerman, Bruce Bassett, John Marriner, Edvard Mörtsell, Robert C. Nichol, Michael W. Richmond, Masao Sako, Donald P. Schneider, Mathew Smith, The effect of peculiar velocities on supernova cosmology. APJ **741**(1), 67 (2011). (November)
40. Dragan Huterer, Daniel L. Shafer, Fabian Schmidt, No evidence for bulk velocity from type Ia supernovae. JCAP **2015**(12), 033 (2015). (December)
41. W.L. Freedman, B.F. Madore, D. Hatt, T.J. Hoyt, I. Sung Jang, R.L. Beaton, C.R. Burns, M. Gyoon Lee, A.J. Monson, J.R. Neeley, M.M. Phillips, J.A. Rich, M. Seibert, The Carnegie-Chicago Hubble program. VIII. An independent determination of the Hubble constant based on the tip of the red giant branch. APJ **882**(1), 34 (2019)

42. K. Krisciunas, C. Contreras, C.R. Burns, M.M. Phillips, M.D. Stritzinger, N. Morrell, M. Hamuy, J. Anais, L. Boldt, L. Busta, A. Campillay, S. Castellón, G. Folatelli, W.L. Freedman, C. González, E.Y. Hsiao, W. Krzeminski, S.E. Persson, M. Roth, F. Salgado, J. Serón, N.B. Suntzeff, S. Torres, A.V. Filippenko, W. Li, B.F. Madore, D.L. DePoy, J.L. Marshall, J.-P. Rheault, S. Villanueva, The Carnegie supernova project. I. Third photometry data release of low-redshift type Ia supernovae and other white dwarf explosions. AJ **154**(5), 211 (2017)
43. D.W. Pesce, J.A. Braatz, M.J. Reid, A.G. Riess, D. Scolnic, J.J. Condon, F. Gao, C. Henkel, C.M.V. Impellizzeri, C.Y. Kuo, K.Y. Lo, The Megamaser cosmology project XIII. Combined Hubble constant constraints. ApJL **891**(1), L1 (2020). (March)
44. R.O. Marzke, M.J. Geller, L.N. da Costa, J.P. Huchra, Pairwise velocities of galaxies in the CfA and SSRS2 redshift surveys. AJ **110**, 477 (1995)
45. M. Betoule, R. Kessler, J. Guy, J. Mosher, D. Hardin, R. Biswas, P. Astier, P. El-Hage, M. Konig, S. Kuhlmann, J. Marriner, R. Pain, N. Regnault, C. Balland, B.A. Bassett, P.J. Brown, H. Campbell, R.G. Carlberg, F. Cellier-Holzem, D. Cinabro, A. Conley, C.B. D'Andrea, D.L. DePoy, M. Doi, R.S. Ellis, S. Fabbro, A.V. Filippenko, R.J. Foley, J.A. Frieman, D. Fouchez, L. Galbany, A. Goobar, R.R. Gupta, G.J. Hill, R. Hlozek, C.J. Hogan, I.M. Hook, D.A. Howell, S.W. Jha, L. Le Guillou, G. Leloudas, C. Lidman, J.L. Marshall, A. Möller, A.M. Mourão, J. Neveu, R. Nichol, M.D. Olmstead, N. Palanque-Delabrouille, S. Perlmutter, J.L. Prieto, C.J. Pritchet, M. Richmond, A.G. Riess, V. Ruhlmann-Kleider, M. Sako, K. Schahmaneche, D.P. Schneider, M. Smith, J. Sollerman, M. Sullivan, N.A. Walton, C.J. Wheeler, Improved cosmological constraints from a joint analysis of the SDSS-II and SNLS supernova samples. Astron. Astrophys. **568**, A22 (2014). (Aug)
46. J. Carrick, S.J. Turnbull, G. Lavaux, M.J. Hudson, Cosmological parameters from the comparison of peculiar velocities with predictions from the 2M++ density field. Monthly Notices of the Royal Astron. Soc. **450**(1), 317–332 (2015)
47. D.M. Scolnic, D.O. Jones, A. Rest, Y.C. Pan, R. Chornock, R.J. Foley, M.E. Huber, R. Kessler, G. Narayan, A.G. Riess, S. Rodney, E. Berger, D.J. Brout, P.J. Challis, M. Drout, D. Finkbeiner, R. Lunnan, R.P. Kirshner, N.E. Sanders, E. Schlafly, S. Smartt, C.W. Stubbs, J. Tonry, W.M. Wood-Vasey, M. Foley, J. Hand, E. Johnson, W.S. Burgett, K.C. Chambers, P.W. Draper, K.W. Hodapp, N. Kaiser, R.P. Kudritzki, E.A. Magnier, N. Metcalfe, F. Bresolin, E. Gall, R. Kotak, M. McCrum, K.W. Smith, The complete light-curve sample of spectroscopically confirmed SNe Ia from Pan-STARRS1 and cosmological constraints from the combined Pantheon sample. APJ **859**(2), 101 (2018). (June)
48. J. Jasche, G. Lavaux, Physical Bayesian modelling of the non-linear matter distribution: new insights into the nearby universe. A&A **625**, A64 (2019). (May)
49. Robert Lilow, Adi Nusser, Constrained realizations of 2MRS density and peculiar velocity fields: growth rate and local flow. MNRAS **507**(2), 1557–1581 (2021). (October)
50. E.R. Peterson, W. D'Arcy Kenworthy, D. Scolnic, A.G. Riess, D. Brout, A. Carr, Hélène Courtois, T. Davis, A. Dwomoh, D.O. Jones, B. Popovic, B.M. Rose, K. Said, The Pantheon+ analysis: evaluating peculiar velocity corrections in cosmological analyses with nearby type Ia supernovae. APJ **938**(2), 112 (2022)
51. W. D'Arcy Kenworthy, A.G. Riess, D. Scolnic, W. Yuan, J. Luis Bernal, D. Brout, S. Casertano, D.O. Jones, L. Macri, E.R. Peterson, Measurements of the Hubble constant with a two-rung distance ladder: two out of three ain't bad. APJ **935**(2), 83 (2022)
52. J.R. Mould, J.P. Huchra, W.L. Freedman, Jr. R.C. Kennicutt, L. Ferrarese, H.C. Ford, B.K. Gibson, J.A. Graham, S.M.G. Hughes, G.D. Illingworth, D.D. Kelson, L.M. Macri, B.F. Madore, S. Sakai, K.M. Sebo, N.A. Silbermann, P.B. Stetson, The Hubble space telescope key project on the extragalactic distance scale. XXVIII. Combining the constraints on the Hubble constant. APJ **529**(2), 786–794 (2000)
53. W.L. Freedman, B.F. Madore, B.K. Gibson, L. Ferrarese, D.D. Kelson, S. Sakai, J.R. Mould, Jr. R.C. Kennicutt, H.C. Ford, J.A. Graham, J.P. Huchra, S.M.G. Hughes, G.D. Illingworth, L.M. Macri, P.B. Stetson, Final results from the Hubble space telescope key project to measure the Hubble constant. APJ **553**(1), 47–72 (2001)

54. John P. Huchra, Lucas M. Macri, Karen L. Masters, Thomas H. Jarrett, Perry Berlind, Michael Calkins, Aidan C. Crook, Roc Cutri, Pirin Erdoğdu, Emilio Falco, Teddy George, Conrad M. Hutcheson, Ofer Lahav, Jeff Mader, Jessica D. Mink, Nathalie Martimbeau, Stephen Schneider, Michael Skrutskie, Susan Tokarz, Michael Westover, The 2MASS redshift survey description and data release. ApJS **199**(2), 26 (2012). (April)
55. L.M. Macri, R.C. Kraan-Korteweg, T. Lambert, M. Victoria Alonso, P. Berlind, M. Calkins, P. Erdoğdu, E.E. Falco, T.H. Jarrett, J.D. Mink, The 2MASS redshift survey in the zone of avoidance. ApJS **245**(1), 6 (2019)
56. D. Heath Jones, M.A. Read, W. Saunders, M. Colless, T. Jarrett, Q.A. Parker, A.P. Fairall, T. Mauch, E.M. Sadler, F.G. Watson, D. Burton, L.A. Campbell, P. Cass, S.M. Croom, J. Dawe, K. Fiegert, L. Frankcombe, M. Hartley, J. Huchra, D. James, E. Kirby, O. Lahav, J. Lucey, G.A. Mamon, L. Moore, B.A. Peterson, S. Prior, D. Proust, K. Russell, V. Safouris, K.-I. Wakamatsu, E. Westra, M. Williams, The 6dF galaxy survey: final redshift release (DR3) and southern large-scale structures. MNRAS **399**(2), 683–698 (2009)
57. T. de Jaeger, L. Galbany, A.G. Riess, B.E. Stahl, B.J. Shappee, A.V. Filippenko, W. Zheng, A 5 percent measurement of the Hubble-Lemaître constant from type II supernovae. MNRAS **514**(3), 4620–4628 (2022). (August)
58. Adam G. Riess, Weidong Li, Peter B. Stetson, Alexei V. Filippenko, Saurabh Jha, Robert P. Kirshner, Peter M. Challis, Peter M. Garnavich, Ryan Chornock, Cepheid calibrations from the Hubble space telescope of the luminosity of two recent type Ia supernovae and a redetermination of the Hubble constant. APJ **627**(2), 579–607 (2005). (July)
59. H.M. Courtois, Y. Hoffman, R. Brent Tully, S. Gottlöber, Three-dimensional velocity and density reconstructions of the local universe with Cosmicflows-1. APJ **744**(1), 43 (2012)
60. R. Brent Tully, H.M. Courtois, A.E. Dolphin, J. Richard Fisher, P. Héraudeau, B.A. Jacobs, I.D. Karachentsev, D. Makarov, L. Makarova, S. Mitronova, L. Rizzi, E.J. Shaya, J.G. Sorce, P.-F. Wu, Cosmicflows-2: the data. AJ **146**(4), 86 (2013)
61. R. Brent Tully, H.M. Courtois, J.G. Sorce. Cosmicflows-3. AJ **152**(2), 50 (2016)
62. R. Brent Tully, E. Kourkchi, H.M. Courtois, G.S. Anand, J.P. Blakeslee, D. Brout, T. de Jaeger, A. Dupuy, D. Guinet, C. Howlett, J.B. Jensen, D. Pomarède, L. Rizzi, D. Rubin, K. Said, D. Scolnic, B.E. Stahl. Cosmicflows-4. APJ **944**(1), 94 (2023)
63. R.W. Pike, M.J. Hudson, Cosmological parameters from the comparison of the 2MASS gravity field with peculiar velocity surveys. APJ **635**(1), 11–21 (2005)
64. Supranta S. Boruah, Michael J. Hudson, Guilhem Lavaux, Peculiar velocities in the local universe: comparison of different models and the implications for H_0 and dark matter. MNRAS **507**(2), 2697–2713 (2021). (October)
65. W. Rahman, R. Trotta, S.S. Boruah, M.J. Hudson, D.A. van Dyk, New constraints on anisotropic expansion from supernovae type Ia. Monthly Notices of the Royal Astron. Soc. **514**(1), 139–163 (2022)
66. Aidan C. Crook, John P. Huchra, Nathalie Martimbeau, Karen L. Masters, Tom Jarrett, Lucas M. Macri, Groups of galaxies in the two micron all sky redshift survey. APJ **655**(2), 790–813 (2007). (February)
67. R. Brent Tully, Galaxy groups. AJ **149**(2), 54 (2015)
68. S.H. Lim, H.J. Mo, Y. Lu, H. Wang, X. Yang, Galaxy groups in the low-redshift universe. MNRAS **470**(3), 2982–3005 (2017)
69. T.S. Lambert, R.C. Kraan-Korteweg, T.H. Jarrett, L.M. Macri, The 2MASS redshift survey galaxy group catalogue derived from a graph-theory based friends-of-friends algorithm. MNRAS **497**(3), 2954–2973 (2020)
70. M. Levi, L.E. Allen, A. Raichoor, C. Baltay, S. BenZvi, F. Beutler, A. Bolton, F.J. Castander, C.-H. Chuang, A. Cooper, J.-G. Cuby, A. Dey, D. Eisenstein, X. Fan, B. Flaugher, C. Frenk, A.X. Gonzalez-Morales, O. Graur, J. Guy, S. Habib, K. Honscheid, S. Juneau, J.-P. Kneib, O. Lahav, D. Lang, A. Leauthaud, B. Lusso, A. de la Macorra, M. Manera, P. Martini, S. Mao, J.A. Newman, N. Palanque-Delabrouille, W.J. Percival, C. Allende Prieto, C.M. Rockosi, V. Ruhlmann-Kleider, D. Schlegel, H.-J. Seo, Y.-S. Song, G. Tarle, R. Wechsler, D. Weinberg, C. Yeche, Y. Zu, The dark energy spectroscopic instrument (DESI). Bull. Am. Astron. Soc. **51**, 57 (2019)

71. Luca Amendola, Stephen Appleby, Anastasios Avgoustidis, David Bacon, Tessa Baker, Marco Baldi, Nicola Bartolo, Alain Blanchard, Camille Bonvin, Stefano Borgani, Enzo Branchini, Clare Burrage, Stefano Camera, Carmelita Carbone, Luciano Casarini, Mark Cropper, Claudia de Rham, Jörg. P. Dietrich, Cinzia Di Porto, Ruth Durrer, Anne Ealet, Pedro G. Ferreira, Fabio Finelli, Juan García-Bellido, Tommaso Giannantonio, Luigi Guzzo, Alan Heavens, Lavinia Heisenberg, Catherine Heymans, Henk Hoekstra, Lukas Hollenstein, Rory Holmes, Zhiqi Hwang, Knud Jahnke, Thomas D. Kitching, Tomi Koivisto, Martin Kunz, Giuseppe La Vacca, Eric Linder, Marisa March, Valerio Marra, Carlos Martins, Elisabetta Majerotto, Dida Markovic, David Marsh, Federico Marulli, Richard Massey, Yannick Mellier, Francesco Montanari, David F. Mota, Nelson J. Nunes, Will Percival, Valeria Pettorino, Cristiano Porciani, Claudia Quercellini, Justin Read, Massimiliano Rinaldi, Domenico Sapone, Ignacy Sawicki, Roberto Scaramella, Constantinos Skordis, Fergus Simpson, Andy Taylor, Shaun Thomas, Roberto Trotta, Licia Verde, Filippo Vernizzi, Adrian Vollmer, Yun Wang, Jochen Weller, Tom Zlosnik, Cosmology and fundamental physics with the Euclid satellite. Living Rev. Relativity **21**(1), 2 (2018). (April)
72. C. Saulder, C. Howlett, K.A. Douglass, K. Said, S. BenZvi, S. Ahlen, G. Aldering, S. Bailey, D. Brooks, T.M. Davis, A. de la Macorra, A. Dey, A. Font-Ribera, J.E. Forero-Romero, S. Gontcho, A. Gontcho, K. Honscheid, A.G. Kim, T. Kisner, A. Kremin, M. Landriau, M.E. Levi, J. Lucey, A.M. Meisner, R. Miquel, J. Moustakas, A.D. Myers, N. Palanque-Delabrouille, W. Percival, C. Poppett, F. Prada, F. Qin, M. Schubnell, G. Tarlé, M. Vargas Magaña, B. Alan Weaver, R. Zhou, Z. Zhou, H. Zou, Target selection for the DESI peculiar velocity survey. MNRAS **525**(1), 1106–1125 (2023)
73. Adam Andrews, Jens Jasche, Guilhem Lavaux, Fabian Schmidt, Bayesian field-level inference of primordial non-Gaussianity using next-generation galaxy surveys. MNRAS **520**(4), 5746–5763 (2023). (April)

Chapter 19
The Impact of Dust on Cepheid and Type Ia Supernova Distances

Dillon Brout and Adam Riess

19.1 Introduction

Cosmic dust particles play a crucial role in cosmological observations because they absorb, scatter, and re-emit starlight, leading to extinction, attenuation, and an overall reddening of light from distant objects. The combination of Milky-Way dust and extragalactic dust affect nearly all measurements in cosmology. From measurements of the cosmic microwave background, to Cepheids and Type Ia Supernovae (SNe Ia), to weak lensing and large-scale structure, to optical counterpart identification of gravitational-wave multimessenger phenomena. In this chapter we focus on the impact of dust specifically for two of the rungs of the SH_0ES [1] cosmic distance ladder: Cepheids and SNe Ia.

Cosmic dust grains are composed of tiny solid particles, ranging in size from a few nanometers to several micrometers, that are generated by a variety of processes, including stellar nucleosynthesis, supernova explosions, and collisions between solid bodies. While the structure and composition of dust particles depend on the specific astrophysical environment in which they are formed, in general, dust grains are made up of silicon, carbon, and large hydrocarbon molecules (see Fig. 19.1 Top), each of which have slightly different effects due to their composition, shape, and reflectivity. The particles can have a wide range of shapes, from spherical to elongated, though their shape and size distributions are not measured directly, but rather are instead inferred through their effects on extinction, polarization, and albedo.

To measure for dust extinction in the Milky Way, specific backlights such as stars and quasars serve as reference sources with well-characterized intrinsic spectral

D. Brout (✉)
Boston University, 685 Commonwealth Ave, Boston, MA 02215, USA
e-mail: dbrout@bu.edu

A. Riess
Space Telescope Science Institute, 3700 San Martin Drive, Baltimore, MD 21218, USA

Department of Physics and Astronomy, Johns Hopkins University, Baltimore, MD 21218, USA

E. Di Valentino and D. Brout (eds.), *The Hubble Constant Tension*, Springer Series in Astrophysics and Cosmology, https://doi.org/10.1007/978-981-99-0177-7_19

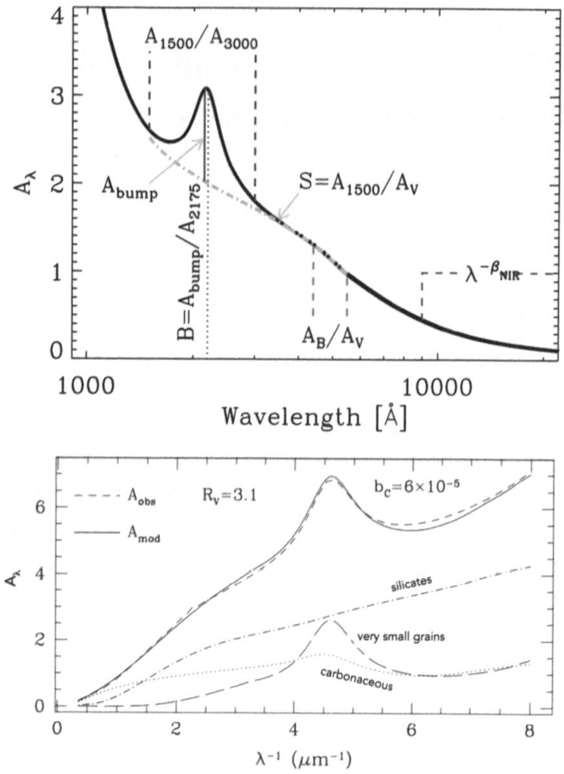

Fig. 19.1 **Top**: Schematic representation [2] of an example non-normalized attenuation or extinction curve. **Bottom**: Schematic representation [3] of the different effects of grain compositions

energy distributions (SEDs). By comparing the observed attenuated SEDs to the unattenuated counterparts, one can estimate the extinction as a function of wavelength. In this chapter, we will explore the mathematical relationships between relevant parameters describing the physics and impact of cosmic dust, including extinction, reddening, and dust laws, with a particular focus on their impact on the cosmic distance ladder and measurements of cosmological parameters.

19.2 Dust Formalism

Dust extinction and attenuation are processes that cause dimming and reddening of light. While they arise from different but related physical mechanisms, they have the same effects on cosmological observables. **Extinction** is due to dust along the line of sight to objects that causes absorption or scattering away from the line of sight. **Attenuation** includes the effects of extinction but also includes scattering back into the line of sight and the contribution to the light by unobscured stars. It is the net loss of light from a complex, spatially extended population and depends on both extinction and the star-dust environment geometry. The wavelength dependent extinction or attenuation can be expressed mathematically as:

$$A_\lambda = m_\lambda - m_{\lambda_0} \quad (19.1)$$

where m_{λ_0} represents the unextinguished magnitude of an astronomical object at a given wavelength, m_λ represents the observed magnitude, and A_λ describes the amount of extinction.

The attenuation or extinction *curve* describes a normalized shape of A_λ as a function of wavelength (see Fig. 19.1 Bottom). The overall normalization (in most cases set by A_V) of the attenuation or extinction curve is set by the column density of grains and the overall shape is set by the mixture of grain types and grain size distribution, and grain optical properties along the line of sight. It is difficult to relate the extinction curve directly to dust physics, so such curves are largely a measured quantity using known background sources and for the purposes of cosmology this is sufficient.

There are several observed features of a typical dust extinction curve (shown in Fig. 19.1). The Ultra-violet (UV)—optical slope (A_{1500}/A_{3000}) and the strength of the UV bump (B) is seen to vary significantly within and across galaxies and provides a large lever arm to constrain the overall dust properties as it provides information about the ratio of the star formation rate (SFR) and the dust column density. The optical slope (A_B/A_V) is important to account for in many cosmological constraints that are observed in the optical, e.g. SNe Ia. The near-IR power-law slope ($\beta_{\rm NIR}$) which is the smallest of these features as dust has the least impact and variation at larger wavelengths.

The color excess $E(B-V)$ characterizes the reddening of an object's spectrum due to dust extinction. It is defined (Eq. 19.2) as the difference between the extinction in two bands (B-V).

$$E(B-V) = A_B - A_V \quad (19.2)$$

By comparing the magnitudes in different bands, one can estimate the amount of reddening caused by dust and gain insights into the properties of the intervening material. Differential measurements of E(B-V) are extremely useful, though to know the true color excess one must have some knowledge of the intrinsic color of the source object. Thus, differential maps of E(B-V) across the Milky-Way galaxy are quite accurate, though knowing the absolute value/normalization of the color excess is a more difficult task (see Sect. 19.3).

Lastly, a useful quantity that is often used in analyses of distance indicators is the 'total-to-selective extinction ratio' R:

$$R_V \equiv A_V/E(B-V) \quad (19.3)$$

where this represents the linear slope of the extinction curve at a given band (e.g. R_V). These ratios provide quantitative information about the differential extinction in the relevant (V-optical) bands/wavelengths and are used in cosmological constraints of Sects. 19.5.1 and 19.5.2.

19.3 Milky Way Dust

Accurate estimates of the Milky Way dust are exceptionally important in cosmology as all electromagnetic observations must contend with it at some level to correct for the observed brightnesses and colors of extra-galactic sources and to understand spatially dependent selection effects.

Reddening: Dust maps can be derived in two main ways: emission-based maps and extinction-based maps. Emission-based maps utilize the observed far-infrared thermal emission from dust to estimate the dust column density. The Schlegel-Finkbeiner-Davis map (SFD: [4]) is based on 100μm data from the Infrared Astronomy Satellite (IRAS), with corrections to remove zodiacal light and the cosmic IR background based on maps from the Diffuse Infrared Background Experiment (DIRBE). This SFD 100μm map has a resolution of 6.1 arcminutes FWHM. Because hotter dust emits more 100μm radiation per column density, the SFD 100μm map is modulated by a temperature correction based on the DIRBE 100μm/240μm color temperature. This modulation is on a 1.3 degree scale where the dust is sufficiently bright, and transitions to a 7 degree scale at high Galactic latitude.

The overall calibration of the SFD map was updated in 2011 [5] based on spectra of main sequence turn-off stars. This followed [6], which used Sloan Digital Sky Survey data of the "blue tip" of the stellar locus (a prominent blue edge in the stellar color distribution) that can be used to estimate extinction along the line of sight and is sensitive to ~10–20 mmag color changes. Using this method [6] find a 14% scaling of the overall SFD map normalization and a preference for the Fitzpatrick 1999 [7] color law over O'Donnell [8] and Cardelli et al. [9] (see next section). [5] follow up on this work using a different technique that uses stellar spectra and modeling codes to find an additional 4% calibration normalization relative to [6] and re-confirm a preference for the Fitzpatrick color law in the optical wavelengths. This 4% difference between modern methodologies has been adopted for the systematic uncertainty in numerous cosmology analyses. Though it is important to emphasize that these uncertainty estimates were best constrained from data within the galactic plane and are not guaranteed to hold true out of the plane of the galaxy where data is more sparse and relative constraints are worse (though with lower reddening amplitude).

After the SFD work, maps based on Planck [10] became available. Planck observations of dust at several frequencies (217, 353, 545, 857 GHz) are used, but in order to constrain the high-frequency side of the thermal emission spectrum, they used the SFD 100μm. In other words, the same deficiencies in the SFD zodiacal light correction and zero point affect the Planck maps. Furthermore, the Planck analysis attempts to infer the emissivity power-law index β along with dust temperature and optical depth τ, introducing additional noise at high latitude where the emission is to faint to constrain 3 parameters. So although the Planck data are superior to IRAS in general, the derived dust map shares some shortcomings with SFD.

Another method for mapping dust involves utilizing inferred stellar extinctions while considering a presumed or fixed reddening law. Empirical constraints on extinction require some knowledge of the unextinguished spectrum, either via models or empirical pairs of more/less reddened stars. Estimations of reddening along the line of sight to these stars is somewhat challenging, but not an overwhelming burden. Furthermore, since each star provides information about the extinction up to its distance, these stellar reddening-based maps allow for the reconstruction of dust extinction in three dimensions [11–13]. However, these maps may suffer from contamination due to incomplete removal of galaxies and quasi-stellar objects (QSOs) from stellar catalogs [14]. This contamination can lead to systematic errors, causing both overestimation and underestimation of extinction at the positions of galaxies and QSOs. Thus, these maps are not perfect and there is the need to motivate further progress for the next generation of experiments.

Dust Laws: For use in broadband photometry fits like that of SN Ia and Cepheid luminosity measurements, it is important that attenuation curves are parameterized for use in fitting. Common parameterizations distill attenuation to single parameter characterizing the slope of of a predefined curve, though some include additional parameters describing the UV bump (see Fig. 19.1). Such parameterizations do not need to be unique to the Milky Way galaxy and can extend to extra-galactic dust measurements for comparison. While the variations in attenuation curves has thus far been difficult to relate to dust physics, for the purposes of precision cosmological constraints the empirical relations described below are adequate.

Fitzpatrick [7] provide a single-family parameterization for extinction curves (originally intended for the Milky-Way, though now extended beyond) from the optical to near-IR. They used a comprehensive dataset of reddened stars with known intrinsic colors and distances, coupled with detailed spectroscopic measurements to compute reddening effects across a wide range of wavelengths and characterize R(V). Fitzpatrick [7] showed significant variations in the extinction law along different sight lines and several notable studies have extensively confirmed and characterized these variations [5, 6, 15]. The impact of this variation is discussed in Sect. 19.5.2.

19.4 Extragalactic Dust

For many cosmic distance probes, it is also important to understand the dust in the galaxy to which distances are being measured. As R_V varies for different dust grain sizes and composition, and galaxies have different dust properties, it is well known that different galaxies and different regions within galaxies exhibit a wide range of

R_V values. In fact, while the Milky Way galaxy has an R_V on average $\sim$3.1, it has a distribution of at least $\sigma_{R_V} = 0.2$ [16]. Additionally, different parts of the LMC and SMC have been found to have R_V values with a range of $R_V \sim 2 - 5$ [17, 18]. The origin of these dramatically different extinction curves is still debated as it could be driven by metallicity, radiation field strength, or both.

Furthermore, [19] study the dust attenuation curves of 230,000 individual galaxies in the local universe, using GALEX, SDSS, and WISE photometry calibrated on the Herschel ATLAS, and they find quiescent galaxies, which are typically high-mass, have a mean $R_V = 2.61$ and star-forming galaxies, which are lower-mass on average, have a mean $R_V = 3.15$.

R_V has also been measured through large SN sample statistics and detailed studies of individual SNe, though often with varying sets of assumptions. [20] compiled 13 various studies of SN Ia samples from the literature which determined a range of R_V values from $\sim$1 to $\sim$3.5. [20] itself determined R_V from nearby SNe and for 21 SNe Ia observed in Sab-Sbp galaxies and 34 SNe in Sbc-Scp they find $R_V = 2.71 \pm 1.58$ and $R_V = 1.70 \pm 0.38$ respectively. While so many past analyses have recovered $R_V < 2$ for studies of individual SNe (e.g. [21, 22]), these were often SNe Ia with high $E(B - V)$, and it was postulated R_V may decrease with $E(B - V)$. However, [23] found from a sample of modestly reddened ($E(B - V) < 0.25$ mag) SNe Ia, a small value of $R_V \sim 1$ and more recently, [24] analyzed high-quality UV-NIR spectra of 6 SNe and found that SNe with high reddening indicated R_V's ranging from $\sim$1.4 to $\sim$2.8 and SNe with low amounts of reddening also indicated R_V's of $\sim$1.4 and $\sim$2.8. Importantly, [24] stressed that the observed diversity in R_V is not accounted for in analyses that measure the cosmological expansion of the universe.

Since the low R_V values (< 2) are not found in studies of the Milky Way, this has motivated various SN Ia studies to ascribe the dust to circumstellar dust around the progenitor at the time of the explosion [25, 26]. However, an alternative interpretation could be that the low R_V values are caused by dust in the interstellar medium [27]. This understanding has been supported by [28, 29], which constrained the location of the dust that caused the reddening in the SN Ia spectra to be, for the majority of the SNe that they observed, on scales of the interstellar medium, rather than circumstellar surroundings.

19.5 Impact on Distance Indicators

19.5.1 Impact on Cepheid Magnitudes

While the impact of dust on measurements of galactic and extra-galactic Cepheids is minimal due to the fact that in the distance ladder approaches they are observed in the Hubble Space Telescope (HST) Near Infrared (NIR) bands, we detail here the possible impact and its relevance for the inference of the Hubble Constant. For the case of nearby first and second rung Cepheids in the SH_0ES distance ladder, they are all at extremely low redshift ($z < 0.01$), and therefore both Milky Way and extra-galactic dust affect the same range of wavelength and their effects are treated simultaneously below.

Extra-galactic Cepheids are observed in regions of recent star formation in late-type galaxies and are found to have a mean reddening of $E(V-I) \sim 0.3$mag. The SH_0ES [1] observations in HST NIR F160W (H-Band) reduces the impact of and correct ('deredden') for the effects of interstellar dust reddening.

For the distance ladder, the goal is to measure the relative distance between two sets of Cepheids (rungs) free from reddening which by definition is $(m_{H,i} - A_{H,i}) - (m_{H,j} - A_{H,j})$ for the ith and jth samples where A_H is the extinction from dust and comes from the color excess, $E(V-I)$ multiplied by a reddening ratio, R_H, so $A_H = R_H \times E(V-I)$. However, in practice we do not directly measure $E(V-I)$ which would require knowledge of the intrinsic color, $(m_V - m_I)_0$. Rather, we observe the apparent color, $(m_V - m_I)_0$. A useful, shortcut to measuring the distance difference using only the apparent quantities is to use so-called "Wesenheit" magnitudes [30] (m_H^W) following:

$$m_H^W = m_H - R_H \times (m_V - m_I) \tag{19.4}$$

This measure is based only on directly observed quantities. Below we denote the components of the Cepheid model that make up $(m_V - m_I)$. Wesenheit magnitudes m_H^W are not true magnitudes in the sense that they do not represent the brightness. Rather, their utility is in the comparison between two sets of such magnitudes, which results in the reddening-free magnitude difference.

Their utility can be seen when we decompose the apparent color $(m_V - m_I)$ into its constituents. The main component is $(m_V - m_I)_0$ which would be the "unreddened" color at the middle of the Cepheid instability strip (typically this term is $V - I \sim 1$). The displacement in color from the midline of the instability strip is given by $\Delta(m_V - m_I)$ which corresponds to an intrinsic property of each Cepheid. Finally, the color excess due to dust is given by $E(V-I)$ which is reddening by dust:

$$m_V - m_I = (m_V - m_I)_0 + \Delta_{V,I} + E(V-I) \tag{19.5}$$

Thus, multiplying $(m_V - m_I)$ of Eq. 19.4 by a single R_H for two samples i and j (i.e. two rungs of the distance ladder) results in the following when applied to the decomposition in 19.5. The terms $(m_V - m_I)_0$ multiplied by R_H will be the same for different galaxies, this being an intrinsic property of Cepheids, and it is important to use the same value of R_H (whatever that is) where multiplying by the intrinsic color. When multiplying R_H on the $\Delta_{V,I}$ term, because the mean of $\Delta_{V,I}$ is at or near zero (by definition) for large samples of Cepheids that fill the instability strip, $R_H \Delta_{V,I}$ will also average to zero for a statistically large sample. However, because Cepheids, like most stars, are dimmer when they are redder (cooler), a positive value of R_H can also reduce the apparent dispersion of the instability strip. In fact, observations of stars show the change in intrinsic color with brightness is of the same order as the reddening law, R, a happy coincidence, so that the term R_H reduces variance in two ways. Therefore, it is common to choose the value of R_H to be the reddening ratio between the H-band and the $V - I$ color, as derived from a reddening law. Lastly, the only term that does not cancel or we expect to not be consistent across galaxies is $R_H \times E(V - I)$ because $E(V - I)$ varies within and across Cepheid samples. Thus, we can now rewrite the observable of Eq. 19.4 as $m_H - R_H \times E(V - I)$. Finally, taking the differential between two samples leaves us with our desired result $m^W_{H,i} - m^W_{H,j} = (m_{H,i} - R_H \times E(V - I)_i) - (m_{H,j} - R_H \times E(V - I)_j)$.

We get the exact difference in distance when $R_{W,i} = R_{W,j}$ (and with a large Cepheid sample, $\Delta_{V,I} = 0$). We note here that the "shortcut" Eq. 19.4 only works if we require $R_{W,i} = R_{W,j}$. If we do not want to require $R_{W,i} = R_{W,j}$, then we must decompose 19.4 into the terms in 19.5 and then we need to define and model the intrinsic standardizable color of Cepheids from another sample and only allow for R_W to vary across for the dust excess component [1, 31, 32].

In the case of correcting NIR magnitudes (m_H) following Eq. 19.4, we have thus far assumed a single R_H. Fortunately, R_H is well-characterized to be small ($\sim$0.4) for Milky Way-like extinction. This value is much smaller than for correcting optical magnitudes ($R_V \sim 3$) and demonstrates the comparatively low sensitivity to dust in the SH_0ES analysis. This dust law insensitivity is also important when considering potential variation in the dust curve across the population of Cepheid host galaxies. While R_V can range from 2 to 6, R_H correspondingly ranges from 0.35 to 0.5 with a definitive lower bound at $\sim$0.2 due to the Rayleigh scattering limit or 0.25 for any known reddening law.

While R_H has been determined independently, R_H can also be determined from the Cepheids themselves as the value that minimizes the observed scatter in the observed Period-Luminosity relation. From the highest quality Milky Way Cepheids $R_H = 0.363 \pm 0.038$ [33] which matches independently determined values in the literature. Additionally, R_H determined from the entire distance ladder of Cepheids is $R_H = 0.34 \pm 0.02$ [1]. For this value, H_0 increases by 0.3.

It is possible to consider different values of the reddening ratio for each individual host. However, because with different R_H the intrinsic color term $(m_V - m_I)_0$ no longer cancels across rungs, the intrinsic term loses its meaning, and defies the interpretation of Cepheids as standard candles. So this cannot be done directly with Wesenheit magnitudes. Therefore, to vary the reddening ratio, one must first subtract the color excess using an empirical period-color relation, such as $(V - I)_0 = 0.5 + 0.25 \log P$ (empirical, Milky Way or Large Magellanic Cloud) from $(V - I)$, or this can lead to a spurious result [32]. When R_H is fit individually for each host, H_0 has been shown to change at the 0.8 km/s/Mpc level, as shown in [31]. In practice, the Cepheid data is not very constraining of individual reddening ratios in the SN Ia hosts, so an alternative and better approach is to assign individual host reddening ratios based on the host star formation rate and mass which predict $2.8 < R_V < 4.2$ [1].

Observing in the NIR to mitigate the impact of dust is not without its own challenges. On HST, the resolution in NIR is 2–3 times worse than that of the optical and in the NIR the background of ubiquitous red giant stars is an order of magnitude higher. However, with the advent of the James Webb Space Telescope (JWST), we have been further reassured that there is not a confounding issue when it comes to the impact of dust for Cepheids. Both [34] and [35] show that JWST provides significantly improved resolution, improved sensitivity in the NIR, and even lower impact of interstellar dust in comparison to HST. They find good agreement between the HST and JWST systems and no evidence for bias beyond the 0.04mag level, which is significantly better than the 0.2mag size of the H_0 tension and is expected to further improve with future observations.

19.5.2 Impact on Type Ia Supernova Distance Moduli

In order to compute distances, Type Ia Supernova (SN Ia) photometric light curves are fit to an underlying spectral time-series model (e.g. SALT3 [36]). These fits output the following summary statistics that characterize each light curve: the peak B-band brightness (m_B), the duration 'stretch' (x_1), and the color (c). The latter of which, c, characterizes the combination of intrinsic SN Ia color and dust reddening because at the photometric light curve level such properties are degenerate. SN Ia light-curve fit parameters are then combined empirically to compute distances ($\mu_{\rm obs}$) where the nuisance parameters (α, β, γ) are single numbers that correlate light-curve and host properties with an intrinsic luminosity and are determined such that they minimize the scatter in distances of SNe Ia [37]. The distance moduli are computed following a modified version of the Tripp [38] equation:

$$\mu_{\rm obs} = m_B + \alpha x_1 - \beta c + \gamma G_{\rm host} - \mathscr{M}_B + \mu_{\rm bias}, \tag{19.6}$$

where $G_{\rm host}$ is each host-galaxy's mass that has been shown to correlate with SN Ia brightness [39–42], $\mu_{\rm bias}$ is a correction for instrumental and survey selection effects for each SN based on careful survey simulations [43], and $\mathscr{M}_B$ is a single number representing the intrinsic magnitude of a SN Ia as determined by the SH_0ES distance ladder.

For SNe Ia, galactic and extra-galactic dust both reddens and obscures light in the rest-frame UV to optical wavelengths. With the exception of the future NASA Roman Space telescope, large area supernova surveys are typically observed from the ground in the optical wavelengths (due to atmospheric transmission). The analysis of Pantheon+ [45] and SH_0ES [1] account for Milky Way galactic dust by applying the Schlafly et al. 2011 stellar locus map [5]. A second correction is applied with a 4% absolute scale offset following [6] found using the stellar spectral modeling approach. This difference is evaluated in the Pantheon+ SN Ia covariance matrix in order to characterize the systematic uncertainty.

Dust in the host galaxy of the supernova is degenerate with the light-curve color and is hypothesized to be the major cause of the so called intrinsic scatter of standardized SN Ia brightnesses (the excess scatter of SN Ia distance residuals compared to a best-fit cosmology after accounting for measurement noise). Recent promising

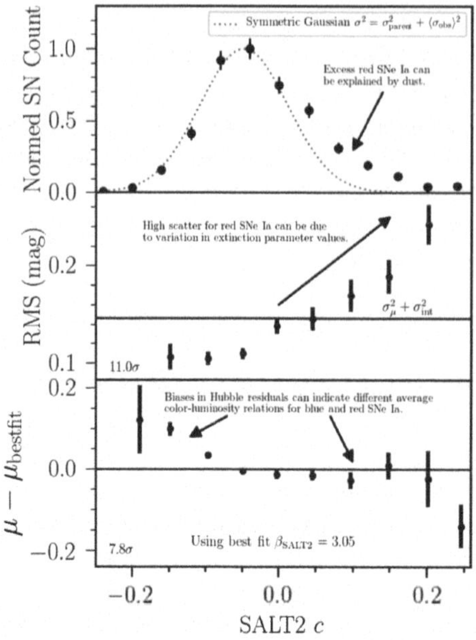

Fig. 19.2 **Top**: Observed SALT light curve fit parameter color (c) from the full data sample of [44], with a symmetric Gaussian overlaid (dashed) for comparison. Skewness in the observed dataset to the red is evident and indicative of dust. **Middle**: RMS scatter of Hubble Diagram residuals as a function of color. The RMS is calculated after Tripp standardization and after subtracting the mean Hubble residual bias. SNe Ia with bluer colors exhibit significantly better standardizability than their redder SN Ia counterparts. This is indicative of unmodeled variation in dust properties for each SN Ia or host galaxy. **Bottom**: Binned Hubble diagram residuals as a function of color, after Tripp standardization using a single best fit β. The bluest SNe Ia exhibit a different color-luminosity relation than the reddest; indicative of an intrinsic SN Ia distribution and a secondary distribution characterized by dust. In the bottom two panels, the significance of the deviation from a flat line is show in the bottom corner

models have now explained the majority of SN Ia intrinsic scatter using dust. The size of SN Ia intrinsic scatter has recently been found to depend on SN Ia color and host galaxy properties (BS21: [44]). Recently, BS21 introduced dust extinction to the SN Ia spectral model by forward modeling simultaneously 1) an intrinsic SN color population, 2) extrinsic dust populations, 3) variation in the dust properties from galaxy to galaxy, and 4) correlations between the dust properties of galaxies with different host properties (G_{host}). BS21 marks a major improvement over previous intrinsic scatter models commonly used in SN Ia cosmology analyses (e.g. grey scatter: Guy et al. 2010 [46] and spectral scatter: Conley 2011 [47]. While the Tripp Eq. 19.6 does not disentangle intrinsic color and extrinsic dust properties, by forward modeling this process and analyzing the forward modeled simulations with the Tripp equation, the model can be compared with and fit to the data in a hierarchical approach (see Fig. 19.2).

The major successes of the BS21 model are that solves two key problems with one simple solution. (1) It allows for R_V's in high and low mass host galaxies which explains the observed correlation with host properties. (2) It explains the observed color dependent intrinsic scatter with variation in dust properties. It also makes key predictions, that the typical SN Ia dust (whether it be local to the SN or global to the host galaxy) corresponds to an $R_V \sigma 2$ and with a large inter-galaxy dispersion ~ 1 (BS21 and [48]) which has similarly been seen in numerous followup studies [49–51]. Continuing upon the improved modeling developed in BS21, [52] implement the dust model used in cosmological analyses. [52] correct the Pantheon+ distance moduli on a population level and [48] address the computational limitations in the hierarchical model fitting process in order to provide robust model uncertainties that include correlations in uncertainties between dust parameters and SN intrinsic parameters.

Note that alternative models for the correlation between SN Ia luminosity and host-galaxy properties have been proposed that involve progenitor differences [53] or different subclasses [54] instead of dust. However, in light of recent findings in Pantheon+ for the very strong color dependence of intrinsic scatter and β, such alternative scenarios must explain now a three path mechanism to explain the observations of SN Ia color skewness to the red, SN Ia one-sided color-dependant scatter, and both scatter and β dependencies on host properties (not depicted in Fig. 19.2). The large statistics of the Pantheon+ sample illuminating these dependencies makes such alternative approaches challenging.

A new SN-host correlations model was recently explored in the Dark Energy Survey Supernova Program that accurately and realistically models physically motivated relations between SN age, SN host galaxy properties, and SN stretch (x_1). Vincenzi et al. (in prep) find that this model does not explain the host-mass correlation with luminosity, so they use this model in conjunction with a BS21 dust based scatter model and find that inferred distances to SN Ia remain relatively unchanged relative to the approach taken in Pantheon+. Finally, further evidence for BS21 is substantiated by the fact that dust properties are also found to correlate with other physical parameters of interest in galaxy surveys. [19] find lower mass galaxies tend to have steeper attenuation curves (R_V).

19.6 Impact on Cosmological Constraints

19.6.1 Dust Impact (or Lack-Thereof) on the H_0 Tension

The SH_0ES methodology of obtaining the Hubble constant is one that leverages differential measurements between the first and second rungs (for Cepheids), and the second and third rungs (for SNe Ia). Here we emphasized that only differences in dust properties between rungs is what could potentially contribute to biases in H_0.

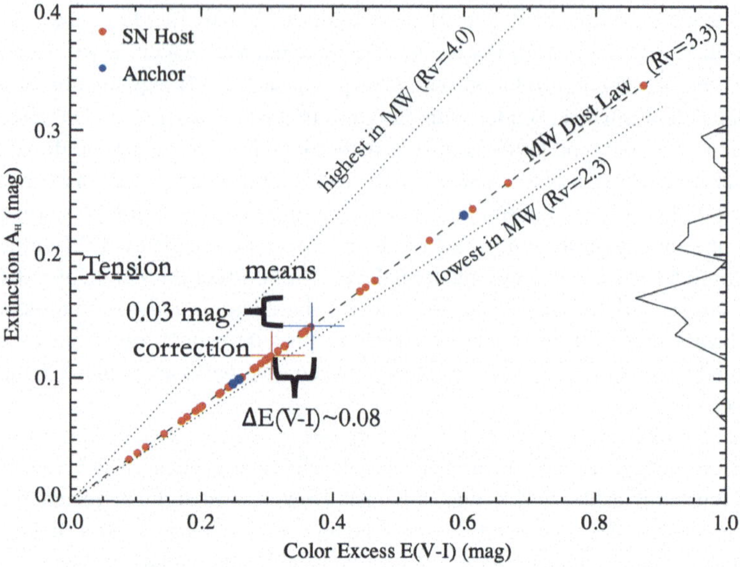

Fig. 19.3 Difference between the observed color excess of the Cepheids for the first (Anchor) and second (SN Host) rungs of the distance ladder for different underlying dust law (R_V). H_0 Tension of 0.2 mag is shown for comparison to the 0.03 mag differences observed in the sample

This is because the SH_0ES likelihood and methodology is a differential measurement for Cepheids and SNe Ia, meaning that offsets that are consistent across rungs cancel.

Cepheids:
Figure 19.3 visualizes the impact of dust on the Cepheids in the first two rungs of the distance ladder (Geometric "Anchor" Cepheids and "SN Host" Cepheids). For the Cepheids in both rungs (red and blue points), color excess $E(V-I)$ is measured relative to the intrinsic Cepheid color at the center of the instability strip and for a typical Milky Way dust law ($R_V = 3.3$) the corresponding extinction (A_V) values are inferred from the color excess. As discussed in Sect. 19.4, there are numerous extinction laws not only within a galaxy but also across galaxies. Therefore, Fig. 19.3 also visualizes other possible dust laws ranging from R_V of 2–4.

The difference in mean color excess and inferred extinction is the most relevant quantity for assessing the impact on the H_0 tension. Thus, we show the Cepheid color excess for each rung (First: blue, Second: red) and the average difference ($\Delta E(V-I) \sim 0.08$) in Fig. 19.3. If the means between rungs are identical, H_0 is completely insensitive to dust by nature of the distance ladder methodology, as only differences in the means of each rung can propagate to changes in H_0.

The entire H_0 tension between SH_0ES and Planck (6km/s/Mpc) can be expressed as 0.18mag in ΔA_V (shown for reference on Fig. 19.3). However, because the mean color excess between the first and second rungs is $\Delta E(V-I) \sim 0.08$ and because a Milky Way dust law of R_V=3.3 corresponds to an R_H=0.4 in the infrared (because the

Cepheid luminosities are computed in HST H-band), this corresponds to a 0.03 A_H difference for 'Milky Way-like' dust. If the typical dust law is not Milky Way-Like, but rather R_V=2.3 ($R_H = 0.35$) or in a worse case scenario of R_V=4.0 ($R_H = 0.35$), the difference in extinction A_H could range from 0.025 to 0.04, however, in no case does the expected A_H approach the 0.18mag needed to resolve the H_0 tension. Additionally, analyses such as [31] and [32] have allowed for R_H to vary from galaxy to galaxy and find that, despite the extra model freedom and parameters, because R_H is small and the color excess are small, variations in R_H across Cepheid galaxies cannot explain the H_0 tension. Finally, in Table 19.1 are depicted the SH_0ES variants; SH_0ES explore numerous attenuation laws (e.g. Cardelli et al. [9] and Fitzpatrick [7]), attenuation law slopes and dependencies, and even remove the dust treatment altogether and show that they all have either minimal impact on H_0 or move the central value up to make the tension worse.

SNe Ia:
As shown in Fig. 13 of Pantheon+, the impact of a 4% absolute scaling of galactic dust maps on H_0 is negligible. This is because, first, SNe Ia are typically discovered in low extinction regions of the sky. Second, the average Milky Way $E(B - V)$ of the second rung is 0.050 and of the third rung is 0.046; multiplying the difference of 0.004 by the typical Milky Way color law of $\sim$ 3 results in an average brightness difference between the rungs of 0.012 mag. Given that the H_0 difference is $\sim$0.20 mag, this suggests very minimal impact on H_0 from differences in Milky Way $E(B - V)$ across rungs (See Table 19.1 for ΔH_0). Lastly, it is important to note that small errors in the dust maps will result in different inferred SALT color parameter (c) of each light curve. Because the SALT color-luminosity relation $\beta \sim 3$ is very similar to that Milky Way dust attenuation law $\sim$3, dust map differences are fortuitously 'absorbed' by the Tripp estimator Eq. 19.6.

Pantheon+ incorporate distance uncertainties in a covariance built from numerous systematic perturbations to the analysis, including perturbations characterizing the uncertainty in the underlying model of dust. The inferred H_0 for each of the systematic perturbations is shown in Fig. 13 of Pantheon+ and shows a 0.2 km/s/Mpc scatter in H_0 resulting from the uncertainty in the dust population parameters, which is well below the level of the H_0 tension. This is bolstered by the fact that because the Cepheids are only visible in edge-on late type/spiral galaxies in the second rung, and robust determinations of H_0 select for the similar types of host galaxies in the third rung of the distance ladder. While the underlying dust properties and differences between spiral galaxies and other galaxy types remain an open question, restricting the third rung of the distance ladder in SH_0ES to only late type galaxies mitigates possible differences in galaxy properties between rungs. The SH_0ES analysis explore an additional variant in which they allow all types of Hubble flow host galaxies (and therefore dust properties) and in this case find a $\Delta H_0 = +0.3$, exacerbating the tension.

The ΔH_0 relative to the BS21 dust model is shown in Table 19.1 for the Guy et al. [46] and Conley [47] intrinsic scatter models. Instead of dust, the Guy et al. [46] model describes SN Ia scatter with a grey luminosity scatter. While this model

Table 19.1 Variants to the Pantheon+ and SH_0ES analysis that assess the impact of dust. Negative ΔH_0 is in the direction of reducing tension

Dust-Related Systematic	ΔH_0(km/s/Mpc)
Cepheids	
Milky Way (MW) Color Law from Fitzpatrick (1999), $R_V = 2.5$	+0.20
MW Color Law from Cardelli, Clayton, & Mathis (1989), $R_V = 3.1$	+0.05
R_W free global	+0.20
Intrinsic color subtracted & MW Color Law from Fitzpatrick (1999) & $R_V =$ 3.3	+0.09
Intrinsic color subtracted & MW Color Law from Fitzpatrick (1999) & $R_V =$ free	+0.30
Intrinsic color subtracted & R_V (host mass-SFR)	+0.81
No dust treatment (A_H values assumed to cancel)	+0.74
SNe Ia	
Milky Way (MW) Dust Map Absolute Scaling (4%)	−0.02
MW Color Law from Cardelli, Clayton, & Mathis (1989)	+0.05
Grey Scatter Model from Conley (2010)	−0.20
Spectral Scatter Model from Guy et al. (2011)	−0.22
SN Ia/Host Dust Model Parameter Uncertainty 1	−0.02
SN Ia/Host Dust Model Parameter Uncertainty 2	−0.12
SN Ia/Host Dust Model Parameter Uncertainty 3	−0.13
All host galaxy morphologies/types in third rung	+0.28
Hubble flow high mass hosts only $log_{10}(M_{\text{stellar}}) > 10$	−0.07

does not perform as well on the most recent Pantheon+ data (see Fig. 5 of BS21), nonetheless while this model is agnostic to host-galaxy dust it only achieves a ΔH_0 of -0.2 km/s/Mpc. Likewise, Conley (2011) characterizes SN Ia scatter with empirical spectral variations and finds ΔH_0 of -0.2 km/s/Mpc.

There have also been numerous independent crosschecks on the SNe Ia and Pantheon+ dust treatment. First, [55] derive SN Ia distances using infrared light curves thus reducing the impact of dust. They find reduced correlations with host-galaxy properties and SN Ia luminosity, they suggest that dust drives the correlations and their infrared observations explain the lack of correlation. Ultimately [55] report a value of the $H_0 = 73.2 \pm 2.3$ km/s/Mpc.

Most recently, [56] assessed the impact of dust on a subset of the SNe Ia for which additional wavelengths from UV to NIR were available and with an improved SN Ia model were able to disentangle the dust and SN properties for 37 calibrator SNe and 67 Hubble flow SNe. While only for a subset of the total Pantheon+ SNe, this importantly meant [56] could simultaneously compute an independent check on the Pantheon+ dust modeling and find $H_0 = 74.82 \pm 1.28$.

A third major crosscheck has been made in [57] on the SH_0ES methodology of restricting SN Ia hosts in the Hubble flow to match that of the second rung of

Cepheid hosts. To avoid the SN Ia dust systematics imposed by restricting to late-type star-forming galaxies in both the second and third rungs of the distance ladder, [57] use what is colloquially called a 4-rung distance ladder. Using galaxy groups, [57] transfer the Cepheid calibration late-type galaxies to elliptical galaxies using the surface brightness fluctuations (SBF) technique of determining relative distances of early-type galaxies. These galaxies have different distributions of mass and dust and also host different populations of SNe Ia. While the introduction of an additional and less precise SBF rung of the distance ladder results in a larger uncertainty, the recovered central value is an important crosscheck on many host dust related systematics. [57] find $H_0 = 74.6 \pm 2.85$.

19.6.2 Dust Impact on SN Ia Constraints of Dark Energy and Dark Matter

While the above has shown that impact on inference of H_0 from Milky-Way and host-galaxy dust is minimal, the same cannot be said for inference of other cosmological parameters: Matter+Dark Matter Density (Ω_M) and Equation of State of Dark Energy (w). The H_0 insensitivity is largely due to similarities between the second and third rung SNe Ia and hosts, but it is also attributable to the small redshift range in the Hubble flow (third) rung of the distance ladder $0.023 < z < 0.15$, which corresponds to a relatively small span of <1.5 billion years of cosmic history over which dust properties have little time to evolve.

For inference of Ω_M and w, a much larger span of cosmic history is used to compile sufficient SN Ia statistics and the constraints of those models are also best performed at relatively higher redshifts. Samples of SNe Ia covering a very large span of cosmic history (currently 11 billion years) opens up the possibility for the evolution of dust parameters or the environments in which SNe Ia are found. Both [44] and [48] show that if uncertainties in dust populations of SN Ia hosts can evolve with redshift and this results in the largest systematic uncertainty in SN Ia cosmology today (on par with photometric calibration). Though, if correlations between dust properties and host-galaxy photometric properties (i.e. stellar mass, star formation rate, color) can be well constrained at low-redshift, then potential population evolution can be traced photometrically for future high redshift surveys. We note here that while there is much effort on survey calibration for future telescopes, there is comparatively less effort being placed on Milky Way and host-galaxy dust modeling, both of which are tractable issues with external datasets and effort.

Future surveys such as The Vera Rubin Observatory Legacy Survey of Space and Time (LSST) and The Nancy Grace Roman Space Telescope (Roman) will be forced to tackle this challenge. LSST observes in optical wavelengths and will be more susceptible to the impact of host-galaxy dust and population evolution similarly to that of the Dark Energy Survey analysis. However, unlike the Dark Energy Survey, LSST will survey the entire southern sky (at much larger values of Milky Way

$E(B - V)$) and will have to contend with increased systematics due to Milky Way dust. It will be important to improve uncertainties on Milky Way dust maps in regions of the sky where extinction is larger. Comparatively, Roman will observe in the infrared, which will make it less sensitive to dust for SNe Ia at moderate redshifts, though Roman will contend with the same dust population evolution systematics as other surveys when observing SNe Ia at redshift 3 (rest-frame optical).

19.7 Conclusion

In conclusion, we have explored the impact of dust on Cepheid and Type Ia Supernova distance indicators and the resulting impact on cosmological constraints, focusing on the measurement of the Hubble Constant (H_0). We show that the effects of dust propagate to changes in H_0 only if there are significant differences between different rungs of the distance ladder. We discuss the impact of Milky-Way dust and host-galaxy dust and the associated uncertainties in these properties. In both cases, the differential measurements employed in the distance ladder approach for both Cepheids and SNe Ia largely cancel out the effects of dust, making them robust tools for measuring H_0.

For Cepheids, which are observed in the Hubble Space Telescope (HST) Near Infrared (NIR) bands, the impact of dust is minimal. The use of Wesenheit magnitudes allows for the correction of dust effects based on directly observed quantities and the small value of the reddening ratio (R_H) in the NIR reduces the sensitivity to dust, which has been well-characterized for Milky Way-like extinction. For SNe Ia, we discuss the application of dust corrections to SNe Ia as a function of host-galaxy properties and as an important component of accounting for SN Ia intrinsic scatter. These dust-based corrections are applied consistently across rungs and uncertainties in the dust model parameters are characterized and accounted for and do not account for the observed Hubble tension.

Overall, the analysis presented in this paper reaffirms the robustness of current H_0 measurements and suggests that dust-related effects are not a significant source of tension in the determination of the Hubble Constant. However, we emphasize the importance of continued refinement in our understanding of dust properties and dust variation for cosmological studies beyond H_0 measurements.

Acknowledgements Brout and Riess would like to thank Doug Finkbeiner for his useful comments on this chapter and discussions on the topic in general. This is a preprint of the following chapter: Dillon Brout and Adam Riess, The Impact of Dust on Cosmic Distance Indicators, published in The Hubble Constant Tension, edited by Eleonora DiValentino and Dillon Brout, expected 2024, Springer Nature reproduced with permission of Springer Nature Singapore Pte Ltd. The final authenticated version will be available online at: https://doi.org/10.48550/arXiv.2311.08253.

References

1. A.G. Riess, W. Yuan, L.M. Macri, D. Scolnic, D. Brout, S. Casertano, D.O. Jones, Y. Murakami, G.S. Anand, L. Breuval, T.G. Brink, A.V. Filippenko, S. Hoffmann, S.W. Jha, W. D'arcy Kenworthy, J. Mackenty, B.E. Stahl, W.K. Zheng, A comprehensive measurement of the local value of the Hubble constant with 1 km s^{-1} Mpc^{-1} uncertainty from the Hubble space telescope and the SH_0ES team. ApJ **934**(1), L7 (2022)
2. S. Salim, D. Narayanan, The dust attenuation law in galaxies. Ann. Rev. Astron. Astrophys. **58**, 529–575 (2020)
3. J.C. Weingartner, B.T. Draine, Dust grain–size distributions and extinction in the milky way, large magellanic cloud, and small magellanic cloud. The Astrophys. J. **548**(1), 296–309 (2001)
4. D.J. Schlegel, D.P. Finkbeiner, M. Davis, Maps of dust infrared emission for use in estimation of reddening and cosmic microwave background radiation foregrounds. The Astrophys. J. **500**(2), 525–553 (1998)
5. E.F. Schlafly, D.P. Finkbeiner, Measuring reddening with Sloan digital sky survey stellar spectra and recalibrating SFD. ApJ **737**(2), 103 (2011)
6. E.F. Schlafly, D.P. Finkbeiner, D.J. Schlegel, M. Jurić, Željko Ivezić, R.R. Gibson, G.R. Knapp, B.A. Weaver, The blue tip of the stellar locus: measuring reddening with the Sloan digital sky survey. The Astrophys. J. **725**(1), 1175–1191 (2010)
7. E.L. Fitzpatrick, Correcting for the effects of interstellar extinction. Publications of the Astronomical Society of the Pacific **111**(755), 63–75 (1999)
8. J.E. O'Donnell, R v-dependent optical and near-ultraviolet extinction. ApJ **422**, 158 (1994)
9. J.A. Cardelli, G.C. Clayton, J.S. Mathis, The relationship between infrared, optical, and ultraviolet extinction. ApJ **345**, 245 (1989)
10. Planck Collaboration, N. Aghanim, M. Ashdown, J. Aumont, C. Baccigalupi, M. Ballardini, A.J. Banday, R.B. Barreiro, N. Bartolo, S. Basak, K. Benabed, J.-P. Bernard, M. Bersanelli, P. Bielewicz, L. Bonavera, J.R. Bond, J. Borrill, F.R. Bouchet, F. Boulanger, C. Burigana, E. Calabrese, J.-F. Cardoso, J. Carron, H.C. Chiang, L.P.L. Colombo, B. Comis, F. Couchot, A. Coulais, B.P. Crill, A. Curto, F. Cuttaia, P. de Bernardis, G. de Zotti, J. Delabrouille, E. Di Valentino, C. Dickinson, J.M. Diego, O. Doré, M. Douspis, A. Ducout, X. Dupac, S. Dusini, F. Elsner, T.A. Enßlin, H.K. Eriksen, E. Falgarone, Y. Fantaye, F. Finelli, F. Forastieri, M. Frailis, A.A. Fraisse, E. Franceschi, A. Frolov, S. Galeotta, S. Galli, K. Ganga, R.T. Génova-Santos, M. Gerbino, T. Ghosh, Y. Giraud-Héraud, J. González-Nuevo, K.M. Górski, A. Gruppuso, J.E. Gudmundsson, F.K. Hansen, G. Helou, S. Henrot-Versillé, D. Herranz, E. Hivon, Z. Huang, A.H. Jaffe, W.C. Jones, E. Keihänen, R. Keskitalo, K. Kiiveri, T.S. Kisner, N. Krachmalnicoff, M. Kunz, H. Kurki-Suonio, J.-M. Lamarre, M. Langer, A. Lasenby, M. Lattanzi, C.R. Lawrence, M. Le Jeune, F. Levrier, P.B. Lilje, M. Lilley, V. Lindholm, M. López-Caniego, Y.-Z. Ma, J.F. Macías-Pérez, G. Maggio, D. Maino, N. Mandolesi, A. Mangilli, M. Maris, P.G. Martin, E. Martínez-González, S. Matarrese, N. Mauri, J.D. McEwen, A. Melchiorri, A. Mennella, M. Migliaccio, M.-A. Miville-Deschênes, D. Molinari, A. Moneti, L. Montier, G. Morgante, A. Moss, P. Natoli, C.A. Oxborrow, L. Pagano, D. Paoletti, G. Patanchon, O. Perdereau, L. Perotto, V. Pettorino, F. Piacentini, S. Plaszczynski, L. Polastri, G. Polenta, J.-L. Puget, J.P. Rachen, B. Racine, M. Reinecke, M. Remazeilles, A. Renzi, G. Rocha, C. Rosset, M. Rossetti, G. Roudier, J.A. Rubiño-Martín, B. Ruiz-Granados, L. Salvati, M. Sandri, M. Savelainen, D. Scott, C. Sirignano, G. Sirri, J.D. Soler, L.D. Spencer, A.-S. Suur-Uski, J.A. Tauber, D. Tavagnacco, M. Tenti, L. Toffolatti, M. Tomasi, M. Tristram, T. Trombetti, J. Valiviita, F. Van Tent, P. Vielva, F. Villa, N. Vittorio, B.D. Wandelt, I.K. Wehus, A. Zacchei, A. Zonca, Planck intermediate results—xlviii. Disentangling galactic dust emission and cosmic infrared background anisotropies. A&A **596**, A109 (2016)
11. Gregory M. Green, Edward Schlafly, Catherine Zucker, Joshua S. Speagle, Douglas Finkbeiner, A 3d dust map based on Gaia, Pan-STARRS 1, and 2MASS. The Astrophys. J. **887**(1), 93 (2019)
12. J.L. Vergely, R. Lallement, N.L.J. Cox, Three-dimensional extinction maps: Inverting intercalibrated extinction catalogues. Astron. Astrophys. **664**, A174 (2022)

13. G. Edenhofer, C. Zucker, P. Frank, A.K. Saydjari, J.S. Speagle, D. Finkbeiner, T. Enßlin, A parsec-scale galactic 3d dust map out to 1.25 kpc from the sun (2023)
14. Y.-K. Chiang, B. Mé nard, Extragalactic imprints in galactic dust maps. The Astrophys. J. **870**(2), 120 (2019)
15. B.T. Draine. Scattering by interstellar dust grains. I. Optical and ultraviolet. The Astrophys. J. **598**(2), 1017 (2003)
16. E.F. Schlafly, A.M. Meisner, A.M. Stutz, J. Kainulainen, J.E.G. Peek, K. Tchernyshyov, H.W. Rix, D.P. Finkbeiner, K.R. Covey, G.M. Green, E.F. Bell, W.S. Burgett, K.C. Chambers, P.W. Draper, H. Flewelling, K.W. Hodapp, N. Kaiser, E.A. Magnier, N.F. Martin, N. Metcalfe, R.J. Wainscoat, C. Waters, The optical-infrared extinction curve and its variation in the milky way. ApJ **821**(2), 78 (2016)
17. J. Gao, B.W. Jiang, A. Li, M.Y. Xue, The mid-infrared extinction law in the large Magellanic cloud. ApJ **776**(1), 7 (2013)
18. P.Y.Merica-Jones, K.M. Sandstrom, L.C. Johnson, J. Dalcanton, A.E. Dolphin, K. Gordon, J. Roman-Duval, D.R. Weisz, B.F. Williams, The small Magellanic cloud investigation of dust and gas evolution (SMIDGE): the dust extinction curve from red clump stars. ApJ **847**(2), 102 (2017)
19. S. Salim, M. Boquien, J.C. Lee, Dust attenuation curves in the local universe: demographics and new laws for star-forming galaxies and high-redshift analogs. ApJ **859**(1), 11 (2018)
20. A. Cikota, S. Deustua, F. Marleau, Determining type Ia supernova host galaxy extinction probabilities and a statistical approach to estimating the absorption-to-reddening ratio R_V. ApJ **819**(2), 152 (2016)
21. X. Wang, L. Wang, Y.-Q. Lou, X. Zhou, Z. Li, A novel color parameter as a luminosity calibrator for type Ia supernovae. ApJ **620**(2), L87–L90 (2005)
22. K. Krisciunas, J.L. Prieto, P.M. Garnavich, J.-L. Riley, A. Rest, C. Stubbs, R. McMillan, Photometry of the type Ia supernovae 1999cc, 1999cl, and 2000cf. AJ **131**(3), 1639–1647 (2006)
23. S. Nobili, A. Goobar, The colour-lightcurve shape relation of type ia supernovae and the reddening law. Astron. Astrophys. **487**(1), 19–31 (2008)
24. R. Amanullah, J. Johansson, A. Goobar, R. Ferretti, S. Papadogiannakis, T. Petrushevska, P.J. Brown, Y. Cao, C. Contreras, H. Dahle, N. Elias-Rosa, J.P.U. Fynbo, J. Gorosabel, L. Guaita, L. Hangard, D.A. Howell, E.Y. Hsiao, E. Kankare, M. Kasliwal, G. Leloudas, P. Lundqvist, S. Mattila, P. Nugent, M.M. Phillips, A. Sand berg, V. Stanishev, M. Sullivan, F. Taddia, G. Östlin, S. Asadi, R. Herrero-Illana, J.J. Jensen, K. Karhunen, S. Lazarevic, E. Varenius, P. Santos, S. Seethapuram Sridhar, S.H.J. Wallström, J. Wiegert, Diversity in extinction laws of type Ia supernovae measured between 0.2 and 2 μm. Mon. Not. Roy. Astron. Soc. **453**(3), 3300–3328 (2015)
25. L. Wang, Dust around type Ia supernovae. ApJ **635**(1), L33–L36 (2005)
26. Ariel Goobar, Low R_V from Circumstellar Dust around Supernovae. ApJ **686**(2), L103 (October 2008)
27. M.M. Phillips, J.D. Simon, N. Morrell, C.R. Burns, N.L.J. Cox, R.J. Foley, A.I. Karakas, F. Patat, A. Sternberg, R.E. Williams, A. Gal-Yam, E.Y. Hsiao, D.C. Leonard, S.E. Persson, M. Stritzinger, I.B. Thompson, A. Campillay, C. Contreras, G. Folatelli, W.L. Freedman, M. Hamuy, M. Roth, G.A. Shields, N.B. Suntzeff, L. Chomiuk, I.I. Ivans, B.F. Madore, B.E. Penprase, D. Perley, G. Pignata, G. Preston, A.M. Soderberg, On the source of the dust extinction in type Ia supernovae and the discovery of anomalously strong Na I absorption. ApJ **779**(1), 38 (2013)
28. M. Bulla, A. Goobar, R. Amanullah, U. Feindt, R. Ferretti, Estimating dust distances to type Ia supernovae from colour excess time evolution. Mon. Not. Roy. Astron. Soc. **473**(2), 1918–1929 (2018)
29. M. Bulla, A. Goobar, S. Dhawan, Shedding light on the type Ia supernova extinction puzzle: dust location found. Mon. Not. Roy. Astron. Soc. **479**(3), 3663–3674 (2018)
30. B.F. Madore, The period-luminosity relation. IV. Intrinsic relations and reddenings for the large Magellanic cloud cepheids. ApJ **253**, 575–579 (1982)

31. B. Follin, L. Knox, Insensitivity of the distance ladder Hubble constant determination to cepheid calibration modelling choices. Monthly Notices of the Royal Astron. Soc. **477**(4), 4534–4542 (2018)
32. E. Mörtsell, A. Goobar, J. Johansson, S. Dhawan, Sensitivity of the Hubble constant determination to Cepheid calibration. The Astrophys. J. **933**(2), 212 (2022)
33. A.G. Riess, S. Casertano, W. Yuan, J.B. Bowers, L. Macri, J.C. Zinn, D. Scolnic. Cosmic distances calibrated to 1% precision with gaia EDR3 parallaxes and Hubble space telescope photometry of 75 milky way Cepheids confirm tension with λCDM. The Astrophys. J. Lett. **908**(1), L6 (2021)
34. W. Yuan, A.G. Riess, S. Casertano, L.M. Macri, A first look at Cepheids in a type ia supernova host with JWST. The Astrophys. J. Lett. **940**(1), L17 (2022)
35. A.G. Riess, G.S. Anand, W. Yuan, S. Casertano, A. Dolphin, L.M. Macri, L. Breuval, D. Scolnic, M. Perrin, R.I. Anderson, Crowded no more: the accuracy of the Hubble constant tested with high resolution observations of Cepheids by JWST (2023)
36. W.D. Kenworthy, D.O. Jones, M. Dai, R. Kessler, D. Scolnic, D. Brout, M.R. Siebert, J.D.R. Pierel, K.G. Dettman, G. Dimitriadis, R.J. Foley, S.W. Jha, Y.C. Pan, A. Riess, S. Rodney, C. Rojas-Bravo, SALT3: an improved type Ia supernova model for measuring cosmic distances. ApJ **923**(2), 265 (2021)
37. J. Marriner, J.P. Bernstein, R. Kessler, H. Lampeitl, R. Miquel, J. Mosher, R.C. Nichol, M. Sako, D.P. Schneider, M. Smith. A more general model for the intrinsic scatter in type Ia supernova distance moduli. The Astrophys. J. **740**(2), 72 (2011)
38. R. Tripp, A two-parameter luminosity correction for type IA supernovae. A&A **331**, 815–820 (1998)
39. M. Hicken, P. Challis, S. Jha, R.P. Kirshner, T. Matheson, M. Modjaz, A. Rest, W.M. Wood-Vasey, G. Bakos, E.J. Barton, P. Berlind, A. Bragg, C. Briceño, W.R. Brown, N. Caldwell, M. Calkins, R. Cho, L. Ciupik, M. Contreras, K.-C. Dendy, A. Dosaj, N. Durham, K. Eriksen, G. Esquerdo, M. Everett, E. Falco, J. Fernandez, A. Gaba, P. Garnavich, G. Graves, P. Green, T. Groner, C. Hergenrother, M.J. Holman, V. Hradecky, J. Huchra, B. Hutchison, D. Jerius, A. Jordan, R. Kilgard, M. Krauss, K. Luhman, L. Macri, D. Marrone, J. McDowell, D. McIntosh, B. McNamara, T. Megeath, B. Mochejska, D. Munoz, J. Muzerolle, O. Naranjo, G. Narayan, M. Pahre, W. Peters, D. Peterson, K. Rines, B. Ripman, A. Roussanova, R. Schild, A. Sicilia-Aguilar, J. Sokoloski, K. Smalley, A. Smith, T. Spahr, K.Z. Stanek, P. Barmby, S. Blondin, C.W. Stubbs, A. Szentgyorgyi, M.A.P. Torres, A. Vaz, A. Vikhlinin, Z. Wang, M. Westover, D. Woods, P. Zhao, CfA3: 185 type Ia supernova light curves from the CfA. ApJ **700**, 331–357 (2009)
40. M. Sullivan, A. Conley, D.A. Howell, J.D. Neill, P. Astier, C. Balland, S. Basa, R.G. Carlberg, D. Fouchez, J. Guy, D. Hardin, I.M. Hook, R. Pain, N. Palanque-Delabrouille, K.M. Perrett, C.J. Pritchet, N. Regnault, J. Rich, V. Ruhlmann-Kleider, S. Baumont, E. Hsiao, T. Kronborg, C. Lidman, S. Perlmutter, E.S. Walker, The dependence of type Ia supernovae luminosities on their host galaxies. Mon. Not. Roy. Astron. Soc. **406**, 782–802 (2010)
41. H. Lampeitl, M. Smith, R.C. Nichol, B. Bassett, D. Cinabro, B. Dilday, R.J. Foley, J.A. Frieman, P.M. Garnavich, A. Goobar, M. Im, S.W. Jha, J. Marriner, R. Miquel, J. Nordin, L. Östman, A.G. Riess, M. Sako, D.P. Schneider, J. Sollerman, M. Stritzinger, The effect of host galaxies on type Ia supernovae in the SDSS-II supernova survey. ApJ **722**, 566–576 (2010)
42. M. Childress, G. Aldering, P. Antilogus, C. Aragon, S. Bailey, C. Baltay, S. Bongard, C. Buton, A. Canto, F. Cellier-Holzem, N. Chotard, Y. Copin, H.K. Fakhouri, E. Gangler, J. Guy, E.Y. Hsiao, M. Kerschhaggl, A.G. Kim, M. Kowalski, S. Loken, P. Nugent, K. Paech, R. Pain, E. Pecontal, R. Pereira, S. Perlmutter, D. Rabinowitz, M. Rigault, K. Runge, R. Scalzo, G. Smadja, C. Tao, R.C. Thomas, B.A. Weaver, C. Wu, Host galaxies of type Ia supernovae from the nearby supernova factory. ApJ **770**, 107 (2013)
43. R. Kessler, D. Brout, S. Crawford et al., First cosmology results using type Ia supernova from the dark energy survey: simulations to correct supernova distance biases. Mon. Not. Roy. Astron. Soc. **485**, 1171–1187 (2019)

44. Dillon Brout, Daniel Scolnic, It's dust: solving the mysteries of the intrinsic scatter and host-galaxy dependence of standardized type Ia supernova brightnesses. Astrophys. J. **909**(1), 26 (2021)
45. D. Brout, D. Scolnic, B. Popovic, A.G. Riess, A. Carr, J. Zuntz, R. Kessler, T.M. Davis, S. Hinton, D. Jones, W. D'Arcy Kenworthy, E.R. Peterson, K. Said, G. Taylor, N. Ali, P. Armstrong, P. Charvu, A. Dwomoh, C. Meldorf, A. Palmese, H. Qu, B.M. Rose, B. Sanchez, C.W. Stubbs, M. Vincenzi, C.M. Wood, P.J. Brown, R. Chen, K. Chambers, D.A. Coulter, M. Dai, G. Dimitriadis, A.V. Filippenko, R.J. Foley, S.W. Jha, L. Kelsey, R.P. Kirshner, A. Möller, J. Muir, S. Nadathur, Y.-C. Pan, A. Rest, C. Rojas-Bravo, M. Sako, M.R. Siebert, M. Smith, B.E. Stahl, P. Wiseman, The Pantheon+ analysis: cosmological constraints. ApJ **938**(2), 110 (2022)
46. J. Guy, M. Sullivan, A. Conley, N. Regnault, P. Astier, C. Balland, S. Basa, R.G. Carlberg, D. Fouchez, D. Hardin, I.M. Hook, D.A. Howell, R. Pain, N. Palanque-Delabrouille, K.M. Perrett, C.J. Pritchet, J. Rich, V. Ruhlmann-Kleider, D. Balam, S. Baumont, R.S. Ellis, S. Fabbro, H.K. Fakhouri, N. Fourmanoit, S. González-Gaitán, M.L. Graham, E. Hsiao, T. Kronborg, C. Lidman, A.M. Mourao, S. Perlmutter, P. Ripoche, N. Suzuki, E.S. Walker, The supernova legacy survey 3-year sample: type Ia supernovae photometric distances and cosmological constraints. A&A **523**, A7 (2010)
47. A. Conley, J. Guy, M. Sullivan, N. Regnault, P. Astier, C. Balland, S. Basa, R.G. Carlberg, D. Fouchez, D. Hardin et al., Supernova constraints and systematic uncertainties from the first three years of the supernova legacy survey. The Astrophys. J. Suppl. Series **192**(1), 1 (2010)
48. B. Popovic, D. Brout, R. Kessler, D. Scolnic, The pantheon+ analysis: forward-modeling the dust and intrinsic colour distributions of type Ia supernovae, and quantifying their impact on cosmological inferences (2021)
49. R. Chen, D. Scolnic, E. Rozo, E. S. Rykoff, B. Popovic, R. Kessler, M. Vincenzi, T. M. Davis, P. Armstrong, D. Brout, L. Galbany, L. Kelsey, C. Lidman, A. Möller, B. Rose, M. Sako, M. Sullivan, G. Taylor, P. Wiseman, J. Asorey, A. Carr, C. Conselice, K. Kuehn, G.F. Lewis, E. Macaulay, M. Rodriguez-Monroy, B.E. Tucker, T.M.C. Abbott, M. Aguena, S. Allam, F. Andrade-Oliveira, J. Annis, D. Bacon, E. Bertin, S. Bocquet, D. Brooks, D.L. Burke, A. Carnero Rosell, M. Carrasco Kind, J. Carretero, R. Cawthon, M. Costanzi, L.N. da Costa, M.E.S. Pereira, S. Desai, H.T. Diehl, P. Doel, S. Everett, I. Ferrero, B. Flaugher, D. Friedel, J. Frieman, J. Garcí a-Bellido, M. Gatti, E. Gaztanaga, D. Gruen, S.R. Hinton, D.L. Hollowood, K. Honscheid, D.J. James, O. Lahav, M. Lima, M. March, F. Menanteau, R. Miquel, R. Morgan, A. Palmese, F. Paz-Chinchón, A. Pieres, A.A. Plazas Malagón, J. Prat, A.K. Romer, A. Roodman, E. Sanchez, M. Schubnell, S. Serrano, I. Sevilla-Noarbe, M. Smith, M. Soares-Santos, E. Suchyta, G. Tarle, D. Thomas, C. To, D.L. Tucker, T.N. Varga, Measuring cosmological parameters with type Ia supernovae in redMaGiC galaxies. The Astrophys. J. **938**(1), 62 (2022)
50. C. Meldorf, A. Palmese, D. Brout, R. Chen, D. Scolnic, L. Kelsey, L. Galbany, W.G. Hartley, T.M. Davis, A. Drlica-Wagner, M. Vincenzi, J. Annis, M. Dixon, O. Graur, C. Lidman, A. Möller, P. Nugent, B. Rose, M. Smith, S. Allam, D.L. Tucker, J. Asorey, J. Calcino, D. Carollo, K. Glazebrook, G.F. Lewis, G. Taylor, B.E. Tucker, A.G. Kim, H. T Diehl, M. Aguena, F. Andrade-Oliveira, D. Bacon, E. Bertin, S. Bocquet, D. Brooks, D.L. Burke, J. Carretero, M. Carrasco Kind, F.J. Castander, M. Costanzi, L.N. da Costa, S. Desai, P. Doel, S. Everett, I. Ferrero, D. Friedel, J. Frieman, J. Garcí a-Bellido, M. Gatti, D. Gruen, J. Gschwend, G. Gutierrez, S.R. Hinton, D.L. Hollowood, K. Honscheid, D.J. James, K. Kuehn, M. March, J.L. Marshall, F. Menanteau, R. Miquel, R. Morgan, F. Paz-Chinchón, M.E.S. Pereira, A.A. Plazas Malagón, E. Sanchez, V. Scarpine, I. Sevilla-Noarbe, E. Suchyta, G. Tarle, T.N. Varga, The dark energy survey supernova program results: type ia supernova brightness correlates with host galaxy dust. Monthly Notices of the Royal Astron. Soc. **518**(2), 1985–2004 (2022)
51. S. Thorp, K.S. Mandel, Constraining the SN Ia host galaxy dust law distribution and mass step: hierarchical Bayesn analysis of optical and near-infrared light curves. Monthly Notices of the Royal Astron. Soc. **517**(2), 2360–2382 (2022)

52. B. Popovic, D. Brout, R. Kessler, D. Scolnic, L. Lu, Improved treatment of host-galaxy correlations in cosmological analyses with type Ia supernovae (2021)
53. M. Rigault, Y. Copin, G. Aldering, P. Antilogus, C. Aragon, S. Bailey, C. Baltay, S. Bongard, C. Buton, A. Canto, F. Cellier-Holzem, M. Childress, N. Chotard, H.K. Fakhouri, U. Feindt, M. Fleury, E. Gangler, P. Greskovic, J. Guy, A.G. Kim, M. Kowalski, S. Lombardo, J. Nordin, P. Nugent, R. Pain, E. Pécontal, R. Pereira, S. Perlmutter, D. Rabinowitz, K. Runge, C. Saunders, R. Scalzo, G. Smadja, C. Tao, R.C. Thomas, B.A. Weaver, Evidence of environmental dependencies of type Ia supernovae from the nearby supernova factory indicated by local Hα. A&A **560**, A66 (2013)
54. A. Polin, P. Nugent, D. Kasen, Observational predictions for sub-chandrasekhar mass explosions: Further evidence for multiple progenitor systems for type Ia supernovae. The Astrophys. J. **873**(1), 84 (2019)
55. C.R. Burns, E. Parent, M.M. Phillips, M. Stritzinger, K. Krisciunas, N.B. Suntzeff, E.Y. Hsiao, C. Contreras, J. Anais, L. Boldt, L. Busta, A. Campillay, S. Castelló n, G. Folatelli, W.L. Freedman, C. González, M. Hamuy, P. Heoflich, W. Krzeminski, B.F. Madore, N. Morrell, S.E. Persson, M. Roth, F. Salgado, J. Serón, S. Torres, The Carnegie supernova project: absolute calibration and the Hubble constant. The Astrophys. J. **869**(1), 56 (2018)
56. S. Dhawan, S. Thorp, K.S. Mandel, S.M. Ward, G. Narayan, S.W. Jha, T. Chant, A BayeSN distance ladder: H_0 from a consistent modelling of type ia supernovae from the optical to the near-infrared. Monthly Notices of the Royal Astron. Soc. **524**(1), 235–244 (2023)
57. P. Garnavich, C.M. Wood, P. Milne, J.B. Jensen, J.P. Blakeslee, P.J. Brown, D. Scolnic, B. Rose, D. Brout, Connecting infrared surface brightness fluctuation distances to type Ia supernova hosts: testing the top rung of the distance ladder. The Astrophys. J. **953**(1), 35 (2023)

Part IV
Theoretical Proposed Solutions

Chapter 20
(Introduction to the Second Part of the Book) What About the Solutions?

Eleonora Di Valentino

Out of various cosmological models proposed in literature, the Lambda cold dark matter (ΛCDM) stands out as the standard model, chosen for its simplicity and remarkable ability to accurately describe a wide array of astrophysical and cosmological observations.

However, despite its success, ΛCDM grapples with numerous unknowns, struggling to elucidate fundamental concepts pertaining to the structure and evolution of the universe. These conceptual challenges arise from three elusive ingredients, unsupported by theoretical first principles or laboratory experiments, but rather inferred from cosmological and astrophysical observations.

These three enigmatic components are: inflation, dark matter (DM), and dark energy (DE). In the realm of ΛCDM, inflation is given by a single, slow-rolling scalar field; DM is postulated to interact only through gravity, be cold and pressureless, and lack direct evidence of its existence; DE is represented by the cosmological constant term Λ, offering little in the way of a robust physical explanation.

While ΛCDM stands as the preferred model for its adeptness in accurately describing observed phenomena, it is not immune to theoretical shortcomings. The model, characterized by its six parameters, lacks a foundation in deep-rooted physical principles and should, at best, be regarded as an approximation of an as-yet-undiscovered underlying physical theory.

As observations multiply and attain greater accuracy, it is anticipated that deviations from the ΛCDM model will come to light [1, 2]. Indeed, discrepancies in pivotal cosmological parameters, such as H_0 [3–6], have already emerged across various observations, each carrying distinct statistical significance.

E. Di Valentino (✉)
School of Mathematics and Statistics, University of Sheffield, Hounsfield Road, Sheffield S3 7RH, Sheffield, UK
e-mail: e.divalentino@sheffield.ac.uk

E. Di Valentino and D. Brout (eds.), *The Hubble Constant Tension*, Springer Series in Astrophysics and Cosmology, https://doi.org/10.1007/978-981-99-0177-7_20

While certain tensions observed in cosmological measurements may originate from systematic errors, their persistent manifestation across multiple probes suggests potential flaws in the conventional cosmological framework. This raises the intriguing possibility that new physics might be essential to elucidate these observational discrepancies. Therefore, the enduring nature of these tensions raises questions about the adequacy of the established ΛCDM model.

Considering the diverse array of the Hubble constant measurements detailed in prior chapters, where late universe observations are consistently higher than early ones and vice versa, pinpointing a singular systematic error to consistently explain the observed disparities becomes challenging. This complexity makes it difficult to envisage a single source of error that uniformly accounts for the discrepancies across the broad spectrum of measurements encountered earlier, therefore addressing the Hubble constant tension.

Persisting at a significance level of 5–6.3 σ [7–9], even after eliminating measurements related to specific object types, teams, or calibrations, the Hubble constant tension remains robust. Identifying a singular error to reconcile this tension becomes an intricate task. While the possibility of multiple independent systematic errors offers some flexibility in resolving the tension, such occurrences are inherently less probable.

Lastly, recognizing that the indirect early universe constraints are model-dependent, there exists an opportunity to explore the prospect of expanding the cosmological framework. Examining potential extensions and their capacity to reconcile the observed discrepancies among various cosmological probes becomes a promising avenue for further investigation.

In the subsequent chapters of this book, it will become evident that while the seemingly straightforward resolution-such as the existence of a local void to address the H_0 tension-is discredited at a significance level of approximately 20σ [10], alternative propositions can be categorized into two broad groups: those involving modifications to the expansion history during early times [11–14] (Chap. 22) (prior to recombination) and those occurring during late times [15] (post-recombination). Additionally, there is the option of considering adjustments to the recombination epoch itself [16, 17] or the inflationary period (Chap. 34). Regardless of the chosen avenue, it is essential to note that, at present, no specific proposal has emerged as a compelling frontrunner, demonstrating a significantly higher likelihood or superiority over others. The exploration of these possibilities in subsequent chapters will shed light on the nuanced complexities of each proposal and their implications for resolving the persistent challenges posed by the H_0 tension.

Indeed, while late-time solutions exhibit an increase in H_0 by reducing the expansion history at intermediate redshifts, they fall short in providing a satisfactory fit to low-redshift measurements, such as Baryon Acoustic Oscillations (BAO) or Type Ia Supernovae data [18]. Conversely, early-time solutions exhibit promising adjustments to both parameters, but they are unable to completely resolve the H_0 tension between Planck Cosmic Microwave Background (CMB) data [19] and the SH_0ES collaboration measurement [20] without introducing an additional Hubble constant prior. These solutions tend to enhance the consistency between data sets primarily due to a volume effect but exacerbate the S_8 tension [22] between the CMB and weak lensing data [21]. While the subsequent chapters explore also intriguing proposals involving negative dark energy densities (Chap. 28), interactions between different components [23], and modified gravity theories (Chaps. 30 and 31), it becomes increasingly evident that relying solely on early or late-time solutions proves insufficient [24]. The complexities of the observed tensions necessitate a more comprehensive approach that may involve a combination of these strategies or the exploration of alternative theoretical frameworks.

Adding to the intricacies, caution is warranted when delving into exotic models. The correlation between parameters in such models may yield artificial signals that are not present in nature, introducing the risk of false detections for new physics [25]. Additionally, the hidden model dependence of certain datasets must be considered, as some datasets assume the ΛCDM model in various steps of the data reduction process, exemplified in techniques like BAO [26]. Notably, the use of a H_0 prior for late-time transitions resembling a hockey stick shape is under scrutiny, with suggestions for its replacement by the M_B value [27]. Further complexity arises from anomalies in CMB data, including the Planck A_{Lens} problem [28], and inconsistencies between different CMB experiments that may produce significantly disparate results in extended cosmologies [29].

The Hubble constant tension not only emphasizes the need for new observations but also serve as catalysts for probing alternative theoretical models and solutions. The current state of the field suggests that while the game is still open, a paradigm shift in cosmology is not beyond the realm of possibility in the near future. The ongoing investigation into these challenges holds the potential to reshape our understanding of the universe.

Acknowledgements EDV is supported by a Royal Society Dorothy Hodgkin Research Fellowship.

References

1. E. Abdalla, G. Franco Abellán, A. Aboubrahim, A. Agnello, O. Akarsu, Y. Akrami, G. Alestas, D. Aloni, L. Amendola, L.A. Anchordoqui et al., JHEAp **34**, 49–211 (2022). https://doi.org/10.1016/j.jheap.2022.04.002 [arXiv:2203.06142 [astro-ph.CO]]
2. L. Perivolaropoulos, F. Skara, New Astron. Rev. **95**, 101659 (2022). https://doi.org/10.1016/j.newar.2022.101659 [arXiv:2105.05208 [astro-ph.CO]]
3. L. Verde, T. Treu, A.G. Riess, Nature Astron. **3**, 891. https://doi.org/10.1038/s41550-019-0902-0 [arXiv:1907.10625 [astro-ph.CO]]
4. E. Di Valentino, L.A. Anchordoqui, O. Akarsu, Y. Ali-Haimoud, L. Amendola, N. Arendse, M. Asgari, M. Ballardini, S. Basilakos, E. Battistelli et al., Astropart. Phys. **131**, 102605 (2021). https://doi.org/10.1016/j.astropartphys.2021.102605 [arXiv:2008.11284 [astro-ph.CO]]
5. E. Di Valentino, O. Mena, S. Pan, L. Visinelli, W. Yang, A. Melchiorri, D. F. Mota, A. G. Riess, J. Silk, Class. Quant. Grav. **38**(15), 153001 (2021). https://doi.org/10.1088/1361-6382/ac086d [arXiv:2103.01183 [astro-ph.CO]]
6. M. Kamionkowski, A.G. Riess, Ann. Rev. Nucl. Part. Sci. **73**, 153–180 (2023) [arXiv:2211.04492 [astro-ph.CO]]

7. A.G. Riess, Nature Rev. Phys. **2**(1), 10-12 (2019). https://doi.org/10.1038/s42254-019-0137-0 [arXiv:2001.03624 [astro-ph.CO]]
8. E. Di Valentino, Mon. Not. Roy. Astron. Soc. **502**(2), 2065–2073 (2021). https://doi.org/10.1093/mnras/stab187 [arXiv:2011.00246 [astro-ph.CO]]
9. E. Di Valentino, Universe **8**(8), 399 (2022). https://doi.org/10.3390/universe8080399
10. D. Huterer, H.Y. Wu, [arXiv:2309.05749 [astro-ph.CO]]
11. F. Niedermann, M.S. Sloth, [arXiv:2307.03481 [hep-ph]]
12. S. Gariazzo, O. Mena, [arXiv:2306.15067 [astro-ph.CO]]
13. A. Nygaard, E.B. Holm, T. Tram, S. Hannestad, [arXiv:2307.00418 [astro-ph.CO]]
14. K. Naidoo, [arXiv:2308.13617 [astro-ph.CO]]
15. M. Raveri, [arXiv:2309.06795 [astro-ph.CO]]
16. K. Jedamzik, L. Pogosian, [arXiv:2307.05475 [astro-ph.CO]]
17. J. Chluba, L. Hart, [arXiv:2309.12083 [astro-ph.CO]]
18. L. Knox, M. Millea, Phys. Rev. D **101**(4), 043533 (2020). https://doi.org/10.1103/PhysRevD.101.043533 [arXiv:1908.03663 [astro-ph.CO]]
19. N. Aghanim et al., [Planck], Astron. Astrophys. **641**, A6 (2020) [erratum: Astron. Astrophys. **652**, C4 (2021)]. https://doi.org/10.1051/0004-6361/201833910 [arXiv:1807.06209 [astro-ph.CO]]
20. A.G. Riess, W. Yuan, L. M. Macri, D. Scolnic, D. Brout, S. Casertano, D.O. Jones, Y. Murakami, L. Breuval, T.G. Brink et al., Astrophys. J. Lett. **934**(1), L7 (2022). https://doi.org/10.3847/2041-8213/ac5c5b [arXiv:2112.04510 [astro-ph.CO]]
21. E. Di Valentino, L.A. Anchordoqui, Ö. Akarsu, Y. Ali-Haimoud, L. Amendola, N. Arendse, M. Asgari, M. Ballardini, S. Basilakos, E. Battistelli et al., Astropart. Phys. **131**, 102604 (2021). https://doi.org/10.1016/j.astropartphys.2021.102604 [arXiv:2008.11285 [astro-ph.CO]]
22. K. Jedamzik, L. Pogosian, G.B. Zhao, Commun. in Phys. **4**, 123 (2021). https://doi.org/10.1038/s42005-021-00628-x [arXiv:2010.04158 [astro-ph.CO]]
23. S. Pan, W. Yang, [arXiv:2310.07260 [astro-ph.CO]]
24. S. Vagnozzi, Universe **9**(9), 393 (2023). https://doi.org/10.3390/universe9090393 [arXiv:2308.16628 [astro-ph.CO]]
25. E. Di Valentino. O. Mena, Mon. Not. Roy. Astron. Soc. **500**(1), L22-L26 (2020) https://doi.org/10.1093/mnrasl/slaa175 [arXiv:2009.12620 [astro-ph.CO]]
26. S. Anselmi, G.D. Starkman, A. Renzi, Phys. Rev. D **107**(12), 123506 (2023). https://doi.org/10.1103/PhysRevD.107.123506 [arXiv:2205.09098 [astro-ph.CO]]
27. D. Camarena, V. Marra, [arXiv:2307.02434 [astro-ph.CO]]
28. G.E. Addison, C.L. Bennett, M. Halpern, G. Hinshaw, J.L. Weiland, [arXiv:2310.03127 [astro-ph.CO]]
29. W. Giarè, [arXiv:2305.16919 [astro-ph.CO]]

Chapter 21
Not Empty Enough: A Local Void Cannot Solve the H_0 Tension

Dragan Huterer and Hao-Yi Wu

21.1 Introduction

In the currently favored flat ΛCDM cosmological model, the local measurement of the Hubble constant H_0 from the distance ladder [1–3] and the global measurement from the CMB anisotropies [4–7] should agree. The fact that they do not may well be a harbinger of new physics—or, at the very least, unexpectedly large systematic errors in either of the two principal measurements.

Hubble tension has sparked huge interest in cosmology, and a dizzying number of explanations for it have been put forth (see Ref. [8], which refers to more than 1000 papers in this regard). However, no single explanation has yet proven to be compelling. Most of the theoretical explanations try to change the global CMB H_0 fit from $\sim$67 km s^{-1}Mpc^{-1} to $\sim$73 km s^{-1}Mpc^{-1}, the value favored by the distance ladder with SNIa. While doing so, these new models have to preserve the excellent fit to CMB, baryon acoustic oscillations, and other cosmological data which—in an unmodified, standard cosmological model—favor $H_0 \sim 67$ km s^{-1}Mpc^{-1}. Even in reasonably successful models, it has proven difficult to raise the Hubble constant all the way to $\sim$73 km s^{-1}Mpc^{-1} (rather than halfway there, to $\sim$70 km s^{-1}Mpc^{-1}) while at the same time improving the goodness of fit to the data to justify the new parameter(s) of the models in question.

In this Chapter, we review arguably the simplest explanation for the Hubble tension—the possibility that we live in a void empty enough to resolve the Hubble tension. The logic of this explanation goes as follows. There are underdensities

D. Huterer (✉)
Department of Physics and Leinweber Center for Theoretical Physics, University of Michigan, 450 Church St, Ann Arbor, MI 48103, USA
e-mail: huterer@umich.edu

H.-Y. Wu (✉)
Department of Physics, Boise State University, Boise, ID 83725, USA
e-mail: hywu@boisestate.edu

E. Di Valentino and D. Brout (eds.), *The Hubble Constant Tension*, Springer Series in Astrophysics and Cosmology, https://doi.org/10.1007/978-981-99-0177-7_21

and overdensities in the universe. If we live in an underdensity, the local Hubble constant, $H_0^{\rm loc}$, will be higher than the global Hubble constant, as less mass in the void implies a higher expansion rate. This would imply a possible mismatch between $H_0^{\rm loc}$ and the global, "true" H_0. In addition, given that the local SNIa sample is highly sparse and inhomogeneous in the 3D spatial distribution, the error bar on the local H_0 might be larger than previously thought. The two effects—local matter density and sample sparsity—constitute the **sample variance** (or cosmic variance) in the local H_0 measurements.

Our goal is to precisely quantify the sample variance, and we do so in Sect. 21.4. First, however, we review the physics behind the void explanation in Sect. 21.2, and briefly review the observational evidence for a void in Sect. 21.3. We conclude in Sect. 21.5.

21.2 The Impact of Underdensity on the Hubble Constant: Theoretical Considerations

The "Hubble bubble" hypothesis refers to the possibility that Milky Way resides in an underdense region in the universe, and the underdensity makes the local expansion rate significantly different from the global expansion rate. If we live in such a void, nearby galaxies would preferentially have positive peculiar velocities (moving away from us faster than expected from the global expansion rate). This effect can cause the local Hubble constant to be higher than the global value.

The effect of the local density of the universe on the locally measured Hubble constant can be quantified as follows. In an underdense region, the expansion rate is higher than the mean (and vice versa in an overdense region) [9–18]. Namely, the relation between the local H_0 shift, ΔH_0, and the local density contrast, δ, is

$$\frac{\Delta H_0}{H_0} = -\frac{1}{3}\delta f(\Omega_M)\Theta(\delta, \Omega_M), \tag{21.1}$$

where $f(\Omega_M) \simeq \Omega_M(a)^{0.55}$ is the growth rate of density perturbations (see Ref. [19] for a review), and $\Theta = 1 - O(\delta)$ is a non-linear correction which is small for underdensities of typical size. Therefore, an underdensity with $\delta < 0$ will automatically lead to a higher measured local expansion rate, that is, $\Delta H_0 > 0$. The effect of an under/overdensity described here is qualitatively *guaranteed*, as the local density alone leads to location-dependent local measurements of H_0. This makes the Hubble bubble an appealing model to potentially explain why local measurements of the Hubble constant disagree with the global measurement given by *Planck*.

Is it possible that sample variance provides a sufficiently large effect to explain the Hubble tension? Over the past decade or so, the answer has been arrived at with increasingly realistic numerical and analytical estimates. The effect of sample variance on the local measurements of H_0 was pioneered by Refs. [20–22] using numerical simulations. These analyses have strongly hinted that the underdensity required to explain the Hubble tension is extremely unlikely. In parallel, theorists have investigated the Hubble bubble scenario, relying on simplified analytical models of overdensity. For example, some work adopted the top-hat spherical model [23] or the Lemaitre–Tolman–Bondi (LTB) metric, a variant of the FLRW metric with a spherically symmetric underdensity or overdensity [24–26].[1] All of these results indicate that it is extremely unlikely that we live in a void that is sufficiently large in size and devoid of matter to explain the Hubble tension.

However, most of these past analyses of sample variance—both numerical and theoretical—have modeled the presence of spherical underdensity but have not attempted to mimic the local H_0 measurements particularly closely. Specifically, these studies have been assuming spherical symmetry in the SNIa measurements. However, merely visualizing the highly inhomogeneous distribution of SNIa in the sky brings into question the accuracy of the homogeneity assumption that underlies most past numerical and theoretical treatments. For example, SNIa distribution is denser in some regions due to historically available observations (e.g. Sloan Digital Sky Survey's Stripe 82; see also, for example, Fig. 1 in Ref. [29]).

Before discussing our work addressing sample inhomogeneity, we briefly review why a direct measurement of the Hubble bubble is challenging.

21.3 Do We Live in a Hubble Bubble? Observations of the Local Density Contrast

As far as the observational evidence for the local density contrast is concerned, the following quote by Arthur Eddington

Never trust an experimental result until it has been confirmed by theory

is perhaps very relevant here. This is because observational evidence for the Hubble bubble must necessarily rely on the estimate of mass density, which is plagued by many observational uncertainties. The principal challenge is twofold: to convert from light in some electromagnetic band (visible, X-ray, radio, etc.) to mass, and then to map out the three-dimensional distribution of the mass (rather than just the 2D projection on the sky). Despite the massive efforts by observers and modelers, we are still not at the level of accuracy where such direct measurements can be made to make robust claims about the Hubble bubble.

Some observations do hint that the Milky Way might be in an underdense region in the universe, but the results have been controversial [30–37]. Several observations use large-area surveys of galaxy number density to map the radial distribution of mass centered around our location in the universe. For example, Ref. [34] uses the 600 deg^2 galaxy catalog from UKIDSS-LAS, finding a significant local underdensity $\delta = -0.3$ within 200 Mpc. Ref. [31] uses the full-sky 2M++ catalog and finds no

[1] Note that some studies combine theoretical and numerical tools [27, 28].

evidence for the local underdensity out to 120 h^{-1}Mpc. Ref. [27] uses the distance–redshift relation derived from ∼1000 SNe from Pantheon, Foundation, and Carnegie Supernova Project and derived that $|\delta| < 27\%$ (5σ constraint) on scales larger than 69 h^{-1}Mpc ($z = 0.023$). In contrast, Ref. [36] find that the local density within $z < 0.075$ is 20% lower than the mean density of the universe, which would lead to a 3% correction of the local Hubble constant.

Clearly, on the front of direct observations, things are unsettled, and there is a wide divergence of interpretations of observational measurements. Nevertheless, in the remainder of this review, we demonstrate that, even if we take these results at face value, none of the claimed observed underdensities can alleviate the Hubble tension. As will be described in the next section, resolving the Hubble tension would require an underdensity $\delta \approx -0.8$ within a radius of 120 h^{-1}Mpc [38]. Such an underdensity is incompatible with any of the observational results and is impossible to achieve within a ΛCDM framework. In sum, quantifying the local density contrast is an interesting research subject per se, but it is unlikely to lead to a viable solution for the Hubble tension.

21.4 Investigating Density Contrasts in Large-Volume N-Body Simulations

We investigate the effect of under/overdensities with a controlled experiment, using an N-body simulation of the large-scale structure of the universe that covers many realizations of the supernova sample volume.

Let us start with some preliminaries. In the presence of a local void, galaxies tend to move away from the center of the void and exhibit positive peculiar velocities. Let's assume we have N standard candles at comoving distances r_i and radial peculiar velocities $v_{r,i}$ (i.e. radial velocities of galaxies with the global expansion rate of the universe taken out). The shift in the local Hubble constant caused by these peculiar velocities is given by

$$\Delta H_0^{\rm loc} = \frac{1}{N}\sum_{i=1}^{N}\frac{v_{r,i}}{r_i}. \tag{21.2}$$

Clearly, positive net radial peculiar velocities will lead to a positive shift in H_0, and vice versa for negative velocities. Equation (21.2) forms a basis for how one can estimate the statistical fluctuations of the observed H_0.[2]

Here we recapitulate the results of Wu and Huterer [38]. We address two aspects of the sample variance: (1) variance caused by the local density contrast, and (2) variance

[2] Note that, in SNIa observations, the peculiar velocities of galaxies are usually removed; for example, Ref. [2] has used the density field reconstructed by the 2M++ survey to correct for the peculiar velocities. Here we discuss the *maximum* impact of the local density by assuming that the peculiar velocities are not corrected for. Our results thus serve as a conservative upper limit of the sample variance.

caused by the inhomogeneous selection of the supernova sample "Supercal" [39, 40], which is used in the Riess et al. [21] analysis. The former effect is irreducible, while the latter effect can be reduced with future larger samples.

To address the contribution of the local density contrast to the Hubble constant, we perform our study on a large N-body simulation. Our general idea is to place the observer at many independent locations in the simulation, mimic the SNIa observations around each observer, and measure the statistical distribution of the inferred Hubble constant with respect to the global value (i.e. in ΔH_0 from Eq. 21.1). The advantage of this approach is that we can mimic the SNIa observations precisely by selecting halos in our simulation that stand for SNIa with their (near-)exact three-dimensional locations around the observer. This enables measurements of ΔH_0 that account for the sample variance of actual SNIa observations very precisely. Given the size of the numerical simulation and our consequent ability to place many independent observed volumes in it, we also have great statistical power at our disposal to quantify the effect of sample variance.

We use the public release of the Dark Sky simulations [41]. Specifically, we use the $(8\,h^{-1}\mathrm{Gpc})^3$ volume with 10^{12} particles and a mass resolution $3.9 \times 10^{10}\ h^{-1}\mathrm{M}_\odot$. The simulation assumes a standard flat ΛCDM model consistent with *Planck* 2016 [42]. We divide this $(8\,h^{-1}\mathrm{Gpc})^3$ volume into 512 subvolumes of $(1\,h^{-1}\mathrm{Gpc})^3$. We then choose the halo with virial mass $M_{\mathrm{vir}} \in [10^{12.3}, 10^{12.4}]\ \mathrm{M}_\odot$ closest to the center of each subvolume as our observer. This choice simulates 512 independent observers living in Milky Way-mass halos, each with a separate subvolume of the large-scale structure out to the distance of interest ($z_{\mathrm{max}} = 0.15$). For the host halos of SNIa, we also use Milky Way-mass halos with $M_{\mathrm{vir}} \in [10^{12.3}, 10^{12.4}]\ \mathrm{M}_\odot$, and we have explicitly checked that using all halos above $10^{13}\ \mathrm{M}_\odot$ does not change the result. We then simulate the Supercal supernova observation in each subvolume.

Our procedure then goes as follows (see also Fig. 21.1):

1. Divide the $8\,h^{-1}$Gpc Dark Sky box into 512 subvolumes as mentioned above; each Milky Way-mass halo in every subvolume is a possible host of an SNIa. Then, for each subvolume.
2. Select a random orientation of the overall SNIa sky coordinates with respect to the simulation's Cartesian coordinates.
3. Assign each SNIa to the closest halo in the subvolume based on its 3D coordinates reported in the Supercal dataset release [39, 40].
4. Calculate the $\Delta H_0^{\mathrm{loc}}$ from the radial velocities of SNIa host halos using the relation

$$\Delta H_0^{\mathrm{loc}} = \frac{1}{N}\sum_{i=1}^{N}\frac{v_{r,i}}{r_i}\,, \tag{21.3}$$

where r_i and $v_{r,i}$ are the comoving distance to, and the peculiar velocity of, the halo that hosts the ith SNIa. We average over the $\Delta H_0^{\mathrm{loc}}$ of all supernovae from this particular orientation for this observer. (The actual analysis includes additional details; see Ref. [38].)

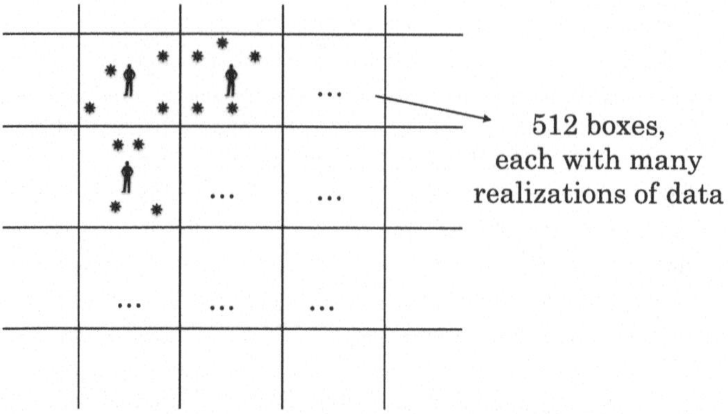

Fig. 21.1 Sketch of the simulation procedure in Ref. [38]. We divide the 8 h^{-1}Gpc N-body simulation into 512 subvolumes. In each subvolume, we place an observer at the center, use ~3000 rotations of the global coordinates of the observed SNIa (not shown in this sketch), and find the closest Milky Way-mass halo to each SNIa. We then use the known properties of that halo—its distance from the observer and its peculiar velocity—as proxies for those of SNIa, and use Eq. 21.2 to estimate the Hubble constant shift

5. Go to step 2., repeat the measurements for many different coordinate-system orientations and obtain the histogram of $\Delta H_0^{\rm loc}$ of different orientations from this particular subvolume.
6. Go to step 1., repeat the measurements for the 512 non-overlapping subvolumes and obtain the distribution of $\Delta H_0^{\rm loc}$ from all subvolumes and all orientations.

The 512 subvolumes account for the variance in the local density in a ΛCDM universe, and the ~3000 SNIa coordinate-system orientations in each subvolume account for the skewed redshift distribution and the sparse angular distribution of the SNIa sample. The full procedure therefore corresponds to ~1.5 million simulations of inferring the Hubble constant from local SNIa based on the Supercal sample ($0.023 < z < 0.15$).

Figure 21.2 shows the histogram of the distributions of ΔH_0 for all subvolumes and SNIa coordinate-system orientations. The overall distribution in the Hubble constant shifts, shown with the dark blue histogram, indicates that the sample variance in the local $H_0^{\rm loc}$ measurements is much smaller than the amount that could explain the Hubble tension (i.e. one that would reach out to the Riess et al. value, shown as the vertical beige bar). For example, even a void that is underdense at the $\sim 2\sigma$ level relative to the mean (the green histogram) would lead to the locally measured Hubble constant, $H_0^{\rm loc}$, that is only slightly larger than the mean value, assumed here to be *Planck*'s $H_0 \simeq 67\,{\rm km\ s^{-1}Mpc^{-1}}$.

We have, in fact, been able to measure the standard deviation of the measured Hubble constant very precisely,

$$\sigma_{\rm sample\ variance} = 0.31\,{\rm km\ s^{-1}Mpc^{-1}}. \tag{21.4}$$

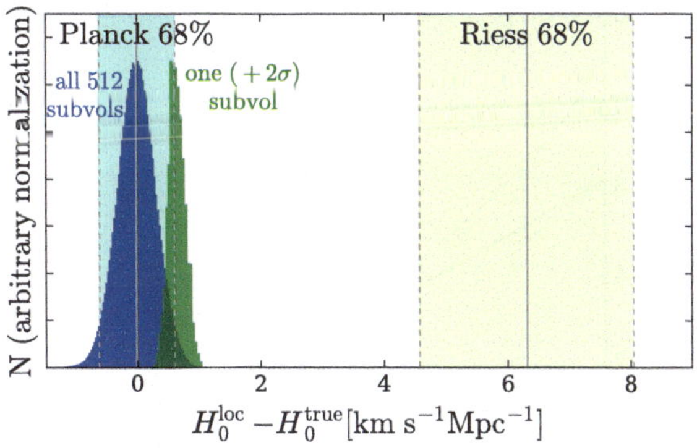

Fig. 21.2 Sample variance in ΔH_0^{loc} from the simulations of Wu and Huterer [38], compared to the *Planck* and distance-ladder ([2], "Riess" in the plot) error bars, assuming *Planck*'s H_0 [6] to be the true global value. The blue histogram shows 3240 rotations of the SNIa coordinate system from 512 subvolumes derived from the 8 h^{-1}Gpc Dark Sky box, corresponding to ~1.5 million SN-to-halo coordinate-system configurations. The green histogram shows a particularly underdense subvolume with a high ΔH_0^{loc} at the 2-σ level relative to the mean of all subvolumes. The sample variance in H_0^{loc} is much smaller than the difference between SNIa and CMB measurements. Adopted from Ref. [38]

In other words, $|H_0^{\rm Planck} - H_0^{\rm loc}|/\sigma_{\rm sample\ variance} \simeq 20$. Therefore, **sample variance is about 20 times too small to explain the Hubble tension**. Interestingly, using completely different approaches based on perturbation theory, Refs. [23, 28] find a very similar result: a sample variance of ~ 0.4 km s^{-1}Mpc^{-1} for the Pantheon sample (also see [43]).

How "empty" should a local void be to resolve the Hubble tension? Fig. 21.3 shows the relation between the change of the local Hubble constant, ΔH_0^{loc}, and the local density contrast, δ. For each of the 512 subvolumes, we measure the dark matter density contrast at $z < 0.04$ ($120\,h^{-1}$Mpc) and perform mock measurements of H_0^{loc} using the Supercal supernova sample ($z < 0.15$). The small green error bars correspond to the 512 simulation subvolumes, and the black line is the linear fit. The dashed curve includes the non-linear correction. This slope is different from that expected from the perturbation theory because we explicitly consider the sample selection, which simple perturbation-theory calculations do not account for.

The horizontal line shows that to account for the 6 km s^{-1}Mpc^{-1} shift requires an underdensity of $\delta \approx -0.8$ with a radius of 120 h^{-1}Mpc, which is incompatible with the density fluctuations in the ΛCDM model. The blue arrows correspond to observational results from Refs. [31, 34, 35], and none of them can cause a 6 km s^{-1}Mpc^{-1} shift.

Why does the sample variance of local measurements with SNIa have such a small effect? The reason is in the fact the Supercal SNIa sample goes out to $z = 0.15$ (corresponding to distances of about $\sim 500\,h^{-1}$Mpc). Averaged over such a large volume, overdensities and underdensities are not very pronounced, and the variations in the locally determined Hubble constant are correspondingly small.

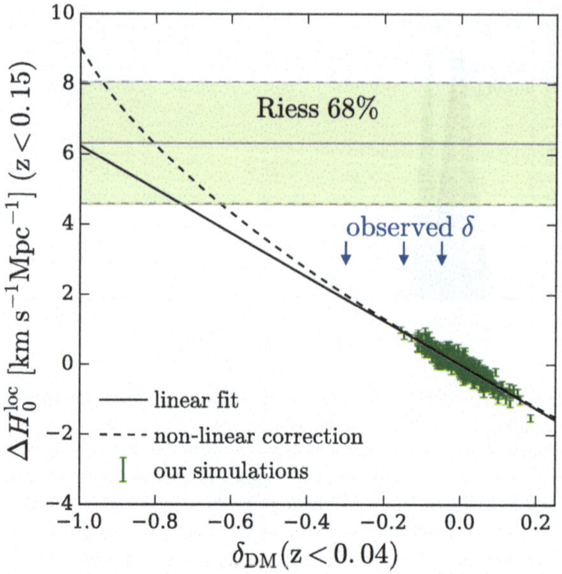

Fig. 21.3 Shift in local Hubble constant versus local density contrast derived from the mock Supercal sample created from the 8 h^{-1}Gpc Dark Sky box (green error bars). The blue arrows mark the local density inferred from Refs. [31, 34, 35]. The black solid line shows the linear fit, and the dashed curve shows the non-linear correction. A local Hubble constant shift of $\Delta H_0^{\rm loc} \simeq$ 6 km s^{-1}Mpc^{-1} required to explain the Hubble tension would consequently require an underdensity of $\delta \approx -0.8$ with a radius of 120 h^{-1}Mpc, which is incompatible with the cosmic structure in a ΛCDM universe. Adopted from Ref. [38]

This result, and the related ones cited above, put a nail in the coffin of the sample-variance explanations for the Hubble tension. Because the sample-variance (or void) explanation was arguably the simplest one, it leads to an exciting situation that the true explanation, whatever it is, is likely to only be more "exotic" and, overall, more unexpected.

21.5 Summary

We have reviewed what is logically the simplest explanation for the Hubble tension—the notion that we might be simply seeing the effect of sample variance in the distance-ladder-based measurements. Sample variance affects the last rung of the distance ladder, whose distance–redshift relation can be impacted by local density fluctuations. In the sample-variance explanation scenario, (1) we live in an underdense region of the universe, which has an enhanced locally measured Hubble constant, and (2) the local supernova sample is sparse and highly inhomogeneous so that the error bars for the local Hubble constant might be larger than previously estimated.

We have first illustrated the formalism that links the local density to the measured Hubble constant and have reviewed the methodology adopted by the community over the past decade. We have then reviewed recent observational results of the local matter density contrast, which have been inconclusive because it is difficult to infer the three-dimensional distribution of mass density from galaxies.

Proceeding to discuss our own work on the subject [38], we set out our goal: to use an N-body simulation to mimic the actual SNIa observations used for H_0 measurements. In the simulation, we place an observer in many independent subvolumes, each containing the 3D distribution of SNIa that precisely matches that used in actual observations. This automatically takes into account the effect of over/underdensities, as well as the inhomogeneous distribution of supernovae. Moreover, each SNIa is represented by a host halo, whose peculiar velocity is known in the simulation. Therefore, we have access to a simulated distance-ladder procedure that takes full account of sample variance, including how it mimics actual SNIa observations. The procedure is also statistically powerful—we use a total of 1.5 million realizations of SNIa observations; see Sect. 21.4.

Combining large-volume N-body ΛCDM simulations with recent SNIa data, we have found that the effect of the sample variance is small. Specifically, the resulting standard deviation in the Hubble constant (the square root of sample variance) is

$$\sigma_{\text{sample variance}} = 0.31 \text{ km s}^{-1}\text{Mpc}^{-1}. \tag{21.5}$$

To explain the 6 km s^{-1}Mpc^{-1} Hubble tension would require a 20σ effect in sample variance. Sample variance is therefore an extremely unlikely explanation for the Hubble tension.

In addition, we have found that resolving the 6 km s^{-1}Mpc^{-1} Hubble tension requires an underdensity $\delta = -0.8$ out to a radius 120 h^{-1}Mpc. The underdensity required to resolve the Hubble tension is so low that it is incompatible with the large-scale structure in a ΛCDM universe. Although the observational evidence for local underdensity is still controversial, even the most extreme observational claims for the local underdensity could not explain the Hubble tension.

Ruling out the local void is an exciting development in the nascent field of Hubble tension, as it removes this very simple explanation and a “guaranteed" effect in cosmology, leaving more exotic or surprising solutions. Future research should focus on the systematics of supernova and CMB measurements, as well as the exploration of new physics beyond the standard model of cosmology.

Acknowledgements DH is supported by NASA grant under contract 19-ATP19-0058 and DOE under contract DE-FG02-95ER40899. HW is supported by the DOE grant under contract DE-SC0021916.

References

1. W.L. Freedman, B.F. Madore, V. Scowcroft, C. Burns, A. Monson, S.E. Persson, M. Seibert, J. Rigby, Carnegie Hubble program: a mid-infrared calibration of the Hubble constant. APJ **758**, 24 (2012)
2. A.G. Riess, L.M. Macri, S.L. Hoffmann, D. Scolnic, S. Casertano, A.V. Filippenko, B.E. Tucker, M.J. Reid, D.O. Jones, J.M. Silverman, R. Chornock, P. Challis, W. Yuan, P.J. Brown, R.J. Foley. A 2.4% determination of the local value of the Hubble constant. APJ **826**, 56 (2016)
3. A.G. Riess et al., A comprehensive measurement of the local value of the Hubble constant with 1 km s^{-1} Mpc^{-1} uncertainty from the Hubble space telescope and the SH_0ES team. Astrophys. J. Lett. **934**(1), L7 (2022)
4. N. Aghanim et al., Planck 2018 results. VI. Cosmological parameters. Astron. Astrophys., **641**, A6 (2020) [Erratum: Astron.Astrophys. 652, C4 (2021)]
5. G. Hinshaw, D. Larson, E. Komatsu, D.N. Spergel, C.L. Bennett, J. Dunkley, M.R. Nolta, M. Halpern, R.S. Hill, N. Odegard, L. Page, K.M. Smith, J.L. Weiland, B. Gold, N. Jarosik, A. Kogut, M. Limon, S.S. Meyer, G.S. Tucker, E. Wollack, E.L. Wright, Nine-year Wilkinson microwave anisotropy probe (WMAP) observations: cosmological parameter results. ApJS **208**, 19 (2013)
6. X.I.I. Planck Collaboration, Planck 2015 results. XIII. Cosmological parameters. A&A **594**, A13 (2016)
7. X.V.I. Planck Collaboration, Planck 2013 results. XVI. Cosmological parameters. A&A **571**, A16 (2014)
8. E. Di Valentino, O. Mena, S. Pan, L. Visinelli, W. Yang, A. Melchiorri, D.F. Mota, A.G. Riess, J. Silk, In the realm of the Hubble tension—a review of solutions. Class. Quant. Grav. **38**(15), 153001 (2021)
9. I. Ben-Dayan, R. Durrer, G. Marozzi, D.J. Schwarz, Value of H_0 in the inhomogeneous universe. Phys. Rev. Lett. **112**(22), 221301 (2014)
10. A. Cooray, R.R. Caldwell, Large-scale bulk motions complicate the Hubble diagram. PhRvD **73**(10), 103002 (2006)
11. H.M. Courtois, D. Pomarède, R.B. Tully, Y. Hoffman, D. Courtois, Cosmography of the local universe. AJ **146**, 69 (2013)
12. P. Fleury, C. Clarkson, R. Maartens, How does the cosmic large-scale structure bias the Hubble diagram? JCAP **3**, 062 (2017)
13. L. Hui, P.B. Greene, Correlated fluctuations in luminosity distance and the importance of peculiar motion in supernova surveys. PhRvD **73**(12), 123526 (2006)
14. L.A. Martinez-Vaquero, G. Yepes, Y. Hoffman, S. Gottlöber, M. Sivan. Constrained simulations of the local universe—II. The nature of the local Hubble flow. MNRAS **397**, 2070–2080 (2009)
15. X. Shi, M.S. Turner, Expectations for the difference between local and global measurements of the Hubble constant. APJ **493**, 519–522 (1998)
16. B. Sinclair, T.M. Davis, T. Haugbølle, Residual Hubble-bubble effects on supernova cosmology. APJ **718**, 1445–1455 (2010)
17. E.L. Turner, R. Cen, J.P. Ostriker, The relation of local measures of Hubble's constant to its global value. AJ **103**, 1427–1437 (1992)
18. L. Wang, P.J. Steinhardt, Cluster abundance constraints for cosmological models with a time-varying, spatially inhomogeneous energy component with negative pressure. APJ **508**, 483–490 (1998)
19. D. Huterer. Growth of cosmic structure **12** (2022)
20. V. Marra, L. Amendola, I. Sawicki, W. Valkenburg, Cosmic variance and the measurement of the local Hubble parameter. Phys. Rev. Lett. **110**(24), 241305 (2013)
21. I. Odderskov, S. Hannestad, T. Haugbølle, On the local variation of the Hubble constant. JCAP **10**, 028 (2014)
22. R. Wojtak, A. Knebe, W.A. Watson, I.T. Iliev, S. Heß, D. Rapetti, G. Yepes, S. Gottlöber, Cosmic variance of the local Hubble flow in large-scale cosmological simulations. MNRAS **438**, 1805–1812 (2014)

23. Z. Zhai, W.J. Percival, Sample variance for supernovae distance measurements and the Hubble tension. PhRvD **106**(10), 103527 (2022)
24. D. Camarena, V. Marra, Z. Sakr, C. Clarkson, A void in the Hubble tension? The end of the line for the Hubble bubble. Class. Quantum Gravity **39**(18), 184001 (2022)
25. S. Castello, M. Högås, E. Mörtsell, A cosmological underdensity does not solve the Hubble tension. JCAP **2022**(7), 003 (2022)
26. V. Marra, T. Castro, D. Camarena, S. Borgani, A. Ragagnin, The BEHOMO project: Λ Lemaître-Tolman-Bondi N-body simulations. A&A **664**, A179 (2022)
27. W. D'Arcy Kenworthy, D. Scolnic, A. Riess, The local perspective on the Hubble tension: local structure does not impact measurement of the Hubble constant. Astrophys. J. **875**(2), 145 (2019)
28. Z. Zhai, W.J. Percival, The effective volume of supernovae samples and sample variance. arXiv e-prints (2023) [arXiv:2303.05717]
29. J. Soltis, A. Farahi, D. Huterer, C. Michael Liberato, Percent-level test of isotropic expansion using type Ia supernovae. Phys. Rev. Lett. **122**(9), 091301 (2019)
30. H. Böhringer, G. Chon, C. A. Collins, Observational evidence for a local underdensity in the universe and its effect on the measurement of the Hubble constant. A&A **633**, A19 (2020)
31. J. Carrick, S.J. Turnbull, G. Lavaux, M.J. Hudson, Cosmological parameters from the comparison of peculiar velocities with predictions from the 2M++ density field. MNRAS **450**, 317–332 (2015)
32. W.J. Frith, G.S. Busswell, R. Fong, N. Metcalfe, T. Shanks, The local hole in the galaxy distribution: evidence from 2MASS. MNRAS **345**(3), 1049–1056 (2003)
33. S. Jha, A.G. Riess, R.P. Kirshner, Improved distances to type Ia supernovae with multicolor light-curve shapes: MLCS2k2. APJ **659**, 122–148 (2007)
34. R.C. Keenan, A.J. Barger, L.L. Cowie, Evidence for a ~300 Megaparsec Scale Under-density in the local galaxy distribution. APJ **775**, 62 (2013)
35. J.R. Whitbourn, T. Shanks, The local hole revealed by galaxy counts and redshifts. MNRAS **437**, 2146–2162 (2014)
36. J. H. Wong, T. Shanks, N. Metcalfe, J.R. Whitbourn, The local hole: a galaxy underdensity covering 90 per cent of sky to ≈200 Mpc. MNRAS **511**(4), 5742–5755 (2022)
37. I. Zehavi, A.G. Riess, R.P. Kirshner, A. Dekel, A local Hubble bubble from type IA supernovae? APJ **503**, 483–491 (1998)
38. W. Hao-Yi, D. Huterer, Sample variance in the local measurements of the Hubble constant. Mon. Not. Roy. Astron. Soc. **471**(4), 4946–4955 (2017)
39. D. Scolnic, S. Casertano, A. Riess, A. Rest, E. Schlafly, R.J. Foley, D. Finkbeiner, C. Tang, W.S. Burgett, K.C. Chambers, P.W. Draper, H. Flewelling, K.W. Hodapp, M.E. Huber, N. Kaiser, R.P. Kudritzki, E.A. Magnier, N. Metcalfe, C.W. Stubbs, Supercal: cross-calibration of multiple photometric systems to improve cosmological measurements with type Ia supernovae. APJ **815**, 117 (2015)
40. D. Scolnic, A. Rest, A. Riess, M.E. Huber, R.J. Foley, D. Brout, R. Chornock, G. Narayan, J.L. Tonry, E. Berger, A.M. Soderberg, C.W. Stubbs, R.P. Kirshner, S. Rodney, S.J. Smartt, E. Schlafly, M.T. Botticella, P. Challis, I. Czekala, M. Drout, M.J. Hudson, R. Kotak, C. Leibler, R. Lunnan, G.H. Marion, M. McCrum, D. Milisavljevic, A. Pastorello, N.E. Sanders, K. Smith, E. Stafford, D. Thilker, S. Valenti, W.M. Wood-Vasey, Z. Zheng, W.S. Burgett, K.C. Chambers, L. Denneau, P.W. Draper, H. Flewelling, K.W. Hodapp, N. Kaiser, R.-P. Kudritzki, E.A. Magnier, N. Metcalfe, P.A. Price, W. Sweeney, R. Wainscoat, C. Waters, Systematic uncertainties associated with the cosmological analysis of the first Pan-STARRS1 type Ia supernova sample. APJ **795**, 45 (2014)
41. S.W. Skillman, M.S. Warren, M.J. Turk, R.H. Wechsler, D.E. Holz, P.M. Sutter, Dark sky simulations: early data release. ArXiv e-prints (2014)
42. Planck Collaboration Int. XLVI. Planck intermediate results. XLVI. Reduction of large-scale systematic effects in HFI polarization maps and estimation of the reionization optical depth. A&A **596**, A107 (2016)
43. D. Camarena, V. Marra, Impact of the cosmic variance on H_0 on cosmological analyses. PhRvD **98**(2), 023537 (2018)

Chapter 22
Resolving the Hubble Tension with Early Dark Energy

Vivian Poulin and Tristan L. Smith

22.1 Introduction: Why Early Dark Energy?

The notion of "Early Dark Energy" was originally introduced as an attempt at tackling the cosmic coincidence problem [1, 2], i.e., the question of why dark matter and dark energy have such similar densities, in particular since the vacuum energy prediction from quantum field theory for the value of the cosmological constant is so far off compared to the observation. If dark energy is a scalar-field, it may have an equation of state different from a cosmological constant, and it may even leave a small contribution already at the last-scattering surface [3–5].

Over the last few years, with the rise of the Hubble tension, the interest for models of EDE has dramatically increased, as it can be used to boost the value of the expansion rate in the decade of redshift prior to recombination to reduce the size of the sound horizon $r_s(z_*)$. Indeed, a naive increase in H_0 would decrease the angular diameter distance and in turn increase the angular size of the sound horizon θ_s. This must be compensated for by a decrease in $r_s(z_*)$, in order to keep θ_s fixed to the very precise value measured by *Planck*.

The Hubble tension is, in fact, more than just a tension in the value of H_0: it also requires an increase in the physical matter density. Under the assumption of a flat ΛCDM Universe, the expansion history measurements derived from Supernovae

V. Poulin (✉)
Laboratoire Univers and Particules de Montpellier (LUPM), CNRS and Université de Montpellier (UMR-5299), Place Eugène Bataillon 34095 Montpellier Cedex 05, France
e-mail: vivian.poulin@umontpellier.fr

T. L. Smith
Department of Physics and Astronomy, Swarthmore College, 500 College Ave., Swarthmore, PA 19081, USA
e-mail: tsmith2@swarthmore.edu

Center for Cosmology and Particle Physics, Department of Physics, New York University, New York, NY 10003, USA

E. Di Valentino and D. Brout (eds.), *The Hubble Constant Tension*, Springer Series in Astrophysics and Cosmology, https://doi.org/10.1007/978-981-99-0177-7_22

Type Ia (SNIa) and Baryon Acoustic Oscillations (BAO) impose a tight constraint on Ω_m (the ratio of matter density ρ_m to the critical density[1] $\rho_{\rm crit}$). When combined with the SH_0ES determination of H_0, they lead to a physical matter density ω_m ($\equiv \Omega_m h^2$) of 0.1603 ± 0.0063, a value that is approximately 3σ larger than the one inferred from *Planck* observations assuming the ΛCDM model ($\omega_m = 0.14228^{+0.00079}_{-0.00091}$). Adjusting cosmological data therefore requires an increase in ω_m that would spoil the amplitude of the peaks measured by *Planck*.

Remarkably, with a single mechanism, EDE manages to reduce the size of the sound horizon and compensate for the effect of an increase of $\omega_{\rm cdm}$ on the CMB. Indeed, the rise in $H(z)$ induced by EDE contributes to an augmentation of Hubble friction, resulting not only in a smaller sound horizon but also in a slowdown of the growth of the Cold Dark Matter (CDM) density contrast. EDE therefore manages to maintain a good fit to the combination of CMB, BAO and SN1a data, while adjusting the value of H_0 (or rather the SN1a magnitude) measured by SH_0ES.

In the following, we review the results of analyses of EDE, focusing on the toy-model of axion-like EDE as a representative, well-studied, example of EDE. Sect. 22.2 introduces the model and present results of analyses of EDE with a suite of cosmological data (in particular *Planck* ACT and SPT data). Sect. 22.3 discusses the impact of EDE beyond the Hubble tension, from observational impacts on large-scale structure (LSS) data in particular, to theoretical "fine-tuning" issues and model-building efforts. We present our conclusions in Sect. 22.4.

22.2 Resolving the Hubble Tension with Early Dark Energy

22.2.1 Modelling Early Dark Energy

Initially suggested as a potential mechanism to explain the Hubble tension in Refs. [6, 7], EDE has since then undergone various refinements in the modelling and additional studies by numerous authors [8–38].

EDE models typically propose the existence of a scalar field, and may differ in the choice of scalar field potential, the number of scalar field involve in the model, and whether or not the scalar field is minimally coupled. In this chapter, we focus on an axion-like potential of the form[2] [7, 8]

$$V(\theta) = m^2 f^2 [1 - \cos(\theta)]^n, \qquad (22.1)$$

where m represents the axion mass, f the axion decay constant, and $\theta \equiv \phi/f$ is a re-normalized field variable defined such that $-\pi \leq \theta \leq \pi$. The next chapter focuses on an alternative realization of an Early Dark Energy model named 'New Early

[1] Here $\rho_{\rm crit} \equiv 3H_0^2/(8\pi G)$, and h is given by $h \equiv H_0/(100\ {\rm km/s/Mpc})$.

[2] This potential was implemented in a modified version of the CLASS code [39] available at https://github.com/PoulinV/AxiCLASS.

Dark Energy' [14, 15]. The scalar field background dynamics[3] can be understood by looking at the homogeneous Klein-Gordon equation of motion:

$$\ddot{\phi} + 3H\dot{\phi} + \frac{dV(\phi)}{d\phi} = 0. \tag{22.2}$$

Due to the initially large Hubble friction, the background field is held fixed at some value, resulting in a constant background energy density. When $H \sim \partial_\theta^2 V(\theta)$ the field becomes dynamical, rolling down its potential to eventually oscillate about the minimum. Although this potential is not tied to a fundamental theory, it provides a flexible EDE model with a rate of dilution set by the value of the exponent n, and this is what we are most interested in here. Taken at face-value, this potential may be generated by higher-order instanton corrections [41–43] that require fine-tuning to cancel the leading contributions [30, 35], or through the interaction of an axion and a dilaton [11] and is therefore non-generic [44]. Related constructions of this potential in the context of 'Natural Inflation' were also proposed in Refs. [45–47]. Interestingly, recent works have paved the way to embed this potential in a string-theory framework [30, 48], showing that the theory challenges for building an EDE with such a potential are not particularly different from that of other cosmological models, such as inflation, dark energy, or fuzzy dark matter [48].

In full generality, this model therefore has 4 free parameters. Yet, most studies trade the 'theory parameters', m and f, for 'phenomenological parameters' that are better constrained by the data. These are the critical redshift z_c at which the field becomes dynamical and the fractional energy density $f_{\rm EDE}(z_c)$ contributed by the field at the critical redshift. The third parameter, n, controls the equation of state $w \equiv (n-1)/(n+1)$ when the field dilutes. Finally, the last parameter is the initial field value θ_i, which has two main impacts: it changes the asymmetry of the energy injection and controls the effective sound speed c_s^2 and thus the dynamics of the perturbations (mostly). Note that it is assumed that the field always starts in slow-roll (as enforced by the very high value of the Hubble rate at early times), and without loss of generality one can restrict $0 \leq \theta_i \leq \pi$. We illustrate the effect of each parameter on the energy injection $f_{\rm EDE}(z)$ in Fig. 22.1.

22.2.2 *EDE in Light of Planck Data*

We now review results of analyses of the axion-like EDE model in light of *Planck* CMB power spectra. The baseline analysis includes the full *Planck* 2018 TT,TE,EE and lensing power spectra [49], BAO measurements from BOSS DR12 [50], SDSS DR7 [51] and 6dFGS [52], and a compilation of uncalibrated luminosity distances to SN1a from Pantheon [53]. Reconstructed parameters are provided in Table 22.1. We refer to this dataset as "BaseTTTEEE". All the analyses we present follow the *Planck*

[3] For a discussion about EDE perturbed dynamics, we refer to the review [40].

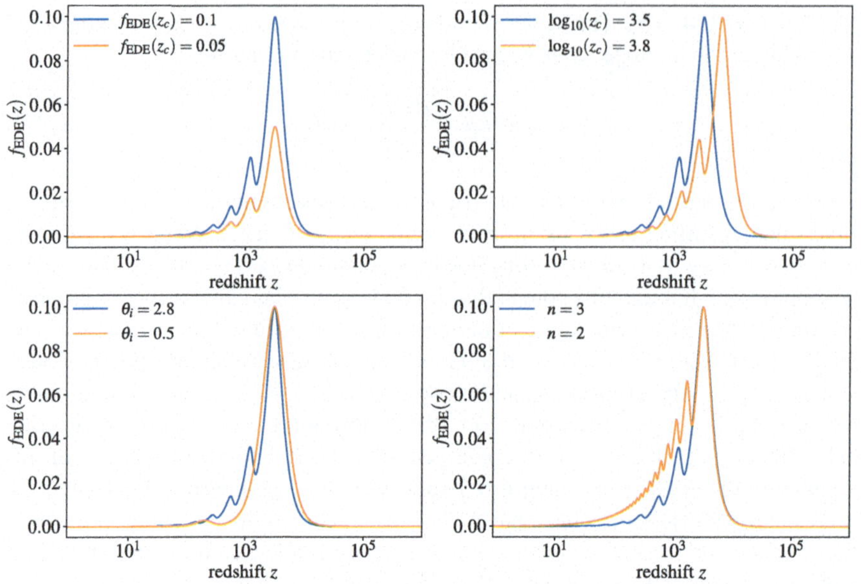

Fig. 22.1 Effect of varying the different EDE parameters on the energy injection $f_{\rm EDE}(z)$. The fiducial model has $\{f_{\rm EDE}(z_c) = 0.1, \log_{10}(z_c) = 3.5, \theta_i = 2.8, n = 3\}$

Table 22.1 Mean (best-fit) $\pm 1\sigma$ (or 2σ for one-sided bounds) of reconstructed parameters in the EDE model confronted to *Planck*TTTEEE+Lensing+BAO+Pantheon. We report constraints that include and leave out the SH_0ES prior, and compute the residual tension using the $Q_{\rm DMAP}$ statistics.

	Planck+BAO+Pantheon	+SH_0ES
$f_{\rm EDE}(z_c)$	$< 0.091(0.088)$	$0.109(0.122)^{+0.030}_{-0.024}$
$\log_{10}(z_c)$	unconstrained (3.55)	$3.599(3.568)^{+0.029}_{-0.081}$
θ_i	unconstrained (2.8)	$2.65(2.73)^{+0.22}_{-0.025}$
h	$0.688(0.706)^{+0.006}_{-0.011}$	$0.715(0.719) \pm 0.009$
S_8	$0.831(0.839)^{+0.011}_{-0.013}$	$0.839(0.843) \pm 0.012$
Ω_m	$0.3084(0.3041) \pm 0.0058$	$0.3008(0.3005) \pm 0.0048$
$\omega_{\rm cdm}$	$0.1227(0.1281)^{+0.0018}_{-0.0036}$	$0.1303(0.1319) \pm 0.0035$
$10^2\omega_b$	$2.258(2.266)^{+0.018}_{-0.020}$	$2.283(2.303) \pm 0.020$
$10^9 A_s$	$2.122(2.135) \pm 0.032$	$2.153(2.145) \pm 0.032$
n_s	$0.9734(0.9823)^{+0.0053}_{-0.0076}$	$0.9883(0.9895) \pm 0.0060$
$\tau_{\rm reio}$	$0.0570(0.0574)^{+0.0069}_{-0.0076}$	$0.0582(0.0579) \pm 0.0075$
Total $\chi^2_{\rm min}$	3799.2	3802.9
$\Delta\chi^2_{\rm min}$	-3.8	-23.7
$Q_{\rm DMAP}$	1.9σ	

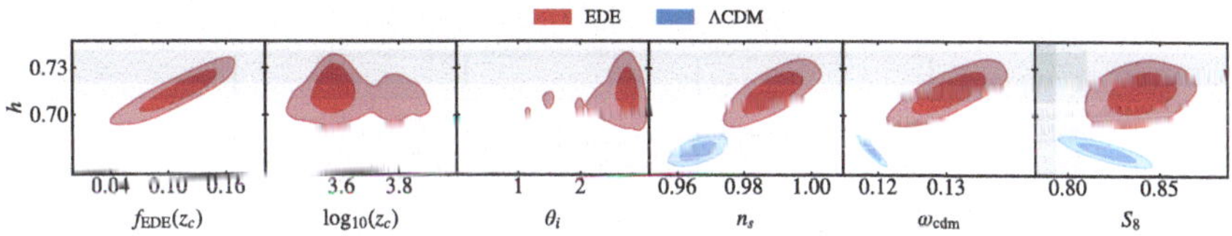

Fig. 22.2 2D posteriors for h vs $\{f_{\rm EDE}(z_c), \log_{10}(z_c), \theta_i, n_s, \omega_{\rm cdm}, S_8\}$

conventions for the treatment of neutrinos and inclue two massless and one massive species with with $m_\nu = 0.06$ eV [49]. For the ΛCDM parameters, large flat priors on the dimensionless baryon energy density ω_b, the dimensionless cold dark matter energy density $\omega_{\rm cdm}$, the Hubble parameter today H_0, the logarithm of the variance $\ln(10^{10}\mathcal{A}_s)$ of curvature perturbations centered around the pivot scale $k_p = 0.05$ Mpc^{-1} (according to the *Planck* convention), the scalar spectral index n_s, and the reionization optical depth $\tau_{\rm reio}$ were imposed. For the EDE parameters, a logarithmic prior on z_c, and flat priors for $f_{\rm EDE}(z_c)$ and θ_i as $\log_{10}(z_c) \in [3, 4]$, $f_{\rm EDE}(z_c) \in [0, 0.3]$, $\theta_i \in [0, \pi]$ were considered. For most of our discussion, the exponent n (or equivalently the equation of state once the field rolls $w_f = (n-1)/(n+1)$) is fixed to $n = 3$ to reduce the dimensionality, without affecting conclusions.

We start by presenting results of analyses that include a direct measurements of H_0 by SH_0ES as modeled by a Gaussian prior on the Hubble parameter $h = 0.7304 \pm 0.001$ where $h \equiv H/(100\ \text{km/s/Mpc})$ [54]. We show the posterior distribution of h vs $\{f_{\rm EDE}(z_c), \log_{10}(z_c), \theta_i, n_s, \omega_{\rm cdm}, S_8\}$ in the EDE and ΛCDM model (Fig. 22.2). In this analysis, we find $f_{\rm EDE}(z_c) = 0.109^{+0.030}_{-0.024}$, around the critical redshift $\log_{10}(z_c) = 3.599(3.568)^{+0.029}_{-0.081}$ with $h = 0.715 \pm 0.009$, and a $\Delta\chi^2 \equiv \chi^2_{\rm EDE} - \chi^2_{\Lambda\rm CDM} = -23.7$, which is a strong preference in favor of the EDE model (a simple estimate gives $\sim 4\sigma$ preference, assuming $\Delta\chi^2$ is χ^2−distributed with 3 degrees of freedom). We stress that it is not correct to estimate the tension with SH_0ES by using the reconstructed value of h from this analysis, given that the SH_0ES prior was included in the analysis. Rather, we report the following tension metric [55, 56]

$$Q_{\rm DMAP} \equiv \sqrt{\chi^2(\text{w/ S}H_0\text{ES}) - \chi^2(\text{w/ S}H_0\text{ES})}\,, \tag{22.3}$$

where χ^2(w/o SH_0ES) is obtained from a fit to the same data set but removing the SH_0ES prior. This metric can easily be interpreted intuitively, as for a single additional data point (SH_0ES here), the χ^2 value of a "good" model should not degrade by more than $\sim$1 (for a rigorous discussion, see Ref. [55]). Applying this to EDE, one finds the tension metric $Q_{\rm DMAP} = 1.9\sigma$, while the metric gives 4.8σ in ΛCDM. Therefore, the EDE model is able to accommodate large values of H_0, while providing a good fit to *Planck*+BAO+SN1a. From these results, one concludes that a contribution of 10% of EDE at a redshift close to matter-radiation equality (recalling $z_{\rm eq} \sim 3500$) *can resolve the Hubble tension.*

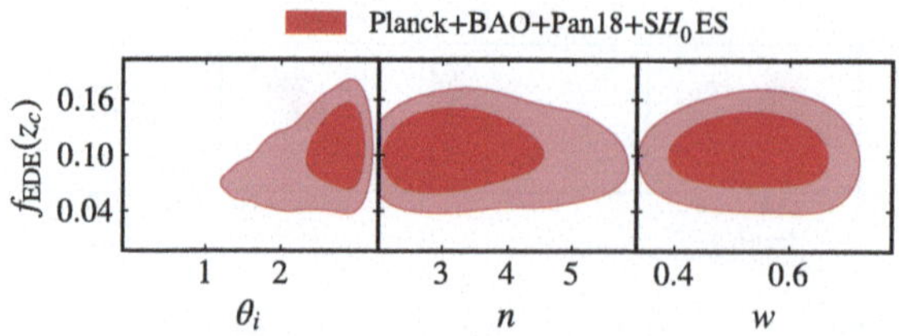

Fig. 22.3 Constraints on EDE "microphysics": n controls the EDE equation of state when it dilutes (displayed as $w = (n-1)/(n+1)$) and θ_i controls the asymmetry of the injection and the dynamics of perturbations

22.2.3 *Constraints on EDE "Microphysics"*

We show in Fig. 22.3, that on top of constraining the peak and location of the energy injection, *Planck* data also provide constraints on the "microphysics" of the energy injection. When n is let free to vary,[4] one finds $n = 3.37^{+0.41}_{-0.99}$ (68% C.L.) [8, 40], with a fairly sharp constraint $n \lesssim 5.2$ (95% C.L.). This translates into a constraint on $w = (n-1)/(n+1) = 0.535 \pm 0.086$ (68% C.L.) and a limit $w < 0.7$ (95% C.L.), i.e., a very fast dilution with a phase of kinetic energy domination ($w = 1$) is disfavored by the data in the axion model. In addition, they impose tight constraints on the initial field value θ_i (in units of f, the axion decay constant). As shown in Sect. 22.2.1, θ_i at the background level controls the shape of the energy injection, which in turn affects the sound horizon and damping scales. At the perturbation level, its dominant impact is setting the effective sound speed of propagation of EDE perturbations.

For an oscillating EDE with a potential of the form $V \propto \phi^{2n}$ around the minimum, an approximate expression[5] for the sound speed can be derived by "cycle-averaging" the sound-speed [57]:

$$c_s^2(k,a) = \frac{\langle \delta P \rangle}{\langle \delta \rho \rangle} \simeq \frac{2a^2(n-1)\varpi^2 + k^2}{2a^2(n+1)\varpi^2 + k^2}, \tag{22.4}$$

where ϖ is the angular frequency of the oscillating background field [8, 57, 58]

$$\varpi(a) \simeq 3\,H(z_c)\frac{\sqrt{\pi}\,\Gamma(\frac{1+n}{2n})}{\Gamma\left(1+\frac{1}{2n}\right)} 2^{-(1+n)/2} \frac{\theta_{\rm env}^{n-1}(a)}{\sqrt{|E_{n,\theta\theta}(\theta_i)|}},$$

[4] For numerical reason, a lower bound $n > 2$ is imposed in this analysis. Part of the parameter space below $n = 2$ may be allowed, although as one gets closer to the matter case $n = 1$, it is expected that constraints will become strong.

[5] Note that this approximate sound speed only applies when $\varpi/H \gg 1$–i.e. when the field oscillates several times per Hubble time.

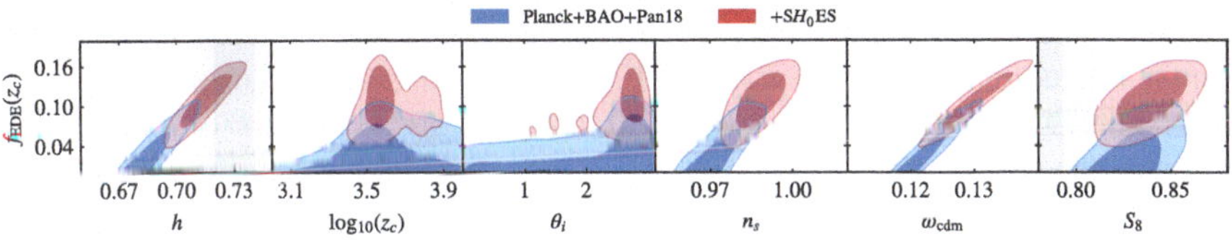

Fig. 22.4 2D posteriors for $f_{\text{EDE}}(z_c)$ vs $\{h, \log_{10}(z_c), \theta_i, n_s, \omega_{\text{cdm}}, S_8\}$ when analyzing *Planck*TTTEEE+BAO+Pantheon with and without the SH_0ES prior ($H_0 = 73.04 \pm 1.04$ [54]), as analyzed in Ref. [60]

where $\Gamma(x)$ is the Euler Gamma function. The envelope of the background field ($\theta_{\text{env}} \equiv \phi_{\text{env}}/f$) once it is oscillating is roughly $\phi_{\text{env}}(a) = \phi_c(a_c/a)^{3/(n+1)}$ and we have written the scalar field potential as $V_n(\phi) = m^2 f^2 E_n(\theta = \phi/f)$. The angular frequency ϖ defines a scale that governs the sound speed that affects a given mode: For $a \gg a_c$, $\varpi = 0$, so that $c_s^2 = 1$; when $a \ll a_c$ and $k \ll a\varpi$, $c_{s,\text{eff}}^2 = (n-1)/(n+1) = w$; when $a \ll a_c$ and $k \gg a\varpi$, $c_s^2 = 1$.

Data prefer an EDE for which modes within the horizon around z_c have effective sound speed $\lesssim 0.9$ [8, 10]. This can be achieved if the potential is flat around the initial field value θ_i such that the term $\theta_i^{n-1}/\sqrt{E_{n,\theta\theta}(\theta_i)} \gg 1$ in Eq. 22.4. Since the range of $k-$modes within the horizon is a sharp function of the initial field value θ_i, data provide a fairly strong constraint on $\theta_i \sim \pi$ (see Fig. 22.4).

These constraints on the dynamics, and as a result on the shape of the potential, explain why simpler power-law potentials (as in the Rock'n'Roll model [9]) fair less favorably in resolving the tension.

Finally, translating the phenomenological parameters back into the theory parameters, the EDE model suggests the existence of an axion-like field with mass $m \simeq 10^6 H_0 = 2.13\ h \times 10^{-27}$ eV, decay constant $f \simeq 0.15 M_{\text{pl}}$, and initial field value $\phi_i/f \sim \pi$. This prompted studies to understand whether the near-Planckian values of the decay constant and field excursion can lead to additional phenomenology (as in the EDS model [27, 59]) and constraints, as it might violate the axion weak gravity conjecture [35, 44].

22.2.4 "Barefoot Analyses": Bayesian or Frequentist?

Concerns often arise regarding the inclusion of SH_0ES data for EDE constraints and for comparison to ΛCDM: ideally, a model that resolves the Hubble tension when analyzed with CMB data alone should predict a higher H_0, improve the χ^2 more than the additional degrees of freedom and be preferred over ΛCDM. It is therefore interesting to look at "barefoot" analyses of EDE, i.e., analyses without the SH_0ES prior.

We show in Fig. 22.4 the 2D reconstructed posteriors for the parameters $f_{\text{EDE}}(z_c)$ vs $\{h, \log_{10}(z_c), \theta_i, n_s, \omega_{\text{cdm}}, S_8\}$ when analyzing *Planck*TTTEEE+BAO+Pantheon with and without the SH_0ES prior [60]. In the absence of a prior on H_0, it is striking that the data combination *Planck*TTTEEE+BAO+Pan18 provides an apparent upper limit on the EDE contribution $f_{\rm EDE}(z_c) < 0.091$ (at 95% C.L.), with $h = 0.688^{+0.006}_{-0.011}$. In fact, taken at face value, this constraint seems to exclude the value of $f_{\rm EDE}(z_c) \sim 0.12$ favored in the analysis that includes SH_0ES.

This apparent inconsistency is the sign of 'prior-volume effects' [56, 61–67]: when the parameter $f_{\rm EDE} \sim 0$, the other EDE parameters (z_c, θ_i) lose all influence on the dynamics and have therefore no impact on the model predictions, which become identical to ΛCDM regardless of the value of these two parameters. This leads to a large prior volume at $f_{\rm EDE} \simeq 0$, which impacts the results of the MCMCs and skews the posterior distributions towards $f_{\rm EDE} \to 0$ [63–65]. These posteriors however do not trace the true data likelihood in the extended ΛCDM+EDE parameter space. To test for the impact of prior-volume effects, one can compute the "profile likelihood" of a given parameter by successively optimizing χ^2 while keeping the parameter of interest fixed at different values within a given range. Following the Frequentist approach, confidence intervals are then extracted by computing the $\Delta\chi^2$ with respect to the bestfit point, with $\Delta\chi^2 \simeq 1$ giving the 1σ region in the case of a Gaussian profile far from the parameter prior boundaries [68].[6]

In Fig. 22.5, we compare the MCMC 1D posteriors for H_0 and $f_{\rm EDE}$ in ΛCDM and EDE to the profile likelihoods of the same parameters obtained when analyzing *Planck* 2018 CMB TTTEEE data. First, it is clear that there is little difference between the posteriors and the likelihood profiles of ΛCDM from the two approaches [70]. This shows that for ΛCDM, conclusions derived from the Bayesian analyses are perfectly robust. For the EDE model however, the profile likelihood shows much wider constraints to both H_0 and $f_{\rm EDE}$ than the Bayesian posteriors. Strikingly, while one would conclude from the MCMC analysis that $f_{\rm EDE}$ is consistent with 0, the profile likelihood analysis shows preference for $f_{\rm EDE} > 0$ at over 2σ, without including the SH_0ES H_0 prior.

22.2.5 Imprints of EDE in the CMB: Target for Future Surveys

The predictions for the residuals of the CMB TT and EE power spectra with respect to ΛCDM in the best-fit EDE cosmology are shown with the solid black line. To emphasize the irreducible effects imprinted by EDE in data (i.e. those that cannot be absorbed by a re-shuffling of ΛCDM parameters) that future surveys can target, we show in Fig. 22.6 the residuals of 100 samples randomly drawn from the MCMC chains, normalized to the ΛCDM best-fit to *Planck*. The color code indicates the

[6] In a more generic (non-Gaussian) case, one can use the Feldman-Cousins prescription to define confidence intervals [69]. For EDE, these were found to match the simple Gaussian prescription [65].

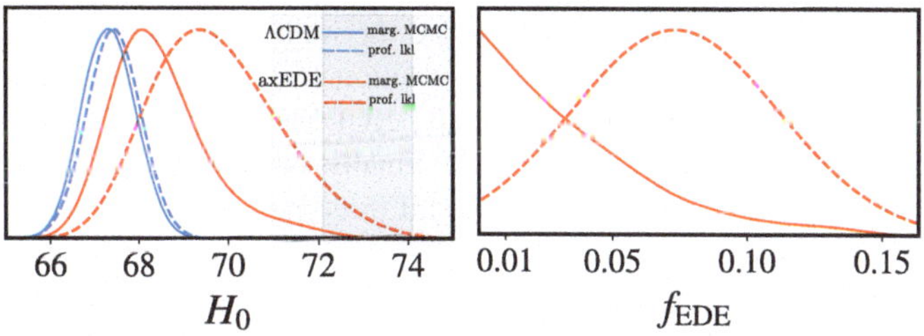

Fig. 22.5 Posteriors obtained from a profile likelihood (dashed) and MCMC methods (solid) are shown for ΛCDM (blue) and EDE (red) for H_0 and $f_{\rm EDE}$, for *Planck* 2018 CMB TTTEEE data. For the profile likelihoods, we plot the normalised curves of $e^{-\chi^2/2}$. While the posterior and profile likelihood match under ΛCDM, the profile likelihood indicates both a higher H_0 and $f_{\rm EDE}$ under the EDE model

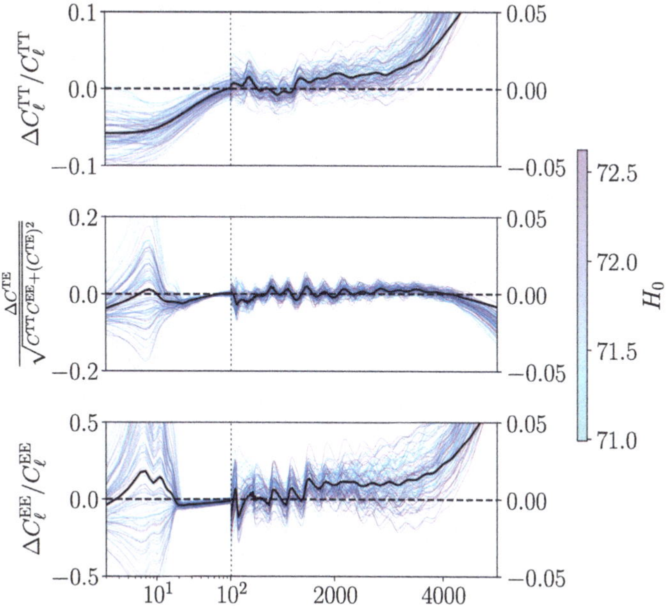

Fig. 22.6 Residuals of 100 samples randomly drawn from the MCMC chains in the EDE model (fit to *Planck*+BAO+Pantheon+SH_0ES), normalized to the ΛCDM best-fit to *Planck*. The color code indicates the value of H_0 in each sample

value of H_0 in each sample. Note that we use a combination of logarithmic and linear spacing in order to clearly show both large and small scale features.

First, the spectra show a bump-like feature around $\ell \sim 1500$ with increased power toward large ℓ in TT and EE that is present in essentially all the samples, and is due to the larger n_s. This feature can be further probed with high-resolution ground-based measurements, such as ACT and SPT, and in the future, the Simons Observatory (SO) [71] and CMB-S4 [72]. Multipoles $\ell \sim 100 - 500$ are particularly sensitive to details of the EDE dynamics, as these are the multipoles entering the horizon concurrently with the largest fractional contribution of the EDE energy density. Yet, the data require no strong departure from the ΛCDM prediction in that multipole range, and the samples oscillate around zero. Finally, one can also see large residual patterns at large angular scales ($\ell < 30$) in all three power spectra. The increased depth of the Sachs-Wolfe plateau in TT is (mostly) a consequence of slightly different values of A_s and n_s, and is common to all samples, albeit at different amplitudes. On the other hand, the large bump/dip in TE and EE, which is due to small differences in τ and ω_b, should not be overinterpreted, as the amplitudes of these vary greatly within the random samples of 100 points. It is therefore possible to easily adjust any forthcoming large-scale EE measurements with a slight shift in parameters and this signal is not a constraining feature for EDE. Note also that the correlation between the value of H_0 and the size of the residuals is stronger in polarization than in temperature, with the EE and TE spectra showing sharper oscillations when H_0 increases. On the other hand, in TT it is possible to find samples of points with large H_0 values that show only small deviations from ΛCDM (and vice versa).

22.2.6 Detecting EDE in the CMB

At the moment, one can conservatively conclude that *Planck* data alone do not strongly favor, nor disfavor, EDE over ΛCDM. This raises a couple of questions: should we have expected these data to be able to detect $f \sim 0.12$, $h \sim 0.72$, as the best-fit model extracted from analyses that include SH_0ES suggest? If not, can future CMB experiment unambiguously detect such a signal?

These questions can be addressed by running mock analyses that include an EDE signal, as performed in Ref. [8], and illustrated in Fig. 22.7. On the left panel, it shows that *Planck* data alone *cannot* detect even a 11.5% contribution of EDE at $z_c \sim 3500$ with $h = 0.72$ and only lead to upper limits on the EDE contribution. However, an experiment like CMB-S4 would detect this signal at more than 10σ. In addition, the dashed lines on the right panel shows that the biases introduced by analysing the mock *Planck* data assuming ΛCDM leads to parameters that are in good agreement with what is found in real data, with a degradation in χ^2 of the same order as to what is found in real data as well. Looking forward, when analysing data like CMB-S4, the ΛCDM model should provide a bad fit to the data, with a fit improved by ~ 500 under EDE, such that any statistical measure should unambiguously favor EDE. ΛCDM parameters should become strongly biased (solid lines on the right panel), with big differences in parameter values compared to what is currently found with *Planck*. Let us stress that the current disagreement between Bayesian and Frequentist approach is purely due to the lack of precision of current CMB data. In the case of CMB-S4, the data likelihood overwhelmingly dominates over the prior, and both approaches agree.

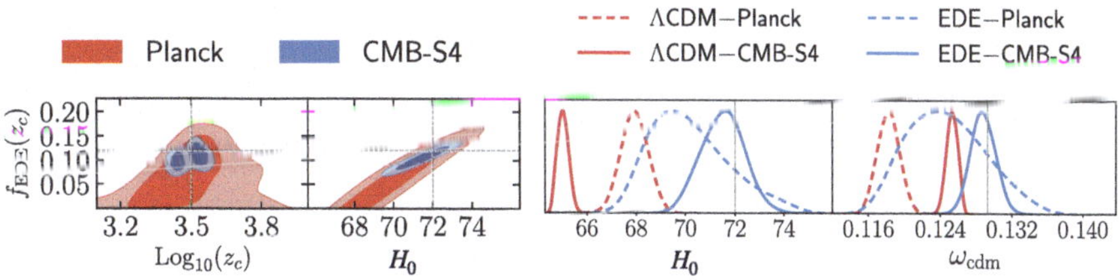

Fig. 22.7 Reconstruction from mock data with fiducial model $\{h=0.72, f_{\rm EDE}(z_c)=0.115, \log_{10}(z_c)=3.53\}$. *Left panel:* 2D Posterior distributions in the EDE model from mock *Planck* data and CMB-S4. *Right panel:* 1D Posterior distributions of H_0 and $\omega_{\rm cdm}$ from mock *Planck* data (dashed lines) and CMB-S4 (full lines) in either the ΛCDM (blue) or EDE (red) model. Taken from Ref. [8]

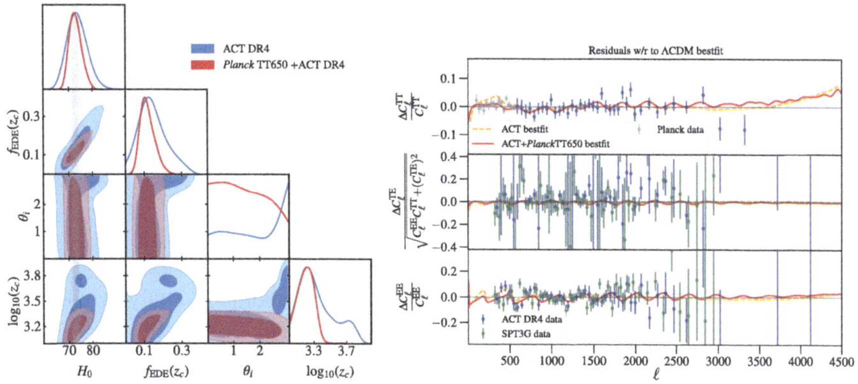

Fig. 22.8 *Left panel:* Posterior distributions of $\{H_0, f_{\rm EDE}(z_c), \theta_i, \log_{10}(z_c)\}$ reconstructed from ACT data (with and without *Planck*TT data upto $\ell \leq 650$). *Right panel:* Residuals between EDE and ΛCDM when fit to ACT DR4 data alone and in combination with *Planck*TT650. We also show the residuals of ACT DR4, SPT3G and *Planck*TT650 in the ΛCDM model

22.2.7 *Early Dark Energy in Light of ACT and SPT Data*

The next-generation surveys, the Atacama Cosmology Telescope (ACT) [73] and the South Pole Telescope (SPT) [74, 75] can already provide additional constraining power on EDE. Interestingly, recent analyses have shown that the ACT DR4 data favor EDE at 2–3 σ over ΛCDM, with H_0 values that are in good agreement with the SH_0ES determination. We review the results from ground-based telescope data analyses, the hints of EDE, and contrasts it with the results from the *Planck* satellite just presented.

We show in Fig. 22.8 left panel the posterior distributions of $\{H_0, f_{\rm EDE}(z_c), \theta_i, \log_{10}(z_c)\}$ in the EDE cosmology, with ACT alone (in blue) and in combination with *Planck*TT restricted data, that is essentially equivalent to the data from WMAP [76]. ACT data favor a non-zero contribution of EDE at $\gtrsim 2\sigma$, with $f_{\rm EDE}(z_c) = 0.152^{+0.055}_{-0.092}$ and $h = 0.742^{+0.036}_{-0.049}$, improving the $\Delta\chi^2 = -9.3$ with

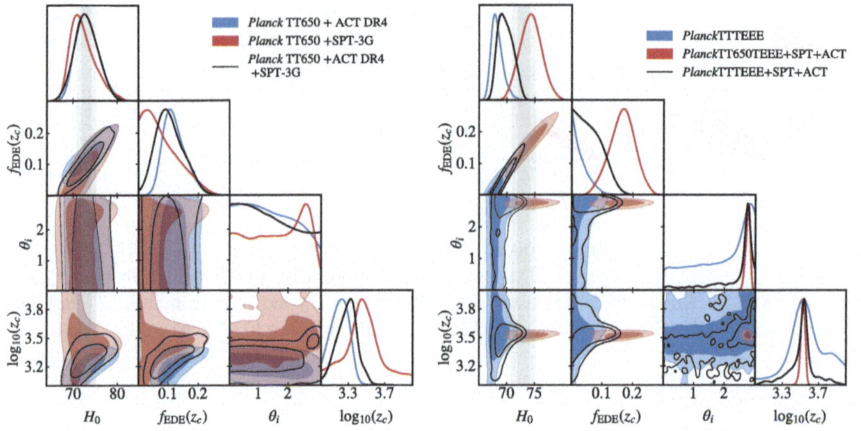

Fig. 22.9 *Left panel:* Posterior distributions of $\{H_0, f_{\rm EDE}(z_c), \theta_i, \log_{10}(z_c)\}$ reconstructed from the analysis of ACT and SPT data with *Planck*TT data up to $\ell \leq 650$. *Right panel:* Same as left panel but now comparing the use of *Planck*TT650TEEE or full *Planck*TTTEEE. From Ref. [79] *Left panel:* Posterior distributions of $\{H_0, f_{\rm EDE}(z_c), \theta_i, \log_{10}(z_c)\}$ reconstructed from the analysis of ACT and SPT data with *Planck*TT data up to $\ell \leq 650$. *Right panel:* Same as left panel but now comparing the use of *Planck*TT650TEEE or full *Planck*TTTEEE. From Ref. [79]

respect to ΛCDM (see also Ref. [77] for the ACT collaboration analysis yielding similar results with a different analysis pipeline).

Ref. [77] established that the preference for EDE in ACT data is due to a residual pattern at a multipole value of around 500 in the EE spectrum. This can be observed in the right panel of Fig. 22.8, which shows the residual between EDE and the best-fit ΛCDM model to the ACT DR4 data. One can see that the pattern is accompanied by a similar bump at $\ell \sim 350$ in TT, that is not observed in *Planck* data. As a result, when the large-scale *Planck*TT650 data is included, the residual features around the multipole value of 500 in both the EE and TT spectra become less prominent. Instead, a residual pattern of oscillations is observed in all spectra, which still provides a significantly better fit to the ACT data, equally supported by the TT, TE, and EE spectra, and in fact the preference for EDE increases to $\sim 3\sigma$, with $\Delta\chi^2 = -15.4$. The inclusion of SPT TEEE data is shown in Fig. 22.9 left panel. SPT data prefer a small deficit of power in EE around ~ 350 where ACT prefers a small increase, and thereby slighlty reduce the preference for EDE. Analyses with full SPT TT data have yet to appear in the literature.

Finally, we consider separately the role of *Planck*TT and *Planck*TEEE on constraining the hints for EDE seen in ACT. Remarkably, the right panel of Fig. 22.9 shows that the analyses of ACT+SPT with *Planck*TT650TEEE data detects $f_{\rm EDE}(z_c) = 0.163(0.179)^{+0.047}_{-0.04}$ at $\log_{10}(z_c) = 3.526(3.528)^{+0.028}_{-0.024}$ and $\theta_i = 2.784(2.806)^{+0.098}_{-0.093}$, with a $\Delta\chi^2 = -16.2$ (3.3σ preference) in favor of EDE over ΛCDM [78]. In addition, the model predicts the value of $H_0 = 74.2^{+1.9}_{-2.1}$ km/s/Mpc, in good agreement with direct determinations.

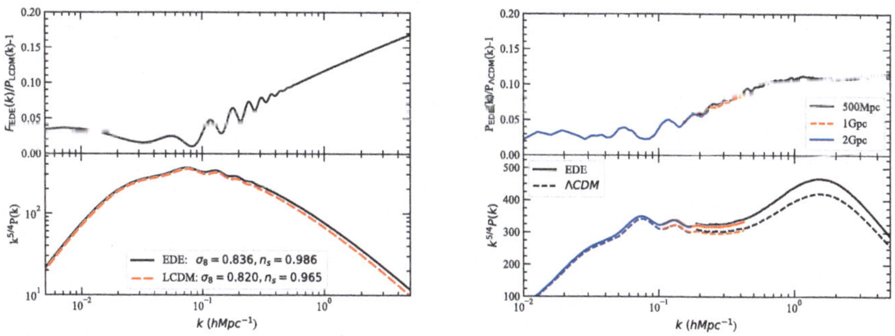

Fig. 22.10 The linear (left) and non linear (right) matter power spectra in an EDE cosmology and in ΛCDM. Taken from Ref. [82]

However, these results are in some tension with the analysis of full *Planck* data presented earlier, which showed only upper limits (in the MCMC analyses) in the absence of a prior on H_0. One can see in the right panel of Fig. 22.9 that the reconstructions from *Planck*TTTEEE (in blue) and *Planck*TT650TEEE+ACT+SPT (in red) show a mild inconsistency, though compatible with a statistical fluke at $\sim 2\sigma$[7]. Consequently, the combination of all datasets *Planck*TTTEEE+ACT+SPT (in black) leads to the constraint $f_{\rm EDE} < 0.128$, significantly weaker than that obtained from analyzing *Planck*TTTEEE alone ($f_{\rm EDE} < 0.091$). Future data from ACT, and later on from SO and CMB-S4, are eagerly awaited to shed light on the situation.

22.3 Consequences of Early Dark Energy

22.3.1 Constraining EDE with Galaxy Clustering Data

We show in Fig. 22.10 the linear (left) and non linear (right) matter power spectra in the EDE cosmology and in ΛCDM as computed in Ref. [82] for a fiducial analysis that includes *Planck*+BAO+Pantheon+SH_0ES data. One can see that the EDE cosmology results in a clear increase of power at scales $k \gtrsim 0.1\ h$/Mpc, although the differences are smaller in the non-linear power spectrum. This is mostly due to the reshuffling of the ΛCDM parameters necessary to preserve the fit to CMB data, in particular the increase in $\omega_{\rm cdm}$ and n_s [83]. As stressed in introduction, the increase in $\omega_{\rm cdm}$, that is the main driver of the larger power at small scales, is necessary in order to keep $\Omega_m \equiv \omega_{\rm cdm}/h^2$ fixed and adjust the late-time measurements of the shape of the expansion rate from BAO and SN1a. The effect of EDE partly compensates that increase, though not perfectly, resulting in a larger value of σ_8. As we discuss below, the change in the matter power spectrum and in σ_8 can be probed with LSS data.

[7] See Refs. [80, 81] for further examination of the statistical agreement between these three CMB data sets.

The use of BOSS data analyzed using the one-loop prediction of the galaxy power spectrum in redshift space from the Effective Field Theory of Large-Scale Structures[8] [84–89] has been suggested to constrain EDE [90, 91]. When analyzing *Planck*+BAO+Pantheon+EFTBOSS, Refs. [92, 93] originally derived $f_{\rm EDE}(z_c) < 0.08$ and $f_{\rm EDE}(z_c) < 0.072$ respectively, with two different implementations of the EFT. However, the apparent constraining power from the BOSS full-shape analysis is partly amplified by the impact of the prior volume artificially favoring ΛCDM in the Bayesian context. This was explicitly verified with a profile-likelihood approach [65, 94] which indicated that EDE is favored by the combination of *Planck*+BOSS at the $\sim 2\sigma$ level, with $f_{\rm EDE}(z_c) = 0.072 \pm 0.036$ (see discussion in Sect. 22.2.4).

The most recent analysis, that corrects an issue in the normalization of BOSS data window function [95], showed that the inclusion of EFTBOSS leads to $f_{\rm EDE}(z_c) < 0.083$, which is a marginal $\sim 5\%$ improvement over the constraints with conventional BAO and redshift space distortion data [60]. The Hubble tension is reduced to the 2.1σ level in the EDE cosmology (1.9σ without EFTBOSS) compared to 4.8σ in the ΛCDM model, and one finds $f_{\rm EDE}(z_c) = 0.103^{+0.027}_{-0.023}$ at $z_c = 3970^{+255}_{-205}$ when the SH_0ES prior is included.

Nevertheless, Fig. 22.11 right panel shows that EFTBOSS data however do further restrict the preference for EDE seen in ACT data, leading to a mild constraint on $f_{\rm EDE}(z_c) < 0.172$ with $\Delta\chi^2({\rm EDE} - \Lambda{\rm CDM}) = -11.1$, to be compared with $f_{\rm EDE}(z_c) = 0.128^{+0.064}_{-0.039}$ without EFTBOSS, with $\Delta\chi^2({\rm EDE} - \Lambda{\rm CDM}) = -14.6$. When full *Planck* data are included, the constraints become $f_{\rm EDE}(z_c) < 0.110$, and one can estimate that the Hubble tension is reduced to 1.5σ in the EDE model, to be compared with 4.7σ in ΛCDM. The inclusion of the SH_0ES prior leads to $f_{\rm EDE}(z_c) = 0.108^{+0.028}_{-0.021}$ at $z_c = 3565^{+220}_{-495}$. Therefore, EFTBOSS data do not rule out EDE as a resolution to the Hubble tension, although they do constrain very high EDE fractions, as seen from analyzing ACT DR4 data. Future data from DESI, Euclid and the Vera Rubin Observatory will be pivotal in testing for the presence of EDE [93].

22.3.2 EDE and the S_8 Tension

Inferences of S_8 from weak lensing surveys such as the CFHTLenS [96], DES [97], KiDS [98], and HSC [99] as well as from *Planck* SZ cluster abundances [49, 100] are about 2–3 σ smaller than that from the CMB. We have already argued that a generic feature of early-universe models proposed to resolve the Hubble tension is to predict a greater S_8 and increase this tension, simply because one cannot keep Ω_m fixed if h increases without increasing $\omega_{\rm m}$, and therefore yielding a larger onset of matter domination and a larger σ_8 [101, 102].

Due to the impact on S_8, Ref. [83] found that the combination of DES data, with priors on S_8 as measured by KIDS and HSC, leads to the constraint $f_{\rm EDE}(z_c) < 0.06$ without SH_0ES, and $f_{\rm EDE}(z_c) = 0.062^{+0.032}_{-0.033}$ with SH_0ES. However, the profile

[8] We refer to this dataset as ‘EFTBOSS’ for brevity.

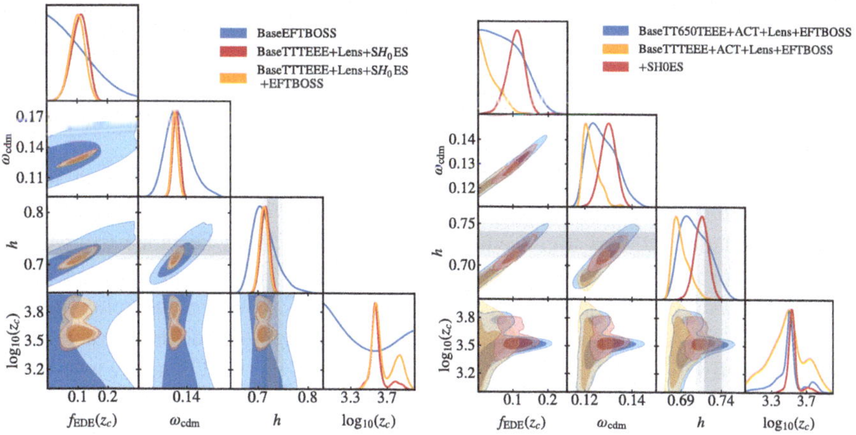

Fig. 22.11 *Left panel:* 2D posterior distributions from BaseEFTBOSS and BaseTTTTEEE+*Planck* Lensing+SH_0ES, with and without EFTBOSS data. BaseTTTTEEE refers to *Planck* TTTEE+BAO+Pan18, while BaseEFTBOSS refers to EFTBOSS+BBN+*Planck* Lensing+BAO+Pan18. *Right Panel:* 2D posterior distributions from ACT+Lens+EFTBOSS in combination with either BaseTT650TEEE, or BaseTTTEEEE with and without SH_0ES

likelihood analysis of *Planck*+DES leads to $f_{\rm EDE}(z_c) = 0.061^{+0.035}_{-0.034}$, $H_0 = 70.28 \pm 1.33$ km/s/Mpc [63, 103], showing that prior-volume effects still play a role in the apparent constraints from weak lensing data.

It is worth noting that the most recent cosmic shear data that combines both KiDS and DES are in less tension with *Planck* [104], and therefore it is expected that up-to-date constraints on EDE would be weaker than reported here, though a dedicated analysis is still lacking. Nevertheless, it is clear that low-S_8 measurements cannot be explained by the presence of EDE [105], and future LSS surveys (e.g. DESI, Euclid, LSST) will be decisive in testing for the existence of EDE.

If EDE does not explain a low S_8, one may wonder whether one (or more) additional degrees of freedom that reduces the amplitude of fluctuations at small scales could help further, whether it is connected to the EDE dynamics itself, or independent of it. For instance, massive neutrinos lead to a power suppression at small scales which decreases the value of σ_8. However, the sum of neutrino masses $\sum m_\nu \sim 0.3$eV required to resolve the tension is excluded by *Planck* data (e.g. [49, 106]), although constraints on the sum of neutrino masses are substantially broadened in extended cosmologies (e.g. [107, 108]). In the context of EDE, simply adding the neutrino mass as a free parameter does not change the reconstructed value of S_8 [63, 109].

It was shown however that, when EDE is combined to another proposal that is able to alleviate the S_8 tension, one can explain both tension independently. For instance, the existence of a dark sector, where DM is composed of $\sim$5% ultra-light axions (ULAs), can help in relieving the S_8 tension without affecting the success of the EDE solution in explaining a high-H_0 [110]. Such a dark sector could be the manifestation of the presence of numerous scalar-fields that are ubiquitous in string theory [111,

112]. Ref. [113] obtained a similar result considering a decaying DM model instead of the ULA component. Though these models may seem 'ad hoc', they demonstrate that H_0 and S_8 solutions can come from different (potentially disconnected) sectors and therefore that one should not dismiss EDE models on the basis that they cannot simultaneously explain both tensions.

Efforts have been made to establish a connection between the resolutions of S_8 and H_0 within a unified theoretical framework [22, 26, 27, 29]. Examples include EDE models coupled to DM as a consequence of the Swampland Distance Conjecture (SDC) [27], or in a chameleon-inspired EDE in Ref. [26], that can induce a "fifth force" that has the potential of reducing the growth of structure and relieving the tension with LSS data. Let us also mention the recent progress made in the framework of the Cold NEDE model, discussed in more details in the next chapter, where the presence of an axion-like field with mass $\sim 10^{-27}$ eV that resolves the S_8 tension also triggers a phase-transition in the EDE field at precisely the right time to explain the Hubble tension [114].

22.3.3 Impact on Higher-Redshift LSS Probes

The impact of EDE on the matter power spectrum can profoundly affect the predicted halo mass function. By running N-body simulations, Ref. [82] established that EDE predicts only a small (1–10%) increase in the number of halos at $z = 0$ for very massive clusters, but at larger redshifts, that difference can increase tremendously. For example, the EDE model predicts about 50% more massive clusters of mass $M = (3 - 5) \times 10^{14} h^{-1} M_\odot$ at $z = 1$, and the difference is expected to keep increasing with increasing redshifts, such that differences will soon be tested by JWST observations [82, 115]. Although a dedicated analysis of the halo mass function in EDE model at such large reshift is still lacking, it has been pointed out that the first publicly available data show a somewhat unexpectedly high number of massive galaxies at high redshift that may be better explained when invoking the presence of EDE [116].

It has also been shown that Lyman-α data from eBOSS [117], the XQ-100 [118] and MIKE/HIRES [119] quasar samples are in tension with the EDE prediction, leading to the strongest constraints to date on $f_{\rm EDE}$ [120]. It is however important to bare in mind that even under ΛCDM, Lyman-α data favor significantly different value (at the 3σ level) of the power spectrum tilt and amplitude at small scales ($k \sim 1$ h/Mpc) and redshift $z = 3$ than reconstructed from analysis of CMB data, which (barring a simple statistical fluctuation) may indicate either a systematic effect, or some new physics is at play such as the running of the spectral tilt [121] or a new type of DM properties [122].

22.3.4 A Second Cosmic Coincidence Problem

The fact that the EDE contribution become significant precisely close to matter-radiation equality when it is most impactful on the CMB, leads to an obvious 'coincidence problem' known as the 'why-then' problem. This echoes the 'why now' or 'cosmic coincidence' problem, that arises from the observation that the densities of dark energy and matter are similar today, when one would naturally expect very different energy scale. Interestingly, this new 'why then' problem may provide insight into the long-standing 'why now' problem.

This new coincidence problem motivates a model-building effort towards dynamically explaining the fact that z_{eq} is so close to matter-radiation equality. An interesting starting point is 'tracking dark energy' [1, 123], which explains the 'why now' problem by postulating that the density of a scalar field is always a fixed fraction of the energy density of the dominant component. The models provide an interesting way to connect various eras of accelerated expansion in the universe. Assisted quintessence [2], which realizes tracking models through a collection of scalar fields with the same potential, was recently applied in the EDE context. However, authors of Ref. [21] found that it could not alleviate the tension.

Other models connect the dynamics of the EDE model to the onset of DM domination by coupling the EDE scalar field to DM. These include the chameleon EDE [26] and trigger EDS [59] models. In these model, the dilution of EDE is precisely triggered around matter-radiation equality, and EDE can affect the growth of matter perturbations, with interesting implications for the S_8 tension. Alternatively, EDE may be coupled to another species experiencing a non-relativistic transition, such as neutrinos. Refs. [12, 124] showed that a conformal coupling between EDE and neutrinos lead to a modified KG equation, that now counts an additional source term proportional to the trace of the neutrino energy-momentum tensor. The effect of the neutrino coupling is to kick the scalar out of its minimum and up its potential right when the neutrinos become non-relativistic, $T_\nu \sim m_\nu$, independently of the value of the scalar field mass and its initial condition. This model sill lacks a dedicated analysis in light of up-to-date cosmological data. Interestingly it requires individual masses of $\mathcal{O}(0.1 - 0.5)$ eV, and it is expected that it can be probed with current and near-future CMB and LSS observations.

Finally, the last avenue is that this coincidence appears from mere chance. For instance, the "Axiverse" scenario [112] suggests that many scalar fields with similar potentials exist in the universe. In Ref. [125], it was argued that the probability that one of such fields explains DE today while leaving observables unaffected is not small, thereby solving the "why now" problem. In this context, it is therefore possible that there exist additional EDE-like fields in the Universe that may have been relevant (or not) at different eras, with a non-negligible probability that one such field becomes dynamical close to matter-radiation equality. The EDE field responsible for the Hubble tension would therefore be a member of a family of such fields, while other member may play the role of Dark Energy, inflation, and more of such fields may be detected in the future, depending on whether they left an imprint

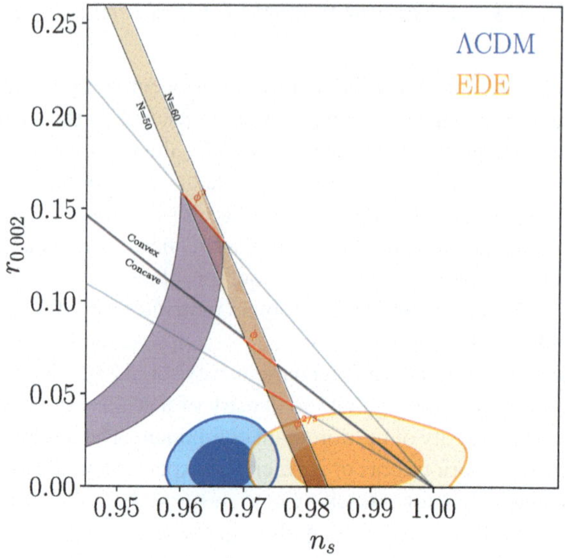

Fig. 22.12 Constraints on inflation in the $n_s - r$ plane from the combination of *Planck*+Bicep/Keck+BAO data, in the ΛCDM and EDE models. constraints are shifted towards larger $n_s \sim 1$ in the EDE model, but the bound on r is unaffected

on observables (see also Refs. [19, 123, 126–128]). In that context, it is important to note that it is not unexpected that that an EDE field acting around matter-radiation equality was detected first *precisely because* our cosmological data gains most of their sensitivity right around recombination. If the existence of EDE is confirmed with future data, the question of connecting EDE with DE and inflation will become particularly relevant.

22.3.5 *Impact of EDE for Inflation*

The presence of EDE has important consequences for inflation, as the value of n_s reconstructed from the fit to CMB data is significantly larger than in ΛCDM. Under the ΛCDM model, current *Planck* data which favor $n_s \sim 0.96$ and $r \lesssim 0.1$ are in perfect agreement with predictions from Starobinsky (or R^2) inflation, while they disfavor power low (or convex) potentials [129]. However, if the Hubble tension persists, and the preference for EDE models with $n_s \simeq 1$ increases, it can reshape our understanding of cosmic inflation [130, 131]. We illustrate bounds on $n_s - r$ in the ΛCDM and EDE model on Fig. 22.12. All constraints in the standard $n_s - r$ plane are shifted towards larger n_s, while the bound on r is essentially unaffected by the presence of EDE [131]. In the EDE case, convex potentials can generally be rescued, with the important consequence that Starobinsky inflation is disfavored compared to power-law potential, or the curvaton model.

It is important to note that the preference for $n_s \sim 1$ is very generic to models that resolve the Hubble tension by increasing the expansion rate pre-recombination

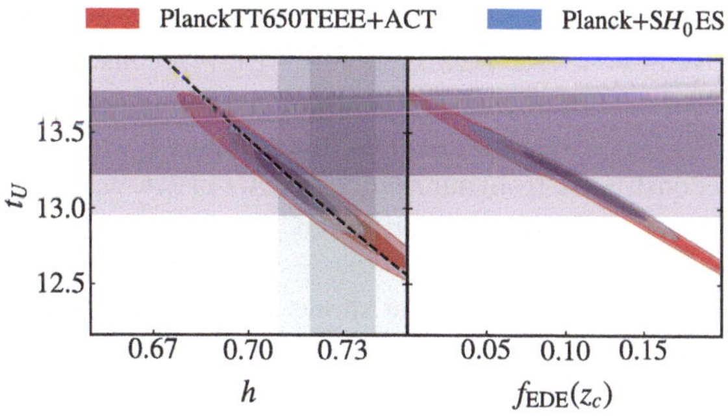

Fig. 22.13 The age of the universe t_U vs $\{h, f_{\rm EDE}(z_c)\}$ in the EDE cosmology. The purple bands mark the recent measurements from GCs, $t_U = 13.5 \pm 0.27$ Gyrs [133–135]. The dashed black line shows the prediction from Eq. 22.6, accounting for the small difference in Ω_m between EDE and ΛCDM

(such as an extra radiation species $\Delta N_{\rm eff}$). It follows from the fact that the angular scale θ_D corresponding to damping roughly depends on

$$\theta_D(z_*) \sim \frac{H_0}{\sqrt{\dot{\tau}(z_*)H(z_*)}}, \tag{22.5}$$

where $\dot{\tau}(z_*) = n_e(z_*)x_e(z_*)\sigma_T/(1+z_*)$, n_e is the electron density, x_e is the ionization fraction, and σ_T is the Thomson cross section. If the indirect H_0 increases by a fraction f, with $\theta_s(z_*)$ fixed, $\theta_D(z_*)$ increases by $f^{1/2}$, leading to a suppression of power, which can be compensated for by increasing n_s. Although details of specific models will change by how much the value of n_s must increase, this degeneracy is very generic, and should therefore be considered as an important caveat to our current understanding of cosmic inflation (regardless of the cosmological nature of the Hubble tension).

22.3.6 The Age of the Universe

Models that resolve the Hubble tension by modifying the pre-recombination era, such as Early Dark Energy, generically predict a significantly smaller age of the universe than ΛCDM. Indeed, assuming flat-ΛCDM and radiation is sub-dominant, the age of the Universe today is approximately [132]

$$t_U \equiv t(a=1) \simeq t_{\rm pl}\left(\frac{\Omega_m}{\Omega_{m,\rm pl}}\right)^{-0.28}\left(\frac{h_{\rm pl}}{h}\right). \tag{22.6}$$

The freedom to adjust Ω_m is limited given the tight constraints on Ω_m from BAO and Pantheon data. Therefore, one generically expects that increasing H_0 without changing Ω_m leads to a universe $\sim$7.5% younger than ΛCDM predicts. This is illustrated in Fig. 22.13 in the context of EDE.

This raises a potential issue for EDE and other "early-universe" models, as they can run into tension with the measured ages of old objects such as globular clusters of stars (GCs) [132, 135, 136]. More precisely, recent measurements from GCs give $t_{\rm U} = 13.5 \pm 0.27$ Gyrs [133–135] (see also Ref. [137–139] for a discussion of other measurements), while the EDE cosmology fit to *Planck*+SH_0ES predicts $t_U = 13.17^{+0.14}_{-0.16}$ Gyr. We illustrate this potential tension in Fig. 22.13, for the EDE model reconstructed from our *Planck*TT650TEEE+ACT and *Planck*+SH_0ES analyses, with the purple band marking the GC measurement that lies somewhat larger than the model predicts.

The dashed black line shows the prediction from Eq. 22.6, accounting for the small difference in Ω_m in the EDE model compared to ΛCDM. Any model that only modifies the early universe would follow a similar degeneracy line, and therefore can be tested with more accurate measurements of t_U. Resolving the Hubble tension while adjusting both measurements of the age of the universe and measurements of the shape of the expansion rate may require both an early and late-time modification [135].

Similar considerations apply to cosmic chronometers (e.g. Ref. [137, 140]) which are currently too imprecise to probe the EDE scenarios [56], but could provide interesting tests in the future. In fact, Ref. [132] points out that the entire time-redshift relation is affected by the EDE model, differing from the base ΛCDM model by at least ∼4% at all z and t that can have significant implications for the era of reionization.

22.4 Conclusions

Early Dark Energy, a new form of dark energy active at early times, has emerged as a promising way to explain the Hubble tension. In this chapter, we have reviewed the status of this solution, paying particular attention to the axion-like EDE model, a toy-model whose flexibility enables us to extract the generic properties required to explain the Hubble tension. EDE is motivated by two observations: first, that one must decrease the sound horizon in order to keep the angular-size of the sound horizon fixed in the CMB and BAO data and predict a magnitude of SN1a in agreement with the measurements from SH_0ES. Second, that a higher h, at fix Ω_m (as imposed by the BAO and SN1a measurements of the shape of the late-time expansion) requires an increase in $\omega_{\rm cdm}$ which, if uncompensated, would spoil the fit to CMB data. By briefly contributing to the Hubble rate in the pre-recombination era, EDE precisely manages to "kill two birds with one stone", and provide a good fit to combinations of up-to-date cosmological data from CMB, BAO and SH_0ES-calibrated SN1a.

Focusing on the "axion-like" EDE model as a representative example, we have reviewed current analyses in details in Sect. 22.2, showing that EDE can reduce the Hubble tension to below the 2σ level. In addition, *Planck* data constraint the specific 'microphysics' of EDE, ultimately tied to the shape of the energy injection, and the dynamics of perturbations in the EDE component. We have also discussed recent indications of EDE in ACT data, that are further supported by *Planck* polarization data. However, *Planck* high-ℓ temperature data are mildly inconsistent with ACT and restricts the preference. Forthcoming data from ACT and SPT will be crucial in confirming or dismissing the current indications of EDE. Looking ahead, it is anticipated that high-precision measurements at high-ℓ values ($\gtrsim$1000) will be sensitive to the distinct increase in power caused by EDE (primarily due to the reshuffling of ΛCDM parameters), while improvements at intermediate ℓ values ($\sim$50–500) can provide insights into the detailed dynamics of EDE perturbations.

There are many interesting implications of EDE that we have reviewed in Sect. 22.3. First, the EDE cosmology shows an increase of power in the matter power spectrum (at the $5-10\%$ level in the range $k \sim 0.01 - 1$ h/Mpc) compared to ΛCDM. This gives an opportunity to probe the model with LSS data such as BOSS, although current data are not constraining enough to rule out EDE. Nevertheless, EDE cannot explain the low-S_8 measured by current galaxy weak-lensing surveys such as KiDS, DES and HSC, triggering a large model-building effort towards embedding a solution to the S_8 tension within the EDE framework. Second, EDE predicts a much younger universe, which may also require some late-time modification if measurements of the age of universe become more precise. Third, the existence of an EDE phase may have important consequences for inflation, as the value of n_s reconstructed in EDE models that resolve the Hubble tension favor curvaton models over Starobinsky inflation [130, 131]. At the theoretical level, EDE also triggers some fine-tuning issues in particular a 'why-then' problem that resembles the 'why-now' problem and suggests that EDE may shed light on this long-standing problem. In fact, connections between inflation, late DE and early DE are still largely unexplored, and may provide an interesting path forward for model-building with undetermined consequences for EDE models.

In this chapter, we have argued that EDE is a promising mechanism to resolve the H_0 tension, but still lacks some missing pieces, for various observational and theoretical reasons. Future work will be of utmost important in order to find an unambiguous solutions to current cosmic tensions, in particular given the numerous surveys to come within the next decade (DESI, Euclid, SO, among others), and hopefully shed light on the unknown nature of dark matter and dark energy.

Acknowledgements This project has received support from the European Union's Horizon 2020 research and innovation program under the Marie Skodowska-Curie grant agreement No 860881-HIDDeN. This project has received funding from the European Research Council (ERC) under the European Union's HORIZON-ERC-2022 (Grant agreement No. 101076865).

References

1. S. Dodelson, M. Kaplinghat, E. Stewart, Solving the coincidence problem?: tracking oscillating energy. Phys. Rev. Lett. **85**, 5276–5279 (2000)
2. S.A. Kim, A.R. Liddle, S. Tsujikawa, Dynamics of assisted quintessence. Phys. Rev. D **72**, 043506 (2005)
3. M. Doran, M.J. Lilley, J. Schwindt, C. Wetterich, Quintessence and the separation of CMB peaks. Astrophys. J. **559**, 501–506 (2001)

4. C. Wetterich, Phenomenological parameterization of quintessence. Phys. Lett. B **594**, 17–22 (2004)
5. M. Doran, G. Robbers, Early dark energy cosmologies. JCAP **06**, 026 (2006)
6. T. Karwal, M. Kamionkowski, Dark energy at early times, the Hubble parameter, and the string axiverse. Phys. Rev. D **94**(10), 103523 (2016)
7. V. Poulin, T.L. Smith, T. Karwal, M. Kamionkowski, Early dark energy can resolve the Hubble tension. Phys. Rev. Lett. **122**(22), 221301 (2019)
8. T.L. Smith, V. Poulin, M.A. Amin, Oscillating scalar fields and the Hubble tension: a resolution with novel signatures. Phys. Rev. D **101**(6), 063523 (2020)
9. P. Agrawal, F.-Y. Cyr-Racine, D. Pinner, L. Randall, Rock 'n' roll solutions to the Hubble tension (2019)
10. M.-X. Lin, G. Benevento, W. Hu, M. Raveri, Acoustic dark energy: potential conversion of the Hubble tension. Phys. Rev. D **100**(6), 063542 (2019)
11. S. Alexander, E. McDonough, Axion-Dilaton destabilization and the Hubble tension. Phys. Lett. B **797**, 134830 (2019)
12. J. Sakstein, M. Trodden, Early dark energy from massive neutrinos as a natural resolution of the Hubble tension. Phys. Rev. Lett. **124**(16), 161301 (2020)
13. A. Gogoi, R. Kumar Sharma, P. Chanda, S. Das, Early mass-varying neutrino dark energy: nugget formation and Hubble anomaly. Astrophys. J. **915**, 132 (2021)
14. F. Niedermann, M.S. Sloth, New early dark energy. Phys. Rev. D **103**(4), L041303 (2021)
15. F. Niedermann, M.S. Sloth, Resolving the Hubble tension with new early dark energy. Phys. Rev. D **102**(6), 063527 (2020)
16. F. Niedermann, M.S. Sloth, Hot new early dark energy. Phys. Rev. D **105**(6), 063509 (2022)
17. G. Ye, Y.-S. Piao, Is the Hubble tension a hint of AdS phase around recombination? Phys. Rev. D **101**(8), 083507 (2020)
18. K.V. Berghaus, T. Karwal, Thermal friction as a solution to the Hubble tension. Phys. Rev. D **101**(8), 083537 (2020)
19. K. Freese, M.W. Winkler, Chain early dark energy: a proposal for solving the Hubble tension and explaining today's dark energy. Phys. Rev. D **104**(8), 083533 (2021)
20. M. Braglia, W.T. Emond, F. Finelli, A.E. Gumrukcuoglu, K. Koyama, Unified framework for early dark energy from α-attractors. Phys. Rev. D **102**(8), 083513 (2020)
21. V.I. Sabla, R.R. Caldwell, No H_0 assistance from assisted quintessence. Phys. Rev. D **103**(10), 103506 (2021)
22. V.I. Sabla, R.R. Caldwell, Microphysics of early dark energy. Phys. Rev. D **106**(6), 063526 (2022)
23. A. Gómez-Valent, Z. Zheng, L. Amendola, V. Pettorino, C. Wetterich, Early dark energy in the pre- and postrecombination epochs. Phys. Rev. D **104**(8), 083536 (2021)
24. A. Moss, E. Copeland, S. Bamford, T. Clarke, A model-independent reconstruction of dark energy to very high redshift **9** (2021)
25. E. Guendelman, R. Herrera, D. Benisty, Unifying inflation with early and late dark energy with multiple fields: Spontaneously broken scale invariant two measures theory. Phys. Rev. D **105**(12), 124035 (2022)
26. T. Karwal, M. Raveri, B. Jain, J. Khoury, M. Trodden, Chameleon early dark energy and the Hubble tension. arXiv e-prints, arXiv:2106.13290 (2021)
27. E. McDonough, M.-X. Lin, J.C. Hill, W. Hu, S. Zhou, The early dark sector, the Hubble tension, and the swampland **12** (2021)
28. H. Wang, Y.-S. Piao, A fraction of dark matter faded with early dark energy? **9** (2022)
29. S. Alexander, H. Bernardo, M.W. Toomey, Addressing the Hubble and S_8 tensions with a kinetically mixed dark sector **7** (2022)
30. E. McDonough, M. Scalisi, Towards early dark energy in string theory **8** (2022)
31. S. Nakagawa, F. Takahashi, W. Yin, Early dark energy by dark Higgs, and axion-induced non-thermal trapping **9** (2022)
32. A. Gómez-Valent, Z. Zheng, L. Amendola, C. Wetterich, V. Pettorino, Coupled and uncoupled early dark energy, massive neutrinos and the cosmological tensions **7** (2022)

33. H. Mohseni Sadjadi, V. Anari, Early dark energy and the screening mechanism **5** (2022)
34. K. Kojima, Y. Okubo, Early dark energy from a higher-dimensional gauge theory. Phys. Rev. D **106**(6), 063540 (2022)
35. T. Rudelius, Constraints on early dark energy from the axion weak gravity conjecture **3** (2022)
36. V.K. Oikonomou, Unifying inflation with early and late dark energy epochs in axion $F(R)$ gravity. Phys. Rev. D **103**(4), 044036 (2021)
37. S.X. Tian, Z.-H. Zhu, Early dark energy in k-essence. Phys. Rev. D **103**(4), 043518 (2021)
38. M. Maziashvili, Inflaton-driven early dark energy. Astropart. Phys. **145**, 102792 (2023)
39. D. Blas, J. Lesgourgues, T. Tram, The cosmic linear anisotropy solving system (CLASS) II: approximation schemes. JCAP **1107**, 034 (2011)
40. V. Poulin, T.L. Smith, T. Karwal, The ups and downs of early dark energy solutions to the Hubble tension: a review of models, hints and constraints circa 2023 **2** (2023)
41. H. Abe, T. Kobayashi, H. Otsuka, Natural inflation with and without modulations in type IIB string theory. JHEP **04**, 160 (2015)
42. K. Choi, H. Kim, Aligned natural inflation with modulations. Phys. Lett. B **759**, 520–527 (2016)
43. R. Kappl, H.P. Nilles, M.W. Winkler, Modulated natural inflation. Phys. Lett. B **753**, 653–659 (2016)
44. N. Kaloper, Dark energy, H_0 and weak gravity conjecture. Int. J. Mod. Phys. D **28**(14), 1944017 (2019)
45. M. Czerny, T. Higaki, F. Takahashi, Multi-natural inflation in supergravity and BICEP2. Phys. Lett. B **734**, 167–172 (2014)
46. D. Croon, V. Sanz, Saving natural inflation. JCAP **02**, 008 (2015)
47. T. Higaki, F. Takahashi, Elliptic inflation: interpolating from natural inflation to R^2-inflation. JHEP **03**, 129 (2015)
48. M. Cicoli, M. Licheri, R. Mahanta, E. McDonough, F.G. Pedro, M. Scalisi, Early dark energy in type IIB string theory **3** (2023)
49. N. Aghanim et al., Planck 2018 results. VI. Cosmological parameters. Astron. Astrophys. **641**, A6 (2020) [Erratum: Astron.Astrophys. 652, C4 (2021)]
50. S. Alam et al., The clustering of galaxies in the completed SDSS-III Baryon oscillation spectroscopic survey: cosmological analysis of the DR12 galaxy sample. Mon. Not. Roy. Astron. Soc. **470**(3), 2617–2652 (2017)
51. A.J. Ross, L. Samushia, C. Howlett, W.J. Percival, A. Burden, M. Manera, The clustering of the SDSS DR7 main Galaxy sample—I. A 4 percent distance measure at $z = 0.15$. Mon. Not. Roy. Astron. Soc. **449**(1), 835–847 (2015)
52. F. Beutler, C. Blake, M. Colless, D.H. Jones, L. Staveley-Smith, L. Campbell, Q. Parker, W. Saunders, F. Watson, The 6dF galaxy survey: Baryon acoustic oscillations and the local Hubble constant. Mon. Not. Roy. Astron. Soc. **416**, 3017–3032 (2011)
53. D.M. Scolnic et al., The complete light-curve sample of spectroscopically confirmed SNe Ia from Pan-STARRS1 and cosmological constraints from the combined pantheon sample. Astrophys. J. **859**(2), 101 (2018)
54. A.G. Riess et al., A comprehensive measurement of the local value of the Hubble constant with 1 km/s/Mpc uncertainty from the Hubble space telescope and the SH_0ES team **12** (2021)
55. M. Raveri, W. Hu, Concordance and discordance in cosmology (2018)
56. N. Schöneberg, G. Franco Abellán, A. Pérez Sánchez, S.J. Witte, V. Poulin, J. Lesgourgues, The H_0 Olympics: a fair ranking of proposed models **7** (2021)
57. V. Poulin, T.L. Smith, D. Grin, T. Karwal, M. Kamionkowski, Cosmological implications of ultralight axionlike fields. Phys. Rev. D **98**(8), 083525 (2018)
58. M.C. Johnson, M. Kamionkowski, Dynamical and gravitational instability of oscillating-field dark energy and dark matter. Phys. Rev. D **78**, 063010 (2008)
59. M.-X. Lin, E. McDonough, J.C. Hill, W. Hu, A dark matter trigger for early dark energy coincidence **12** (2022)
60. T. Simon, P. Zhang, V. Poulin, T.L. Smith, Updated constraints from the effective field theory analysis of BOSS power spectrum on early dark energy. Phys. Rev. D. **In press**, 8 (2022)

61. A. Lewis, S. Bridle, Cosmological parameters from CMB and other data: A Monte Carlo approach. Phys. Rev. D **66**, 103511 (2002)
62. B. Audren, J. Lesgourgues, K. Benabed, S. Prunet, Conservative constraints on early cosmology: an illustration of the Monte Python cosmological parameter inference code. JCAP **02**, 001 (2013)
63. R. Murgia, G.F. Abellán, V. Poulin, Early dark energy resolution to the Hubble tension in light of weak lensing surveys and lensing anomalies. Phys. Rev. D **103**(6), 063502 (2021)
64. T.L. Smith, V. Poulin, J.L. Bernal, K.K. Boddy, M. Kamionkowski, R. Murgia, Early dark energy is not excluded by current large-scale structure data. Phys. Rev. D **103**(12), 123542 (2021)
65. L. Herold, E.G.M. Ferreira, E. Komatsu, New constraint on early dark energy from Planck and BOSS data using the profile likelihood **12** (2021)
66. A. Gómez-Valent, Fast test to assess the impact of marginalization in Monte Carlo analyses and its application to cosmology. Phys. Rev. D **106**(6), 063506 (2022)
67. B. Hadzhiyska, K. Wolz, S. Azzoni, D. Alonso, C. García-García, J. Ruiz-Zapatero, A. Slosar, Cosmology with 6 parameters in the stage-IV era: efficient marginalisation over nuisance parameters **1** (2023)
68. J. Neyman, Outline of a theory of statistical estimation based on the classical theory of probability. Phil. Trans. Roy. Soc. Lond. A **236**(767), 333–380 (1937)
69. G.J. Feldman, R.D. Cousins, A unified approach to the classical statistical analysis of small signals. Phys. Rev. D **57**, 3873–3889 (1998)
70. P.A.R. Ade et al., Planck intermediate results. XVI. Profile likelihoods for cosmological parameters. Astron. Astrophys. **566**, A54 (2014)
71. P. Ade et al., The Simons observatory: science goals and forecasts. JCAP **02**, 056 (2019)
72. K.N. Abazajian et al., *CMB-S4 Science Book*, vol. 10, 1st edn. (2016)
73. S. Aiola et al., The Atacama cosmology telescope: DR4 maps and cosmological parameters. JCAP **12**, 047 (2020)
74. D. Dutcher et al., Measurements of the E-mode polarization and temperature-E-mode correlation of the CMB from SPT-3G 2018 data. arXiv e-prints, arXiv:2101.01684 (2021)
75. L. Balkenhol et al., A measurement of the CMB temperature power spectrum and constraints on cosmology from the SPT-3G 2018 TT/TE/EE data set **12** (2022)
76. Y. Huang, G.E. Addison, J.L. Weiland, C.L. Bennett, Assessing consistency between WMAP 9-year and Planck 2015 temperature power spectra. Astrophys. J. **869**(1), 38 (2018)
77. J.C. Hill et al., The Atacama cosmology telescope: constraints on pre-recombination early dark energy **9** (2021)
78. T.L. Smith, V. Poulin, T. Simon, Assessing the robustness of sound horizon-free determinations of the Hubble constant **8** (2022)
79. T.L. Smith, M. Lucca, V. Poulin, G.F. Abellan, L. Balkenhol, K. Benabed, S. Galli, R. Murgia, Hints of early dark energy in Planck, SPT, and ACT data: new physics or systematics? **2** (2022)
80. W. Handley, P. Lemos, Quantifying the global parameter tensions between ACT, SPT and Planck. Phys. Rev. D **103**(6), 063529 (2021)
81. E. Di Valentino, W. Giarè, A. Melchiorri, J. Silk, A health checkup test of the standard cosmological model in view of recent Cosmic Microwave Background Anisotropies experiments **9** (2022)
82. A. Klypin, V. Poulin, F. Prada, J. Primack, M. Kamionkowski, V. Avila-Reese, A. Rodriguez-Puebla, P. Behroozi, D. Hellinger, T.L. Smith, Clustering and halo abundances in early dark energy cosmological models. Mon. Not. Roy. Astron. Soc. **504**(1), 769–781 (2021)
83. J.C. Hill, E. McDonough, M.W. Toomey, S. Alexander, Early dark energy does not restore cosmological concordance. Phys. Rev. D **102**(4), 043507 (2020)
84. D. Baumann, A. Nicolis, L. Senatore, M. Zaldarriaga, Cosmological non-linearities as an effective fluid. JCAP **07**, 051 (2012)
85. J.J.M. Carrasco, M.P. Hertzberg, L. Senatore, The effective field theory of cosmological large scale structures. JHEP **09**, 082 (2012)

86. L. Senatore, M. Zaldarriaga, The IR-resummed effective field theory of large scale structures. JCAP **1502**(02), 013 (2015)
87. L. Senatore, Bias in the effective field theory of large scale structures. JCAP **1511**(11), 007 (2015)
88. L. Senatore, M. Zaldarriaga, Redshift space distortions in the effective field theory of large scale structures (2014)
89. A. Perko, L. Senatore, E. Jennings, R.H. Wechsler, Biased tracers in redshift space in the EFT of large-scale structure (2016)
90. G. D'Amico, J. Gleyzes, N. Kokron, D. Markovic, L. Senatore, P. Zhang, F. Beutler, H. Gil-Marín, The cosmological analysis of the SDSS/BOSS data from the effective field theory of large-scale structure. JCAP **05**, 005 (2020)
91. M.M. Ivanov, M. Simonovi, M. Zaldarriaga, Cosmological parameters from the BOSS galaxy power spectrum (2019)
92. G. D'Amico, L. Senatore, P. Zhang, H. Zheng, The Hubble tension in light of the full-shape analysis of large-scale structure data. JCAP **05**, 072 (2021)
93. M.M. Ivanov, E. McDonough, J.C. Hill, M. Simonović, M.W. Toomey, S. Alexander, M. Zaldarriaga, Constraining early dark energy with large-scale structure. Phys. Rev. D **102**(10), 103502 (2020)
94. A. Reeves, L. Herold, S. Vagnozzi, B.D. Sherwin, E.G.M. Ferreira, Restoring cosmological concordance with early dark energy and massive neutrinos? **7** (2022)
95. T. Simon, G. Franco Abellán, P. Du, V. Poulin, Y. Tsai, Constraining decaying dark matter with BOSS data and the effective field theory of large-scale structures. Phys. Rev. D. **106**(2), 023516 (2022)
96. C. Heymans et al., CFHTLenS: the Canada-France-Hawaii telescope lensing survey. Mon. Not. Roy. Astron. Soc. **427**, 146 (2012)
97. T.M.C. Abbott et al., Dark energy survey year 3 results: cosmological constraints from galaxy clustering and weak lensing. Phys. Rev. D **105**(2), 023520 (2022)
98. C. Heymans et al., KiDS-1000 cosmology: multi-probe weak gravitational lensing and spectroscopic galaxy clustering constraints **7** (2020)
99. C. Hikage et al., Cosmology from cosmic shear power spectra with Subaru hyper suprime-cam first-year data. Publ. Astron. Soc. Jap. **71**(2), Publications of the Astronomical Society of Japan, vol. 71, no. 2, April 2019, p. 43. https://doi.org/10.1093/pasj/psz010 (2019)
100. P.A.R. Ade et al., Planck 2015 results. XXIV. cosmology from Sunyaev-Zeldovich cluster counts. Astron. Astrophys. **594**, A24 (2016)
101. K. Jedamzik, L. Pogosian, G.-B. Zhao, Why reducing the cosmic sound horizon alone can not fully resolve the Hubble tension. Commun. in Phys. **4**, 123 (2021)
102. S. Vagnozzi, Consistency tests of ΛCDM from the early ISW effect: implications for early-time new physics and the Hubble tension. arXiv e-prints, arXiv:2105.10425 (2021)
103. L. Herold, E.G.M. Ferreira, Resolving the Hubble tension with early dark energy **10** (2022)
104. T.M.C. Abbott et al., DES Y3 + KiDS-1000: consistent cosmology combining cosmic shear surveys **5** (2023)
105. L.F. Secco, T. Karwal, W. Hu, E. Krause, The role of the Hubble scale in the weak lensing versus CMB tension **9** (2022)
106. G.F. Abellan, R. Murgia, V. Poulin, J. Lavalle, Hints for decaying dark matter from S_8 measurements **8** (2020)
107. Z. Chacko, A. Dev, P. Du, V. Poulin, Y. Tsai, Cosmological limits on the neutrino mass and lifetime. JHEP **04**, 020 (2020)
108. I.M. Oldengott, G. Barenboim, S. Kahlen, J. Salvado, D.J. Schwarz, How to relax the cosmological neutrino mass bound. JCAP **04**, 049 (2019)
109. E. Fondi, A. Melchiorri, L. Pagano, No evidence for EDE from Planck data in extended scenarios. Astrophys. J. Lett. **931**, L18 (2022)
110. I.J. Allali, M.P. Hertzberg, F. Rompineve, Dark sector to restore cosmological concordance. Phys. Rev. D **104**(8), L081303 (2021)
111. P. Svrcek, E. Witten, Axions in string theory. JHEP **06**, 051 (2006)

112. A. Arvanitaki, S. Dimopoulos, S. Dubovsky, N. Kaloper, J. March-Russell, String axiverse. Phys. Rev. D **81**, 123530 (2010)
113. S.J. Clark, K. Vattis, J. Fan, S.M. Koushiappas, The H_0 and S_8 tensions necessitate early and late time changes to ΛCDM **10** (2021)
114. J.S. Cruz, F. Niedermann, M.S. Sloth, Cold new early dark energy pulls the trigger on the H_0 and S_8 tensions: a simultaneous solution to both tensions without new ingredients **5** (2023)
115. R. Endsley, P. Behroozi, D.P. Stark, C.C. Williams, B.E. Robertson, M. Rieke, S. Gottlöber, G. Yepes, Clustering with JWST: constraining galaxy host halo masses, satellite quenching efficiencies, and merger rates at z= 4-10. Monthly Notices of the Royal Astron. Soc. **493**, 1178–1196, 02 (2020)
116. M. Boylan-Kolchin, Stress testing ΛCDM with high-redshift galaxy candidates **8** (2022)
117. S. Chabanier et al., The one-dimensional power spectrum from the SDSS DR14 Lyα forests. JCAP **07**, 017 (2019)
118. V. Iršič, M. Viel, T.A.M. Berg, V. D'Odorico, M.G. Haehnelt, S. Cristiani, G. Cupani, T.-S. Kim, S. López, S. Ellison, G.D. Becker, L. Christensen, K.D. Denney, G. Worseck, J.S. Bolton, The Lyman ? forest power spectrum from the XQ-100 legacy survey. Monthly Notices of the Royal Astron. Soc. **466**, 4332–4345, 12 (2016)
119. M. Viel, G.D. Becker, J.S. Bolton, M.G. Haehnelt, Warm dark matter as a solution to the small scale crisis: new constraints from high redshift Lyman-α forest data. Phys. Rev. D **88**, 043502 (2013)
120. S. Goldstein, J.C. Hill, V. Iršič, B.D. Sherwin, Canonical Hubble-tension-resolving early dark energy cosmologies are inconsistent with the Lyman-α forest **3** (2023)
121. N. Palanque-Delabrouille, C. Yèche, N. Schöneberg, J. Lesgourgues, M. Walther, S. Chabanier, E. Armengaud, Hints, neutrino bounds and WDM constraints from SDSS DR14 Lyman-α and Planck full-survey data. JCAP **04**, 038 (2020)
122. D.C. Hooper, M. Lucca, Hints of dark matter-neutrino interactions in Lyman-α data. Phys. Rev. D **105**(10), 103504 (2022)
123. K. Griest, Toward a possible solution to the cosmic coincidence problem. Phys. Rev. D **66**, 123501 (2002)
124. M. Carrillo González, Q. Liang, J. Sakstein, M. Trodden, Neutrino-assisted early dark energy: theory and cosmology. JCAP **04**, 063 (2021)
125. M. Kamionkowski, J. Pradler, D.G.E. Walker, Dark energy from the string axiverse. Phys. Rev. Lett. **113**(25), 251302 (2014)
126. E.V. Linder, T.L. Smith, Dark before light: testing the cosmic expansion history through the cosmic microwave background. JCAP **1104**, 001 (2011)
127. J. Samsing, E.V. Linder, T.L. Smith, Model independent early expansion history and dark energy. Phys. Rev. D **86**, 123504 (2012)
128. A. Hojjati, E.V. Linder, J. Samsing, New constraints on the early expansion history of the universe. Phys. Rev. Lett. **111**(4), 041301 (2013)
129. Y. Akrami et al., Planck 2018 results. X. Constraints on inflation. Astron. Astrophys. **641**, A10 (2020)
130. F. Takahashi, W. Yin, Cosmological implications of ns 1 in light of the Hubble tension. Phys. Lett. B **830**, 137143 (2022)
131. J.S. Cruz, F. Niedermann, M.S. Sloth, A grounded perspective on new early dark energy using ACT, SPT, and BICEP/Keck **9** (2022)
132. M. Boylan-Kolchin, D.R. Weisz, Uncertain times: the redshift-time relation from cosmology and stars. Mon. Not. Roy. Astron. Soc. **505**(2), 2764–2783 (2021)
133. D. Valcin, J.L. Bernal, R. Jimenez, L. Verde, B.D. Wandelt, Inferring the age of the universe with globular clusters. JCAP **12**, 002 (2020)
134. D. Valcin, R. Jimenez, L. Verde, J.L. Bernal, B.D. Wandelt, The age of the universe with globular clusters: reducing systematic uncertainties. JCAP **08**, 017 (2021)
135. J.L. Bernal, L. Verde, R. Jimenez, M. Kamionkowski, D. Valcin, B.D. Wandelt, The trouble beyond H_0 and the new cosmic triangles. Phys. Rev. D **103**(10), 103533 (2021)

136. S. Vagnozzi, F. Pacucci, A. Loeb, Implications for the Hubble tension from the ages of the oldest astrophysical objects. JHEAp **36**, 27–35 (2022)
137. M. Moresco et al., Unveiling the universe with emerging cosmological probes. Living Rev. Rel. **25**(1), 6 (2022)
138. A. Cimatti, M. Moresco, Revisiting oldest stars as cosmological probes: new constraints on the Hubble constant **2** (2023)
139. J.M. Ying, B. Chaboyer, E.M. Boudreaux, C. Slaughter, M. Boylan-Kolchin, D. Weisz, The absolute age of M92. Astron. J. **166**(1), 18 (2023)
140. S. Vagnozzi, A. Loeb, M. Moresco, Eppur è piatto? The cosmic chronometers take on spatial curvature and cosmic concordance. Astrophys. J. **908**(1), 84 (2021)

Chapter 23
New Early Dark Energy as a Solution to the H_0 and S_8 Tensions

Florian Niedermann and Martin S. Sloth

23.1 Introduction

New Early Dark Energy (NEDE) is a promising framework for resolving the Hubble and the S8 tension using well-motivated standard particle physics methods. Within the NEDE framework, the Hubble tension is resolved by a fast-triggered phase transition around the eV energy scale just before matter-radiation equality and the subsequent recombination time. The fast-triggered phase transition, distinct from other EDE-like models (see [1] and for reviews [2, 3]), leads to the quick decay of an early dark energy component, resolving the Hubble tension.

The trigger of the phase transition can be realized by different microphysical degrees of freedom in different NEDE models. NEDE is, therefore, not a single model but an entire framework for addressing the Hubble tension, depending on the trigger. Simple cases discussed so far are the cases where the trigger of the phase transition is a second scalar field (Cold NEDE [4, 5]) or a non-vanishing temperature of the dark sector (Hot NEDE [6, 7]). Since the CMB and LSS are very precise probes of perturbations, different NEDE models differ at the phenomenological level, as well as being distinguishable from other EDE-type models, and we expect that we will be able to identify the correct model with future data.

F. Niedermann (✉)
Nordita, KTH Royal Institute of Technology and Stockholm University, Hannes Alfvéns väg 12, 106 91 Stockholm, Sweden
e-mail: florian.niedermann@su.se

M. S. Sloth
Universe-Origins, University of Southern Denmark, Campusvej 55, 5230 Odense M, Denmark
e-mail: sloth@sdu.dk

E. Di Valentino and D. Brout (eds.), *The Hubble Constant Tension*, Springer Series in Astrophysics and Cosmology, https://doi.org/10.1007/978-981-99-0177-7_23

The physics of NEDE has some similarities with old inflation models[1] [8], in which inflation ends in a first-order phase transition, as opposed to the second-order slow-roll over phase transition of new inflation [9, 10]. The "new" in NEDE, therefore, refers to the similar but opposite role of NEDE versus other EDE-like models when compared to old versus new inflation. The NEDE framework differs as much from other EDE-like models as old inflation differs from new inflation.

First-order phase transitions happen in nature at virtually all energy scales, like the boiling or freezing of water or the QCD phase transition. Since we know that we have an extended dark sector consisting of at least both dark energy and dark matter, it is compelling to think that similar phase transitions can occur in the dark sector. This makes NEDE conceptually appealing at the theoretical level. In fact, in the Cold NEDE model, the two fields are naturally realized as axions [11], which might reside in the axiverse of string theory [12, 13], one with a mass at the eV scale, around the mass scale of also the QCD axion, and the other an ultralight axion with a mass of order 10^{-27} eV playing the role of a small fuzzy Dark Matter (DM) component [14–19]. In the landscape [20], or in a dynamical solution to the cosmological constant problem, phase transitions triggered by multiple axions may indeed be expected and be consistent [21] with expectations based on swampland conjectures, which requires the DE sector to have non-trivial dynamics [22, 23]. On the other hand, in the Hot NEDE model, an eV phase transition could explain how the neutrinos get their mass in a Higgs-like mechanism, where the NEDE boson plays a similar role that the Higgs boson plays in the electroweak phase-transition in giving mass to all the other Standard Model (SM) particles [6, 7].

On the phenomenological level, NEDE distinguishes itself from other EDE-like models, not only by its fast-triggered phase transition on the background level but particularly at the perturbation level. The discontinuous change in the equation of state at the background level, characteristic of a first-order phase transition, is also felt by the perturbations. Since the transition happens at different times in different places due to the unavoidable adiabatic perturbations in the trigger field, the evolution of perturbations across the phase transition is non-trivial and distinct for the NEDE model. As these perturbations get imprinted in the CMB and LSS today, measuring the detailed perturbations is our primary way to test different NEDE models against each other and other EDE-like models (see [2, 5, 24–27] for a first phenomenological model comparison).

One prominent way the differences between models belonging to the NEDE framework and other EDE-like models manifest themselves at the perturbation level is through their ability to address the S8 problem in tandem with the Hubble tension. In the Cold NEDE model, it was originally assumed that the energy density of the light trigger field is completely sub-dominant (in that case the S_8 tension remained similar significant as in ΛCDM [27, 28]). Still, if one allows it to contribute just 0.5% of the energy density of the universe, it can resolve the S8 tension while also improving the ability of NEDE to solve the H_0 tension [11]. This enables Cold NEDE to solve both the Hubble tension and the S8 tension with only four new degrees of free-

[1] Although a fast triggered second order phase transition, called Hybrid NEDE, is also possible [5].

dom compared to ΛCDM (AxiEDE model also has four extra parameters but cannot address the S8 problem without adding more [29, 30]). In the Hot NEDE model, the non-vanishing dark temperature, different from the temperature of the visible sector, requires an extended dark sector. The extended thermal dark sector allows for non-trivial Dark Matter (DM) and Dark Radiation (DR) evolution and interactions, which can also lead to resolutions of the S8 problem. In both Cold NEDE and Hot NEDE, the trigger required for solving the Hubble tension is also central to resolving the S8 tension. This sets NEDE apart from other EDE-like models, which do not have a trigger mechanism like NEDE.

A novel feature of NEDE at the more technical level is that the phase transition happens when the smallest length scales relevant to the CMB have already entered the horizon. In other cosmological phase transitions, like the QCD phase transition, all relevant cosmological scales are super-horizon at the time of the phase transition, and the matching of the perturbations across the discontinuous phase transition is trivial as the co-moving curvature perturbation is conserved on super-horizon scales. In the case of NEDE, the matching of perturbations on sub-horizon scales across the phase transition has therefore required novel theoretical developments of the perturbation theory, which have been implemented in a numerical Boltzmann code called `TriggerCLASS`.[2]

The treatment of perturbations through a fast-triggered phase transition is what we describe as a phenomenological model below. The most important input for determining the phenomenological NEDE model is the choice of trigger (scalar field in Cold NEDE versus temperature in Hot NEDE), as well as the details of the NEDE fluid after the transition. It is the phenomenological model that has been implemented into the Boltzmann solver. It is crucial that the phenomenological model has its motivation in a fundamental microphysical realization, and the relation of NEDE to particle physics is through the microphysical description. In a minimal microscopic model, there is a one-to-one mapping between the fundamental parameters of the microphysical description and the degrees of freedom in the phenomenological model fitted to data (more generally, at least a surjective mapping). Using cosmological observations and constraints on the phenomenological model, this mapping enables us also to constrain the underlying fundamental particle physics. From particle physics, we will have separate constraints on the microscopic particle physics parameters, which makes the NEDE framework highly predictive, promising to take precision particle cosmology to a new level.

23.2 A Phenomenological Model

We start by outlining a purely phenomenological model describing a fast-triggered decay of a new phase of dark energy before discussing two possible microscopic realizations in Sects. 23.3 and 23.4. As we will argue in more detail later, the presence of the trigger is crucial for having a *fast* phase transition that completes quickly

[2] https://github.com/NEDE-Cosmo/TriggerCLASS.

on cosmological time scales set by $1/H$. In particular, this avoids large inhomogeneities that could be produced in a slow phase transition and would conflict with the homogeneity of the CMB.

23.2.1 Fluid Description

Our phenomenological model introduces an ideal fluid, which, on background level, is characterized by an equation of state parameter $w_{\rm NEDE}(\tau)$, relating its pressure and energy density, $\bar{p}_{\rm NEDE}(\tau) = w_{\rm NEDE}(\tau)\bar{\rho}_{\rm NEDE}(\tau)$. General for a first-order phase transition is a discontinuous change in the properties of the microscopic state at the phase boundary, reflected in our case in the sharp change in the equation of state.

Initially, NEDE behaves as vacuum energy or a cosmological constant equivalently and, accordingly, $w_{\rm NEDE} = -1$ or $\bar{\rho}_{\rm NEDE} = \text{const}$. After the phase transition at (conformal) time τ_*, we require the fluid to subside at least as fast as radiation. This corresponds to an equation of state parameter[3] $w_{\rm NEDE}(\tau) \geq 1/3$, which, at this stage, can be a general time-dependent function. Integrating the energy conservation equation $\bar{\rho}'_{\rm NEDE} + 3\mathscr{H}(\bar{\rho}_{\rm NEDE} + \bar{p}_{\rm NEDE}) = 0$, fixes the time-dependence of $\rho_{\rm NEDE}$,

$$\bar{\rho}_{\rm NEDE}(\tau) = \bar{\rho}_{\rm NEDE}(\tau_*) \exp\left[-3\int \mathrm{d}\tilde{\tau}\left[1 + w_{\rm NEDE}(\tilde{\tau})\right]\mathscr{H}(\tilde{\tau})\right], \tag{23.1}$$

where $\mathscr{H} = a'(\tau)/a(\tau) = aH$ is the conformal Hubble parameter and $a(\tau)$ is the scale factor. We thus model the phase transition as a discontinuity in $w_{\rm NEDE}$ at time τ_*, explicitly

$$\begin{aligned} w_{\rm NEDE} &= -1 && \text{for} \quad \tau < \tau_*\,, \\ 1/3 \leq & w_{\rm NEDE}(\tau) \leq 1 && \text{for} \quad \tau \geq \tau_*\,, \end{aligned} \tag{23.2}$$

where the upper bound on $w_{\rm NEDE} \leq 1$ ensures that the fluid is compatible with the dominant energy condition. The abrupt change of $w_{\rm NEDE}(\tau)$ at τ_*, which is a distinctive feature of NEDE, underlies the assumption that the phase transition completes quickly on cosmological time scales. With these choices, the NEDE fluid never dominates the energy budget and instead gives rise to a short energy injection that peaks at the time of the phase transition. The larger $w_{\rm NEDE}$, the quicker the fall-off of $\rho_{\rm NEDE}$ and the more localized the energy injection becomes. As a rule of thumb, the NEDE phenomenology favors $w_{\rm NEDE} \approx 2/3$ just after the phase transition (this is consistent with the findings for EDE [1]). We plot the fractional energy density of NEDE as the red curve in Fig. 23.1 for a typical NEDE cosmology. The maximal fraction is defined as

$$f_{\rm NEDE} = \bar{\rho}_{\rm NEDE}(\tau_*)/\bar{\rho}_{\rm tot}(\tau_*) \tag{23.3}$$

[3] We note that this condition can be relaxed after the onset of matter domination to $w_{\rm NEDE}(\tau) \geq 0$.

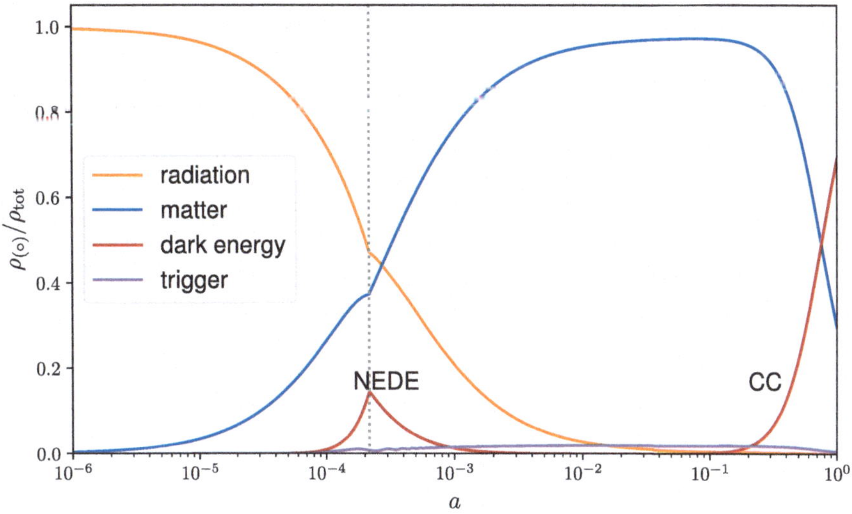

Fig. 23.1 NEDE background evolution in the case of Cold NEDE for one of the best-fit cosmologies (using CMB, BAO, supernovae, and LSS data) in [11]. NEDE leads to a sharp energy injection shortly before matter-radiation equality. The phase transition at time τ_* (dotted vertical line) is modeled as a discontinuity in the equation of state parameter. The trigger is sub-dominant (yet not negligible) throughout the evolution

and corresponds to the peak labeled "NEDE" in Fig. 23.1. The second peak, labeled "CC", corresponds to the (final) cosmological constant contribution to the energy budget.

Another distinctive feature in our setting is the presence of a sub-dominant trigger, plotted in purple in Fig. 23.1. It is the physical agent that fixes the time of the phase transition. Without committing to a particular microphysical implementation, we can describe it in terms of a function $q(t, \mathbf{x})$. The decay time $\tau_*(\mathbf{x})$, or transition surface equivalently, is then defined implicitly through

$$q(t_*(\mathbf{x}), \mathbf{x}) \equiv q_*(= \text{const}) , \tag{23.4}$$

where q_* is the threshold value at which the phase transition becomes efficient. Due to the spatial variations in $q(t, \mathbf{x})$, also $\tau_*(\mathbf{x})$ is in general position dependent. To make this more concrete, we will later discuss the case where $q(\tau, \mathbf{x})$ is identified with either a scalar field (Cold NEDE) or a temperature (Hot NEDE). Then the threshold value q_* corresponds to a particular field value or temperature, respectively.

23.2.2 *Perturbations*

Except for the sharper transition and presence of the trigger field, the above model shares many similarities with the original EDE model proposed in [1]. However, more differences arise on the perturbation level. In short, NEDE density perturbations $\delta\rho_{\text{NEDE}}(\tau, \mathbf{x}) = \rho_{\text{NEDE}}(\tau, \mathbf{x}) - \bar{\rho}_{\text{NEDE}}(\tau)$ are generated at the time of the phase transition τ_* (for earlier times, $\tau < \tau_*$, $\rho_{\text{NEDE}}(\tau, \mathbf{x}) = \text{const}$ with vanishing adiabatic perturbations). To be precise, their initial amplitude is determined by adiabatic perturbations in the trigger $\delta q(\tau_*, \mathbf{x}) = q(\tau_*, \mathbf{x}) - \bar{q}(\tau_*)$, where $\bar{q}(\tau)$ is the trigger's background value. The picture is simple: Without trigger perturbations, the phase transition would be an instantaneous event on large scales. As will become clearer later, tiny differences in its timing could still arise on short length scales that characterize the microphysics of the phase transition. For example, in a first-order transition, such a scale is the typical size of vacuum bubbles when they collide. Still, on much larger cosmological scales the transition would occur at constant time τ_*. Once we consider perturbations in the trigger—we know there always will be adiabatic perturbations—this is no longer true and the transition time acquires a small spatial dependence, i.e. $\tau_* \to \tau_*(\mathbf{x}) = \bar{\tau}_* + \delta\tau_*(\mathbf{x})$, where $\delta\tau_*(\mathbf{x}) = -\delta q(\tau_*, \mathbf{x})/\bar{q}'(\tau_*)$. The last formula can be understood intuitively: If $\bar{q}'(\tau_*) > 0$, at a fixed position $\mathbf{x}$, the threshold value q_* is reached earlier if $\delta q(\tau_*, \mathbf{x}) > 0$ and later if $\delta q(\tau_*, \mathbf{x}) < 0$, corresponding to $\delta\tau_*(\mathbf{x}) < 0$ and $\delta\tau_*(\mathbf{x}) > 0$, respectively. Now, these time variations imply that the decay of NEDE will start at slightly different times at different positions in space, which, in turn, creates perturbations $\delta\rho_{\text{NEDE}}(t_*)$. Simply speaking, an earlier decay corresponds to an underdensity $\delta\rho_{\text{NEDE}}(\bar{\tau}_*, \mathbf{x}) < 0$ and a later decay to an overdensity $\delta\rho_{\text{NEDE}}(\bar{\tau}_*, \mathbf{x}) > 0$. A more detailed matching calculation, based on work by Deruelle and Mukhanov [31], that matches the cosmological perturbations across the transition surface yields (in momentum space) [5]

$$\frac{\delta\rho_{\text{NEDE}}(\bar{\tau}_*, \mathbf{k})}{\bar{\rho}_{\text{NEDE}}(\bar{\tau}_*)} = 3\mathcal{H}(\bar{\tau}_*)\left[1 + w_{\text{NEDE}}(\bar{\tau}_*)\right]\delta\tau_*(\mathbf{k}) \,, \tag{23.5a}$$

$$\theta_{\text{NEDE}}(\bar{\tau}_*) = -k^2\delta\tau_*(\mathbf{k}) \,. \tag{23.5b}$$

We defined the divergence of the NEDE fluid velocity θ as $ik^i(\delta T_{\text{NEDE}})^0_i \equiv (1 + w_{\text{NEDE}})\bar{\rho}_{\text{NEDE}}\,\theta_{\text{NEDE}}$ and its density contrast as $\delta_{\text{NEDE}} \equiv \delta\rho_{\text{NEDE}}(\bar{\tau}_*, \mathbf{k})\bar{\rho}_{\text{NEDE}}(\bar{\tau}_*)$. The derivation is based on Israel's matching equations that relate the extrinsic curvature on both sides of the transition surface [32]. The expressions in (23.5) provide the initial conditions for the subsequent evolution of the fluid perturbations, described by the usual dynamical equations for an interacting fluid with vanishing viscosity parameter. In synchronous gauge they are [33, 34]

$$\begin{aligned}\delta'_{\text{NEDE}} &= -(1 + w_{\text{NEDE}})\left(\theta_{\text{NEDE}} + \frac{h'}{2}\right) - 3(c_s^2 - w_{\text{NEDE}})\mathcal{H}\delta_{\text{NEDE}} \\ &\quad - 9(1 + w_{\text{NEDE}})(c_s^2 - c_a^2)\mathcal{H}^2\frac{\theta_{\text{NEDE}}}{k^2}\,,\end{aligned} \tag{23.6a}$$

$$\theta'_{\text{NEDE}} = -(1 - 3c_s^2)\mathcal{H}\theta_{\text{NEDE}} + \frac{c_s^2k^2}{1 + w_{\text{NEDE}}}\delta_{\text{NEDE}} \tag{23.6b}$$

where h is the trace of spatial metric perturbation, and $c_a(\tau)$ and $c_s(\tau, \mathbf{k})$ are the adiabatic and the rest-frame sound speed of the decaying NEDE fluid, respectively.

In summary, the full dynamical system is given by the background solution in (23.1) and the perturbation equations in (23.6), subject to the boundary conditions (23.5).

The above description is very general and requires fixing $f_{\rm NEDE}$ [which determines $\rho_{\rm NEDE}(t_*)$], $w_{\rm NEDE}(\tau)$, $c_s(\tau, \mathbf{k})$ and $q(\tau_*, \mathbf{k})$ [or $\delta\tau_*(\mathbf{k})$] to close the dynamical system. At this stage, it is clear that we need further input to reduce the amount of freedom we have. After all, $w_{\rm NEDE}(\tau)$ alone gives infinite freedom for the background evolution. Therefore, in the next section, we propose a set of simple assumptions to close the system.

23.2.3 A Simple Model for Extending ΛCDM

The aim of this section is to highlight the phenomenological prospects of a triggered decay scenario. To that end, we close the dynamical system by imposing a set of simple assumptions. We will present a microscopic scenario to motivate them later.

1. The trigger is given by an ultra-light scalar field $\phi(\tau, \mathbf{k})$ with mass m and adiabatic perturbations $\delta\phi(\tau, \mathbf{k})$, explicitly $q(\tau, \mathbf{k}) \equiv \phi(\tau, \mathbf{k})$. We then assume that the decay of NEDE is triggered when ϕ drops out of slow-roll, corresponding to $H(\tau_*)/m \approx 0.2$. This then fixes the decay time τ_* as a function of m. In other words, the lighter the mass, the later the phase transition is triggered. A microscopic realization in terms of a tunneling field ψ that couples to ϕ will be presented in Sect. 23.3. In any event, this introduces the trigger's initial field value, $\phi_{\rm ini}$, as a new parameter. Moreover, if $\phi_{\rm ini}$ is sufficiently large, the trigger field makes a contribution to dark matter $\Omega_\phi = \rho_{0,\phi}/(3M_{\rm pl}^2 H_0^2)$ through its coherent oscillations at late times. It can be related to the other model parameters through [11]

$$\Omega_\phi \simeq 0.4 \times \left(\frac{1+z_*}{5000}\right)\left(\frac{\phi_{\rm ini}}{M_{\rm Pl}}\right)^2 (1 - f_{\rm NEDE}) \,. \tag{23.7}$$

   where $z_* = 1/a(\tau_*) - 1$ is the decay redshift. The trigger field can therefore make a sizeable contribution to the energy budget provided its initial field value is near (yet still below) the Planck scale.
2. The equation of state parameter after the phase transition is constant and, in accordance with (23.2), lies in the range $1/3 \leq w_{\rm NEDE} \leq 1$. This can be understood as keeping the first term in a Taylor expansion, $w_{\rm NEDE}(\tau) = w^*_{\rm NEDE} + \frac{{\rm d}w^*_{\rm NEDE}}{{\rm d}\tau}(\tau - \tau_*) + \ldots$, where τ_* is the decay time. This is a reasonable truncation provided $\frac{1}{\mathscr{H}_*}\frac{{\rm d}w^*_{\rm NEDE}}{{\rm d}\tau} \ll 1$, which ensures that $\rho_{\rm NEDE}$ has become negligible before cosmological observables become sensitive to the running of $w_{\rm NEDE}$. Otherwise, higher-order corrections need to be included.
3. The rest-frame sound speed equals the adiabatic sound speed, explicitly

$$\begin{aligned} c_s^2(\tau, \mathbf{k}) = c_a^2(\tau) &= \frac{\bar{p}'_{\rm NEDE}}{\bar{\rho}'_{\rm NEDE}} \\ &= w_{\rm NEDE} - \frac{1}{3(1+w_{\rm NEDE})\mathscr{H}} \frac{\mathrm{d}w_{\rm NEDE}}{\mathrm{d}\tau} \\ &\approx w^*_{\rm NEDE} \,. \end{aligned} \tag{23.8}$$

In other words, this assumes that the NEDE fluid does not produce sizeable entropy perturbations on cosmological scales. We will formulate a corresponding necessary condition on the microscopic model in the next section.

These three assumptions are enough to reduce the NEDE framework to a 4-parameter extension, supplementing the six ΛCDM parameters with $f_{\rm NEDE}$, m, $w^*_{\rm NEDE}$ and $\phi_{\rm ini}$, where m can be traded for the decay redshift z_* and $\phi_{\rm ini}$ for $\Omega_\phi = \rho_{0,\phi}/(3M_{\rm pl}^2 H_0^2)$, quantifying the trigger contribution to the energy budget today. This model has been implemented in the state-of-the-art Boltzmann code `TriggerCLASS`, which extends the Cosmic Linear Anisotropy Solving System (`CLASS` [37]). It can be fitted to CMB (Planck 2018 temperature, polarization, and lensing [38]), baryonic acoustic oscillations (6dFGS [39], MGS [40], and BOSS [41]), and supernovae data (Pantheon [42]). The result of a recent analysis performed in [11] is shown in Fig. 23.2. The blue contour depicts the marginalized ΛCDM constraints, which is discrepant with the distance ladder result of H_0 obtained by the SH$_0$ES team (grey vertical band [35]) and at a milder level also with the value of S_8 inferred from weak lensing measurements of KiDS, VIKING, and DES (colored vertical band [36]). Specifically, for ΛCDM the Gaussian tensions are 4.8σ for H_0 and 2.3σ for S_8.

The main observation is then that within NEDE (red contour), both the H_0 and S_8 tension are reduced below 2σ. This makes a combined analysis possible that includes the weak lensing and the distance ladder constraint implemented as a Gaussian prior on S_8 and H_0 (dotted contour). In that case, we obtain $f_{\rm NEDE} = 0.134^{+0.032}_{-0.025}$ (68% C.L.), which corresponds to a Gaussian evidence for a non-vanishing fraction of NEDE that exceeds 5σ. We also report the new concordance values $H_0 = 71.71 \pm 0.88\,\mathrm{km\,sec^{-1}\,Mpc^{-1}}$ (68% C.L.) and $S_8 = 0.793 \pm 0.018$ (68% C.L.). A selection of cosmological parameters for the four data set combinations is presented in Table 23.1.

The NEDE phenomenology can be best understood in terms of different parameter correlations depicted in Fig. 23.3:

- There is the usual anticorrelation between $f_{\rm NEDE}$ and the sound horizon r_s typical for EDE-type models caused by the energy injection in the primordial fluid (see the NEDE feature in Fig. 23.1 and first panel in Fig. 23.3) [1]. The angular scale of r_s (as measured by BAO and CMB observations at high precision) is held constant by raising H_0. This is the core mechanism responsible for addressing the H_0 tension (second panel).
- There is a positive correlation between $f_{\rm NEDE}$ and the amount of cold dark matter $w_{\rm cdm} \equiv \Omega_c h^2$ (third panel). It is generically expected for models that reduce the sound horizon [45, 46]. In terms of the CMB power spectrum, this increase is

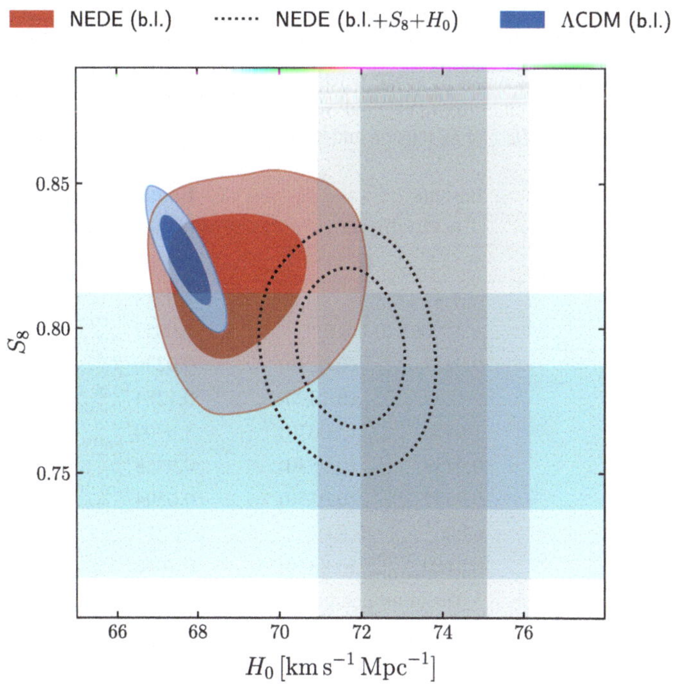

Fig. 23.2 NEDE addresses the H_0 and S_8 tension simultaneously. The red and blue contours were obtained for the baseline (b.l.) datasets (Planck, BAO, uncalibrated supernovae). The vertical (gray) and horizontal (magenta) shaded region corresponds to the measured value $H_0 = 73.04 \pm 1.04\,\mathrm{kms^{-1}Mpc^{-1}}$ (SH_0ES [35]) and $S_8 = 0.762^{+0.025}_{-0.024}$ (KiDS+VIKING+DES [36]), respectively. The dotted contour was obtained by including these values as Gaussian priors in the MCMC analysis. Here and henceforth, the light and dark-shaded regions correspond to the 68% and 95% C.L., respectively

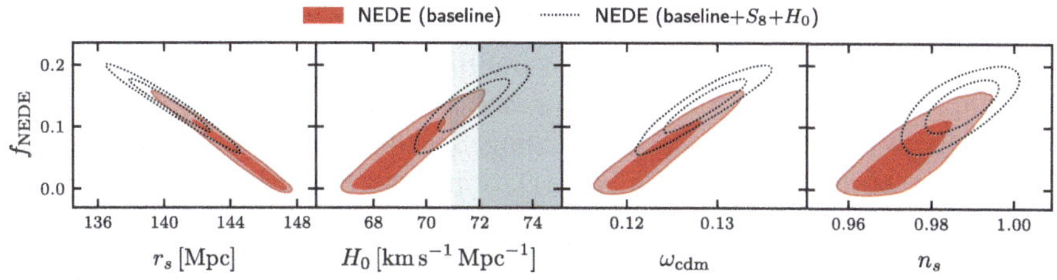

Fig. 23.3 The NEDE model introduces approximate degeneracies between f_{NEDE} and (r_s (at matter drag), H_0, ω_{cdm}, n_s). The gray band represents the H_8 constraint from SH_0ES

Table 23.1 Constraints on all (6 + 4) NEDE parameters (means with 1σ error bars and best-fit in parenthesis) obtained from an analysis with the baseline datasets (CMB, BAO, uncalibrated supernovae) and one supplemented by weak lensing and distance ladder priors on S_8 and H_0, respectively. The values are cited from [11], and S_8 has been added as a derived parameter. The last three rows quantify the H_0 and S_8 tension and state the overall χ^2 improvement compared to ΛCDM

	Baseline	Baseline + H_0 + S_8
$f_{\rm NEDE}$	< 0.130 (0.1140)	$0.1340^{+0.0320}_{-0.0250}$ (0.1460)
$z_{\rm decay}$	4911^{+1000}_{-2000} (4441)	4414^{+500}_{-800} (4626)
$3\omega^*_{\rm NEDE}$	> 1.41 (2.09)	$2.05^{+0.12}_{-0.18}$ (2.03)
Ω_ϕ	< 0.0068 (0.0010)	$0.0057^{+0.0025}_{-0.0029}$ (0.0060)
$\Omega_{\rm b}h^2$	$0.0227^{+0.0002}_{-0.0003}$ (0.0230)	$0.0230^{+0.0002}_{-0.0002}$ (0.0230)
$\Omega_{\rm c}h^2$	$0.1232^{+0.0023}_{-0.0044}$ (0.1290)	$0.1284^{+0.0030}_{-0.0030}$ (0.1300)
$\log(10^{10}A_{\rm s})$	$3.0570^{+0.0160}_{-0.0160}$ (3.0670)	$3.0700^{+0.0160}_{-0.0160}$ (3.0690)
$n_{\rm s}$	$0.9739^{+0.0060}_{-0.0090}$ (0.9840)	$0.9868^{+0.0058}_{-0.0058}$ (0.9880)
$\tau_{\rm reio}$	$0.0572^{+0.0074}_{-0.0074}$ (0.0550)	$0.0594^{+0.0070}_{-0.0080}$ (0.0560)
H_0 [km/s/Mpc]	$69.06^{+0.78}_{-1.40}$ (70.84)	$71.71^{+0.88}_{-0.88}$ (71.74)
S_8	$0.8160^{+0.0180}_{-0.0150}$ (0.8330)	$0.7930^{+0.0180}_{-0.0180}$ (0.7990)
H_0 tension[a]	3.05σ (1.4σ)	–
S_8 tension	1.9σ	–
$\Delta\chi^2$	-3.7	-20.1

[a] The Gaussian tension measure is unreliable in the case of H_0 due to non-Gaussian posteriors caused by prior volume effects [5, 24]. A way out is offered by the DMAP criterion (in parenthesis), which is prior-independent [43]. This was also supported by the recent maximum likelihood analysis in [27] (see also [44] for EDE)

compensated by the presence of dark sector acoustic oscillations, whose positive pressure leads to an excess decay of the gravitational potential [5, 47].

- Lowering r_s generically leads to enhanced diffusion damping on small scales [5]. This is balanced by increasing the spectral tilt n_s and amplitude A_s. As argued in [48], the correlation between $f_{\rm NEDE}$ and n_s (fourth panel) has far-reaching implications for inflationary model-building. Specifically, it makes simple models of inflation such as the curvaton [49–51], or perhaps power-law inflation, viable again (see f.ex. [48] for a more detailed discussion of this point).
- The presence of the ultralight trigger field allows for a sub-percent fraction of non-thermal dark matter Ω_ϕ. It suppresses the matter power spectrum on scales relevant for determining the S_8 parameter due to acoustic oscillations in its perturbations. The anti-correlation between Ω_ϕ and S_8 is then responsible for resolving the S_8 tension (see Fig. 23.4 and [11] for more details).

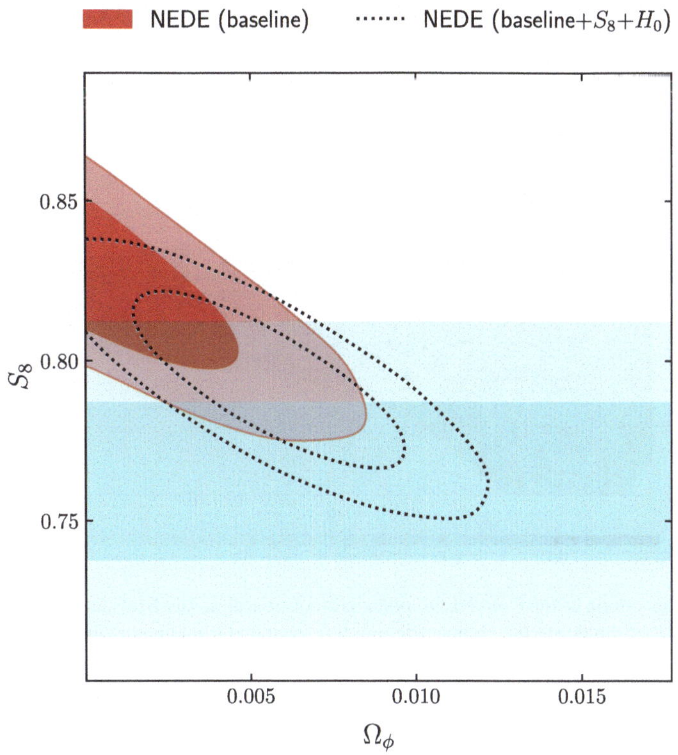

Fig. 23.4 The trigger field acts like axion dark matter with a mass $m \sim 10^{-27}$eV and suppresses the matter power spectrum on scales relevant for the determination of S_8 due to the Jeans stabilization of its perturbations. The magenta band represents the S_8 constraint from weak lensing

23.3 Cold New Early Dark Energy

The model of Cold New Early Dark Energy (Cold NEDE) relies on a two-field mechanism to explain the sudden transition in the equation of state proposed in (23.2). Here, the trigger field ϕ, which we introduced before as defining the transition surface $q(\tau, \mathbf{k})$, initiates the vacuum tunneling of a scalar field ψ when it drops out of slow-roll. The tunneling corresponds to a first-order phase transition that converts vacuum energy into a new form of energy that decays with redshift (for another first-order model, see [52]). A schematic illustration is provided in Fig. 23.5. A similar model has been proposed in the early 90's at higher energies as a mechanism to exit from inflation in a first-order phase transition and yet avoid typical problems with old inflation [53–56]. In any event, this idea can be realized in terms of the renormalizable two-field potential

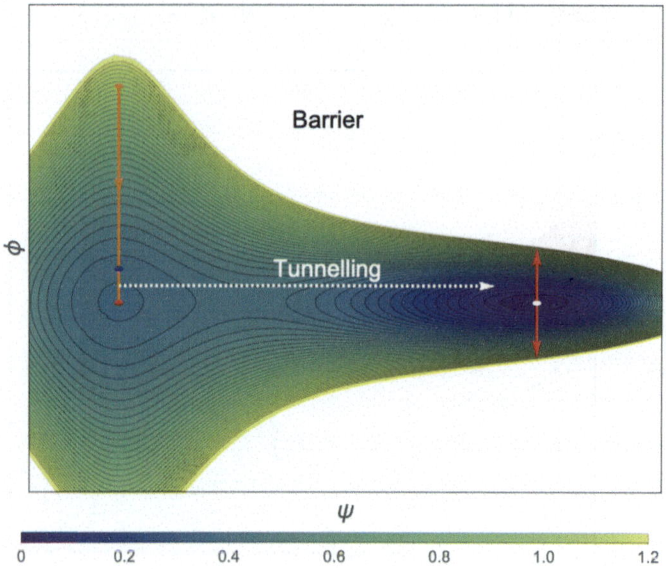

Fig. 23.5 Illustration of the Cold NEDE potential (not-to-scale). The field starts at the orange dot but is initially frozen due to the Hubble friction. It is separated from the true minimum (white dot) through a high potential barrier. Shortly before matter-radiation equality, when $m \sim H \sim 10^{-27}$eV, the field starts rolling along the orange line. As a result, the barrier between the field and the true minimum shrinks, and sub-barrier tunneling becomes admissible after the field passes the blue dot. The tunneling rate increases quickly, leading to the percolation of space with the true vacuum. In its final state, the field oscillates around the true minimum (red arrow)

$$V(\psi,\phi) = \frac{\lambda}{4}\psi^4 + \frac{\beta}{2}M^2\psi^2 - \frac{1}{3}\alpha M\psi^3 + \frac{1}{2}m^2\phi^2 + \frac{1}{2}\tilde{\lambda}\phi^2\psi^2 \ldots, \tag{23.9}$$

where λ, α, and $\tilde{\lambda}$ are dimensionless constants, and the ellipses denote higher terms in an effective field theory expansion that are suppressed at low energies. This is a two-scale potential with $M \sim$ eV being of the order of the critical energy density at the same time of matter-radiation equality and $m \sim M^2/M_{\rm pl} \sim 10^{-27}$eV an ultralight-scale set by the value of the Hubble parameter at the same time. This could have a UV completion in a model with two axions, one with a mass close to the QCD axion mass at the eV scale and an Ultra-Light Axion (ULA) with a mass of 10^{-27} eV [11]. For $\lambda < 2\alpha^2/(9\beta)$, this potential admits a false minimum at $(\psi,\phi)_{\rm False} = (0,0)$ and a true one at

$$(\psi,\phi)_{\rm True} = \left(\frac{M}{2\lambda}\left[\alpha + \sqrt{\alpha^2 - 4\lambda\beta}\right], 0\right). \tag{23.10}$$

Initially the field is frozen at $(\psi,\phi)_{\rm False} = (0,\phi_{\rm ini})$ due to a large Hubble friction. In particular, it is blocked from tunneling to the true minimum (23.10) by a huge potential barrier created by the term $\tilde{\lambda}\phi^2\psi^2$ in (23.9) (which requires $\phi_{\rm ini}$ to be sufficiently large). Eventually, when $H \sim m$ during the CMB epoch close to matter-radiation equality, ϕ starts rolling towards zero. This gradually removes the barrier and increases the tunneling probability per spacetime volume denoted as Γ. In our case, it can be approximated as [57]

$$\Gamma \sim M^4 \, e^{-S_E} \tag{23.11}$$

where

$$S_E \approx \frac{4\pi^2}{3\lambda} \left(2 - \delta_{\text{eff}}\right)^{-3} \left(\alpha_1 \delta_{\text{eff}} + \alpha_2 \delta_{\text{eff}}^2 + \alpha_3 \delta_{\text{eff}}^3\right) , \tag{23.12}$$

is the Euclidian action corresponding to the so-called "bounce solution" [58, 59] with numerically determined coefficients [60] $\alpha_1 = 13.832$, $\alpha_2 = -10.819$, and $\alpha_3 = 2.0765$. Its time dependence is captured through

$$\delta_{\text{eff}}(\tau) = 9 \frac{\lambda}{\alpha^2} \left(\beta + \tilde{\lambda} \frac{\phi^2(\tau)}{M^2} \right) . \tag{23.13}$$

Tunneling becomes efficient when Γ outperforms the Hubble expansion, corresponding to $\Gamma \gtrsim H^4$. For $M \sim$ eV and $H \sim m \sim 10^{-27}$eV, this happens when $S_E \lesssim 250$ (this condition is always met as $\phi \to 0$, provided $9\beta/\alpha^2 \lesssim 0.1$ and $\lambda < 1$), which in turn fixes the decay time τ_* (or z_* alternatively). A more quantitative argument finds [5] $H(\tau_*)/m \approx 0.2$, which will be used as a reference value below. Moreover, the fact that Γ is exponentially sensitive to the value of ϕ (through δ_{eff}) justifies that the phase transition occurs across a surface of constant ϕ.

For $\tau > \tau_*$, the space is quickly filled with bubbles of the true vacuum. These are spherical configurations of the scalar field ψ with $\psi \approx \psi_{\text{True}}$ in their interior and $\psi \approx \psi_{\text{False}}$ in their exterior. Both phases are separated by a bubble wall across which the field interpolates between both vacua. The percolation time scale is given by the inverse of $\bar{\beta} = -\mathrm{d}S_E/\mathrm{d}t$, where t denotes cosmological time ($\mathrm{d}t = a\mathrm{d}\tau$). After a time $\Delta t \sim \bar{\beta}^{-1}$, the tunneling rate has increased so much that all of space has been converted to the true vacuum. Since the vacuum bubbles expand with almost the speed of light, $\bar{\beta}^{-1}$ also sets the typical size of bubbles at the time when they collide. These bubbles introduce large anisotropies in the cosmic fluid (corresponding to a density contrast of the order of f_{NEDE}). Requiring that we cannot resolve them in our cosmological probes, which have an angular resolution $\theta_{\min}$, sets an upper bound on $\bar{\beta}^{-1}$ [5]:

$$H_* \bar{\beta}^{-1} < 0.3 \times \left(\frac{\theta_{\min}}{0.14^\circ}\right) \left(\frac{z_*}{5000}\right) \left(\frac{g_*}{3.4}\right)^{1/2} \left(\frac{0.7}{h}\right) \left(1 - f_{\text{NEDE}}\right)^{-1/2} , \tag{23.14}$$

where g_* is the number of relativistic degrees of freedom at the time of the transition and $H_* = H(\tau_*)$. We normalized the expression with respect to a typical CMB resolution of $\theta_{\min} \sim 2\pi/2500 \simeq 0.14^\circ$. By relating $\bar{\beta}$ to the microscopic parameters, it can be shown that λ can be used as a dial to suppress $H_*\bar{\beta}^{-1}$. A sufficient condition to satisfy (23.14) is

$$\lambda < 2.1 \times \left(\frac{\theta_{\rm min}}{0.14^\circ}\right)\left(\frac{z_*}{5000}\right)\left(\frac{g_*}{3.4}\right)^{1/2}\left(\frac{0.7}{h}\right)\sqrt{\frac{\delta^*_{\rm eff}-\delta}{1-f_{\rm NEDE}}}\left[\frac{d(\lambda S_E)}{d\delta_{\rm eff}}\right], \tag{23.15}$$

where $\delta^*_{\rm eff} = \delta_{\rm eff}(\tau_*)$ and $\delta = \delta_{\rm eff}|_{\phi=0} = 9\lambda\beta/\alpha^2$. This is also a necessary condition to justify our effective fluid description introduced in Sect. 23.2.3, which relies on the assumption that the scalar field condensate after the phase transition is homogenous and isotropic on cosmological scales. Moreover, this is a necessary condition for the absence of entropy perturbations in the large-scale fluid, which would be created if the phase transition were too slow and structures emerged at cosmological scales.

The liberated vacuum energy after the phase transition is given as the difference in potential energy between the true and false vacuum, $\rho_{\rm NEDE}(\tau_*) = V(0, \phi(\tau_*)) - V(\psi_{\rm True}, 0)$, amounting to

$$f_{\rm NEDE} = \frac{\rho_{\rm NEDE}(\tau_*)}{\rho_{\rm tot}(\tau_*)} = \frac{c_\delta}{36}\frac{\alpha^4}{\lambda^3}\frac{M^4}{M_{\rm Pl}^2 H_*^2}, \tag{23.16}$$

where we introduced the function

$$c_\delta = \frac{1}{216}\left(3+\sqrt{9-4\delta}\right)^2\left(3-2\delta+\sqrt{9-4\delta}\right), \tag{23.17}$$

which is of order unity for $\delta < 1$. We finally provide a dictionary between the microscopic and the phenomenological parameters ($f_{\rm NEDE}$, z_* and Ω_ϕ):

$$M^4 \simeq (0.4\,{\rm eV})^4\,\frac{1}{c_\delta}\left(\frac{\lambda^3\alpha^{-4}}{0.01}\right)\left(\frac{f_{\rm NEDE}/(1-f_{\rm NEDE})}{0.1}\right)\left(\frac{g_*}{3.4}\right)\left(\frac{z_*}{5000}\right)^4, \tag{23.18a}$$

$$m \simeq 1.7\times 10^{-27}\,{\rm eV}\,(1-f_{\rm NEDE})^{-1/2}\left(\frac{g_*}{3.4}\right)^{1/2}\left(\frac{z_*}{5000}\right)^2\left(\frac{0.2}{H_*/m}\right), \tag{23.18b}$$

$$\left(\frac{\phi_{\rm ini}}{M_{\rm Pl}}\right)^2 \simeq 0.13\times\left(\frac{1+z_*}{5000}\right)^{-1}\left(\frac{\Omega_\phi}{0.005}\right)(1-f_{\rm NEDE})^{-1}. \tag{23.18c}$$

Here, (23.18a) and (23.18c) follow directly from (23.16) and (23.7), respectively, and we assumed radiation domination to obtain (23.18b). Different terms are normalized with respect to their phenomenologically relevant values (cf. Table 23.1). The upshot from this is that Cold NEDE requires an ultralight trigger field and an eV-scale tunneling field. Moreover, the initial field value for the trigger field $\phi_{\rm ini}$ remains sub-

Planckian even if we require that it makes a contribution to dark matter large enough to resolve the S_8 tension.

Relating w_{NEDE} to the parameters in (23.9) is more complicated, and more work is needed to achieve this. In general terms, however, there are at least two contributions to ρ_{NEDE} that have the potential to explain the observational fact that $w_{\mathrm{NEDE}} > 1/3$. First, the small-scale anisotropic stress, which corresponds to the colliding bubble walls, sources tensor, vector, and scalar shear. While tensor shear is equivalent to gravitational radiation and hence associated with $w = 1/3$, the vector and scalar shear are known to behave like a stiff fluid; in particular, their energy density averaged over large scales decays as $1/a^6$ [61]. This possibility requires the gravitational sourcing to be very efficient, which corresponds to the regime where $H_*\bar{\beta}^{-1}$ is close to its upper bound in (23.14). Second, the scalar field condensate after the decay, despite being highly fragmented, oscillates in an effective field theory potential that contains higher order terms such as $\propto \psi^6/M^2$ or $\propto \psi^8/M^4$ [indicated by the ellipses in (23.9)]. If these operators are probed within the perturbative regime, the expectation is that this will increase w_{NEDE} initially (similar to the fact that coherent oscillations in higher order potentials lead to a stiffer fluid when averaged over cycles).

An entirely different possibility has been prosed in [6] as "scenario B". Here the idea is that the vacuum bubbles during the collision phase quickly dissipate into relativistic light particles that become non-relativistic shortly after (before recombination). The net effect is a reduction of the dark matter density at early times which is then replenished at later times. In particular, at background level, this was shown to reproduce the shape of the energy injection in Fig. 23.1 without relying on a stiff fluid component.

Another question regards the ultraviolet completion of (23.9). A successful phenomenology requires a large separation of scales where $m \ll M$. For this choice to be radiative stable at low energies, the coupling $\tilde{\lambda}$ has to be very weak, approximately [5] $\tilde{\lambda} \leq \mathcal{O}(1) \times 10^3 m^2/(\beta M^2)$. This raises the question of how this can be achieved in a more fundamental theory. As mentioned earlier, a preliminary answer was given in [11] recently. The idea is to explain the smallness of m and $\tilde{\lambda}$ through the breaking of a continuous shift symmetry, $\phi \to \phi + \mathrm{const}$, down to a discrete symmetry, $\phi \to \phi + 2\pi f$, which occurs at an energy scale $\Lambda \ll f$. To be specific, in this axion-like setup, we find $m = \Lambda^2/f$, which for $\Lambda \sim \mathrm{eV}$ and $f \sim M_{\mathrm{Pl}}$ recovers the right value of the trigger mass.

23.4 Hot New Early Dark Energy

Since the cosmological electroweak and QCD phase transitions are both known to be triggered by the temperature of the primordial plasma dropping below some critical temperature, it is natural to consider the possibility that a temperature triggers the NEDE phase transition. In that case, when the ultralight scalar field trigger is replaced by a dark sector temperature T_d, there is only one energy scale, the eV energy scale, controlling the phase transition.

This scenario has been introduced as Hot NEDE in [6, 7]. Correspondingly, in (23.4), we identify $q(\tau, \mathbf{x}) \equiv T_d(\tau, \mathbf{x})$. Moreover, the dark sector is assumed to be colder than the visible sector, $T_{\rm vis} > T_d$. In fact, we require $\xi \equiv T_d/T_{\rm vis} < 0.6/g_{\rm rel,d}^{1/4}$, where $g_{\rm rel,d}$ is the effective number of relativistic degrees of freedom in the dark sector. This avoids bounds on the equivalent number of neutrino species[4], which receives a contribution [64] $\Delta N_{\rm eff} = \frac{4}{7}(\frac{11}{4})^{4/3} g_{\rm rel,d}\xi^4 < 0.29$. As we will see, it also makes sure that for $f_{\rm NEDE} \sim 0.1$, the corresponding dark radiation fluid is subdominant compared to the released vacuum energy measured by $\rho_{\rm NEDE}$, implying a strong first-order phase transition in agreement with the sharp energy injection in Fig. 23.1.

On a microphysical level, the idea is that a complex NEDE field Ψ is charged under a local $U(1)_D$ with gauge coupling $g_{\rm NEDE}$ and self-coupling λ. The corresponding gauge bosons form a plasma of temperature T_d and the NEDE field receives thermal corrections of the form (valid in the perturbative regime where [6, 65] $\lambda \lesssim g_{\rm NEDE}^3$)

$$\delta V(\psi; T_d) = 3T_d^4\, K\left(g_{\rm NEDE}\psi/T_d\right) \mathrm{e}^{-g_{\rm NEDE}\psi/T_d}\,, \tag{23.19}$$

where $\psi = \sqrt{2}|\Psi|$ and $K(a)$ can be approximated within the range $0 < a \equiv g_{\rm NEDE}\psi/T_d < 30$ as

$$K(a) = -0.1134\,(1+a) - 0.0113\,a^2 \\ + 4.32\times 10^{-6}\ln(a)\,a^{3.58} + 0.0038\,\mathrm{e}^{-a(a-1)}\,. \tag{23.20}$$

For small argument, we can expand $K(a)\mathrm{e}^{-a} \simeq -\pi^2/90 + a^2/24 + \ldots$ and recover the usual result in the literature valid for gauge boson masses $g_d\psi \ll T_d$ [66]. The full potential then reads

$$V(\psi; T_d) = -\frac{1}{8}\, g_{\rm NEDE}^2 T_\circ^2\psi^2 + \frac{\lambda}{4}\psi^4 + \delta V(\psi; T_d) + V_{1-\rm loop} + \ldots\,, \tag{23.21}$$

where the first two terms correspond to the vacuum potential, and $T_\circ$ is a characteristic temperature scale, which is related to the mass scale μ through $T_\circ = 2\mu/g_{\rm NEDE}$. The term $V_{1-\rm loop}$ denotes 1-loop corrections to the vacuum potential [67, 68]. While they can introduce changes to the precise shape of the potential, we do not expect them to change the conclusion about the character and strength of the phase transition (for a more detailed discussion see [69]). The tachyonic character of the quadratic term reflects the fact that the symmetric state with $\psi = 0$ is unstable at zero temperature. To describe the temperature dependence of $V(\psi; T_d)$, it is useful to introduce the parameter

$$\tilde{\delta}_{\rm eff}(T_d) = \pi\left(1 - \frac{T_\circ^2}{T_d^2}\right)\,. \tag{23.22}$$

---

[4] To be specific, big bang nucleosynthesis and the CMB give rise to the bounds $\Delta N_{\rm eff} < 0.39$ [62] and $\Delta N_{\rm eff} < 0.3$ [63], respectively.

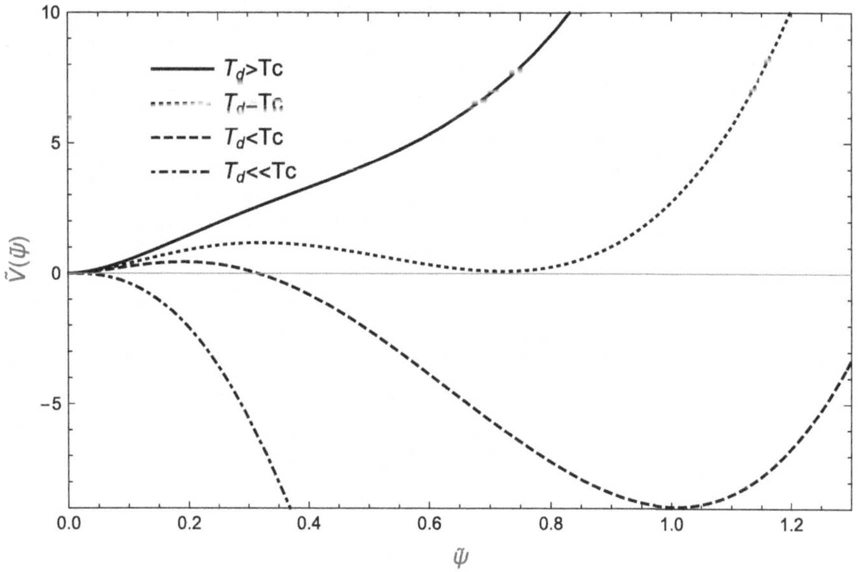

Fig. 23.6 The NEDE scalar field potential $\tilde{V} = 4\pi V/T_d^4$ as a function of T_d and $\tilde{\psi} = g_{\text{NEDE}}\sqrt{\gamma}\psi/T_d$ for $\gamma = 0.02$. Initially, $T_d > T_c$, and the symmetric vacuum at $\tilde{\psi} = 0$ minimizes the potential energy. As the dark sector cools further, the potential develops a second minimum at $\tilde{\psi} > 0$. It becomes degenerate with the symmetric vacuum at $T_d = T_c$. At this point, the tunneling rate $\Gamma(T_d)$ becomes non-vanishing. Percolation becomes possible later on when $\Gamma(T_d) \approx H^4$ is met, which implicitly defines the temperature T_d^* at which the NEDE phase transition occurs

We depict different stages of the potential's evolution as a function of T_d in Fig. 23.6. Initially, for $T_d \gg T_0$ (or $\tilde{\delta}_{\text{eff}} \approx \pi$ equivalently), we are in the symmetric phase where the global minimum lies at $\psi = 0$. However, as the temperature cools, the potential develops a second minimum. Both minima become degenerate at a critical temperature T_c. Finally, at an even lower temperature $T_d^* < T_c$, the phase transition occurs. We are mostly interested in the supercooled regime for which $T_d^* \ll T_c$. In that case, we have a strong first-order phase transition where the difference in vacuum energy dominates over the energy density of the radiation fluid, a necessary requirement to have a localized energy injection. It can be shown that this regime is realized for [6] [requiring the general form of the thermal corrections in (23.19)]

$$\gamma \equiv \frac{4\pi\lambda}{g_{\text{NEDE}}^4} \lesssim 1\,. \tag{23.23}$$

As before, we define $\rho_{\text{NEDE}}(\tau_*) = V(0) - V(\psi_{\text{True}})$, where $\psi_{\text{True}} \simeq \mu/\sqrt{\lambda} \equiv v_\Psi$. The maximal fraction of NEDE then evaluates to

$$f_{\rm NEDE} = \frac{\pi}{16\gamma}\left(1 - \frac{\tilde{\delta}^*_{\rm eff}}{\pi}\right)^2 \frac{T_d^{*4}}{\rho_{\rm tot}(\tau_*)}\,, \tag{23.24}$$

where $\delta^*_{\rm eff} \equiv \delta_{\rm eff}(\tau_*)$. In accordance with Fig. 23.6, and as explicitly shown in [6], strong supercooling corresponds to the regime where $\tilde{\delta}^*_{\rm eff}/\pi \ll 1$ (or $T_d^* \ll T_c$ equivalently). Together with $\rho_{\rm tot}(\tau_*) \approx 1.1 \times T_{\rm vis}^{*4}/(1 - f_{\rm NEDE})$, this implies

$$\xi_*^4 \equiv (T_d/T_{\rm vis})^4|_{\tau=\tau_*} \simeq 0.6 \times \gamma \left[\frac{f_{\rm NEDE}/(1 - f_{\rm NEDE})}{0.1}\right]. \tag{23.25}$$

We note that for $\gamma \ll 1$, $f_{\rm NEDE}$ can be sizeable, although we assume $\xi_* < 0.6/g_{\rm rel,d}^{1/4}$. As mentioned before, this regime avoids bounds on $\Delta N_{\rm eff}$ and ensures that the dark radiation energy density, $\propto T_d^4$, is smaller than $\rho_{\rm NEDE}$. In other words, like in Cold NEDE, the trigger remains sub-dominant. In absolute terms, we derive the dark sector temperature to be

$$T_d^{*4} \simeq (0.7 {\rm eV})^4 \gamma \left[\frac{f_{\rm NEDE}/(1 - f_{\rm NEDE})}{0.1}\right]\left[\frac{1 + z_*}{5000}\right]^4. \tag{23.26}$$

This then constitutes a central relation between the phenomenological parameters, $f_{\rm NEDE}$ and z_*, and the more fundamental parameters, T_d^* and γ. As with Cold NEDE, the phase transition is required to complete quickly on cosmological time scales. The percolation time sale $\bar{\beta}^{-1}$ was derived to be [6]

$$H_* \bar{\beta}^{-1} \sim 10^{-2} g_{\rm NEDE}^2 \qquad (\text{valid if } \gamma \lesssim 1 \text{ and } g_{\rm NEDE} \lesssim 0.1)\,, \tag{23.27}$$

which for a sufficiently weak gauge coupling can easily satisfy the phenomenological bound in (23.14).

To summarize, Hot NEDE can provide us with a sizeable fraction of (false) vacuum energy that, in the supercooled regime, dominates over the dark radiation component. While the former provides us with a sizeable amount of (decaying) early dark energy, the latter plays the role of a subdominant trigger component in agreement with Fig. 23.1. In particular, we can work in a regime where $\xi = T_d/T_{\rm vis} \ll 1$. This property sets Hot NEDE apart from dark radiation approaches to the Hubble tension—see, for example, the models that introduce a mass-threshold [70–74]—which require ξ to be of order unity in order to be able to address the H_0 tension. This typically leads to challenges with big bang nucleosynthesis (BBN), which gives rise to an upper bound on [64] $\Delta N_{\rm eff}$ (and hence ξ) or requires a post-BBN thermalization of the dark sector, which comes with its own model-building challenges [75, 76].

Phenomenological differences with Cold NEDE will arise due to the fact that the NEDE perturbations in (23.5) are now seeded by dark sector temperature perturbations, $\delta\tau_* = -\delta T_d(\tau_*, \mathbf{x})/T_d'(\tau_*)$, as opposed to perturbations of a slowly rolling scalar field in the Cold NEDE case. A detailed phenomenological study is planned for future work.

23.4.1 Neutrino Mass Generation

It has been pointed out long ago in the context of the inverse seesaw mechanism [77–80] that the active neutrino masses can be created through a mass mixing with a set of sterile neutrinos ν_s and right-handed neutrinos ν_R. Besides two high-energy Dirac mixings $\nu_s \leftrightarrow \nu_R$ and $\nu_R \leftrightarrow \nu_{e,\mu\tau}$, the crucial ingredient is a Majorana mass term $\propto m_s \overline{(\nu_s)^c}(\nu_s)$, where m_s is of order (or larger) than $\mathcal{O}(\text{eV})$. The mechanism is theoretically compelling because the active neutrino masses are protected against radiative corrections through a Lepton symmetry that is recovered in the limit of vanishing neutrino mass. This is similar to the electron mass m_e, which is radiatively stable due to a chiral symmetry that is recovered in the limit $m_e \to 0$ [81]. However, in contrast to the electron, which receives its mass in the electroweak phase transition at a temperature $T_{\text{vis}} \sim 160\,\text{GeV}$ [82] through the Higgs mechanism, the inverse seesaw mechanism does not provide a dynamical explanation for how the Majorana mass term is created. It is, therefore, natural to ask if the sterile mass m_s was generated during the NEDE phase transition when $T_d \sim \text{eV}$. In other words, we can ask if there was as dark sector analog of the electroweak phase transition.

To make these statements more concrete, we first review the inverse seesaw mechanism in its original form (following [80]). For simplicity, we only consider one neutrino generation. Besides one sterile neutrino ν_s, we introduce the Dirac spinor of a right-handed neutrino ν_R, and denote the left-handed neutrino as ν_L (this could be ν_e, ν_μ, or ν_τ). Defining $N \equiv (\nu_L, \nu_R^c, \nu_s)^T$, we can introduce the mass mixing

$$\mathcal{L}_\nu = -\frac{1}{2} N^T C M N + \text{h.c.} \tag{23.28}$$

where

$$M = \begin{pmatrix} 0 & d & 0 \\ d & 0 & n \\ 0 & n & m_s \end{pmatrix} \tag{23.29}$$

is a (complex-valued) mass matrix, and C denotes charge conjugation. Using this notation, off-diagonal and diagonal elements correspond to Dirac and Majorana mass terms, respectively. For concreteness, we assume that $m_s = \mathcal{O}(100\text{eV})$, $d = \mathcal{O}(100\text{GeV})$ and $n = \mathcal{O}(10\text{TeV})$ (other choices are possible too). The crucial observation is that this mass matrix, upon diagonalization, admits a light eigenmode with mass

$$m_{\text{light}} = \frac{|d|^2}{|d|^2 + |n|^2}|m_s| + \mathcal{O}(|m_s|^2)\,, \tag{23.30}$$

which, for the numerical example provided above, amounts to $m_{\text{light}} = \mathcal{O}(0.05\text{eV})$ and thus can play the role of one of the active neutrinos. In addition, there is a pair of super-TeV pseudo-Dirac fermions, which are irrelevant for the low-energy dynamics. As described in [80], this mechanism can reproduce the observed mixing angles and mass spectrum when generalized to three generations. Moreover, only the Majorana mass term $\propto m_s$ breaks Lepton number, and thus the symmetry gets restored for $m_s \to 0$. Consequently, m_s and, due to (23.30) also m_{light}, can be small in a technically natural way.

Now, the idea is to introduce an interaction between the sterile neutrino and the NEDE scalar Ψ with Yukawa coupling g_s,

$$\mathcal{L} \supset -\frac{1}{\sqrt{2}} \sum_{ij} g_s \Psi \overline{(\nu_s)^c}(\nu_s) \;\; + \text{h.c.}\,, \tag{23.31}$$

which is invariant under a global Lepton symmetry $U(1)_L$ if we assign a charge $L = -2$ to Ψ and $L = 1$ to ν_s. In addition, we assume appropriate charges under our local $U(1)_D$ to preserve gauge invariance. In any event, when $\Psi \to v_\Psi/\sqrt{2}$ in the NEDE phase transition, we indeed generate the Majorana mass entry in (23.29); more explicitly, we have $m_s = g_s v_\Psi$. In particular, the vacuum expectation value v_Ψ can be related to the phenomenological NEDE parameters through

$$m_s \approx (1.2\,\text{eV}) \times \frac{g_s}{(4\pi\lambda)^{1/4}} \times \left[\frac{f_{\text{NEDE}}/(1 - f_{\text{NEDE}})}{0.1}\right]^{1/4} \left[\frac{1 + z_*}{5000}\right]. \tag{23.32}$$

As a result, we find that we can create a (super-) eV-scale mass entry m_s as required by the inverse seesaw mechanism, provided the NEDE self-coupling is sufficiently small. To be precise, we need $4\pi\lambda < g_s^4$. Due to the mixing between the active neutrino ν_L and ν_s of order of $\mathcal{O}(d^2/n^2)$, there is a bound $g_s < 10^{-7}\mathcal{O}(n^2/d^2)$ [83], which makes sure that the Boltzmann evolution of the active neutrinos is preserved. It translates to $\lambda < 10^{-20}(10^{-2}\text{eV}/m_{\text{light}})^4$. As a consistency check, such a small value of λ is consistent with the supercooling bound in (23.23), if the gauge coupling is not to small ($g_{\text{NEDE}} > 10^{-5}$). There is one concern, though, with λ being that small. It can lead to a large vacuum expectation value because $v_\Psi = \mu/\sqrt{\lambda}$ and hence a heavy gauge boson mass $\propto g_{\text{NEDE}} v_\Psi$, suppressing the gauge boson number density in the broken phase. However, for the thermal description to be applicable, the gauge field and Ψ must be in thermal equilibrium. As argued in [7], this translates into a lower bound $\xi_* > 0.3$, ensuring the dark sector is populated strongly enough to maintain thermal equilibrium. This, in turn, makes the neutrino scenario more constrained, leading to the finite range $0.3 < \xi_* < 0.6/g_{\text{rel,d}}^{1/4}$.

Before ending this section, we note that this mechanism only explains the eV-scale entry in (23.29). A more complete mechanism was suggested in [6, 7]. It creates the

d entry in the electroweak phase transition and postulates a new, super-TeV phase transition to generate the $n(> d)$ term[5] . The latter phase transition takes place again in the dark sector and corresponds to a breaking of a $SU(2)_D \times U(1)_{Y_D} \rightarrow U(1)_D$. If the higher breaking scale is sufficiently large, the same phase transition could give mass to a super-TeV dark matter, which, due to its gravitational interactions, could be produced at early times through the freeze-in mechanism [85, 86]. Overall, the Hot NEDE proposal should be understood as an attempt to develop a more fundamental and comprehensive dark sector model guided by the aim of resolving the cosmic tension.

23.5 Conclusion

New Early Dark Energy is a framework for a triggered low-scale phase transition in the dark sector. As discussed in Sect. 23.2, it admits a very general phenomenological description in terms of an abruptly decaying early dark energy component. The corresponding sharp energy injection into the cosmic fluid can resolve the Hubble tension, making use of the well-established degeneracy between H_0 and the sound horizon r_s (see Fig. 23.3). Its unique feature is a trigger sector. It sets off the phase transition and seeds a new type of dark sector acoustic oscillations carried by the decaying NEDE fluid. The trigger leads to unique signatures in the CMB and matter power spectra, which are different from other EDE models and is crucial for the model's phenomenological success in reducing the H_0 and S_8 tension below the two-sigma level. When identified with an ultralight scalar field, it was recently shown that the trigger could resolve the S_8 tension when it makes a small contribution to dark matter through its coherent oscillations at late times [11]. When testing the model against CMB, BAO, SNe data and including a local prior on the value of S_8 and H_0 (which has become statistically viable within the NEDE framework), we obtain $f_{\mathrm{NEDE}} = 0.134^{+0.032}_{-0.025}$, which corresponds to a Gaussian evidence for a non-vanishing fraction of NEDE that exceeds 5σ along with the new concordance values $H_0 = 71.71 \pm 0.88\,\mathrm{km\,sec^{-1}\,Mpc^{-1}}$ and $S_8 = 0.793 \pm 0.018$.

As mentioned before, NEDE admits different microscopical descriptions with their own unique signatures. Here we have presented two possibilities. One where the trigger is an ultralight scalar field (Cold NEDE) and one where it is identified with a dark sector temperature (Hot NEDE)—each of which comes with its own opportunities and challenges. In Cold NEDE, the primary theoretical tasks involve linking the equation of state parameter w_{NEDE} to the effective field theory parameters in the potential and explaining the hierarchy between the trigger scale m and the tunneling scale $M \gg m$. While we believe the latter can be achieved within a multi-

[5] As mentioned above, this is a generic choice of energy scales, which avoids any additional tuning of the d entry relative to the electroweak scale. It is, however, possible that both the d and the n energy scales are lower, bringing the new UV dark phase transition, generating the n scale, within an energy scale relevant for a possible interpretation of the recent NANOgrav observation [69, 84].

axion framework, the former might require more detailed numerical studies. On a phenomenological level, new CMB data, such as future ACT releases [87] and the CMB-S4 experiment [88], as well as additional LSS data from observations of galaxy abundances and clustering or spectroscopic galaxy surveys, such as the James Web Telescope [89] or Euclid [90], will help discriminate the model from its competitors and constrain its unique signatures (for first studies including ground-based CMB and full-shape matter power spectrum data see [27, 28, 48, 91]).

For Hot NEDE, a full cosmological data extraction is still outstanding. Here, we expect changes arising from the difference in trigger perturbations. Such an analysis is important because it can reveal how predictive the physics of the trigger sector is. Moreover, in the case where we include the neutrino mass generation, we could look for additional signatures related to the physics of the inverse seesaw mechanism (for a detailed list see [7]).

Although the preferred microscopic scenario remains uncertain at the moment—in fact, other triggered scenarios different from cold and Hot NEDE, such as hybrid NEDE, are possible [5]—the crucial aspect is that, due to the low-energy scale of the transition, any scenario will be highly constrained by our cosmological probes. So, NEDE is a promising and natural framework for resolving the Hubble tension, and if the Hubble tension really is the signature of a triggered phase transition in the dark sector, we will soon know.

Acknowledgements M. S. Sloth is supported by Independent Research Fund Denmark grant 0135-00378B. The work of F. Niedermann was supported by VR Starting Grant 2022-03160 of the Swedish Research Council.

References

1. V. Poulin, T.L. Smith, T. Karwal, M. Kamionkowski, Early dark energy can resolve the Hubble tension. Phys. Rev. Lett. **122**, 221301 (2019). https://doi.org/10.1103/PhysRevLett.122.221301 [arXiv:1811.04083]
2. V. Poulin, T.L. Smith, T. Karwal, The ups and downs of early dark energy solutions to the Hubble tension: a review of models, hints and constraints circa 2023. arXiv:2302.09032
3. M. Kamionkowski, A.G. Riess, The Hubble tension and early dark energy. arXiv:2211.04492
4. F. Niedermann, M.S. Sloth, New early dark energy. Phys. Rev. D **103**, L041303 (2021). https://doi.org/10.1103/PhysRevD.103.L041303 [arXiv:1910.10739]
5. F. Niedermann, M.S. Sloth, Resolving the Hubble tension with new early dark energy. Phys. Rev. D **102**, 063527 (2020). https://doi.org/10.1103/PhysRevD.102.063527 [arXiv:2006.06686]
6. F. Niedermann, M.S. Sloth, Hot new early dark energy. Phys. Rev. D **105**, 063509 (2022). https://doi.org/10.1103/PhysRevD.105.063509 [arXiv:2112.00770]
7. F. Niedermann, M.S. Sloth, Hot new early dark energy: towards a unified dark sector of neutrinos, dark energy and dark matter. Phys. Lett. B **835**, 137555 (2022). https://doi.org/10.1016/j.physletb.2022.137555 [arXiv:2112.00759]
8. A.H. Guth, The inflationary universe: a possible solution to the horizon and flatness problems. Phys. Rev. D **23**, 347 (1981). https://doi.org/10.1103/PhysRevD.23.347

9. A.D. Linde, A new inflationary universe scenario: a possible solution of the horizon, flatness, homogeneity, isotropy and primordial monopole problems. Phys. Lett. B **108**, 389 (1982). https://doi.org/10.1016/0370-2693(82)91219-9
10. A. Albrecht, P.J. Steinhardt, Cosmology for grand unified theories with radiatively induced symmetry breaking. Phys. Rev. Lett. **48**, 1220 (1982). https://doi.org/10.1103/PhysRevLett.48.1220
11. J.S. Cruz, F. Niedermann, M.S. Sloth, Cold new early dark energy pulls the trigger on the H_0 and S_8 tensions: a simultaneous solution to both tensions without new ingredients. arXiv:2305.08895
12. P. Svrcek, E. Witten, Axions in string theory. JHEP **06**, 051 (2006). https://doi.org/10.1088/1126-6708/2006/06/051 [arXiv:hep-th/0605206]
13. A. Arvanitaki, S. Dimopoulos, S. Dubovsky, N. Kaloper, J. March-Russell, String axiverse. Phys. Rev. D **81**, 123530 (2010). https://doi.org/10.1103/PhysRevD.81.123530 [arXiv:0905.4720]
14. J. Preskill, M.B. Wise, F. Wilczek, Cosmology of the invisible axion. Phys. Lett. B **120**, 127 (1983) . https://doi.org/10.1016/0370-2693(83)90637-8
15. M.S. Turner, Coherent scalar field oscillations in an expanding universe. Phys. Rev. D **28**, 1243 (1983). https://doi.org/10.1103/PhysRevD.28.1243
16. R.H. Brandenberger, Cosmological perturbations in a universe dominated by a coherent scalar field. Phys. Rev. D **32**, 501 (1985). https://doi.org/10.1103/PhysRevD.32.501
17. B. Ratra, P.J.E. Peebles, Cosmological consequences of a rolling homogeneous scalar field. Phys. Rev. D **37**, 3406 (1988). https://doi.org/10.1103/PhysRevD.37.3406
18. J.E. Kim, Light pseudoscalars, particle physics and cosmology. Phys. Rept. **150**, 1 (1987). https://doi.org/10.1016/0370-1573(87)90017-2
19. D.J.E. Marsh, Axion cosmology. Phys. Rept. **643**, 1 (2016). https://doi.org/10.1016/j.physrep.2016.06.005 [arXiv:1510.07633]
20. L. Susskind, The anthropic landscape of string theory. arXiv:hep-th/0302219
21. L.F. Abbott, A mechanism for reducing the value of the cosmological constant. Phys. Lett. B **150**, 427 (1985). https://doi.org/10.1016/0370-2693(85)90459-9
22. P. Agrawal, G. Obied, P.J. Steinhardt, C. Vafa, On the cosmological implications of the string swampland. Phys. Lett. B **784**, 271 (2018). https://doi.org/10.1016/j.physletb.2018.07.040 [arXiv:1806.09718]
23. E. Palti, The swampland: introduction and review. Fortsch. Phys. **67**, 1900037 (2019). https://doi.org/10.1002/prop.201900037 [arXiv:1903.06239]
24. N. Schöneberg, G. Franco Abellán, A. Pérez Sánchez, S.J. Witte, V. Poulin, J. Lesgourgues, The H_0 Olympics: a fair ranking of proposed models. Phys. Rept. **984**, 1 (2022). https://doi.org/10.1016/j.physrep.2022.07.001 [arXiv:2107.10291]
25. V. Poulin, T.L. Smith, A. Bartlett, Dark energy at early times and ACT data: a larger Hubble constant without late-time priors. Phys. Rev. D **104**, 123550 (2021). https://doi.org/10.1103/PhysRevD.104.123550 [arXiv:2109.06229]
26. B.S. Haridasu, H. Khoraminezhad, M. Viel, Scrutinizing early dark energy models through CMB lensing. arXiv:2212.09136
27. J.S. Cruz, S. Hannestad, E.B. Holm, F. Niedermann, M.S. Sloth, T. Tram, Profiling cold new early dark energy. arXiv:2302.07934
28. F. Niedermann, M.S. Sloth, New early dark energy is compatible with current LSS data. Phys. Rev. D **103**, 103537 (2021). https://doi.org/10.1103/PhysRevD.103.103537 [arXiv:2009.00006]
29. R. Murgia, G.F. Abellán, V. Poulin, Early dark energy resolution to the Hubble tension in light of weak lensing surveys and lensing anomalies. Phys. Rev. D **103**, 063502 (2021). https://doi.org/10.1103/PhysRevD.103.063502 [arXiv:2009.10733]
30. T.L. Smith, V. Poulin, J.L. Bernal, K.K. Boddy, M. Kamionkowski, R. Murgia, Early dark energy is not excluded by current large-scale structure data. Phys. Rev. D **103**, 123542 (2021). https://doi.org/10.1103/PhysRevD.103.123542 [arXiv:2009.10740]

31. N. Deruelle, V.F. Mukhanov, On matching conditions for cosmological perturbations. Phys. Rev. D **52**, 5549 (1995). https://doi.org/10.1103/PhysRevD.52.5549 [arXiv:gr-qc/9503050]
32. W. Israel, Singular hypersurfaces and thin shells in general relativity. Nuovo Cim. B **44S10**, 1 (1966). https://doi.org/10.1007/BF02710419
33. C.-P. Ma, E. Bertschinger, Cosmological perturbation theory in the synchronous and conformal Newtonian gauges. Astrophys. J. **455**, 7 (1995). https://doi.org/10.1086/176550 [arXiv:astro-ph/9506072]
34. W. Hu, Structure formation with generalized dark matter. Astrophys. J. **506**, 485 (1998). https://doi.org/10.1086/306274 [arXiv:astro-ph/9801234]
35. A.G. Riess et al., A comprehensive measurement of the local value of the Hubble constant with 1 km s^{-1} Mpc $^{-1}$ uncertainty from the Hubble space telescope and the SH_0ES team. Astrophys. J. Lett. **934**, L7 (2022). https://doi.org/10.3847/2041-8213/ac5c5b [arXiv:2112.04510]
36. S. Joudaki et al., KiDS+VIKING-450 and DES-Y1 combined: cosmology with cosmic shear. Astron. Astrophys. **638**, L1 (2020). https://doi.org/10.1051/0004-6361/201936154[arXiv:1906.09262]
37. D. Blas, J. Lesgourgues, T. Tram, The cosmic linear anisotropy solving system (CLASS) II: approximation schemes. JCAP **07**, 034 (2011). https://doi.org/10.1088/1475-7516/2011/07/034 [arXiv:1104.2933]
38. Planck collaboration, Planck 2018 results. VI. Cosmological parameters. Astron. Astrophys. **641**, A6 (2020). https://doi.org/10.1051/0004-6361/201833910 [arXiv:1807.06209]
39. F. Beutler, C. Blake, M. Colless, D.H. Jones, L. Staveley-Smith, L. Campbell et al., The 6dF galaxy survey: Baryon acoustic oscillations and the local Hubble constant. Mon. Not. R. Astron Soc. **416**, 3017 (2011). https://doi.org/10.1111/j.1365-2966.2011.19250.x
40. A.J. Ross, L. Samushia, C. Howlett, W.J. Percival, A. Burden, M. Manera, The clustering of the SDSS DR7 main galaxy sample—I. A 4 percent distance measure at $z = 0.15$. Mon. Not. Roy. Astron. Soc. **449**, 835 (2015). https://doi.org/10.1093/mnras/stv154 [arXiv:1409.3242]
41. BOSS collaboration, The clustering of galaxies in the completed SDSS-III Baryon oscillation spectroscopic survey: cosmological analysis of the DR12 galaxy sample. Mon. Not. Roy. Astron. Soc. **470**, 2617 (2017). https://doi.org/10.1093/mnras/stx721 [arXiv:1607.03155]
42. Pan-STARRS1 collaboration, The complete light-curve sample of spectroscopically confirmed SNe Ia from Pan-STARRS1 and cosmological constraints from the combined Pantheon sample. Astrophys. J. **859**, 101 (2018). https://doi.org/10.3847/1538-4357/aab9bb [arXiv:1710.00845]
43. M. Raveri, W. Hu, Concordance and discordance in cosmology. Phys. Rev. D **99**, 043506 (2019). https://doi.org/10.1103/PhysRevD.99.043506 [arXiv:1806.04649]
44. L. Herold, E.G.M. Ferreira, E. Komatsu, New constraint on early dark energy from Planck and BOSS data using the profile likelihood. Astrophys. J. Lett. **929**, L16 (2022). https://doi.org/10.3847/2041-8213/ac63a3 [arXiv:2112.12140]
45. K. Jedamzik, L. Pogosian, G.-B. Zhao, Why reducing the cosmic sound horizon alone can not fully resolve the Hubble tension. Commun. Phys. **4**, 123 (2021). https://doi.org/10.1038/s42005-021-00628-x [arXiv:2010.04158]
46. S. Vagnozzi, Consistency tests of Λ CDM from the early integrated Sachs-Wolfe effect: implications for early-time new physics and the Hubble tension. Phys. Rev. D **104**, 063524 (2021). https://doi.org/10.1103/PhysRevD.104.063524 [arXiv:2105.10425]
47. M.-X. Lin, G. Benevento, W. Hu, M. Raveri, Acoustic dark energy: potential conversion of the Hubble tension. Phys. Rev. D **100**, 063542 (2019). https://doi.org/10.1103/PhysRevD.100.063542 [arXiv:1905.12618]
48. J.S. Cruz, F. Niedermann, M.S. Sloth, A grounded perspective on new early dark energy using ACT, SPT, and BICEP/Keck. JCAP **02**, 041 (2023). https://doi.org/10.1088/1475-7516/2023/02/041 [arXiv:2209.02708]
49. K. Enqvist, M.S. Sloth, Adiabatic CMB perturbations in pre-big bang string cosmology. Nucl. Phys. B **626**, 395 (2002). https://doi.org/10.1016/S0550-3213(02)00043-3 [arXiv:hep-ph/0109214]
50. D.H. Lyth, D. Wands, Generating the curvature perturbation without an inflaton. Phys. Lett. B **524**, 5 (2002). https://doi.org/10.1016/S0370-2693(01)01366-1 [arXiv:hep-ph/0110002]

51. T. Moroi, T. Takahashi, Effects of cosmological moduli fields on cosmic microwave background. Phys. Lett. B **522**, 215 (2001). https://doi.org/10.1016/S0370-2693(01)01295-3 [arXiv:hep-ph/0110096]
52. K. Freese, M.W. Winkler, Chain early dark energy: a proposal for solving the Hubble tension and explaining today's dark energy. Phys. Rev. D **104**, 083533 (2021). https://doi.org/10.1103/PhysRevD.104.083533 [arXiv:2102.13655]
53. A.D. Linde, Eternal extended inflation and graceful exit from old inflation without Jordan-Brans-Dicke. Phys. Lett. B **249**, 18 (1990). https://doi.org/10.1016/0370-2693(90)90521-7
54. F.C. Adams, K. Freese, Double field inflation. Phys. Rev. D **43**, 353 (1991). https://doi.org/10.1103/PhysRevD.43.353 [arXiv:hep-ph/0504135]
55. E.J. Copeland, A.R. Liddle, D.H. Lyth, E.D. Stewart, D. Wands, False vacuum inflation with Einstein gravity. Phys. Rev. D **49**, 6410 (1994). https://doi.org/10.1103/PhysRevD.49.6410 [arXiv:astro-ph/9401011]
56. M. Cortês, A.R. Liddle, Viable inflationary models ending with a first-order phase transition. Phys. Rev. D **80**, 083524 (2009). https://doi.org/10.1103/PhysRevD.80.083524 [arXiv:0905.0289]
57. A.D. Linde, Decay of the false vacuum at finite temperature. Nucl. Phys. B **216**, 421 (1983). https://doi.org/10.1016/0550-3213(83)90072-X
58. S.R. Coleman, The fate of the false vacuum. 1. Semiclassical theory. Phys. Rev. D **15**, 2929 (1977). https://doi.org/10.1103/PhysRevD.16.1248
59. C.G. Callan, Jr., S.R. Coleman, The fate of the false vacuum. 2. First quantum corrections. Phys. Rev. D **16**, 1762 (1977). https://doi.org/10.1103/PhysRevD.16.1762
60. F.C. Adams, General solutions for tunneling of scalar fields with quartic potentials. Phys. Rev. D **48**, 2800 (1993). https://doi.org/10.1103/PhysRevD.48.2800 [arXiv:hep-ph/9302321]
61. B. Xue, P.J. Steinhardt, Evolution of curvature and anisotropy near a nonsingular bounce. Phys. Rev. D **84**, 083520 (2011). https://doi.org/10.1103/PhysRevD.84.083520 [arXiv:1106.1416]
62. B.D. Fields, K.A. Olive, T.-H. Yeh, C. Young, Big-Bang nucleosynthesis after Planck. JCAP **03**, 010 (2020). https://doi.org/10.1088/1475-7516/2020/03/010 [arXiv:1912.01132]
63. Planck collaboration, Planck 2018 results. VI. Cosmological parameters. Astron. Astrophys. **641**, A6 (2020). https://doi.org/10.1051/0004-6361/201833910 [arXiv:1807.06209]
64. M.A. Buen-Abad, G. Marques-Tavares, M. Schmaltz, Non-Abelian dark matter and dark radiation. Phys. Rev. D **92**, 023531 (2015). https://doi.org/10.1103/PhysRevD.92.023531 [arXiv:1505.03542]
65. P.B. Arnold, O. Espinosa, The effective potential and first order phase transitions: beyond leading-order. Phys. Rev. D **47**, 3546 (1993). https://doi.org/10.1103/PhysRevD.47.3546 [arXiv:hep-ph/9212235]
66. M. Dine, R.G. Leigh, P.Y. Huet, A.D. Linde, D.A. Linde, Towards the theory of the electroweak phase transition. Phys. Rev. D **46**, 550 (1992). https://doi.org/10.1103/PhysRevD.46.550 [arXiv:hep-ph/9203203]
67. S.R. Coleman, E.J. Weinberg, Radiative corrections as the origin of spontaneous symmetry breaking (1888). Phys. Rev. D **7** (1973). https://doi.org/10.1103/PhysRevD.7.1888
68. D.A. Kirzhnits, A.D. Linde, Symmetry behavior in Gauge theories. Annals Phys. **101**, 195 (1976). https://doi.org/10.1016/0003-4916(76)90279-7
69. J.S. Cruz, F. Niedermann, M.S. Sloth, NANOGrav meets hot new early dark energy and the origin of neutrino mass. arXiv:2307.03091
70. D. Aloni, A. Berlin, M. Joseph, M. Schmaltz, N. Weiner, A step in understanding the Hubble tension. Phys. Rev. D **105**, 123516 (2022). https://doi.org/10.1103/PhysRevD.105.123516 [arXiv:2111.00014]
71. N. Schöneberg, G. Franco Abellán, A step in the right direction? Analyzing the Wess Zumino dark radiation solution to the Hubble tension. JCAP **12**, 001 (2022). https://doi.org/10.1088/1475-7516/2022/12/001 [arXiv:2206.11276]
72. N. Schöneberg, G. Franco Abellán, T. Simon, A. Bartlett, Y. Patel, T.L. Smith, The weak, the strong and the ugly—a comparative analysis of interacting stepped dark radiation. arXiv:2306.12469

73. M.A. Buen-Abad, Z. Chacko, C. Kilic, G. Marques-Tavares, T. Youn, Stepped partially acoustic dark matter: likelihood analysis and cosmological tensions. arXiv:2306.01844
74. I.J. Allali, F. Rompineve, M.P. Hertzberg, Dark sectors with mass thresholds face cosmological datasets. arXiv:2305.14166
75. A. Berlin, N. Blinov, S.W. Li, Dark sector equilibration during nucleosynthesis. Phys. Rev. D **100**, 015038 (2019). https://doi.org/10.1103/PhysRevD.100.015038 [arXiv:1904.04256]
76. D. Aloni, M. Joseph, M. Schmaltz, N. Weiner, Dark radiation from neutrino mixing after big bang nucleosynthesis. arXiv:2301.10792
77. R.N. Mohapatra, J.W.F. Valle, Neutrino mass and Baryon number nonconservation in superstring models. Phys. Rev. D **34**, 1642 (1986). https://doi.org/10.1103/PhysRevD.34.1642
78. M.C. Gonzalez-Garcia, J.W.F. Valle, Fast decaying neutrinos and observable flavor violation in a new class of Majoron models. Phys. Lett. B **216**, 360 (1989). https://doi.org/10.1016/0370-2693(89)91131-3
79. F. Deppisch, J.W.F. Valle, Enhanced lepton flavor violation in the supersymmetric inverse seesaw model. Phys. Rev. D **72**, 036001 (2005). https://doi.org/10.1103/PhysRevD.72.036001 [arXiv:hep-ph/0406040]
80. A. Abada, M. Lucente, Looking for the minimal inverse seesaw realisation. Nucl. Phys. B **885**, 651 (2014). https://doi.org/10.1016/j.nuclphysb.2014.06.003 [arXiv:1401.1507]
81. G. 't Hooft, Naturalness, chiral symmetry, and spontaneous chiral symmetry breaking. NATO Sci. Ser. B **59**, 135 (1980). https://doi.org/10.1007/978-1-4684-7571-5_9
82. M. D'Onofrio, K. Rummukainen, Standard model cross-over on the lattice. Phys. Rev. D **93**, 025003 (2016). https://doi.org/10.1103/PhysRevD.93.025003 [arXiv:1508.07161]
83. M. Archidiacono, S. Hannestad, Updated constraints on non-standard neutrino interactions from Planck. JCAP **07**, 046 (2014). https://doi.org/10.1088/1475-7516/2014/07/046 [arXiv:1311.3873]
84. NANOGrav collaboration, The NANOGrav 15-year data set: search for signals from new physics. Astrophys. J. Lett. **951** (2023). https://doi.org/10.3847/2041-8213/acdc91 [arXiv:2306.16219]
85. M. Garny, M. Sandora, M.S. Sloth, Planckian interacting massive particles as dark matter. Phys. Rev. Lett. **116**, 101302 (2016). https://doi.org/10.1103/PhysRevLett.116.101302 [arXiv:1511.03278]
86. M. Garny, A. Palessandro, M. Sandora, M.S. Sloth, Theory and phenomenology of Planckian interacting massive particles as dark matter. JCAP **02**, 027 (2018). https://doi.org/10.1088/1475-7516/2018/02/027 [arXiv:1709.09688]
87. ACT collaboration, The Atacama cosmology telescope: DR4 maps and cosmological parameters. J. Cosmol. Astropart. Phys. **12**, 047 (2020). https://doi.org/10.1088/1475-7516/2020/12/047 [arXiv:2007.07288]
88. K. Abazajian et al., CMB-S4 science case, reference design, and project plan. arXiv:1907.04473
89. J.P. Gardner et al., The James Webb space telescope. Space Sci. Rev. **123**, 485 (2006). https://doi.org/10.1007/s11214-006-8315-7 [arXiv:astro-ph/0606175]
90. L. Amendola et al., Cosmology and fundamental physics with the Euclid satellite. Living Rev. Rel. **21**, 2 (2018). https://doi.org/10.1007/s41114-017-0010-3 [arXiv:1606.00180]
91. T.L. Smith, M. Lucca, V. Poulin, G.F. Abellan, L. Balkenhol, K. Benabed et al., Hints of early dark energy in Planck, SPT, and ACT data: New physics or systematics?. Phys. Rev. D **106**, 043526 (2022). https://doi.org/10.1103/PhysRevD.106.043526 [arXiv:2202.09379]

Chapter 24
On the Dark Radiation Role in the Hubble Constant Tension

Stefano Gariazzo and Olga Mena

24.1 Introduction

Dark radiation in our universe provides a unique window to new mechanisms able to solve many open questions in particle physics and cosmology nowadays. The particle content of the minimal Standard Model (SM) scenario may be enlarged with additional species that feebly couple to ordinary matter, i.e. dark radiation. The putative existence of these species is strongly motivated, both from the pure theoretical and phenomenological/experimental perspectives. The explanation of neutrino masses within the SM and the short-baseline neutrino oscillation anomalies may need the existence of additional sterile neutrino species. The so-called strong CP problem may imply the existence of axions in our universe, which could contribute significantly to the (observationally required) dark matter component. Last, but not least, Majoron models provide very appealing scenarios where lepton number may be spontaneously broken and rare events such as neutrinoless double beta decay are possible.

Currently, there are several open anomalies that can not be fully understood in the minimal cosmological constant plus cold dark matter (ΛCDM) scenario. The most significant, 5σ disagreement is related to the value of the Hubble constant H_0 extracted from model-independent, direct measurements of local distances and redshifts in the nearby Universe ($H_0 = 73.04 \pm 1.04$ km s^{-1} Mpc^{-1}) [1] and the one indirectly inferred from model-dependent Cosmic Microwave Background (CMB) observations ($H_0 = 67.4 \pm 0.5$ km s^{-1} Mpc^{-1}) [2], see also [1, 3–6]. The goal of this manuscript is to analyze the role of dark radiation models in the Hubble constant tension. We start in Sect. 24.2 with a brief introduction which summarizes the main properties of the dark radiation component, such as its thermal abundance and its impact on the cosmological observables, focusing mostly on the CMB. Sect. 24.3 describes the most simple models involving dark radiation species ad their role in

S. Gariazzo (✉)
Istituto Nazionale di Fisica Nucleare (INFN), Sezione di Torino, Via P. Giuria 1, 10125 Turin, Italy
e-mail: gariazzo@to.infn.it

O. Mena
Instituto de Física Corpuscular (CSIC-Universitat de València), 46980 Paterna, Spain
e-mail: omena@ific.uv.es

E. Di Valentino and D. Brout (eds.), *The Hubble Constant Tension*, Springer Series in Astrophysics and Cosmology, https://doi.org/10.1007/978-981-99-0177-7_24

the H_0 conundrum. Sect. 24.4 presents more sophisticated and refined dark radiation scenarios, which may provide a resolution of the H_0 issue. We conclude in Sect. 24.5.

24.2 Dark Radiation in the Early Universe

One of the most important phases of the Universe's evolution is radiation domination. During this epoch, the largest contribution to the energy density of the Universe comes from relativistic particles, i.e. mainly photons, neutrinos and electrons by the time the temperature of the fluid is below 100 MeV. During radiation domination, several interesting processes take place: (i) neutrinos decouple when the temperature of the plasma is approximately 2 MeV; (ii) electrons become non-relativistic shortly after and transfer their energy density to photons while annihilating away, and (iii) light nuclei are produced in the Big-Bang Nucleosynthesis (BBN) process until the photon temperature drops below $\sim$0.05 MeV. Shortly after the end of radiation domination (at $T \sim 1$ eV), the last scattering of photons occurs ($T \sim 0.3$ eV) and the CMB radiation is produced. The expansion speed at all these times, proportional to the square root of the radiation energy density, is crucial because it affects the observables we measure today.

After electron-positron pairs annihilate into photons, neutrinos and photons are the only SM particles behaving as radiation. Other relativistic particles may exist, as we will discuss in the following. Such non-standard particles are normally grouped under the name "dark radiation", since they do not take part into electroweak interactions. The amount of the dark radiation energy density, ρ_R, can be conveniently parameterized by means of the effective number of relativistic neutrino-like species, $N_{\rm eff}$, by

$$\rho_R = \rho_\gamma \left(1 + \frac{7}{8}\left(\frac{4}{11}\right)^{4/3} N_{\rm eff}\right), \qquad (24.1)$$

where ρ_γ represents the photon energy density. If only three standard neutrinos which underwent instantaneous decoupling exist, $N_{\rm eff}$ would be equal to 3. In case the neutrino energy density is different from the instantaneous decoupling one, $N_{\rm eff}$ may deviate from 3 even in presence of only three standard neutrino families. The presence of additional contributions to the radiation energy density, moreover, would correspond to an increased value of $N_{\rm eff}$. In other words, $N_{\rm eff} \neq 3$ might originate either because of new degrees of freedom which have nothing to do with standard neutrinos, or from a non-standard momentum distribution for the three neutrinos.

Our theoretical knowledge of their decoupling, however, states that standard neutrinos did not undergo instantaneous decoupling and that their momentum distribution function slightly deviates from a pure Fermi-Dirac one. When taking into account the momentum dependence of neutrino-electron interactions and the evolution of the thermal plasma around electron-positron annihilation, the standard neutrino energy density is computed to be $N_{\rm eff}^{\rm std} = 3.044$ [7–9], see also [10]. This number was previously claimed to be a bit higher [11, 12], but state-of-the-art calculations confirm that the (theoretical and numerical) error on this value is at the level of 10^{-4}. The presence of new physics related to the three standard neutrinos, such as non-standard interactions (NSI, see e.g. [13]) between neutrinos and electrons [14, 15], or a non-unitary neutrino mixing matrix [16] can lead to deviations from $N_{\rm eff}^{\rm std}$. These effects are rather small and do not significantly impact the H_0 value, although they may be tested by next-generation CMB measurements [17, 18]. As we shall see, the situation changes significantly if we consider the presence of additional particles (axions, sterile neutrinos, and so on) or more complicated neutrino interactions, for example with dark matter. In such cases, it is common to have much larger contributions to $N_{\rm eff}$, usually denoted as $\Delta N_{\rm eff} \equiv N_{\rm eff} - N_{\rm eff}^{\rm std}$. An exhaustive description of the effect of $\Delta N_{\rm eff}$ on the different cosmological observables can be found in Ref. [19]. Current limits on $\Delta N_{\rm eff}$ arise primarily from observables at two epochs: (i) at the BBN period, and, (ii) at the Recombination epoch. Here we briefly summarize the main impact of this parameter on both BBN light element abundances and on the CMB power spectrum, as the impact on the matter power spectrum is subdominant.

BBN refers to the formation of the first light nuclei (heavier than the lightest isotope of hydrogen) in the very first three minutes of our universe's lifetime. BBN abundances are computed by employing a solid understanding of the nuclear interactions involved in the production of elements, providing a natural laboratory where to test extensions of the SM of elementary particles that involve additional relativistic species $\Delta N_{\rm eff}$. Any extra contribution to the dark radiation of our universe will increase the expansion rate $H(z)$ and will shift towards higher temperatures the freeze-out epoch of the weak interactions, implying a higher neutron-to-proton ratio and therefore a larger fraction of primordial Helium and Deuterium (as well as a higher fraction of other primordial elements) with respect to hydrogen. This makes BBN a powerful tool for constraining the total amount of relativistic species and other beyond-the-SM physics frameworks: given a concrete model, we can solve numerically the set of differential equations that regulate the nuclear interactions in the primordial plasma, see e.g. [20–22], compute the light element abundances and compare the results to the values inferred by astrophysical and cosmological observations. Given current uncertainties, the standard BBN predictions show a good agreement with direct measurements of primordial abundances [23–26] limiting $\Delta N_{\rm eff} \lesssim 0.3 - 0.4$ at 95% CL. Notice also that the BBN predictions for the Helium abundance can impact the CMB angular spectra because they can be used to estimate the baryon energy density through a simple formula [27]:

$$\Omega_b h^2 = \frac{1 - 0.007125\, Y_p^{\rm BBN}}{273.279} \left(\frac{T_{\rm CMB}}{2.7255\ {\rm K}} \right)^3 \eta_{10}\,. \tag{24.2}$$

Here, $\eta_{10} \equiv 10^{10} n_b / n_\gamma$ is the photon-baryon ratio today, $T_{\rm CMB}$ is the CMB temperature at the present time and $Y_p^{\rm BBN} \equiv 4 n_{\rm He}/n_b$ is the Helium *nucleon fraction*, defined as the ratio of the 4-Helium number density to the total baryon one.

Concerning the CMB temperature power spectrum, first of all, varying $N_{\rm eff}$ changes the redshift of the matter radiation equivalence, $z_{\rm eq}$, inducing an enhancement of the early Integrated Sachs Wolfe (ISW) effect which increases the CMB spectrum around the first acoustic peak. Nevertheless, this is a sub-dominant effect. The authors of Ref. [28] carefully explained that the most relevant impact of changing $N_{\rm eff}$ is located at high multipoles, i.e. at the damping tail. If $\Delta N_{\rm eff}$ increases, the Hubble parameter H during radiation domination will increase as well. The delay in matter radiation equality will also modify the sound speed and the comoving sound horizon, proportional to $1/H$, which reads as:

$$r_{\rm s} = \int_0^{\tau'} d\tau c_{\rm s}(\tau) = \int_0^a \frac{da}{a^2 H} c_{\rm s}(a) \, ,$$

and it is proportional to the inverse of the expansion rate $r_{\rm s} \propto 1/H$. The consequence is a reduction in the angular scale of the acoustic peaks $\theta_{\rm s} = r_{\rm s}/D_{\rm A}$, where $D_{\rm A}$ is the angular diameter distance, causing a horizontal shift of the peak positions towards higher multipoles. Furthermore, a vertical shift also affects the amplitude of the peaks at high multipoles, where the ISW effect is negligible, and it is related to the so-called Silk damping. This results from the fact that the baryon-photon decoupling is not an instantaneous process, leading to a diffusion damping of oscillations in the plasma. If decoupling starts at $\tau_{\rm d}$ and ends at $\tau_{\rm ls}$, during $\Delta\tau$ the radiation free streams on scale $\lambda_{\rm d} = (\lambda\Delta\tau)^{1/2}$ where λ is the photon mean free path and $\lambda_{\rm d}$ is shorter than the thickness of the last scattering surface. As a consequence, temperature fluctuations on scales smaller than λ_d are damped, because on such scales photons can spread freely both from overdensities and underdensities. The overall result is that the damping angular scale $\theta_{\rm d} = r_{\rm d}/D_{\rm A}$ is proportional to the square root of the expansion rate $\theta_{\rm d} \propto \sqrt{H}$ and consequently it increases with $\Delta N_{\rm eff}$ which therefore induces a suppression of the peaks and a smearing of the oscillations that intensifies at the CMB damping tail.

The three aforementioned effects caused by a non-zero $\Delta N_{\rm eff}$ (redshift of equivalence, the size of the sound horizon at recombination, and the damping tail suppression) can be easily compensated by varying other cosmological parameters, including the Hubble constant H_0. Notice that the horizontal shift towards smaller angular scales caused by an increased value of $N_{\rm eff}$ can be compensated by decreasing $D_{\rm A}$, which can be automatically satisfied by increasing H_0. The effect of $N_{\rm eff}$ on the damping tail is however more difficult to mimic via H_0, as it is mostly degenerated with the helium fraction which enters directly in $r_{\rm d}$, i.e. the mean square diffusion distance at recombination via n_e, the number density of free electrons.

There is however one effect induced by $N_{\rm eff}$ which cannot be mimicked by other cosmological parameters: the neutrino anisotropic stress [29, 30], related to the fact that neutrinos are free-streaming particles propagating at the speed of light, faster than the sound speed in the photon fluid, suppressing the oscillation amplitude of CMB modes that entered the horizon in the radiation epoch. The effect on the CMB power spectrum is therefore located at scales that cross the horizon before the matter-radiation equivalence by an increase in power of $5/(1+\frac{4}{15}f_\nu)$ [31], where f_ν is the fraction of radiation density contributed by free-streaming particles.

All in all, our current knowledge confirms that $N_{\rm eff}$ is close to 3 as measured by CMB observations ($N_{\rm eff}=2.99^{+0.34}_{-0.33}$ at 95% confidence level (CL) [2]) or BBN abundances (e.g. $N_{\rm eff}=2.87^{+0.24}_{-0.21}$ at 68% CL [21]) independently.

24.3 *Tree-level* Dark Radiation Solutions to the H_0 Tension

Let us now analyse the simplest dark radiation candidate that can be used to extend the standard three-neutrino picture and discuss their effect on the H_0 tension.

24.3.1 Sterile Neutrinos

One simple dark radiation candidate is the sterile neutrino, intended as a right-handed fermion which cannot take part into SM interactions but that participates in neutrino oscillations. In order to be considered as dark radiation, the sterile neutrino must be relativistic in the early universe, indicatively at the time of BBN or CMB epochs. This excludes sterile neutrinos with masses much larger than the MeV scale, which are usually proposed in order to build a seesaw model and explain the smallness of active neutrino masses, see e.g. [32–34]. Sterile neutrinos at the keV scale can play a significant role in the universe as warm dark matter [35, 36], but since they become non-relativistic during BBN, they are not considered as dark radiation components. When considering masses at the eV scale, sterile neutrinos would become non-relativistic approximately at the time of matter-radiation equality, thus being relativistic during the entire radiation-dominated era, although already non-relativistic at CMB decoupling ($T\sim 0.3$ eV). Sterile neutrinos corresponding to much smaller mass splittings with respect to active neutrinos (mass splittings smaller than the solar or atmospheric ones) are also considered in quasi-Dirac or pseudo-Dirac scenarios, see e.g. [37, 38]. In this case, they may remain relativistic until today, depending on the mass of the lightest neutrino. In the following, however, we will focus primarily on the eV-scale sterile neutrino.

Sterile neutrinos at the eV scale have been introduced to give a possible neutrino-oscillation explanation to the so-called Short-Baseline (SBL) anomalies [39–42], which originally included the excess appearance of electron antineutrinos measured by LSND [43], later confirmed also by MiniBooNE [44, 45], and the anomalous disappearance of electron neutrinos at Gallium experiments [46, 47] and of electron antineutrinos from nuclear reactors [48]. The original anomalies have been studied at different experimental probes and at the theoretical level for more than ten years. Here we will briefly summarize the present status.

Concerning the appearance channel, the LSND results have been tested by several experiments apart from MiniBooNE. None of these experiments has observed an anomalous signal, including KARMEN [49] and OPERA [50], which therefore put only upper bounds on active-sterile mixing angles. In more recent years, the MicroBooNE experiment is testing the MiniBooNE anomaly, and the first results seem to indicate that the excess of events is not compatible with active-sterile neutrino oscillations [51, 52].

In the disappearance channel, we can separate the electron and muon (anti)neutrino channels. In the former case, the Gallium and reactor anomalies have been studied both experimentally and theoretically. Recent revisitations of the Gallium cross sections appear to reduce the significance of the Gallium anomaly with respect to the original calculations [53, 54], but the BEST experiment observed a disappearance with stronger significance [55, 56] with respect to previous GALLEX and SAGE results, see also [57]. If neutrino oscillations are responsible for the BEST results, part of the preferred region is in tension with SBL reactor antineutrino experiments [58–61]. On the other hand, although most of the reactor experiments do not confirm the original reactor anomaly [58–63], the Neutrino-4 experiment claims a best-fit at a high mass splitting and mixing angle [64], whose significance may be overestimated due to the treatment of the χ^2 [65]. Neutrino-4 oscillation results are perfectly compatible with BEST constraints, which however have a much broader allowed range for the new mass splitting. Recent theoretical calculations of the reactor antineutrino flux [66–68] differ slightly from the two that produced the original reactor anomaly [69, 70], which is therefore significantly reduced or even disappeared if the most updated flux estimations are considered [68, 71].

Finally, concerning muon (anti)neutrino disappearance, most of the experiments, for example MINOS/MINOS+ [72], only put upper limits on the mixing parameters because no SBL oscillations are observed. Only in recent years, IceCube reported a mild preference in favor of active-sterile neutrino oscillations by analysing 8 years of muon neutrino events [73], with a broad uncertainty on the preferred mass splitting and mixing angles.

Assuming that all the above-mentioned anomalies are generated by active-sterile neutrino mixing, when one attempts to combine all these appearance and disappearance results in order to produce a global fit of active-sterile oscillation parameters, it seems impossible to reconcile the strong upper bounds from disappearance experiments with the observations at appearance probes. The tension was already strong a few years ago [40, 74, 75], at the point that a combination of appearance and disappearance constraints has no statistical meaning. In summary, the sterile neutrino searches from different probes seem unable to pin out a single preferred region for the active-sterile oscillation parameters, and possibly a consistent explanation of the anomalies requires different kinds of new physics. In the following, therefore, we will consider the potential impact of eV-scale sterile neutrinos on the cosmological environment and ignore the problems in terrestrial searches of sterile neutrinos.

The thermalization of sterile neutrinos in the early universe has been studied for a long time, see e.g. [76–82], or [83] for a review of early studies. The most important result in these early calculations is to show that the production of sterile neutrinos occurs mostly in a non-resonant way through neutrino oscillations. Reference [81] provides a rather simple formula that determines the relation between active sterile mixing parameters and the approximated value of $\Delta N_{\rm eff}$ they generate. In more recent years, precise numerical calculations have been developed in order to determine the contribution of eV-scale sterile neutrinos to $\Delta N_{\rm eff}$. Calculations have been first implemented in a "one active plus one sterile neutrino" scenario but keeping the full momentum dependence, see e.g. [84–86], and then extended to include multiple flavors (two active and one sterile neutrino) [87–89]. Finally, more accurate calculations at the numerical level have been proposed in recent years [90, 91].

The amount of sterile neutrinos produced through oscillations in the early universe depends significantly on the values of the active-sterile mixing parameters. Because the dense plasma blocks neutrino oscillations at very early times by maintaining neutrinos into a flavor state, oscillations cannot take place until rather late. Since this effect depends on the oscillation frequency, higher mass splittings correspond to an earlier start of oscillations, that in turn means that there is more time to produce sterile neutrinos. Lower mass splittings, instead, generate oscillations that are blocked for a longer time, and are less efficient in producing sterile states. On the other hand, a large mixing angle allows to have a faster conversion between active and sterile flavors. Finally, it has been noticed that the new active-sterile mixing angles are equivalent at the time of producing neutrino oscillations in the early universe. More specifically, for a $\Delta m^2_{41} \sim 1$ eV, it is sufficient to have any of the mixing angles larger than $\sim 10^{-3}$ to generate a sterile state in full thermal equilibrium with the active neutrinos, or $\Delta N_{\rm eff} \simeq 1$ [90].

This is of course in tension with the constraints we obtain from CMB and BBN observables. Cosmological analyses normally parameterize the presence of sterile neutrinos by means of its contribution $\Delta N_{\rm eff}$ to the effective number of neutrinos when relativistic, and its contribution to the matter energy density, proportional to the effective mass $m_{\rm eff,sterile}$, when it becomes non-relativistic. In case of non-resonant production of the sterile state [92], the relation $m_{\rm eff,sterile} = m_s \Delta N_{\rm eff}$ applies, where m_s is the sterile neutrino physical mass. From the sterile neutrino search, Planck obtains upper limits $N_{\rm eff} < 3.30$ and $m_{\rm eff,sterile} < 0.652$ eV at 95% CL (TTTEEE+lensing+BAO) [2]. Notice that in this case, the analysis also yields $H_0 = 67.8^{+1.3}_{-1.2}$ km s^{-1} Mpc^{-1} at 95% CL, meaning that the Hubble tension is not alleviated. The constraints are particularly strong thanks to the combined effect of $N_{\rm eff}$ and H_0 on both temperature and polarization. If one only considers the TT+lensing+BAO dataset, Planck reports $N_{\rm eff} < 3.52$, $m_{\rm eff,sterile} < 0.551$ eV and $H_0 = 68.2^{+2.1}_{-1.7}$ km s^{-1} Mpc^{-1}, which is still in tension with a fully thermalized fourth neutrino, but with a slightly reduced significance. Notice that the bounds on $N_{\rm eff}$ being smaller than 4 as reported by Planck are also confirmed by ACT and SPT, which respectively yield $N_{\rm eff} < 3.38$, $m_{\rm eff,sterile} < 0.561$ eV and $N_{\rm eff} < 3.86$, $m_{\rm eff,sterile} < 0.232$ eV when analysed in combination with WMAP 9-year data and other low-redshift probes [93]. Notice that in the SPT case, the bound on $m_{\rm eff,sterile}$ is much stronger, while the bound on $N_{\rm eff}$ is relaxed. This arises from the fact that heavier states are allowed only if their contribution to $N_{\rm eff}$ is rather small (otherwise their contribution to the non-relativistic energy density, proportional to $m_{\rm eff,sterile}$, would be too high), while for lighter states a larger value of $N_{\rm eff}$ is allowed because these particles mostly act as radiation. Even in the SPT case, however, these results show that none of the available cosmological data sets currently allows for the presence of a fully thermalized sterile neutrino, and the allowed parameter ranges do not permit a full solution of the H_0 tension.

24.3.2 Neutrino Asymmetries

In the standard big-bang model, the neutrino asymmetry $\eta_{\nu_\alpha} \equiv (n_{\nu_\alpha} - n_{\bar{\nu}_\alpha})/n_\gamma$ is assumed to be negligibly small, comparable to the tiny-but-crucial baryon asymmetry. That is, while standard baryogenesis models involving sphalerons suggest that η_{ν_α} should be of the order of the baryon asymmetry $\eta_B = n_B/n_\gamma \simeq 6.1 \times 10^{-10}$ [94, 95], other proposed models [96–98] manage to combine a large lepton asymmetry with the value of η_B. The initial asymmetry in a given neutrino flavour α, defined as the difference between the neutrino and antineutrino comoving densities, is related to the neutrino chemical potential μ_α. It is possible to compute the constraints on this parameter, or, as it is usually done in the literature, on the dimensionless degeneracy parameter $\xi_\alpha \equiv \mu_\alpha/T_\nu$. A non-vanishing value of the electron neutrino chemical potential ξ_e affects the neutron/proton freeze-out, modifying BBN predictions [99]. Notice that a priori, the individual neutrino flavour chemical potentials can have different values. Depending on the neutrino mixing angles as well as on the initial values of the neutrino chemical potentials there could be or not an equilibration of the different ξ_α [100].[1] A significant lepton asymmetry in the early universe, represented by a nonvanishing chemical potential ξ_{α_i}, for thermal distributions of two light mass eigenstates ν_i, will contribute to the dark radiation of the universe as [100]

$$\Delta N_{\rm eff} = \sum_{i=1,2} \frac{15}{7}\left(\frac{\xi}{\pi}\right)^2 \left(2 + \left(\frac{xi}{\pi}\right)^2\right) . \tag{24.3}$$

The authors of Ref. [100] show that CMB data alone lead to a 95% CL limit of $\xi < 0.77$, associated to $\Delta N_{\rm eff} = 0.3$ and $H_0 = 67.71 \pm 0.95$ km s^{-1} Mpc^{-1}, and therefore not alleviating the H_0 tension. However, a combination with Supernovae I data provides weak evidence for a non negligible value of ξ, together with a 95% CL lower bound for the Hubble parameter of $H_0 > 69.8$ km s^{-1} Mpc^{-1}, larger than the value without Supernovae observations and in a better agreement with local measurements of the Hubble constant.

[1] When assuming a unique common degeneracy parameter ξ for the three neutrino degeneracy parameters due to neutrino oscillations [101–103], one can derive bounds from BBN observations [94, 104], from CMB data [2, 105], or from a combination of the former two measurements, providing $\xi = 0.001 \pm 0.016$ [106, 107].

24.3.3 Axions

Axions are the dynamical pseudo-scalar fields providing one of the most compelling solutions to the strong CP problem [108, 109]. In addition, axions also may be considered as a natural candidate for the dark matter component in our universe. Apart from non-thermal mechanisms, axions can be copiously produced in the early Universe also via scattering and decays of particles belonging to the primordial bath [110–118], contributing to the radiation energy density $N_{\rm eff}$ in the same way as massive neutrinos do when they are still relativistic. Eventually, axions become non-relativistic and provide a hot and sub-dominant dark matter component. The presence of such a cosmic axion background can leave distinct and detectable imprints, and current cosmological data put bounds on the axion mass and interactions [119–132]. Axion couplings are proportional to the mass of the axion itself, and thermal production channels are efficient only if the axion mass is large enough. On the contrary, the cold axion dark matter density is a decreasing function of the mass. Therefore, a significant thermal population of axions would be possible only if cold axions provide a sub-dominant component to the observed cosmic dark matter abundance. In terms of f_a, known as the axion decay constant, the axion mass reads as [133–135]

$$m_a \simeq 5.7\,\mu\text{eV}\left(\frac{10^{12}\,\text{GeV}}{f_a}\right) . \tag{24.4}$$

The landscape of axion models is broad [136, 137]. Nevertheless, it is possible to divide them into two main classes according to the origin of the color anomaly: the Kim-Shifman-Vainshtein-Zakharov (KSVZ) [138, 139] and Dine-Fischler-Srednicki-Zhitnitsky (DFSZ) [140, 141] frameworks, leading to the very same axion mass, see Eq. (24.4). However, the interactions with electroweak gauge bosons and SM fermions are different. Scatterings are the only thermal production channel available for the minimal KSVZ and DFSZ frameworks due to the flavor-conserving axion interactions. If flavor-violating couplings are allowed then decays also contribute [118]. Initially, axions may belong to the thermal bath if during or after inflationary reheating there are efficient mechanisms to produce them. Even if they are not present at the onset of the radiation domination epoch, they are produced by thermal scatterings and their interaction strength could bring them to equilibrium. A quick method to estimate the axion relic density is via the *instantaneous decoupling approximation*, assuming that axions at some point reach thermal equilibrium, and that they suddenly decouple when the bath had a temperature T_D identified by the relation

$$\Gamma_a(T_D) = H(T_D) . \tag{24.5}$$

The asymptotic axion comoving density results in

$$Y_a^\infty = Y_a(T \lesssim T_D) = \frac{45\zeta(3)}{2\pi^4 g_{\star s}(T_D)} , \tag{24.6}$$

with $\zeta(3) \simeq 1.2$ the Riemann zeta function. The asymptotic comoving density depends on the decoupling temperature only through the factor $g_{\star s}(T_D)$ in the denominator. Axions decoupling at later times are a more significant fraction of the thermal bath and therefore their comoving density is larger. Notice that the criterion in Eq. (24.5) leads to a convenient estimate of the decoupling epoch but it is far from being rigorous. As it is manifest from Eq. (24.6), the final axion abundance is sensitive to the detailed value of the decoupling temperature only if $g_{\star s}$ is changing around the decoupling time, see Ref. [130], which provides a robust computation of the axion relic density calculation in terms of the Boltzmann equation, in order to find solid cosmological bounds on the axion mass, finding that the actual axion relic density is substantially larger than previously estimated. Once the relic density is computed, it is possible to evaluate the axion contribution to the effective number of additional neutrino species via the relation $\Delta N_{\rm eff} \simeq 75.6\,\left(Y_a^{\infty}\right)^{4/3}$, assuming that the phase-space distribution is thermal. Therefore, as long as thermal axions remain relativistic particles ($T_a \gg m_a$), they behave as radiation in the early Universe and their cosmological effects are those produced via their contribution to the effective number of neutrino species $N_{\rm eff}$. By means of BBN light element abundances data, Ref. [130] finds for the KSVZ axion $\Delta N_{\rm eff} < 0.33$ and an axion mass bound $m_a < 0.53$ eV (i.e., a bound on the axion decay constant $f_a > 1.07 \times 10^7$ GeV), both at 95% CL. These BBN bounds are improved to $\Delta N_{\rm eff} < 0.14$ and $m_a < 0.16$ eV ($f_a > 3.56 \times 10^7$ GeV) if a prior on the baryon energy density from CMB data is assumed. When cosmological observations from the CMB temperature, polarization and lensing from the Planck satellite are combined with large scale structure data limits of $\Delta N_{\rm eff} < 0.23$ and $m_a < 0.28$ eV are found, both at 95% CL. Very similar results are quoted for the DFSZ axion. The authors of Ref. [114] showed that a model in which hot axions are produced from the coupling with muons leads to an alleviation of the Hubble tension at 3σ.

24.4 *Higher Order* Dark Radiation Corrections to the H_0 Tension

In the following, we shall present proposed scenarios which imply extra interactions beyond the SM paradigm, involving also in some cases exotic particles, such as sterile neutrino species, the dark matter component, or Majorons.

24.4.1 Interacting Scenarios

Both neutrinos and dark matter provide evidence for physics beyond the SM of elementary particles and possible interactions among them can induce important changes in the canonical evolution within the minimal ΛCDM model, see Refs. [142–150]. In particular, in these models, the value of $N_{\rm eff}$ will generically be increased at photon decoupling while will remain unchanged during the BBN period. Since the effective number of relativistic degrees of freedom is proportional to the ratio of the neutrino and photon temperatures, if the photon temperature T_γ is reduced, or, alternatively, the neutrino temperature is increased, $N_{\rm eff}$ will be larger than in the canonical scenario. If there exists a dark matter particle that remains in thermal equilibrium with neutrinos after their weak decoupling processes, these interactions will effectively reheat the neutrino sector with respect to the electromagnetic plasma. This neutrino reheating will take place after standard neutrino decoupling once the dark matter particles are no longer relativistic, implying they should have a mass of tens of MeV [151, 152]. It has been explored if such a scenario could alleviate the H_0 tension [148, 153]. Including Planck polarization data, however, dilutes such a possibility, although this very same scenario is able to significantly relax the lower bounds on the value of the clustering parameter σ_8 inferred in the context of ΛCDM from the Planck data, leading to agreement within $1-2\sigma$ with weak lensing estimates of the σ_8 parameter.

24.4.2 Self-interacting Active Neutrinos

Free-streaming neutrinos travel supersonically through the photon-baryon plasma at early times, inducing a net phase shift in the CMB power spectra towards larger scales (smaller multipoles), leading to a physical size of the photon sound horizon at last scattering that is slightly larger [29, 154–157]. Self-interacting neutrinos, see e.g. [158, 159] shift the power spectrum towards smaller scales and boost their fluctuation amplitude, reducing the physical size of photon sound horizon at last scattering: a smaller value of the angular diameter distance would be required, implying a higher value of H_0 [160]. If the mass of the mediator is heavy, the process can be expressed in terms of a four-fermion interaction $G_{\rm eff}\nu\bar{\nu}\nu\bar{\nu}$ [160], with $G_{\rm eff} \gg G_{\rm F}$, the former governing weak-interaction processes. Neutrinos experience self-scatterings after weak decoupling, and increasing $G_{\rm eff}$ delays neutrino free-streaming. The authors of Ref. [160] found a strongly interacting mode with $G_{\rm eff} = 2.5^{+0.8}_{-0.5} \times 10^4\ \mathrm{GeV}^{-2}$ and associated it to a much larger value of H_0 than in the standard ΛCDM model, $H_0 = 72.3 \pm 1.4$ km s^{-1} Mpc^{-1}. This very same strongly interacting neutrino cosmology prefers $N_{\rm eff} = 4.02 \pm 0.29$. Despite being a very interesting result, the former analysis was restricted to Planck observations of temperature power spectrum data: the authors of Ref. [161] realized that when high polarisation data from the Planck 2018 release are included in the fit, the strongly interacting neutrino mode was disfavoured, even if it could not be completely excluded. As a consequence, strong neutrino self-interactions do not lead to a high value of the Hubble constant and therefore such a model is not a viable solution to the current H_0 discrepancy when considering the full Planck 2018 data, see also Refs. [159, 162–168].

Among the classes of theoretical neutrino model building, the self-interacting scenario discussed here includes an important class of models related to the explanation of the smallness of neutrino masses. Why are the neutrino masses within the SM much smaller than those of the charged fermions in the very same family? One of the most elegant and complete benchmarks is the so-called seesaw mechanism [32–34], in which additional right handed neutrinos are added via a Majorana mass term, naturally suppressing the (light, active) neutrino mass scale. These models imply a lepton number spontaneous symmetry breaking, with an associated pseudo-Goldstone boson, the *Majoron* [169, 170], a light scalar with very weak interactions with visible matter [171, 172]. Nevertheless, Majoron-neutrino interactions will change the cosmological observables at both the BBN and CMB periods, as they contribute to both ΔN_{eff} and to the neutrino anistotropic stress, reducing neutrino free-streaming [29, 173–176]. With Planck 2015 data, the Majoron models were shown to predict a higher value of the H_0 parameter, especially when N_{eff} is free in the analysis [177]. Updated studies demonstrated that the flexibility in explaining the H_0 tension was reduced when considering non-CMB data together with Planck [178], as the H_0 values obtained in the combined analysis of Planck 2015 TTTEEE plus external data is 68.13 ± 0.48 km s^{-1} Mpc^{-1}. Considering Planck 2018 data and an updated local measurement of H_0, instead, the authors of Ref. [179] show that Majorons could alleviate the Hubble constant tension: as these particles contribute naturally to the dark radiation component of the universe, with $\Delta N_{\text{eff}} \sim 0.11$, they are able to reduce by a significant level the H_0 discrepancy. The combined fit of CMB, BAO and H_0 measurements improves by $\Delta\chi^2 = -12.2$ [179] when considering the self-interacting scenario with respect to the standard ΛCDM model, see also Refs. [180–182], although the CMB χ^2 alone is slightly worsened. Exploiting the very same mechanism, i.e. a very light scalar Majoron coupled to neutrinos, the authors of Ref. [183] found a better agreement among high redshift estimates and local, direct measurements of H_0.

24.4.3 Self-interacting Sterile Neutrinos

Another class of self-interacting scenarios involves the coupling between a sterile neutrino and a new pseudo-scalar particle, see e.g. [42] and references therein. This secret interaction would induce a large matter potential that suppresses active-sterile oscillations in the early universe. During the universe expansion, when the matter potential becomes similar to the vacuum oscillation frequency, the sterile neutrino may encounter a resonance [184]. Depending on the mass of the mediator, the resonance can occur at different times. If the new particle is much lighter than the sterile neutrino, the resonance occurs much later than BBN, leaving the abundances unaffected, and the sterile state is produced through vacuum neutrino oscillations. At some point of the evolution, moreover, the sterile neutrino can decay into the new mediator and the mass bounds are avoided [185–187]. If the mediator is heavier than the MeV scale [188, 189], instead, the sterile neutrino might be produced through a resonant process. If this happens before BBN, the light element abundances would be significantly altered. If the sterile neutrino production occurs after BBN, even ignoring the resonant mechanism, a copious abundance of sterile neutrinos is generated by active-sterile conversions. In such case, the scenario would be disfavored by cosmological bounds on the neutrino energy density.

When considering a light pseudoscalar, the most recent analyses performed with Planck data [190] show that the self-interacting model naturally corresponds to a high value for H_0, namely $H_0 = 71.6^{+1.1}_{-1.6}$ km s^{-1} Mpc^{-1}, which is very close to the local observed value. This result corresponds to a slightly better fit ($\Delta\chi^2 = -1.0$) of PlanckTTTEEE + H_0 constraints than a simple sterile neutrino scenario, although the CMB χ^2 alone is slightly worsened with respect to the non-interacting case. When fitting PlanckTTTEEE+lensing+BAO data, the model allows to obtain $H_0 = 70.0^{+0.7}_{-1.1}$ km s^{-1} Mpc^{-1}, which is still much higher than the CMB prediction in the ΛCDM model. When considering high-multipoles CMB data, the light pseudoscalar scenario is instead preferred over the ΛCDM model by ACT data alone ($\Delta\chi^2 = -5.3$), moreover with a preferred value of $H_0 = 72.7^{+2.0}_{-2.8}$ km s^{-1} Mpc^{-1}, although the combination of Planck+ACT prefers the simpler ΛCDM model with a $\Delta\chi^2 = 17.4$ against the pseudoscalar case and a lower value of $H_0 = 70.6^{+0.7}_{-0.9}$ km s^{-1} Mpc^{-1} [191].

Concerning secret sterile neutrino interactions with a heavy mediator, the situation for the H_0 tension is instead worsened [184]. When the sterile neutrino mass is enforced to obey SBL constraints and be close to 1 eV, the study reports indeed that H_0 is even smaller in the self-interacting scenario than in the ΛCDM model, because of the lower value of $N_{\rm eff}$ enforced by the interactions and the high value of m_s. Since higher neutrino masses are correlated with lower H_0 values and a small $N_{\rm eff}$ also corresponds to a decrease in H_0, both effects drive H_0 even down to 59.56 ± 0.88 km s^{-1} Mpc^{-1} [184], which of course increases the already strong H_0 tension. This class of model is therefore ruled out by cosmological constraints on the sum of neutrino masses and on free-streaming of active neutrinos, unless additional phenomena are considered to evade the neutrino mass bound, for example a fast decay mode from the sterile state into active neutrinos [192].

24.5 Summary

Dark radiation models provide a natural environment where to explain many theoretical (QCD strong CP problem, origin of neutrino masses) and observational (short baseline neutrino oscillation anomalies) open problems in the Standard Model of elementary particles. Some of these new particles can also play a very important role in astrophysics and cosmology, e.g. contributing to the hot dark matter component of the universe, stellar evolution and more. The question we have explored throughout this review is whether the extra dark radiation component can also play a relevant role in resolving the so-called Hubble constant tension.

The standard and commonly exploited way of parameterizing the mass-energy density in the new sector makes use of the $N_{\rm eff}$ parameter, the effective number of relativistic degrees of freedom. A larger value of $N_{\rm eff}$ can be obtained in very simple models, such as those with sterile neutrinos or with non-vanishing neutrino asymmetries. However, to solve the H_0 tension, more elaborated models leading to a non-zero $\Delta N_{\rm eff}$ are required. Interacting neutrino scenarios with a Majoron or with extra sterile species are examples in which the Hubble constant tension is alleviated. In addition, some of these models are also able to explain some of the aforementioned open questions within the SM, and therefore should be regarded as very appealing scenarios, both from the theoretical and observational perspectives. Upcoming data, not only from future cosmological probes, but also from laboratory, man-made particle beams and/or from astroparticle physics searches (neutrino, gamma-ray and cosmic ray telescopes in the case of axions and Majorons, for instance) may shed light on the role of extra dark radiation and the Hubble constant problem.

Acknowledgements This work has been supported by the MCIN/AEI/10.13039/501100011033 of Spain under grant PID2020-113644GB-I00, by the Generalitat Valenciana of Spain under grant PROMETEO/2019/083 and by the European Union's Framework Programme for Research and Innovation Horizon 2020 (2014-2020) under grant agreement 754496 (FELLINI) and 860881 (HIDDeN).

References

1. A.G. Riess et al., A comprehensive measurement of the local value of the Hubble constant with 1 km s^{-1} Mpc^{-1} uncertainty from the Hubble space telescope and the SH_0ES team. Astrophys. J. Lett. **934**, L7 (2022). https://doi.org/10.3847/2041-8213/ac5c5b [https://arxiv.org/abs/2112.04510 2112.04510]
2. Planck collaboration, Planck 2018 results. VI. Cosmological parameters. Astron. Astrophys. **641**, A6 (2020). https://doi.org/10.1051/0004-6361/201833910 [https://arxiv.org/abs/1807.06209 1807.06209]
3. L. Verde, T. Treu, A. Riess, Tensions Between the Early and the Late Universe, vol. 3, p. 891 (2019). https://doi.org/10.1038/s41550-019-0902-0 [https://arxiv.org/abs/1907.10625 1907.10625]
4. E. Di Valentino et al., Snowmass2021—letter of interest cosmology intertwined II: the Hubble constant tension. Astropart. Phys. **131**, 102605 (2021). https://doi.org/10.1016/j.astropartphys.2021.102605 [https://arxiv.org/abs/2008.11284 2008.11284]
5. E. Di Valentino et al., In the realm of the Hubble tension—a review of solutions. Class. Quant. Grav. **38**, 153001 (2021). https://doi.org/10.1088/1361-6382/ac086d [https://arxiv.org/abs/2103.01183 2103.01183]
6. N. Schöneberg et al., The H_0 Olympics: a fair ranking of proposed models. Phys. Rep. **984**, 2228 (2022). https://doi.org/10.1016/j.physrep.2022.07.001 [https://arxiv.org/abs/2107.10291 2107.10291]
7. K. Akita, M. Yamaguchi, A precision calculation of relic neutrino decoupling. JCAP **08**, 012 (2020). https://doi.org/10.1088/1475-7516/2020/08/012 [https://arxiv.org/abs/2005.07047 005.07047]
8. J. Froustey, C. Pitrou, M.C. Volpe, Neutrino decoupling including flavour oscillations and primordial nucleosynthesis. JCAP **12**, 015 (2020). https://doi.org/10.1088/1475-7516/2020/12/015 [https://arxiv.org/abs/2008.01074 2008.01074]
9. J.J. Bennett et al., Towards a precision calculation of $N_{\rm eff}$ in the standard model II: neutrino decoupling in the presence of flavour oscillations and finite-temperature QED. JCAP **04**, 073

(2021). https://doi.org/10.1088/1475-7516/2021/04/073 [https://arxiv.org/abs/2012.02726 2012.02726]
10. M. Cielo, M. Escudero, G. Mangano, O. Pisanti, Neff in the standard model at NLO is 3.043. https://arxiv.org/abs/2306.05460 2306.05460
11. G. Mangano, G. Miele, S. Pastor, T. Pinto, O. Pisanti, P.D. Serpico, Relic neutrino decoupling including flavor oscillations. Nucl. Phys. **B729**, 221 (2005). https://doi.org/10.1016/j.nuclphysb.2005.09.041 [https://arxiv.org/abs/hep-ph/0506164 hep-ph/0506164]
12. P.F. de Salas, S. Pastor, Relic neutrino decoupling with flavour oscillations revisited. JCAP **07**, 051 (2016). https://doi.org/10.1088/1475-7516/2016/07/051 [https://arxiv.org/abs/1606.06986 1606.06986]
13. Y. Farzan, M. Tortola, Neutrino oscillations and non-standard interactions. Front. Phys. **6**, 10 (2018). https://doi.org/10.3389/fphy.2018.00010 [https://arxiv.org/abs/1710.09360 1710.09360]
14. Y. Du, J.-H. Yu, Neutrino non-standard interactions meet precision measurements of N_{eff}. JHEP **05**, 058 (2021). https://doi.org/10.1007/JHEP05(2021)058 [https://arxiv.org/abs/2101.10475 2101.10475]
15. P.F. de Salas, S. Gariazzo, P. Martínez-Miravé, S. Pastor, M.Tórtola, Cosmological radiation density with non-standard neutrino-electron interactions. Phys. Lett. B **820**, 136508 (2021). https://doi.org/10.1016/j.physletb.2021.136508 [https://arxiv.org/abs/2105.08168 2105.08168]
16. S. Gariazzo, P. Martínez-Miravé, O. Mena, S. Pastor, M. Tórtola, Non-unitary three-neutrino mixing in the early universe. JCAP **03**, 046 (2023). https://doi.org/10.1088/1475-7516/2023/03/046 [https://arxiv.org/abs/2211.10522 2211.10522]
17. Simons observatory collaboration, The Simons observatory: science goals and forecasts. JCAP **02**, 056 (2019). https://doi.org/10.1088/1475-7516/2019/02/056 [https://arxiv.org/abs/1808.07445 1808.07445]
18. CMB-S4 collaboration, *CMB-S4 Science Book*, 1st edn. https://arxiv.org/abs/1610.02743 1610.02743
19. M. Archidiacono, E. Giusarma, S. Hannestad, O. Mena, Cosmic dark radiation and neutrinos. Adv. High Energy Phys. **2013**, 191047 (2013). https://doi.org/10.1155/2013/191047 [https://arxiv.org/abs/1307.0637 1307.0637]
20. O. Pisanti et al., PArthENoPE: public algorithm evaluating the nucleosynthesis of primordial elements. Comput. Phys. Commun. **178**, 956 (2008). https://doi.org/10.1016/j.cpc.2008.02.015 [https://arxiv.org/abs/0705.0290 0705.0290]
21. R. Consiglio, P.F. de Salas, G. Mangano, G. Miele, S. Pastor, O. Pisanti, PArthENoPE reloaded. Comput. Phys. Commun. **233**, 237 (2018). https://doi.org/10.1016/j.cpc.2018.06.022 [https://arxiv.org/abs/1712.04378 1712.04378]
22. S. Gariazzo, P.F. de Salas, O. Pisanti, R. Consiglio, PArthENoPE revolutions. Comput. Phys. Commun. **271**, 108205 (2022). https://doi.org/10.1016/j.cpc.2021.108205 [https://arxiv.org/abs/2103.05027 2103.05027]
23. C. Pitrou, A. Coc, J.-P. Uzan, E. Vangioni, A new tension in the cosmological model from primordial deuterium? Mon. Not. Roy. Astron. Soc. **502**, 2474 (2021). https://doi.org/10.1093/mnras/stab135 [https://arxiv.org/abs/2011.11320 2011.11320]
24. V. Mossa et al., The Baryon density of the Universe from an improved rate of deuterium burning. Nature **587**, 210 (2020). https://doi.org/10.1038/s41586-020-2878-4
25. O. Pisanti, G. Mangano, G. Miele, P. Mazzella, Primordial deuterium after LUNA: concordances and error budget. JCAP **04**, 020 (2021). https://doi.org/10.1088/1475-7516/2021/04/020 [https://arxiv.org/abs/2011.11537 2011.11537]
26. T.-H. Yeh, K.A. Olive, B.D. Fields, The impact of new $d(p,\gamma)3$ rates on Big Bang nucleosynthesis. JCAP **03**, 046 (2021). https://doi.org/10.1088/1475-7516/2021/03/046[https://arxiv.org/abs/2011.13874 2011.13874]
27. P.D. Serpico et al., Nuclear reaction network for primordial nucleosynthesis: a detailed analysis of rates, uncertainties and light nuclei yields. JCAP **12**, 010 (2004). https://doi.org/10.1088/1475-7516/2004/12/010 [https://arxiv.org/abs/astro-ph/0408076 astro-ph/0408076]

28. Z. Hou, R. Keisler, L. Knox, M. Millea, C. Reichardt, How Massless neutrinos affect the cosmic microwave background damping tail. Phys. Rev. D **87**, 083008 (2013). https://doi.org/10.1103/PhysRevD.87.083008 [https://arxiv.org/abs/1104.2333 1104.2333]
29. S. Bashinsky, U. Seljak, Neutrino perturbations in CMB anisotropy and matter clustering. Phys. Rev. D **69**, 083002 (2004). https://doi.org/10.1103/PhysRevD.69.083002 [https://arxiv.org/abs/astro-ph/0310198 astro-ph/0310198]
30. S. Hannestad, Structure formation with strongly interacting neutrinos–implications for the cosmological neutrino mass bound. JCAP **02**, 011 (2005). https://doi.org/10.1088/1475-7516/2005/02/011
31. W. Hu, N. Sugiyama, Small scale cosmological perturbations: an analytic approach. Astrophys. J. **471**, 542 (1996). https://doi.org/10.1086/177989 [https://arxiv.org/abs/astro-ph/9510117 astro-ph/9510117]
32. S.F. King, *Neutrino Mass and Mixing in the Seesaw Playground*, vol. 908, pp. 456–466 (2016). https://doi.org/10.1016/j.nuclphysb.2015.12.015DOI [https://arxiv.org/abs/1511.03831 1511.03831]
33. O.G. Miranda, J.W.F. Valle, Neutrino oscillations and the seesaw origin of neutrino mass. Nucl. Phys. **B908**, 436 (2016). https://doi.org/10.1016/j.nuclphysb.2016.03.027 https://arxiv.org/abs/1602.00864 1602.00864]
34. Z.-Z. Xing, Z.-H. Zhao, The minimal seesaw and leptogenesis models. Rep. Prog. Phys. **84**, 066201 (2021). https://doi.org/10.1088/1361-6633/abf086 [https://arxiv.org/abs/2008.12090 2008.12090]
35. A. Boyarsky, M. Drewes, T. Lasserre, S. Mertens, O. Ruchayskiy, Sterile neutrino dark matter. Prog. Part. Nucl. Phys. **104**, 1 (2019). https://doi.org/10.1016/j.ppnp.2018.07.004 [https://arxiv.org/abs/1807.07938 1807.07938]
36. R. Adhikari et al., A white paper on keV sterile neutrino dark matter. JCAP **01**, 025 (2017). https://doi.org/10.1088/1475-7516/2017/01/025 [https://arxiv.org/abs/1602.04816 1602.04816]
37. G. Anamiati, V. De Romeri, M. Hirsch, C.A. Ternes, M. Tórtola, Quasi-Dirac neutrino oscillations at DUNE and JUNO. Phys. Rev. D **100**, 035032 (2019). https://doi.org/10.1103/PhysRevD.100.035032 [https://arxiv.org/abs/1907.00980 1907.00980]
38. K. Carloni, I. Martinez-Soler, C.A. Arguelles, K. Babu, P.B. Dev, Probing pseudo-dirac neutrinos with astrophysical sources at IceCube. https://arxiv.org/abs/2212.00737 2212.00737
39. S. Gariazzo, C. Giunti, M. Laveder, Y.F. Li, E.M. Zavanin, Light sterile neutrinos. J. Phys. G **43**, 033001 (2016). https://doi.org/10.1088/0954-3899/43/3/033001 [https://arxiv.org/abs/1507.08204 1507.08204]
40. C. Giunti, T. Lasserre, eV-scale sterile neutrinos. Ann. Rev. Nucl. Part. Sci. **69**, 163 (2019). https://doi.org/10.1146/annurev-nucl-101918-023755 [https://arxiv.org/abs/1901.08330 1901.08330]
41. S. Böser et al., Status of light sterile neutrino searches. Prog. Part. Nucl. Phys. **111**, 103736 (2020). https://doi.org/10.1016/j.ppnp.2019.103736 [https://arxiv.org/abs/1906.01739 1906.01739]
42. M. Archidiacono, S. Gariazzo, Two sides of the same coin: sterile neutrinos and dark radiation, status and perspectives. Universe **8**, 175 (2022). https://doi.org/10.3390/universe8030175 [https://arxiv.org/abs/2201.10319 2201.10319]
43. LSND collaboration, Evidence for neutrino oscillations from the observation of $\bar{\nu}_e$ appearance in a $\bar{\nu}_\mu$ beam. Phys. Rev. D **64**, 112007 (2001). https://doi.org/10.1103/PhysRevD.64.112007 [https://arxiv.org/abs/hep-ex/0104049 hep-ex/0104049]
44. MiniBooNE collaboration, Unexplained excess of electron-like events from a 1-GeV neutrino beam. Phys. Rev. Lett. **102**, 101802 (2009). https://doi.org/10.1103/PhysRevLett.102.101802 [https://arxiv.org/abs/0812.2243 0812.2243]
45. MiniBooNE collaboration, Significant excess of electron like events in the MiniBooNE short-baseline neutrino experiment. Phys. Rev. Lett. **121**, 221801 (2018). https://doi.org/10.1103/PhysRevLett.121.221801 [https://arxiv.org/abs/1805.12028 1805.12028]

46. J. Abdurashitov et al., Measurement of the response of a Ga solar neutrino experiment to neutrinos from an Ar-37 source. Phys. Rev. C **73**, 045805 (2006). https://doi.org/10.1103/PhysRevC.73.045805 [https://arxiv.org/abs/nucl-ex/0512041 nucl-ex/0512041]
47. C. Giunti, M. Laveder, Statistical significance of the gallium anomaly. Phys. Rev. C **83**, 065504 (2011). https://doi.org/10.1103/PhysRevC.83.065504 [https://arxiv.org/abs/1006.3244 1006.3244]
48. G. Mention, M. Fechner, T. Lasserre, T.A. Mueller, D. Lhuillier, M. Cribier et al., The reactor antineutrino anomaly. Phys. Rev. D **83**, 073006 (2011). https://doi.org/10.1103/PhysRevD.83.073006 [https://arxiv.org/abs/1101.2755 1101.2755]
49. KARMEN collaboration, Upper limits for neutrino oscillations muon-anti-neutrino → electron-anti-neutrino from muon decay at rest. Phys. Rev. D **65**, 112001 (2002). https://doi.org/10.1103/PhysRevD.65.112001 [https://arxiv.org/abs/hep-ex/0203021 hep-ex/0203021]
50. OPERA collaboration, Final results of the search for $\nu_\mu \to \nu_e$ oscillations with the OPERA detector in the CNGS beam. JHEP **06**, 151 (2018). https://doi.org/10.1007/JHEP06(2018)151 [https://arxiv.org/abs/1803.11400 1803.11400]
51. MicroBooNE collaboration, Search for neutrino-induced neutral-current Δ radiative decay in MicroBooNE and a first test of the MiniBooNE low energy excess under a single-photon hypothesis. Phys. Rev. Lett. **128**, 111801 (2022). https://doi.org/10.1103/PhysRevLett.128.111801 [https://arxiv.org/abs/2110.00409 2110.00409]
52. MicroBooNE collaboration, First constraints on light sterile neutrino oscillations from combined appearance and disappearance searches with the MicroBooNE detector. Phys. Rev. Lett. **130**, 011801 (2023). https://doi.org/10.1103/PhysRevLett.130.011801 [https://arxiv.org/abs/2210.10216 2210.10216]
53. C. Giunti, Y. Li, C. Ternes, O. Tyagi, Z. Xin, Gallium anomaly: critical view from the global picture of ν_e and $\overline{\nu}_e$ disappearance. JHEP **10**, 164 (2022). https://doi.org/10.1007/JHEP10(2022)164 [https://arxiv.org/abs/2209.00916 2209.00916]
54. C. Giunti, Y. Li, C. Ternes, Z. Xin, Inspection of the detection cross section dependence of the Gallium anomaly. Phys. Lett. B **842**, 137983 (2023). https://doi.org/10.1016/j.physletb.2023.137983 [https://arxiv.org/abs/2212.09722 2212.09722]
55. V. Barinov et al., Results from the Baksan experiment on sterile transitions (BEST). Phys. Rev. Lett. **128**, 232501 (2022). https://doi.org/10.1103/PhysRevLett.128.232501 [https://arxiv.org/abs/2109.11482 2109.11482]
56. V. Barinov et al., Search for electron-neutrino transitions to sterile states in the BEST experiment. Phys. Rev. C **105**, 065502 (2022). https://doi.org/10.1103/PhysRevC.105.065502 [https://arxiv.org/abs/2201.07364 2201.07364]
57. S.R. Elliott, V. Gavrin, W. Haxton, The gallium anomaly. https://arxiv.org/abs/2306.03299 2306.03299
58. STEREO collaboration, Improved sterile neutrino constraints from the STEREO experiment with 179 days of reactor-on data. Phys. Rev. D **102**, 052002 (2020). https://doi.org/10.1103/PhysRevD.102.052002 [https://arxiv.org/abs/1912.06582 1912.06582]
59. PROSPECT collaboration, Improved short-baseline neutrino oscillation search and energy spectrum measurement with the PROSPECT experiment at HFIR. Phys. Rev. D **103**, 032001 (2021). https://doi.org/10.1103/PhysRevD.103.032001 [https://arxiv.org/abs/2006.11210 2006.11210]
60. DANSS collaboration, Search for sterile neutrinos at the DANSS experiment. Phys. Lett. B **787**, 56 (2018). https://doi.org/10.1016/j.physletb.2018.10.038 [https://arxiv.org/abs/1804.04046 1804.04046]
61. RENO, NEOS collaboration, Search for sterile neutrino oscillations using RENO and NEOS data. Phys. Rev. D **105**, L111101 (2022). https://doi.org/10.1103/PhysRevD.105.L111101 https://arxiv.org/abs/2011.00896 2011.00896]
62. DANSS collaboration, *Recent Results from DANSS*, vol. NOW2022, p. 017 (2023). https://doi.org/10.22323/1.421.0017 [https://arxiv.org/abs/2305.07417 2305.07417]

63. NEOS collaboration, Sterile neutrino search at the NEOS experiment. Phys. Rev. Lett. **118**, 121802 (2017). https://doi.org/10.1103/PhysRevLett.118.121802 [https://arxiv.org/abs/1610.05134 1610.05134]
64. A. Serebrov, R. Samoilov, M. Chaikovskii, O. Zherebtsov, The result of the Neutrino-4 experiment, sterile neutrinos and dark matter, the fourth neutrino and the Hubble constant. https://arxiv.org/abs/2302.09958 2302.09958
65. C. Giunti, Y. Li, C. Ternes, Y. Zhang, Neutrino-4 anomaly: oscillations or fluctuations? Phys. Lett. B **816**, 136214 (2021). https://doi.org/10.1016/j.physletb.2021.136214 [https://arxiv.org/abs/2101.06785 2101.06785]
66. M. Estienne et al., Updated summation model: an improved agreement with the Daya Bay antineutrino fluxes. Phys. Rev. Lett. **123**, 022502 (2019). https://doi.org/10.1103/PhysRevLett.123.022502 [https://arxiv.org/abs/1904.09358 1904.09358]
67. V. Kopeikin, Y.N. Panin, A. Sabelnikov, *Measurement of the Ratio of Cumulative Spectra of Beta Particles from* ^{235}U and ^{239}*Pu Fission Products for Solving Problems of Reactor-Antineutrino Physics*, vol. 84, pp. 1–10 (2021). https://doi.org/10.1134/S1063778821010129
68. L. Perissé et al., A comprehensive revision of the summation method for the prediction of reactor antineutrino fluxes and spectra. https://arxiv.org/abs/2304.14992 2304.14992
69. T.A. Mueller et al., Improved predictions of reactor antineutrino spectra. Phys. Rev. C **83**, 054615 (2011). https://doi.org/10.1103/PhysRevC.83.054615 https://arxiv.org/abs/1101.2663 1101.2663]
70. P. Huber, On the determination of anti-neutrino spectra from nuclear reactors. Phys. Rev. C **84**, 024617 (2011). https://doi.org/10.1103/PhysRevC.84.024617 [https://arxiv.org/abs/1106.0687 1106.0687]
71. C. Giunti, Y. Li, C. Ternes, Z. Xin, Reactor antineutrino anomaly in light of recent flux model refinements. Phys. Lett. B **829**, 137054 (2022). https://doi.org/10.1016/j.physletb.2022.137054 [https://arxiv.org/abs/2110.06820 2110.06820]
72. MINOS+ collaboration, Search for sterile neutrinos in MINOS and MINOS+ using a two-detector fit. Phys. Rev. Lett. **122**, 091803 (2019). https://doi.org/10.1103/PhysRevLett.122.091803 [https://arxiv.org/abs/1710.06488 1710.06488]
73. IceCube collaboration, eV-scale sterile neutrino search using eight years of atmospheric muon neutrino data from the IceCube neutrino observatory. Phys. Rev. Lett. **125**, 141801 (2020). https://doi.org/10.1103/PhysRevLett.125.141801 [https://arxiv.org/abs/2005.12942 2005.12942]
74. S. Gariazzo, C. Giunti, M. Laveder, Y.F. Li, Updated global 3+1 analysis of short-BaseLine neutrino oscillations. JHEP **06**, 135 (2017). https://doi.org/10.1007/JHEP06(2017)135 [https://arxiv.org/abs/1703.00860 1703.00860]
75. M. Dentler et al., Updated global analysis of neutrino oscillations in the presence of eV-scale sterile neutrinos. JHEP **08**, 010 (2018). https://doi.org/10.1007/JHEP08(2018)010 [https://arxiv.org/abs/1803.10661 1803.10661]
76. R. Barbieri, A. Dolgov, Bounds on sterile-neutrinos from nucleosynthesis. Phys. Lett. B **237**, 440 (1990). https://doi.org/10.1016/0370-2693(90)91203-N
77. R. Barbieri, A. Dolgov, Neutrino oscillations in the early universe. Nucl. Phys. B **349**, 743 (1991). https://doi.org/10.1016/0550-3213(91)90396-F
78. K. Kainulainen, Light singlet neutrinos and the primordial nucleosynthesis. Phys. Lett. B **244**, 191 (1990). https://doi.org/10.1016/0370-2693(90)90054-A
79. K. Enqvist, K. Kainulainen, J. Maalampi, Resonant neutrino transitions and nucleosynthesis. Phys. Lett. B **249**, 531 (1990). https://doi.org/10.1016/0370-2693(90)91030-F
80. K. Enqvist, K. Kainulainen, M.J. Thomson, Stringent cosmological bounds on inert neutrino mixing. Nucl. Phys. B **373**, 498 (1992). https://doi.org/10.1016/0550-3213(92)90442-E
81. A.D. Dolgov, F.L. Villante, BBN bounds on active sterile neutrino mixing. Nucl. Phys. **B679**, 261 (2004). https://doi.org/10.1016/j.nuclphysb.2003.11.031 [https://arxiv.org/abs/hep-ph/0308083 hep-ph/0308083]
82. M. Cirelli, G. Marandella, A. Strumia, F. Vissani, Probing oscillations into sterile neutrinos with cosmology, astrophysics and experiments. Nucl. Phys. B **708**,

215 (2005). https://doi.org/10.1016/j.nuclphysb.2004.11.056 [https://arxiv.org/abs/hep-ph/0403158 hep-ph/0403158]
83. A.D. Dolgov, Neutrinos in cosmology. Phys. Rep. **370**, 333 (2002). https://doi.org/10.1016/S0370-1573(02)00139-4 [https://arxiv.org/abs/hep-ph/0202122 hep-ph/0202122]
84. S. Hannestad, I. Tamborra, T. Tram, Thermalisation of light sterile neutrinos in the early universe. JCAP **07**, 025 (2012)[. https://doi.org/10.1088/1475-7516/2012/07/025 https://arxiv.org/abs/1204.5861 1204.5861]
85. S. Hannestad, R.S. Hansen, T. Tram, Can active-sterile neutrino oscillations lead to chaotic behavior of the cosmological lepton asymmetry? JCAP **04**, 032 (2013). https://doi.org/10.1088/1475-7516/2013/04/032 [https://arxiv.org/abs/1302.7279 1302.7279]
86. S. Hannestad, R.S. Hansen, T. Tram, Y.Y.Y. Wong, Active-sterile neutrino oscillations in the early universe with full collision terms. JCAP **08**, 019 (2015). https://doi.org/10.1088/1475-7516/2015/08/019 https://arxiv.org/abs/1506.05266 1506.05266]
87. A. Mirizzi, N. Saviano, G. Miele, P.D. Serpico, Light sterile neutrino production in the early universe with dynamical neutrino asymmetries. Phys. Rev. D **86**, 053009 (2012). https://doi.org/10.1103/PhysRevD.86.053009 [https://arxiv.org/abs/1206.1046 1206.1046]
88. A. Mirizzi, G. Mangano, N. Saviano, E. Borriello, C. Giunti, G. Miele et al., The strongest bounds on active-sterile neutrino mixing after Planck data. Phys. Lett. **B726**, 8 (2013). https://doi.org/10.1016/j.physletb.2013.08.015 [https://arxiv.org/abs/1303.5368 1303.5368]
89. N. Saviano, A. Mirizzi, O. Pisanti, P.D. Serpico, G. Mangano, G. Miele, Multi-momentum and multi-flavour active-sterile neutrino oscillations in the early universe: role of neutrino asymmetries and effects on nucleosynthesis. Phys. Rev. D **87**, 073006 (2013). https://doi.org/10.1103/PhysRevD.87.073006 [https://arxiv.org/abs/1302.1200 1302.1200]
90. S. Gariazzo, P.F. de Salas, S. Pastor, Thermalisation of sterile neutrinos in the early universe in the 3+1 scheme with full mixing matrix. JCAP **07**, 014 (2019). https://doi.org/10.1088/1475-7516/2019/07/014 [https://arxiv.org/abs/1905.11290 1905.11290]
91. L. Mastrototaro, P.D. Serpico, A. Mirizzi, N. Saviano, Massive sterile neutrinos in the early Universe: from thermal decoupling to cosmological constraints. Phys. Rev .D **104**, 016026 (2021). https://doi.org/10.1103/PhysRevD.104.016026 [https://arxiv.org/abs/2104.11752 2104.11752]
92. S. Dodelson, L.M. Widrow, Sterile-neutrinos as dark matter. Phys. Rev. Lett. **72**, 17 (1994). https://doi.org/10.1103/PhysRevLett.72.17 https://arxiv.org/abs/hep-ph/9303287 hep-ph/9303287]
93. E. Di Valentino, S. Gariazzo, W. Giarè, O. Mena, Weighing neutrinos at the damping tail.https://arxiv.org/abs/2305.12989 2305.12989
94. B.D. Fields, K.A. Olive, T.-H. Yeh, C. Young, Big-bang nucleosynthesis after Planck. JCAP **03**, 010 (2020). https://doi.org/10.1088/1475-7516/2020/03/010 [https://arxiv.org/abs/1912.01132 1912.01132]
95. Particle Data Group collaboration, Review of particle physics. PTEP **2022**, 083C01 (2022). https://doi.org/10.1093/ptep/ptac097
96. P.-H. Gu, Large Lepton asymmetry for small Baryon asymmetry and warm dark matter. Phys. Rev. D **82**, 093009 (2010). https://doi.org/10.1103/PhysRevD.82.093009 [https://arxiv.org/abs/1005.1632 1005.1632]
97. J. March-Russell, H. Murayama, A. Riotto, The small observed Baryon asymmetry from a large lepton asymmetry. JHEP **11**, 015 (1999). https://doi.org/10.1088/1126-6708/1999/11/015 [https://arxiv.org/abs/hep-ph/9908396 hep-ph/9908396]
98. J. McDonald, Naturally large cosmological neutrino asymmetries in the MSSM. Phys. Rev. Lett. **84**, 4798 (2000). https://doi.org/10.1103/PhysRevLett.84.4798 [https://arxiv.org/abs/hep-ph/9908300 hep-ph/9908300]
99. J. Froustey, C. Pitrou, Incomplete neutrino decoupling effect on big bang nucleosynthesis. Phys. Rev. D **101**, 043524 (2020). https://doi.org/10.1103/PhysRevD.101.043524 [https://arxiv.org/abs/1912.09378 1912.09378]
100. G. Barenboim, W.H. Kinney, W.-I. Park, Flavor versus mass eigenstates in neutrino asymmetries: implications for cosmology. Eur. Phys. J. C **77**, 590 (2017). https://doi.org/10.1140/epjc/s10052-017-5147-4 [https://arxiv.org/abs/1609.03200 1609.03200]

101. G. Mangano, G. Miele, S. Pastor, O. Pisanti, S. Sarikas, Updated BBN bounds on the cosmological lepton asymmetry for non-zero θ_{13}. Phys. Lett. B **708**, 1 (2012). https://doi.org/10.1016/j.physletb.2012.01.015 [https://arxiv.org/abs/1110.4335 1110.4335]
102. A. Dolgov, S. Hansen, S. Pastor, S. Petcov, G. Raffelt, D. Semikoz, Cosmological bounds on neutrino degeneracy improved by flavor oscillations. Nucl. Phys. B **632**, 363 (2002). https://doi.org/10.1016/S0550-3213(02)00274-2
103. Y.Y.Y. Wong, Analytical treatment of neutrino asymmetry equilibration from flavor oscillations in the early universe. Phys. Rev. D **66**, 025015 (2002). https://doi.org/10.1103/PhysRevD.66.025015 [https://arxiv.org/abs/hep-ph/0203180 hep-ph/0203180]
104. V. Simha, G. Steigman, Constraining the universal Lepton asymmetry. JCAP **08**, 011 (2008). https://doi.org/10.1088/1475-7516/2008/08/011 [https://arxiv.org/abs/0806.0179 0806.0179]
105. I.M. Oldengott, D.J. Schwarz, Improved constraints on lepton asymmetry from the cosmic microwave background. Europhys. Lett. **119**, 29001 (2017). https://doi.org/10.1209/0295-5075/119/29001 [https://arxiv.org/abs/1706.01705 1706.01705]
106. C. Pitrou, A. Coc, J.-P. Uzan, E. Vangioni, Precision big bang nucleosynthesis with improved Helium-4 predictions. Phys. Rep. **04**, 005 (2018). https://doi.org/10.1016/j.physrep.2018.04.005 [https://arxiv.org/abs/1801.08023 1801.08023]
107. J. Froustey, C. Pitrou, Primordial neutrino asymmetry evolution with full mean-field effects and collisions. JCAP **03**, 065 (2022). https://doi.org/10.1088/1475-7516/2022/03/065 [https://arxiv.org/abs/2110.11889 2110.11889]
108. R.D. Peccei, H.R. Quinn, CP conservation in the presence of instantons. Phys. Rev. Lett. **38**, 1440 (1977). https://doi.org/10.1103/PhysRevLett.38.1440
109. R.D. Peccei, H.R. Quinn, Constraints imposed by CP conservation in the presence of instantons. Phys. Rev. D **16**, 1791 (1977). https://doi.org/10.1103/PhysRevD.16.1791
110. M.S. Turner, Thermal production of not SO invisible axions in the early universe. Phys. Rev. Lett. **59**, 2489 (1987). https://doi.org/10.1103/PhysRevLett.59.2489
111. Z.G. Berezhiani, A.S. Sakharov, M.Y. Khlopov, Primordial background of cosmological axions. Sov. J. Nucl. Phys. Series **55**, 1063 (1992)
112. C. Brust, D.E. Kaplan, M.T. Walters, New light species and the CMB. JHEP **12**, 058 (2013). https://doi.org/10.1007/JHEP12(2013)058 [https://arxiv.org/abs/1303.5379 1303.5379]
113. D. Baumann, D. Green, B. Wallisch, New target for cosmic axion searches. Phys. Rev. Lett. **117**, 171301 (2016). https://doi.org/10.1103/PhysRevLett.117.171301 [https://arxiv.org/abs/1604.08614 1604.08614]
114. F. D'Eramo, R.Z. Ferreira, A. Notari, J.L. Bernal, Hot axions and the H_0 tension. JCAP **11**, 014 (2018). https://doi.org/10.1088/1475-7516/2018/11/014 [https://arxiv.org/abs/1808.07430 1808.07430]
115. F. Arias-Aragón, F. D'eramo, R.Z. Ferreira, L. Merlo, A. Notari, Cosmic imprints of XENON1T axions. JCAP **11**, 025 (2020). https://doi.org/10.1088/1475-7516/2020/11/025 https://arxiv.org/abs/2007.06579 2007.06579]
116. F. Arias-Aragón, F. D'Eramo, R.Z. Ferreira, L. Merlo, A. Notari, Production of thermal axions across the ElectroWeak phase transition. JCAP **03**, 090 (2021). https://doi.org/10.1088/1475-7516/2021/03/090 [https://arxiv.org/abs/2012.04736 2012.04736]
117. D. Green, Y. Guo, B. Wallisch, Cosmological implications of axion-matter couplings. JCAP **02**, 019 (2022). https://doi.org/10.1088/1475-7516/2022/02/019 [https://arxiv.org/abs/2109.12088 2109.12088]
118. F. D'Eramo, S. Yun, Flavor violating axions in the early universe. Phys. Rev. D **105**, 075002 (2022). https://doi.org/10.1103/PhysRevD.105.075002 [https://arxiv.org/abs/2111.12108 2111.12108]
119. S. Hannestad, A. Mirizzi, G. Raffelt, New cosmological mass limit on thermal relic axions. JCAP **07**, 002 (2005). https://doi.org/10.1088/1475-7516/2005/07/002 https://arxiv.org/abs/hep-ph/0504059 hep-ph/0504059]
120. A. Melchiorri, O. Mena, A. Slosar, An improved cosmological bound on the thermal axion mass. Phys. Rev. D **76**, 041303 (2007). https://doi.org/10.1103/PhysRevD.76.041303 [https://arxiv.org/abs/0705.2695 0705.2695]

121. S. Hannestad, A. Mirizzi, G.G. Raffelt, Y.Y. Wong, Cosmological constraints on neutrino plus axion hot dark matter. JCAP **08**, 015 (2007). https://doi.org/10.1088/1475-7516/2007/08/015 [https://arxiv.org/abs/0706.4198 0706.4198]
122. S. Hannestad, A. Mirizzi, G.G. Raffelt, Y.Y. Wong, Cosmological constraints on neutrino plus axion hot dark matter: update after WMAP-5. JCAP **04**, 019 (2008). https://doi.org/10.1088/1475-7516/2008/04/019 [https://arxiv.org/abs/0803.1585 0803.1585]
123. S. Hannestad, A. Mirizzi, G.G. Raffelt, Y.Y. Wong, Neutrino and axion hot dark matter bounds after WMAP-7. JCAP **08**, 001 (2010). https://doi.org/10.1088/1475-7516/2010/08/001 [https://arxiv.org/abs/1004.0695 1004.0695]
124. M. Archidiacono, S. Hannestad, A. Mirizzi, G. Raffelt, Y.Y. Wong, Axion hot dark matter bounds after Planck. JCAP **10**, 020 (2013). https://doi.org/10.1088/1475-7516/2013/10/020 https://arxiv.org/abs/1307.0615 1307.0615]
125. E. Giusarma, E. Di Valentino, M. Lattanzi, A. Melchiorri, O. Mena, Relic neutrinos, thermal axions and cosmology in early 2014. Phys. Rev. D **90**, 043507 (2014). https://doi.org/10.1103/PhysRevD.90.043507 [https://arxiv.org/abs/1403.4852 1403.4852]
126. E. Di Valentino, S. Gariazzo, E. Giusarma, O. Mena, Robustness of cosmological axion mass limits. Phys. Rev. D D **91**, 123505 (2015). https://doi.org/10.1103/PhysRevD.91.123505 [https://arxiv.org/abs/1503.00911 1503.00911]
127. E. Di Valentino, E. Giusarma, M. Lattanzi, O. Mena, A. Melchiorri, J. Silk, Cosmological Axion and neutrino mass constraints from Planck 2015 temperature and polarization data. Phys. Lett. B **752**, 182 (2016). https://doi.org/10.1016/j.physletb.2015.11.025 https://arxiv.org/abs/1507.08665 1507.08665]
128. M. Archidiacono, T. Basse, J. Hamann, S. Hannestad, G. Raffelt, Y.Y.Y. Wong, Future cosmological sensitivity for hot dark matter axions. JCAP **05**, 050 (2015). https://doi.org/10.1088/1475-7516/2015/05/050 [https://arxiv.org/abs/1502.03325 1502.03325]
129. W. Giarè, E. Di Valentino, A. Melchiorri, O. Mena, New cosmological bounds on hot relics: axions and neutrinos. Mon. Not. Roy. Astron. Soc. **505**, 2703 (2021). https://doi.org/10.1093/mnras/stab1442 [https://arxiv.org/abs/2011.14704 2011.14704]
130. F. D'Eramo et al., Cosmological bound on the QCD axion mass, redux. JCAP **09**, 022 (2022). https://doi.org/10.1088/1475-7516/2022/09/022 [https://arxiv.org/abs/2205.07849 2205.07849]
131. W. Giarè, F. Renzi, A. Melchiorri, O. Mena, E. Di Valentino, Cosmological forecasts on thermal axions, relic neutrinos, and light elements. Mon. Not. Roy. Astron. Soc. **511**, 1373 (2022). https://doi.org/10.1093/mnras/stac126 [https://arxiv.org/abs/2110.00340 2110.00340]
132. E. Di Valentino et al., Novel model-marginalized cosmological bound on the QCD axion mass. Phys. Rev. D **107**, 103528 (2023). https://doi.org/10.1103/PhysRevD.107.103528 [https://arxiv.org/abs/2212.11926 2212.11926]
133. W.A. Bardeen, S.H.H. Tye, J.A.M. Vermaseren, Phenomenology of the new light higgs Boson search. Phys. Lett. B **76**, 580 (1978). https://doi.org/10.1016/0370-2693(78)90859-6
134. G. Grilli di Cortona, E. Hardy, J. Pardo Vega, G. Villadoro, The QCD axion, precisely. HEP **01**, 034 (2016). https://doi.org/10.1007/JHEP01(2016)034 J [https://arxiv.org/abs/1511.02867 1511.02867]
135. M. Gorghetto, G. Villadoro, Topological susceptibility and QCD axion mass: QED and NNLO corrections. JHEP **03**, 033 (2019). https://doi.org/10.1007/JHEP03(2019)033 [https://arxiv.org/abs/1812.01008 1812.01008]
136. J.E. Kim, G. Carosi, Axions and the strong CP problem. Rev. Mod. Phys. **82**, 557 (2010). https://doi.org/10.1103/RevModPhys.82.557 [https://arxiv.org/abs/0807.3125 0807.3125]
137. L. Di Luzio, M. Giannotti, E. Nardi, L. Visinelli, The landscape of QCD axion models. Phys. Rep. **870**, 1 (2020). https://doi.org/10.1016/j.physrep.2020.06.002 [https://arxiv.org/abs/2003.01100 2003.01100]
138. J.E. Kim, Weak interaction singlet and strong CP invariance. Phys. Rev. Lett. **43**, 103 (1979). https://doi.org/10.1103/PhysRevLett.43.103
139. M.A. Shifman, A.I. Vainshtein, V.I. Zakharov, Can confinement ensure natural CP invariance of strong interactions? Nucl. Phys. B **166**, 493 (1980). https://doi.org/10.1016/0550-3213(80)90209-6

140. M. Dine, W. Fischler, M. Srednicki, A simple solution to the strong CP problem with a harmless axion. Phys. Lett. B **104**, 199 (1981). https://doi.org/10.1016/0370-2693(81)90590-6
141. A.R. Zhitnitsky, On possible suppression of the axion hadron interactions. Sov. J. Nucl. Phys. **31**, 260 (1980). ((in Russian))
142. G. Mangano, A. Melchiorri, P. Serra, A. Cooray, M. Kamionkowski, Cosmological bounds on dark matter-neutrino interactions. Phys. Rev. D **74**, 043517 (2006). https://doi.org/10.1103/PhysRevD.74.043517 [https://arxiv.org/abs/astro-ph/0606190 astro-ph/0606190]
143. P. Serra, F. Zalamea, A. Cooray, G. Mangano, A. Melchiorri, Constraints on neutrino—dark matter interactions from cosmic microwave background and large scale structure data. Phys. Rev. D **81**, 043507 (2010). https://doi.org/10.1103/PhysRevD.81.043507 [https://arxiv.org/abs/0911.4411 0911.4411]
144. R. Diamanti, E. Giusarma, O. Mena, M. Archidiacono, A. Melchiorri, Dark radiation and interacting scenarios. Phys. Rev. D **87**, 063509 (2013). https://doi.org/10.1103/PhysRevD.87.063509 [https://arxiv.org/abs/1212.6007 1212.6007]
145. R.J. Wilkinson, C. Boehm, J. Lesgourgues, Constraining dark matter-neutrino interactions using the CMB and large-scale structure. JCAP **05**, 011 (2014). https://doi.org/10.1088/1475-7516/2014/05/011 [https://arxiv.org/abs/1401.7597 1401.7597]
146. A. Olivares-Del Campo, C. Bœhm, S. Palomares-Ruiz, S. Pascoli, Dark matter-neutrino interactions through the lens of their cosmological implications. Phys. Rev. D **97**, 075039 (2018). https://doi.org/10.1103/PhysRevD.97.075039 [https://arxiv.org/abs/1711.05283 1711.05283]
147. J. Stadler, C. Bœhm, O. Mena, First numerical study of neutrino-dark matter mixed damping. JCAP **08**, 014 (2019). https://doi.org/10.1088/1475-7516/2019/08/014 [https://arxiv.org/abs/1903.00540 1903.00540]
148. M.R. Mosbech et al., The full Boltzmann hierarchy for dark matter-massive neutrino interactions. JCAP **03**, 066 (2021). https://doi.org/10.1088/1475-7516/2021/03/066 [https://arxiv.org/abs/2011.04206 2011.04206]
149. M. Escudero, Neutrino decoupling beyond the standard model: CMB constraints on the dark matter mass with a fast and precise $N_{\rm eff}$ evaluation. JCAP **02**, 007 (2019). https://doi.org/10.1088/1475-7516/2019/02/007 [https://arxiv.org/abs/1812.05605 1812.05605]
150. M. Escudero, Precision early universe thermodynamics made simple: $N_{\rm eff}$ and neutrino decoupling in the standard model and beyond. JCAP **05**, 048 (2020). https://doi.org/10.1088/1475-7516/2020/05/048 https://arxiv.org/abs/2001.04466 2001.04466]
151. C. Boehm, M.J. Dolan, C. McCabe, Increasing Neff with particles in thermal equilibrium with neutrinos. JCAP **12**, 027 (2012). https://doi.org/10.1088/1475-7516/2012/12/027 [https://arxiv.org/abs/1207.0497 1207.0497]
152. C. Boehm, M.J. Dolan, C. McCabe, A lower bound on the mass of cold thermal dark matter from Planck. JCAP **08**, 041 (2013). https://doi.org/10.1088/1475-7516/2013/08/041 [https://arxiv.org/abs/1303.6270 1303.6270]
153. E. Di Valentino, C.B.e. hm, E. Hivon, F.R. Bouchet, Reducing the H_0 and σ_8 tensions with dark matter-neutrino interactions. Phys. Rev. D **97**, 043513 (2018). https://doi.org/10.1103/PhysRevD.97.043513 [https://arxiv.org/abs/1710.02559 1710.02559]
154. B. Follin, L. Knox, M. Millea, Z. Pan, First detection of the acoustic oscillation phase shift expected from the cosmic neutrino background. Phys. Rev. Lett. **115**, 091301 (2015). https://doi.org/10.1103/PhysRevLett.115.091301 [https://arxiv.org/abs/1503.07863 1503.07863]
155. D. Baumann, D. Green, J. Meyers, B. Wallisch, Phases of new physics in the CMB. JCAP **01**, 007 (2016). https://doi.org/10.1088/1475-7516/2016/01/007 [https://arxiv.org/abs/1508.06342 1508.06342]
156. D. Baumann, D. Green, M. Zaldarriaga, Phases of new physics in the BAO spectrum. JCAP **11**, 007 (2017). https://doi.org/10.1088/1475-7516/2017/11/007 [https://arxiv.org/abs/1703.00894 1703.00894]
157. G. Choi, C.-T. Chiang, M. LoVerde, Probing decoupling in dark sectors with the cosmic microwave background. JCAP **06**, 044 (2018). https://doi.org/10.1088/1475-7516/2018/06/044 [https://arxiv.org/abs/1804.10180 1804.10180]

158. J.M. Berryman et al., *Neutrino Self-interactions: A White Paper*, vol. 42, p. 101267, 3 (2023). https://doi.org/10.1016/j.dark.2023.101267 [https://arxiv.org/abs/2203.01955 2203.01955]
159. P. Taule, M. Escudero, M. Garny, Global view of neutrino interactions in cosmology: The free streaming window as seen by Planck. Phys. Rev. D **106**, 063539 (2022). https://doi.org/10.1103/PhysRevD.106.063539 [https://arxiv.org/abs/2207.04062 2207.04062]
160. C.D. Kreisch, F.-Y. Cyr-Racine, O. Doré, Neutrino puzzle: anomalies, interactions, and cosmological tensions. Phys. Rev. D **101**, 123505 (2020). https://doi.org/10.1103/PhysRevD.101.123505 [https://arxiv.org/abs/1902.00534 1902.00534]
161. S. Roy Choudhury, S. Hannestad, T. Tram, Updated constraints on massive neutrino self-interactions from cosmology in light of the H_0 tension. JCAP **03**, 084 (2021). https://doi.org/10.1088/1475-7516/2021/03/084 [https://arxiv.org/abs/2012.07519 2012.07519]
162. T. Brinckmann, J.H. Chang, M. LoVerde, Self-interacting neutrinos, the Hubble parameter tension, and the cosmic microwave background. Phys. Rev. D **104**, 063523 (2021). https://doi.org/10.1103/PhysRevD.104.063523 [https://arxiv.org/abs/2012.11830 2012.11830]
163. A. Das, S. Ghosh, Self-interacting neutrinos as a solution to the Hubble tension?, in *European Physical Society Conference on High Energy Physics 2021*, vol. EPS-HEP2021, p. 124, 9 (2022). https://doi.org/10.22323/1.398.0124 [https://arxiv.org/abs/2109.03263 2109.03263]
164. T. Brinckmann, J.H. Chang, P. Du, M. LoVerde, Confronting interacting dark radiation scenarios with cosmological data. https://arxiv.org/abs/2212.13264 2212.13264
165. S. Sandner, M. Escudero, S.J. Witte, Precision CMB constraints on eV-scale bosons coupled to neutrinos. https://arxiv.org/abs/2305.01692 2305.01692
166. H.G. Escudero, J.-L. Kuo, R.E. Keeley, K.N. Abazajian, Early or phantom dark energy, self-interacting, extra, or massive neutrinos, primordial magnetic fields, or a curved universe: an exploration of possible solutions to the H_0 and σ_8. Phys. Rev. D **106**, 103517 (2022). https://doi.org/10.1103/PhysRevD.106.103517 [https://arxiv.org/abs/2208.14435 2208.14435]
167. C.D. Kreisch et al., The Atacama cosmology telescope: the persistence of neutrino self-interaction in cosmological measurements. https://arxiv.org/abs/2207.03164 2207.03164
168. J. Venzor, G. Garcia-Arroyo, A. Pérez-Lorenzana, J. De-Santiago, Resonant neutrino self-interactions and the H_0 tension. https://arxiv.org/abs/2303.12792 2303.12792
169. Y. Chikashige, R.N. Mohapatra, R. Peccei, Spontaneously Broken Lepton Number and Cosmological Constraints on the Neutrino Mass. Spectrum (1926). Phys. Rev. Lett. Series **45** (1980). https://doi.org/10.1103/PhysRevLett.45.1926
170. Y. Chikashige, R.N. Mohapatra, R. Peccei, Are there real goldstone bosons associated with broken Lepton number? Phys. Lett. B **98**, 265 (1981). https://doi.org/10.1016/0370-2693(81)90011-3
171. V.D. Barger, W.-Y. Keung, S. Pakvasa, Majoron emission by neutrinos. Phys. Rev. D **25**, 907 (1982). https://doi.org/10.1103/PhysRevD.25.907
172. K. Akita, M. Niibo, Updated constraints and future prospects on majoron dark matter. https://arxiv.org/abs/2304.04430 2304.04430
173. Z. Chacko, L.J. Hall, T. Okui, S.J. Oliver, CMB signals of neutrino mass generation. Phys. Rev. D **70**, 085008 (2004). https://doi.org/10.1103/PhysRevD.70.085008 [https://arxiv.org/abs/hep-ph/0312267 hep-ph/0312267]
174. M. Lattanzi, J.W.F. Valle, Decaying warm dark matter and neutrino masses. Phys. Rev. Lett. **99**, 121301 (2007). https://doi.org/10.1103/PhysRevLett.99.121301 [https://arxiv.org/abs/0705.2406 0705.2406]
175. M. Lattanzi, S. Riemer-Sorensen, M. Tortola, J.W.F. Valle, Updated CMB and x- and γ-ray constraints on Majoron dark matter. Phys. Rev. D **88**, 063528 (2013). https://doi.org/10.1103/PhysRevD.88.063528 [https://arxiv.org/abs/1303.4685 1303.4685]
176. C. Biggio, L. Calibbi, T. Ota, S. Zanchini, *Type-II Majoron Dark Matter*. https://arxiv.org/abs/2304.12527 2304.12527
177. F. Forastieri, M. Lattanzi, P. Natoli, Constraints on secret neutrino interactions after Planck. JCAP **07**, 014 (2015). https://doi.org/10.1088/1475-7516/2015/07/014 [https://arxiv.org/abs/1504.04999 1504.04999]

178. F. Forastieri, M. Lattanzi, P. Natoli, Cosmological constraints on neutrino self-interactions with a light mediator. Phys. Rev. D **100**, 103526 (2019). https://doi.org/10.1103/PhysRevD.100.103526 [https://arxiv.org/abs/1904.07810 1904.07810]
179. M. Escudero, S.J. Witte, A CMB search for the neutrino mass mechanism and its relation to the Hubble tension. Eur. Phys. J. C **80**, 294 (2020). https://doi.org/10.1140/epjc/s10052-020-7854-5 [https://arxiv.org/abs/1909.04044 1909.04044]
180. F. Arias-Aragon, E. Fernandez-Martinez, M. Gonzalez-Lopez, L. Merlo, Neutrino masses and Hubble tension via a Majoron in MFV. Eur. Phys. J. C **81**, 28 (2021). https://doi.org/10.1140/epjc/s10052-020-08825-8 [https://arxiv.org/abs/2009.01848 2009.01848]
181. G.-Y. Huang, W. Rodejohann, Solving the Hubble tension without spoiling Big Bang Nucleosynthesis. Phys. Rev. D **103**, 123007 (2021). https://doi.org/10.1103/PhysRevD.103.123007 https://arxiv.org/abs/2102.04280 2102.04280]
182. M. Escudero, S.J. Witte, The Hubble tension as a hint of leptogenesis and neutrino mass generation. Eur. Phys. J. C **81**, 515 (2021). https://doi.org/10.1140/epjc/s10052-021-09276-5 [https://arxiv.org/abs/2103.03249 2103.03249]
183. G. Barenboim, U. Nierste, Modified majoron model for cosmological anomalies. Phys. Rev. D **104**, 023013 (2021). https://doi.org/10.1103/PhysRevD.104.023013 [https://arxiv.org/abs/2005.13280 2005.13280]
184. F. Forastieri et al., Cosmic microwave background constraints on secret interactions among sterile neutrinos. JCAP **07**, 038 (2017). https://doi.org/10.1088/1475-7516/2017/07/038 [https://arxiv.org/abs/1704.00626 1704.00626]
185. M. Archidiacono, S. Hannestad, R.S. Hansen, T. Tram, Cosmology with self-interacting sterile neutrinos and dark matter—a pseudoscalar model. Phys. Rev. D **91**, 065021 (2015). https://doi.org/10.1103/PhysRevD.91.065021 [https://arxiv.org/abs/1404.5915 1404.5915]
186. M. Archidiacono, S. Hannestad, R.S. Hansen, T. Tram, Sterile neutrinos with pseudoscalar self-interactions and cosmology. Phys. Rev. D **93**, 045004 (2016). https://doi.org/10.1103/PhysRevD.93.045004 [https://arxiv.org/abs/1508.02504 1508.02504]
187. M. Archidiacono, S. Gariazzo, C. Giunti, S. Hannestad, R. Hansen, M. Laveder et al., Pseudoscalar–sterile neutrino interactions: reconciling the cosmos with neutrino oscillations. JCAP **08**, 067 (2016). https://doi.org/10.1088/1475-7516/2016/08/067 [https://arxiv.org/abs/1606.07673 1606.07673]
188. N. Saviano, O. Pisanti, G. Mangano, A. Mirizzi, Unveiling secret interactions among sterile neutrinos with big-bang nucleosynthesis. Phys. Rev. D **90**, 113009 (2014). https://doi.org/10.1103/PhysRevD.90.113009 https://arxiv.org/abs/1409.1680 1409.1680]
189. A. Mirizzi, G. Mangano, O. Pisanti, N. Saviano, Collisional production of sterile neutrinos via secret interactions and cosmological implications. Phys. Rev. D **91**, 025019 (2015). https://doi.org/10.1103/PhysRevD.91.025019 [https://arxiv.org/abs/1410.1385 1410.1385]
190. M. Archidiacono, S. Gariazzo, C. Giunti, S. Hannestad, T. Tram, Sterile neutrino self-interactions: H_0 tension and short-baseline anomalies. JCAP **12**, 029 (2020). https://doi.org/10.1088/1475-7516/2020/12/029 [https://arxiv.org/abs/2006.12885 2006.12885]
191. M.A. Corona et al., Pseudoscalar sterile neutrino self-interactions in light of Planck, SPT and ACT data. JCAP **06**, 010 (2022). https://doi.org/10.1088/1475-7516/2022/06/010 [https://arxiv.org/abs/2112.00037 2112.00037]
192. X. Chu, B. Dasgupta, M. Dentler, J. Kopp, N. Saviano, Sterile neutrinos with secret interactions—cosmological discord? JCAP **11**, 049 (2018). https://doi.org/10.1088/1475-7516/2018/11/049 https://arxiv.org/abs/1806.10629 1806.10629]

Chapter 25
Decaying Dark Matter and the Hubble Tension

Andreas Nygaard, Emil Brinch Holm, Thomas Tram, and Steen Hannestad

25.1 Introduction

Although the nature of dark matter remains unknown, a brief look at the Standard Model contents of the Universe reveals that a majority of the known particles are unstable and decay [1, 2]. By analogy, a natural question to ask is whether dark matter may decay on cosmological timescales. Decays of dark matter into electromagnetically interacting particles are strongly constrained by CMB observations [3]. Decays into a dark sector, so-called *invisible decays*, on the other hand, are much less constrained because no direct observation channel exists. Nonetheless, there are strong constraints on models that assume *all* of dark matter to decay on cosmological timescales (e.g., the simple observation that we observe it today) [4, 5]. However, these constraints may always be evaded by considering a scenario where only a fraction of the dark matter decays invisibly. It is this class of models we study in this chapter.

There exist several phenomenological models of invisibly decaying dark matter, largely varying in their assumptions on the decaying particle (cold or warm) and on the decay products (massive or massless, two- or many-body decays). In this chapter, we review constraints on the three most studied models:

DCDM→DR: Decaying cold dark matter (DCDM) decaying into dark radiation (DR). Presented in Sect. 25.2.

DCDM→DR+WDM: Decaying cold dark matter decaying into warm dark matter (WDM) *and* dark radiation. Presented in Sect. 25.3.

A. Nygaard (✉) · E. B. Holm · T. Tram · S. Hannestad
Department of Physics and Astronomy, Aarhus University, 8000 Aarhus C, Denmark
e-mail: andreas@phys.au.dk

E. B. Holm
e-mail: ebholm@phys.au.dk

T. Tram
e-mail: thomas.tram@phys.au.dk

S. Hannestad
e-mail: steen@phys.au.dk

E. Di Valentino and D. Brout (eds.), *The Hubble Constant Tension*, Springer Series in Astrophysics and Cosmology, https://doi.org/10.1007/978-981-99-0177-7_25

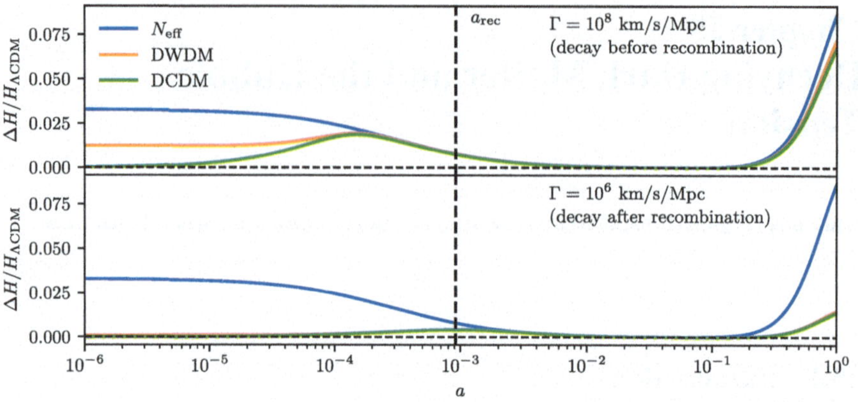

Fig. 25.1 Hubble parameter H as a function of scale factor a, relative to its value in the ΛCDM model, for models of decaying cold dark matter (DCDM), decaying warm dark matter (DWDM), and additional relativistic degrees of freedom, $\Delta N_{\rm eff}$. Taken from Ref. [1]

DWDM→DR: Decaying *warm* dark matter (DWDM) decaying into dark radiation. Presented in Sect. 25.4.

Why are these models relevant to the Hubble tension? Fig. 25.1 shows the relative change in the Hubble parameter $H(a)$ as a function of the scale factor a for the DWDM→DR and DCDM→DR models, as well as a model with additional relativistic degrees of freedom, $\Delta N_{\rm eff}$. The abundances of the species have been fixed such that they all contribute $\Delta N_{\rm eff} = 0.5$ to the number of relativistic degrees of freedom at $a = 1$, the acoustic scale is fixed (thus letting H_0 vary) to simulate observational constraints, and the DWDM and DCDM models have the same decay rate Γ.

In the top panel, Γ is such that both models decay before recombination. In this case, the energy density is greater than that of ΛCDM, increasing the value of the Hubble parameter. As the species decay, their values of H converge to that of the $N_{\rm eff}$ model. In all three models, the increase in $H(a)$ prior to recombination decreases the sound horizon at recombination, $r_s(a_*)$. Since observations essentially fix the acoustic scale to $\theta_s = r_s(a_*)/D_A(a_*)$, where $D_A(a_*)$ is the angular diameter distance to recombination, this results in a decrease of $D_A(a_*)$, which manifests in an increased value of H_0, as can be seen in the figure.[1] The model of extra relativistic degrees of freedom obtains the largest increase in H_0 but is known to have too strong an impact on the CMB spectrum to satisfactorily solve the Hubble tension [6]. The motivation for the decaying models, then, is that they may be able to circumvent these constraints by virtue of injecting the radiation energy density more locally around recombination.

[1] However, in the bottom panel, where Γ is such that both models decay after recombination, only a comparatively negligible increase in H_0 is seen.

25.2 DCDM→DR

The most simple model involving decaying dark matter is decaying *cold* dark matter (DCDM), where a cold mother particle decays into massless particles denoted as *dark radiation* (DR), which is a type of radiation that does not interact electromagnetically. We can simply write this process as

$$X_{\rm dcdm} \rightarrow \gamma_{\rm dr}\,.$$

As mentioned in the introduction, the scenario where all dark matter is decaying is heavily constrained, so we consider a model where just a fraction,

$$f_{\rm dcdm} \equiv \frac{\Omega^{\rm ini}_{\rm dcdm}}{\Omega^{\rm ini}_{\rm dcdm} + \Omega_{\rm scdm}}\,, \tag{25.1}$$

of the dark matter is unstable. Here $\Omega_{\rm scdm}$ is the density parameter of the stable cold dark matter, and $\Omega^{\rm ini}_{\rm dcdm}$ is the density parameter of the decaying component as it would be today if none of it had decayed [7]. This partial decay can also be interpreted as all CDM decaying into a stable CDM particle and DR [7].

The equations for the homogeneous and isotropic energy densities in this model resemble the normal background equations for CDM and radiation but with additional source terms dependent on the decay rate, $\Gamma_{\rm dcdm}$ (w.r.t. proper time) [7],

$$\begin{aligned} \rho'_{\rm dcdm} &= -3\frac{a'}{a}\rho_{\rm dcdm} - a\Gamma_{\rm dcdm}\rho_{\rm dcdm}\,, \\ \rho'_{\rm dr} &= -4\frac{a'}{a}\rho_{\rm dr} + a\Gamma_{\rm dcdm}\rho_{\rm dcdm}\,, \end{aligned} \tag{25.2}$$

where the prime denotes derivatives w.r.t. conformal time.

The equations in first-order general relativistic perturbation theory for the cold mother particle are exactly the same as for normal CDM, where only the density perturbation is non-trivial. For DR, the Boltzmann hierarchy resembles that of massless neutrinos but with additional source terms in the lowest modes [4].

It is useful to distinguish between two regions of the parameter space when dealing with this model, since the two limits $\Gamma_{\rm dcdm} \rightarrow 0$ and $\Gamma_{\rm dcdm} \rightarrow \infty$ correspond to the ΛCDM model, with the first having no constraints on the fractional amount of DCDM, $f_{\rm dcdm}$, and the latter having a very weak constraint of $f_{\rm dcdm} \neq 1$. This leads to a large parameter space volume of good fits to data in both limiting cases, and this funnel-shaped likelihood is difficult to sample simultaneously with an MCMC algorithm. We therefore split the parameter space into a *long-lived* region, where $\Gamma_{\rm dcdm} < 10^3\ {\rm Gyr}^{-1}$, and a *short-lived* region, where $\Gamma_{\rm dcdm} > 10^3\ {\rm Gyr}^{-1}$. The long-lived regime corresponds to a very late decay long after recombination and the short-lived regime corresponds to an early decay before recombination. The latter is probably the most interesting in relation to the Hubble tension, since it can impact the

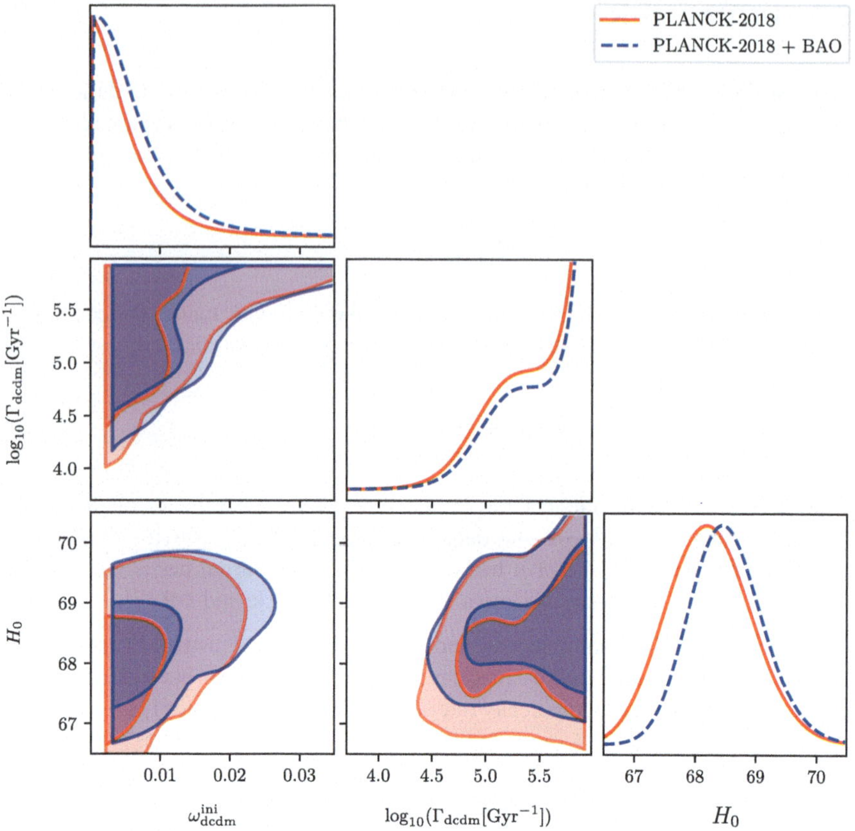

Fig. 25.2 Triangle plot of posteriors in the short-lived regime using Planck 2018 data with and without BAO peak data. The figure shows the model-specific parameters, $\omega^{\rm ini}_{\rm dcdm}$ and $\Gamma_{\rm dcdm}$, along with H_0, and has been produced using the same MCMC runs used in Ref. [7]

physics around recombination, and hence the formation of the CMB. The constraints on H_0 from Ref. [7] are different for the two regimes, with the long-lived regime yielding a 95.45% credible interval of $H_0 = 67.7^{+1.0}_{-0.9}$ km s^{-1} Mpc^{-1} and the short-lived one yielding one of $H_0 = 68.6^{+1.2}_{-1.4}$ km s^{-1} Mpc^{-1}. As we might expect, the short-lived regime alleviates the Hubble tension better of the two with almost one standard deviation (compared to a value inferred from local measurements of $H_0 = 73.04 \pm 1.04$ km s^{-1} Mpc^{-1} from Ref. [2]). We will therefore only consider the short-lived regime from hereon.

The DCDM model has been investigated in numerous papers with Bayesian statistics [8–15] and recently also with frequentist statistics and profile likelihoods [16]. Figure 25.2 shows the posteriors for the short-lived regime in the model-specific parameters (along with H_0) using high-ℓ and low-ℓ temperature and polarization as well as lensing data from Planck 2018 [17] (from now on just referred to as Planck

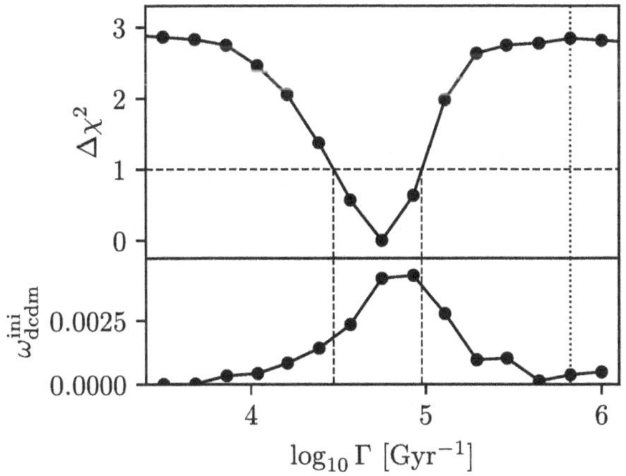

Fig. 25.3 Top panel: One-dimensional profile likelihood of $\log_{10}\Gamma/\mathrm{Gyr}$. Bottom panel: The abundance of decaying cold dark matter $\omega^{\rm ini}_{\rm dcdm}$ associated with every point in the profile above. Dashed lines indicate the $\Delta\chi^2 = 1$ intersections, giving the 1σ CIs. Taken from Ref. [16]

2018 data) with and without the BOSS DR12 BAO peak data [18] (from now on referred to as BAO peak data). The posterior of the decay rate, $\Gamma_{\rm dcdm}$, shows a volume effect towards higher values, where a large part of the parameter space fits the data well. Because of this, it is not possible to assign a reasonable credible interval to the decay rate since any such interval would be prior dependent. It is, however, also apparent that a region around $\log_{10}(\Gamma/\mathrm{Gyr}) \in [4.5, 5.5]$ includes a separate effect in the posterior, visualized as a small "bump" or plateau, which marks this as a region of interest. By investigating the same region in the decay rate with profile likelihoods, we find that a peak in the likelihood (a "well" in $\chi^2(\Gamma) \equiv -2\log(\mathcal{L}(\Gamma)/\max(\mathcal{L}(\Gamma)))$ arises here, as shown in Fig. 25.3. This study has been performed with the same data (Planck 2018 data and BAO peak data) along with low redshift BAO data from the 6dF survey [19] and the BOSS main galaxy sample [20] (the inclusion of which will be referred to as the full BOSS DR12 data set). The approximate 68% confidence interval obtained is $\log_{10}\Gamma\ \mathrm{Gyr}^{-1} = 4.763^{+0.214}_{-0.290}$, while it is unconstrained at 95%. The global best-fit of the model lies in this region at $\Delta\chi^2 = 2.8$ relative to the ΛCDM model, hinting at a mild preference for the DCDM model over ΛCDM. Interestingly, the best-fitting parameters $\log_{10}\Gamma\ \mathrm{Gyr}^{-1} = 4.763$ and $\omega^{\rm ini}_{\rm dcdm} = 0.00429$ correspond to a scenario where about 3% of the cold dark matter decays just prior to recombination, in support of the tendency of early time solutions to the Hubble tension to preferentially modify physics temporally close to recombination [6, 21]. Despite the stronger signature of DCDM in the frequentist analysis, the resulting 68% constraint $H_0 = 68.14^{+0.54}_{-0.49}$ km s^{-1} Mpc^{-1} solidifies the conclusion that the DCDM→DR model is unable to solve the H_0 tension (Fig. 25.4).

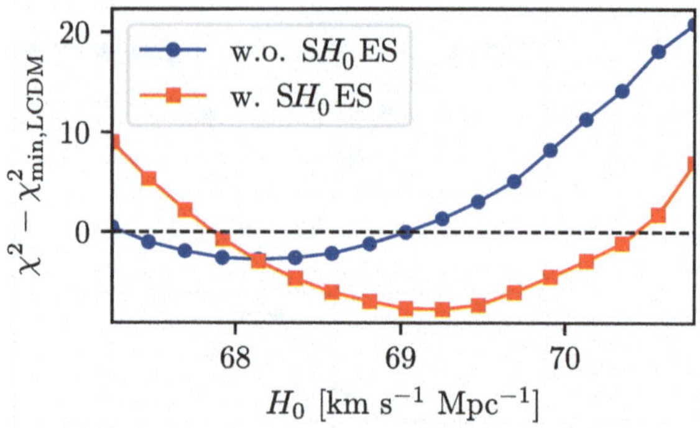

Fig. 25.4 One-dimensional profile likelihoods of H_0 using Planck 2018 data and the full BOSS DR12 data set (see text). The two profiles are with and without an additional SH_0ES prior. Taken from Ref. [16]

25.3 DCDM→DR+WDM

A reasonable increase in model complexity is to allow one of the daughter particles to be massive. We still have a cold mother particle decaying into dark radiation, but now also accompanied by a massive daughter particle acting as *warm dark matter* (WDM). We can write this process as

$$X_{\rm dcdm} \rightarrow \gamma_{\rm dr} + Y_{\rm wdm} \,.$$

This model has the same model parameters as the DCDM→DR model (abundance and decay rate), but it also has a new parameter: The mass ratio of the WDM particle to the DCDM particle, $\widetilde{m}$. Using conservation of energy and momentum, one can relate this mass ratio to the WDM velocity and the fraction of energy transferred to the massless DR particle in the decay, ϵ, through the following two equations [22],

$$\epsilon = \frac{1}{2}(1-\widetilde{m}^2), \qquad \beta^2_{\rm wdm} = \frac{\epsilon^2}{(1-\epsilon)^2}\,, \tag{25.3}$$

where $\beta_{\rm wdm}$ is the velocity of the massive daughter in natural units.

The background and perturbation equations are the same for DCDM and DR as in the DCDM→DR model, except for an additional factor of ϵ on the source term in the background equation for DR. The equations for the massive daughter can be expressed in terms of the equation-of-state parameter, $w_{\rm wdm}(a)$, which can be shown to have the following form [22],

$$w_{\rm wdm}(a) = \frac{1}{3}\frac{\Gamma\beta^2}{1-{\rm e}^{-\Gamma t}}\int_{a_{\rm ini}}^{a(t)} \frac{{\rm e}^{-\Gamma t_D}\, {\rm d}ln(a_D)}{H_D[(a/a_D)^2(1-\beta^2)+\beta^2]}, \tag{25.4}$$

where $a_{\rm ini}$ refers to some initial value of a where our numerical integration begins, and the subscript "D" refers to the integration variable. The background equation for the massive daughter particle is then

$$\dot{\rho}_{\rm wdm} = -3\big(1 + w_{\rm wdm}(a)\big)\frac{a'}{a}\rho_{\rm wdm} + (1-\epsilon)a\Gamma\rho_0\,. \tag{25.5}$$

The perturbation equations resemble those of massive neutrinos but can be rewritten in terms of $w_{\rm wdm}$ in a similar manner [23]. An accurate implementation of the full Boltzmann hierarchy fast enough for MCMC runs to be feasible is still missing, but results have been produced using a fluid approximation [5, 6, 23–25].

Figure 25.5 shows the posteriors for the model-specific parameters, Γ and ϵ, along with H_0. These results assume that all dark matter is decaying, i.e., the abundance ratio, $f_{\rm dcdm}$, is fixed to unity. Because of this, the results are within the very long-lived regime, and the energy transfer ratio, ϵ, is rather low as well, resulting in a heavy WDM daughter particle acting much like CDM. In particular, if $\log_{10}(\epsilon) \lesssim -2.7$, the decay rate becomes unconstrained because the massive daughter particle becomes virtually indistinguishable from the cold mother particle. The 68.27% credible interval for the Hubble constant from this analysis is $H_0 = 67.71^{+0.42}_{-0.43}$ km s^{-1} Mpc^{-1}, which is similar to the result from the long-lived regime in the simple DCDM→DR model. We would expect that an analysis with a fractional decay in the short-lived regime would yield a higher value of H_0. In order to do this accurately, the full hierarchy should be solved for WDM instead of using a fluid approximation.

25.4 DWDM→DR

Another extension of the simplest decaying dark matter model in Sect. 25.2 is to allow the decaying particle a non-negligible momentum, making it a warm dark matter species. The model of a decaying warm dark matter (DWDM) species decaying to dark radiation was studied in Refs. [1, 26]. This model has several interesting particle physics realizations, such as decaying neutrinos [27] and majoron decays [28]. In particular, in [29] it was used to constrain the lifetime of the active neutrino species.

The fundamental characteristic that separates the DWDM model from the other decaying dark matter models is that the non-negligible momentum increases the lifetime by a factor of E/m through time dilation, where E and m denote the energy and mass of the decaying particle, respectively. In the general case, the model contains three parameters: The decay constant, Γ (or, equivalently, the lifetime $\tau \equiv 1/\Gamma$), the DWDM mass, m, and the abundance of the DWDM species, specified either at final or initial time. The background equations are [1]

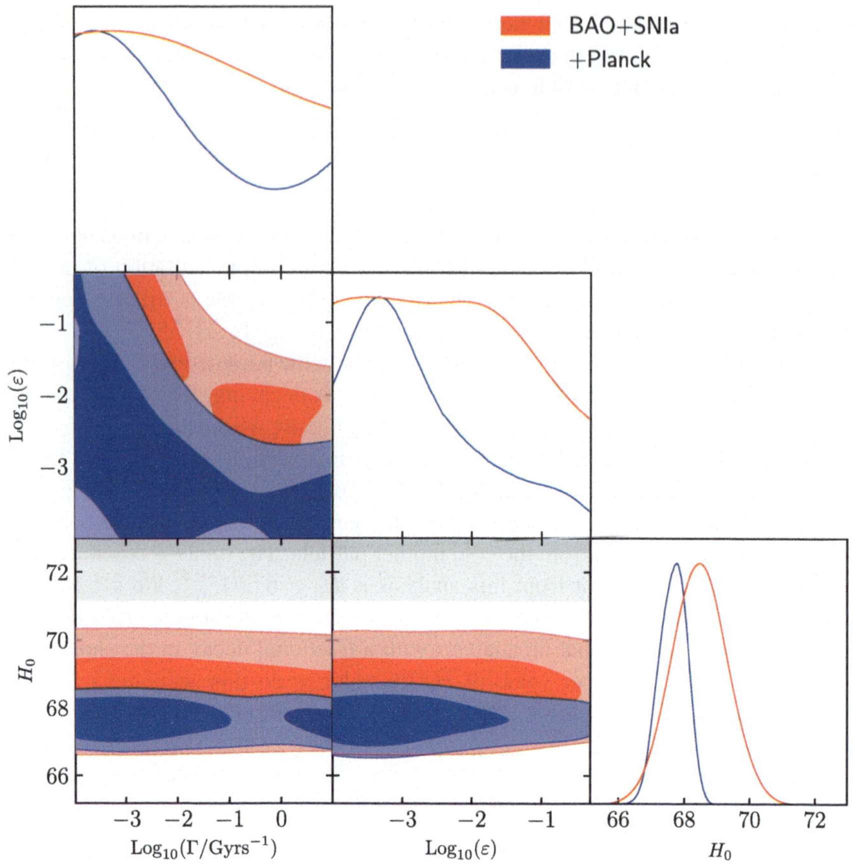

Fig. 25.5 Triangle plot of the decay rate, the energy transfer ratio, and the Hubble constant. The red lines and contours are from a background-only MCMC run with the full BOSS DR12 data set and a SH_0ES prior, while the blue lines and contours are from an MCMC run including perturbations (with the fluid approximation) and also the Planck 2018 data in addition to the other data sets. The gray contours are the H_0 measurement by the SH_0ES collaboration. The figure is taken from Ref. [23]

$$\begin{aligned} \rho'_{\text{dwdm}} &= -3\frac{a'}{a}(\rho_{\text{dwdm}} + p_{\text{dwdm}}) - a\Gamma m n_{\text{dwdm}}, \\ \rho'_{\text{dr}} &= -4\frac{a'}{a}\rho_{\text{dr}} + a\Gamma_{\text{dcdm}} m n_{\text{dwdm}}, \end{aligned} \tag{25.6}$$

where ρ_i and p_i denote the energy and pressure density of the i'th species and n_{dwdm} the number density of the decaying species. Apart from the addition of the non-zero DWDM pressure, p_{dwdm}, these equations are identical to the case of a cold decaying species (25.2) up to the substitution $\rho \to mn$, i.e., it is the rest mass and not the total energy that drives the decay.

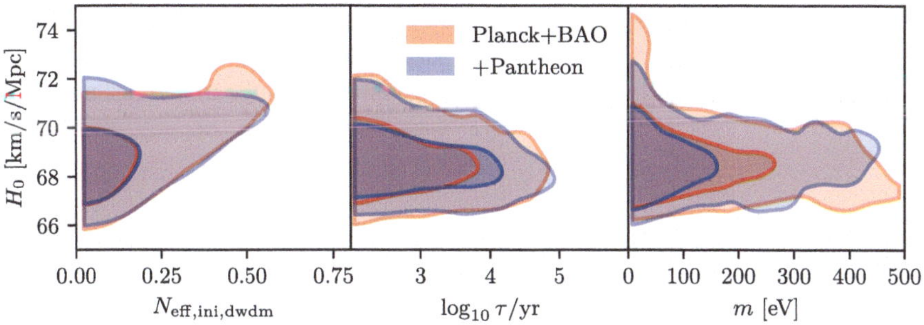

Fig. 25.6 Two-dimensional marginalized posteriors for the Hubble constant, H_0, and the three parameters of the DWDM→DR model: The initial abundance, $N_{\mathrm{eff,ini,dwdm}}$, parametrized as its contribution to the effective number of relativistic degrees of freedom, the DWDM lifetime, τ, and the DWDM mass, m. Taken from Ref. [1]

As in the DCDM→DR model, the perturbations of the DWDM species are identical to those of massive neutrinos [30]. Moreover, the decay product perturbations are influenced by a collision term, which captures the fact that the DWDM species decays preferentially at low momenta [1].

Figure 25.6, adopted from [1], shows contours of the two-dimensional marginalized posterior distributions for H_0 and three parameters of the DWDM→DR model using Planck 2018 data combined with the full BOSS DR12 data set. Here, the initial abundance is parametrized as $N_{\mathrm{eff,ini,dwdm}}$, its contribution to the effective number of relativistic degrees of freedom. The prior on the lifetime studied here is roughly equivalent to the short-lived regime explored earlier.

Evidently, although the maximum a posteriori estimate favors the ΛCDM limit, a considerable amount of the DWDM species is permitted by the data. Ref. [1] finds a 68% credible upper bound of $N_{\mathrm{eff,ini,dwdm}} < 1.05$. Furthermore, as seen in Fig. 25.6, there is a mild correlation between H_0 and the DWDM abundance, indicating that the DWDM→DR model may admit a larger value of H_0 than ΛCDM. A 68% credible interval of $H_0 = 68.73^{+0.81}_{-1.3}$ km s^{-1} Mpc^{-1}, at a 2.7σ Gaussian tension with the representative local value $H_0 = 73.04 \pm 1.04$ km s^{-1} Mpc^{-1} [2], is presented in this study. Thus, the broad conclusion is that there is no evidence from CMB data that the DWDM→DR model solves the Hubble tension.

Although the constraints on the lifetime, τ, and mass, m, are strongly driven by the bounds of the uniform prior chosen, in broad terms, the data prefers small masses and small lifetimes. Unfortunately, in all treatments of the DWDM→DR model to date, the effects of *inverse decays* have been neglected. The inverse decay process is kinematically allowed when the DWDM species decays while still relativistic, which is exactly the scenario in the preferred region of parameter space [1]. Thus, a complete understanding of the DWDM→DR model inevitably requires a numerical implementation of the inverse decay processes and their quantum statistical corrections [27]. Finally, we also note that the Bayesian constraints on the DWDM→DR model are expected to be strongly influenced by prior volume effects [1] since it reduces to the ΛCDM model in the limit of vanishing abundance, making the lifetime and mass unconstrained and thereby storing a large probability mass in the posterior distribution. At this time, a frequentist analysis of the model has not been conducted.

25.5 Discussion and Conclusion

The three decay models presented in this chapter all show an ability to slightly alleviate the Hubble tension. The simplest of the models (DCDM→DR), has been exhaustively studied, and it appears that it has reached its limits regarding how much it can alleviate the tension. The other two models, however, still have potential for further investigation. The DCDM→DR+WDM model still needs a fast and accurate implementation of the full Boltzmann hierarchy, and with a fractional decay, this model can also be studied in the short-lived regime. The DWDM→DR model also prefers slightly larger values of H_0, but a study including the inverse decay process is needed for a definite conclusion. Further investigation of the latter two models, along with improvements, could potentially alleviate the Hubble tension further. Ultimately, at the time of writing, we cannot definitively say which decaying dark matter model has the strongest alleviation of the Hubble tension, although there is a slight preference for the DWDM→DR model with the 68% C.I. $H_0 = 68.73^{+0.81}_{-1.3}$ km s^{-1} Mpc^{-1} at a 2.7σ Gaussian tension with the representative local value $H_0 = 73.04 \pm 1.04$ km s^{-1} Mpc^{-1} [2]. Thus, although decaying dark matter models are somewhat more natural than several other proposed solutions to the H_0 discrepancy, they only accomplish a mild alleviation of the latter [6].

The next natural step to take in the hierarchy of numerical complexity would be a fully general two-body decay from a mother particle with a definite mass to two daughter particles with different masses, i.e., DWDM→ WDM$_{(1)}$+WDM$_{(2)}$. This is a challenging task since it combines all the most difficult aspects of the previous models while also introducing new ones. Although the full set of perturbation equations has been derived [27], an implementation in a numerical Einstein–Boltzmann solver code still remains. Furthermore, we expect the full model to be very computationally expensive to evaluate, possibly making it unfeasible for immediate inference purposes. Nevertheless, a fully general decay scheme like this would, apart from other use cases like neutrino decays, possibly be able to affect the physics around recombination in just the right way for the Hubble tension to be relieved.

Acknowledgements A. Nygaard, E. B. Holm, and T. Tram were supported by a research grant (29337) from VILLUM FONDEN. We would like to thank Guillermo Franco Abellán (first author of Ref. [23]) for letting us use Fig. 25.5 in this chapter.

References

1. E.B. Holm, T. Tram, S. Hannestad, Decaying warm dark matter revisited. JCAP **08**(08), 044 (2022). https://doi.org/10.1088/1475-7516/2022/08/044, http://arxiv.org/abs/2205.13628 [astro-ph.CO]
2. A.G. Riess et al., A comprehensive measurement of the local value of the hubble constant with 1 km s^{-1} Mpc^{-1} uncertainty from the hubble space telescope and the SH_0ES team. Astrophys. J. Lett. **934**(1), L7(2022). https://doi.org/10.3847/2041-8213/ac5c5b, http://arxiv.org/abs/2112.04510 [astro-ph.CO]
3. L. Zhang, X. Chen, M. Kamionkowski, Z.-G. Si, Z. Zheng, Constraints on radiative dark-matter decay from the cosmic microwave background. Phys. Rev. D **76**, 061301 (2007). https://doi.org/10.1103/PhysRevD.76.061301, http://arxiv.org/abs/0704.2444 [astro-ph]
4. B. Audren, J. Lesgourgues, G. Mangano, P.D. Serpico, T. Tram, Strongest model-independent bound on the lifetime of dark matter. JCAP **12**, 028 (2014). https://doi.org/10.1088/1475-7516/2014/12/028, http://arxiv.org/abs/1407.2418 [astro-ph.CO]
5. T. Simon, G. Franco Abellán, P. Du, V. Poulin, Y. Tsai, Constraining decaying dark matter with BOSS data and the effective field theory of large-scale structuresPhys. Rev. D **106** (2), 023516 (2022). https://doi.org/10.1103/PhysRevD.106.023516, http://arxiv.org/abs/2203.07440 [astro-ph.CO]
6. N. Schöneberg, G. Franco Abellán, A. Pérez Sánchez, S. J. Witte, V. Poulin, J. Lesgourgues, The H_0 Olympics: A fair ranking of proposed models. Phys. Rept. **984**, 1–55 (2022). https://doi.org/10.1016/j.physrep.2022.07.001, http://arxiv.org/abs/2107.10291 [astro-ph.CO]
7. A. Nygaard, T. Tram, S. Hannestad, Updated constraints on decaying cold dark matter. JCAP **05**, 017 (2021). https://doi.org/10.1088/1475-7516/2021/05/017, http://arxiv.org/abs/2011.01632 [astro-ph.CO]
8. K. Ichiki, M. Oguri, K. Takahashi, WMAP constraints on decaying cold dark matter. Phys. Rev. Lett. **93**, 071302 (2004). https://doi.org/10.1103/PhysRevLett.93.071302, https://arxiv.org/abs/astro-ph/0403164
9. Z. Berezhiani, A.D. Dolgov, I.I. Tkachev, Reconciling Planck results with low redshift astronomical measurements. Phys. Rev. D **92**(6), 061303 (2015). https://doi.org/10.1103/PhysRevD.92.061303, http://arxiv.org/abs/1505.03644 [astro-ph.CO]
10. A. Chudaykin, D. Gorbunov, I. Tkachev, Dark matter component decaying after recombination: Lensing constraints with Planck data. Phys. Rev. D **94**, 023528 (2016). https://doi.org/10.1103/PhysRevD.94.023528, http://arxiv.org/abs/1602.08121 [astro-ph.CO]
11. V. Poulin, P.D. Serpico, J. Lesgourgues, A fresh look at linear cosmological constraints on a decaying dark matter component. JCAP **08**, 036 (2016). https://doi.org/10.1088/1475-7516/2016/08/036, http://arxiv.org/abs/1606.02073 [astro-ph.CO]
12. I.M. Oldengott, D. Boriero, D.J. Schwarz, Reionization and dark matter decay. JCAP **08**, 054 (2016). https://doi.org/10.1088/1475-7516/2016/08/054, http://arxiv.org/abs/1605.03928 [astro-ph.CO]
13. A. Chudaykin, D. Gorbunov, I. Tkachev, Dark matter component decaying after recombination: sensitivity to baryon acoustic oscillation and redshift space distortion probes. Phys. Rev. D **97**(8), 083508 (2018). https://doi.org/10.1103/PhysRevD.97.083508, http://arxiv.org/abs/1711.06738 [astro-ph.CO]
14. K.L. Pandey, T. Karwal, S. Das, Alleviating the H_0 and σ_8 anomalies with a decaying dark matter model. JCAP **07**, 026 (2020). https://doi.org/10.1088/1475-7516/2020/07/026, http://arxiv.org/abs/1902.10636 [astro-ph.CO]
15. L. Xiao, L. Zhang, R. An, C. Feng, B. Wang, Fractional Dark Matter decay: cosmological imprints and observational constraints. JCAP **01**, 045 (2020). https://doi.org/10.1088/1475-7516/2020/01/045, http://arxiv.org/abs/1908.02668 [astro-ph.CO]
16. E.B. Holm, L. Herold, S. Hannestad, A. Nygaard, T. Tram, Decaying dark matter with profile likelihoods. Phys. Rev. D **107**(2), L021303 (2023). https://doi.org/10.1103/PhysRevD.107.L021303, http://arxiv.org/abs/2211.01935 [astro-ph.CO]

17. **Planck** Collaboration, N. Aghanim et al., Planck 2018 results. VI. Cosmological parameters. Astron. Astrophys. **641**, A6 (2020). https://doi.org/10.1051/0004-6361/201833910, http://arxiv.org/abs/1807.06209 [astro-ph.CO]. [Erratum: Astron.Astrophys. 652, C4 (2021)]
18. **BOSS** Collaboration, S. Alam et al., The clustering of galaxies in the completed SDSS-III Baryon oscillation spectroscopic survey: cosmological analysis of the DR12 galaxy sample. Mon. Not. Roy. Astron. Soc. **470**(3), 2617–2652 (2017). https://doi.org/10.1093/mnras/stx721, http://arxiv.org/abs/1607.03155 [astro-ph.CO]
19. **BOSS** Collaboration, S. Alam et al., The clustering of galaxies in the completed SDSS-III Baryon oscillation spectroscopic survey: cosmological analysis of the DR12 galaxy sample. Mon. Not. Roy. Astron. Soc. **470**(3), 2617–2652 (2017). https://doi.org/10.1093/mnras/stx721, http://arxiv.org/abs/1607.03155 [astro-ph.CO]
20. A.J. Ross, L. Samushia, C. Howlett, W.J. Percival, A. Burden, M. Manera, The clustering of the SDSS DR7 main galaxy sample – I. A 4 per cent distance measure at $z = 0.15$. Mon. Not. Roy. Astron. Soc. **449**(1), 835–847 (2015). https://doi.org/10.1093/mnras/stv154, http://arxiv.org/abs/1409.3242 [astro-ph.CO]
21. L. Knox, M. Millea, Hubble constant hunter's guide. Phys. Rev. D **101**(4), 043533 (2020). https://doi.org/10.1103/PhysRevD.101.043533, http://arxiv.org/abs/1908.03663 [astro-ph.CO]
22. G. Blackadder, S.M. Koushiappas, Dark matter with two- and many-body decays and supernovae type Ia. Phys. Rev. D **90**(10), 103527 (2014). https://doi.org/10.1103/PhysRevD.90.103527, http://arxiv.org/abs/1410.0683 [astro-ph.CO]
23. G. Blackadder, S.M. Koushiappas, Dark matter with two- and many-body decays and supernovae type Ia. Phys. Rev. D **90**(10), 103527 (2014). https://doi.org/10.1103/PhysRevD.90.103527, http://arxiv.org/abs/1410.0683 [astro-ph.CO]
24. G. Franco Abellán, R. Murgia, V. Poulin, J. Lavalle, Implications of the S_8 tension for decaying dark matter with warm decay products. Phys. Rev. D **105**(6), 063525 (2022). https://doi.org/10.1103/PhysRevD.105.063525, http://arxiv.org/abs/2008.09615 [astro-ph.CO]
25. L. Fuß, M. Garny, Decaying Dark Matter and Lyman-α forest constraints. http://arxiv.org/abs/2210.06117 [astro-ph.CO]
26. N. Blinov, C. Keith, D. Hooper, Warm decaying dark matter and the hubble tension. JCAP **06**, 005 (2020). https://doi.org/10.1088/1475-7516/2020/06/005, http://arxiv.org/abs/2004.06114 [astro-ph.CO]
27. G. Barenboim, J.Z. Chen, S. Hannestad, I.M. Oldengott, T. Tram, Y.Y.Y. Wong, Invisible neutrino decay in precision cosmology. JCAP **03**, 087 (2021). https://doi.org/10.1088/1475-7516/2021/03/087, http://arxiv.org/abs/2011.01502 [astro-ph.CO]
28. M. Escudero, S.J. Witte, The hubble tension as a hint of leptogenesis and neutrino mass generation. Eur. Phys. J. C **81**(6), 515 (2021). https://doi.org/10.1140/epjc/s10052-021-09276-5, http://arxiv.org/abs/2103.03249 [hep-ph]
29. G. Franco Abellán, Z. Chacko, A. Dev, P. Du, V. Poulin, Y. Tsai, Improved cosmological constraints on the neutrino mass and lifetime. JHEP **08**, 076 (2022). https://doi.org/10.1007/JHEP08(2022)076, http://arxiv.org/abs/2112.13862 [hep-ph]
30. C.-P. Ma, E. Bertschinger, Cosmological perturbation theory in the synchronous and conformal Newtonian gauges. Astrophys. J. **455**, 7–25 (1995). https://doi.org/10.1086/176550, https://arxiv.org/abs/astro-ph/9506072

Chapter 26
Signs of a Non-zero Equation-of-State for Dark Matter

Krishna Naidoo

26.1 Tensions and Anomalies

The standard model of cosmology ΛCDM has been extremely successful in describing cosmological observations, including the cosmic microwave background [15], galaxy clustering [2] and distance ladder [18] measurements. This simple model is comprised of roughly 5% normal matter, 25% dark matter and 70% dark energy. Despite it's many successes the model remains incomplete. Almost 95% of our current universe is made of stuff we have never directly measured, with few signs that it's properties will be detected in the near future. In ΛCDM we assume dark matter interacts exclusively through gravity and is cold, meaning it's particles have low velocities that allow them to sink and form large clumps. On the other hand, dark energy is assumed to be a cosmological constant, a scalar field which has a single value across all of space-time. While general relativity allows such a solution, fundamental physics has found explaining its small value extremely difficult. This latter point is often the premise for considering models proposing modifications to gravity.

Over the last decade tensions have emerged between the determination of parameters from the early universe and the late universe. The most egregious of these discrepancies is the Hubble tension, a tension between the expansion rate of the universe today measured from cepheid variables [16] and those derived from CMB measurements [15]. Typically measurements from the late universe find a larger Hubble constant than those derived from CMB and early universe physics (including Baryonic Acoustic Oscillations from large scale structure). Combined with the S_8 tension—a discrepancy between measurements of the clustering amplitude from the early universe and late-time measurements of weak lensing have placed intense scrutiny on the standard model of cosmology. Are these signs that the simple but

K. Naidoo (✉)
University College London, Gower Street, London WC1E 6BT, UK
e-mail: k.naidoo@ucl.ac.uk

E. Di Valentino and D. Brout (eds.), *The Hubble Constant Tension*, Springer Series in Astrophysics and Cosmology, https://doi.org/10.1007/978-981-99-0177-7_26

powerful assumptions of ΛCDM are starting to crack, are we seeing symptoms of known limitations or are these just the result of systematics?

A lesser known discrepancy is the Integrated Sachs-Wolfe (ISW) void anomaly. The ISW is a secondary anisotropy which impacts the light from the CMB and is caused by the decay of gravitational potentials, which either draw or impart energy to photons traversing a gravitational potential well (for an overdensity) or trough (for an underdensity). This effect means large scale structure distorts the CMB we observe. There are a number of ways to measure this effect, the simplest approach is to cross-correlate large scale structure observations with the CMB, but this signal is typically quite weak and difficult to measure (see for e.g. [14]). Another approach is to use stacking, typically carried out with voids (see [5, 9]). From the latter, cosmologists have measured a surprising anomaly, in that the measurements of void ISW is often significantly larger than what is expected from ΛCDM. This discrepancy is strongest when measured from photometric voids, which are elongated and preferentially aligned with the line-of-sight. What makes this result particularly puzzling is observations of the lensing signal show no large discrepancies, and therefore no obvious systematics. Instead these measurements are either in agreement with ΛCDM or slightly lower.

26.2 The Dark Matter Solution

ΛCDM assumes the presence of cold dark matter, massive uninteractive particles that seed structure formation through their gravitational effects. However, little is known about what dark matter really is; is it a single particle or many, is the microphysics simple or complicated, and how does it fit in with the standard model of particle physics? Its presumed simplicity has proved to be extremely powerful in making predictions but have we reached the limit of these simplistic assumptions, i.e. has precision cosmology highlighted our ignorance—could cosmological tensions actually be manifestations of unknown dark matter physics?

Extensions to dark matter often only depart from ΛCDM on very small scales, when the mass of the dark matter particle starts to become relevant. For example 'warm' dark matter models [3], made of moderately light particles, prevent the growth of structures on the very smallest scales due to their larger streaming velocities. Other extensions such a fuzzy dark matter [7] similarly depart from ΛCDM on small scales, but in this case are caused by extremely light particles with de Broglie wavelengths on kiloparsec scales, causing quantum mechanical interference patterns on the matter distribution field. The tendency for dark matter extensions to be only relevant on small scales has created a perception that dark matter extensions are irrelevant for any solutions to cosmological tensions. However, since we are yet to detect dark matter directly there is plenty of room for unknown physics—physics that could be relevant on large scales. For example what if dark matter is not a single particle but many, with complicated interactions and decays? What if a subset of the particles are unstable and decay, what if dark matter interacts with other dark sector components—

i.e. dark energy and/or neutrinos? What approach do we take to explore these exotic dark matter properties? One approach is to take each extension and explore them individually, however this is a highly ineffective exploration of the parameter space. An alternative approach is to explore this with phenomenological descriptions. This allows us to explore the parameter space more effectively but comes at the cost of losing specificity, since the dark matter properties will be generally defined. The benefit of such an approach is that we can rely on data to dictate what dark matter models are most viable and relevant for observations. With this in mind we turn to the phenomenological *generalised* dark matter model [6]. In this model dark matter is modelled as a fluid with free functions for the Equation-of-State (EoS), speed of sound and shear viscosity. Since we are interested in the role this type of model can have on cosmological tensions we will use previous works, in particular [8, 19], to limit the parameters and functional forms considered. In both [19] and [8] a constant EoS for dark matter is constrained extremely tightly to $< 10^{-4}$ from CMB experiments. For this reason we will assume at very early times dark matters EoS is effectively zero. While we will assume the speed of sound is adiabatic and shear viscosity is zero.

26.2.1 A Non-zero Equation-of-State

With this in mind we will be exploring the consequence of a dark matter model with an EoS that begins at zero at early times and at some scale factor $a_{\rm nz}$ transitions to a monotonic function over cosmic time (specifically, linear as a function of the scale factor a);

$$w_{\rm dm}(a) = \begin{cases} w_{\rm dm,0}\left(\frac{a-a_{\rm nz}}{1-a_{\rm nz}}\right), & \text{for } a \geq a_{\rm nz}, \\ 0, & \text{otherwise}, \end{cases} \tag{26.1}$$

where $w_{\rm dm}(a)$ is the EoS, $w_{\rm dm,0}$ the EoS today, and $a_{\rm nz}$ is the scale factor at which the EoS transitions from zero to non-zero. In ΛCDM dark matter's EoS is $w_{\rm dm} = 0$. From the continuity equation we can define the time evolution of the density as

$$\rho_{\rm dm}(a) = \frac{\rho_{\rm dm,0}}{a^3} W(a), \tag{26.2}$$

where

$$W(a) = \exp\left(3\int_a^1 \frac{w_{\rm dm}(a')}{a'} da'\right). \tag{26.3}$$

The total matter distribution is then defined as

$$\rho_{\rm m}(a) = \frac{\rho_{\rm m,0}}{a^3}\Gamma(a), \tag{26.4}$$

$$\Gamma(a) = f_{\rm b} + f_\nu + f_{\rm dm} W(a), \tag{26.5}$$

where $f_{\rm i}$ is the fraction of component i to the matter density today (b for baryons, ν for non-relativistic neutrinos and dm for dark matter). The addition of the time-varying function $W(a)$ changes the time evolution of gravitational potentials in the Poisson equation, which becomes

$$\nabla^2 \Phi(x, a) = \frac{3}{2} H_0^2 a^2 \Omega_{\rm m}(a) \delta(x, a). \tag{26.6}$$

In the linear regime the time evolution and spatial components of the density contrast can be separated to $\delta(x, a) = D(a)\delta(x, 1)$ where $D(a)$ is the linear growth function. This means the time evolution of the gravitational potential is proportional to

$$\Phi(a) \propto \frac{D(a)}{a} \Gamma(a), \tag{26.7}$$

where as in ΛCDM $\Gamma(a) = 1$. This means the time-derivative of the gravitational potential is given by

$$\dot{\Phi}(a) = H(a) \left[f(a) - 1 + \Upsilon(a) \right] \Phi(a), \tag{26.8}$$

where $H(a)$ is the Hubble expansion rate, $f(a) = {\rm d} \ln D / {\rm d} \ln a$ the growth rate and

$$\Upsilon(a) = \frac{{\rm d}\Gamma(a)}{{\rm d} \ln a} = 3 f_{\rm dm} \frac{w_{\rm dm}(a) W(a)}{\Gamma(a)}. \tag{26.9}$$

In ΛCDM, or any cosmological model where $\Omega_{\rm m} \propto a^{-3}$, this last term ($\Upsilon$) is zero. A non-zero value is a unique signature of the non-zero EoS for dark matter.

26.2.2 *Impact on the Integrated Sachs-Wolfe and Lensing Signals*

Consider now what effect this will have on the ISW and lensing convergence. The ISW is defined as

$$\frac{T_{\rm ISW}(\hat{\eta})}{T_{\rm CMB}} = \frac{2}{c^3} \int_0^{\chi_{\rm LS}} \dot{\Phi}(\chi \hat{\eta}, a) a \, {\rm d}\chi, \tag{26.10}$$

and is sensitive to departures from the ΛCDM prediction that $\Upsilon(a) = 0$ (see Eq. 26.8). While the lensing convergence is defined as

$$\kappa(\hat{\eta}) = \frac{3 H_0^2}{2c^2} \Omega_{\rm m,0} \int_0^{\chi_s} \frac{\chi}{\chi_{\rm s}} (\chi_s - \chi) \frac{D(\chi)}{a(\chi)} \Gamma(\chi) \delta(\chi) {\rm d}\chi. \tag{26.11}$$

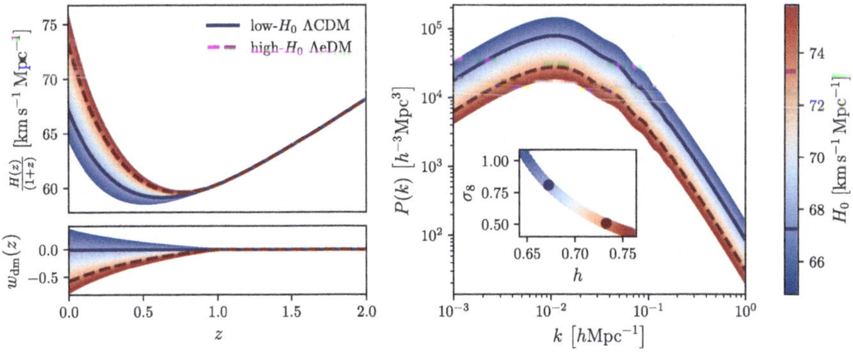

Fig. 26.1 The Hubble function $H(z)$, dark matter EoS $w_{\rm dm}(z)$ and power spectrum $P(k)$ are displayed for models of ΛeDM as a function of the Hubble constant H_0. Early universe physics are fixed to CMB constraints. A negative dark mater EoS results in a larger H_0 and a lower σ_8 (shown in the inset plot on the right). Therefore, ΛeDM naturally couples the two tensions, providing an explanation for both the H_0 and σ_8 tensions simultaneously. Profiles for ΛCDM (assuming early universe CMB constraints) are shown in dark blue (solid lines) and ΛeDM (assuming late universe constraints on H_0) are shown in red (dashed lines). The figure was originally presented in [11]

and is therefore sensitive to departures from the ΛCDM prediction $\Gamma(a) = 1$.

By construction, ΛeDM is identical to ΛCDM at early times, since in both cases dark matter EoS is zero, $w_{\rm dm} = 0$. We can use this property to fix the early universe to the conditions of ΛCDM inferred from Planck [15]. This allows us to focus on the impact of $w(a > a_{\rm nz}) \neq 0$. In Fig. 26.1 we fix $a_{\rm nz} = 0.5$ (equivalent to redshift $z = 1$) and vary $w_{\rm dm,0}$ showing a negative EoS for dark matter results in a larger H_0 and smaller σ_8. The higher H_0 is the results of a slower decay in the dark matter density, while the lower σ_8 is the result of suppressed structure growth. The coupling of these two effects shows the model can provide an explanation for both cosmological tensions simultaneously.

In a similar fashion we can explore the effect of ΛeDM on the ISW and the lensing profiles from voids. We look at three types of voids: spectroscopic voids—spherical and relatively small [12], photometric voids—large and elongated along the line-of-sight [9] and the Eridanus voids—voids founds along the line-of-sight of the cold spot anomaly [10] (see [11] for specific void profile properties). In Fig. 26.2 we show the effect of the same models considered in Fig. 26.1, showing a negative EoS leads to much deeper ISW signals from voids, an effect which is particularly strong for large elongated photometric and Eridanus voids. Rather interestingly this is coupled with equal or slightly lower lensing signals. This is perfectly in keeping with observations, where void ISW signals have been found to be much larger than predicted by ΛCDM but with lensing signals either lower or in agreement with ΛCDM. This unique prediction of ΛeDM is able to explain this long standing anomaly, which makes this model of extreme interest should these anomalies persists in the future.

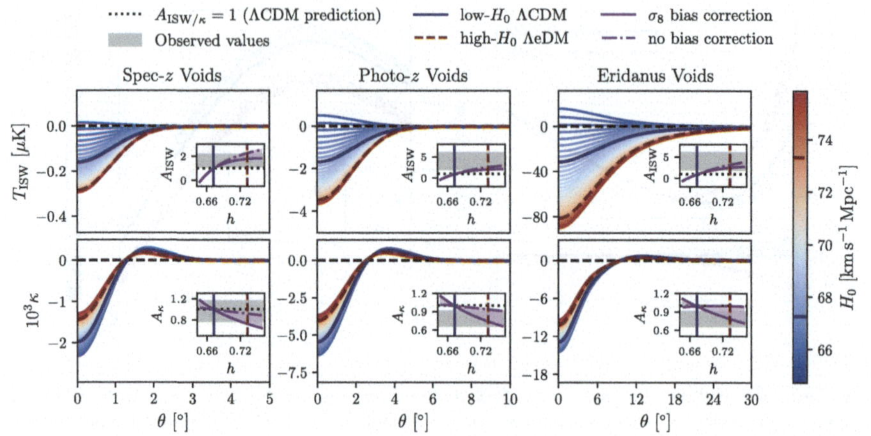

Fig. 26.2 The ISW (top panels) and lensing convergence κ (bottom panels) angular profiles for three void types (spec-z on the left, photo-z in the middle and Eridanus voids on the right) are shown as a function of H_0 in ΛeDM with all other parameters fixed to early universe CMB constraints. In the inset plots we show the amplitude of the ISW and κ profiles with respect to ΛCDM, shown with a σ_8 bias correction (solid purple lines) and without a bias correction (dashed purple lines). Measured values are indicated with horizontal grey bands. A higher H_0 resolves the void-ISW anomaly and recovers lower κ amplitudes. For reference the profiles for ΛCDM assuming early universe CMB constraints are shown with dark blue solid lines and the profiles for ΛeDM assuming late universe constraints on H_0 are shown with red dashed lines. The figure was originally presented in [11]

26.2.3 Observational Constraints

In the prior sections we have described how an evolution of the dark matter EoS at late-times can effect cosmological observables, with a keen focus on those relevant to cosmological tensions and the ISW-void anomaly. We can now take this a step further and constrain cosmological parameters with the EoS for dark matter allowed to vary. To avoid exploring largely degenerate parameters we fix $a_{\rm nz} = 0.5$ and allow $w_{\rm dm,0}$ to vary between -1 and 0 (the CDM limit). In Fig. 26.3 we show the constraints on the standard 6 parameter model with the addition of $w_{\rm dm,0}$ using observation of the CMB from Planck [15], supernova from Pantheon [18] with SH_0ES [16] Cepheid constraints on the absolute magnitude M_B and BAO and redshift space distortions from 6dF, SDSS DR7 and BOSS [1, 2, 17]. The constraints on the base ΛCDM parameters are virtually identical since these are mostly set at early redshift where the models are equivalent. Note, in the case of the dark matter fractional density, we define it according to its asymptotic value at early times, which can be evaluated by calculating

$$\Omega_{\rm dm}^{\rm init} = \Omega_{\rm dm,0}\, W(a_{\rm nz}), \tag{26.12}$$

where $\Omega_{\rm dm,0}$ is the fractional density of dark matter today. This is a measure of what the dark matter fractional density would be today in ΛCDM and is much easier to sample as this removes a strong degeneracy between $\Omega_{\rm dm,0}$ and $w_{\rm dm,0}$. With Planck alone $\Omega_{\rm dm,0}$ is poorly constrained, unsurprisingly since this model considers a late change to ΛCDM, CMB measurements can only constrain this model through their impact on the sound horizon scale. With the addition of supernovae and BAO the model is better constrained, with SN constraints preferring higher values of H_0, pushing constraints to higher $w_{\rm dm,0}$. However, this preference is not definitive and is still consistent with ΛCDM. In Fig. 26.4 we show the constraints on ΛeDM for the derived parameters H_0 and σ_8. This figure shows that there is a strong anti-correlation for high-H_0 and low-σ_8. In other words a small increase in H_0 is balanced with a significantly larger shift in σ_8. So while this model does correlate both tensions, raising $H_0 > 70\,{\rm Km\,s^{-1}\,Mpc^{-1}}$, as measured by [16], are strongly disfavoured due to the resulting low σ_8.

A negative EoS for dark matter at late-time is a promising solution to tensions and the ISW-void anomaly. Currently, cosmological data shows a marginal preference for ΛCDM [11], with more precise measurements from void ISW and lensing (or more typical galaxy and weak lensing auto and cross-correlations) required to make a more definitive statement. Current best-fit constraints on ΛeDM, with a negative EoS for dark matter at late-times, predict a dampening of the void lensing signal by a factor of ∼0.9 with respect to ΛCDM and an increase in the void ISW signal by a factor of ∼2 with respect to ΛCDM. These predictions fit perfectly with observational measurements from voids. To our knowledge, this is the only model that can simultaneously provide a solution to the Hubble and S_8 tension, and the void-ISW anomaly.

26.2.4 Implications: An Interacting Dark Sector

A negative EoS for dark matter is a significant departure from the cold dark matter paradigm. So what does a negative EoS for dark matter actually mean? Generally, the EoS for the different constituents in the universe are as follows, radiation has an EoS = 1/3, cold dark matter and baryons have an EoS = 0 and a cosmological constant has an EoS = −1. Imagine now that dark matter was unstable and decayed into radiation, over time its *effective* EoS would be slightly positive, since it would describe a combination of dark matter and radiation decay products. Similarly, a slightly negative EoS can be thought of as being an effective description of dark matter that decays or interacts with dark energy. For this reason the model is degenerate with models of interacting dark matter and dark energy [4]. To follow the behaviour of ΛeDM, the interactions would need to be relevant only at very late-times. This could be achieved if this interaction/decay only occurred at low energies or densities, or if the dark matter particle was unstable and decayed with a very long half-life. This behaviour is similar to the properties of the evanescent matter model of [13]. Should this turn out to be true, this would mark a significant departure from ΛCDM and imply that dark energy is not a cosmological constant but rather something else. Furthermore it would mean that cosmology could provide a window into new

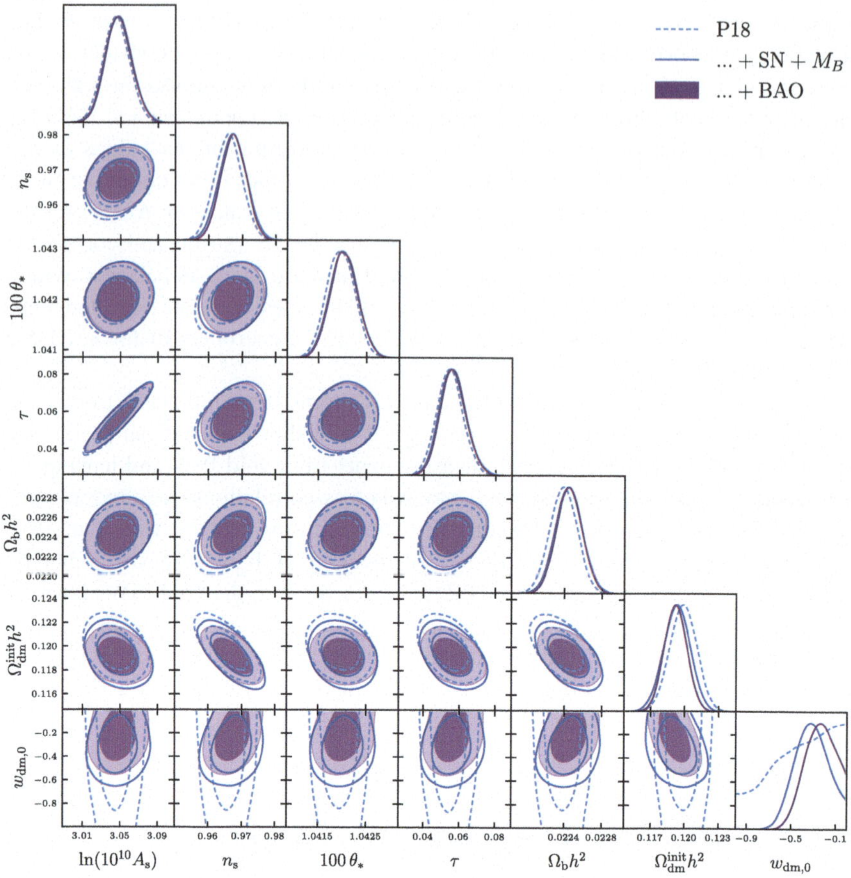

Fig. 26.3 Constraints on the standard 6 parameter cosmological model with the addition of the dark matter EoS today ($w_{dm,0}$). Constraints on the standard 6 parameters are identical to those of ΛCDM, since by construction ΛeDM does not alter the early universe. We see that Planck is very unconstraining on the additional $w_{dm,0}$ parameter. The addition of Pantheon supernovae (with SH_0ES constraints on the absolute magnitude M_B) and BAO, significantly constrain $w_{dm,0}$, and while the value is consistent with a zero there is a slight preference for a negative EoS of $w_{dm,0} = -0.2$

physics in the dark sector, physics which has remained elusive to direct detection from particle physics experiments.

26.3 Future Work and Predictions for Experiment

A non-zero EoS for dark matter can alleviate tensions in cosmology and provide an explanation for the ISW-void anomaly. Whether this models holds up against

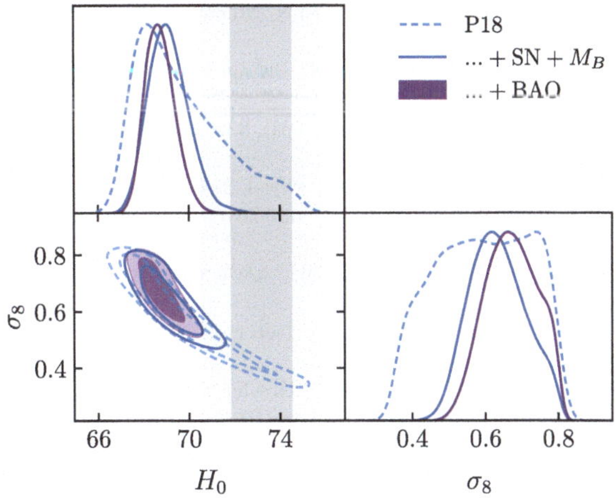

Fig. 26.4 Constraints on ΛeDM on the Hubble constants H_0 and σ_8 parameters. This shows a direct anti-correlation between high-H_0 and low-σ_8

the test of time depends on a number of factors: (1) that tensions and the ISW-void anomaly remain, without clear systematics in observations and/or methodology being discovered and (2) that limitations to the model are resolved.

Generalised dark matter is a good way to explore exotic dark matter models, however the model is limited to linear perturbation theory and has not been explored in great detail in high-resolution simulations and on small scales. Exploring these model in simulations will allow us to test whether such a model is compatible with observations and demonstrate how these models behave on small scales.

A unique consequence of this model is its prediction of larger ISW at low redshift. Improved ISW measurements from voids, including better understanding of void galaxy bias, will help to clearly establish the strength of the void-ISW anomaly—it's persistent will be the clearest sign of a negative EoS for dark matter at late times.

References

1. F. Beutler, C. Blake, M. Colless et al., MNRAS **416**, 3017 (2011). https://doi.org/10.1111/j.1365-2966.2011.19250.x
2. S. Alam et al., MNRAS **470**, 2617–2652 (2017). https://doi.org/10.1093/mnras/stx721
3. S. Colombi, S. Dodelson, L.M. Widrow, ApJ **458**, 1 (1996). https://doi.org/10.1086/176788
4. G.R. Farrar, P.J.E. Peebles, ApJ **604**, 1 (2004)
5. B.R. Granett, M.C. Neyrinck, I. Szapudi, ApJ **683**, L99 (2008). https://doi.org/10.1086/591670
6. W. Hu, ApJ **506**, 485 (1998). https://doi.org/10.1086/306274
7. W. Hu, R. Barkana, A. Gruzinov, Phys. Rev. Lett. **85**, 1158–1161 (2000). https://doi.org/10.1103/PhysRevLett.85.1158
8. M. Kopp, C. Skordis, D.B. Thomas et al., Phys. Rev. Lett. **120**, 221102 (2018). https://doi.org/10.1103/PhysRevLett.120.221102
9. A. Kovács, C. Sánchez, J. García-Bellido et al., MNRAS **484**, 5267 (2019). https://doi.org/10.1093/mnras/stz341

10. R. Mackenzie, T. Shanks, M.N. Bremer et al., MNRAS **470**, 2328 (2017). https://doi.org/10.1093/mnras/stx931
11. K. Naidoo, M. Jaber, W.A. Hellwing, M. Bilicki, (2022). https://doi.org/10.48550/arXiv.2209.08102, arXiv:2209.08102
12. S. Nadathur, R. Crittenden, ApJL **830**, L19 (2016). https://doi.org/10.3847/2041-8205/830/1/L19
13. P.J.E. Peebles, Annalen der Physik **524**(9), 591 (2012)
14. Planck Collaboration, et al., 2016. A&A, 594. doi:https://doi.org/10.1051/0004-6361/201525831
15. Planck Collaboration et al., A&A **641**, A6 (2020). https://doi.org/10.1051/0004-6361/201833910
16. A.G. Riess, W. Yuan, L.M. Macri et al., ApJL **934**, L7 (2022). https://doi.org/10.3847/2041-8213/ac5c5b
17. A.J. Ross, L. Samushia, C. Howlett, W.J. Percival, A. Burden, M. Manera, MNRAS **449**, 835–847 (2015). https://doi.org/10.1093/mnras/stv154
18. Scolnic, D. M., et al., 2018, ApJ, 859. doi:https://doi.org/10.3847/1538-4357/aab9bb
19. L. Xu, Y. Chang, Phys. Rev. D **88**, 127301 (2013). https://doi.org/10.1103/PhysRevD.88.127301

Chapter 27
Resolving the Hubble Tension at Late Times with Dark Energy

Marco Raveri

27.1 Introduction

In this chapter, we review attempts at solving the Hubble constant tension with non-standard properties of Dark Energy (DE) at late times. The empirical evidence for the existence of DE during the late cosmic epochs has been well-established through numerous observational studies, starting from the discovery of cosmic acceleration [1, 2]. Little progress has been made towards knowing its physical nature and properties, that, to this day, remain unknown. So far, the best description we have for DE is a cosmological constant, Λ, which is one of the cornerstones of the standard ΛCDM cosmological model.

DE is a natural candidate to try to explain existing cosmological tensions between the early and late time Universe. All physical components of the standard ΛCDM model, but DE, are showcased in the early universe, in particular at recombination. Dark matter, baryons, radiation, and neutrinos are all relevant at times when anisotropies in the Cosmic Microwave Background (CMB) form, and these observations can be used to study their properties. DE, on the other hand, is, at least in the standard model, completely irrelevant at those times. At late times, in turn, DE becomes relevant and is the only difference between the physical components of the early and late time universe. This is the empirical motivation to try to tie the physical properties of DE, beyond a cosmological constant, to the resolution of cosmological tensions.

In this chapter, we review these attempts, and the discussion is structured as follows: we first show, in Sect. 27.2, how DE enters the cosmological observables that are relevant for the Hubble constant tension; in Sect. 27.3 we discuss how DE can

M. Raveri (✉)
Department of Physics, INFN and INAF at the University of Genova, Via Dodecaneso 33, 16146 Genova, Italy
e-mail: marco.raveri@unige.it

E. Di Valentino and D. Brout (eds.), *The Hubble Constant Tension*, Springer Series in Astrophysics and Cosmology, https://doi.org/10.1007/978-981-99-0177-7_27

be used to relieve the tension by changing the calibration of low redshift observables; in Sect. 27.4 we discuss attempts at solving the Hubble tension by altering the late times expansion history of the Universe with non-standard DE properties.

27.2 Dark Energy and the Hubble Constant

The observable universe is characterized by large-scale homogeneity and isotropy, properties well-captured by the flat Friedmann-Lemaître-Robertson-Walker (FLRW) metric with the line element $ds^2 = -dt^2 + a(t)^2 d\mathbf{x}^2$. Here, $a(t)$ represents the scale factor governing the background expansion of the universe, with t being cosmic time. The scale factor is directly related to redshift by $z = 1/a(t) - 1$.

The temporal evolution of the scale factor is governed by the Friedmann equation, given by

$$H^2(a) = H_0^2 \left[\Omega_\mathrm{r}\, a^{-4} + \Omega_\mathrm{m}\, a^{-3} + \Omega_\mathrm{de}(a)\right], \tag{27.1}$$

where $H(a) \equiv \dot{a}/a$ represents the time-dependent Hubble expansion rate, in cosmic time. In this equation, H_0 denotes the present-day Hubble constant, while Ω_r and Ω_m correspond to the current relative energy densities of relativistic and non-relativistic particle species, respectively. The term $\Omega_\mathrm{de}(a) \equiv \rho_\mathrm{de}(a)/\rho^\mathrm{critical}_\mathrm{today}$ describes the effective density contribution of DE, ρ_de, relative to present day critical density. Note that, from Eq. (27.1), at present ($a = 1$), the constraint $\Omega_\mathrm{r} + \Omega_\mathrm{m} + \Omega_\mathrm{de} = 1$ has to be satisfied.

In the context of the ΛCDM model, DE is described with a cosmological constant, Λ, resulting in a constant value for Ω_de. However, in a broader context, $\Omega_\mathrm{de}(a)$ could encompass the cumulative contribution from all components other than radiation and matter densities. This includes possible modifications to gravity, which would lead to a modified Friedmann equation, as well as the potential existence of a non-zero curvature term, $\Omega_k\, a^{-2}$. In all these cases $\Omega_\mathrm{de}(a)$ becomes explicitly time dependent. A crucial test for the standard ΛCDM model involves verifying whether Ω_de remains constant throughout cosmic history.

Friedman equations written as in Eq. (27.1) are very general and can in principle describe any density contribution. For this reason, this approach is especially useful to probe empirical discrepancies and span different families of models.

Numerous investigations in literature [3, 4] focus on the equation of state of DE, which is given by

$$w_\mathrm{de}(a) = \frac{1}{3}\frac{d \ln \Omega_\mathrm{de}}{d \ln a} - 1, \tag{27.2}$$

looking for deviations from the standard model value of $w_\mathrm{de} = -1$, which characterizes a cosmological constant. We note here that this approach significantly restricts the family of models that can be studied in the framework, as Eq. (27.2) is well defined only for positive $\Omega_{\rm de}$. This constraint is not necessarily satisfied in the context of Modified Gravity (MG) theories [5], in which $\Omega_{\rm de}$ is an effective quantity—a placeholder—for structural modifications of Friedman equations. This constraint is also not satisfied by curvature effective density contributions and a negative cosmological constant. For these reasons, we phrase, whenever possible, the discussion in this chapter in terms of DE effective density.

The Hubble rate determines the redshift dependence of observable cosmological distances, and in particular luminosity distance that is defined as:

$$D_L(z) \equiv \sqrt{\frac{L}{4\pi S}} = (1+z) \int_0^z \frac{dz}{H(z)} \tag{27.3}$$

where L and S are the observed luminosity and flux of a source and we have assumed spatial flatness, from here onward.

Luminosity distances in the universe at late times can be measured with Supernovae (SN) observations that, once standardized, give us the apparent magnitude of an event, as a function of luminosity distance:

$$m(z) \equiv 5 \log_{10} \left(\frac{D_L(z)}{10\, pc} \right) + M_0 \tag{27.4}$$

where M_0 is a calibration parameter that, crucially, is the same for all (standardized) events of a given type and is redshift independent. This procedure is discussed in detail, for a state-of-the-art SN survey, in [6]. Note that in Eq. (27.4), the value of the Hubble constant enters as an additive parameter and, as such, is completely degenerate with the determination of M_0.

Distances can also be inferred, at late times, and in full analogy with the CMB, by measuring the angular size of the Baryon Acoustic Oscillation (BAO) feature in the galaxy correlation function. The reference physical scale of the BAO feature is the sound horizon, r_s, of the Thompson scattering coupled photo-baryon fluid, at the time when baryons stopped being dragged by photons. Crucially, at late times, this quantity does not evolve with redshift and is fully determined by pre-recombination physics. At a given redshift this scale projects on a feature at an angular separation of:

$$\theta_{\rm BAO} = \frac{r_s}{D_A(z)} \tag{27.5}$$

where $D_A(z)$ is the redshift-dependent angular diameter distance, related to luminosity distance by $D_L = (1+z)^2 D_A$. The value of the Hubble constant enters multiplicatively in Eq. (27.5), and, as such, is completely degenerate with the determination of the sound horizon.

Both SN and BAO measurements have to be calibrated, in different ways, to yield distance measurements. This is what we show in Fig. 27.1 where we can see: SN measurements from the Pantheon compilation [11], with the redshift binning

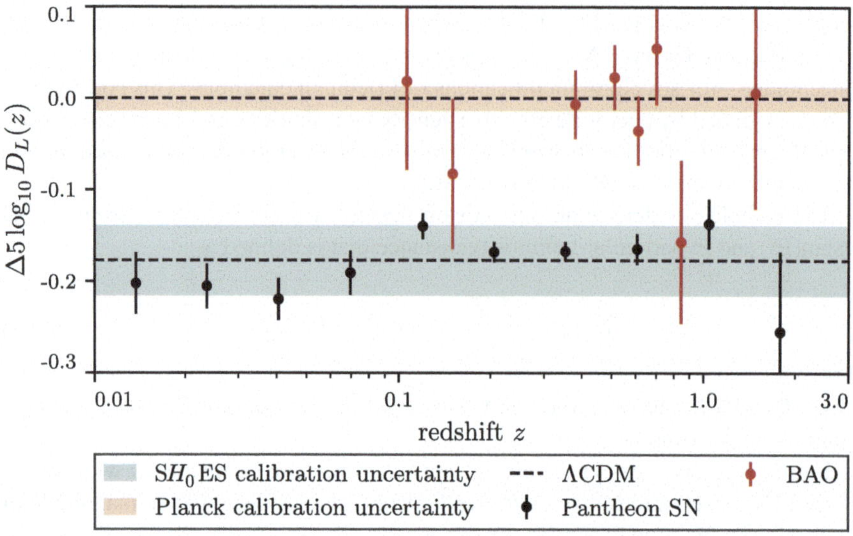

Fig. 27.1 The luminosity distance residuals with respect to the ΛCDM model best fit to CMB observations from Planck [7]. BAO [8–10] data points, in red, are calibrated by the Planck determination of the physical scale of the sound horizon in the ΛCDM model. Pantheon SN [11] data points, in black, are calibrated by the SH_0ES [12] measurement of absolute SN magnitude, M_0. The upper colored band shows the uncertainty in the sound horizon calibration while the lower band shows the errors in the measurements of M_0. The separation between the two bands shows in full display the Hubble tension. The gap between the data points of the two data sets shows the difficulty of solving the tension with late times modifications to the expansion history: SN and BAO data overlap in redshift, but, once calibrated, do not overlap in distance space. Figure adapted from [13]

used in [13, 14], calibrated by the measurement of M_0 trough the distance ladder, from [12], which is often phrased as a measurement of H_0; BAO measurements from the eBOSS [8], MGS [9] and 6dF [10] surveys, calibrated by the measurement of the physical scale of the sound horizon, inferred from CMB observations of the Planck satellite [7]. For both data sets, we show luminosity distance residuals, with respect to the CMB best fit ΛCDM cosmological model for visual clarity. BAO angular diameter distances are converted to luminosity distances using the effective redshift of the survey tomographic bins. The gray/red bands show the uncertainty in the calibration of the two data sets respectively and the difference between the two showcases the Hubble tension. Figure 27.1 illustrates the core issue in addressing the Hubble constant tension at late times. The calibrated SN and BAO data sets share a common redshift range, but their distance measurements do not coincide. Any given cosmological model, that preserves the behavior of either calibrator, gives a unique distance-redshift relation—is a line in Fig. 27.1—and cannot be in two different positions at the same time.

For this reason achieving a complete resolution of the Hubble tension at late times seems unattainable, although several models have been proposed to, at least, alleviate the tension.

A crucial aspect highlighting the challenge of fully addressing the Hubble constant tension at late times arises from the joint use of H_0/M_0 measurements, SN, BAO, and CMB data. In the context of this chapter, we assume that all four datasets have no residual systematic effects when we discuss DE model performances at solving the tension. It might be possible to achieve a better resolution of the Hubble constant tension by excluding some parts of these data sets.

27.3 Dark Energy and Low Redshift Calibration

In this Section, we review attempts at solving the Hubble constant tension by using DE to alter the low redshift calibration of distances.

This calibration is usually reported as a constraint on H_0 and, in the standard approach, the local value of H_0 is obtained using the SN Ia magnitude-redshift relation, as in Eq. 27.3, calibrated with Cepheid variables [15], with Tip of the Red Giant Branch (TRGB) measurements [16], and others, see [17] for a review. To strike a balanced tradeoff between sensitivity to expansion history, at high redshift, and contamination from peculiar velocities, at low redshift, only a subset of SN Ia is used in this determination of H_0. For example, in [15], only SN with $z > 0.023$ and $z < 0.15$ are used. Across this narrow redshift range the sensitivity to the shape of the distance-redshift relation is limited and, in the standard analysis of [15], it is fit with a template, by fixing the value of the deceleration parameter.

The H_0 calibration has then some residual dependency on the low redshift expansion rate and its impact on the determination of the Hubble constant was studied in detail in [18]. Several models altering the low redshift expansion history were studied, with an emphasis on DE models. None of these models were found to give significant changes to the inferred value of H_0, with a maximum difference of $\Delta H_0 = 0.47$ Km s^{-1} Mpc^{-1}, or a shift of 0.6%, with an expansion history constrained to be very close to the ΛCDM one.

In retrospect, this conclusion is compatible with residuals in Fig. 27.1. The SN calibrator sample includes only the first few low redshift data points and, once the best fitting ΛCDM expansion history is subtracted, there is no indication of redshift-dependent residuals that could significantly change the estimate of the residuals mean.

We mention here the special case of a very late transition in the DE density, as was studied in [19]. If such a transition occurs very close to the present epoch (i.e. $z < 0.1$ or lower) it would go unnoticed in observations at higher redshifts, like the CMB [20]. For these types of models it is crucial to treat the H_0 measurement properly, as a constraint on SN calibration and in [19, 21] we discussed how to convert back a constraint on H_0 from calibrated SN to a constraint on M_0. A late DE transition, before the first SN at $z < 0.01$ can increase the local expansion rate, pushing H_0 to values compatible with local measurements of [12] and even higher. However, just as in void scenarios, the problem with such a solution is that raising H_0 does not resolve the origin of the Hubble tension, as shown in Fig. 27.1, which resides in the calibration of SN in the Hubble flow. In other words, at redshift lower than the Hubble flow SN sample, $z < 0.023$, the expansion history is never relevant as the distance-redshift relation is never used in the fit to derive the value of the Hubble constant, which is effectively extrapolated to $z = 0$.

Another, radically different, way of changing the SN Ia calibration is to change the behavior of a step in the distance ladder calibration, especially Cepheids, by altering gravity through DE, as discussed in [22]. A modification of gravity through the introduction of a fifth force is a common outcome of theories that couple the degree of freedom of DE with matter, propose novel interactions between objects, or attempt to dynamically explain the nature of dark energy [5, 23–25]. Given the absence of any observed fifth force in tests of gravity within the solar system, these theories require a mechanism to maintain cosmological relevance while remaining hidden within the solar system. The concept of screening—which involves weakening the strength of the fifth force in regions of strong gravitational fields—was initially introduced to ensure the consistency of scalar-tensor theories of gravity with tests of the equivalence principle, the inverse square law, and post-Newtonian gravity tests within the confines of the solar system [26, 27]. Known screening mechanisms include: chameleon [28, 29], symmetron [30], dilaton [31], K-mouflage [32] and Vainshtein [33] screening.

Since screening mechanisms generally depend on the local environment they could alter the period-luminosity relation of Cepheids in different host galaxies. In particular, this could happen between the Milky Way and N4258 and Cepheids in the SN Ia calibrator sample, changing the inferred value of the Hubble constant. In [22], the authors assume that the Cepheids used for calibrating the period-luminosity relationship are subject to screening, and compute the expected change in H_0, as a function of the residual strength of the fifth force. Overall changes are not sufficient to fully resolve the Hubble tension: models in which the cores of Cepheids are unscreened (leading to increased luminosities) bring the results of the distance ladder closer to, at most, 2–3 σ from the Planck value when local variations in the effective gravitational constant, $\Delta G/G$ are about 5%; models that exclusively unscreen Cepheid envelopes (like chameleon models) require a higher 30–40% variations in the gravitational constant to achieve a similar level of improvement.

27.4 Dark Energy and the Expansion History

In this Section, we review attempts at solving the Hubble constant tension by using DE to change the expansion history at low to intermediate redshifts. This means in particular that we consider changes in the expansion history at $z > 0.01$, which allows us to treat the calibration of SN Ia as a constraint on the Hubble constant.

As previously discussed, all these models incur in the no-go argument shown in Sect. 27.2 so we generally expect to be able to somewhat relieve the Hubble tension, without resolving it fully.

27.4.1 Models for the de Equation of State

The simplest DE model we can consider, called wCDM, has a constant equation of state, $w_{\rm de} = w_0$, in analogy with matter and radiation components. In this case, the value of the equation of state fixes the time dependence of the DE density as $\Omega_{\rm de} \propto a^{-3(1+w_0)}$. To increase the expansion rate at low redshifts, with respect to constant $\Omega_{\rm de}$, we need $\Omega_{\rm de}$ to increase in time, rather than diluting, and hence we would need $w_0 < -1$. As discussed in [34], because of the CMB geometric degeneracy, this might look like an empirically viable model when considering CMB-only measurements. Once the degeneracy is broken though, by the use of either SN or BAO data, the model cannot solve the Hubble constant tension, leaving it qualitatively unchanged.

Similar conclusions apply to other phenomenological models for $w_{\rm de}$, as reviewed in [34]. If we consider the combination of Planck 2018, SN and BAO measurements, as in Fig. 27.1, the baseline ΛCDM value for the Hubble constant is, $H_0 = 67.4 \pm 0.8$ Km s^{-1} Mpc^{-1}, in $\sim 4\sigma$ tension with results of $H_0 = 73.04 \pm 1.04$ in [15]. The wCDM model achieves, when fit to the same data combination $H_0 = 68.34 \pm 0.82$ Km s^{-1} Mpc^{-1}, as reported in [34], which is a $\sim 0.5\sigma$ reduction in the tension.

We could consider as an extension of the wCDM model the Chevallier—Polarski—Linder parameterization (CPL) [35, 36] for the equation of state, in which $w_{\rm de} = w_0 + w_a(1-a)$. When fitting our benchmark combination of data this model yields $H_0 = 68.35 \pm 0.84$ Km s^{-1} Mpc^{-1}, which is virtually identical to the wCDM case. Other parametrizations have been tested in [34], in particular: the Jassal-Bagla-Padmanabhan parameterization (JBP) [37] model; the Logarithmic DE equation of state proposed by [38]; the BA parameterization put forward by Barboza and Alcaniz [39]. All these models would not qualitatively change the results with respect to the wCDM model. Some of these models might slightly change quantitatively the tension, by either slightly shifting the central value of the H_0 estimate or inflating the error bars of the final estimate. None would be on target and we refer the reader to the discussion in [34] to gauge different model performances.

In this section, we have discussed only phenomenological approaches and not concrete physical model constructions. On the one hand, it is always possible to build a model that is physically viable and reproduces any given expansion history [40]. On the other hand, considering specific models would impose requirements of physical viability on phenomenological quantities [41], hence restricting the available parameter space for solutions and necessarily degrading the performances of a model, towards the solution of the Hubble tension. We refer the reader to [34, 42] for an overview of DE model constraints in the context of the Hubble constant tension.

27.4.2 Reconstructed Dark Energy Models

While we have seen several models that alleviate the Hubble constant tension, it is natural to ask what are the ultimate performances that can be achieved by DE models at this task. This requires going beyond testing specific model realizations or simple parametric forms for $w_{\mathrm{de}}(z)$ and estimating $\Omega_{\mathrm{de}}(z)$ directly from data. Indeed, employing simpler constructions can introduce a bias into the results and potentially lead to the loss of important information about ways in which tensions can be relieved.

Multiple strategies exist for the non-parametric reconstruction of cosmological functions like $w_{\mathrm{de}}(z)$ or $\Omega_{\mathrm{de}}(z)$. Widely used techniques encompass discretizing the functions at multiple redshifts, adopting Gaussian Processes (GP) [43–45], and employing the correlated prior method [3, 46–48]. Employing a binning strategy, where the function is assumed to be constant and independent within each bin or smoothly interpolated between redshift nodes, leads to results that rely on an unphysical implicit assumption of smoothness. Opting for a limited number of bins could potentially introduce bias into the reconstruction while increasing their number risks fitting noise features. On the other hand, the GP approach doesn't impose restrictions on bin numbers but enforces Gaussianity of the space where it is performed. This space, incorporates a Gaussian prior that correlates the function across neighboring redshifts. The selection of the GP prior is fundamentally empirical, devoid of any connection to a physical theory. Moreover, the prior's parameters are typically marginalized, which can obscure the Bayesian interpretation of the resulting reconstruction. The correlated prior method similarly introduces a correlation between neighboring redshifts, but unlike the GP approach, it employs a fixed prior covariance matrix derived from theoretical considerations. This explicit prior enables a clear articulation of the extent to which the data improves upon the prior and facilitates computation of Bayesian evidence that can be compared to that of the ΛCDM model, as discussed in [14].

In this section, we show what happens to the Hubble constant tension when we opt for approaches 1 and 3, following [13, 14]. We do not pursue the GP strategy as we want to reconstruct $\Omega_{\mathrm{de}}(z)$. Distance measurements, for both SN and BAO, are Gaussian distributed in data space but their measurements are non-linearly related to $\Omega_{\mathrm{de}}(z)$, which means that we generally expect the posterior distribution of $\Omega_{\mathrm{de}}(z)$ to be non-Gaussian. In the reconstruction $\Omega_{\mathrm{de}}(z)$ is represented by its value at 11 discrete points (nodes) in redshift, with a cubic spline used to interpolate between them. The first 10 nodes are distributed uniformly in the interval $a \in [1, 0.25]$ while the last node is positioned at $a = 0.2$. At higher redshift $\Omega_{\mathrm{de}}(z)$ was made to approach its ΛCDM value, even though it is possible to study, with the same framework earlier times deviations from the standard model [49].

On its own, the cubic spline introduces an implicit smoothing (and correlation) between values of $\Omega_{\mathrm{de}}(z)$ at different redshifts, as shown in [14]. In addition the time correlation of $\Omega_{\mathrm{de}}(z)$ is set from the theoretical correlation matrix obtained in [50, 51] for the general family of Horndeski DE models [52]. This theory prior has a much longer correlation time, with respect to the spline model, and acts as a filter, discouraging but not completely prohibiting fast variations in $\Omega_{\mathrm{de}}(z)$. Since the correlation time is much longer than the spline model the theory prior reconstruction is independent of the number of bins used in the reconstruction.

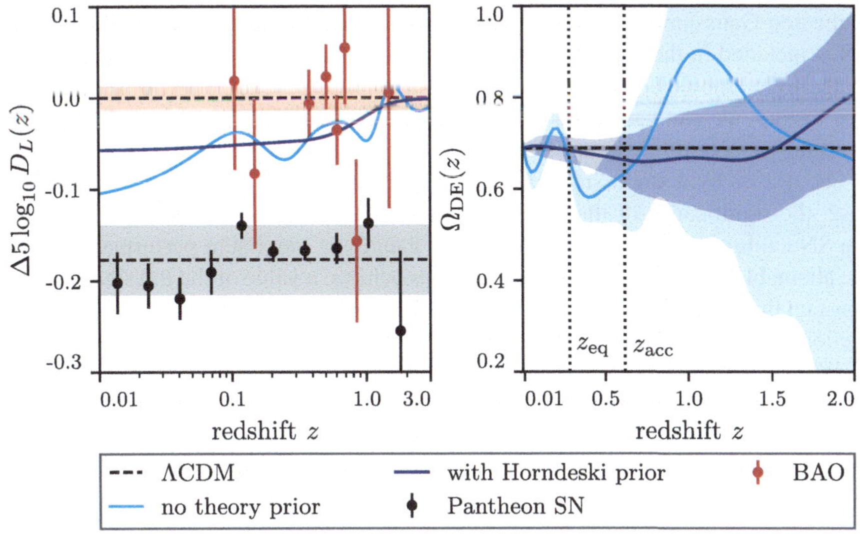

Fig. 27.2 *Left panel:* luminosity distance residuals, as in Fig. 27.1, with the two best fitting reconstructed DE models, with and without theoretical (Horndeski) priors. *Right panel:* reconstructed $\Omega_{\rm de}(z)$ for the data sets discussed in this chapter, with and without theory priors. The bands show the 68% confidence level regions. The two vertical lines show the redshifts of equality between the matter and DE densities, $z_{\rm eq}$, and the beginning of cosmic acceleration, $z_{\rm acc}$, in the best fit ΛCDM model. Figure adapted from [13]

We note that the discussion of the Hubble constant tension, in Sect. 27.2, relies only on cosmological background arguments. For this reason, we show results for $\Omega_{\rm de}(z)$ and do not discuss the behavior of inhomogeneities beyond the standard model. We refer the reader to [13, 14] for an in-depth discussion and how it relates to other tensions.

As in [13, 14] we jointly fit all data discussed in Sect. 27.2, and the resulting reconstruction of $\Omega_{\rm de}(z)$ is shown in Fig. 27.2. The reconstruction is performed with and without theoretical priors and the difference between the two showcases the important role played by the theory priors. Without theory priors one can see oscillations, in the redshift range $0 < z < 0.6$, in the reconstructed $\Omega_{\rm de}(z)$, which are mainly driven by the scatter of BAO and SN data. These oscillations do not significantly improve the fit and are hence suppressed by the theory prior. Figure 27.2 demonstrates the general difficulty of solving the Hubble tension by changing the expansion history at late times. The CMB anchors the expansion rate at high redshift and then the reconstruction struggles to strike a balance between fitting jointly SN and BAO data points.

For the spline reconstruction, which deploys the largest fitting freedom we consider, the inferred value for the Hubble constant, in the reconstructed DE model, is $H_0 = (69.9)\, 69.3 \pm 1.4\,{\rm Km\,s^{-1}\,Mpc^{-1}}$. In parentheses, hereafter, we have indicated the value for the best-fitting model, which is significantly higher than the mean due to the non-Gaussianity of the posterior that we had anticipated. If the calibration of SN is included in the fit, to force the reconstruction to struggle to increase the estimate of the Hubble constant, then we obtain $H_0 = (71.34)\, 70.9 \pm 1.4$. From these two results, we can gauge the extent to which it is possible to relieve the Hubble constant tension at late times. In the first case, the significance is slightly lowered, to about 2 σ, by a slight shift and significantly enlarged error bars. In the second case, the significance of the tension is not changed—as the joint dataset contains the SN calibration [53] but it is helpful to gauge the maximum performances that are attainable. Note that both reconstructions achieve a value of the inferred Hubble constant that is significantly higher than the DE models we discussed in the previous section, because of the much larger fitting freedom.

When we include the DE theoretical priors in the reconstruction the ability of the model to relieve the Hubble constant tension is significantly lowered. When fitting only to CMB, BAO, and SN we can obtain at most $H_0 = (67.7)67.8^{+1.1}_{-1.4}$ Km s^{-1} Mpc^{-1}, a slight improvement over the standard ΛCDM result of $H_0 = 67.4 \pm 0.8$ Km s^{-1} Mpc^{-1}.

27.5 Summary

In this chapter, we have discussed how DE at late times could be invoked to solve the tension over the value of the Hubble constant, between early and late time Universe measurements. DE is a natural candidate to provide a resolution to this conundrum: its physical properties are unknown; it's the only physical constituent that is irrelevant at early times and important at late times.

We have seen how DE at late times enters into the discussion over the Hubble constant tension, how it influences the cosmological observables that are relevant to the tension, and how this limits the capability of DE models to fully solve it. Changing the expansion history, or the laws of gravity, at very late times, as we have discussed in Sect. 27.3, could help alleviate the tension by changing the very way in which the distance to Hubble flow SN is calibrated. Changing the expansion history at intermediate redshifts, in the range $0.01 < z < 3$, as we have reviewed in Sect. 27.4, can also help alleviate the tension. For this family of models, the main limitation to a full tension resolution is that the distance redshift relation is tomographically probed by both SN and BAO observations. Both probes constrain the expansion history, in most of this redshift interval, better than what is needed to solve the tension. We have seen how this translates to the impossibility of achieving a full tension resolution, even when deploying very general techniques that reconstruct, rather than constrain, DE models at late times. These approaches are so general that they provide an upper bound to the performances that can be achieved, which in this case are at about $H_0 = (71.34)70.9 \pm 1.4$, thus solving the tension only halfway.

Acknowledgements We thank Levon Pogosian for helpful discussions and comments. M.R. acknowledges financial support from the INFN InDark initiative.

References

1. A.G. Riess et al., Observational evidence from supernovae for an accelerating universe and a cosmological constant. Astron. J. **116**, 1009–1038 (1998)
2. S. Perlmutter et al., Measurements of Ω and Λ from 42 high redshift supernovae. Astrophys. J. **517**, 565–586 (1999)
3. G.-B. Zhao et al., Dynamical dark energy in light of the latest observations. Nature Astron. **1**(9), 627–632 (2017)
4. L.A. Escamilla, W. Giarè, E. Di Valentino, R.C. Nunes, S. Vagnozzi, The state of the dark energy equation of state circa 2023 **7** (2023)
5. T. Clifton, P.G. Ferreira, A. Padilla, C. Skordis, Modified gravity and cosmology. Phys. Rept. **513**, 1–189 (2012)
6. D. Brout et al., The Pantheon+ analysis: cosmological constraints. Astrophys. J. **938**(2), 110 (2022)
7. N. Aghanim et al., Planck 2018 results. V. CMB power spectra and likelihoods. Astron. Astrophys. **641**, A5 (2020)
8. S. Alam et al., Completed SDSS-IV extended Baryon oscillation spectroscopic survey: cosmological implications from two decades of spectroscopic surveys at the Apache point observatory. Phys. Rev. D **103**(8), 083533 (2021)
9. A.J. Ross, L. Samushia, C. Howlett, W.J. Percival, A. Burden, M. Manera, The clustering of the SDSS DR7 main Galaxy sample—I. a 4 percent distance measure at $z = 0.15$. Mon. Not. Roy. Astron. Soc. **449**(1), 835–847 (2015)
10. F. Beutler, C. Blake, M. Colless, D.H. Jones, L. Staveley-Smith, L. Campbell, Q. Parker, W. Saunders, F. Watson, The 6dF Galaxy survey: Baryon acoustic oscillations and the local Hubble constant. Mon. Not. Roy. Astron. Soc. **416**, 3017–3032 (2011)
11. D.M. Scolnic et al., The complete light-curve sample of spectroscopically confirmed SNe Ia from Pan-STARRS1 and cosmological constraints from the combined Pantheon sample. Astrophys. J. **859**(2), 101 (2018)
12. A.G. Riess, S. Casertano, W. Yuan, J.B. Bowers, L. Macri, J.C. Zinn, D. Scolnic, Cosmic distances calibrated to 1% precision with Gaia EDR3 parallaxes and Hubble space telescope photometry of 75 milky way Cepheids confirm tension with ΛCDM. Astrophys. J. Lett. **908**(1), L6 (2021)
13. L. Pogosian, M. Raveri, K. Koyama, M. Martinelli, A. Silvestri, G.-B. Zhao, J. Li, S. Peirone, A. Zucca, Imprints of cosmological tensions in reconstructed gravity. Nature Astron. **6**(12), 1484–1490 (2022)
14. M. Raveri, L. Pogosian, M. Martinelli, K. Koyama, A. Silvestri, G.-B. Zhao, J. Li, S. Peirone, A. Zucca, Principal reconstructed modes of dark energy and gravity. JCAP **02**, 061 (2023)
15. A.G. Riess et al., A comprehensive measurement of the local value of the Hubble constant with 1 km s^{-1} Mpc^{-1} uncertainty from the Hubble space telescope and the SH_0ES team. Astrophys. J. Lett. **934**(1), L7 (2022)
16. S.A. Uddin et al., Carnegie supernova project-I and -II: measurements of H_0 using Cepheid, TRGB, and SBF distance calibration to type Ia supernovae **8** (2023)
17. A.G. Riess, L. Breuval, The local value of H_0 **8** (2023)
18. S. Dhawan, D. Brout, D. Scolnic, A. Goobar, A.G. Riess, V. Miranda, Cosmological model insensitivity of local H_0 from the Cepheid distance ladder. Astrophys. J. **894**(1), 54 (2020)
19. G. Benevento, W. Hu, M. Raveri, Can late dark energy transitions raise the Hubble constant? Phys. Rev. D **101**(10), 103517 (2020)
20. M. Mortonson, W. Hu, D. Huterer, Hiding dark energy transitions at low redshift. Phys. Rev. D **80**, 067301 (2009)
21. M. Raveri, G. Zacharegkas, W. Hu, Quantifying concordance of correlated cosmological data sets. Phys. Rev. D **101**(10), 103527 (2020)
22. H. Desmond, B. Jain, J. Sakstein, Local resolution of the Hubble tension: the impact of screened fifth forces on the cosmic distance ladder. Phys. Rev. D **100**(4), 043537 (2019) [Erratum: Phys.Rev.D 101, 069904 (2020), Erratum: Phys.Rev.D 101, 129901 (2020)]

23. K. Koyama, Cosmological tests of modified gravity. Rept. Prog. Phys. **79**(4), 046902 (2016)
24. A. Joyce, L. Lombriser, F. Schmidt, Dark energy versus modified gravity. Ann. Rev. Nucl. Part. Sci. **66**, 95–122 (2016)
25. C. Burrage, J. Sakstein, Tests of chameleon gravity. Living Rev. Rel. **21**(1), 1 (2018)
26. J. Khoury, Theories of dark energy with screening mechanisms **11** (2010)
27. B. Jain, J. Khoury, Cosmological tests of gravity. Annals Phys. **325**, 1479–1516 (2010)
28. J. Khoury, A. Weltman, Chameleon fields: awaiting surprises for tests of gravity in space. Phys. Rev. Lett. **93**, 171104 (2004)
29. J. Khoury, A. Weltman, Chameleon cosmology. Phys. Rev. D **69**, 044026 (2004)
30. K. Hinterbichler, J. Khoury, Symmetron fields: screening long-range forces through local symmetry restoration. Phys. Rev. Lett. **104**, 231301 (2010)
31. P. Brax, C. van de Bruck, A.-C. Davis, D. Shaw, The Dilaton and modified gravity. Phys. Rev. D **82**, 063519 (2010)
32. E. Babichev, C. Deffayet, R. Ziour, k-Mouflage gravity. Int. J. Mod. Phys. D **18**, 2147–2154 (2009)
33. A.I. Vainshtein, To the problem of nonvanishing gravitation mass. Phys. Lett. B **39**, 393–394 (1972)
34. E. Di Valentino, O. Mena, S. Pan, L. Visinelli, W. Yang, A. Melchiorri, D.F. Mota, A.G. Riess, J. Silk, In the realm of the Hubble tension—a review of solutions. Class. Quant. Grav. **38**(15), 153001 (2021)
35. M. Chevallier, D. Polarski, Accelerating universes with scaling dark matter. Int. J. Mod. Phys. D **10**, 213–224 (2001)
36. E.V. Linder, Exploring the expansion history of the universe. Phys. Rev. Lett. **90**, 091301 (2003)
37. H.K. Jassal, J.S. Bagla, T. Padmanabhan, Observational constraints on low redshift evolution of dark energy: how consistent are different observations? Phys. Rev. D **72**, 103503 (2005)
38. G. Efstathiou, Constraining the equation of state of the universe from distant type Ia supernovae and cosmic microwave background anisotropies. Mon. Not. Roy. Astron. Soc. **310**, 842–850 (1999)
39. E.M. Barboza Jr., J.S. Alcaniz, A parametric model for dark energy. Phys. Lett. B **666**, 415–419 (2008)
40. N. Frusciante, L. Perenon, Effective field theory of dark energy: a review. Phys. Rept. **857**, 1–63 (2020)
41. S. Peirone, M. Martinelli, M. Raveri, A. Silvestri, Impact of theoretical priors in cosmological analyses: the case of single field quintessence. Phys. Rev. D **96**(6), 063524 (2017)
42. N. Schöneberg, G. Franco Abellán, A. Pérez Sánchez, S.J. Witte, V. Poulin, J. Lesgourgues, The H_0 Olympics: a fair ranking of proposed models. Phys. Rept. **984**, 1–55 (2022)
43. A. Shafieloo, A.G. Kim, E.V. Linder, Gaussian process cosmography. Phys. Rev. D **85**, 123530 (2012)
44. M. Seikel, C. Clarkson, M. Smith, Reconstruction of dark energy and expansion dynamics using Gaussian processes. JCAP **06**, 036 (2012)
45. F. Gerardi, M. Martinelli, A. Silvestri, Reconstruction of the dark energy equation of state from latest data: the impact of theoretical priors. JCAP **07**, 042 (2019)
46. R.G. Crittenden, G.-B. Zhao, L. Pogosian, L. Samushia, X. Zhang, Fables of reconstruction: controlling bias in the dark energy equation of state. JCAP **02**, 048 (2012)
47. G.-B. Zhao, R.G. Crittenden, L. Pogosian, X. Zhang, Examining the evidence for dynamical dark energy. Phys. Rev. Lett. **109**, 171301 (2012)
48. Y. Wang, L. Pogosian, G.-B. Zhao, A. Zucca, Evolution of dark energy reconstructed from the latest observations. Astrophys. J. Lett. **869**, L8 (2018)
49. M.-X. Lin, M. Raveri, W. Hu, Phenomenology of modified gravity at recombination. Phys. Rev. D **99**(4), 043514 (2019)
50. M. Raveri, P. Bull, A. Silvestri, L. Pogosian, Priors on the effective dark energy equation of state in scalar-tensor theories. Phys. Rev. D **96**(8), 083509 (2017)
51. J. Espejo, S. Peirone, M. Raveri, K. Koyama, L. Pogosian, A. Silvestri, Phenomenology of large scale structure in scalar-tensor theories: joint prior covariance of w_{DE}, Σ and μ in Horndeski. Phys. Rev. D **99**(2), 023512 (2019)

52. G.W. Horndeski, Second-order scalar-tensor field equations in a four-dimensional space. Int. J. Theor. Phys. **10**, 363–384 (1974)
53. M. Raveri, W. Hu, Concordance and discordance in cosmology. Phys. Rev. D **99**(4), 043506 (2019)

Chapter 28
Cosmological Models with Negative Λ

Anjan A. Sen

28.1 Introduction

The evolution of Universe always presents us with many surprises and puzzles. Till date, cosmologists are still trying to find answers to two biggest puzzles related to our Universe: the dark matter (DM) and dark energy (DE). For dark matter, the answer may lie in some form of extension of Standard model of particle physics. But for dark energy, we currently have no clue about its actual nature originating from a field theory prescription, instead we are working with different kind of phenomenological characterisations of dark energy. This is due to the fact that the energy scale of the dark energy is so abnormally low compared to any reasonable quantum field theory scale [33, 34, 40], it is extremely difficult to model dark energy within realm of standard model of particle physics and its various extensions.

Given this, the simplest candidate for dark energy is the cosmological constant Λ, the ground state or vacuum energy for the fields. Although we can not get the correct value for Λ (as needed to be consistent with observations) from any reasonable field theory set up, still it is the best option due to its simple form (especially while considering the perturbed Universe) and its unmatched consistency with almost all the cosmological observations. This confirms the presence of a field (in the simplest case, a canonical scalar field) at the non zero positive minimum of its potential which confirms the presence of a stable or meta stable de-Sitter (dS) vacuum depending on the nature of the minimum of the potential. We can also construct scenarios where the scalar field is away from the minimum of the potential and is slowly rolling at the positive part of the potential. This scenario gives rise to a DE behaviour, which unlike Λ, is not a constant but evolves with time giving rise to an evolving equation of state $w_{de} = p_{de}/\rho_{de}$ for the DE (for Λ, $w_{de} = -1$). Such evolving DE component is termed as "Quintessence" [17, 45]. Most of the quintessence scalar fields, discussed in the context of DE, have a dS minima. Hence whether DE is Λ or an evolving scalar field, the Universe always settles in a dS phase in asymptotic future.

A. A. Sen (✉)
Centre for Theoretical Physics, Jamia Millia Islamia, New Delhi 110025, India
e-mail: aasen@jmi.ac.cin

E. Di Valentino and D. Brout (eds.), *The Hubble Constant Tension*, Springer Series in Astrophysics and Cosmology, https://doi.org/10.1007/978-981-99-0177-7_28

But this is not always true. A scalar field rolling over the positive part of a potential which is sufficiently flat at present energy scale and also contains an Anti de-Sitter (AdS) minima is fully consistent with a late time accelerating Universe. In fact, constructing such potentials with AdS minima is not difficult in string theory context as compared to that with dS minima [16, 18, 19, 43], although the right value for the energy scale consistent with the late time Universe, has not been obtained as yet. But still there are not many attempts in the literature for cosmological studies with scalar fields having AdS minima.

In recent years, cosmological observations have unveiled tensions in the value of present day Hubble constant H_0 [6, 24, 36], the amplitude of matter fluctuations $S_8 := \sigma_8\sqrt{\Omega_m/0.3}$ that is related to the galaxy shear measurements using weak lensing [1, 6, 10, 28, 44] as well as anomalies in the lensing parameter $A_{\rm lens}$ and/or curvature Ω_k [3, 6, 25, 46]. All these tensions and anomalies are primarily present in the concordance flat ΛCDM model. The recent observations of very massive high redshift galaxies by JWST are also difficult to explain under the assumption of a concordance flat ΛCDM model as our observable Universe [2, 13, 29].

All these indicate that the flat ΛCDM model needs modifications. In this regard, in recent years, cosmological models in the presence of a tiny negative cosmological constant (with the energy density at the same order as dark energy) has been tested with cosmological observations, particularly in the context of Hubble tension. It has been shown that not only such models are consistent with cosmological observations, in some cases they can be better in statistical sense, compared to flat ΛCDM model. Moreover, they may also help to alleviate different cosmological tensions.

With such motivation, in this chapter, we shall review the recent studies related to cosmological models containing a negative cosmological constant and the scope of alleviate some of the cosmological tensions in such models.

28.2 Universe with Negative Λ

28.2.1 Possibility of Accelerating Universe with a Negative Λ

In contrast to positive cosmological constant, a negative cosmological constant results an attractive gravitational effect and hence can not result in an accelerated expanding Universe on its own. But it was first shown by Hartle, Hawking and Hertog [26] that a fundamental theory of quantum cosmology with negative cosmological constant can give rise to classical low energy theory with positive cosmological constant that can drive the accelerated expansion in the Universe. This follows from a underlying symmetry of the model called "signature reversal symmetry" which is present in both quantum and classical theories. Later it was argued by Mithani and Vilenkin [31] that such theories are unstable as the mass of the corresponding field becomes tachyonic.

28.2.2 Early Observational Hints for a Negative Dark Energy

The first observational hint that there may be a negative contribution in the Dark Energy behaviour, especially at redshifts $z > 1$ was from the Baryon Acoustic Oscillation measurements in the Lyman-α forest of high redshift quasars by BOSS DR11 [23]. With their measurement of $H(z = 2.34) = 222 \pm 7$ Km/s/Mpc, and using the precise values of r_d and $\Omega_m h^2$ by Planck, they obtained the constraint on dark energy density at $z = 2.34$:

$$\frac{\rho_{de}(z = 2.34)}{\rho_{de}(z = 0)} = -1.2 \pm 0.8, \tag{28.1}$$

which is in tension of around 2.5 σ with a value 1 (that is needed for Λ to be dark energy). If this results persists, it shows that the dark energy density at $z = 2.34$ is lower than that at $z = 0$ and perhaps with an opposite sign.

This negative dark energy behaviour at higher redshifts was further corroborated by another study by Poulin, Boddy, Bird and Kamionkowski (PBBK) [35] who did a model independent reconstruction of the dark energy density using a large set of cosmological data including CMB, LSS, BAO, SnIa as well as Lyman-α BAO measurements at $z = 2.34$. They showed that when one relaxes the prior $\rho_{de} > 0$, the reconstructed ρ_{de} behaviour evolves towards negative values at higher redshifts.

Similar conclusions were arrived by Sahni, Shafieloo and Starobinsky (SSS) [39] using the Om diagnostics. For ΛCDM model $\Omega_m h^2$ is constant and is tightly constrained by Planck: $\Omega_m h^2 = 0.1426 \pm 0.0025$. Using the relation

$$H^2(z) = \kappa[\rho_{de}(z) + \rho_{m0}(1 + z)^3], \tag{28.2}$$

together with the constraint $H(z = 2.34) = 222 \pm 7$ Km/s/Mpc by BOSS-DR11 with Lyman-α measurments, one gets $h^2(z)/(1 + z)^3 = 0.132 \pm 0.008$ (where $h(z) = H(z)/H(z = 0)$). If one now assumes $\Omega_m h^2 = 0.1426 \pm 0.0025$ as constrained by CMB observations by Planck, one can see $\rho_{de}(z = 2.34) < 0$ (for details see [39]) showing the possibility of a negative dark energy behaviour at higher redshifts. We should also mention this may also happen if the matter energy density does not obey the standard conservation equation due to some interactions with other component as well as for modified gravity models.

In a later work, Wang, Pogosian, Zhao and Zucca (WPZZ) [49] applied a different model independent reconstruction method to reconstruct the dark energy evolution using updated data from Planck as well as measurements from SnIa, BAO (including the Lyman-α BAO measurement by BOSS-DR11), cosmic chronometer as well as the more recent H_0 measurement by SH_0ES and showed the same negative dark energy behaviour at higher redshifts.

All these studies independently confirmed the possible hint of a negative dark energy behaviour at higher redshift primarily driven by the Lyman-α BAO measurement by BOSS-DR11.

28.2.3 Recent Observational Hint for the Presence of a Negative Λ

- **Evidence of Negative Cosmological Constant from Reconstructed H(z) using Pade Approximation**
  In recent years, the first observational hint for presence of a negative cosmological constant in the energy budget of the Universe came from the work by Dutta et al. [22]. They studied a model independent but parametrized form for the Hubble parameter $H(z)$ in order to reconstruct the dark energy density. For this, they assumed the Pade approximation $P_{2,2}$ for the Hubble parameter $H(z)$ and used various low-redshift data to constrain the $H(z)$ behaviour. They obtained the similar negative dark energy behaviour at higher redshifts as obtained by earlier works by PBBK, SSS and WPZZ discussed in the previous section. This reconstruction is valid till $z \sim 3$ as there are no low redshift data beyond that redshift. Subsequently they imposed the condition that this low-redshift behaviour of $H(z)$ should match with the Planck constrained $H(z)$ behaviour for ΛCDM at redshift $z \sim 4$ and beyond. The rationale for such assumption is that for models with non interacting dark energy, the Universe should behave as Einstein-de Sitter at higher redshifts (because dark energy is negligible at higher redshifts and Universe is mostly dominated by ordinary matter and radiation) irrespective of the dark energy model. Together with this condition, the SH_0ES prior was also imposed and the $H(z)$ was reconstructed beyond $z > 4$. From the reconstructed $H(z)$, one can now obtain the dark energy redshift dependence as follows:

$$\frac{H^2(z)}{H_0^2} = \Omega_{m0}(1+z)^3 + (1-\Omega_{m0})f(z), \tag{28.3}$$

  where $f(z)$ is the redshift dependence of the dark energy density. With this, one can reconstruct the $f(z)$ and it was shown that $f(z)$ has a negative minima at higher redshifts which can be modelled with the presence of small cosmological constant with a value $\Lambda = 3H_0^2(1-\Omega_{m0})f_{min}$ where $f_{min} < 0$. As evident, the value of this small cosmological constant is of the same order of the dark energy density at present and hence it does not involve any new energy scale. The authors argued that (a) the BAO Lyman-α data at $z = 2.34$, (b) the SH_0ES prior on H_0 (R16) as well as (c) the constraint that the $H(z)$ should reproduce the ΛCDM $H(z)$ behaviour as constrained by Planck-2018 at higher redshifts, all three together play a crucial role for the presence of tiny negative cosmological constant in the energy budget of the Universe.
  Although the study involves a particular parametrization for $H(z)$, namely the Pade approximation, the fact that reconstructed behaviour for $\rho_{de}(z)$ upto $z \sim 3$ using the low redshift data is consistent with earlier low-redshifts reconstructions of $\rho_{de}(z)$ by PBBK and WPZZ, supports the validity of the reconstructed $H(z)$ by Dutta et al.

- **Dark energy with a constant Equation of state on top of negative Cosmological Constant**
  Visinelli, Vagnozzi and Danielsson (VVD) [48] considered a string inspired extension for ΛCDM model where dark energy with a constant equation of state is considered on top of a negative cosmological constant. The equation of state for the dark energy can be phantom in nature which can be possibly modelled through string inspired multiverse scenario where multiple scalar fields with AdS vacua can give an effective phantom equation of state. They used only low-redshift observational data from BAO and SnIA observations together with the two anchors: one is r_d from Planck-2018 and the other is H_0 from SH_0ES measurements.
  There was no definite evidence for the existence of a negative cosmological constant although at 68% confidence level, the constraints are consistent with the presence of a non zero negative cosmological constant.
- **What happens when one includes CMB measurement with low redshift measurements?**
  Recently Sen, Adil and Sen (SAS) [42] have done a comprehensive analysis on cosmological constraint for models with negative cosmological constant. Similar to the work by VVD, they considered dark energy models with constant equation of state top of a negative cosmological constant. They also considered dark energy energy models with CPL parametrization for the equation of state together with a negative cosmological constant. The role plays by negative cosmological constant can be understood in the following way:
  CMB measurement give extremely tight constraint on the angular scale θ_s of sound horizon at last scattering:

$$\theta_s = \frac{r_d(z_d)}{D_A(z*)}, \tag{28.4}$$

where $z* = 1100$ is the redshift at last scattering and $z_d \sim z*$ is the redshift at drag epoch. r_d is the sound horizon at z_d and D_A is the angular diameter distance to the last scattering. As CMB gives tight constraint on θ_s from the peak position, $\frac{r_d(z_d)}{D_A(z*)}$ needs to be very tightly constrained. If there is no new physics at the early Universe, r_d will not change also, and hence D_A is very tightly constrained. The expression of D_A is as follows:

$$D_A = \frac{c}{H_0}\frac{1}{(1+z)}\int_0^{z*}\frac{dz}{E(z)} = \frac{c}{H_0}\frac{1}{(1+z)}F(z*), \tag{28.5}$$

where $E(z) = \frac{H(z)}{H_0}$ and $F(z*) = \int_0^{z*}\frac{dz}{E(z)}$. So with tightly constrained D_A, if we need to increase H_0 without any change in peak locations in CMB, we need to increase $F(z*)$ proportionately. $F(z*)$ contains the dark energy information at late times. With a simple case, where dark energy is parametrized by a constant equation of state, it is not difficult to check that a phantom type equation of state can result the increase in $F(z*)$ in comparison to ΛCDM model. In this regard, the presence of a negative cosmological constant plays a crucial role. As shown by SAS [42], presence of a $-$ve Λ increases the $F(z*)$ for a fixed phantom like w_{de}.

So one can obtain a required increase in $F(z*)$ (in order to increase H_0) with less phantom equation of state but with a certain contribution from $-$ve Λ whereas without the contribution from $-$ve Λ, one needs much larger phantom equation state for the dark energy, which may be difficult to obtain theoretically (see Fig. 1 in SAS [42]).
Subsequently a dark energy model with constant dark energy equation of state together with a negative cosmological constant (WCDMCC) was confronted with various cosmological datasets including CMB likelihood by Planck-2018 (TTTEEE+LowTT+LowEE+Lensing) [6], BAO [9, 11, 37] as well as Pantheon data for SnIA [41] together with SH_0ES-2019 prior (R19) [36]. With CMB+BAO data alone, a dark energy model (with constant equation of state) together with a $-$ve Λ (WCDMCC) results $H_0 = 69.6^{+1.3}_{-1.5}$ Km/s/Mpc which is at much reduced tension ($\sim 2\sigma$) with the SH_0ES R19 prior for H_0. The constraint on the r_d is 147.17 ± 0.22Mpc showing no change in the early Universe physics. The constraint on the dark energy equation of state is $w_{de} = -1.061^{+0.037}_{-0.015}$ showing a very mild phantom behaviour. The constraint on the amount of negative cosmological constant is $\Omega_\Lambda = -0.88^{+1.3}_{-0.46}$. Although the zero value of Ω_Λ is consistent within 68% confidence level, the results shows that the CMB+BAO data is fully consistent with a dark energy model with - ve Λ. Most interestingly, the WCDMCC model has better evidence in terms of both AIC as well as Bayesian Evidence criteria. As shown in the SAS paper [42], there is a reduction of around 16 in AIC for WCDMCC model compared to ΛCDM and an improvement of around 7 in the Bayesian Evidence in WCDMCC model compared to ΛCDM for the CMB+BAO data. This shows that the WCDMCC is surely a viable cosmological model.
Subsequenty the Pantheon data for SnIA measurements as well as the prior on H_0 by R19 were added to CMB+BAO data. This gives $H_0 = 70.34^{+0.82}_{-0.77}$ Km/s/Mpc, $\Omega_\Lambda = -2.48^{+1.90}_{-0.29}$ and $w_{de} = -1.02^{+0.035}_{-0.013}$. For this full data set, the improvement in AIC in WCDMCC model is around 8.15 compared to ΛCDM and corresponding improvement in Bayesian Evidence is 3.79 in WCDMCC compared to ΛCDM. Both of these numbers give a strong preference of WCDMCC model over ΛCDM when considering the full CMB+BAO+SnIA+H_0 data.

- **Graduated Dark Energy**
Inspired by the "Graduated Inflationary Universe" model proposed by Barrow in 1990, recently a "Graduated Dark Energy" model has been proposed by Akarsu, Barrow, Escamilla and Vazquez (ABEV) [8]. The basic idea is to model the inertial mass of the dark energy $\rho_{intertia} = \rho + p$ as $\rho_{inertia} = \gamma\rho_0(\frac{\rho}{\rho_0})^\lambda$. Here γ and λ are two parameters of the model and ρ_0 is positive constant with its value at present. It was claimed that this set up can describe a dark energy model where the cosmological constant switches sign from a negative value in the early Universe to a positive value in late Universe. While confronting with data from Planck (compressed likelihood), SnIa, BAO (including the Lyman-α measurement at $z -$ 2.34) as well as the H_0 measurement by SH_0ES, the sign switching happens at around $z \sim 2.3$. The model results a higher value for H_0 and results a better χ^2 compared to ΛCDM. This is another independent study where it was shown that the

presence of a small negative cosmological constant is consistent with cosmological observations.
This sign switching Cosmological constant mode Λ_sCDM was further studied in another recent work by Akarsu et al. [7] where different cosmological observations were taken into account including the full CMB likelihood from Planck-2018 as well as various BAO measurements and SnIa measurement by Pantheon together with the SH_0ES prior. It was observed that such sign switching cosmological constant model can address several cosmological tensions like H_0, S_8, M_B etc in a single model set up. The model is also better fit to the cosmological observations compared to ΛCDM. This model also gives independent hint of the presence of a negative cosmological constant in the energy budget of the Universe, especiallly in the higher redshifts.

- **Possibility of Negative Cosmological Constant in Dark Sector**
This is another independent study by Calderon, Gannouji, L'Huillier and Polarski [12]. While modelling dark energy density, they considered both the constant equation state as well as CPL parametrization for this purpose. Moreover they also considered various piecewise continuous functions for the equation of state of the dark energy. Together this, they also considered the presence of a negative cosmological constant in the energy budget of the Universe. Using various cosmological observations, they showed that the best models are those with phantom behaviour at $z > 1$ together with a higher evidence for the non-zero negative cosmological constant.
- **Dark Energy with Phantom Crossing: Omnipotent Dark Energy**
Recently Di Valentino, Mukherjee and Sen (DMS) [47] have proposed a parametrization for the dark energy density that involves a phantom crossing behaviour. The parametrization is as follows:

$$\rho_{de}(z) = \rho_{de}(z=0)\frac{1+\alpha(a-a_m)^2+\beta(a-a_m)^3}{1+\alpha(1-a_m)^2+\beta(1-a_m)^3}. \tag{28.6}$$

The model has three parameters: a_m, α and β. The absence of the linear term ensures that the ρ_{de} has an extremum at the scale factor $a = a_m$. This results the phantom crossing at $a = a_m$. In recent work, it have been shown that this parametrization can unify a host of dark energy behaviours and is termed as "Omnipotent Dark Energy".
DMS showed that this model can result higher H_0 for both CMB+BAO as well as CMB+SN data combinations, practically resolving the Hubble tension. Later Chudaykin et al. [14] have shown by adding the SPT-3G [21] and SPTPOL [50] data as well as information from BAO measurements [9, 11, 37] to the Planck-2018 likelihood and Pantheon SnIa data, that this phantom crossing model can resolve both H_0 and S_8 tensions simultaneosly and also can result a better fit to the observational data compared to ΛCDM model.
More recently, Adil et al. [4] have shown that the constrained Omnipotent dark energy behaviour is equivalent to a phantom dark energy model together with a small negative cosmological constant. One should note that the Omnipotent

parametrization for ρ_{de} does not necessarily involve the presence of a negative cosmological constant. But as it is constrained by observational data, the constrained behaviour for ρ_{de} eventually shows the existence of a negative cosmological constant which is very interesting. This is another independent hint for the existence of a negative cosmological constant in the energy budget of the Universe.

- **AdS Early Dark Energy**
  A cosmological phase of Early Dark Energy before recombination can also help to resolve the Hubble Tension. Recently Jiang and Piao [27] have shown that a short AdS period just before recombination can help the early dark energy model to work better to solve the Hubble tension problem. With low-l Planck-TT likelihood together with polarization data from SPTpol and lensing as well as with BAO and Pantheon data (without SHOES) prior, it was shown an AdS-early dark energy model can result $H_0 = 71.9$ Km/s/Mpc which resolves the H_0 tension. Inclusion of galaxy shear and clustering data gives $S_8 = 0.785$, also resolving the S_8 tension.
- **Multiple Transitions Vacuum Dark Energy**
  In another work, Moshafi, Firouzjahi and Talebian [32]has considered multiple transition vacuum dark energy model, where vacuum energy undergoes multiple transition both in early time as well as in late time. The transition can be dS to AdS or vice versa. They showed while constraining the model using various cosmological observations that a transient late time AdS-like vacuum energy can give rise larger H_0 but at the same results a scale invariant Harrison Zeldovich type priomordial power spectrum.

All these models, discussed above give independent hints for the existence of small negative cosmological constant in the Universe. We should emphasis that modelling the presence of a small negative cosmological constant is not difficult. The simplest way to include a negative cosmological constant in the dark energy sector is include an AdS minima in the quintessence field potential that gives rise to the dark energy behaviour. One such potential was obtained by Cicoli et al. [15]. in models with multiple axion:

$$V(\phi) = V_0\left[1 + pcos\left(\frac{\phi}{f}\right)\right]. \tag{28.7}$$

Here for $p > 1$, the potential has a AdS minima. This scalar field potential has been confronted with variety of observational data and it was shown that $p > 1$ case which has an AdS minima in the potential, is consistent with observational data and at the same time is better fit to the data than the ΛCDM model [38]. But this model can not result a phantom type dark energy. In most of the studies that resolve Hubble tension using late time physics, it is known that phantom type dark energy is more suitable. In such cases, this multi-axions scalar field model needs to be interacting with other components in the Universe, like dark matter, to result an effective dark energy equation of state that is phantom.

In another recent work, Malekjani et a. [30] have shown that, considering latest Pantheon+ sample for the SnIa, there is a redshift evolution for the cosmological parameters H_0 and Ω_{m0}. In particular, there is an increasing Ω_{m0} trend with redshift that leads to $\Omega_{m0} > 1$ for $z > 1$. If this is true, this may correspond to the existence of a negative dark energy at $z > 1$.

Finally, in a recent work, Demirtas et al. [20] have constructed a supersymmetric vacuum in string theory where the magnitude of the cosmological constant is $\sim 10^{-123}$ in Planck unit and it is negative. This is consistent with all the recent studies where a small negative cosmological constant is considered to ameliorate different cosmological tensions. This can have interesting consequences in modelling dark energy models with small negative cosmological constant that may be a viable option to resolve different cosmological tensions.

28.3 Summary

To summarize, presence of small negative cosmological constant in the energy budget of the Universe is not unusual and can be modelled with a scalar field rolling over potential that have an AdS minima. This is also very relevant in the context of dark energy model building in string theory. As long as the scalar field is slowly rolling at the positive part of the potential around present time and is also dominating in the energy budget of the Universe, it can always result a late time accelerating Universe. The non-zero AdS minima in the potential will result a small contribution to the energy budget of the Universe in the form of a negative cosmological constant. As shown in the various recent studies, discussed in this chapter, this contribution can have interesting signatures in the cosmological evolution of the Universe. It can help to resolve some of the cosmological tensions. Such models can also be statistically a better fit of the data compared to ΛCDM. Moreover the presence of this small negative cosmological constant (due to its attractive gravitational effect), will reduce the Hubble friction in the evolution of matter density fluctuations, thereby enhancing the growth of matter fluctuations, especially at higher redshifts and the nonlinear regime may start earlier than the usual ΛCDM model. This may have interesting implications in the context of recent JWST observations of very massive galaxies at higher redshifts.

Hence cosmological models where dark energy energy contains the contribution from a small negative cosmological constant, is definitely an interesting option in the current context of different cosmological tensions and anomalies. Already a good number of research works have shown the potential for such models in the context of observational data and such models should be pursued further, especially in the context of structure formation in the Universe.

Acknowledgements The author acknowledges the funding from SERB, Govt of India under the research grant no: CRG/2020/004347.

References

1. T. Abbott, M. Aguena, A. Alarcon, S. Allam, O. Alves, A. Amon, F. Andrade-Oliveira, J. Annis, S. Avila, D. Bacon, E. Baxter, K. Bechtol, M. Becker, G. Bernstein, S. Bhargava, S. Birrer, J. Blazek, A. Brandao-Souza, S. Bridle, D. Brooks, E. Buckley-Geer, D. Burke, H. Camacho, A. Campos, A.C. Rosell, M.C. Kind, J. Carretero, F. Castander, R. Cawthon, C. Chang, A. Chen, R. Chen, A. Choi, C. Conselice, J. Cordero, M. Costanzi, M. Crocce, L. da Costa, M. da Silva Pereira, C. Davis, T. Davis, J.D. Vicente, J. DeRose, S. Desai, E.D. Valentino, H. Diehl, J. Dietrich, S. Dodelson, P. Doel, C. Doux, A. Drlica-Wagner, K. Eckert, T. Eifler, F. Elsner, J. Elvin-Poole, S. Everett, A. Evrard, X. Fang, A. Farahi, E. Fernandez, I. Ferrero, A. Ferté, P. Fosalba, O. Friedrich, J. Frieman, J. García-Bellido, M. Gatti, E. Gaztanaga, D. Gerdes, T. Giannantonio, G. Giannini, D. Gruen, R. Gruendl, J. Gschwend, G. Gutierrez, I. Harrison, W. Hartley, K. Herner, S. Hinton, D. Hollowood, K. Honscheid, B. Hoyle, E. Huff, D. Huterer, B. Jain, D. James, M. Jarvis, N. Jeffrey, T. Jeltema, A. Kovacs, E. Krause, R. Kron, K. Kuehn, N. Kuropatkin, O. Lahav, P.-F. Leget, P. Lemos, A. Liddle, C. Lidman, M. Lima, H. Lin, N. MacCrann, M. Maia, J. Marshall, P. Martini, J. McCullough, P. Melchior, J. Mena-Fernández, F. Menanteau, R. Miquel, J. Mohr, R. Morgan, J. Muir, J. Myles, S. Nadathur, A. Navarro-Alsina, R. Nichol, R. Ogando, Y. Omori, A. Palmese, S. Pandey, Y. Park, F. Paz-Chinchón, D. Petravick, A. Pieres, A.P. Malagón, A. Porredon, J. Prat, M. Raveri, M. Rodriguez-Monroy, R. Rollins, A. Romer, A. Roodman, R. Rosenfeld, A. Ross, E. Rykoff, S. Samuroff, C. Sánchez, E. Sanchez, J. Sanchez, D.S. Cid, V. Scarpine, M. Schubnell, D. Scolnic, L. Secco, S. Serrano, I. Sevilla-Noarbe, E. Sheldon, T. Shin, M. Smith, M. Soares-Santos, E. Suchyta, M. Swanson, M. Tabbutt, G. Tarle, D. Thomas, C. To, A. Troja, M. Troxel, D. Tucker, I. Tutusaus, T. Varga, A. Walker, N. Weaverdyck, R. Wechsler, J. Weller, B. Yanny, B. Yin, Y. Zhang, J.Z And, Dark energy survey year 3 results: cosmological constraints from galaxy clustering and weak lensing. Phys. Rev. D **105**, 2 (2022)
2. N.J. Adams, C.J. Conselice, L.Ferreira, D. Austin, J.A.A. Trussler, I. Juodzbalis, S.M. Wilkins, J. Caruana, P. Dayal, A. Verma, A.P. Vijayan, Discovery and properties of ultra-high redshift galaxies ($9 < z < 12$) in the jwst ero smacs 0723 field. Monthly Notices of the Royal Astron. Soc. 4755–4766
3. G.E. Addison, Y. Huang, D.J. Watts, C.L. Bennett, M. Halpern, G. Hinshaw, J.L. Weiland, Quantifying discordance in the 2015 Planck CMB spectrum. The Astrophys. J. **818**(2), 132 (2016)
4. S.A. Adil, O. Akarsu, E.D. Valentino, R.C. Nunes, E. Ozulker, A.A. Sen, E. Specogna, Omnipotent dark energy: a phenomenological answer to the Hubble tension. arXiv e-prints 2306.08046 (2023)
5. S.A. Adil, U. Mukhopadhyay, A.A. Sen, S. Vagnozzi, Dark energy in light of the early jwst observations: case for a negative cosmological constant? arXiv e-prints 2307.12763 (2023)
6. N. Aghanim, Y. Akrami, M. Ashdown, J. Aumont, C. Baccigalupi, M. Ballardini, A.J. Banday, R.B. Barreiro, N. Bartolo, S. Basak, R. Battye, K. Benabed, J.-P. Bernard, M. Bersanelli, P. Bielewicz, J.J. Bock, J.R. Bond, J. Borrill, F.R. Bouchet, F. Boulanger, M. Bucher, C. Burigana, R.C. Butler, E. Calabrese, J.-F. Cardoso, J. Carron, A. Challinor, H.C. Chiang, J. Chluba, L.P.L. Colombo, C. Combet, D. Contreras, B.P. Crill, F. Cuttaia, P. de Bernardis, G. de Zotti, J. Delabrouille, J.-M. Delouis, E.D. Valentino, J.M. Diego, O. Doré , M. Douspis, A. Ducout, X. Dupac, S. Dusini, G. Efstathiou, F. Elsner, T.A. Enßlin, H.K. Eriksen, Y. Fantaye, M. Farhang, J. Fergusson, R. Fernandez-Cobos, F. Finelli, F. Forastieri, M. Frailis, A.A. Fraisse, E. Franceschi, A. Frolov, S. Galeotta, S. Galli, K. Ganga, R.T. Génova-Santos, M. Gerbino, T. Ghosh, J. González-Nuevo, K.M. Górski, S. Gratton, A. Gruppuso, J.E. Gudmundsson, J. Hamann, W. Handley, F.K. Hansen, D. Herranz, S.R. Hildebrandt, E. Hivon, Z. Huang, A.H. Jaffe, W.C. Jones, A. Karakci, E. Keihänen, R. Keskitalo, K. Kiiveri, J. Kim, T.S. Kisner, L. Knox, N. Krachmalnicoff, M. Kunz, H. Kurki-Suonio, G. Lagache, J.-M. Lamarre, A. Lasenby, M. Lattanzi, C.R. Lawrence, M.L. Jeune, P. Lemos, J. Lesgourgues, F. Levrier, A. Lewis, M. Liguori, P.B. Lilje, M. Lilley, V. Lindholm, M. López-Caniego, P.M. Lubin, Y.-Z. Ma, J.F.

Macías-Pérez, G. Maggio, D. Maino, N. Mandolesi, A. Mangilli, A. Marcos-Caballero, M. Maris, P.G. Martin, M. Martinelli, E. Martínez-González, S. Matarrese, N. Mauri, J.D. McEwen, P.R. Meinhold, A. Melchiorri, A. Mennella, M. Migliaccio, M. Millea, S. Mitra, M.-A. Miville-Deschênes, D. Molinari, L. Montier, G. Morgante, A. Moss, P. Natoli, H.U. Nørgaard-Nielsen, L. Pagano, D. Paoletti, B. Partridge, G. Patanchon, H.V. Peiris, F. Perrotta, V. Pettorino, F. Piacentini, L. Polastri, G. Polenta, J.-L. Puget, J.P. Rachen, M. Reinecke, M. Remazeilles, A. Renzi, G. Rocha, C. Rosset, G. Roudier, J.A. Rubiño-Martín, B. Ruiz-Granados, L. Salvati, M. Sandri, M. Savelainen, D. Scott, E.P.S. Shellard, C. Sirignano, G. Sirri, L.D. Spencer, R. Sunyaev, A.-S. Suur-Uski, J.A. Tauber, D. Tavagnacco, M. Tenti, L. Toffolatti, M. Tomasi, T. Trombetti, L. Valenziano, J. Valiviita, B.V. Tent, L. Vibert, P. Vielva, F. Villa, N. Vittorio, B.D. Wandelt, I.K. Wehus, M. White, S.D.M. White, A. Zacchei, A. Zonca, Planck 2018 results. Vi. Cosmological parameters. Astron. Astrophys. **641**, A6 (2020)

7. O. Akarsu, S. Kumar, E. Ozulker, J.A. Vazquez, A. Yadav, Relaxing cosmological tensions with a sign switching cosmological constant: improved results with Planck, bao and pantheon data (2022). arXiv e-prints 2211.05742
8. Ö. Akarsu, J.D. Barrow, L.A. Escamilla, J.A. Vazquez, Graduated dark energy: Observational hints of a spontaneous sign switch in the cosmological constant. Phys. Rev. D **101**, 6 (2020)
9. S. Alam, M. Ata, S. Bailey, F. Beutler, D. Bizyaev, J.A. Blazek, A.S. Bolton, J.R. Brownstein, A. Burden, C.-H. Chuang, J. Comparat, A.J. Cuesta, K.S. Dawson, D.J. Eisenstein, S. Escoffier, H. Gil-Marín, J.N. Grieb, N. Hand, S. Ho, K. Kinemuchi, D. Kirkby, F. Kitaura, E. Malanushenko, V. Malanushenko, C. Maraston, C.K. McBride, R.C. Nichol, M.D. Olmstead, D. Oravetz, N. Padmanabhan, N. Palanque-Delabrouille, K. Pan, M. Pellejero-Ibanez, W.J. Percival, P. Petitjean, F. Prada, A.M. Price-Whelan, B.A. Reid, S.A. Rodríguez-Torres, N.A. Roe, A.J. Ross, N.P. Ross, G. Rossi, J.A. Rubiño-Martín, S. Saito, S. Salazar-Albornoz, L. Samushia, A.G. Sánchez, S. Satpathy, D.J. Schlegel, D.P. Schneider, C.G. Scóccola, H.-J. Seo, E.S. Sheldon, A. Simmons, A. Slosar, M.A. Strauss, M.E.C. Swanson, D. Thomas, J.L. Tinker, R. Tojeiro, M.V. Magaña, J.A. Vazquez, L. Verde, D.A. Wake, Y. Wang, D.H. Weinberg, M. White, W.M. Wood-Vasey, C. Yèche, I. Zehavi, Z. Zhai, G.-B. Zhao, The clustering of galaxies in the completed SDSS-III Baryon oscillation spectroscopic survey: cosmological analysis of the DR12 galaxy sample. Monthly Notices of the Royal Astron. Soc. **470**(3), 2617–2652 (2017)
10. M. Asgari, C.-A. Lin, B. Joachimi, B. Giblin, C. Heymans, H. Hildebrandt, A. Kannawadi, B. Stölzner, T. Tröster, J.L. van den Busch, A.H. Wright, M. Bilicki, C. Blake, J. de Jong, A. Dvornik, T. Erben, F. Getman, H. Hoekstra, F. Köhlinger, K. Kuijken, L. Miller, M. Radovich, P. Schneider, H. Shan, E. Valentijn, KiDS-1000 cosmology: cosmic shear constraints and comparison between two point statistics. Astron. Astrophys. **645**, A104 (2021)
11. F. Beutler, C. Blake, M. Colless, D.H. Jones, L. Staveley-Smith, L. Campbell, Q. Parker, W. Saunders, F. Watson, The 6dF galaxy survey: Baryon acoustic oscillations and the local Hubble constant. Monthly Notices of the Royal Astron. Soc. **416**(4), 3017–3032 (2011)
12. R. Calderón, R. Gannouji, B. L'Huillier, D. Polarski, Negative cosmological constant in the dark sector? Phys. Rev. D **103**, 2 (2021)
13. M. Castellano, A. Fontana, T. Treu, P. Santini, E. Merlin, N. Leethochawalit, M. Trenti, E. Vanzella, U. Mestric, A. Bonchi, D. Belfiori, M. Nonino, D. Paris, G. Polenta, G. Roberts-Borsani, K. Boyett, Marus **L15**
14. A. Chudaykin, D. Gorbunov, N. Nedelko, Exploring ΛCDM extensions with spt-3g and Planck data: 4σ evidence for neutrino masses and implications of extended dark energy models for cosmological tensions (2023). arXiv e-prints 2203.03666
15. M. Cicoli, S. de Alwis, A. Maharana, F. Muia, F.. De. Quevedo, Sitter versus quintessence in string theory. Fortschritte der Physik **67**(1–2), 1800079 (2018)
16. M. Cicoli, S. de Alwis, A. Maharana, F. Muia, F.. De. Quevedo, Sitter versus quintessence in string theory. Fortschritte der Physik **67**(1–2), 1800079 (2019)
17. E.J. Copeland, M. Sami, S. Tsujikawa, Dynamics of dark energy. Int. J. Modern Phys. D **15**(11), 1753–1935 (2006)
18. U.H. Danielsson, P. Koerber, T. van Riet, Universal de sitter solutions at tree-level. J. High Energy Phys. **2010**, 90 (2010)

19. U.H. Danielsson, G. Shiu, T. Van Riet, T. Wrase, A note on obstinate tachyons in classical dS solutions. J. High Energy Phys. **2013**, 138 (2013)
20. M. Demirtas, M. Kim, L. McAllister, J. Moritz, A. Rios-Tascon, Exponentially small cosmological constant in string theory. Phys. Rev. Lett
21. D. Dutcher, L. Balkenhol, P. Ade, Z. Ahmed, E. Anderes, A. Anderson, M. Archipley, J. Avva, K. Aylor, P. Barry, R.B. Thakur, K. Benabed, A. Bender, B. Benson, F. Bianchini, L. Bleem, F. Bouchet, L. Bryant, K. Byrum, J. Carlstrom, F. Carter, T. Cecil, C. Chang, P. Chaubal, G. Chen, H.-M. Cho, T.-L. Chou, J.-F. Cliche, T. Crawford, A. Cukierman, C. Daley, T. de Haan, E. Denison, K. Dibert, J. Ding, M. Dobbs, W. Everett, C. Feng, K. Ferguson, A. Foster, J. Fu, S. Galli, A. Gambrel, R. Gardner, N. Goeckner-Wald, R. Gualtieri, S. Guns, N. Gupta, R. Guyser, N. Halverson, A. Harke-Hosemann, N. Harrington, J. Henning, G. Hilton, E. Hivon, G. Holder, W. Holzapfel, J. Hood, D. Howe, N. Huang, K. Irwin, O. Jeong, M. Jonas, A. Jones, T. Khaire, L. Knox, A. Kofman, M. Korman, D. Kubik, S. Kuhlmann, C.-L. Kuo, A. Lee, E. Leitch, A. Lowitz, C. Lu, S. Meyer, D. Michalik, M. Millea, J. Montgomery, A. Nadolski, T. Natoli, H. Nguyen, G. Noble, V. Novosad, Y. Omori, S. Padin, Z. Pan, P. Paschos, J. Pearson, C. Posada, K. Prabhu, W. Quan, S. Raghunathan, A. Rahlin, C. Reichardt, D. Riebel, B. Riedel, M. Rouble, J. Ruhl, J. Sayre, E. Schiappucci, E. Shirokoff, G. Smecher, J. Sobrin, A. Stark, J. Stephen, K. Story, A. Suzuki, K. Thompson, B. Thorne, C. Tucker, C. Umilta, L. Vale, K. Vanderlinde, J. Vieira, G. Wang, N. Whitehorn, W. Wu, V. Yefremenko, K. Yoon, M.Y., Measurements of the e-mode polarization and temperature-e-mode correlation of the CMB from SPT-3g 2018 data. Phys. Rev. D **104**, 2 (2021)
22. K. Dutta, A. Roy, A.A. Ruchika Sen, M.M. Sheikh-Jabbari, Beyond λcdm with low and high redshift data: implications for dark energy. General Relativity and Gravitation **52**, 2 (2020)
23. T. Delubac, J.E. Bautista, J. Rich, D. Kirkby, S. Bailey, A. Font-Ribera, A. Slosar, K.G. Lee, M.M. Pieri, J.C. Hamilton, E. Aubourg, Baryon acoustic oscillations in the ly-α forest of BOSS DR11 quasars. Astron. Astrophys. **574**, A59 (2015)
24. W.L. Freedman, Measurements of the Hubble constant: tensions in perspective. The Astrophys. J. **919**(1), 16 (2021)
25. W. Handley, Curvature tension: evidence for a closed universe. Phys. Rev. D **103**, 4 (2021)
26. J.B. Hartle, S.W. Hawking, T. Hertog, Accelerated expansion from negative λ (2012). arXiv e-prints 1205.3807
27. J.-Q. Jiang, Y.-S. Piao, Testing AdS early dark energy with planck, SPTpol, and LSS data. Phys. Rev. D **104**, 10 (2021)
28. S. Joudaki, C. Blake, C. Heymans, A. Choi, J. Harnois-Deraps, H. Hildebrandt, B. Joachimi, A. Johnson, A. Mead, D. Parkinson, M. Viola, L. van Waerbeke, CFHTLenS revisited: assessing concordance with Planck including astrophysical systematics. Monthly Notices of the Royal Astron. Soc. **465**(2), 2033–2052 (2016)
29. I. Labbé, P. van Dokkum, E. Nelson, R. Bezanson, K.A. Suess, J. Leja, G. Brammer, K. Whitaker, E. Mathews, M. Stefanon, B. Wang, A population of red candidate massive galaxies $\sim$ 600 myr after the big bang. Nature **616**(7956), 266–269 (2023)
30. M. Malekjani, R.M. Conville, E.O. Colgain, S. Pourojaghi, M.M. Sheikh-Jabbari, Negative dark energy density from high redshift pantheon+ supernovae (2023). arXiv e-prints 2301.12725
31. A.T. Mithani, A. Vilenkin, Inflation with negative potentials and the signature reversal symmetry. J. Cosmol. Astroparticle Phys. **2013**(04), 024–024 (2013)
32. H. Moshafi, H. Firouzjahi, A. Talebian, Multiple transitions in vacuum dark energy and H_0 tension. The Astrophys. J. **940**(2), 121 (2022)
33. T. Padmanabhan, Cosmological constant-the weight of the vacuum. Phys. Rep. **380**(5–6), 235–320 (2003)
34. P.J.E. Peebles, B. Ratra, The cosmological constant and dark energy. Rev. Mod. Phys. **75**, 559–606 (2003)
35. V. Poulin, K.K. Boddy, S. Bird, M. Kamionkowski, Implications of an extended dark energy cosmology with massive neutrinos for cosmological tensions. Phys. Rev. D **97**, 12 (2018)
36. A.G. Riess, W. Yuan, L.M. Macri, D. Scolnic, D. Brout, S. Casertano, D.O. Jones, Y. Murakami, G.S. Anand, L. Breuval, T.G. Brink, A.V. Filippenko, S. Hoffmann, S.W. Jha, W.D. Kenworthy,

J. Mackenty, B.E. Stahl, W. Zheng, A comprehensive measurement of the local value of the Hubble constant with 1 km/s/mpc uncertainty from the Hubble space telescope and the SH_0ES team. The Astrophys. J. Lett. **934**(1), L7 (2022)
37. A.J. Ross, L. Samushia, C. Howlett, W.J. Percival, A. Burden, M. Manera, The clustering of the SDSS DR7 main galaxy sample—I. A 4 percent distance measure at z = 0.15. Monthly Notices of the Royal Astron. Soc. **449**(1), 835–847 (2015)
38. Ruchika, S.A. Adil, K. Dutta, A. Mukherjee, A.A. Sen, Observational constraints on axion(s) dark energy with a cosmological constant. Phys. Dark Univ. **40**, 101199 (2023)
39. V. Sahni, A. Shafieloo, A.A. Starobinsky, Model-independent evidence for dark energy evolution from Baryon acoustic oscillations. The Astrophys. J. Lett. **793**(2), L40 (2014)
40. V. Sahni, A. Starobinsky, The case for a positive cosmological Λ-term. Int. J. Modern Phys. D **9**(4), 373–443 (2000)
41. D.M. Scolnic, D.O. Jones, A. Rest, Y.C. Pan, R. Chornock, R.J. Foley, M.E. Huber, R. Kessler, G. Narayan, A.G. Riess, S. Rodney, E. Berger, D.J. Brout, P.J. Challis, M. Drout, D. Finkbeiner, R. Lunnan, R.P. Kirshner, N.E. Sanders, E. Schlafly, S. Smartt, C.W. Stubbs, J. Tonry, W.M. Wood-Vasey, M. Foley, J. Hand, E. Johnson, W.S. Burgett, K.C. Chambers, P.W. Draper, K.W. Hodapp, N. Kaiser, R.P. Kudritzki, E.A. Magnier, N. Metcalfe, F. Bresolin, E. Gall, R. Kotak, M. McCrum, K.W. Smith, The complete light-curve sample of spectroscopically confirmed SNe Ia from Pan-STARRS1 and cosmological constraints from the combined Pantheon sample. Astrophys. J. **859**(2), 101 (2018)
42. A.A. Sen, S.A. Adil, S. Sen, Do cosmological observations allow a negative λ? Monthly Notices of the Royal Astron. Soc. **518**(1), 1098–1105 (2022)
43. E. Silverstein, Simple de sitter solutions. Phys. Rev. D **77**, 106006 (2008)
44. M. Troxel, N. MacCrann, J. Zuntz, T. Eifler, E. Krause, S. Dodelson, D. Gruen, J. Blazek, O. Friedrich, S. Samuroff, J. Prat, L. Secco, C. Davis, A. Ferté, J. DeRose, A. Alarcon, A. Amara, E. Baxter, M. Becker, G. Bernstein, S. Bridle, R. Cawthon, C. Chang, A. Choi, J.D. Vicente, A. Drlica-Wagner, J. Elvin-Poole, J. Frieman, M. Gatti, W. Hartley, K. Honscheid, B. Hoyle, E. Huff, D. Huterer, B. Jain, M. Jarvis, T. Kacprzak, D. Kirk, N. Kokron, C. Krawiec, O. Lahav, A. Liddle, J. Peacock, M. Rau, A. Refregier, R. Rollins, E. Rozo, E. Rykoff, C. Sánchez, I. Sevilla-Noarbe, E. Sheldon, A. Stebbins, T. Varga, P. Vielzeuf, M. Wang, R. Wechsler, B. Yanny, T. Abbott, F. Abdalla, S. Allam, J. Annis, K. Bechtol, A. Benoit-Lévy, E. Bertin, D. Brooks, E. Buckley-Geer, D. Burke, A.C. Rosell, M.C. Kind, J. Carretero, F. Castander, M. Crocce, C. Cunha, C. D'Andrea, L. da Costa, D. DePoy, S. Desai, H. Diehl, J. Dietrich, P. Doel, E. Fernandez, B. Flaugher, P. Fosalba, J. García-Bellido, E. Gaztanaga, D. Gerdes, T. Giannantonio, D. Goldstein, R. Gruendl, J. Gschwend, G. Gutierrez, D. James, T. Jeltema, M. Johnson, M. Johnson, S. Kent, K. Kuehn, S. Kuhlmann, N. Kuropatkin, T. Li, M. Lima, H. Lin, M. Maia, M. March, J. Marshall, P. Martini, P. Melchior, F. Menanteau, R. Miquel, J. Mohr, E. Neilsen, R. Nichol, B. Nord, D. Petravick, A. Plazas, A. Romer, A. Roodman, M. Sako, E. Sanchez, V. Scarpine, R. Schindler, M. Schubnell, M. Smith, R. Smith, M. Soares-Santos, F. Sobreira, E. Suchyta, M. Swanson, G. Tarle, D. Thomas, D. Tucker, V. Vikram, A. Walker, J. Weller, Y.Z., Dark energy survey year 1 results: cosmological constraints from cosmic shear. Phys. Rev. D **98**, 4 (2018)
45. S. Tsujikawa, Quintessence: a review. Class. Quantum Gravity **30**(21), 214003 (2013)
46. E.D. Valentino, A. Melchiorri, J. Silk, Planck evidence for a closed universe and a possible crisis for cosmology. Nat. Astron. **4**(2), 196–203 (2019)
47. E.D. Valentino, A. Mukherjee, A.A. Sen, Dark energy with phantom crossing and the H_0 tension. Entropy **23**(4), 404 (2021)
48. L. Visinelli, S. Vagnozzi, U. Danielsson, Revisiting a negative cosmological constant from low-redshift data. Symmetry **11**(8), 1035 (2019)
49. Y. Wang, L. Pogosian, G.-B. Zhao, A. Zucca, Evolution of dark energy reconstructed from the latest observations. The Astrophys. J. **869**(1), L8 (2018)
50. W.L.K. Wu, L.M. Mocanu, P.A.R. Ade, A.J. Anderson, J.E. Austermann, J.S. Avva, J.A. Beall, A.N. Bender, B.A. Benson, F. Bianchini, L.E. Bleem, J.E. Carlstrom, C.L. Chang, H.C. Chiang, R. Citron, C.C. Moran, T.M. Crawford, A.T. Crites, T. de Haan, M.A. Dobbs, W. Everett, J.

Gallicchio, E.M. George, A. Gilbert, N. Gupta, N.W. Halverson, N. Harrington, J.W. Henning, G.C. Hilton, G.P. Holder, W.L. Holzapfel, Z. Hou, J.D. Hrubes, N. Huang, J. Hubmayr, K.D. Irwin, L. Knox, A.T. Lee, D. Li, A. Lowitz, A. Manzotti, J.J. McMahon, S.S. Meyer, M. Millea, J. Montgomery, A. Nadolski, T. Natoli, J.P. Nibarger, G.I. Noble, V. Novosad, Y. Omori, S. Padin, S. Patil, C. Pryke, C.L. Reichardt, J.E. Ruhl, B.R. Saliwanchik, J.T. Sayre, K.K. Schaffer, C. Sievers, G. Simard, G. Smecher, A.A. Stark, K.T. Story, C. Tucker, K. Vanderlinde, T. Veach, J.D. Vieira, G. Wang, N. Whitehorn, V. Yefremenko, A measurement of the cosmic microwave background lensing potential and power spectrum from $500deg^2$ of SPTpol temperature and polarization data. The Astrophys. J. **884**(1), 70 (2019)

Chapter 29
On the Interacting Dark Energy Scenarios—The Case for Hubble Constant Tension

Supriya Pan and Weiqiang Yang

29.1 Introduction

The dark sector of our universe is highly mysterious and it has become a very appealing topic for modern cosmology and cosmologists. According to a large amount of observational evidences, the dark sector of the universe is comprised of two heavy fluids, namely, dark matter (DM) and dark energy (DE) together which occupy nearly 96% of the total energy budget of our universe. The DM sector is responsible for the observed structure formation of our universe while the role of DE is to drive the accelerating expansion of the universe [1, 2]. While the origin, nature and evolution of this dark picture is not fully understood at this moment, but on the report of a large number of astronomical observations, Λ-Cold Dark Matter model, a very simple cosmological model constructed in the context of Einstein's General Relativity (GR) where $\Lambda > 0$ acts as the source of DE and DM being pressure-less, has been quite successful in describing the current observational evidences. In this simplest cosmological picture, DE and DM are conserved independently, that means, evolution of DM does not affect the evolution of DE, and vice-versa. Despite marvellous successes, ΛCDM cosmological model has some limitations. The cosmological constant problem is one of the biggest issues that still needs an answer [3]. On the other hand, the cosmic coincidence problem (the so-called "why now" problem) [4] is another cosmological issue associated with this cosmological model. Apart from the above issues, it was reported that the estimations of some key cosmological parameters using the cosmic microwave background (CMB) observations by Planck (within the ΛCDM paradigm) are in tension at many standard deviations with the local measurements.

S. Pan (✉)
Department of Mathematics, Presidency University, 86/1 College Street, 700073 Kolkata, India
e-mail: supriya.maths@presiuniv.ac.in

W. Yang
Department of Physics, Liaoning Normal University, 116029 Dalian, People's Republic of China
e-mail: d11102004@163.com

E. Di Valentino and D. Brout (eds.), *The Hubble Constant Tension*, Springer Series in Astrophysics and Cosmology, https://doi.org/10.1007/978-981-99-0177-7_29

In particular, the measurements of the Hubble constant by the cosmic microwave background observations by Planck within the ΛCDM paradigm ($H_0 = 67.4 \pm 0.5$ km/s/Mpc at 68% CL) [5] and SH_0ES (Supernovae and H_0 for the Equation of State of dark energy) collaboration ($H_0 = 73.0 \pm 1.0$ km/s/Mpc at 68% CL, labeled as R21) [6] are in tension with $\gtrsim 5\sigma$ confidence. Additionally, some further estimations by the SH_0ES team indicate that this tension can increase up to 5.3σ [7]. Moreover, the S_8 parameter, defined as a combination of the amplitude of the matter power spectrum σ_8 with the matter density at present Ω_m ($S_8 = \sigma_8\sqrt{\Omega_m/0.3}$) exhibits a tension at the level of 2–3 σ between Planck (within ΛCDM paradigm) and low redshift probes, e.g. weak gravitational lensing and galaxy clustering (e.g. [8–17]). Thus, even if ΛCDM is extremely successful with a series of observational data, however, based on the aforementioned limitations, it is argued that perhaps ΛCDM is an approximation of a more realistic scenario that is yet to be discovered. Following this, several modifications of the ΛCDM cosmological scenario were proposed by many authors, see the recent reviews [18–21].

In this chapter we focus on a very generalized cosmological scenario where DE and DM interact with each other in a non-gravitationally way. That means an exchange of energy and/or momentum between DE and DM is allowed. This class of cosmological models is known as Interacting DE (or coupled DE or Interacting DE-DM). According to the historical records, the idea of interaction between gravity and a scalar field with exponential potential was proposed by Wetterich to explain the tiny value of the cosmological constant [22]. However, in Wetterich's proposal [22], explicit DE-DM interaction was not introduced. Soon after the discovery of the late-time accelerating expansion of our universe, when the dynamical DE in the name of quintessence was proposed and the cosmic coincidence problem emerged in the cosmological domain [4], Amendola wrote an article on the "Coupled quintessence" [23] in which he formally introduced an interaction between the matter sector and a scalar field with exponential potential acting as a source of DE where the author pointed out that such an interaction can solve the long debating "cosmic coincidence" problem. This coincidence problem was further investigated by other authors and it was found that interacting DE models can excellently solve the cosmic coincidence problem [24–30]. This was the first achievement of the theory of interacting DE and this influenced various investigators to work with the interacting DE models. As a result, over the last several years a number of interacting DE models have been investigated with some interesting outcomes [31–63, 65, 66] (also see the review articles on the interacting DE scenarios [67, 68]). In particular, it was observed that an interaction between DE-DM can lead to a phantom DE scenario [69–72] without invoking any scalar field theory with negative kinetic term which may lead to instabilities in classical and quantum levels. In the context of the cosmological tensions, interacting DE plays a very promising role. There are many evidences showing that an interaction between DE and DM can increase the H_0 value and thus such scenarios can alleviate the H_0 tension between Planck (within the ΛCDM paradigm) and SH_0ES collaborations [73–82]. It has been also noticed that the interacting DE has the potentiality to alleviate the S_8 tension [74, 78, 83, 84]. Based on the above observations, it is clear that the cosmological scenarios allowing a non-gravitational interaction between DE and DM are highly appealing and they should get more attention in the forthcoming years.

This chapter is organized in the following way. In Sect. 29.2 we give an overview of the interacting DE-DM models and present the evolution equations of a general interaction model at the level of background and perturbations. In Sect. 29.3 we discuss various models of interaction that have been investigated in the literature to alleviate the Hubble tension. In particular, in Sect. 29.3.1 we discuss the interacting models in which the DE equation of state is constant; in Sect. 29.3.2 we present those results where the DE equation of state is dynamical; and in Sect. 29.3.3 we discuss a special kind of interaction models in which the coupling parameter of the interaction scenarios is assumed to be time dependent. In Sect. 29.5 we conclude the chapter with a brief summary of the entire work.

29.2 Interacting DE

In the large scales, our universe is almost homogeneous and isotropic. This homogeneous and isotropic configuration of the universe is well described by the Friedmann-Lemaître-Robertson-Walker (FLRW) line element

$$ds^2 = -dt^2 + a^2(t)\left[\frac{dr^2}{1-kr^2} + r^2(d\theta^2 + \sin^2\theta d\phi^2)\right] \tag{29.1}$$

where (t, r, θ, ϕ) are the co-moving coordinates; $a(t)$ is the expansion scale factor of the universe; k denotes the spatial geometry of the universe which can take three distinct values, namely, $0, +1, -1$ corresponding to three distinct spatial geometries of the universe—flat ($k = 0$), closed ($k = +1$) and open ($k = -1$). We further assume that the gravitational sector of the universe is well described by GR and the matter distribution is minimally coupled to gravity. In the matter distribution, one can consider a variety of components, such as radiation, baryons, neutrinos, dark matter, dark energy. Now, within this framework, the Einstein's gravitational equations $G_{\mu\nu} = 8\pi G\, T_{\mu\nu}$ can be recast as

$$H^2 + \frac{k}{a(t)^2} = \frac{8\pi G}{3}\sum_i \rho_i, \tag{29.2}$$

$$2\dot{H} + 3H^2 + \frac{k}{a(t)^2} = -8\pi G \sum_i p_i, \tag{29.3}$$

where an overhead dot represents the derivative with respect to the cosmic time; $H \equiv \dot{a}(t)/a(t)$ is the Hubble rate of the FLRW universe; p_i and ρ_i are respectively the pressure and energy density of the i-th fluid. Hence $\sum_i \rho_i$ and $\sum_i p_i$ denote the total energy density and total pressure of the matter distribution, respectively. Using the Bianchi's identity $\nabla^{\mu} T_{\mu\upsilon} = 0$, or, alternatively, using Eqs. (29.2) and (29.3), one can derive the conservation equation of the total fluid as follows

$$\sum_i \dot{\rho}_i + 3H \sum_i (\rho_i + p_i) = 0. \quad (29.4)$$

We assume that DM and DE are coupled to each other, but the remaining fluids are independently conserved. That means, in the context of the FLRW universe, the conservation equations for DM and DE become,

$$\dot{\rho}_{\rm dm} + 3H(1 + w_{\rm dm})\rho_{\rm dm} = -Q(t), \quad (29.5)$$

$$\dot{\rho}_{\rm de} + 3H(1 + w_{\rm de})\rho_{\rm de} = +Q(t), \quad (29.6)$$

where $\rho_{\rm dm}$ and $\rho_{\rm de}$ are respectively the energy density of DM and DE; $w_{\rm dm} \geq 0$, $w_{\rm de} < -1/3$, are respectively the barotropic equation of state parameters of DM and DE. The quantity Q denotes the rate of energy or/and momentum exchange between these dark fluids and it is the heart of the theory of interacting dark energy. For $Q(t) > 0$, the energy flows from DM to the DE while for $Q(t) < 0$, energy flow takes place from DE to DM. In the limiting case $Q \to 0$, the non-interacting cosmological scenario as a special case is recovered. Now, given a specific interaction function, one can in principle determine the dynamics of the universe at the background level using the conservation Eqs. (29.5), (29.6) together with the Hubble equation (29.2). The interaction function also affects the evolution of the universe at the perturbation level. However, there is no such guiding principle to select the interaction function. Although several attempts have been made to derive the interaction function from some action principle [85–94], but no decisive conclusion has been made yet. Thus, in most of the cases, the choice of the interaction function is made from the phenomenological ground and hence there is no unique choice of the interaction function. The existing interaction functions are usually divided into two classes: the interaction functions including explicitly the Hubble rate H of the FLRW universe, e.g. $Q(t) = Hf(\rho_{\rm dm}, \rho_{\rm de})$, where f is an analytic function of the energy densities of the dark components. A general form of the interaction function in this category that recovers many interactions is the following $Q = 3H\xi\rho_{dm}^{\alpha}\rho_{de}^{\beta}(\rho_{dm} + \rho_{de})^{\gamma}$, where $\alpha + \beta + \gamma = 1$ and ξ is a dimensionless coupling parameter. Note that for different values of (α, β, γ), one can produce a cluster of interaction functions with only one coupling parameter. The well known interaction function $Q = 3H\xi\rho_{de}$ is obtained for $\alpha = 0 = \gamma$. We stress that the above choice is a generalized interaction function in a specific direction and this does not include other kind of interaction models, e.g. the interaction models with two coupling parameters, see for instance [49, 57], or the interaction functions involving the exponential or trigonometric functions [66]. Secondly, one can consider another class of interaction functions where the Hubble rate does not appear explicitly, i.e. of the form $Q(t) = f(\rho_{\rm dm}, \rho_{\rm de})$. A generalized interaction function can be of the following: $Q = \Gamma\rho_{dm}^{\alpha}\rho_{de}^{\beta}(\rho_{dm} + \rho_{de})^{\gamma}$, where $\alpha + \beta + \gamma = 1$ but here Γ has the dimension equal to the dimension of the

Hubble parameter.[1] It has been argued in many places, see for instance, Ref. [95] that as the interaction between DE and DM is a local phenomenon, therefore, it should depend only on the local quantities and the appearance of the Hubble parameter H, the global expansion rate, in the interaction functions is mainly designed for the mathematical simplicity. While as argued in Ref. [96], explicit appearance of H in the interaction functions can be motivated from some well established cosmological theories. Thus, the mathematical simplicity does not strongly hold in such cases. On the other hand, it has been observed in Ref. [97] that a particular choice of the interaction function may lead to negative energy density of DM or (and) DE component either in the past or in the future. This is a striking issue since the consequences of the negative energy density of the dark components are not clearly understood at this moment.

Notice that depending on the nature of the equation of state parameters, namely, $w_{\rm dm}$, $w_{\rm de}$, various interacting scenarios can be considered. In most of the cases, $w_{\rm dm}$ is set to zero due to the abundances of the cold DM in the universe sector[2], and the nature of $w_{\rm de}$ remains unspecified, that means it can be either constant or dynamical.

Notice that the conservation Eqs. (29.5) and (29.6) can alternatively be expressed as

$$\dot{\rho}_{dm} + 3H(1 + w_{dm}^{\rm eff})\rho_{dm} = 0, \tag{29.7}$$

$$\dot{\rho}_{de} + 3H(1 + w_{de}^{\rm eff})\rho_{de} = 0, \tag{29.8}$$

where $w_{de}^{\rm eff}$ and $w_{dm}^{\rm eff}$ are termed as the effective equation-of-state parameters of DM and DE, and they are given by

$$w_{dm}^{\rm eff} = w_{\rm dm} + \frac{Q(t)}{3H\rho_{\rm dm}}. \tag{29.9}$$

$$w_{de}^{\rm eff} = w_{\rm de} - \frac{Q(t)}{3H\rho_{\rm de}}. \tag{29.10}$$

Thus, we see that even though we started with an interacting scenario between DE and DM quantified through the interaction function $Q(t)$, but this interacting scenario can always be expressed as a non-interacting scenario described by the conservation equations in Eqs. (29.7) and (29.8) in terms of two effective equation of state parameters of DM and DE displayed in Eqs. (29.9) and (29.10). Now, if the DM sector is cold, i.e. $w_{\rm dm} = 0$, then $w_{dm}^{\rm eff} = \frac{Q(t)}{3H\rho_{\rm dm}}$. We remark that the presence of $Q(t)$ in eqns. (29.9) and (29.10) can significantly affect their nature. For example, in an expanding universe ($H > 0$), an interacting quintessence model in which $w_{\rm de} > -1$ can offer an effective phantom-like behaviour ($w_{\rm de} < -1$) provided $Q(t) > 0$. This is indeed

[1] In the statistical analysis, we thus consider the dimensionless parameter Γ/H_0.

[2] However, as the nature of DE and DM is still yet to be discovered, and since the sensitivity of the observational data is increasing over time, therefore, there is certainly no guiding rule to consider $w_{\rm dm} = 0$. In fact, one may consider $w_{\rm dm}$ to be a free parameter in the underlying cosmological scenarios and let the observational data to decide its fate, see for instance, Refs. [98–104].

an interesting observation since a phantom equation-of-state in DE can increase the expansion rate of the universe faster and as a result, the Hubble constant can take higher values [105]. In this context, an interaction which can offer a phantom DE equation of state through (29.10) can do this job.

As we noted, the evolution of the interacting scenario at the background level is affected by presence of the interaction function $Q(t)$. Similarly, the evolution of the interacting scenario at the level of perturbations is also affected due to the presence of $Q(t)$. Following the perturbed FLRW metric in any gauge [106–108], one can determine how the interacting scenario behaves in presence of the interaction function, see for instance Refs. [41, 42, 95].

29.3 Different Interacting Scenarios and Their Roles in Alleviating the H_0 Tension

As already argued, the interaction function, Q, plays the key role in the interacting dynamics. However, as the choice of Q is not unique, therefore, one can freely choose any interaction model and test its viability with the observational data. In the following we describe how the interaction models help in alleviating the Hubble constant tension. Throughout this section we have assumed that the curvature of the universe is zero.

29.3.1 IDE with Constant Equation of State of DE

The interacting DE scenarios with constant equation of state in DE, w_{de}, have received tremendous attention in the community. In this context, the most simplest scenario is the one where the equation of state of DE mimics the cosmological constant, i.e. $w_{de} \sim -1$. On the other hand, one can assume a constant w_{de} other than -1 and investigate the consequences of the underlying interacting scenarios. In this section we shall discuss how different interacting models with constant w_{de} return different values of H_0 and in all cases we consider the spatial flatness of the FLRW universe.

In Ref. [82], an interaction between DE and CDM has been investigated using multiple CMB datasets from various experiments where the DE equation of state has been fixed at $w_{de} = -0.999$ and the interaction function has the following form

$$Q = H\xi\rho_{de}, \tag{29.11}$$

where ξ is the coupling parameter of the interaction function. Using different CMB data from multiple surveys, the constraints on H_0 are found to be as follows [82]: for Planck 2018 alone, $H_0 = 71.6 \pm 2.1$ km/s/Mpc at 68% CL and hence the tension with respect to R21 is reduced down to 0.6σ; $H_0 = 72.6^{+3.4}_{-2.6}$ km/s/Mpc at 68% CL (for ACT)[3] and hence the tension with respect to R21 is reduced down to 0.1σ which means in this case the H_0 tension is solved; for ACT+WMAP, we find $H_0 = 71.3^{+2.6}_{-3.2}$ km/s/Mpc at 68% CL and hence the tension is reduced down to 0.6σ; and for Planck 2018+ACT, $H_0 = 71.4^{+2.5}_{-2.8}$ km/s/Mpc at 68% CL and in this case the tension is reduced down to 0.6σ. We note that in all the cases, we find an evidence of $\xi \neq 0$ at more than 68% CL for all these datasets [82].

In Ref. [64], the authors investigated an interaction scenario between CDM and DE with the same interaction function (29.11), but w_{de} was fixed at -1 which exactly mimics the vacuum energy sector. In this case, Planck 2018 alone leads to $H_0 = 70.84^{+4.26}_{-2.50}$ km/s/Mpc at 68% CL with an evidence of a non-zero coupling ($\xi = 0.132^{+0.142}_{-0.077}$ at 68% CL) at more than 68% CL, and hence with respect to R21, the tension on H_0 is reduced down to 0.6σ. This interacting scenario is quite appealing because even if in presence of massive neutrinos the tension in H_0 is significantly alleviated [64].

In Ref. [65], another interacting scenario between CDM and DE with $w_{de} = -1$ has been investigated for the interaction function

$$Q = \Gamma \rho_{de}, \tag{29.12}$$

which does not involve the Hubble parameter but it depends on the intrinsic nature of the dark sector, e.g. the energy density of the DE density. Here, Γ is the coupling parameter of the interaction function having the dimension of the Hubble parameter and hence Γ/H_0 which is dimensionless. For CMB alone from Planck 2018, this interaction model leads to $H_0 = 70.3^{+3.3}_{-2.0}$ km/s/Mpc at 68% CL which thus alleviates the tension with R21 at 1σ. For Planck 2018+BAO, $H_0 = 69.0^{+1.4}_{-1.8}$ km/s/Mpc at 68% CL, and as a consequence, the tension with R21 is reduced down to 2.1σ.

In Ref. [81], the authors considered the same interaction function (29.11) describing an interaction between CDM and DE where $w_{de} \neq -1$ is a constant and it could lie either in the quintessence regime ($w_{de} > -1$) or in the phantom regime ($w_{de} < -1$). For the quintessence case, Planck 2018 alone leads to $H_0 = 69.8^{+4}_{-2.5}$ km/s/Mpc at 68% CL (hence, the tension with R21 is reduced down to 0.9σ) with an evidence of a non-zero coupling in the dark sector at more than 3σ significance [81]. When w_{de} has a phantom nature, then the 68% CL lower bound on the Hubble constant is found to be, $H_0 > 70.4$ km/s/Mpc with $w_{de} = -1.59^{+0.18}_{-0.33}$ (at 68% CL for Planck 2018 alone). In this case the resolution of the H_0 tension appears due to the strong degeneracy between H_0-w_{de}, not due to the interaction between CDM and DE.

In Ref. [61], a special kind of an interaction scenario between CDM and DE (with constant w_{de}) in which the interaction function may change its sign during the evolution of the universe, known as sign changeable interaction function[4] was investigated for two variants of the interaction function. The interacting scenarios

[3] Here ACT refers to the full Atacama Cosmology Telescope temperature and polarization DR4 likelihood [109], assuming a conservative Gaussian prior on $\tau = 0.065 \pm 0.015$ as in [110].

[4] For sign changeable interaction functions we refer to Refs. [111–115].

with $w_{de} \neq -1$ are found to increase the H_0 values and therefore here we shall report the constraints for this particular case. Considering the following interaction function

$$Q = 3H\xi(\rho_{dm} - \rho_{de}), \tag{29.13}$$

where ξ is the coupling parameter of the interaction function, it was found that Planck 2015+BAO returns $H_0 = 69.12^{+0.93}_{-1.39}$ km/s/Mpc at 68% CL [61] which thus alleviates the tension with R21 at 2.5σ. For a generalized interaction function of this type

$$Q = 3H(\alpha\rho_{dm} - \beta\rho_{de}), \tag{29.14}$$

where α, β ($\alpha \neq \beta$) are the coupling parameters of this interaction function and they will have the same sign, Planck 2015+BAO gives $H_0 = 69.81^{+1.18}_{-1.39}$ km/s/Mpc at 68% CL, and hence the tension with R21 is reduced down to 2σ. The sign changeable interaction functions did not get much attention in the literature without any specific reason, however, they deserve further attention.

In Ref. [66], an interacting scenario between CDM and DE with a constant w_{de} other than -1, has been constrained with the following interaction function

$$Q = 3H\xi\rho_{de}\sin\left(\frac{\rho_{de}}{\rho_{dm}} - 1\right), \tag{29.15}$$

where ξ is the coupling parameter of the interaction function. The analysis with Planck 2018 gives $H_0 = 72.67^{+5.43}_{-8.26}$ km/s/Mpc at 68% CL and Planck 2018+BAO gives $H_0 = 69.17^{+1.52}_{-1.71}$ km/s/Mpc at 68% CL [66]. Hence, for Planck 2018 and Planck 2018+BAO, the tension on H_0 with respect to R21, is reduced down to 0.05σ, 2σ, respectively. That means, for Planck 2018 alone case, the tension is completely solved. While it is important to mention that the resolution of the tension in this case is also influenced by the uncertainties.

In Ref. [104], an interacting scenario between a non-cold DM characterized by a non-null equation of state of DM, w_{dm} (≥ 0), and a DE component mimicking the vacuum energy sector, i.e. $w_{de} = -1$, has been explored for the interaction function $Q = 3H\xi\rho_{de}$ of (29.11). Unlike other interacting scenarios in which DM is assumed to be pressureless or cold, here, w_{dm} is considered to be a free-to-vary parameter in [0, 1] and the cosmological probes are allowed to decide its fate. This generalized interacting scenario is labeled as **IWDM**. We present the results in Table 29.1 and Fig. 29.1 for different datasets. We notice a mild evidence of $w_{dm} \neq 0$ for all the datasets at 68% CL, while at 95% CL, this evidence goes away. In addition, an evidence of $\xi \neq 0$ is also found from Planck 2018 and Planck 2018+Lensing datasets together with higher values of H_0 in both the datasets: $H_0 = 70.6^{+4.3}_{-2.4}$ km/s/Mpc at 68% CL for Planck 2018 and $H_0 = 70.3^{+4.3}_{-2.8}$ km/s/Mpc at 68% CL for Planck 2018+Lensing. Therefore, the tension in H_0 with respect to R21 is reduced down to 0.7σ for both Planck 2018 and Planck 2018+Lensing.

Table 29.1 Summary of the observational constraints on the interacting scenario between a non-cold DM and vacuum considering various observational datasets. Here, the free parameters of this interacting scenario are as follows: $\Omega_b h^2$ is the physical baryon density; $\Omega_{dm} h^2$ is the non-cold DM density; θ_{MC} is ratio of the sound horizon to the angular diameter distance; τ refers to the optical depth; n_s is the spectral index; the amplitude of the primordial scalar perturbations is denoted by A_s. Among the derived parameters, H_0, S_8 are already defined; Ω_m refers to the density parameter of matter (non-cold DM+baryons) at present time and $r_{\rm drag}$ refers to the comoving size of the sound horizon at $z_{\rm drag}$ (redshift at which baryon-drag optical depth equals unity). The table has been taken from Ref. [104]

Parameters	Planck 2018	Planck 2018+lensing	Planck 2018+BAO	Planck 2018+Pantheon	Planck 2018+Lensing+BAO+Pantheon
$\Omega_b h^2$	$0.02248^{+0.00017+0.00035}_{-0.00017-0.00033}$	$0.02249^{+0.00016+0.00032}_{-0.00016-0.00032}$	$0.02250^{+0.00016+0.00032}_{-0.00016-0.00031}$	$0.02247^{+0.00016+0.00033}_{-0.00017-0.00032}$	$0.02252^{+0.00015+0.00031}_{-0.00015-0.00030}$
$\Omega_{dm} h^2$	$0.077^{+0.033}_{-0.056} < 0.14$	$0.082^{+0.055}_{-0.036} < 0.14$	$0.104^{+0.023+0.036}_{-0.015-0.040}$	$0.109^{+0.013+0.022}_{-0.011-0.024}$	$0.112^{+0.011+0.020}_{-0.009-0.020}$
$100\theta_{MC}$	$1.0436^{+0.0024+0.0049}_{-0.0036-0.0046}$	$1.0433^{+0.0020+0.0051}_{-0.0037-0.0043}$	$1.0419^{+0.0008+0.0024}_{-0.0014-0.0021}$	$1.04147^{+0.00065+0.0015}_{-0.00076-0.0014}$	$1.04137^{+0.00054+0.0012}_{-0.00056-0.0012}$
τ	$0.0534^{+0.0075+0.015}_{-0.0074-0.015}$	$0.0526^{+0.0072+0.015}_{-0.0072-0.014}$	$0.0541^{+0.0077+0.016}_{-0.0075-0.015}$	$0.0533^{+0.0075+0.015}_{-0.0074-0.015}$	$0.0539^{+0.0070+0.01[illegible]}_{-0.0071-0.01[illegible]}$
n_s	$0.9628^{+0.0046+0.0093}_{-0.0046-0.0088}$	$0.9636^{+0.0042+0.0082}_{-0.0042-0.0082}$	$0.9637^{+0.0042+0.0083}_{-0.0042-0.0083}$	$0.9623^{+0.0044+0.0088}_{-0.0044-0.0089}$	$0.9646^{+0.0039+0.0078}_{-0.0039-0.0078}$
$\ln(10^{10} A_s)$	$3.046^{+0.015+0.030}_{-0.016-0.030}$	$3.043^{+0.014+0.029}_{-0.014-0.028}$	$3.046^{+0.016+0.032}_{-0.016-0.031}$	$3.046^{+0.015+0.031}_{-0.015-0.030}$	$3.044^{+0.014+0.028}_{-0.014-0.027}$
w_{dm}	$0.00122^{+0.00053}_{-0.00097} < 0.0025$	$0.00112^{+0.00047}_{-0.00091} < 0.0023$	$0.00115^{+0.00045}_{-0.00098} < 0.0025$	$0.00123^{+0.00052}_{-0.00099} < 0.0031$	$0.00108^{+0.00040}_{-0.00096} < 0.0023$
ξ	$0.11^{+0.14+0.17}_{-0.07-0.20}$	$0.10^{+0.14+0.17}_{-0.08-0.19}$	$0.048^{+0.053+0.12}_{-0.064-0.11}$	$0.036^{+0.036+0.072}_{-0.036-0.072}$	$0.025^{+0.032+0.063}_{-0.032-0.063}$
Ω_m	$0.21^{+0.07+0.19}_{-0.14-0.16}$	$0.22^{+0.08+0.18}_{-0.14-0.17}$	$0.270^{+0.057+0.10}_{-0.046-0.11}$	$0.285^{+0.033+0.066}_{-0.032-0.066}$	$0.289^{+0.028+0.055}_{-0.028-0.055}$
H_0 [Km/s/Mpc]	$70.6^{+4.3+5.2}_{-2.4-6.0}$	$70.3^{+4.3+5.4}_{-2.8-5.6}$	$68.8^{+1.3+3.0}_{-1.6-2.7}$	$68.2^{+1.0+2.1}_{-1.0-2.1}$	$68.33^{+0.82+1.6}_{-0.81-1.5}$
S_8	$1.01^{+0.08+0.41}_{-0.23-0.27}$	$0.98^{+0.06+0.40}_{-0.21-0.24}$	$0.880^{+0.032+0.13}_{-0.071-0.11}$	$0.869^{+0.028+0.070}_{-0.038-0.066}$	$0.849^{+0.022+0.056}_{-0.031-0.052}$
$r_{\rm drag}$ [Mpc]	$146.80^{+0.33+0.64}_{-0.33-0.65}$	$146.91^{+0.29+0.56}_{-0.29-0.57}$	$146.90^{+0.31+0.61}_{-0.31-0.63}$	$146.78^{+0.33+0.64}_{-0.33-0.65}$	$146.98^{+0.27+0.53}_{-0.27-0.53}$

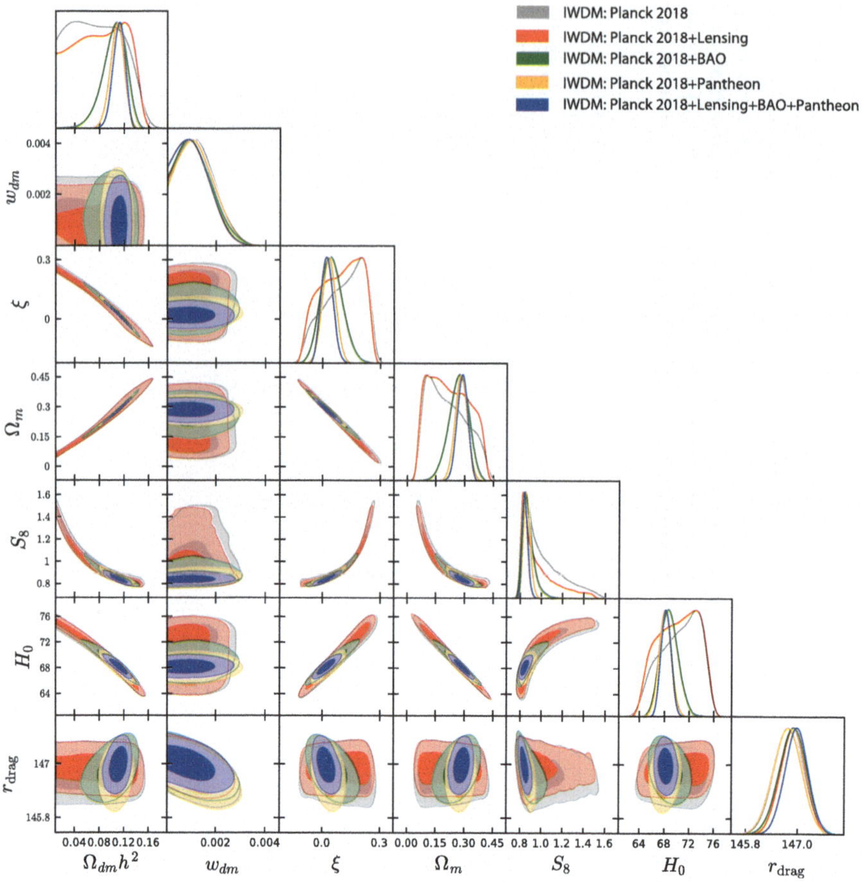

Fig. 29.1 One dimensional marginalized posterior distributions of some of the free and derived parameters and the two dimensional joint contours for the interacting scenario between a non-cold DM and vacuum have been displayed considering various observational datasets. This figure has been taken from Ref. [104]

29.3.2 IDE with Dynamical Equation of State of DE

The interacting scenarios in which the DE equation of state is dynamical, are the most generalized scenarios since such scenarios can recover the interacting scenarios with constant w_{de} models as a special case. From both the theoretical and observational grounds, such scenarios are quite appealing. One can investigate such scenarios following two approaches—either one can parametrize the DE equation of state in terms of some dynamical variable, e.g. the scale factor of the universe, see for instance Refs. [36, 37, 49, 51, 54, 77, 79], or, one can assume that DE is described by a scalar field ϕ with potential $V(\phi)$ where the equation of state of DE is dynamical being dependent on the scalar field and its potential, see Refs. [116–120]. Surprisingly, according to the existing literature, only a few works have been done considering the parametrized forms for w_{de} [36, 37, 49, 51, 54, 77, 79] without any particular reasons, while these scenarios and quite simple and they can lead to higher values of H_0. We suspect that one of the possible reasons could be the behaviour of the interacting scenarios at the level of perturbations which is certainly affected by the parametrized dynamical DE equation of state and an arbitrary choice of the dynamical w_{de} may invite instabilities in the large scale. However, with a suitable choice of a dynamical w_{de}, one can cure the instability issues from the picture. In this section we shall discuss the impact of these scenarios in the context of the H_0 tension.

In Ref. [79], an interacting scenario between CDM and DE has been investigated with the following interaction function

$$Q = 3H\xi(1 + w_{de})\rho_{de}, \tag{29.16}$$

where ξ is the coupling parameter of the interaction model and w_{de} has a dynamical nature which can take one of the expressions below

$$\text{Model I}: \quad w_{de}(a) = w_0\, a[1 - \log(a)], \tag{29.17}$$

$$\text{Model II}: \quad w_{de}(a) = w_0\, a \exp(1 - a), \tag{29.18}$$

$$\text{Model III}: \quad w_{de}(a) = w_0\, a[1 + \sin(1 - a)], \tag{29.19}$$

$$\text{Model IV}: \quad w_{de}(a) = w_0\, a[1 + \arcsin(1 - a)]. \tag{29.20}$$

where w_0 in every model refers to the present value of w_{de}. In Table 29.2, we present the estimated values of H_0 in each interacting scenario for Planck 2015 and Planck 2015+BAO datasets [79]. From Table 29.2, it is clearly seen that for CMB data alone, the estimated values of H_0, for all dynamical equation of state parameters (29.17)—(29.20), are significantly large compared to R21 by SH_0ES collaboration [6], however, due to the large error bars, the tension with R21 is reduced down to less than 1σ. However, for the Planck 2015+BAO dataset, the situation becomes quite interesting. As shown in Table 29.2, the H_0 values are very close to its local estimation by SH_0ES [6] and the uncertainties on H_0 are significantly decreased compared yo Planck 2015. Thus, with respect to R21, the tension is reduced down to 1.1σ (for Model I), 0.7σ (for Model II), 0.3σ (for Model III), 0.1σ (for Model IV). We thus conclude that the last two models are quite sound since they can completely resolve the H_0 tension for both Planck 2018 and Planck 2015+BAO datasets.

In the context of interacting scalar field models, the determination of the cosmological parameters is highly dependent on the choice of the scalar field potential and also on the interaction function as well. As argued in Ref. [118], such scalar field interacting models can also relieve the H_0 tension if a suitable choice of the potential is made.

However, considering both the scalar field interacting scenarios and the parametrized w_{de} interacting scenarios, we believe that both the scenarios are quite appealing in the light of the H_0 tension, and hence, they need further attention in the upcoming years.

Table 29.2 The table presents the constraints on H_0 for CMB and CMB+BAO datasets where CMB data have been taken from Planck 2015 considering various interacting scenarios in which the interaction function has the form (29.16) and the equation of state for DE is dynamical

Equation of state	Hubble constant H_0 (km/s/Mpc)	
	Planck 2015	Planck 2015+BAO
$w_{\rm DE}(a) = w_0 a[1 - \log(a)]$	81^{+13}_{-14}	$71.0^{+1.5}_{-1.5}$
$w_{\rm DE}(a) = w_0 a \exp(1 - a)$	84^{+14}_{-7}	$71.7^{+1.5}_{-1.7}$
$w_{\rm DE}(a) = w_0 a[1 + \sin(1 - a)]$	84^{+12}_{-5}	$73.5^{+1.6}_{-1.7}$
$w_{\rm DE}(a) = w_0 a[1 + \arcsin(1 - a)]$	82^{+14}_{-17}	$72.8^{+1.5}_{-1.8}$

29.3.3 *Interacting Scenarios with Time-Dependent Coupling Parameter*

In most of the interaction functions explored in the literature, the coupling parameter is usually assumed to be constant. However, there is no such guiding principle available yet which dictates that the coupling parameter of the interaction functions should be constant instead of dynamical. Even though the interaction scenarios with constant coupling parameter are relatively easier to handle, but the interacting functions with time dependent coupling represent more generalized scenarios [113, 114, 121–125]. In the following we shall discuss that one can reconcile the Hubble constant tension within these generalized frameworks.

Let us focus on an interacting scenario between CDM and DE with constant equation-of-state w_{de} and the interaction function has the following form

$$Q = 3H\xi(\mathcal{X}) f(\rho_{dm}, \rho_{de}), \tag{29.21}$$

where f is any analytic function of ρ_{dm}, ρ_{de}; $\xi(\mathcal{X})$ is a dimensionless time-dependent coupling and $\mathcal{X}$ could be any cosmological variable. We consider $\mathcal{X} = a/a_0 = a$ (a_0 is the present value of the scale factor and we set $a_0 = 1$) and consider the first two terms of the Taylor series expansion of $\xi(a)$ around $a_0 = 1$ as follows

$$\xi(a) = \xi_0 + \xi_a\,(1 - a)\,, \tag{29.22}$$

where ξ_0, ξ_a are constants and ξ_0 refers to the present day value of $\xi(a)$. Now, depending on the choice of $f(\rho_{dm}, \rho_{de})$, and the DE equation of state parameter, w_{de}, one could explore the dynamics of the underlying interacting scenario and estimate the constraints on the key cosmological parameters. We focus on the following two well known interaction functions of the form

$$Q = 3\xi(\mathcal{X})\ H\ \rho_{de}, \tag{29.23}$$

$$Q = 3\xi(\mathcal{X})\ H\ \frac{\rho_{dm}\rho_{de}}{\rho_{dm} + \rho_{de}}, \tag{29.24}$$

29.3.3.1 The Case for $w_{de} = -1$

The simplest interacting scenario is the one when CDM interacts with the vacuum energy corresponding to the DE equation of state $w_{de} = -1$, known as the Interacting Vacuum Scenario (IVS). The interacting scenarios driven by the interaction functions in Eq. (29.23) [labeled as IVS1], (29.24) [labeled as IVS2] together with the dynamical coupling $\xi(a)$ as in (29.22), were found to increase the Hubble constant value compared to the ΛCDM based Planck's estimation, see for instance, Ref. [122]. In particular, for the interaction function $Q = 3H[\xi_0 + \xi_a(1-a)]\rho_{de}$ of Eq. (29.23), one finds [122] $H_0 = 69.2 \pm 2.8$ km/sec/Mpc at 68% CL (Planck 2015), and $H_0 = 68.5^{+1.1}_{-1.5}$ km/sec/Mpc at 68% CL (Planck 2015+BAO) [122], and hence, with respect to R21, the tension is alleviated at 1.3σ and 2.7σ, respectively.

On the other hand, for the interaction function $Q = 3H[\xi_0 + \xi_a(1-a)]\frac{\rho_{dm}\rho_{de}}{\rho_{dm}+\rho_{de}}$ of Eq. (29.24), one finds $H_0 = 68.3 \pm 3.5$ km/sec/Mpc at 68% CL (Planck 2015) [122] and $H_0 = 68.0^{+1.3}_{-1.5}$ km/sec/Mpc at 68% CL (Planck 2015+BAO) [122], and hence with respect to R21, the tension is alleviated at 1.3σ and 2.9σ, respectively.

Now, referring to Tables 2 and 3 of [122], one can find that even though the strong evidence for non-null values of ξ_0 and ξ_a is not found from both Planck 2015 and Planck 2015+BAO, however, we cannot readily exclude the possibility of their non-null values. This can be evident from Figs. 29.2 and 29.3 where we show the evolution of $\xi(z)$ $[1 + z = a_0/a = 1/a]$ considering the 68% CL values of ξ_0 and ξ_a estimated from Planck 2015 and Planck 2015+BAO. One can notice from Figs. 29.2 and 29.3 that $\xi(z)$ has a transition from $\xi(z) > 0$ to $\xi(z) < 0$ which is prominent in the right graph of each Figs. 29.2 and 29.3. This is an interesting observation which tells that for $\xi(z) > 0$ [equivalently, $Q(t) > 0$], the energy flow occurs from the cold DM to vacuum sector while for $\xi(z) < 0$ [equivalently, $Q(t) < 0$] energy flow occurs from the vacuum energy sector to cold DM. Finally, as the choice of the dynamical coupling $\xi(a)$ is not unique, therefore, investigations with various choices for $\xi(a)$ will be certainly interesting.

29.3.3.2 The Case for w_{de} = Constant ($\neq -1$)

One can extend the interacting scenarios of the previous Sect. 29.3.3.1 by a constant DE equation of state w_{de} ($\neq -1$) which is allowed to vary in a certain region. Such extensions can lead to many interesting outcomes and also alleviate the Hubble constant tension [123]. As reported in Ref. [123], if the DE equation of state w_{de} is allowed to vary in the phantom regime (i.e. $w_{de} < -1$), then for the interaction function $Q = 3H[\xi_0 + \xi_a(1-a)]\rho_{de}$, the H_0 is constrained to be $H_0 = 81.8^{+10.1}_{-11.7}$ km/sec/Mpc at 68% CL for Planck 2018 together with a mild evi-

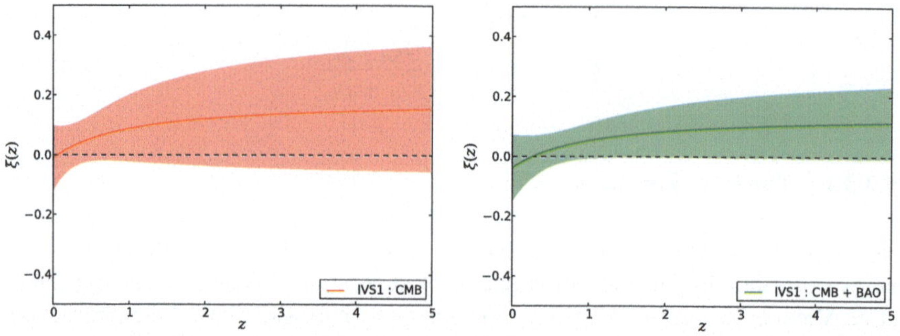

Fig. 29.2 Evolution of the coupling parameter $\xi(z)$ $[1+z = a_0/a = 1/a]$ of Eq. (29.22) for IVS1 for CMB (left graph) and CMB+BAO (right graph). The red (green) solid curve corresponds to the best fit curve for CMB (CMB+BAO) with the corresponding 68% CL region. Figure has been drawn taking the values of (ξ_0, ξ_a) from Table 3 of Ref. [122]

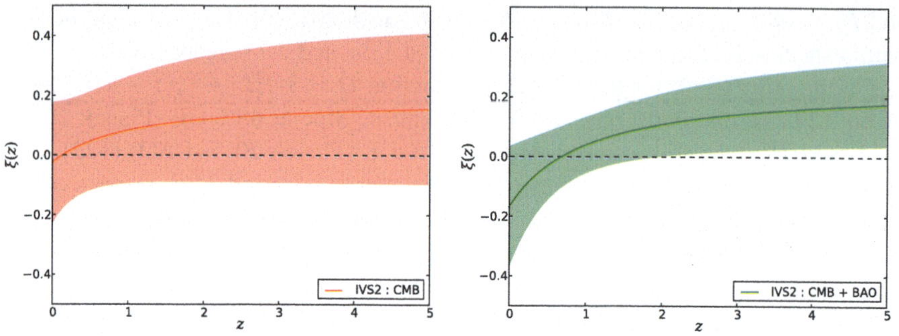

Fig. 29.3 Evolution of the coupling parameter $\xi(z)$ $[1+z = a_0/a = 1/a]$ of (29.22) for IVS2 for CMB (left graph) and CMB+BAO (right graph). The red (green) solid curve corresponds to the best fit curve for CMB (CMB+BAO) with the corresponding 68% CL region. Figure has been drawn taking the values of (ξ_0, ξ_a) from Table 3 of Ref. [122]

dence for $\xi_a = -0.077^{+0.064}_{-0.032}$ at 68% CL and a strong preference for a phantom DE ($w_{de} - 1.800^{+0.494}_{-0.387}$ at 68% CL), thus, the tension with R21 is alleviated within 1σ. For Planck 2018+BAO dataset, the same scenario leads to $H_0 = 69.4^{+1.4}_{-1.9}$ km/sec/Mpc at 68% CL together with a strong preference for a phantom DE ($w_{de} = -1.207^{+0.196}_{-0.208}$ at 95% CL) but here the evidence for ξ_a goes away. In this case, the tension with R21 is reduced down to 1.9σ. On the other hand, for the quintessence DE ($w_{de} > -1$), the interaction function $Q = 3H[\xi_0 + \xi_a(1-a)]\rho_{de}$ can lead to a higher value of the Hubble constant. For Planck 2018 alone, one finds that $H_0 = 70.2^{+4.1}_{-3.1}$ km/sec/Mpc at 68% CL, and thus the tension with R21 is resolved within 0.7σ. On the other hand, for Planck 2018+BAO, this interaction function leads to $H_0 = 68.4^{+1.3}_{-1.4}$ km/sec/Mpc at 68% CL which thus reduces the tension with R21 down to $2.7\,\sigma$.

For the interaction function $Q = 3H[\xi_0 + \xi_a(1-a)]\frac{\rho_{dm}\rho_{de}}{\rho_{dm}+\rho_{de}}$ of Eq. (29.24), if w_{de} is allowed to vary in the phantom regime, then H_0 tension can be alleviated [123]. For

Planck 2018 alone, this interacting scenario returns $H_0 > 72.5$ km/sec/Mpc at 95% CL with a strong evidence of phantom DE ($w_{de} = -1.66652146^{+0.485}_{-0.370}$ at 68% CL) and for Planck 2018+BAO, $H_0 = 69.4^{+1.0}_{-1.5}$ with a mild indication of the phantom DE ($w_{de} = -1.080^{+0.067}_{-0.031}$ at 68% CL). Therefore, the tension in H_0 with R21 is reduced down to $2\,\sigma$ and $2.3\,\sigma$, respectively, for Planck 2018 alone and Planck 2018+BAO dataset.

29.4 Interacting DE in a Non-flat Universe

In almost every works on the interacting scenarios, the curvature of our universe is usually assumed to be zero. This is usually motivated from some earlier investigations reporting a very small value of the curvature density parameter [126–129], and, on the other hand, in presence of the curvature of the universe, the analysis of the underlying cosmological model becomes complicated a bit compared to the analysis of the models with no curvature. However, with the growing sensitivity in the astronomical data, as the constraints on the cosmological parameters are getting improved, therefore, there is no physical motivation to exclude the curvature parameter from the analysis. In fact, one can include the curvature parameter in the analysis of the cosmological models and finally allow the observational data to decide its fate. Some recent investigations argued that the observational evidences are in favour of a closed universe scenario [130, 131]. And it was also contended in [130, 131] that the preference of a closed universe may worsen the H_0 tension. Thus, the question arises, can interacting scenarios in the context of a non-flat universe mitigate or solve the H_0 tension? This issue was investigated in Refs. [132, 133].

In Ref. [132], an interaction between a CDM and DE with $w_{de} = -0.999$ was investigated for the interaction function (29.11). As discussed in [132], different datasets return different values of H_0. For Planck 2018 alone, we find a lower value of H_0 with an evidence of a closed universe ($\Omega_k = -0.036^{+0.017}_{-0.013}$ at 68% CL for Planck 2018) but no evidence of an interaction, however, from Planck 2018+BAO, while Ω_k is consistent to the flat universe, but we obtain an evidence of an interaction ($\xi = -0.32^{+0.31}_{-0.09}$ at 68%) within 68% CL together with $H_0 = 69.7^{+1.2}_{-1.6}$ km/s/Mpc at 68% CL which thus alleviates the tension with R21 at 1.9σ.

Further in Ref. [133], an interaction between the CDM and DE with $w_{de} \neq -1$ has been investigated with the interaction function $Q = 3H\xi\rho_{de}$.[5] For the large scale stability of the interacting scenario, the DE equation of state w_{de} was divided into two separate regions, namely, $w_{de} < -1$ (phantom DE) and $w_{de} > -1$ (quintessence DE). We find that for $w_{de} < -1$, $H_0 = 69.0^{+1.2}_{-1.7}$ km/s/Mpc at 68% CL for Planck 2018+BAO with an evidence of an interaction at 68% CL ($\xi = -0.052^{-0.045}_{-0.025}$ at 68% CL), however, the curvature parameter is consistent to a flat universe [133]. Therefore,

[5] Note that the interaction model (29.11) is same with the interaction function $Q = 3H\xi\rho_{de}$. The presence of the factor 3 has some mathematical importance, but no physics is associated with this choice.

in this case, the tension with R21 is alleviated at 2.3σ for Planck 2018+BAO. For $w_{de} > -1$, one leads to $H_0 = 68.5 \pm 1.4$ km/s/Mpc at 68% CL for Planck 2018+BAO [133], and hence, the tension with R21 is reduced down to 2.6σ.

The interaction functions presented in this section are not the only choices. One can investigate other kind of interaction functions considering the fact that w_{de} could be dynamical. We anticipate that the interacting scenarios in the context of a non-flat universe are promising and these models need further investigations with the upcoming observational data from various astronomical surveys.

29.5 Summary and Conclusions

The theme of the present chapter is the theory of non-gravitational interaction between two heavy dark components of the universe, namely, DM and DE. The interaction between them is characterized by the transfer or energy between them. Over the last couple of years, interacting DE-DM scenarios have gained remarkable attention in the scientific community for many appealing consequences, such as, alleviation of the cosmic coincidence problem, crossing of the phantom divide line, and addressing the cosmological tensions. The heart of this theory is the interaction function $Q(t)$ that governs the transfer of energy between the dark sectors, and, as a consequence, this interaction function modifies the evolution equations for DE and DM at the level of background and perturbations. Thus, it is natural to expect that the cosmological parameters should be affected due to the presence of this interaction function. However, there is no such fundamental principle available yet for deriving the interaction function. Even though some attempts have been made to formulate the interaction function, but no conclusive statement has been made so far in this direction. Nevertheless, due to the existing features available in this particular theory, interacting DE has been proved as a phenomenologically rich cosmological theory.

In this chapter we have discussed a variety of interacting scenarios between DE and DM characterized by the triplet $(Q(t),\ w_{dm},\ w_{de})$ where $Q(t)$ has a wide variations; w_{dm}, representing the equation of state of the DM sector could be either zero or some non-negative constant, however, the DE equation of state, w_{de}, could be either constant or time dependent. Initially we have focused on the interacting scenarios assuming that the curvature of the universe is zero. In this set up, we have found that depending on the nature of w_{de} and the interaction function $Q(t)$, the 5σ tension on H_0 between Planck [5] and SH_0ES collaboration [6] can be significantly reduced. In particular, in some models, H_0 tension is reduced down to $< 3\sigma$; in some scenarios this tension is reduced down to $< 2\sigma$; in some scenarios it is reduced down to $< 1\sigma$; and in some scenarios, one can completely solve this tension. On the other hand, assuming that our universe may not be spatially flat, we have discussed the impact of the interacting scenarios on the H_0 tension.

Based on the discussions presented in this chapter, it is pretty clear that the interacting scenarios between DE and DM are really promising for reconciling the Hubble constant tension. In addition, we would like to stress that there are many gaps in this domain which need to be explored crucially with the upcoming cosmological probes from several astronomical surveys.

Acknowledgements S. Pan acknowledges the financial support from the Department of Science and Technology (DST), Govt. of India under the Scheme "Fund for Improvement of S&T Infrastructure (FIST)" [File No. SR/FST/MS-I/2019/41]. W. Yang acknowledges the support by the National Natural Science Foundation of China under Grants No. 12175096 and No. 11705079, and Liaoning Revitalization Talents Program under Grant no. XLYC1907098.

References

1. A.G. Riess et al., Supernova search team. Astron. J. **116**, 1009–1038 (1998) [arXiv:astro-ph/9805201 [astro-ph]]
2. S. Perlmutter et al., Supernova cosmology project. Astrophys. J. **517**, 565–586 (1999) [arXiv:astro-ph/9812133 [astro-ph]]
3. S. Weinberg, Rev. Mod. Phys. **61**, 1–23 (1989)
4. I. Zlatev, L.M. Wang, P.J. Steinhardt, Phys. Rev. Lett. **82**, 896–899 (1999) [arXiv:astro-ph/9807002 [astro-ph]]
5. N. Aghanim et al., Planck. Astron. Astrophys. **641**, A6 (2020) [erratum: Astron. Astrophys. **652**, C4 (2021)] [arXiv:1807.06209 [astro-ph.CO]]
6. A.G. Riess, W. Yuan, L.M. Macri, D. Scolnic, D. Brout, S. Casertano, D.O. Jones, Y. Murakami, L. Breuval, T.G. Brink et al., Astrophys. J. Lett. **934**(1), L7 (2022) [arXiv:2112.04510 [astro-ph.CO]]
7. A.G. Riess, L. Breuval, W. Yuan, S. Casertano, L.M. Macri, J.B. Bowers, D. Scolnic, T. Cantat-Gaudin, R.I. Anderson, M.C. Reyes, Astrophys. J. **938**(1), 36 (2022) [arXiv:2208.01045 [astro-ph.CO]]
8. M. Asgari, T. Tröster, C. Heymans, H. Hildebrandt, J.L. van den Busch, A.H. Wright, A. Choi, T. Erben, B. Joachimi, S. Joudaki et al., Astron. Astrophys. **634**, A127 (2020) [arXiv:1910.05336 [astro-ph.CO]]
9. M. Asgari et al., KiDS. Astron. Astrophys. **645**, A104 (2021) [arXiv:2007.15633 [astro-ph.CO]]
10. S. Joudaki, H. Hildebrandt, D. Traykova, N.E. Chisari, C. Heymans, A. Kannawadi, K. Kuijken, A.H. Wright, M. Asgari, T. Erben et al., Astron. Astrophys. **638**, L1 (2020) [arXiv:1906.09262 [astro-ph.CO]]
11. T.M.C. Abbott et al., DES. Phys. Rev. D **105**(2), 023520 (2022) [arXiv:2105.13549 [astro-ph.CO]]
12. A. Amon et al., DES. Phys. Rev. D **105**(2), 023514 (2022) [arXiv:2105.13543 [astro-ph.CO]]
13. L.F. Secco et al., DES. Phys. Rev. D **105**(2), 023515 (2022) [arXiv:2105.13544 [astro-ph.CO]]
14. A. Loureiro et al., KiDS and Euclid. Astron. Astrophys. **665**, A56 (2022) [arXiv:2110.06947 [astro-ph.CO]]
15. C. Heymans, T. Tröster, M. Asgari, C. Blake, H. Hildebrandt, B. Joachimi, K. Kuijken, C.A. Lin, A.G. Sánchez, J.L. van den Busch et al., Astron. Astrophys. **646**, A140 (2021) [arXiv:2007.15632 [astro-ph.CO]]
16. T.M.C. Abbott et al., DES. Phys. Rev. D **102**(2), 023509 (2020) [arXiv:2002.11124 [astro-ph.CO]]
17. O.H.E. Philcox, M.M. Ivanov, Phys. Rev. D **105**(4), 043517 (2022) [arXiv:2112.04515 [astro-ph.CO]]
18. E. Di Valentino, O. Mena, S. Pan, L. Visinelli, W. Yang, A. Melchiorri, D.F. Mota, A.G. Riess, J. Silk, Class. Quant. Grav. **38**(15), 153001 (2021) [arXiv:2103.01183 [astro-ph.CO]]

19. L. Perivolaropoulos, F. Skara, New Astron. Rev. **95**, 101659 (2022) [arXiv:2105.05208 [astro-ph.CO]]
20. E. Abdalla, G. Franco Abellán, A. Aboubrahim, A. Agnello, O. Akarsu, Y. Akrami, G. Alestas, D. Aloni, L. Amendola, L.A. Anchordoqui et al., JHEAp **34**, 49–211 (2022) [arXiv:2203.06142 [astro-ph.CO]]
21. M. Kamionkowski, A.G. Riess [arXiv:2211.04492 [astro-ph.CO]]
22. C. Wetterich, Astron. Astrophys. **301**, 321–328 (1995) [arXiv:hep-th/9408025 [hep-th]]
23. L. Amendola, Coupled quintessence. Phys. Rev. D **62**, 043511 (2000) [astro-ph/9908023]
24. L.P. Chimento, A.S. Jakubi, D. Pavon, W. Zimdahl, Phys. Rev. D **67**, 083513 (2003) [arXiv:astro-ph/0303145 [astro-ph]]
25. R.G. Cai, A. Wang, JCAP **0503**, 002 (2005) [hep-th/0411025]
26. D. Pavón, W. Zimdahl, Phys. Lett. B **628**, 206 (2005) [arXiv:gr-qc/0505020]
27. S. del Campo, R. Herrera, G. Olivares, D. Pavon, Phys. Rev. D **74**, 023501 (2006) [arXiv:astro-ph/0606520 [astro-ph]]
28. M.S. Berger, H. Shojaei, Phys. Rev. D **73**, 083528 (2006) [arXiv:gr-qc/0601086 [gr-qc]]
29. S. del Campo, R. Herrera, D. Pavón, Phys. Rev. D **78**, 021302 (2008) [arXiv:0806.2116 [astro-ph]]
30. S. del Campo, R. Herrera, D. Pavón, JCAP **0901**, 020 (2009) [arXiv:0812.2210 [gr-qc]]
31. J.D. Barrow, T. Clifton, Phys. Rev. D **73**, 103520 (2006) [gr-qc/0604063]
32. L. Amendola, G. Camargo Campos, R. Rosenfeld, Phys. Rev. D **75**, 083506 (2007) [astro-ph/0610806]
33. T. Koivisto, D.F. Mota, Phys. Rev. D **75**, 023518 (2007) [hep-th/0609155]
34. J.H. He, B. Wang, JCAP **0806**, 010 (2008) [arXiv:0801.4233 [astro-ph]]
35. M.B. Gavela, D. Hernandez, L. Lopez Honorez, O. Mena, S. Rigolin, JCAP **0907**, 034 (2009) [arXiv:0901.1611 [astro-ph.CO]]
36. E. Majerotto, J. Väliviita, R. Maartens, Mon. Not. Roy. Astron. Soc. **402**, 2344 (2010) [arXiv:0907.4981 [astro-ph.CO]]
37. J. Valiviita, R. Maartens, E. Majerotto, Mon. Not. Roy. Astron. Soc. **402**, 2355–2368 (2010) [arXiv:0907.4987 [astro-ph.CO]]
38. T. Clemson, K. Koyama, G.B. Zhao, R. Maartens, J. Väliviita, Phys. Rev. D **85**, 043007 (2012) [arXiv:1109.6234 [astro-ph.CO]]
39. V. Salvatelli, N. Said, M. Bruni, A. Melchiorri, D. Wands, Phys. Rev. Lett. **113**(18), 181301 (2014) [arXiv:1406.7297 [astro-ph.CO]]
40. W. Yang, L. Xu, JCAP **1408**, 034 (2014) [arXiv:1401.5177 [astro-ph.CO]]
41. W. Yang, L. Xu, Phys. Rev. D **89**(8), 083517 (2014) [arXiv:1401.1286 [astro-ph.CO]]
42. Y. Wang, D. Wands, G.B. Zhao, L. Xu, Phys. Rev. D **90**(2), 023502 (2014). https://doi.org/10.1103/PhysRevD.90.023502 [arXiv:1404.5706 [astro-ph.CO]]
43. V. Faraoni, J.B. Dent, E.N. Saridakis, Phys. Rev. D **90**(6), 063510 (2014) [arXiv:1405.7288 [gr-qc]]
44. S. Pan, S. Bhattacharya, S. Chakraborty, Mon. Not. Roy. Astron. Soc. **452**(3), 3038 (2015) [arXiv:1210.0396 [gr-qc]]
45. W. Yang, L. Xu, Phys. Rev. D **90**(8), 083532 (2014) [arXiv:1409.5533 [astro-ph.CO]]
46. Y.H. Li, J.F. Zhang, X. Zhang, Phys. Rev. D **93**(2), 023002 (2016) [arXiv:1506.06349 [astro-ph.CO]]
47. R.C. Nunes, S. Pan, E.N. Saridakis, Phys. Rev. D **94**(2), 023508 (2016) [arXiv:1605.01712 [astro-ph.CO]]
48. W. Yang, H. Li, Y. Wu, J. Lu, JCAP **1610**(10), 007 (2016) [arXiv:1608.07039 [astro-ph.CO]]
49. S. Pan, G.S. Sharov, Mon. Not. Roy. Astron. Soc. **472**(4), 4736 (2017) [arXiv:1609.02287 [gr-qc]]
50. A. Mukherjee, N. Banerjee, Class. Quant. Grav. **34**(3), 035016 (2017) [arXiv:1610.04419 [astro-ph.CO]]
51. G.S. Sharov, S. Bhattacharya, S. Pan, R.C. Nunes, S. Chakraborty, Mon. Not. Roy. Astron. Soc. **466**(3), 3497 (2017) [arXiv:1701.00780 [gr-qc]]

52. R.Y. Guo, Y.H. Li, J.F. Zhang, X. Zhang, JCAP **1705**(05), 040 (2017) [arXiv:1702.04189 [astro-ph.CO]]
53. R.G. Cai, N. Tamanini, T. Yang, JCAP **1705**(05), 031 (2017) [arXiv:1703.07323 [astro-ph.CO]]
54. W. Yang, N. Banerjee, S. Pan, Phys. Rev. D **95**(12), 123527 (2017) [arXiv:1705.09278 [astro-ph.CO]]
55. W. Yang, S. Pan, D.F. Mota, Phys. Rev. D **96**(12), 123508 (2017) [arXiv:1709.00006 [astro-ph.CO]]
56. W. Yang, S. Pan, J.D. Barrow, Phys. Rev. D **97**(4), 043529 (2018) [arXiv:1706.04953 [astro-ph.CO]]
57. S. Pan, A. Mukherjee, N. Banerjee, Mon. Not. Roy. Astron. Soc. **477**(1), 1189 (2018) [arXiv:1710.03725 [astro-ph.CO]]
58. W. Yang, S. Pan, A. Paliathanasis, Mon. Not. Roy. Astron. Soc. **482**(1), 1007 (2019) [arXiv:1804.08558 [gr-qc]]
59. W. Yang, S. Pan, L. Xu, D.F. Mota, Mon. Not. Roy. Astron. Soc. **482**(2), 1858 (2019) [arXiv:1804.08455 [astro-ph.CO]]
60. A. Paliathanasis, S. Pan, W. Yang, Int. J. Mod. Phys. D **28**(12), 1950161 (2019) [arXiv:1903.02370 [gr-qc]]
61. S. Pan, W. Yang, C. Singha, E.N. Saridakis, Phys. Rev. D **100**(8), 083539 (2019) [arXiv:1903.10969 [astro-ph.CO]]
62. W. Yang, S. Pan, E. Di Valentino, B. Wang, A. Wang, JCAP **05**, 050 (2020) [arXiv:1904.11980 [astro-ph.CO]]
63. W. Yang, S. Vagnozzi, E. Di Valentino, R.C. Nunes, S. Pan, D.F. Mota, JCAP **1907**, 037 (2019) [arXiv:1905.08286 [astro-ph.CO]]
64. W. Yang, S. Pan, R.C. Nunes, D.F. Mota, JCAP **04**, 008 (2020). https://doi.org/10.1088/1475-7516/2020/04/008 [arXiv:1910.08821 [astro-ph.CO]]
65. W. Yang, E. Di Valentino, S. Pan, S. Basilakos, A. Paliathanasis, Phys. Rev. D **102**(6), 063503 (2020). https://doi.org/10.1103/PhysRevD.102.063503 [arXiv:2001.04307 [astro-ph.CO]]
66. S. Pan, W. Yang, A. Paliathanasis, Mon. Not. Roy. Astron. Soc. **493**(3), 3114–3131 (2020). https://doi.org/10.1093/mnras/staa213[arXiv:2002.03408 [astro-ph.CO]]
67. Y.L. Bolotin, A. Kostenko, O.A. Lemets, D.A. Yerokhin, Int. J. Mod. Phys. D **24**(03), 1530007 (2015) [arXiv:1310.0085 [astro-ph.CO]]
68. B. Wang, E. Abdalla, F. Atrio-Barandela, D. Pavón, Rept. Prog. Phys. **79**(9), 096901 (2016) [arXiv:1603.08299 [astro-ph.CO]]
69. B. Wang, Y.G. Gong, E. Abdalla, Phys. Lett. B **624**, 141 (2005) [hep-th/0506069]
70. S. Das, P.S. Corasaniti, J. Khoury, Phys. Rev. D **73**, 083509 (2006) [astro-ph/0510628]
71. H.M. Sadjadi, M. Honardoost, Phys. Lett. B **647**, 231 (2007) [gr-qc/0609076]
72. S. Pan, S. Chakraborty, Int. J. Mod. Phys. D **23**(11), 1450092 (2014) [arXiv:1410.8281 [gr-qc]]
73. S. Kumar, R.C. Nunes, Phys. Rev. D **94**(12), 123511 (2016) [arXiv:1608.02454 [astro-ph.CO]]
74. S. Kumar, R.C. Nunes, Phys. Rev. D **96**(10), 103511 (2017) [arXiv:1702.02143 [astro-ph.CO]]
75. E. Di Valentino, A. Melchiorri, O. Mena, Phys. Rev. D **96**(4), 043503 (2017) [arXiv:1704.08342 [astro-ph.CO]]
76. W. Yang, S. Pan, E. Di Valentino, R.C. Nunes, S. Vagnozzi, D.F. Mota, JCAP **1809**(09), 019 (2018) [arXiv:1805.08252 [astro-ph.CO]]
77. W. Yang, A. Mukherjee, E. Di Valentino, S. Pan, Phys. Rev. D **98**(12), 123527 (2018) [arXiv:1809.06883 [astro-ph.CO]]
78. S. Kumar, R.C. Nunes, S.K. Yadav, Eur. Phys. J. C **79**(7), 576 (2019) [arXiv:1903.04865 [astro-ph.CO]]
79. S. Pan, W. Yang, E. Di Valentino, E.N. Saridakis, S. Chakraborty, Phys. Rev. D **100**(10), 103520 (2019) arXiv:1907.07540 [astro-ph.CO]
80. E. Di Valentino, A. Melchiorri, O. Mena, S. Vagnozzi, Phys. Dark Univ. **30**, 100666 (2020) [arXiv:1908.04281 [astro-ph.CO]]
81. E. Di Valentino, A. Melchiorri, O. Mena, S. Vagnozzi, Phys. Rev. D **101**(6), 063502 (2020) [arXiv:1910.09853 [astro-ph.CO]]

82. Y. Zhai, W. Giarè, C. van de Bruck, E. Di Valentino, O. Mena, R.C. Nunes, JCAP **07**, 032 (2023) [arXiv:2303.08201 [astro-ph.CO]]
83. A. Pourtsidou, T. Tram, Phys. Rev. D **94**(4), 043518 (2016) [arXiv:1604.04222 [astro-ph.CO]]
84. R. An, C. Feng, B. Wang, JCAP **1802**(02), 038 (2018) [arXiv:1711.06799 [astro-ph.CO]]
85. J. Gleyzes, D. Langlois, M. Mancarella, F. Vernizzi, JCAP **1508**, 054 (2015) [arXiv:1504.05481 [astro-ph.CO]]
86. C.G. Böehmer, N. Tamanini, M. Wright, Phys. Rev. D **91**(12), 123002 (2015) [arXiv:1501.06540 [gr-qc]]
87. C.G. Böehmer, N. Tamanini, M. Wright, Phys. Rev. D **91**(12), 123003 (2015) [arXiv:1502.04030 [gr-qc]]
88. C. van de Bruck, J. Morrice, JCAP **1504**, 036 (2015) [arXiv:1501.03073 [gr-qc]]
89. L. Xiao, R. An, L. Zhang, B. Yue, Y. Xu, B. Wang, Phys. Rev. D **99**(2), 023528 (2019) [arXiv:1807.05541 [astro-ph.CO]]
90. G. D'Amico, T. Hamill, Nemanja Kaloper, Phys. Rev. D **94**, 103526 (2016) [arXiv:1605.00996 [hep-th]]
91. M.C.D. Marsh, Phys. Rev. Lett. **118**, 011302 (2017) [arXiv:1606.01538 [astro-ph.CO]]
92. R. Kase, S. Tsujikawa, Phys. Rev. D **101**(6), 063511 (2020) [arXiv:1910.02699 [gr-qc]]
93. S. Alexander, M. Cortês, A.R. Liddle, J. Magueijo, R. Sims, L. Smolin, Phys. Rev. D **100**(8), 083507 (2019) [arXiv:1905.10382 [gr-qc]]
94. S. Pan, G.S. Sharov, W. Yang, Phys. Rev. D **101**(10), 103533 (2020) [arXiv:2001.03120 [astro-ph.CO]]
95. J. Valiviita, E. Majerotto, R. Maartens, JCAP **07**, 020 (2008) [arXiv:0804.0232 [astro-ph]]
96. S. Pan, J. de Haro, W. Yang, J. Amorós, Phys. Rev. D **101**(12), 123506 (2020) [arXiv:2001.09885 [gr-qc]]
97. W. Yang, S. Pan, L. Aresté Saló, J. de Haro, Phys. Rev. D **103**(8), 083520 (2021) [arXiv:2104.04505 [astro-ph.CO]]
98. C.M. Muller, Phys. Rev. D **71**, 047302 (2005) [arXiv:astro-ph/0410621 [astro-ph]]
99. S. Kumar, L. Xu, Phys. Lett. B **737**, 244–247 (2014) [arXiv:1207.5582 [gr-qc]]
100. C. Armendariz-Picon, J.T. Neelakanta, JCAP **03**, 049 (2014) [arXiv:1309.6971 [astro-ph.CO]]
101. M. Kopp, C. Skordis, D.B. Thomas, S. Ilić, Phys. Rev. Lett. **120**(22), 221102 (2018) [arXiv:1802.09541 [astro-ph.CO]]
102. S. Ilić, M. Kopp, C. Skordis, D.B. Thomas, Phys. Rev. D **104**(4), 043520 (2021) [arXiv:2004.09572 [astro-ph.CO]]
103. K. Naidoo, M. Jaber, W.A. Hellwing, M. Bilicki [arXiv:2209.08102 [astro-ph.CO]]
104. S. Pan, W. Yang, E. Di Valentino, D.F. Mota, J. Silk, JCAP **07**, 064 (2023) [arXiv:2211.11047 [astro-ph.CO]]
105. E. Di Valentino, A. Melchiorri, J. Silk, Phys. Lett. B **761**, 242–246 (2016) [arXiv:1606.00634 [astro-ph.CO]]
106. V.F. Mukhanov, H.A. Feldman, R.H. Brandenberger, Phys. Rept. **215**, 203–333 (1992)
107. C.P. Ma, E. Bertschinger, Astrophys. J. **455**, 7–25 (1995) [arXiv:astro-ph/9506072 [astro-ph]]
108. K.A. Malik, D. Wands, Phys. Rept. **475**, 1–51 (2009) [arXiv:0809.4944 [astro-ph]]
109. S.K. Choi et al., ACT. JCAP **12**, 045 (2020) [arXiv:2007.07289 [astro-ph.CO]]
110. S. Aiola et al., ACT. JCAP **12**, 047 (2020) [arXiv:2007.07288 [astro-ph.CO]]
111. H. Wei, Nucl. Phys. B **845**, 381–392 (2011). https://doi.org/10.1016/j.nuclphysb.2010.12.010 [arXiv:1008.4968 [gr-qc]]
112. H. Wei, Commun. Theor. Phys. **56**, 972–980 (2011) [arXiv:1010.1074 [gr-qc]]
113. Y.H. Li, X. Zhang, Eur. Phys. J. C **71**, 1700 (2011) [arXiv:1103.3185 [astro-ph.CO]]
114. J.J. Guo, J.F. Zhang, Y.H. Li, D.Z. He, X. Zhang, Sci. China Phys. Mech. Astron. **61**(3), 030011 (2018) [arXiv:1710.03068 [astro-ph.CO]]
115. F. Arevalo, A. Cid, L.P. Chimento, P. Mella, Eur. Phys. J. C **79**(4), 355 (2019) [arXiv:1901.04300 [gr-qc]]
116. E.M. Teixeira, R. Daniel, N. Frusciante, C. van de Bruck [arXiv:2309.06544 [astro-ph.CO]]
117. A.Gómez-Valent, Z. Zheng, L. Amendola, C. Wetterich, V. Pettorino, Phys. Rev. D **106**(10), 103522 (2022) [arXiv:2207.14487 [astro-ph.CO]]

118. A. Gómez-Valent, V. Pettorino, L. Amendola, Phys. Rev. D **101**(12), 123513 (2020) [arXiv:2004.00610 [astro-ph.CO]]
119. C. Van De Bruck, J. Mifsud, Phys. Rev. D **97**(2), 023506 (2018) [arXiv:1709.04882 [astro-ph.CO]]
120. C. van de Bruck, J. Mifsud, J. Morrice, Phys. Rev. D **95**(4), 043513 (2017) [arXiv:1609.09855 [astro-ph.CO]]
121. Y. Wang, G.B. Zhao, Astrophys. J. **869**(1), 26 (2018) [arXiv:1805.11210 [astro-ph.CO]]
122. W. Yang, O. Mena, S. Pan, E. Di Valentino, Phys. Rev. D **100**, 083509 (2019) [arXiv:1906.11697 [astro-ph.CO]]
123. W. Yang, E. Di Valentino, O. Mena, S. Pan, Phys. Rev. D **102**(2), 023535 (2020) [arXiv:2003.12552 [astro-ph.CO]]
124. Y.H. Yao, X.H. Meng [arXiv:2207.05955 [astro-ph.CO]]
125. W. Yang, S. Pan, O. Mena, E. Di Valentino [arXiv:2209.14816 [astro-ph.CO]]
126. E. Gaztanaga, R. Miquel, E. Sanchez, Phys. Rev. Lett. **103**, 091302 (2009) [arXiv:0808.1921 [astro-ph]]
127. M.J. Mortonson, Phys. Rev. D **80**, 123504 (2009) [arXiv:0908.0346 [astro-ph.CO]]
128. S.H. Suyu, T. Treu, S. Hilbert, A. Sonnenfeld, M.W. Auger, R.D. Blandford, T. Collett, F. Courbin, C.D. Fassnacht, L.V.E. Koopmans et al., Astrophys. J. Lett. **788**, L35 (2014) [arXiv:1306.4732 [astro-ph.CO]]
129. B. L'Huillier, A. Shafieloo, JCAP **01**, 015 (2017) [arXiv:1606.06832 [astro-ph.CO]]
130. E. Di Valentino, A. Melchiorri, J. Silk, Nature Astron. **4**(2), 196–203 (2019) [arXiv:1911.02087 [astro-ph.CO]]
131. W. Handley, Phys. Rev. D **103**(4), L041301 (2021) [arXiv:1908.09139 [astro-ph.CO]]
132. E. Di Valentino, A. Melchiorri, O. Mena, S. Pan, W. Yang, Mon. Not. Roy. Astron. Soc. **502**(1), L23–L28 (2021) [arXiv:2011.00283 [astro-ph.CO]]
133. W. Yang, S. Pan, E. Di Valentino, O. Mena, A. Melchiorri, JCAP **10**, 008 (2021) [arXiv:2101.03129 [astro-ph.CO]]

Chapter 30
Modified Gravity

Emmanuel N. Saridakis

30.1 Introduction

As we discussed in the Introduction of this book, there are many directions that one can follow in order to alleviate the H_0 tension. In particular, since $\theta_s = \frac{r_s}{D_A}$, where r_s is the sound horizon and D_A is the angular diameter distance, one could try to alter either r_s or D_A or both. Solutions affecting r_s are usually called "early-time" solutions, while solutions affecting D_A are usually called "late-time" solutions. Incorporating gravitational modifications [1] is probably the only framework in the H_0-literature which is so broad that could be used in all directions, namely one could have early-time modified-gravity solutions, late-time modified-gravity solutions, or modified-gravity solutions that alter both regimes. Since early modified-gravity scenarios will be investigated in the next chapter, in the present chapter we will focus on modified-gravity solutions that alter the late-time evolution of the Universe.

The effect of modified gravity on the late-time universe evolution is two-fold. The first is that it induces new terms in the Friedmann equations, which can collectively be absorbed into an effective dark-energy sector. The second is that it typically leads to a modified Newton's constant (actually this is the first thing one expects that modified gravity should do). Hence, in every cosmology governed by a modified theory of gravity one typically obtains Friedmann equations of the form [1]

$$H^2 = \frac{8\pi G_{eff}}{3}\left(\rho_m + \rho_{DE}^{eff}\right), \tag{30.1}$$

$$\dot{H} = -4\pi G_{eff}\left(\rho_m + \rho_{DE}^{eff} + p_m + p_{DE}^{\text{eff}}\right), \tag{30.2}$$

E. N. Saridakis (✉)
National Observatory of Athens, Lofos Nymfon 11852, Athens, Greece
e-mail: msaridak@noa.gr

CAS Key Laboratory for Researches in Galaxies and Cosmology, Department of Astronomy, University of Science and Technology of China, Hefei, Anhui 230026, People's Republic of China

E. Di Valentino and D. Brout (eds.), *The Hubble Constant Tension*, Springer Series in Astrophysics and Cosmology, https://doi.org/10.1007/978-981-99-0177-7_30

where ρ_{DE}^{eff} and p_{DE}^{eff} are respectively the effective dark energy density and pressure, and G_{eff} the effective Newton's constant, all depending on the parameters of the theory. Hence, qualitatively, we deduce that in order to alleviate the H_0 tension in this framework, i.e. obtain a higher H_0 than standard lore predicts, we have two ways [2, 3]. (i) One could either try to obtain smaller effective Newton's constant, since "weaker" gravity is reasonable to induce faster expansion. (ii) One could try to obtain suitable modified-gravity-oriented extract terms in the effective dark-energy sector, which could lead to faster expansion, e.g. obtain and effective dark-energy equation-of-state parameter $w_{DE} := p_{DE}/\rho_{DE}$ lying in the phantom regime.

In this chapter we will present three broad classes of modified gravity that can fulfill the above qualitative requirements in the correct quantitative way and alleviate the H_0 tension. The first is Horndeski gravity [4] (or equivalently Generalized Galileon theories [5, 6]), the second is $f(T)$ gravity [7], and the third is $f(R)$ gravity [8].

30.2 Horndeski/Generalized Galileon Theories

In this section we focus on Horndeski gravity [4], which coincides with generalized Galileon theory [5, 6]. The most general Lagrangian with curvature terms and with one extra scalar degree of freedom ϕ, giving rise to second-order field equations, is [4, 9, 10]

$$\mathcal{L} = \sum_{i=2}^{5} \mathcal{L}_i, \tag{30.3}$$

with

$$\mathcal{L}_2 = K(\phi, X), \tag{30.4}$$
$$\mathcal{L}_3 = -G_3(\phi, X)\Box\phi, \tag{30.5}$$
$$\mathcal{L}_4 = G_4(\phi, X)\, R + G_{4,X}\, [(\Box\phi)^2 - (\nabla_\mu\nabla_\nu\phi)\,(\nabla^\mu\nabla^\nu\phi)], \tag{30.6}$$
$$\begin{aligned}\mathcal{L}_5 = G_5(\phi, X)\, G_{\mu\nu}\,(\nabla^\mu\nabla^\nu\phi) - \frac{1}{6}\, G_{5,X}\, [(\Box\phi)^3 - 3(\Box\phi)\,(\nabla_\mu\nabla_\nu\phi)\,(\nabla^\mu\nabla^\nu\phi) \\ + 2(\nabla^\mu\nabla_\alpha\phi)\,(\nabla^\alpha\nabla_\beta\phi)\,(\nabla^\beta\nabla_\mu\phi)]\,.\end{aligned} \tag{30.7}$$

As usual, R is the Ricci scalar, $G_{\mu\nu}$ is the Einstein tensor, the functions K and G_i ($i = 3, 4, 5$) depend on ϕ and its kinetic energy $X = -\partial^\mu\phi\partial_\mu\phi/2$, and $G_{i,X} := \partial G_i/\partial X$, $G_{i,\phi} := \partial G_i/\partial\phi$. Focusing on Friedmann-Robertson-Walker (FRW) geometry with metric $ds^2 = -dt^2 + a^2(t)\,\delta_{ij}dx^i dx^j$, and adding the matter Lagrangian $\mathcal{L}_m$ corresponding to a perfect fluid with energy density ρ_m and pressure p_m, and performing variation we obtain the two generalized Friedmann equations:

$$
\begin{aligned}
&2XK_{,X} - K + 6X\dot{\phi}HG_{3,X} - 2XG_{3,\phi} - 6H^2G_4 + 24H^2X(G_{4,X} + XG_{4,XX}) \\
&-12HX\dot{\phi}G_{4,\phi X} - 6H\dot{\phi}G_{4,\phi} + 2H^3X\dot{\phi}(5G_{5,X} + 2XG_{5,XX}) \\
&-6H^2X(3G_{5,\phi} + 2XG_{5,\phi X}) = -\rho_m,
\end{aligned}
\tag{30.8}
$$

$$
\begin{aligned}
&K - 2X(G_{3,\phi} + \ddot{\phi}G_{3,X}) + 2(3H^2 + 2\dot{H})G_4 - 8\dot{H}XG_{4,X} - 12H^2XG_{4,X} \\
&-4H\dot{X}G_{4,X} - 8HX\dot{X}G_{4,XX} + 2(\ddot{\phi} + 2H\dot{\phi})G_{4,\phi} + 4XG_{4,\phi\phi} \\
&+4X(\ddot{\phi} - 2H\dot{\phi})G_{4,\phi X} - 4H^2X^2\ddot{\phi}G_{5,XX} - 2X(2H^3\dot{\phi} + 2H\dot{H}\dot{\phi} + 3H^2\ddot{\phi})G_{5,X} \\
&+4HX(\dot{X} - HX)G_{5,\phi X} + 4HX\dot{\phi}G_{5,\phi\phi} \\
&+2[2(\dot{H}X + H\dot{X}) + 3H^2X]G_{5,\phi} = -p_m,
\end{aligned}
\tag{30.9}
$$

with dots marking derivatives with respect to t. Moreover, variation with respect to $\phi(t)$ gives

$$
\frac{1}{a^3}\frac{d}{dt}(a^3J) = P_\phi, \tag{30.10}
$$

with

$$
\begin{aligned}
J :=\ &\dot{\phi}K_{,X} + 6HXG_{3,X} - 2\dot{\phi}G_{3,\phi} - 12HXG_{4,\phi X} + 6H^2\dot{\phi}(G_{4,X} + 2XG_{4,XX}) \\
&+2H^3X(3G_{5,X} + 2XG_{5,XX}) + 6H^2\dot{\phi}(G_{5,\phi} + XG_{5,\phi X}),
\end{aligned}
\tag{30.11}
$$

$$
\begin{aligned}
P_\phi :=\ &K_{,\phi} - 2X(G_{3,\phi\phi} + \ddot{\phi}G_{3,\phi X}) + 6(2H^2 + \dot{H})G_{4,\phi} \\
&+6H(\dot{X} + 2HX)G_{4,\phi X} - 6H^2XG_{5,\phi\phi} + 2H^3X\dot{\phi}G_{5,\phi X}.
\end{aligned}
\tag{30.12}
$$

Lastly, as usual we consider the matter conservation equation $\dot{\rho}_m + 3H(\rho_m + p_m) = 0$.

In the following subsections we will present various sub-cases of Horndeski theories that can alleviate H_0 tension. Since Horndeski theory recovers ΛCDM cosmology for $G_4 = 1/(16\pi G)$, $K = -2\Lambda = const$, and $G_3 = G_5 = 0$, our strategy is to introduce deviations which are negligible at high redshifts, where the CMB structure is formed, but become significant at low redshifts, in which local Hubble measurements take place.

30.2.1 Horndeski Theories with Non-minimal Derivative Coupling

One can first examine one interesting subclass of Horndeski gravity that contains the G_5 term, which is called "non-minimal derivative coupling". In particular, one

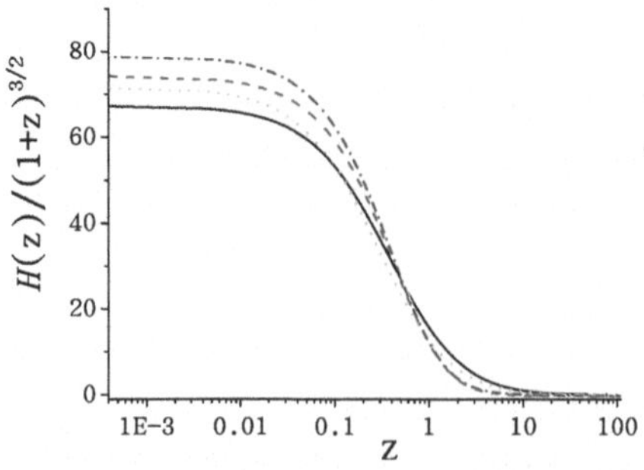

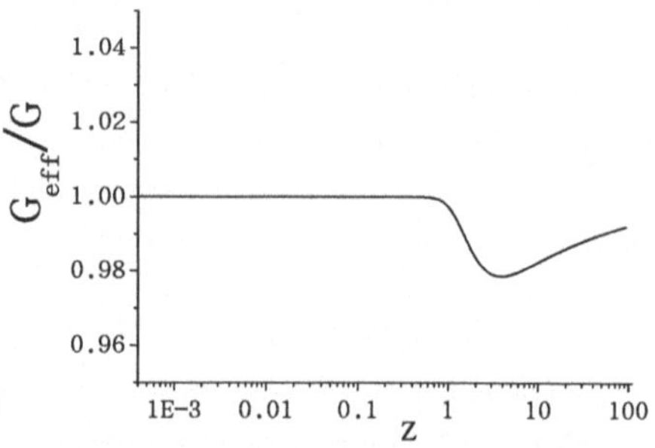

Fig. 30.1 **Left Graph**: The normalized $H(z)/(1+z)^{3/2}$ in units of km/s/Mpc as a function of the redshift, for ΛCDM cosmology (black–solid) and for Model I with $V_0 = 0.08$ and with $G_5(X) = \xi X^2$, for $\xi = 1.5$ (green–dotted), $\xi = 1.3$ (red–dashed) and $\xi = 1$ (blue–dashed-dotted), in H_0 units. We have imposed $\Omega_{m0} \approx 0.31$. **Right graph**: The corresponding normalized effective Newton's constant $\frac{G_{eff}}{G}$ given in (30.15) as a function of the redshift. The figure is from [11]

can consider models with $G_4 = 1/(16\pi G)$ and $G_3 = 0$, which is the case in ΛCDM cosmology, and impose a simple scalar-field potential and standard kinetic term, namely $K = -V(\phi) + X$ [11]. Additionally, since G_5 affects the friction terms of the scalar field [12–14], one chooses the G_5 term to depend only on X, i.e. $G_5(\phi, X) = G_5(X)$ [11]. Inserting into (30.8), (30.9) gives the effective dark energy density and pressure [11]

$$\rho_{DE} = 2X - K + 2H^3 X\dot{\phi}(5G_{5,X} + 2XG_{5,XX}), \tag{30.13}$$
$$p_{DE} = K - 2XG_{5,X}\left(2H^3\dot{\phi} + 2H\dot{H}\dot{\phi} + 3H^2\ddot{\phi}\right) - 4H^2X^2\dot{\phi}G_{5,XX}. \tag{30.14}$$

One can choose $G_5(X)$ suitably in order for $H(z)$ to coincide with $H_{\Lambda\text{CDM}}(z) := H_0\sqrt{\Omega_{m0}(1+z)^3 + 1 - \Omega_{m0}}$ at $z = z_{\text{CMB}} \approx 1100$, namely $H(z \to z_{\text{CMB}}) \approx H_{\Lambda\text{CDM}}(z \to z_{\text{CMB}})$, but satisfy $H(z \to 0) > H_{\Lambda\text{CDM}}(z \to 0)$. For simplicity, we focus on dust matter (i.e. $p_m = 0$), and without loss of generality we consider $K = -V_0\phi + X$.

- Model I: $G_5(X) = \xi X^2$
  A first model is the one with $G_5(X) = \xi X^2$, i.e where G_5 has a quadratic dependence on the field's kinetic energy. In the left graph of Fig. 30.1 we depict the normalized $H(z)/(1+z)^{3/2}$ for various choices of ξ. As we observe, the scenario coincides with ΛCDM one at high and intermediate redshifts, while at small redshifts it leads to higher values. In particular, the present-day value H_0 depends on the model parameter ξ, and it can be around $H_0 \approx 74$ km/s/Mpc for $\xi = 1.3$ (note that since ξ has dimensions of $[M]^{-9}$ and since $H_0 \approx 10^{-61}$ in Planck units, this implies that $\xi^{1/9} \sim 10^{40}\text{GeV}^{-1}$). As we see, H_0 tension can be alleviated at 3σ if $1.2 < \xi < 1.7$. The mechanism behind the alleviation is the decreased effective Newton's constant, brought about in turn by the friction term. Since in Horndeski theories we have [15, 16]

$$\frac{G_{eff}}{G} = \frac{1}{2}\left(G_4 - 2XG_{4,X} + XG_{5,\phi} - \dot{\phi}HXG_{5,X}\right)^{-1}, \tag{30.15}$$

one can easily see that in the present Model I this exhibits a decrease as can be observed in the right graph of Fig. 30.1. Thus, although w_{DE} in this scenario does not evolve into the phantom regime, the decrease in G_{eff} is adequate to alleviate the tension. Finally, one can check that the scenario at hand is stable at the perturbative level too [11].

- Model II: $G_5(X) = \lambda X^4$

  In the case $G_5(X) = \lambda X^4$, one obtains a similar behavior with that of Fig. 30.1, and one can obtain $H_0 \approx 74$ km/s/Mpc for $\lambda = 1$ in H_0 units (since λ has dimensions of $[M]^{-17}$ we acquire $\lambda^{1/17} \sim 10^{30}\text{GeV}^{-1}$), and the tension can be alleviated at 3σ if $0.5 < \lambda < 1.2$ in H_0 units. Similarly to the previous model, the mechanism behind the tension alleviation is the decreased G_{eff} at intermediate redshifts [11].

In summary, the above particular sub-class of Horndeski/generalized Galileon gravity can alleviate the H_0 tension due to the effect of the kinetic-energy-dependent G_5 term on decreasing G_{eff}. Nevertheless, note that models with $G_5 \neq 0$ need some fine-tuning in order to obtain gravitational wave speed equal to the speed of light [11].

30.2.2 Shift-Symmetric Horndeski Theories

Let us now focus on shift-symmetric Horndeski theories, namely models that are invariant under the transformation $\phi \to \phi + c$ where c is constant. The cubic-order shift-symmetric action reads as [17]

$$S = \int d^4x\sqrt{-g}\left[\frac{M_{\rm Pl}^2}{2}R + G_2(X) + G_3(X)\Box\phi\right], \tag{30.16}$$

where $M_{\rm Pl}$ is the reduced Planck mass. In this case we obtain the effective dark density and pressure as

$$\rho_{DE} = -G_2 - 2\dot{\phi}^2 G_{2,X} + 6H\dot{\phi}^3 G_{3,X}, \tag{30.17}$$

$$p_{DE} = G_2 - 2\ddot{\phi}\dot{\phi}^2 G_{3,X}. \tag{30.18}$$

This subclass of Horndeski gravity can also alleviate the H_0 tension. In particular, one can choose:

- Galileon Ghost Condensate (GGC)
  One can consider the Galileon Ghost Condensate (GGC) model, having [18, 19]

$$G_2(X) = a_1X + a_2X^2, \qquad G_3(X) = 3a_3X, \tag{30.19}$$

with a_1, a_2, a_3 constants. This model deviates from the Cubic Galileon scenario [20] due to the term $a_2 X^2$, which is responsible for altering the late-time cosmic evolution, since it prevents the Universe from approaching the tracker solution. We mention here that the mechanism behind the H_0 tension alleviation is the fact that dark-energy equation-of-state parameter can lie in the phantom regime ($w_{DE} < -1$) without exhibiting ghosts [19, 21]. Note also that this model suppresses the large-scale CMB temperature anisotropy comparing to ΛCDM paradigm. These features imply statistical preference of GGC over ΛCDM when data from CMB, BAO, SNIa, and RSDs are used [21]. In summary, one obtains $H_0 = \left(69.3^{+3.6}_{-3.0}\right)$ km s^{-1} Mpc^{-1} at 95% confidence level with best fit $H_0^{bf} = 70$ km s^{-1} Mpc^{-1} [21], consistent within 2σ with the direct measurements.

- Generalized Cubic Covariant Galileon (GCCG):
A different shift-symmetric model is the Generalized Cubic Covariant Galileon (GCCG) one [9], having

$$G_2(X) = -c_2 M_2^{4(1-p)}(-X/2)^p, \qquad G_3(X) = -c_3 M_3^{1-4p_3}(-X/2)^{p_3}, \quad (30.20)$$

with M_2, M_3 constants with dimension $[M]$, and where c_2, c_3, p, p_3 are dimensionless constants (one can also re-define them as $q = p_3 - p + \frac{1}{2}$ and $s = p/q$). The GCCG model is an extension of the Covariant Galileon ($s = 2, q = 1/2$), and similarly to the latter it also accepts a tracker solution with $H\dot{\phi}^{2q} =$ constant for $q > 0$. In such a case one acquires $H_0 = 72^{+8}_{-5}$ km s^{-1} Mpc^{-1} at 95% CL, finding a consistency with Cepheids measurements at 1σ [22], while it is also compatible with the estimation obtained with the tip of the red giant branch in LMC, $H_0 = (72.4 \pm 2)$ km s^{-1} Mpc^{-1} [23]. Nevertheless, note that for the GCCG model if the tension between CMB data and low redshift H_0 measurements disappears, another tension between the latter and BAO data appears [22, 24].

30.2.3 Brans-Dicke Theories

Brans-Dicke (BD) theory [25] was the first scalar-tensor theory that appeared in the literature, which later on was discovered that it falls within Horndeski class. It is characterized by the action

$$S = \int dx^4 \sqrt{-g}\left[\frac{1}{16\pi}\left(\varphi R - \frac{\omega_{BD}}{\varphi} g^{\mu\nu}\partial_\mu\varphi\partial_\nu\varphi\right) - \rho_\Lambda\right] + \int dx^4 \sqrt{-g}\mathcal{L}_m, \quad (30.21)$$

where ρ_Λ is the usual vacuum energy and $\omega_{\rm BD}$ (or equivalently $\epsilon_{\rm BD} := 1/\omega_{\rm BD}$) is a dimensionless parameter. Note that one can generalize the original BD theory by introducing a potential $V(\varphi)$ for the scalar field. Moreover, the Klein-Gordon equation is $\Box\varphi = 8\pi GT/(3 + 2\omega_{\rm BD})$, with T the trace of the total energy-momentum tensor. As expected, General Relativity is recovered by considering the simultaneous

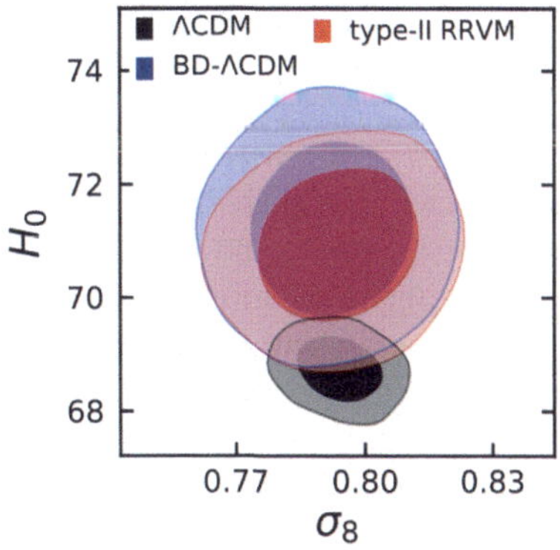

Fig. 30.2 The 1σ and 2σ confidence level regions in the (σ_8–H_0)-plane for the GR ΛCDM, type-II RRVM and BD-ΛCDM (see text), together with the data points used in [27]. The type-II RRVM and BD-ΛCDM are able to alleviate the H_0 tension without worsening the σ_8 one. H_0 is expressed in units of km s^{-1} Mpc^{-1}. The figure is from [28]

limits $\varphi \to 1$ and $\epsilon_{\rm BD} \to 0$, namely by suppressing the dynamics of φ and matching its constant value with the Newton's constant G, and actually the first condition is needed to obtain GR+Λ, while the second condition guarantees that BD theory recovers the weak-field regime of GR, but not for higher orders in the parametrization of post-Newtonian expansion [26].

Let us call BD-ΛCDM the Brans-Dicke model with the usual vacuum energy, a theory that has been recently confronted with a large string of cosmological data [28], such as SNIa+BAO+$H(z)$+CMB+LSS, where $H(z)$ may include or not the SH_0ES prior on H_0. Focusing only on the results obtained including the latter, in Fig. 30.2 we depict the results of the fittings. As we observe, BD-ΛCDM scenario is capable of reducing the H_0 (and σ_8) tension to $\sim 1.5\,\sigma$ level (when the high-l polarization and lensing data from Planck 2018 are not considered) [28]. This is similar to the case of type-II Ricci-Running-Vacuum models (RRVM), which the BD-ΛCDM model mimics [27]. Additionally, we mention that, as confirmed by various information criteria, the model is favoured comparing to standard ΛCDM scenario [27, 28]. One should add that in order not to spoil the correct fit of the CMB temperature data the model needs to accommodate values of the spectral index n_s larger than those of the GR-ΛCDM, nevertheless this does not cause problems since the dynamics of φ can compensate the corresponding changes in the matter power spectrum [28]. Finally, note that the H_0 tension is alleviated for $|\epsilon_{\rm BD}| \lesssim \mathcal{O}(10^{-3})$, and since this parameter has an impact on the LSS the compatibility is possible provided a screening mechanism is invoked.

30.2.4 Minimal Scalar-Tensor Theories

Let us now focus on a different subclass of Horndeski modified gravity that can correspond to extended versions of Brans-Dicke teories. In particular, one considers the action is [29, 30]

$$S = \int d^4x\sqrt{-g}\left[\frac{F(\phi)}{2}R - \frac{g^{\mu\nu}}{2}\partial_\mu\phi\partial_\nu\phi - V(\phi) + \mathscr{L}_m\right], \tag{30.22}$$

where $F(\phi)$ is a generic function of ϕ. In such cases one can easily attain values for H_0 that are larger than the ΛCDM one, due to a degeneracy between the coupling to the Ricci curvature and H_0, and is connected to the rolling of the scalar field which regulates the gravitational strength.

In the case where $F(\phi) = \xi\phi^2$, with $\xi > 0$ the coupling, namely of an effectively massless scalar field at rest in the radiation era and with adiabatic initial condition for the scalar field perturbations [31], the current constraints on the Hubble parameter from Planck 2018 data at 68% confidence level are $H_0 = \left(69.6^{+0.8}_{-1.7}\right)$ km s^{-1} Mpc^{-1} [32]. When BAO data from BOSS DR12 are added, the constraint at 68% confidence level becomes $H_0 = \left(68.78^{+0.53}_{-0.78}\right)$ km s^{-1} Mpc^{-1}, while once a Gaussian likelihood based on the determination of the Hubble constant from the SH_0ES team is also included one obtains $H_0 = (70.1 \pm 0.8)$ km s^{-1} Mpc^{-1}. Note that these constraints have been obtained by fixing the value of the current value of the scalar field $\phi_0 := \phi(z = 0)$ to

$$\frac{1}{8\pi F_0}\frac{2F_0 + 4F_{0,\phi}^2}{2F_0 + 3F_{0,\phi}^2} = G, \tag{30.23}$$

where $F_0 := F(\phi_0)$ and $F_{0,\phi} = \partial F/\partial\phi|_{\phi=\phi_0}$, in order to guarantee that the current effective Newton's constant corresponds to the standard one [33] (see [34–36] for studies of these scenarios without imposing such condition).

30.3 $f(T)$ Gravity

In this section we will show how the H_0 tension can be alleviated in a different class of gravitational modification, namely in the framework of torsional gravity. In torsional formulation one uses as dynamical variables the vierbeins fields, which form an orthonormal basis at a manifold point. In a coordinate basis they are related to the metric through $g_{\mu\nu}(x) = \eta_{AB}\ ,e^A_\mu(x)\ ,e^B_\nu(x)$, with Greek and Latin indices respectively used for the coordinate and tangent space. One introduces the Weitzenböck connection $W^\lambda_{\nu\mu} := e^\lambda_A\,\partial_\mu e^A_\nu$ [37], and thus the corresponding torsion tensor is $T^\lambda_{\mu\nu} := W^\lambda_{\nu\mu} - W^\lambda_{\mu\nu} = e^\lambda_A\,(\partial_\mu e^A_\nu - \partial_\nu e^A_\mu)$. The torsion tensor incorporates all the information of the gravitational field, and the torsion scalar arises from its contraction as

$$T := \frac{1}{4}T^{\rho\mu\nu}T_{\rho\mu\nu} + \frac{1}{2}T^{\rho\mu\nu}T_{\nu\mu\rho} - T_{\rho\mu}{}^{\rho}T^{\nu\mu}{}_{\nu}\,. \tag{30.24}$$

This forms the Lagrangian of teleparallel gravity (similarly to general relativity where the Lagrangian is the Ricci scalar), and since variation in terms of the vierbeins gives the same equations with general relativity, the constructed theory was named teleparallel equivalent of general relativity (TEGR).

In its most simple formulation, it is possible to consider $f(T)$ extensions of teleparallel gravity, intended as corrections to TEGR where only the torsion scalar T is considered. The action takes the form of

$$S = \frac{1}{16\pi G} \int \mathrm{d}^4 x\, e \left[T + f(T)\right] + S_m, \tag{30.25}$$

where $f(T)$ is a generic function of the torsion scalar T, S_m is the action of matter fields, and $e = \det(e^i_{\ \mu}) = \sqrt{-g}$.

For a flat FLRW background, it is possible to derive the relation between the torsion T and the Hubble parameter, $T = 6H^2$. The sign of such a torsion element depends on the signature of the metric, which is identified by the vierbien fields $e^a_\mu = \mathrm{diag}(-1, a, a, a)$. Assuming that matter sector is described by a perfect fluid with energy density ρ and pressure p, the field equations give

$$H^2 = \frac{8\pi G}{3}\rho_m - \frac{f(T)}{6} + \frac{Tf_T}{3} \tag{30.26}$$

$$\dot{H} = -\frac{4\pi G\,(\rho_m + p_m)}{1 + f_T + 2\,Tf_{TT}}. \tag{30.27}$$

Moreover, the set of equations is closed with the equation of continuity for the matter sector $\dot{\rho}_m + 3H(\rho_m + p_m) = 0$. Equations (30.26), (30.27) can be rewritten in terms of the effective energy density ρ_T and pressure p_T arising from $f(T)$,

$$\rho_{f(T)} = \frac{M_{\rm Pl}^2}{2}\left[2\,Tf_T - f(T)\right] \tag{30.28}$$

$$p_{f(T)} = \frac{M_{\rm Pl}^2}{2}\left[\frac{f(T) - Tf_T + 2\,T^2 f_{TT}}{1 + f_T + 2\,Tf_{TT}}\right], \tag{30.29}$$

and define the effective torsion equation-of-state

$$w := \frac{p_{f(T)}}{\rho_{f(T)}} = \frac{f(T) - Tf_T + 2\,T^2 f_{TT}}{\left[1 + f_T + 2\,Tf_{TT}\right]\left[2Tf_T - f(T)\right]}. \tag{30.30}$$

These effective models are hence responsible for the accelerated phases of the early or/and late Universe [7].

In order to study the background evolution, we can rewrite the first FLRW equation, making explicit the form of the torsional energy density [38–40]

$$\frac{H(a)^2}{H_0} := E(a)^2 = \left[\Omega_m a^{-3} + \Omega_r a^{-4} + \frac{1}{T_0}(-f + 2Tf')\right], \tag{30.31}$$

where we considered the relation $T = 6H^2$ and with T_0 the present value of T.

Let us now try to suitably extract an $f(T)$ form that could alleviate H_0 tension. As we mentioned in the Introduction, in order to achieve this, one needs either a positive correction to the first Friedmann equation at late times that could yield an increase in H_0 compared to the ΛCDM scenario, or a weaker gravity, quantified by a smaller G_{eff}. In the case of $f(T)$ gravity the effective gravitational coupling is given by [38]

$$G_{eff} = G\left(1 + \frac{\partial f(T)}{\partial T}\right)^{-1}. \tag{30.32}$$

According to the Effective Field Theory approach to $f(T)$ gravity [41–43] one can choose $f(T)$ in order to obtain a Hubble function evolution of the form

$$H(z) = -\frac{d(z)}{4} + \sqrt{\frac{d^2(z)}{16} + H^2_{\Lambda\mathrm{CDM}}(z)}, \tag{30.33}$$

where $H_{\Lambda\mathrm{CDM}}(z) := H_0\sqrt{\Omega_m(1+z)^3 + \Omega_\Lambda}$ is the Hubble rate in ΛCDM cosmology after recombination, Ω_m is the present matter density parameter, and primes denote derivatives with respect to z. The function $d(z)$ can be selected to be $d < 0$ in order to have $H(z \to z_{\mathrm{CMB}}) \approx H_{\Lambda\mathrm{CDM}}(z \to z_{\mathrm{CMB}})$, therefore the H_0 tension is solved. One should choose $|d(z)| < H(z)$, and thus, since $H(z)$ decreases for smaller z, the deviation from ΛCDM will be significant only at low redshifts. Additionally, since G_{eff} becomes smaller than G during the growth of matter perturbations, the σ_8 tension is also solved.

We follow Ref. [42] and we consider the ansatz $f(T) = -[6H_0^2(1-\Omega_m) + F(T)]$, where $F(T)$ describes the deviation from GR (mind the difference in the conventions). Under these assumptions, the first Friedmann equation becomes

$$T(z) + 2\frac{F'(z)}{T'(z)}T(z) - F(z) = 6H^2_{\Lambda\mathrm{CDM}}(z). \tag{30.34}$$

In order to solve the H_0 tension, we need $T(0) = 6H_0^2$, with $H_0 = 74.03\,\mathrm{km\,s^{-1}\,Mpc^{-1}}$ following the results with local measurements [44], while in the early era of $z \gtrsim 1100$ we require the Universe expansion to evolve as in ΛCDM, namely $H(z \gtrsim 1100) \simeq H_{\Lambda\mathrm{CDM}}(z \gtrsim 1100)$. This implies $F(z)|_{z \gtrsim 1100} \simeq cT^{1/2}(z)$ (the value $c = 0$ corresponds to standard GR, while for $c \neq 0$ we obtain ΛCDM too).

By solving Eq. (30.34) with initial and boundary conditions at $z \sim 0$ and $z \sim 1100$, we can find the $f(T)$ form that can alleviate H_0 tension. In particular, we find that we can fit the numerical solutions very efficiently through $F(T) \approx 375.47\,[T/(6H_0^2)]^{-1.65}$, dubbed Model-1, and $F(T) \approx 375.47\,[T/(6H_0^2)]^{-1.65} + 25T^{1/2}$, dubbed Model-2. We examine G_{eff}, and as expected we find that at high redshifts in both models G_{eff} becomes G, recovering the ΛCDM paradigm. We have checked that both models can easily pass the BBN constraints, which demand $|G_{eff}/G - 1| \leq 0.2$ [45], as well as the ones from the Solar System, which demand $|G'_{eff}(z=0)/G| \leq 10^{-3}h^{-1}$ and $|G_{eff}''(z=0)/G| \leq 10^5 h^{-2}$ [46].

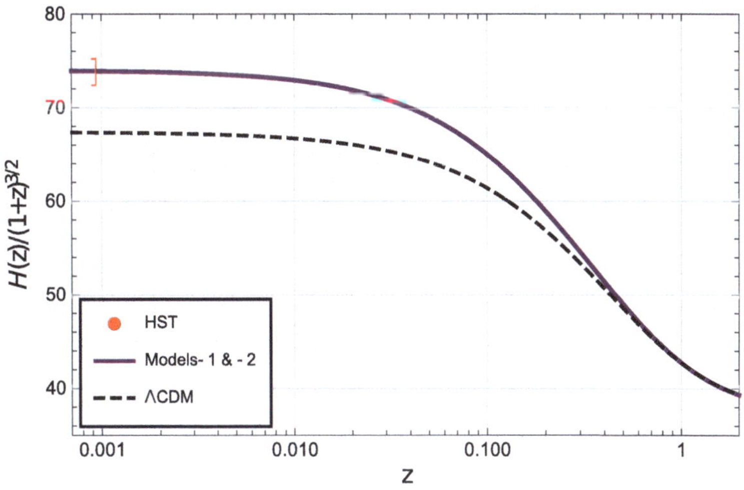

Fig. 30.3 Evolution of the Hubble parameter $H(z)$ in the two $f(T)$ models (purple solid line) and in ΛCDM cosmology (black dashed line). The red point (with error bars) represents the latest data from extragalactic Cepheid-based local measurement of H_0 [44]. The figure is from [42]

In Fig. 30.3 we present the evolution of $H(z)$, for the two $f(T)$ models, and we compare them with ΛCDM. We stress that the H_0 tension can be alleviated as $H(z)$ remains statistically consistent with all CMB and CC measurements at all redshifts. We remind the reader that the two $f(T)$ models differing merely by a term $\propto T^{1/2}$, which does not affect the background as explained before, are degenerate at the background level. However, at the perturbation level, the two models behave differently as their gravitational coupling G_{eff} differs.

In summary, we conclude that the class of $f(T)$ gravity described by

$$f(T) = -2\Lambda/M_{\rm Pl}^2 + \alpha T^{\beta}, \tag{30.35}$$

where only two out of the three parameters Λ, α, and β are independent (the third one is eliminated using Ω_{m0}), can alleviate H_0 tension with suitable parameter choices. And as we mentioned, the mechanism behind this alleviation is the fact that the effective Newton's constant G_{eff} in $f(T)$ gravity can be smaller that G at intermediate times, thus leading to faster expansion. Moreover, these type of $f(T)$ gravity models could also be examined through galaxy-galaxy lensing effects [47], strong lensing effects around black holes [48], and gravitational-wave experiments [49].

For completeness, we proceed to the investigation of different $f(T)$ models too, i.e models that are efficient in successfully passing the confrontation with observations [38, 50].

1. The power-law model [38, 50] (hereafter f_1CDM model), in which

$$f(T) = \alpha(-T)^b, \tag{30.36}$$

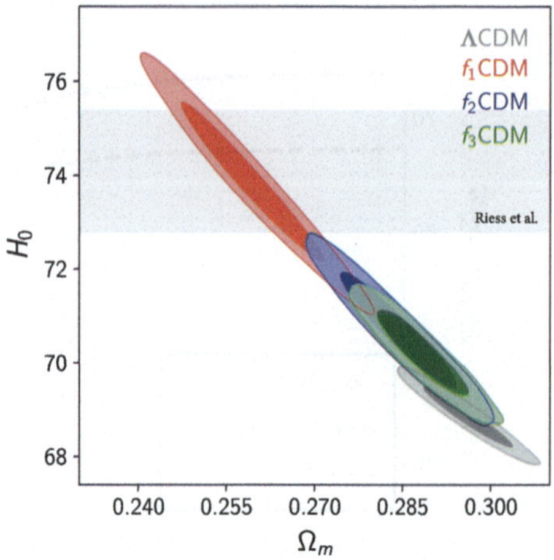

Fig. 30.4 The likelihood contours for the three $f(T)$ models described in the text, in the H_0-Ω_m plane, for CMB+lensing+BAO+R19+Pth+DES data. The grey region indicate the latest constrain on Hubble constant of Riess, i.e $H_0 = 74.03 \pm 1.42$ km/s/Mpc [44]. The figure is from [51]

where $\alpha = (6H_0^2)^{1-b}\frac{\Omega_{F0}}{2b-1}$ and hence the only free parameter is b. Thus, for $b = 0$ the model f_1CDM reduces to ΛCDM cosmology, i.e. $T + f(T) = T - 2\Lambda$, with $\Lambda = 3\Omega_{F0}H_0^2$ and $\Omega_{F0} = \Omega_{\Lambda 0}$.

2. The square-root exponential model (hereafter f_2CDM) [38, 50]

$$f(T) = \alpha T_0(1 - e^{-p\sqrt{T/T_0}}), \tag{30.37}$$

with $\alpha = \frac{\Omega_{F0}}{1-(1+p)e^{-p}}$. This model reduces to ΛCDM paradigm for $p \to +\infty$, and thus we can replace p by $p = 1/b$ which tends to 1 for $b \to 0^+$.

3. The exponential model (hereafter f_3CDM) [38, 50]:

$$f(T) = \alpha T_0(1 - e^{-pT/T_0}), \tag{30.38}$$

with $\alpha = \frac{\Omega_{F0}}{1-(1+2p)e^{-p}}$, where using $p = 1/b$ implies that for $b \to 0^+$ the f_3CDM model reduces to ΛCDM one.

In Fig. 30.4 we present the likelihood contours in the case of the aforementioned three models. As we observe, the power-law, f_1CDM model of (30.36), gives $H_0 = (73.85 \pm 1.05)$ km s^{-1} Mpc^{-1} [51] (see also [52] for an analysis with several other combinations of large-scale data). Hence, this allows to significantly relax the Hubble constant tension, nevertheless it worsens σ_8 tension [53–55] since the correlation of the two parameters does not seem to be removed.

Finally, we close this section mentioning that one can have other classes of torsional gravity, arising from different extensions of TEGR, that can also alleviate the H_0 tension, such as $f(T, B)$ models [56, 57] , where B is the boundary term that quantifies the difference of (Levi-Civita) Ricci scalar R and (Weitzenböck) torsion scalar T [58].

30.4 $f(R)$ Gravity

In this section we investigate the possible alleviation of the H_0 tension in the framework of another class of modified gravity, namely $f(R)$ gravity [8]. In $f(R)$ gravitational theories one extends the Einstein-Hilbert action to

$$S = \int d^4x\sqrt{-g}\,\frac{f(R)}{16\pi G}. \tag{30.39}$$

Following the metric formulation, variation of the action (30.39) with respect to the metric $g_{\mu\nu}$ leads to the field equations

$$FG_{\mu\nu} = -\frac{1}{2}g_{\mu\nu}\,(FR-f) + \nabla_\mu\nabla_\nu F - g_{\mu\nu}\Box F + 8\pi G\, T^{(m)}_{\mu\nu}, \tag{30.40}$$

with $G_{\mu\nu} = R_{\mu\nu} - (1/2)\, g_{\mu\nu}R$ the Einstein tensor, ∇_μ the covariant derivative, $\Box := g^{\mu\nu}\nabla_\mu\nabla_\nu$, and where we have defined $F(R) := f_{,R} = df(R)/dR$. Additionally, $T^{(m)}_{\mu\nu}$ is the energy-momentum tensor for the matter sector.

Applying the above theory in a flat FRW metric, we obtain the modified Friedmann equations as

$$H^2 = 8\pi G\,(\rho_m + \rho_{DE}), \tag{30.41}$$

$$-2\dot{H} = 8\pi G\,(\rho_m + p_m + \rho_{DE} + P_{DE}), \tag{30.42}$$

where we have considered that the matter sector corresponds to a perfect fluid with energy density ρ_m and pressure p_m, while the effective dark energy sector is characterized by

$$\rho_{DE} := \frac{1}{8\pi G}\left[\frac{1}{2}\,(FR-f) - 3H\dot{F} + 3\,(1-F)\,H^2\right], \tag{30.43}$$

$$p_{DE} := \frac{1}{8\pi G}\left[-\frac{1}{2}\,(FR-f) + \ddot{F} + 2H\dot{F} - (1-F)\left(2\dot{H} + 3H^2\right)\right], \tag{30.44}$$

and thus we can introduce the effective dark-energy equation-of-state parameter as $w := p_{DE}/\rho_{DE}$. Finally, note that in flat FRW geometry one obtains the useful relation $R = 6\left(2H^2 + \dot{H}\right)$.

Amongst the various specific models, we consider the exponential one, where [59–61]:

$$f(R) = R - 2\Lambda\left[1 - \exp\left(-\beta\frac{R}{2\Lambda}\right)\right]. \tag{30.45}$$

This model recovers ΛCDM cosmology at the limit $\beta \to \infty$. Additionally, note that at large redshifts the model also comes close to ΛCDM scenario, since the curvature becomes large, namely $R >> \Lambda/\beta$.

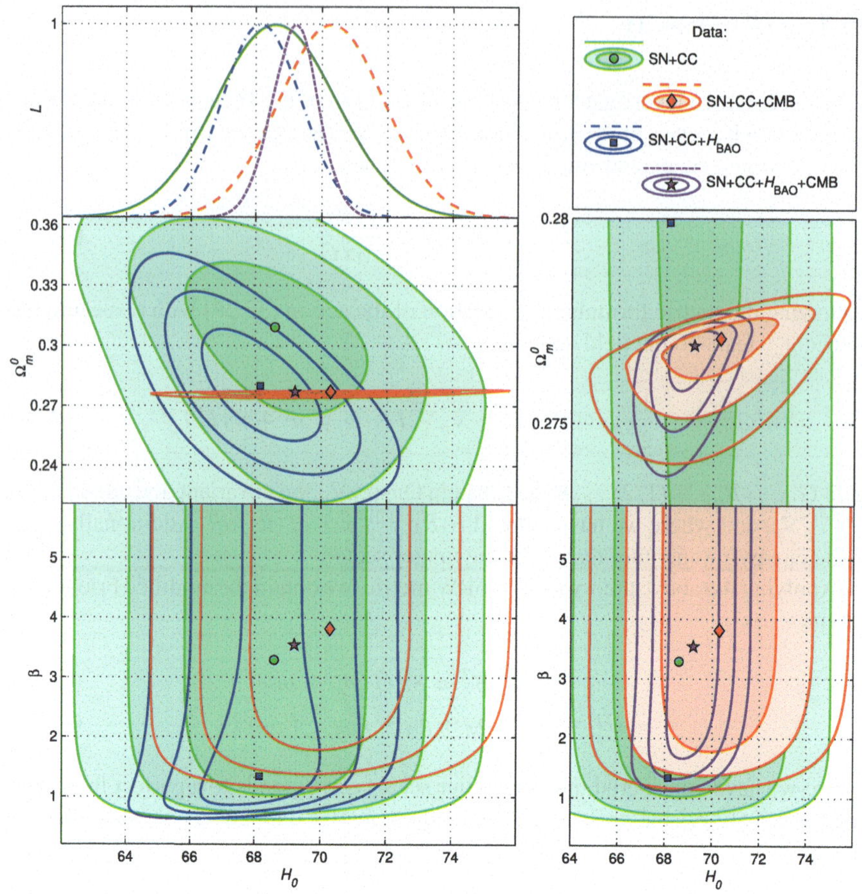

Fig. 30.5 Likelihood contours for the exponential $f(R)$ gravity model (30.45), at 1σ, 2σ and 3σ confidence levels, for SNe Ia + CC + $H_{\rm BAO}$ + CMB observational data. The figure is from [61]

Figure 30.5 depicts the likelihood contours at 1σ, 2σ and 3σ confidence levels for SNe Ia + CC + $H_{\rm BAO}$ + CMB data, in the $H_0 - \Omega_m^0$ and $H_0 - \beta$ planes, while the absolute maximums are described by stars, circles, etc. [61]. The right panels depict the same contour plots but with additional details, e.g. re-scaled along the Ω_m^0 axis. As we observe, the H_0 tension can be partially alleviated in the framework of exponential $f(R)$ gravity.

We close this section by mentioning that, as we saw, H_0 can only be partially alleviated in exponential $f(R)$ gravity, and things are similar in the case of other viable models of the literature such as the Hu-Sawicki one [62].

$$f(R) = R - \frac{c_1 R_{\rm HS}\,(R/R_{\rm HS})^p}{c_2\,(R/R_{\rm HS})^p + 1}, \tag{30.46}$$

where $c_1, c_2, R_{\rm HS}, p > 0$ are the model parameters, or the Starobinsky one [63]

$$f(R) = R - \lambda R_{\rm S}\left[1 - \left(1 + \frac{R^2}{R_{\rm S}^2}\right)^{-n}\right], \tag{30.47}$$

with $\lambda(> 0)$, $R_{\rm S}$, $n > 0$ the model parameters, or the Tsujikawa model [64]

$$f(R) = R - \mu R_{\rm T} \tanh\left(\frac{R}{R_{\rm T}}\right), \tag{30.48}$$

with $\mu(> 0)$ and $R_{\rm T}(> 0)$ the two positive parameters. The reason for this is that in $f(R)$ gravity the effective Newton's constant is [65]

$$\frac{G_{\it eff}}{G} = \frac{1}{f_R}\frac{1 + 4\frac{k^2}{a^2}\frac{f_{RR}}{f_R}}{1 + 3\frac{k^2}{a^2}\frac{f_{RR}}{f_R}}, \tag{30.49}$$

where note that we have a dependence on the wave-number k. On the other hand, perturbation analysis reveals that in order for an $f(R)$ model to be stable one should have $f_{,R} > 0$ for $R \geq R_0$, with R_0 the present value of the Ricci scalar, in order to avoid a ghost state, and $f_{,RR} > 0$ for $R \geq R_0$, in order to avoid the scalar-field degree of freedom to become tachyonic [8]. Additionally, one should have $0 < \frac{Rf_{,RR}}{f_{,R}}(r) < 1$ at $r = -\frac{Rf_{,R}}{f} = -2$, in order to have the presence and stability of a late-time de Sitter solution [8]. One can see that $f(R)$ choices that satisfy the above requirements typically lead to $\frac{G_{\it eff}}{G} > 1$ and thus operate in the opposite direction of the H_0 alleviation. Nevertheless, such models can still alleviate the tension if they lead to strong phantom behavior (according to the requirements of [2, 3]), however it is clear that the full H_0-tension alleviation is hard to be obtained since viable and pathologies-free $f(R)$ models induce effects that simultaneously increase and decrease H_0.

30.5 Bi-scalar Modified Theories of Gravity

As a last modified gravity model that can alleviate the H_0 tension we consider a class of bi-scalar theories, characterized by the action [66, 67]

$$S = \int d^4\sqrt{-g} f\left(R, (\nabla R)^2, \Box R\right), \tag{30.50}$$

with $(\nabla R)^2 = g^{\mu\nu}\nabla_\mu R \nabla_\nu R$. In the following we focus on models with $f(R, (\nabla R)^2, \Box R) = \mathscr{K}((R, (\nabla R)^2) + \mathscr{G}(R, (\nabla R)^2)\Box R$. One can rewrite the above action by converting the Lagrangian using double Lagrange multipliers, resulting to actions of multi-scalar fields coupled minimally to gravity. In order to achieve it, one fixes the dependence of f on $\Box R = \beta$. In particular, the χ and ϕ fields are introduced through the conformal transformations $g_{\mu\nu} = \frac{1}{2}e^{-\sqrt{\frac{2}{3}}\chi}\hat{g}_{\mu\nu}$, $\varphi := f_\beta$, and they enter in a specific combination in a way that the final form of the action is equivalent to the original higher-derivative gravitational action. Doing so, and performing variation, one results in the Friedmann equations [66, 67]

$$H^2 = \frac{1}{3}(\rho_{DE} + \rho_m) \tag{30.51}$$

$$2\dot{H} + 3H^2 = -(p_{DE} + p_m), \tag{30.52}$$

with the effective dark energy and pressure defined as

$$\begin{aligned}\rho_{DE} := \frac{1}{2}\dot{\chi}^2 - \frac{1}{4}e^{-2\sqrt{\frac{2}{3}}\chi}\mathcal{K} - \frac{2}{3}\dot{\phi}^2\left[\dot{\phi}\left(\sqrt{6}\dot{\chi} - 9H\right) - 3\ddot{\phi}\right]\mathcal{G}_B \\ +\frac{1}{2}e^{-\sqrt{\frac{2}{3}}\chi}\left[\dot{B}\dot{\phi}\mathcal{G}_B + \frac{\phi}{2} + \dot{\phi}^2\left(\mathcal{G}_\phi - 2\mathcal{K}_B\right)\right],\end{aligned} \tag{30.53}$$

$$p_{DE} := \frac{1}{2}\dot{\chi}^2 + \frac{1}{4}e^{-2\sqrt{\frac{2}{3}}\chi}\mathcal{K} + \frac{1}{2}e^{-\sqrt{\frac{2}{3}}\chi}\left(\dot{B}\dot{\phi}\mathcal{G}_B + \dot{\phi}^2\mathcal{G}_\phi - \frac{\phi}{2}\right), \tag{30.54}$$

with $\mathcal{K} = \mathcal{K}(\phi, B)$ and $\mathcal{G} = \mathcal{G}(\phi, B)$, with $B = 2e^{\sqrt{\frac{2}{3}}\chi}g^{\mu\nu}\nabla_\mu\phi\nabla_\nu\phi$, and where for simplicity we set the Planck mass to 1. Finally, one can define the effective dark-energy equation-of-state parameter as $w_{DE} := p_{DE}/\rho_{DE}$.

In the following we desire to find specific models which can coincide with ΛCDM cosmology at CMB redhsifts, while at low-redhsifts to deviate from it, giving rise to higher H_0.

- Model I: $\mathcal{K}(\phi, B) = \frac{1}{2}\phi - \frac{\zeta}{2}B$ and $\mathcal{G}(\phi, B) = 0$
  In this case we solve the cosmological equations numerically and in the left graph of Fig. 30.6 we present the normalised combination $H(z)/(1+z)^{3/2}$ as a function of the redshift for ΛCDM cosmology, and for Model I for different values of ζ. We find that the present value of H_0 depends on the model parameter ζ as expected, and for $\zeta = -10$ the present value of the Hubble parameter is around $H_0 \approx 74km/s/Mpc$, which is consistent with the direct measurement of the present Hubble parameter (note that in natural units this corresponds to a typical value $\zeta^{1/4} \sim -10^{-19}$ GeV^{-1}). Additionally, in the right graph of Fig. 30.6 we plot the evolution of $w_{DE}(z)$, and as we can see it lies in the phantom regime ($w_{DE} < -1$) most of the time. Hence, the mechanism behind H_0 alleviation in the present model is phantom behavior, which is known to be one of the mechanisms since it induces faster expansion [2, 3]. Notice the difference between the single-field, Horndeski theories analyzed in Sect. 30.2.1 above, in which the alleviation mechanism was the smaller G_{eff}.

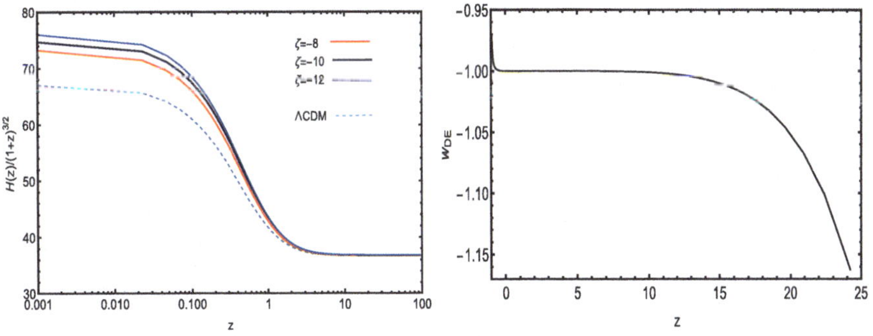

Fig. 30.6 **Left Graph**: The normalized $H(z)/(1+z)^{3/2}$ in units of km/s/Mpc as a function of the redshift, for ΛCDM cosmology (blue dotted line) and for Model I $\zeta = -12$ (solid blue line), for $\zeta = -10$ (solid black line), and for $\zeta = -8$ (solid red line), in Planck units. We have imposed $\Omega_{m0} \approx 0.31$. **Right graph**: The corresponding effective dark-energy equation-of-state parameter w_{DE} as a function of the redshift, for $\zeta = -10$ in Planck units. The figure is form [68]

- Model II: $\mathcal{K}(\phi, B) = \frac{1}{2}\phi$ and $\mathcal{G}(f, B) = xiB$,
  In this case we repeat the same steps as in the previous model and we find that the present Hubble value H_0 depends on the model parameter ξ. Specifically, for $\xi = -10$ it is around $H_0 \approx 74km/s/Mpc$ (in natural units $\xi \sim -10$ corresponds to a typical value $\xi^{1/8} \sim -10^{-19}$ GeV^{-1}). Similarly to the previous case, the mechanism behind H_0-tension alleviation is the phantom behavior.

30.6 Conclusions

Modified gravity is one of the best frameworks that can alleviate H_0 tension, and simultaneously other tensions such as the σ_8 one, due to the huge freedom one has to construct models. In particular, modified gravity can alter either the early-time Universe evolution, in which case one obtains early-time solutions to the H_0 tension which are considered in the next chapter, or it can alter the late-time evolution, and thus it can lead to late-time solutions, or both.

Gravitational modifications affect the late-time evolution of the universe through the new terms they bring in the Friedmann equations, namely in the effective dark-energy sector, as well as through the effective Newton's constant they induce. If these effects lead to weaker gravity (smaller G_{eff}) at suitable redshifts, or to more repulsive effective dark-energy (for instance exhibiting phantom behavior) then they can cause faster expansion comparing to ΛCDM paradigm, and thus lead to an increased H_0 value.

We presented H_0-tension alleviations based on various subclasses of Horndeski/Generalized Galileon theories, such as Horndeski gravity with non-minimal derivative coupling, shift-symmetric Horndeski theories, Brans-Dicke theories, and minimal scalar-tensor theories. Then we presented alleviations based on torsional modifications of gravity, and in particular of $f(T)$ gravity. Finally we presented solutions based on $f(R)$ gravity and bi-scalar modified theories of gravity. Such capabilities of gravitational modifications may be added in their other known advantages, and act as additional indication that modified gravity might be needed for the description of Nature.

Acknowledgements The author acknowledges the contribution of the LISA CosWG, and the COST Actions CA18108 "Quantum Gravity Phenomenology in the multi-messenger approach" and CA21136 "Addressing observational tensions in cosmology with systematics and fundamental physics (CosmoVerse)".

References

1. CANTATA Collaboration, E.N. Saridakis et al., Modified gravity and cosmology: an update by the CANTATA network. arXiv:2105.12582 [gr-qc]
2. L. Heisenberg, H. Villarrubia-Rojo, J. Zosso, Simultaneously solving the H_0 and σ_8 tensions with late dark energy. Phys. Dark Univ. Series **39**, 101163 (2023). (arXiv:2201.11623 [astro-ph.CO])
3. L. Heisenberg, H. Villarrubia-Rojo, J. Zosso, Can late-time extensions solve the H_0 and σ_8 tensions? Phys. Rev. D Series **106**(4), 043503 (2022). (arXiv:2202.01202 [astro-ph.CO])
4. G.W. Horndeski, Second-order scalar-tensor field equations in a four-dimensional space. Int. J. Theor. Phys. Series **10**, 363–384 (1974)
5. A. De Felice, S. Tsujikawa, Generalized Galileon cosmology. Phys. Rev. D Series **84**, 124029 (2011). (arXiv:1008.4236 [hep-th])
6. C. Deffayet, X. Gao, D.A. Steer, G. Zahariade, From k-essence to generalised Galileons. Phys. Rev. D Series **84**, 064039 (2011). (arXiv:1103.3260 [hep-th])
7. Y.-F. Cai, S. Capozziello, M. De Laurentis, E.N. Saridakis, f(T) teleparallel gravity and cosmology. Rept. Prog. Phys. Series **79**(10), 106901 (2016). (arXiv:1511.07586 [gr-qc])
8. A. De Felice, S. Tsujikawa, f(R) theories. Living Rev. Rel. Series **13**, 3 (2010). (arXiv:1002.4928 [gr-qc])
9. A. De Felice, S. Tsujikawa, Conditions for the cosmological viability of the most general scalar-tensor theories and their applications to extended Galileon dark energy models. JCAP series **02**, 007 (2012). (arXiv:1110.3878 [gr-qc])
10. T. Kobayashi, M. Yamaguchi, J. Yokoyama, Generalized G-inflation: inflation with the most general second-order field equations. Prog. Theor. Phys. Series **126**, 511–529 (2011). (arXiv:1105.5723 [hep-th])
11. M. Petronikolou, S. Basilakos, E.N. Saridakis, Alleviating H_0 tension in Horndeski gravity. arXiv:2110.01338 [gr-qc]
12. E.N. Saridakis, S.V. Sushkov, Quintessence and phantom cosmology with non-minimal derivative coupling. Phys. Rev. D Series **81**, 083510 (2010). (arXiv:1002.3478 [gr-qc])
13. G. Koutsoumbas, K. Ntrekis, E. Papantonopoulos, E.N. Saridakis, Unification of dark matter—dark energy in generalized Galileon theories. JCAP Series **02**, 003 (2018). (arXiv:1704.08640 [gr-qc])
14. S. Karydas, E. Papantonopoulos, E.N. Saridakis, Successful Higgs inflation from combined nonminimal and derivative couplings. Phys. Rev. D Series **104**(2), 023530 (2021). (arXiv:2102.08450 [gr-qc])
15. E. Bellini, I. Sawicki, Maximal freedom at minimum cost: linear large-scale structure in general modifications of gravity. JCAP Series **07**, 050 (2014). (arXiv:1404.3713 [astro-ph.CO])

16. S. Peirone, K. Koyama, L. Pogosian, M. Raveri, A. Silvestri, Large-scale structure phenomenology of viable Horndeski theories. Phys. Rev. D Series **97**(4), 043519 (2018). (arXiv:1712.00444 [astro-ph.CO])
17. T. Kobayashi, Horndeski theory and beyond: a review. Rept. Prog. Phys. Series **82**(8), 086901 (2019). (arXiv:1901.07183 [gr-qc])
18. C. Deffayet, O. Pujolas, I. Sawicki, A. Vikman, Imperfect dark energy from kinetic gravity braiding. JCAP Series **10**, 026 (2010). (arXiv:1008.0048 [hep-th])
19. R. Kase, S. Tsujikawa, Dark energy scenario consistent with GW170817 in theories beyond Horndeski gravity. Phys. Rev. D Series **97**(10), 103501 (2018). (arXiv:1802.02728 [gr-qc])
20. C. Deffayet, G. Esposito-Farese, A. Vikman, Covariant Galileon. Phys. Rev. D Series **79**, 084003 (2009). (arXiv:0901.1314 [hep-th])
21. S. Peirone, G. Benevento, N. Frusciante, S. Tsujikawa, Cosmological data favor Galileon ghost condensate over ΛCDM. Phys. Rev. D Series **100**(6), 063540 (2019). (arXiv:1905.05166 [astro-ph.CO])
22. N. Frusciante, S. Peirone, L. Atayde, A. De Felice, Phenomenology of the generalized cubic covariant Galileon model and cosmological bounds. Phys. Rev. D Series **101**(6), 064001 (2020). (arXiv:1912.07586 [astro-ph.CO])
23. W. Yuan, A.G. Riess, L.M. Macri, S. Casertano, D. Scolnic, Consistent calibration of the tip of the red giant branch in the large Magellanic cloud on the Hubble space telescope photometric system and a re-determination of the Hubble constant. Astrophys. J. Series **886**, 61 (2019). (arXiv:1908.00993 [astro-ph.GA])
24. J. Renk, M. Zumalacárregui, F. Montanari, A. Barreira, Galileon gravity in light of ISW, CMB, BAO and H_0 data. JCAP Series **10**, 020 (2017). (arXiv:1707.02263 [astro-ph.CO])
25. C.H. Brans, R.H. Dicke, Mach's principle and a relativistic theory of gravitation. Phys. Rev Series **124**, 925 (1961)
26. V. Faraoni, J. Côté, A. Giusti, Do solar system experiments constrain scalar–tensor gravity? Eur. Phys. J. C Series **80**(2), 132 (2020). (arXiv:1906.05957 [gr-qc])
27. J. Solà Peracaula, A. Gómez-Valent, J. de Cruz Pérez, C. Moreno-Pulido, Running vacuum against the H_0 and σ_8 tensions. EPL **134**(1), 19001 (2021) arXiv:2102.12758 [astro-ph.CO]
28. J. Solà Peracaula, A. Gómez-Valent, J. de Cruz Pérez, C. Moreno-Pulido, Brans–Dicke cosmology with a Λ-term: a possible solution to ΛCDM tensions. Class. Quant. Grav. **37**(24), 245003 (2020) arXiv:2006.04273 [astro-ph.CO]
29. S. Capozziello, R. de Ritis, Noether's symmetries and exact solutions in flat nonminimally coupled cosmological models. Class. Quant. Grav. Series **11**, 107–117 (1994)
30. Y.-F. Cai, E.N. Saridakis, M.R. Setare, J.-Q. Xia, Quintom cosmology: theoretical implications and observations. Phys. Rept. Series **493**, 1–60 (2010). (arXiv:0909.2776 [hep-th])
31. D. Paoletti, M. Braglia, F. Finelli, M. Ballardini, C. Umiltà, Isocurvature fluctuations in the effective Newton's constant. Phys. Dark Univ. Series **25**, 100307 (2019). (arXiv:1809.03201 [astro-ph.CO])
32. M. Ballardini, M. Braglia, F. Finelli, D. Paoletti, A.A. Starobinsky, C. Umiltà, Scalar-tensor theories of gravity, neutrino physics, and the H_0 tension. JCAP Series **10**, 044 (2020). (arXiv:2004.14349 [astro-ph.CO])
33. B. Boisseau, G. Esposito-Farese, D. Polarski, A.A. Starobinsky, Reconstruction of a scalar tensor theory of gravity in an accelerating universe. Phys. Rev. Lett. Series **85**, 2236 (2000). (arXiv:gr-qc/0001066)
34. Ö. Akarsu, N. Katırcı, N. Özdemir, J.A. Vázquez, Anisotropic massive Brans-Dicke gravity extension of the standard ΛCDM model. Eur. Phys. J. C Series **80**(1), 32 (2020). (arXiv:1903.06679 [gr-qc])
35. S. Joudaki, P.G. Ferreira, N.A. Lima, H.A. Winther, Testing gravity on cosmic scales: a case study of Jordan-Brans-Dicke theory. Phys. Rev. D Series **105**(4), 043522 (2022). (arXiv:2010.15278 [astro-ph.CO])
36. M. Ballardini, F. Finelli, D. Sapone, Cosmological constraints on Newton's gravitational constant. arXiv:2111.09168 [astro-ph.CO]
37. R. Weitzenböock, *Invariantentheorie* (Noordhoff, Gronningen, 1923)

38. S. Nesseris, S. Basilakos, E.N. Saridakis, L. Perivolaropoulos, Viable $f(T)$ models are practically indistinguishable from ΛCDM. Phys. Rev. D Series **88**, 103010 (2013). (arXiv:1308.6142 [astro-ph.CO])
39. R.C. Nunes, Structure formation in $f(T)$ gravity and a solution for H_0 tension. JCAP Series **05**, 052 (2018). (arXiv:1802.02281 [gr-qc])
40. R.C. Nunes, S. Pan, E.N. Saridakis, New observational constraints on f(T) gravity from cosmic chronometers. JCAP Series **08**, 011 (2016). (arXiv:1606.04359 [gr-qc])
41. C. Li, Y. Cai, Y.-F. Cai, E.N. Saridakis, The effective field theory approach of teleparallel gravity, $f(T)$ gravity and beyond. JCAP Series **10**, 001 (2018). (arXiv:1803.09818 [gr-qc])
42. S.-F. Yan, P. Zhang, J.-W. Chen, X.-Z. Zhang, Y.-F. Cai, E.N. Saridakis, Interpreting cosmological tensions from the effective field theory of torsional gravity. Phys. Rev. D Series **101**(12), 121301 (2020). (arXiv:1909.06388 [astro-ph.CO])
43. X. Ren, S.-F. Yan, Y. Zhao, Y.-F. Cai, E.N. Saridakis, Gaussian processes and effective field theory of $f(T)$ gravity under the H_0 tension. arXiv:2203.01926 [astro-ph.CO]
44. A.G. Riess, S. Casertano, W. Yuan, L.M. Macri, D. Scolnic, Large Magellanic cloud Cepheid standards provide a 1% foundation for the determination of the Hubble constant and stronger evidence for physics beyond ΛCDM. Astrophys. J. Series **876**(1), 85 (2019). (arXiv:1903.07603 [astro-ph.CO])
45. C.J. Copi, A.N. Davis, L.M. Krauss, A New nucleosynthesis constraint on the variation of G. Phys. Rev. Lett. Series **92**, 171301 (2004). (arXiv:astro-ph/0311334)
46. S. Nesseris, L. Perivolaropoulos, The limits of extended quintessence. Phys. Rev. D Series **75**, 023517 (2007). (arXiv:astro-ph/0611238)
47. Z. Chen, W. Luo, Y.-F. Cai, E.N. Saridakis, New test on general relativity and $f(T)$ torsional gravity from galaxy-galaxy weak lensing surveys. Phys. Rev. D Series **102**(10), 104044 (2020). (arXiv:1907.12225 [astro-ph.CO])
48. S.-F. Yan, C. Li, L. Xue, X. Ren, Y.-F. Cai, D.A. Easson, Y.-F. Yuan, H. Zhao, Testing the equivalence principle via the shadow of black holes. Phys. Rev. Res. Series **2**(2), 023164 (2020). (arXiv:1912.12629 [astro-ph.CO])
49. Y.-F. Cai, C. Li, E.N. Saridakis, L. Xue, $f(T)$ gravity after GW170817 and GRB170817A. Phys. Rev. D Series **97**(10), 103513 (2018). (arXiv:1801.05827 [gr-qc])
50. S. Basilakos, S. Nesseris, F.K. Anagnostopoulos, E.N. Saridakis, Updated constraints on $f(T)$ models using direct and indirect measurements of the Hubble parameter. JCAP Series **08**, 008 (2018). (arXiv:1803.09278 [astro-ph.CO])
51. M. Benetti, S. Capozziello, G. Lambiase, Updating constraints on f(T) teleparallel cosmology and the consistency with Big Bang nucleosynthesis. Mon. Not. Roy. Astron. Soc. Series **500**(2), 1795–1805 (2020). (arXiv:2006.15335 [astro-ph.CO])
52. R. Briffa, C. Escamilla-Rivera, J.L. Said, J. Mifsud, N.L. Pullicino, Impact of H_0 priors on $f(T)$ late time cosmology arXiv:2108.03853 [astro-ph.CO]
53. N. MacCrann, J. Zuntz, S. Bridle, B. Jain, M.R. Becker, Cosmic discordance: are Planck CMB and CFHTLenS weak lensing measurements out of tune? Mon. Not. Roy. Astron. Soc. Series **451**(3), 2877–2888 (2015). (arXiv:1408.4742 [astro-ph.CO])
54. R.A. Battye, T. Charnock, A. Moss, Tension between the power spectrum of density perturbations measured on large and small scales. Phys. Rev. D Series **91**(10), 103508 (2015). (arXiv:1409.2769 [astro-ph.CO])
55. M. Benetti, L.L. Graef, J.S. Alcaniz, The H_0 and σ_8 tensions and the scale invariant spectrum. JCAP Series **07**, 066 (2018). (arXiv:1712.00677 [astro-ph.CO])
56. S. Bahamonde, C.G. Böhmer, M. Wright, Modified teleparallel theories of gravity. Phys. Rev. D Series **92**(10), 104042 (2015). (arXiv:1508.05120 [gr-qc])
57. G. Farrugia, J. Levi Said, V. Gakis, E.N. Saridakis, Gravitational waves in modified teleparallel theories. Phys. Rev. D **97**(12), 124064 (2018) arXiv:1804.07365 [gr-qc]
58. C. Escamilla-Rivera, J. Levi Said, Cosmological viable models in $f(T, B)$ theory as solutions to the H_0 tension. Class. Quant. Grav. **37**(16), 165002 (2020) arXiv:1909.10328 [gr-qc]
59. G. Cognola, E. Elizalde, S. Nojiri, S.D. Odintsov, L. Sebastiani, S. Zerbini, A class of viable modified f(R) gravities describing inflation and the onset of accelerated expansion. Phys. Rev. D Series **77**, 046009 (2008). (arXiv:0712.4017 [hep-th])

60. R.C. Nunes, S. Pan, E.N. Saridakis, E.M.C. Abreu, New observational constraints on $f(R)$ gravity from cosmic chronometers. JCAP Series **01**, 005 (2017). (arXiv:1610.07518 [astro-ph.CO])
61. S.D. Odintsov, D. Sáez-Chillón Gómez, G.S. Sharov, Analyzing the H_0 tension in $F(R)$ gravity models. Nucl. Phys. B **966**, 115377 (2021) arXiv:2011.03957 [gr-qc]
62. W. Hu, I. Sawicki, Models of f(R) cosmic acceleration that evade solar-system tests. Phys. Rev. D Series **76**, 064004 (2007). (arXiv:0705.1158 [astro-ph])
63. A.A. Starobinsky, Disappearing cosmological constant in f(R) gravity. JETP Lett. Series **86**, 157–163 (2007). (arXiv:0706.2041 [astro-ph])
64. S. Tsujikawa, Observational signatures of $f(R)$ dark energy models that satisfy cosmological and local gravity constraints. Phys. Rev. D Series **77**, 023507 (2008). (arXiv:0709.1391 [astro-ph])
65. S. Basilakos, S. Nesseris, L. Perivolaropoulos, Observational constraints on viable f(R) parametrizations with geometrical and dynamical probes. Phys. Rev. D Series **87**(12), 123529 (2013). (arXiv:1302.6051 [astro-ph.CO])
66. A. Naruko, D. Yoshida, S. Mukohyama, Gravitational scalar–tensor theory. Class. Quant. Grav. **33**(9), 09LT01 (2016) arXiv:1512.06977 [gr-qc]
67. E.N. Saridakis, M. Tsoukalas, Cosmology in new gravitational scalar-tensor theories. Phys. Rev. D Series **93**(12), 124032 (2016). (arXiv:1601.06734 [gr-qc])
68. S. Banerjee, M. Petronikolou, E.N. Saridakis, Alleviating the H_0 tension with new gravitational scalar tensor theories. Phys. Rev. D Series **108**(2), 024012 (2023). (arXiv:2209.02426 [gr-qc])

Chapter 31
Early-Time Modified Gravity and the Hubble Tension

Matteo Braglia

31.1 Introduction

As of the current writing of this review, the value of the Hubble parameter obtained through distance ladder calibration of type Ia supernovae (SN Ia) from the Pantheon+ compilation [1] by the SH_0ES collaboration [2] stands at $H_0 = 73.0 \pm 1.0\,\mathrm{km\,s^{-1}\,Mpc^{-1}}$. In contrast, the one *inferred* by the cosmic microwave background (CMB) measurements from the *Planck* DR3 [3] is $H_0 = 67.36 \pm 0.54\,\mathrm{km\,s^{-1}\,Mpc^{-1}}$. This discrepancy reveals a 5σ tension between the two datasets. Importantly, the tension does not appear unique to the two datasets; rather, numerous studies suggest that it may be a generic tension between indirect, or early time, and direct, or late time, measurements of the Hubble parameter [4–9]. This has stimulated intense efforts in the community, on two fronts. On one hand, it is crucial to try and identify possible systematic effects that could plague such measurements. On the other, we should be open to the possibility that the H_0 tension genuinely hints at new Physics beyond the ΛCDM standard cosmological model. Within this review, we embrace the latter perspective, focusing on a particular form of novel Physics potentially indicated by the tension: Early time Modified Gravity (EMG).

Cosmological tests of deviations from General Relativity (GR) have garnered considerable attention (see e.g. the reviews [10, 11]), forming a crucial aspect of many large cosmological collaborations. Throughout scientific history, modified gravity effects have been invoked to account for observations that defy explanation within the standard ΛCDM framework. Consequently, it is unsurprising that various research groups have recently turned to modified gravity as a potential resolution to the Hubble tension. Such deviations from GR can manifest from the early universe to the present day. Attempts to address the H_0 tension have been made at both low and high

M. Braglia (✉)
Center for Cosmology and Particle Physics, New York University, 726 Broadway, New York, NY 10003, USA
e-mail: mb9289@nyu.edu

E. Di Valentino and D. Brout (eds.), *The Hubble Constant Tension*, Springer Series in Astrophysics and Cosmology, https://doi.org/10.1007/978-981-99-0177-7_31

redshifts, involving modifications to gravity around the times of matter-dark energy or radiation-matter equality, respectively. Our focus lies on the latter scenario.

The primary objective of this review is to provide an accessible explanation of how early-time modifications to gravity impact the comoving size of the sound horizon at baryon drag (r_s), creating a degeneracy with the H_0 value inferred from CMB data. Essentially, a modification to gravity can be understood as injecting energy into the cosmic fluid, thereby altering the Hubble parameter during the early universe [12]. This mechanism has been extensively explored in the literature as a potential solution to the Hubble tension, utilizing various phenomenological parameterizations [13–16] and models where both the background and the perturbed cosmological dynamics is fully worked out. The latter include Jordan-Brans-Dicke (JBD) models [17–23], models with nonminimal coupled scalar fields [22, 24–27] (sometimes also denoted as extended quintessence [28, 29]), Galileon-like models within Horndeski theory [30, 31], or the so-called *early modified gravity* (EMG) model presented in Refs. [32, 33],[1] but the list is far from comprehensive.

In this pursuit, we delve into how modifications to gravity during the early universe can either expand or contract the comoving sound horizon (r_s). While the core mechanism is universal across these models, the specific embodiment of the gravity modification holds paramount importance. Modern data possess the sensitivity required to scrutinize the cosmological evolution of different model dynamics, thus making model-building a central aspect.

Our review does not aim for exhaustive coverage of each individual model. Instead, we thoroughly examine the EMG model introduced in Refs. [32, 33], illustrating the aforementioned features. Within this model, a self-interacting scalar field non-minimally coupled to gravity evolves from the radiation era, driven by its coupling to non-relativistic matter. Subsequently, the scalar field enters a phase of coherent oscillations due to a quartic term in its potential, causing its energy density to dilute faster than non-relativistic matter after recombination. The reduction in scalar field amplitude naturally suppresses gravity modifications at late redshifts, aligning excellently with laboratory and Solar System tests of GR without necessitating screening mechanisms.

The review is structured as follows. In Sect. 31.2, we introduce the formalism of Scalar-Tensor (ST) theories, and discuss experimental constraints on deviations from GR. Then, in Sect. 31.3, we we delve into the EMG dynamics and explain the features that make it an appealing candidate to solve the H_0 tension. Finally we present some of the latest cosmological constraints and summarize their status in Sect. 31.4.

[1] Here, we are just adopting the nomenclature used in the literature. In fact, all the models listed posses some kind of non-minimal coupling (NMC) to gravity, as well as produce an early time modification to it.

31.2 Scalar-Tensor Theories and Experimental Constraints

In this paper, we will be dealing with ST theories where a scalar field σ is non-minimally coupled to the Ricci scalar as follows:

$$S = \int d^4x\sqrt{-g}\left[\frac{F(\sigma)}{2}R - \frac{g^{\mu\nu}}{2}\partial_\mu\sigma\,\partial_\nu\sigma - \Lambda - V(\sigma) + \mathscr{L}_m\right], \tag{31.1}$$

where $\mathscr{L}_m$ is the matter Lagrangian, V is the scalar field potential and Λ is the cosmological constant. The field equations derived from the variation of the action with respect to the field σ and the metric are the Klein-Gordon (KG) and Friedmann equations respectively and are given by

$$\ddot{\sigma} + 3H\dot{\sigma} + V_\sigma = \tfrac{1}{2}F_\sigma R, \tag{31.2}$$

$$3H^2F = \rho + \tfrac{\dot{\sigma}^2}{2} + V - 3H\dot{F} \equiv \rho + \rho_{\rm scf}. \tag{31.3}$$

It is clear that the model and its cosmological dynamics are then uniquely specified by the non-minimal coupling $F(\sigma)$, the potential $V(\sigma)$ and the initial conditions on σ deep in the radiation era. The form of $F(\sigma)$ is arbitrary, as long as it satisfy $F > 0$ and $F\,(2\,F + 3\,F_\sigma^2) > 0$, required to avoid negative kinetic terms in the quadratic action of tensor and scalar field perturbations respectively [34].

Comparing to the Einstein-Hilbert action, we immediately note that the Planck mass $M_{\rm pl}^2 \equiv 1/(8\pi G)$ is now a functional of σ, and therefore can vary with time according to the scalar field evolution. Specifically, one can define a *cosmological* Newton constant $G_N = 1/(8\pi F)$ and an effective Newton constant as [34, 35] $G_{\rm eff} = (2F + 4F_\sigma^2)/(2F + 3F_\sigma^2)(8\pi F)$. The latter is the quantity governing the interactions between test bodies.

Deviations from general relativity (GR) can also be parameterized by expanding the metric in powers of the gravitational potential Φ in the weak field regime as $ds^2 = -(1 + 2\Phi - 2\beta_{\rm PN}\Phi^2)dt^2 + (1 - 2\gamma_{\rm PN}\Phi)dx_i dx^i$, where we encapsulated modifications to gravity in the functions $\gamma_{\rm PN}$ and $\beta_{\rm PN}$. They are usually denoted as Post-Newtonian (PN) parameters [36] and, for the Lagrangian in Eq. (31.1), are given by $\gamma_{\rm PN} = 1 - F_{,\sigma}^2/(F + 2F_{,\sigma}^2)$ and $\beta_{\rm PN} = 1 + \frac{d\gamma_{\rm PN}}{d\sigma}FF_{,\sigma}/(8F + 12F_{,\sigma}^2)$. For GR we have $F_\sigma = 0$ so that both $\gamma_{\rm PN}$ and $\beta_{\rm PN}$ reduce to 1.

Before we discuss constraints on this class of theories from large scale cosmological data, let us use quickly summarize other experimental constraints on the quantities we have just introduced.

Big Bang Nucleosynthesis (BBN) constraints. The time at which weak and nuclear processes freeze out is extremely sensitive to the expansion rate H, which is in turn proportional to the *cosmological* Newton constant, see e.g. Eq. (31.3) (and also Eq. (31.4) in the next Section), and is therefore tightly constrained by BBN observations. The most up-to-date constraints on the variation of the gravitational constant G_N are $\Delta G_N = 0.02 \pm 0.06$ at 95% CL based on Ref. [37]. Let us note that this constraint is only valid at very early times, long before recombination.

Laboratory constraints. Constraints on the deviation from GR at later times can be obtained through laboratory experiments. Modern techniques used to constrain $G_{\rm eff}$ include for example atom interferometry and extremely precise torsion balances. Based on such measurements, the current value recommended by CODATA is $G = 6.67430(15) \times 10^{-11}$ N·m$^2 \cdot$ kg^{-2}, with a standard uncertainty of 22 ppm. Different experiments are in tension on the 4th figure after the comma, so there is currently an uncertainty (in units where $G = 1$) of 10^{-4} [36].

Solar System constraints. Finally, solar system tests set stringent bounds on the PN parameters at late times [36]. The tightest constraint $\gamma_{\rm PN} - 1 = (2.1 \pm 2.3) \times 10^{-5}$ was derived in Ref. [38] using the Doppler tracking of the Cassini spacecraft. Observations of the anomalous 43 arcseconds perihelion shift of Mercury's orbit are affected by a combination of $\gamma_{\rm PN}$ and $\beta_{\rm PN}$. Breaking this degeneracy using the bounds on $\gamma_{\rm PN}$ quoted above, Ref. [36] derived $\beta_{\rm PN} - 1 = (4.1 \pm 7.8) \times 10^{-5}$.

Hence, it becomes evident that although deviations from GR are tightly constrained at late times and within the scales of our Solar System, the constraints are notably less stringent when high redshifts. As we will demonstrate, these limitations are lenient to the extent that they permit significant dynamic variations in early times, which could potentially yield intriguing implications for the H_0 tension.

31.3 Dynamics of Early Modified Gravity

Let us delve into the dynamics of the EMG. For the moment being, let us refrain from making any assumptions on the function $F(\sigma)$. We will only require that the scalar field potential has a global minimum at $\sigma = 0$, and its second derivative is much smaller than the Hubble rate at very high redshift, before matter-radiation equality. If the latter condition is not met, the scalar field becomes dynamical at very early times and settles into its minimum long before recombination where the visibility function has no support. As a result there would be no imprints on the CMB anisotropies. Additionally, we will assume that the scalar field never dominates the total energy budget of the Universe. In other words, its effective energy density defined through $\rho_{\rm scf}$ in Eq. (31.3) is always much smaller than $\rho_{\rm crit}$.

With these assumptions in mind, the evolution of σ can be understood intuitively by examining the KG Eq. (31.2). At very early times $z \gg z_{\rm eq}$, the Universe is radiation dominated so that the Ricci scalar is extremely small, and the source term on the right hand side of (31.2) is negligible. Simultaneously, since the field is very light by assumption, the source term V_σ is also very small and the KG equation reduces to $\ddot{\sigma} + 3H\dot{\sigma} \simeq 0$. The Hubble friction thus keeps the scalar field frozen to its initial condition σ_i. This contributes to a shift in the cosmological Newton constant as $G_N(z \gg z_{\rm eq})/G_N(z = 0) \neq 1$. Furthermore, neglecting, the term $3H\dot{F}$ in Eq. (31.3), the NMC induces a relative displacement of the Hubble rate, compared to the standard ΛCDM evolution, as given by:

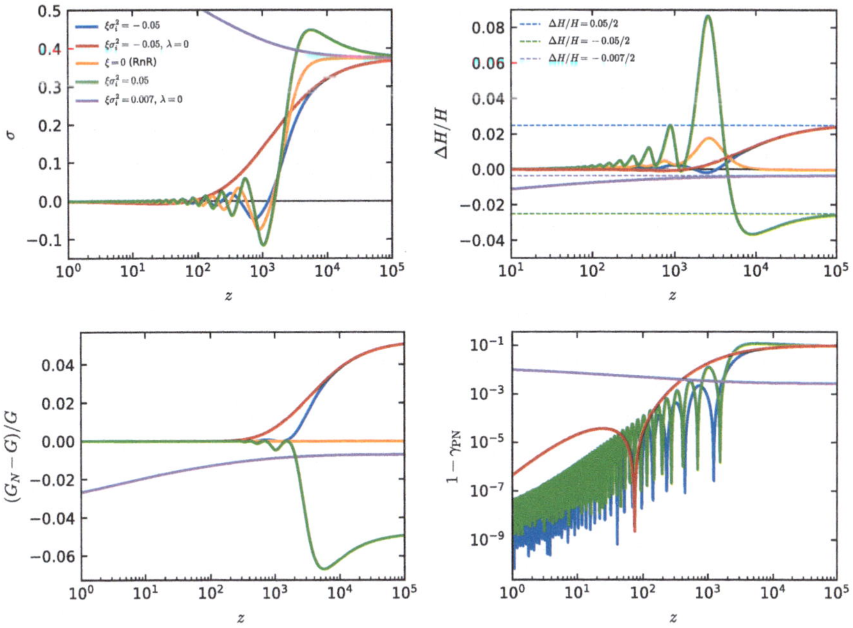

Fig. 31.1 For some illustrative parameters, we plot the evolution of the scalar field [top-left], the relative variation of H with respect to ΛCDM [top-right], the relative variation of the *cosmological* Newton constant [bottom-left] and the post-Newtonian parameter $\gamma_{\rm PN}$ [bottom-right]

$$\frac{H^{\rm EMG} - H^{\rm CDM}}{H^{\rm CDM}}(z \gg z_{\rm eq}) \sim \frac{M_{\rm pl}^2 - F(\sigma_i)}{2}, \tag{31.4}$$

as depicted in the top-right panel of Fig. 31.1. Equation (31.4) is crucial in comprehending the potential of ST models in addressing the H_0 tension. These models have the ability to decrease ($\xi > 0$) or increase ($\xi < 0$) the Hubble rate at early times with a corresponding increase/decrease in the comoving size of the sound horizon $r_s = \int_{z_{\rm dec}}^{\infty} {\rm d}z\, c_s(z)/H(z)$ respectively, which is highly degenerate with H_0 [12].

As we approach matter-radiation equality, the Ricci scalar is no longer negligible, causing the source term $-F_\sigma R$ on the right hand side of the KG equation to gradually activate. This drives σ to larger/smaller values for positive/negative couplings respectively. At some critical redshift, that we denote by z_c, the second derivative of the potential becomes comparable to the Hubble rate, and the potential source term can no longer be neglected, eventually dominating the dynamics of the scalar field.

Thus far, we have presented the general feature of the class of theories in Eq. (31.1). To gain a deeper understanding, let us now specify the functional form for the NMC and the potential. As illustrative example, we consider the EMG model introduced in Ref. [32], corresponding to the choice

$$F(\sigma) = M_{\rm pl}^2 + \xi\sigma^2, \tag{31.5}$$

$$V(\sigma) = \lambda\sigma^4/4, \tag{31.6}$$

where ξ and λ are dimensionless constants[2]. It is noteworthy that the minimally coupled limit, i.e. the $\xi = 0$ limit, of this model corresponds to the RnR EDE model introduced in Ref. [39]. To illustrate the cosmological evolution of our model, we fix the initial conditions on the scalar fields to[3] $\sigma_i = 0.375$, $\dot{\sigma}_i = 0$, and, for a handful of representative combinations of ξ and λ, numerically integrate the background KG and Friedmann equations using the Einstein-Boltzmann code `hi_class` [40–43]. The results are displayed in Fig. 31.1. We now turn to comment each cases separately.

Massless case ($\lambda = 0$), **positive coupling** ($\xi > 0$). This case, represented by the purple line in Fig. 31.1, was examined in Refs. [22, 24, 25]. As we approach $z_{\rm eq}$ the scalar field begins to thaw, and grows until today, leading to a deviation of G_N and the PN parameters from their observed value. This results in a departure from the late-time ΛCDM standard cosmological evolution and fails to adhere to Solar System constraints. In principle, we could get around the former problem by promoting $M_{\rm pl}$ in Eq. (31.5) to a free parameter (see e.g. Ref. [14, 24]), while the second issuc must be tackled by incorporating a screening mechanism. We will not discuss this case further in the following.

Massless case ($\lambda = 0$), **negative coupling** ($\xi < 0$). This case was studied in Refs. [24–27], and is more interesting due to several reasons. Because the potential is absent, the scalar field is entirely driven by the coupling to non- relativistic matter around the radiation-matter equality era as before. However, unlike the positive coupling case, it decreases towards $\sigma = 0$ and its energy density dilutes away as fast or faster than radiation, reducing r_s in a similar way to extra relativistc degrees of freedom [44–46], at least for what concerns the background evolution [25, 26]. Notably, unlike popular EDE models, which requires fine tuning in both the initial condition σ_i and the value of the potential λ (see also next bullet point), here the scalar field starts to roll in a natural way after radiation-matter equality, driven by its NMC to gravity.[4] Another positive feature of the model, as can be seen from Fig. 31.1, is the consistency with late time Solar System constraints, unlike most models of modified gravity that necessitate screening mechanisms [11].

General case ($\lambda \neq 0$). This is the most general case, where the potential intervenes around the critical redshift z_c at which the second derivative of the potential becomes comparable to the Hubble rate. This makes the evolution of σ resemble the one of EDE models ($\xi = 0$ case in the figure) where the scalar fields undergoes damped oscillations around the minimum at $\sigma = 0$ and is diluted much faster than radiation, and injects a large and sharp (in time) amount of energy into the cosmic

---

[2] For convenience, we define V_0 through $\lambda = 10^{2V_0}/(3.516 \times 10^{109})$ as in Ref. [32].

[3] The condition $\dot{\sigma}_i = 0$ is not restricting, as any non-zero $\dot{\sigma}_i$ would be immediately damped by Hubble friction.

[4] As discussed in Refs. [24, 25] and further emphasized in Ref. [27], the invariance of the theory under conformal transformations could also justify setting $\xi = -1/6$, leading to a minimal extension of the ΛCDM with only 1 extra parameter.

fluid. However, the NMC produces some notable differences. First of all, as already discussed, the scalar field starts to roll, driven by the NMC, before the potential kicks in. Furthermore, not only do we have a sharp energy injection around z_c, but also a constant constribution to $\Delta H/H$, as discussed above. Interestingly, the faster dilution results in a more efficient decay of σ, suppressing even more $G_{N,\text{eff}}$ and the PN parameters at late times.

Finally, let us stress that the models considered above provide only a glimpse into early modifications to gravity. The interplay between the NMC to gravity and the comoving sound horizon, and thus H_0, can be observed in various other models [19–23, 30, 31, 47, 48]. However, unilke the model detailed in this Section, not all of these alternatives are consistent with late time Solar System or laboratory constraints.

31.4 Current Constraints on EMG and Relation to the H_0 Tension

In this chapter, we delve into the cosmological constraints that shed light on the modified gravity theories introduced earlier. We will start by offering a detailed account of the constraints applied to the Early Time Modified Gravity (EMG) model, and subsequently, we will juxtapose the inferred value of H_0 with other models discussed in Sects. 31.1–31.3.

Cosmological Constraints on the EMG Model. We present the cosmological constraints on the EMG model using a variety of datasets. The triangle plot in Fig. 31.2 provides a comprehensive view of these constraints. Notably, due to the intrinsic degeneracy between the non-minimal coupling $\xi\sigma_i^2$, which quantifies $\Delta H/H$ prior to recombination (see Eq. (31.4)) at early times, and the comoving sound horizon r_s, the Hubble parameter derived from *Planck* data assumes a value of $H_0 = 68.58^{+0.83}_{-1.3}$, km, s^{-1}, Mpc^{-1} at the 95% confidence level (CL) [33]. This value is higher than the H_0 value inferred from the standard ΛCDM model. However, it is important to note that *Planck* can only establish an upper limit on the $\xi\sigma_i^2$ so that, no definitive detection of modifications to gravity can be made solely from *Planck* data.

An intriguing development arises when ACT data is included. Although this inclusion does not alter the posterior distribution of H_0, it does lead to the detection of a modification to gravity at a 95% CL, specifically $\xi\sigma_i^2 = 0.033 \pm 0.014$. As highlighted in Ref. [33], taking both the data fit and the additional model parameters into account, the EMG model becomes favored over the ΛCDM model at a significance of $\gtrsim 2\sigma$ using the Aikike criterion. This preference is prominently driven by a feature observed in the Temperature-Temperature (TT) angular power spectra of both *Planck* and ACT around $\ell \sim 750$, which the EMG model adeptly fits. Interestingly, while ACT hints at other popular approaches aimed at resolving the H_0 tension, such as Axion Early Dark Energy (EDE) models [57, 58], the addition of SPT data seems to contradict those models, in contrast with the EMG model which is still favored by data.

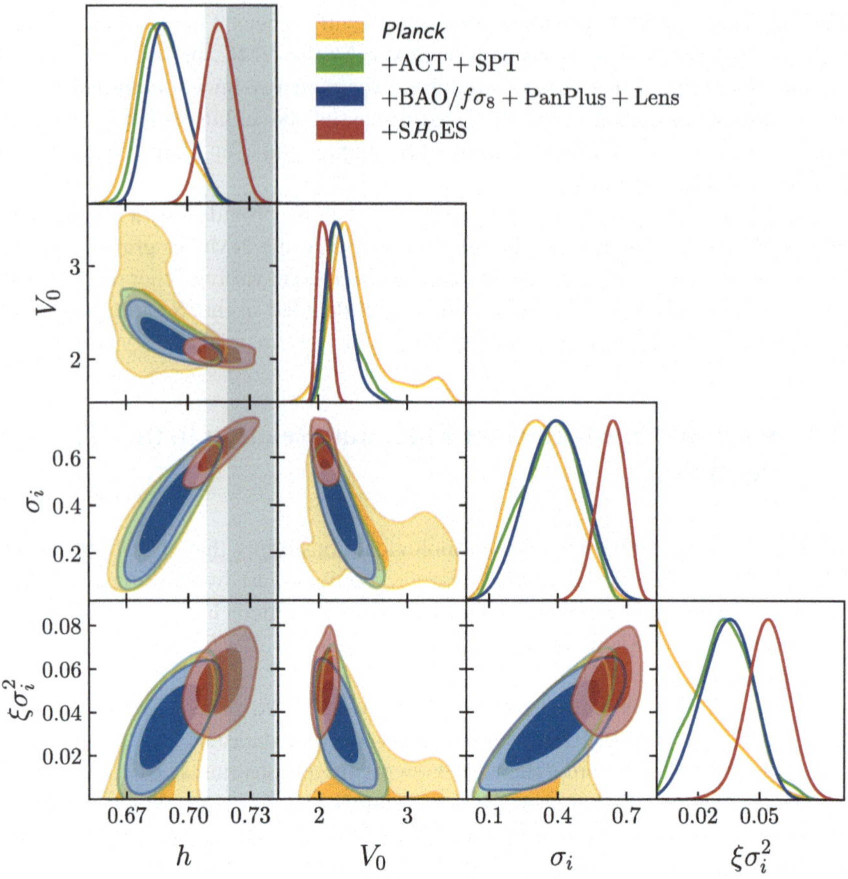

Fig. 31.2 1D and 2D posterior distributions (68% and 95% C.L.) for H_0 and the EMG model parameters. The data used include TT, TE, EE data from *Planck* DR3 [49], ACT DR4 [50] and SPT [51], lensing potential reconstructed from *Planck* DR3 [52], measurements of the BAO and redshift space distortion '$f\sigma_8$' from the CMASS and LOWZ galaxy samples of BOSS DR12 at $z = 0.38, 0.51$, and 0.61 [53] and the BAO measurements from 6dFGS at $z = 0.106$ and SDSS DR7 at $z = 0.15$ [54, 55], information about the luminosity distance to over 1550 SN Ia in the redshift range $0.001 < z < 2.3$ [56], and finally the SH0ES determination of the SN Ia magnitude $M_b = -19.253 \pm 0.027$ [2] modeled as a Gaussian likelihood. See Ref. [33] for a detailed description

The triangle plots in the figure highlight that incorporating external (non-CMB) datasets minimally impacts the constraints on model parameters. Only when a Gaussian prior on the Supernova Type Ia (SN Ia) magnitude (M_b) is included, a significant shift is observed due to the correlation between the EMG energy injection and H_0.

Furthermore, we quantify the residual tension with the SH0ES collaboration. Utilizing the $Q_{\rm DMAP}$ tension metric introduced in [6, 59], we observe a substantial reduction in the tension on H_0, from 6.4σ in the ΛCDM model to 3.3σ in the EMG

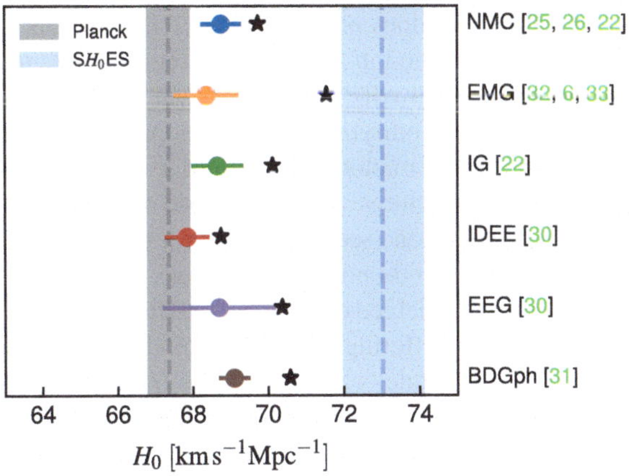

Fig. 31.3 Constraints on H_0 from P18+BAO data (error-bars) and mean values of H_0 (black stars) for some models involving early times modifications to gravity. Constraints are taken from the references cited in the legend. The black stars come from analyses involving priors on H_0 from the SH0ES collaboration. We warn the reader that different models were analyzed with priors on H_0 that reflect the time of publication of the papers cited in the figure

model.[5] While the EMG model does not outright resolve the tension, it does play a role in alleviating it.

The results are intriguing, prompting further investigation and testing of the model's robustness against forthcoming cosmological data releases. This is particularly crucial given the current marginal significance of the detection.

Model Comparison. We depict the estimated H_0 values for the models discussed in this review in Fig. 31.3, and we compare them with the EMG model. As access to constraints from CMB data alone for all models was not available, we present the estimated values for the *Planck*+BAO data combination, which generally yields slightly lower values of H_0 than the *Planck*-only data, as BAO are highly constraining. Evidently, models that introduce modifications to gravity before or around recombination consistently predict higher values of H_0 in comparison to the ΛCDM model, because of the degeneracies outlined in Sect. 31.3.

Additionally, we incorporate information from the SH0ES measurement in the form of a H_0 prior or an M_b prior, represented by the black stars in the figure. However, we must caution the reader about the differing publication dates of the cited papers, leading to distinct priors for each model. Consequently, a direct comparison becomes challenging. Nevertheless, it is evident that the EMG model, as introduced in Ref. [32], exhibits promising potential in addressing the H_0 tension.

In summary, this review has delved into the cosmological constraints of the EMG model and compared its predictions for H_0 with other early-time modified gravity

[5] This estimate of the tension in ΛCDM, quoted in Ref. [33], is larger than what is reported by the SH0ES collaboration [2] due to the inclusion of new CMB data and the use of the full Pantheon+ dataset [56].

models. While the EMG model does not fully resolve the tension, the mechanism addressing the H_0 tension explained in this review offers valuable insights and lays the basis for studying modified gravity models that can finally resolve the H_0 tension. Furthermore, we note that modifying the gravitational strength certainly impact the efficiency of the clustering of matter perturbations as well, with non-trivial consequences on the growth of structures. This could potentially offer a lever arm on addressing the so-called S_8 tension (see e.g. [60] for a review), which the models described here fail to address (but do not worsen either).

In conclusion, this review has delved extensively into the cosmological constraints surrounding the EMG model, offering a thorough understanding of its implications. By comparing the EMG model's predictions for H_0 with those of other early-time modified gravity models, we gain valuable insights into the intricate interplay between gravity modifications and cosmological observations. Although the EMG model does not entirely solve the H_0 tension, the mechanism elucidated in this review lays a foundational understanding for exploring modified gravity models that possess the potential to ultimately resolve this long-standing discrepancy.

References

1. A. Carr, T.M. Davis, D. Scolnic, D. Scolnic, K. Said, D. Brout, E.R. Peterson, R. Kessler, Publ. Astron. Soc. Austral. **39**, e046 (2022). https://doi.org/10.1017/pasa.2022.41 [arXiv:2112.01471 [astro-ph.CO]]
2. A.G. Riess, W. Yuan, L.M. Macri, D. Scolnic, D. Brout, S. Casertano, D.O. Jones, Y. Murakami, L. Breuval, T.G. Brink et al., Astrophys. J. Lett. **934**(1), L7 (2022). https://doi.org/10.3847/2041-8213/ac5c5b [arXiv:2112.04510 [astro-ph.CO]]
3. N. Aghanim et al., [Planck], Astron. Astrophys. **641**, A6 (2020) [erratum: Astron. Astrophys. **652**, C4 (2021)]. https://doi.org/10.1051/0004-6361/201833910[arXiv:1807.06209 [astro-ph.CO]]
4. E. Di Valentino, L.A. Anchordoqui, O. Akarsu, Y. Ali-Haimoud, L. Amendola, N. Arendse, M. Asgari, M. Ballardini, S. Basilakos, E. Battistelli et al., Astropart. Phys. **131**, 102605 (2021). https://doi.org/10.1016/j.astropartphys.2021.102605 [arXiv:2008.11284 [astro-ph.CO]]
5. E. Di Valentino, O. Mena, S. Pan, L. Visinelli, W. Yang, A. Melchiorri, D.F. Mota, A.G. Riess, J. Silk, Class. Quant. Grav. **38**(15), 153001 (2021). https://doi.org/10.1088/1361-6382/ac086d [arXiv:2103.01183 [astro-ph.CO]]
6. N. Schöneberg, G. Franco Abellán, A. Pérez Sánchez, S.J. Witte, V. Poulin, J. Lesgourgues, Phys. Rept. **984**, 1-55 (2022). https://doi.org/10.1016/j.physrep.2022.07.001 [arXiv:2107.10291 [astro-ph.CO]]
7. P. Shah, P. Lemos, O. Lahav, Astron. Astrophys. Rev. **29**(1), 9 (2021). https://doi.org/10.1007/s00159-021-00137-4 [arXiv:2109.01161 [astro-ph.CO]]
8. E. Abdalla, G. Franco Abellán, A. Aboubrahim, A. Agnello, O. Akarsu, Y. Akrami, G. Alestas, D. Aloni, L. Amendola, L.A. Anchordoqui et al., JHEAp **34**, 49-211 (2022).https://doi.org/10.1016/j.jheap.2022.04.002 [arXiv:2203.06142 [astro-ph.CO]]
9. L. Verde, T. Treu, A.G. Riess, Nature Astron. **3**, 891. https://doi.org/10.1038/s41550-019-0902-0 [arXiv:1907.10625 [astro-ph.CO]]
10. T. Clifton, P.G. Ferreira, A. Padilla, C. Skordis, Phys. Rept. **513**, 1-189 (2012). https://doi.org/10.1016/j.physrep.2012.01.001 [arXiv:1106.2476 [astro-ph.CO]]
11. K. Koyama, Rept. Prog. Phys. **79**(4), 046902 (2016). https://doi.org/10.1088/0034-4885/79/4/046902 [arXiv:1504.04623 [astro-ph.CO]]

12. L. Knox, M. Millea, Phys. Rev. D **101**(4), 043533 (2020). https://doi.org/10.1103/PhysRevD.101.043533 [arXiv:1908.03663 [astro-ph.CO]]
13. M.X. Lin, M. Raveri, W. Hu, Phys. Rev. D **99**(4), 043514 (2019). https://doi.org/10.1103/PhysRevD.99.043514 [arXiv:1810.02333 [astro-ph.CO]]
14. M. Ballardini, F. Finelli, D. Sapone, JCAP **06**(06), 004 (2022). https://doi.org/10.1088/1475-7516/2022/06/004 [arXiv:2111.09168 [astro-ph.CO]]
15. Z. Sakr, D. Sapone, JCAP **03**(03), 034 (2022). https://doi.org/10.1088/1475-7516/2022/03/034 [arXiv:2112.14173 [astro-ph.CO]]
16. G. Benevento, J.A. Kable, G.E. Addison, C.L. Bennett, Astrophys. J. **935**(2), 156 (2022). https://doi.org/10.3847/1538-4357/ac80fd [arXiv:2202.09356 [astro-ph.CO]]
17. P. Jordan, Nature **164**, 637–640 (1949). https://doi.org/10.1038/164637a0
18. C. Brans, R.H. Dicke, Phys. Rev. **124**, 925–935 (1961). https://doi.org/10.1103/PhysRev.124.925
19. C. Umiltà, M. Ballardini, F. Finelli, D. Paoletti, JCAP **08**, 017 (2015). https://doi.org/10.1088/1475-7516/2015/08/017 [arXiv:1507.00718 [astro-ph.CO]]
20. M. Ballardini, F. Finelli, C. Umiltà, D. Paoletti, JCAP **05**, 067 (2016). https://doi.org/10.1088/1475-7516/2016/05/067 [arXiv:1601.03387 [astro-ph.CO]]
21. J. Solà Peracaula, A. Gomez-Valent, J. de Cruz Pérez, C. Moreno-Pulido, Astrophys. J. Lett. **886**(1), L6 (2019). https://doi.org/10.3847/2041-8213/ab53e9[arXiv:1909.02554 [astro-ph.CO]]
22. M. Ballardini, M. Braglia, F. Finelli, D. Paoletti, A.A. Starobinsky, C. Umiltà, JCAP **10**, 044 (2020). https://doi.org/10.1088/1475-7516/2020/10/044 [arXiv:2004.14349 [astro-ph.CO]]
23. M. Ballardini, A.G. Ferrari, F. Finelli, JCAP **04**, 029 (2023). https://doi.org/10.1088/1475-7516/2023/04/029 [arXiv:2302.05291 [astro-ph.CO]]
24. M. Rossi, M. Ballardini, M. Braglia, F. Finelli, D. Paoletti, A.A. Starobinsky, C. Umiltà, Phys. Rev. D **100**(10), 103524 (2019). https://doi.org/10.1103/PhysRevD.100.103524[arXiv:1906.10218 [astro-ph.CO]]
25. M. Braglia, M. Ballardini, W.T. Emond, F. Finelli, A.E. Gumrukcuoglu, K. Koyama, D. Paoletti, Phys. Rev. D **102**(2), 023529 (2020). https://doi.org/10.1103/PhysRevD.102.023529[arXiv:2004.11161 [astro-ph.CO]]
26. G. Ballesteros, A. Notari, F. Rompineve, JCAP **11**, 024 (2020). https://doi.org/10.1088/1475-7516/2020/11/024 [arXiv:2004.05049 [astro-ph.CO]]
27. T. Adi, E.D. Kovetz, Phys. Rev. D **103**(2), 023530 (2021). https://doi.org/10.1103/PhysRevD.103.023530 [arXiv:2011.13853 [astro-ph.CO]]
28. F. Perrotta, C. Baccigalupi, S. Matarrese, Phys. Rev. D **61**, 023507 (1999). https://doi.org/10.1103/PhysRevD.61.023507 [arXiv:astro-ph/9906066 [astro-ph]]
29. N. Bartolo, M. Pietroni, Phys. Rev. D **61**, 023518 (2000). https://doi.org/10.1103/PhysRevD.61.023518 [arXiv:hep-ph/9908521 [hep-ph]]
30. M. Zumalacarregui, Phys. Rev. D **102**(2), 023523 (2020). https://doi.org/10.1103/PhysRevD.102.023523 [arXiv:2003.06396 [astro-ph.CO]]
31. A.G. Ferrari, M. Ballardini, F. Finelli, D. Paoletti, N. Mauri, [arXiv:2307.02987 [astro-ph.CO]]
32. M. Braglia, M. Ballardini, F. Finelli, K. Koyama, Phys. Rev. D **103**(4), 043528 (2021). https://doi.org/10.1103/PhysRevD.103.043528 [arXiv:2011.12934 [astro-ph.CO]]
33. G. Franco Abellán, M. Braglia, M. Ballardini, F. Finelli, V. Poulin, [arXiv:2308.12345 [astro-ph.CO]]
34. R. Gannouji, D. Polarski, A. Ranquet, A.A. Starobinsky, JCAP **09**, 016 (2006). https://doi.org/10.1088/1475-7516/2006/09/016 [arXiv:astro-ph/0606287 [astro-ph]]
35. B. Boisseau, G. Esposito-Farese, D. Polarski, A.A. Starobinsky, Phys. Rev. Lett. **85**, 2236 (2000). https://doi.org/10.1103/PhysRevLett.85.2236 [arXiv:gr-qc/0001066 [gr-qc]]
36. C.M. Will, Living Rev. Rel. **17**, 4 (2014). https://doi.org/10.12942/lrr-2014-4 [arXiv:1403.7377 [gr-qc]]
37. J. Alvey, N. Sabti, M. Escudero, M. Fairbairn, Eur. Phys. J. C **80**(2), 148 (2020). https://doi.org/10.1140/epjc/s10052-020-7727-y [arXiv:1910.10730 [astro-ph.CO]]

38. B. Bertotti, L. Iess, P. Tortora, Nature **425**, 374–376 (2003). https://doi.org/10.1038/nature01997
39. P. Agrawal, F.Y. Cyr-Racine, D. Pinner, L. Randall, [arXiv:1904.01016 [astro-ph.CO]]
40. J. Lesgourgues, [arXiv:1104.2932 [astro-ph.IM]]
41. D. Blas, J. Lesgourgues, T. Tram, JCAP **07**, 034 (2011). https://doi.org/10.1088/1475-7516/2011/07/034 [arXiv:1104.2933 [astro-ph.CO]]
42. M. Zumalacárregui, E. Bellini, I. Sawicki, J. Lesgourgues, P.G. Ferreira, JCAP **08**, 019 (2017). https://doi.org/10.1088/1475-7516/2017/08/019 [arXiv:1605.06102 [astro-ph.CO]]
43. E. Bellini, I. Sawicki, M. Zumalacárregui, JCAP **02**, 008 (2020). https://doi.org/10.1088/1475-7516/2020/02/008 [arXiv:1909.01828 [astro-ph.CO]]
44. A.G. Riess, L. Macri, S. Casertano, H. Lampeitl, H.C. Ferguson, A.V. Filippenko, S.W. Jha, W. Li, R. Chornock, Astrophys. J. **730**, 119 (2011) [erratum: Astrophys. J. **732**, 129 (2011)]. https://doi.org/10.1088/0004-637X/732/2/129 [arXiv:1103.2976 [astro-ph.CO]]
45. M. Wyman, D.H. Rudd, R.A. Vanderveld, W. Hu, Phys. Rev. Lett. **112**, no.5, 051302 (2014). https://doi.org/10.1103/PhysRevLett.112.051302 [arXiv:1307.7715 [astro-ph.CO]]
46. J.L. Bernal, L. Verde, A.G. Riess, JCAP **10**, 019 (2016). https://doi.org/10.1088/1475-7516/2016/10/019 [arXiv:1607.05617 [astro-ph.CO]]
47. J. Solà Peracaula, A. Gómez-Valent, J. de Cruz Pérez, C. Moreno-Pulido, Class. Quant. Grav. **37**(24), 245003 (2020). https://doi.org/10.1088/1361-6382/abbc43 [arXiv:2006.04273 [astro-ph.CO]]
48. S. Joudaki, P.G. Ferreira, N.A. Lima, H.A. Winther, Phys. Rev. D **105**(4), 043522 (2022). https://doi.org/10.1103/PhysRevD.105.043522 [arXiv:2010.15278 [astro-ph.CO]]
49. N. Aghanim et al., [Planck], Astron. Astrophys. **641**, A5 (2020). https://doi.org/10.1051/0004-6361/201936386 [arXiv:1907.12875 [astro-ph.CO]]
50. S.K. Choi et al., [ACT], JCAP **12**, 045 (2020). https://doi.org/10.1088/1475-7516/2020/12/045 [arXiv:2007.07289 [astro-ph.CO]]
51. L. Balkenhol et al., [SPT-3G], Phys. Rev. D **108**(2,) 023510 (2023). https://doi.org/10.1103/PhysRevD.108.023510 [arXiv:2212.05642 [astro-ph.CO]]
52. N. Aghanim et al., [Planck], Astron. Astrophys. **641**, A8 (2020). https://doi.org/10.1051/0004-6361/201833886 [arXiv:1807.06210 [astro-ph.CO]]
53. S. Alam et al., [BOSS], Mon. Not. Roy. Astron. Soc. **470**(3), 2617–2652 (2017). https://doi.org/10.1093/mnras/stx721 [arXiv:1607.03155 [astro-ph.CO]]
54. F. Beutler, C. Blake, M. Colless, D.H. Jones, L. Staveley-Smith, L. Campbell, Q. Parker, W. Saunders, F. Watson, Mon. Not. Roy. Astron. Soc. **416**, 3017–3032 (2011). https://doi.org/10.1111/j.1365-2966.2011.19250.x [arXiv:1106.3366 [astro-ph.CO]]
55. A.J. Ross, L. Samushia, C. Howlett, W.J. Percival, A. Burden, M. Manera, Mon. Not. Roy. Astron. Soc. **449**(1), 835-847 (2015). https://doi.org/10.1093/mnras/stv154 [arXiv:1409.3242 [astro-ph.CO]]
56. D. Brout, D. Scolnic, B. Popovic, A.G. Riess, J. Zuntz, R. Kessler, A. Carr, T.M. Davis, S. Hinton, D. Jones et al., Astrophys. J. **938**(2), 110 (2022). https://doi.org/10.3847/1538-4357/ac8e04 [arXiv:2202.04077 [astro-ph.CO]]
57. V. Poulin, T.L. Smith, T. Karwal, M. Kamionkowski, Phys. Rev. Lett. **122**(22), 221301 (2019). https://doi.org/10.1103/PhysRevLett.122.221301 [arXiv:1811.04083 [astro-ph.CO]]
58. T.L. Smith, M. Lucca, V. Poulin, G.F. Abellan, L. Balkenhol, K. Benabed, S. Galli, R. Murgia, Phys. Rev. D **106**(4), 043526 (2022). https://doi.org/10.1103/PhysRevD.106.043526 [arXiv:2202.09379 [astro-ph.CO]]
59. M. Raveri, W. Hu, Phys. Rev. D **99**(4), 043506 (2019). https://doi.org/10.1103/PhysRevD.99.043506 [arXiv:1806.04649 [astro-ph.CO]]
60. E. Di Valentino, L.A. Anchordoqui, Ö. Akarsu, Y. Ali-Haimoud, L. Amendola, N. Arendse, M. Asgari, M. Ballardini, S. Basilakos, E. Battistelli et al., Astropart. Phys. **131**, 102604 (2021). https://doi.org/10.1016/j.astropartphys.2021.102604 [arXiv:2008.11285 [astro-ph.CO]]

Chapter 32
Primordial Magnetic Fields and the Hubble Tension

Karsten Jedamzik and Levon Pogosian

32.1 Introduction

The statistical significance of the Hubble tension, currently just over 5σ, is primarily driven by the difference between the SH_0ES measurement of $H_0 = 73.04 \pm 1.04$ km/s/Mpc using Cepheid calibrated supernova [1] and the $H_0 = 67.36 \pm 0.54$ km/s/Mpc obtained from fitting the ΛCDM model to the Planck CMB data [2]. Other independent measurements tend to re-enforce the tension [3], with a clear trend of all measurements that do not rely on a model of recombination giving a higher H_0, in the 69-73 km/s/Mpc range [4–9], and estimates that use the standard treatment of recombination giving H_0 of around 67–68 km/s/Mpc [10–12]. This trend points to a missing ingredient in the standard description of the universe prior and/or during recombination—something that would help reduce the sound horizon at last scattering, $r_\star$. One such ingredient could be a stochastic primordial magnetic field (PMF) embedded in the plasma and generating inhomogeneities in the baryon distribution, or baryon clumping [13, 14]. The inhomogeneities make the recombination complete faster [13, 15], resulting in a smaller $r_\star$.

At the time of writing this Chapter, it remains an open question whether PMFs will play a key role in resolving the Hubble tension. Preliminary investigations [16–20], based on a toy-model of baryon clumping [13], have shown that cosmological data is broadly compatible with the presence of PMFs, and they could help relieve the tension to a certain extent. Moreover, the required strength of the PMF appears to be of just the right order of magnitude to help explain all observed galactic, cluster and intergalactic magnetic fields. Needless to say that, if confirmed, this would be a truly exciting

K. Jedamzik (✉)
Laboratoire de Univers et Particules de Montpellier, UMR5299-CNRS, Universite de Montpellier, 34095 Montpellier, France
e-mail: karsten.jedamzik@umontpellier.fr

L. Pogosian (✉)
Department of Physics, Simon Fraser University, Burnaby, BC V5A 1S6, Canada
e-mail: levon@sfu.ca

E. Di Valentino and D. Brout (eds.), *The Hubble Constant Tension*, Springer Series in Astrophysics and Cosmology, https://doi.org/10.1007/978-981-99-0177-7_32

development, solving two puzzles at the same time. Achieving a higher level of certainty requires obtaining the exact ionization history from compressible magneto-hydrodynamics (MHD) simulations including the effects of radiation transport that are currently underway.

In what follows, we start by providing an overview of cosmological magnetic fields, their possible primordial origin and evolution, and their observational signatures. We then describe the physics of recombination in the presence of magnetic fields, introducing the simple model of baryon clumping used in the existing studies. After presenting the current observational status of this model, we discuss the current state of the MHD simulations and prospects for the future.

32.2 Primordial Magnetic Fields and Their Observational Signatures

The PMF hypothesis was first proposed as a possible explanation of the origin of galactic magnetic fields [21, 22]. All galaxies, irrespective of their type or age, appear to contain a magnetic field of ~micro-Gauss (μG) strength that is coherent over the extent of the galaxy. Clusters of galaxies are also known to have magnetic fields of similar strength. The origin of the galactic and cluster fields is not entirely understood [23–25]. It may well be that initially small fields of strength of $\sim 10^{-20}$–10^{-18} G excited by the Biermann battery operating at first structure formation are subsequently amplified by the small-scale and/or large-scale dynamo to reach approximate equipartition of magnetic energy with the turbulent energy in collapsed structures (see [26] for a review on astrophysical dynamos). Interestingly, μG strength fields are observed in high redshift galaxies that would be too young to have gone through the number of revolutions necessary for the large-scale dynamo to work [27]. Still, there is a possibility that the supernova explosions in protogalaxies provided the magnetic seed fields that were later amplified by compression, shearing and stochastic motions [28, 29]. An alternative explanation is that magnetic fields emerged from the evolution of the early Universe, having been generated either during cosmic phase transitions (e.g. the electroweak transition) [30] or during inflation [31, 32] (see also [23, 25, 33, 34]) for reviews).

The pre-recombination universe was fully ionized and, if a PMF was ever generated, dissipation due to magnetic diffusion would be negligible. Dissipation due to the kinetic viscosity damping fluid motions induced by the PMF on small scales would, however, significantly drain the energy of the PMF [35, 36] over the many e-folds of cosmic expansion between the very early Universe and the recombination era. Notwithstanding this fact, if only the smallest fraction $\sim 10^{-10}$ of a PMF initially in equipartition with the radiation would survive, it could be sufficient to eliminate the need for dynamo amplification altogether. As discussed below, it could also leave observable signatures in the CMB.

While PMFs have been a subject of continuous study over many decades, there were commonly considered to be unnecessary because less exotic astrophysical explanations could not be ruled out. The interest in PMFs was renewed when evidence emerged of magnetic fields in ∼Mpc size voids in the intergalactic space. Observations of TeV blazars by Hess and Fermi [37–39] have led to the surprising conclusion that magnetic fields exist in the extra-galactic medium between galaxies with a very large volume filling factor [40]. TeV gamma-rays emitted by the blazar may pair-produce on the diffuse extra-galactic background star light. The resulting electron-positron pairs will subsequently Compton scatter on CMB photons converting them to GeV gamma-rays. Observations of blazars have not found these GeV photons with the flux expected. Though other less understood explanations exist [41], the most straightforward explanation is that the electron-positron pairs were deflected by magnetic fields out of the light cone. A lower limit on the field strength of $B \gtrsim 3 \times 10^{-15}$G (assuming a coherence scale of ∼kpc) could be inferred. While adding to the case for PMFs, it is not ruled out that such fields could be the result of outflows from galaxies, filling essentially all space with magnetic fields. The synchrotron emission from a few mega-parsec (Mpc) long ridge connecting two merging clusters of galaxies is another observation that is well-matched by simulations based on the PMF hypothesis [42].

The question of the origin of the magnetic fields in the universe is difficult to settle without the help of further observations. It is hoped that blazar gamma-ray observations by the future Cherenkov Telescope Array mission [43] may raise the lower limit to the ∼3–10 pico-gauss (pG) range [44]. However, the only way to be certain of the primordial origin of the observed fields is to find their signatures in the CMB. Since the PMF survives Silk damping [35, 36], magnetosonic modes on large scales, ∼1–10 Mpc, may lead to additional power in the high ℓ tail of the CMB temperature anisotropy spectrum [45–65]. Dissipation of magnetic fields before recombination may lead to spectral distortions [66–70], and after recombination may lead to an increase of the optical depth [63, 68, 70–73]. The PMF may lead to additional polarization anisotropies due to Faraday rotation [47–49, 65, 74–82] The non-Gaussianity of the PMF may lead to a signal in the bi-spectrum or tri-spectrum of the anisotropies [63, 83–93]. All of the above effects have led to upper limits in the nano-gauss (nG) range. However, as we elaborate in the following subsection, CMB appears to be most sensitive to the generation of small-scale baryonic density fluctuations before recombination, with detectable PMF strength in the 0.01–0.1 nG range [14]. Moreover, as shown in [16], the baryon clumping before recombination may help relieve the Hubble tension.

32.3 Recombination with Primordial Magnetic Fields

A statistically homogeneous and isotropic PMF is characterized by a power spectrum expected to be a smooth function of the Fourier number k. On large scales, the universe is a highly conducting plasma with the magnetic field "frozen-in" and diluting with the expansion as $B = B_0/a^2$, where B_0 is the present day (comoving) strength and a is the scale factor normalized to unity today. On smaller scales, the PMF generates perturbations in the plasma and dissipates its strength. PMFs generated causally in phase transitions generically [94] have a Batchelor spectrum [95] that monotonically increases with k until a peak at a certain small integral scale [96], beyond which it decays. Inflationary magnetogenesis, on the other hand, can produce a range of spectral shapes depending on the particular model (see e.g. [33] for a review), with the simplest models [31, 32] predicting a scale-invariant spectrum. CMB constraints are often quoted in terms of the comoving field strength smoothed on the 1 Mpc scale, $B_{1\mathrm{Mpc}}$, which, in many cases, is not a representative measure of the PMF. The more relevant quantity for the baryon clumping effect is the effective comoving strength, B_{eff}, quantifying the average magnetic total energy density. For scale-invariant fields, the two measures are essentially the same. For causal fields, however, the difference is quite dramatic, with $B_{\mathrm{eff}} \gg B_{1\mathrm{Mpc}}$. Unless specified otherwise, the field strengths quoted in this paper are B_{eff}.

Unlike most of the evolution of the magnetized plasma in the early universe, when it is an incompressible fluid, MHD becomes compressible before recombination for scales well below the photon mean free path $l_\gamma \sim 1$ Mpc[1]. Photons are free-steaming over $\sim$ kpc scales and their sole effect is to introduce a drag force on peculiar motions in the CMB rest frame characterized by a linear drag coefficient α. The Euler equation for the baryon velocity field takes the form of

$$\frac{\partial \mathbf{v}}{\partial t} + (\mathbf{v} \cdot \nabla)\mathbf{v} + c_s^2 \frac{\nabla \rho}{\rho} = -\alpha \mathbf{v} - \frac{1}{4\pi\rho} \mathbf{B} \times (\nabla \times \mathbf{B}) \tag{32.1}$$

where the speed of sound, $c_s \approx 2 \times 10^{-5} c$, is very small on scales well-below l_γ, while being much larger, $c_s = \sqrt{1/3}c$, on length scales larger than l_γ. The first term on the right-hand side (RHS) is the photon drag, while the last term is the Lorentz force due to the PMF, $\mathbf{B} \times (\nabla \times \mathbf{B}) = \nabla B^2/2 - (\mathbf{B} \cdot \nabla)\mathbf{B}$, which pushes baryons into regions of lower magnetic energy density. This force is opposed by the baryon-photon fluid pressure characterized by c_s^2. On scales well-below l_γ, the plasma is free to compress until the baryonic pressure gradients backreact.

Using the continuity equation,

$$\frac{\partial \rho}{\partial t} + \nabla(\rho \mathbf{v}) = 0\,, \tag{32.2}$$

one can estimate the amplitude of the generated density fluctuations in a back-of-the-envelope estimate [13]. Imagine a stochastic magnetic field at uniform baryon density and negligible velocities initially. These are very realistic initial conditions, since at earlier times the evolving l_γ is smaller than the magnetic fluctuations scale L, such that the plasma is incompressible, and photon diffusion suppresses any fluid motion. At redshifts $z \sim 10^4$–10^6, depending on the scale, the photon mean free

[1] All cited length scales are comoving.

path l_γ starts exceeding the scale L. At this point Eq. (32.1) applies and there is a significant drop in c_s. Initially, only the terms on the RHS of Eq. (32.1) are important and the plasma obtains a terminal velocity $\mathrm{v} \sim c_A^2/(\alpha L)$, where $c_A = B/\sqrt{4\pi\rho}$ is the Alfven speed of the plasma. Using Eq. (32.2), it is straigthforward to show that the density fluctuations grow as $\delta\rho/\rho \simeq \mathrm{v}t/L \simeq c_A^2 t/\alpha L^2$ with time t. These density fluctuations become larger with time until either the pressure forces become important in counteracting further compression, or the source magnetic stress term decays. The former happens when the last term on the LHS of Eq. (32.1), $(c_s^2/L)\delta\rho/\rho$, is of the order of the magnetic force term c_A^2/L. That is, density fluctuations cannot become larger than $\delta\rho/\rho \lesssim (c_A/c_s)^2$. It has been shown by numerical simulation [97] that even in the highly dragged viscous MHD regime magnetic fields decay, albeit slower than during turbulent MHD. Magnetic fields excite fluid motions which in turn get dissipated by the photon drag. It was found that a magnetic structure decays when the Eddy turn-over rate $\mathrm{v}/L \simeq c_A^2/\alpha L^2$ equals the Hubble rate $H \simeq 1/t$. Entering this into the expression for $\delta\rho/\rho$ one finds that the average density fluctuation cannot exceed unity by much. This leads to the following estimate for the generated density fluctuations:

$$\frac{\delta\rho}{\rho} \simeq \min\left[1, \left(\frac{c_A}{c_s}\right)^2\right]. \tag{32.3}$$

Since, at recombination,

$$c_A = 4.34\mathrm{km/s}\left(\frac{B}{0.03\mathrm{nG}}\right) \tag{32.4}$$

and $c_s = 6.33\mathrm{km/s}$, even fairly weak fields may generate order-unity density fluctuations on small scales. This simple estimate has subsequently been confirmed by full numerical simulations [14].

As the Universe expands, density fluctuations on ever larger scales are generated and subsequently decay. Causally generated PMFs loose a significant fraction of their power with each e-fold as the peak of the Batchelor spectrum keeps moving to larger scales and weaker fields according to $B(L) = B_0(L/L_0)^{-5/2}$. Scale-invariant fields, on the other hand, are largely unaffected by the evolution. As discussed above, that magnetic fields and the associated density fluctuations decay when the Eddy turnover time equals the Hubble time. This can be used to derive a correlation between the magnetic field strength and the scale of the smallest magnetic structure that survives until just before recombination [97]:

$$B_{\mathrm{rec}} \lesssim 80\,p\mathrm{G}\left(\frac{L}{\mathrm{kpc}}\right), \tag{32.5}$$

where approximate equality is attained for fields which do undergo dissipation in the first place. For fields which are so weak and/or are on such large scales that the Eddy turnover time never reaches the Hubble time, no dissipation ever occurs and no correlation can be derived (hence the inequality sign). Thus, quite generally,

Table 32.1 Different measures (in nG) of a PMF that has an effective comoving magnetic strength of $B_{\rm eff} = 0.05$ nG at the redshift of last-scattering $z_\star$. All the measures are the same for scale-invariant fields, but differ significantly for a causally generated PMF with a Batchelor spectrum

Spectrum	$B_{\rm eff}$ at $z_\star$	$B_{\rm 1Mpc}$ at $z_\star$	$B_{\rm eff}$ at $z = 10^6$	$B_{\rm eff}^{\rm pre-collapse}$
Scale-invariant	0.05	0.05	0.05	0.05
Batchelor	0.05	10^{-9}	5	0.008

for sufficiently strong fields, $B \sim 10 - 100$ pG, inhomogeneities are generated on ~0.1–1 kpc scales shortly before recombination. Further dissipation occurs across recombination as a result of the sudden drop in the free electron density. This leads to a factor of 6.2 drop in the strength of the causal PMF, while, again, a scale-invariant field strength is not affected. After recombination, the PMF evolution is almost entirely halted or at least significantly slowed down as recent numerical simulations show [98]. The post-recombination field strength is the one relevant for the subsequent structure formation and is commonly referred to as "pre-collapse". For causal fields, the resultant pre-collapse correlation is given by $B_0 \simeq 5\, pG(L/\mathrm{kpc})^2$. As the differences between various measures of the PMF are model-dependent and can be a source of confusion, we provide a glossary in Table 32.1. It has been shown by detailed simulations [99, 100] that, irrespective of their coherence length, pre-collapse magnetic fields of ~5 pG lead to final post-collapse cluster fields of $\sim \mu$G with Faraday rotation measures in good agreement with observations.

The magnetically induced inhomogeneities are on scales much too small (i.e. $\ell \sim 10^6$–10^7) to be observed directly as a contribution to CMB anisotropies. However, the baryon clumping has a prominent effect on the rate of recombination, speeding it up and resulting in a smaller sound horizon at photon-baryon decoupling, $r_\star$. The speed up of recombination follows from the recombination rate being proportional to the square of the electron density n_e. Since, generally, $\langle n_e^2 \rangle > \langle n_e \rangle^2$ in a clumpy universe, where $\langle\rangle$ denotes spatial average, the recombination rate is enhanced compared to ΛCDM. A smaller $r_\star$ raises the value of H_0 inferred from the very accurately measured angular size of the sound horizon, helping reduce the Hubble tension.

32.4 Inhomogeneous Recombination

The evolution of the cosmic electron density depends on the full details of the probability distribution function (PDF) $p(\Delta)$ to find baryons at density $\Delta\langle\rho\rangle$, where $\langle\rho\rangle$ is the average baryonic density and Δ is an enhancement factor. The shape of the

[2] It is noted that there is no conflict between the statement that the magnitude of $B_{\rm eff}$ diminishes by a factor 6.2 during recombination, on the one hand, and Eq. 32.5 and this relation, on the other hand, as the coherence length L is increasing during the dissipation.

baryon density PDF is currently unknown, and neither is its evolution before recombination, which could be substantial, particularly for blue Batchelor spectra. In the absence of knowledge of the PDF, [14] and [16] utilized a simple three zone model (see Sect. 32.4.1) to describe the distribution of baryons. In the three-zone model, the amplitude of baryon inhomogeneity is controlled by the clumping factor

$$b \equiv \left(\frac{\langle \rho^2 \rangle - \langle \rho \rangle^2}{\langle \rho \rangle^2} \right) \tag{32.6}$$

which sets the second moment of the PDF, while the higher order moments are effectively determined by the choice of the three-zone model parameters. As we show next, the higher moments, and the third moment in particular, play a very important role in determining the enhancement of the recombination rate.

The evolution of the electron number density around recombination is given by

$$\frac{dn_e}{dt} + 3\,Hn_e = -C\left(\alpha_e n_e^2 - \beta n_{H_0} e^{-\frac{h\nu_\alpha}{T}}\right), \tag{32.7}$$

where n_e is the free electron density, n_H is the total matter number density (protons and hydrogen), and $n_{H_0} = n_H - n_e$ is the density of neutral hydrogen. For simplicity, to illustrate the importance of the higher moment, we consider a universe with only hydrogen (i.e. no helium). Since Eq. (32.7) is quadratic in n_e, and since $\langle n_e^2 \rangle > \langle n_e \rangle^2$ in inhomogeneous universes, we expect the recombination rate to be enhanced when inhomogeneities exist. It would be tempting to solve the above equation by the introduction of a clumping factor given by the underlying density distribution, i.e. $b = (\langle n_H^2 \rangle - \langle n_H \rangle^2)/\langle n_H \rangle^2$, such that the recombination term would be replaced by $\alpha_e(1+b)\langle n_e \rangle^2$ and the effect of inhomogeneities would be entirely described by the first two moments of the distribution, i.e. the average density $\langle n_H \rangle$ and the variance $\langle n_H^2 \rangle$. However, such a procedure would be incorrect as it would assume that $\langle n_e^2 \rangle \sim \langle n_H^2 \rangle$. This is not the case if different density regions have different ionization fractions $\chi_e = n_e/n_H$. Rather, as shown below, the average ionization depends on all the moments of the underlying probability distribution of the density fluctuations $P(n_H)$.

Consider detailed balance, i.e. set the RHS of Eq. (32.7) to be approximately zero. This assumption holds fairly during the middle of recombination because both the recombination and the photoionisation rates are much larger than the Hubble rate, but is increasingly inaccurate towards the end of recombination. The equilibrium ionization fraction can then be derived as

$$\chi_e^2 = \frac{\beta_e}{\alpha_e}\frac{1}{n_H^2}(n_H - n_e)e^{-\frac{h\nu_\alpha}{T}} = \frac{\beta_e}{\alpha_e}\frac{1}{n_H}e^{-\frac{h\nu_\alpha}{T}}(1-\chi_e). \tag{32.8}$$

Let us now allow for inhomogeneities in the total baryon density by defining

$$n_H = \Delta \langle n_H \rangle, \tag{32.9}$$

where ⟨...⟩ denotes the ensemble average, which we will assume to be the same as the average over any sufficiently large volume. Then we can write

$$\chi_e^2 = \frac{\beta_e}{\alpha_e} \frac{1}{\Delta \langle n_H \rangle} e^{-\frac{h\nu_\alpha}{T}} (1 - \chi_e) = \frac{A}{\Delta} (1 - \chi_e), \tag{32.10}$$

where we have defined

$$A \equiv \frac{\beta_e}{\alpha_e} \frac{1}{\langle n_H \rangle} e^{-\frac{h\nu_\alpha}{T}}. \tag{32.11}$$

From (32.10), we can write the local ionized fraction as

$$\chi_e(\Delta) = \frac{A}{2\Delta} \left(1 + \frac{4\Delta}{A}\right)^{1/2} - \frac{A}{2\Delta}. \tag{32.12}$$

In the homogeneous case, $\Delta = 1$, we have

$$\chi_e^0(\Delta = 1) = \frac{A}{2} \left(1 + \frac{4}{A}\right)^{1/2} - \frac{A}{2}. \tag{32.13}$$

Note that in the limit $A \gg 1$ we have a full ionisation, i.e. $\chi_e \to 1$, while in the opposite limit, $A \ll 1$, we have $\chi_e \to 0$.

For consistency, the inhomogeneity parameter $\Delta(x)$ (x are space coordinates) must fulfill

$$\frac{1}{V_{tot}} \int \mathrm{d}V \Delta(x) = 1 \tag{32.14}$$

where the integration is over the total spatial volume V_{tot}. For simplicity, as assumed earlier, we will take averages over this volume to be the same as ensemble averages and the above constraint can be simply written as $\langle \Delta \rangle = 1$. The average electron density is given by

$$\langle n_e \rangle = \langle \chi_e(\Delta) \Delta \langle n_H \rangle \rangle\ , \tag{32.15}$$

and the average ionzation fraction is

$$\langle \chi_e \rangle \equiv \frac{\langle n_e \rangle}{\langle n_H \rangle} = \frac{A}{2} \langle \sqrt{1 + \frac{4\Delta}{A}} - 1 \rangle. \tag{32.16}$$

Note that this is not the same as the average of Eq. (32.10), where $\chi_e = n_e/n_H$. Namely, $\langle n_e \rangle / \langle n_H \rangle \neq \langle n_e / n_H \rangle$. The relevant quantity for us is actually $\langle n_e \rangle$, which is what appears in the CMB calculations, hence the relevant quantity is the one in Eq. (32.16).

One can now show that, for arbitrary distribution functions $P(n_H)$ or $\tilde{P}(\Delta)$ fulfilling the integral constraint

$$\langle \Delta \rangle \equiv \int \mathrm{d}\Delta \Delta \tilde{P}(\Delta) = 1, \tag{32.17}$$

the $\langle \chi_e \rangle$ of Eq. (32.16) is smaller than χ_e^0. Namely, that

$$\langle \sqrt{1 + \frac{4\Delta}{A}} \rangle < \sqrt{1 + \frac{4}{A}} \, . \tag{32.18}$$

This follows from Jensen's inequality [101] which states that the expectation value $\langle g(X) \rangle$ of any concave function $g(X)$ of a random variable X is smaller than $g(\langle X \rangle)$. In our case, the function $\sqrt{1 + 4\Delta/A}$ is certainly concave and hence we have our result.

For smaller density perturbations one can also see this without using Jensen's theorem by introducing $\delta \equiv \Delta - 1$, such that $\langle \delta \rangle = 0$, and performing a Taylor expansion under the assumption that $\delta < 1$. Then,

$$\langle \sqrt{1 + \frac{4}{A} + \frac{4\delta}{A}} \rangle \approx \sqrt{1 + \frac{4}{A}} - \gamma_2 \langle \delta^2 \rangle + \gamma_3 \langle \delta^3 \rangle - ..., \tag{32.19}$$

where all $\gamma_i > 0$. One can see that the leading second order term is negative (the function is concave), as well as all even order moment terms, while the odd moments come with a positive signs. This means, e.g., that PDFs with a large third moment could reduce the impact on the ionized fraction which is a trend we have observed while experimenting numerically with different three-zone models. Ultimately, the full redshift evolution of the baryon PDF and the associated ionized fraction will be determined exactly from MHD simulations that are currently in progress.

32.4.1 The Three-Zone Model

In the absence of the baryon density PDF derived from MHD simulations, when discussing the current observational constraints on the PMF-enhanced recombination, we will use the simple three-zone model introduced in [13]. The model is described by the density parameters Δ_i and volume fractions f_V^i in each zone. Baryon densities in the individual zones are simply given by $n_b^i = \langle n_b \rangle \Delta_i$. Parameters Δ_i and f_V^i have to fulfil the following constraints:

$$\sum_{i=1}^{3} f_V^i = 1, \; \sum_{i=1}^{3} f_V^i \Delta_i = 1, \; \sum_{i=1}^{3} f_V^i \Delta_i^2 = 1 + b \, , \tag{32.20}$$

i.e. the total volume fraction is one and the three-zone model has average density $\langle n_b \rangle$ and clumping factor b. This leads to three constraints for six free parameters f_V^i, Δ_i, such that one may choose three parameters freely. To obtain the average ionization fraction $\langle \chi_e \rangle$, one computes the ionization fraction in each of the zones and take the average, i.e.

$$\langle \chi_e \rangle = \sum_{i=1}^{3} f_V^i \Delta_i \chi_e^i \,. \tag{32.21}$$

In [13], the parameters were chosen to be $f_V^2 = 1/3$, $\Delta_1 = 0.1$ and $\Delta_2 = 1$, which we will subsequently refer to as the M1 model. In [16], for comparison, we introduced a second model, M2, that had $f_V^2 = 1/3$, $\Delta_1 = 0.3$ and $\Delta_2 = 1$. It is not clear at this point if either of these models provides a good representation of the actual baryon PDF, but we note that M2 has a larger third moment and, for reasons discussed earlier, results in a lesser reduction of the average ionized fraction for the same value of the clumping factor b.

32.5 The Impact of Baryon Clumping on CMB Spectra

The discussion in this and the subsequent Sections is largely based on the results obtained in [19], where the impact of clumping on the CMB spectra and, in particular, their effect on CMB polarization were studied. As previously discussed, inhomogeneous recombination completes sooner, which shifts the peak of the visibility function to an earlier epoch. This lowers $r_\star$, defined as the comoving sound horizon at the peak of the visibility function, which has the effect of shifting the acoustic peaks in CMB spectra to smaller angular scales. One can compensate for the shift with a larger value of H_0, which may help in relieving the Hubble tension.

In addition to the shift of the peaks, earlier recombination means that CMB polarization is produced at an earlier epoch and, as a consequence, has a larger overall amplitude. This is because the value of the speed of sound c_s (on scales where the baryon-photon fluid is tightly coupled) was larger at earlier times, and the amplitude of polarization is set by the quadrupole of temperature anisotropy, which is derived from the dipole, which is set by the time derivative of the monopole, which in turn is proportional to c_s [102].

Clumping also broadens the visibility function, which is a consequence of the overdense baryon pockets recombining earlier and underdense baryon pockets recombining later. This broadening tends to further enhance polarization, because of the period of time during which polarization can be generated is longer.

The two effects, the shift of the peak and the broadening of the visibility function, are illustrated in Fig. 32.1, which compares the visibility functions in the ΛCDM model, with an M1 model with $b = 2$ with all other parameters kept the same, and with the M1 model (ΛCDM+b) that best fits the Planck CMB data combined with the SH_0ES prior on H_0. The broadening effect is apparent from the lower peak, since the visibility function is normalized to integrate to unity. We note that, while the general trends in the visibility function are common to all clumping models,

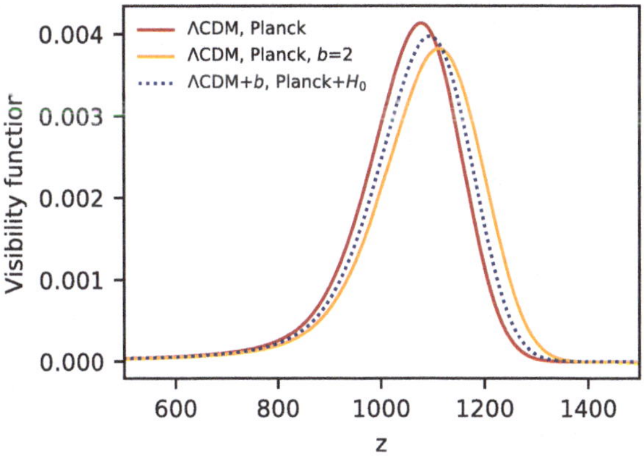

Fig. 32.1 Impact of baryon clumping on the CMB visibility function. We compare the visibility functions in the Planck best-fit ΛCDM model (solid red line), an M1 model with $b = 2$ and all cosmological parameters set to the best-fit ΛCDM (solid orange line), and the ΛCDM+b Planck+SH_0ES best-fit M1 model (dotted blue line). One can see that clumping shifts the peak of the visibility function to earlier times, and increases its width. Figure reproduced from [19]

the quantitative details are dependent on the shape and the evolution of the baryon density PDF.

Another significant effect of clumping on CMB spectra comes from a modification of the Silk damping scale r_D. Here, there are three competing effects. Firstly, r_D decreases due to an overall earlier completion of recombination. This decrease, however, could be negated by the fact that an earlier broad helium recombination in clumping models results in a smaller electron density available at the later stages of recombination, and due to the broadening of the visibility function. Since much of the Silk damping occurs right at recombination, where the visibility function is of order unity, the details of the visibility function play an important role. The first effect, due the overall shift to higher redshifts, would reduce the Silk damping effect on CMB spectra, as it pushes the onset of the damping tail to higher multipoles ℓ. The second and third effects, however, can also be important, and the balance between them is model-dependent and varies with the clumping factor. In the best-fit M1 model, we find that there is less Silk damping compared to ΛCDM. But, in the best-fit M2 model (see Sect. 32.4.1), the Silk damping is virtually identical to that in the best-fit ΛCDM. Also, in M1, at (observationally disallowed) high values of b, the Silk damping is actually enhanced. Additional discussion of the evolution of the damping scale as a function of b in a different clumping models can be found in [18]. The evolution of magnetically induced clumping is unknown at present and is a significant source of uncertainty in observational constraints on the PMF. For example, if clumping was stronger at $z \sim 3000$ than at $z \sim 1200$, the helium recombination could be the dominant effect, inducing more Silk damping.

The reduction in Silk damping present in the M1 model is illustrated in Fig. 32.2, which compares the CMB spectra in the Planck best-fit ΛCDM to those in the

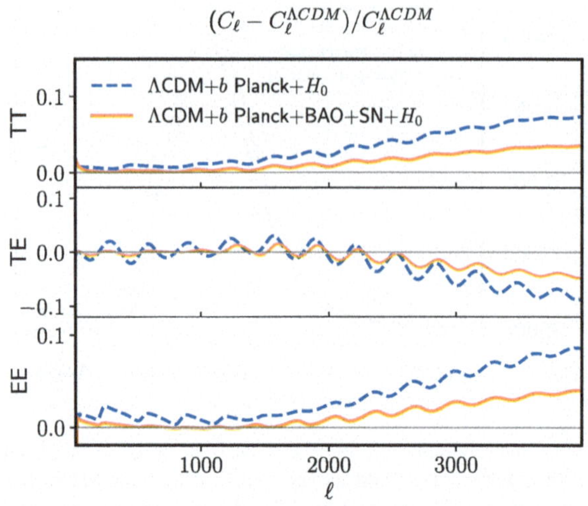

Fig. 32.2 Impact of baryon clumping on CMB power spectra. Shown is the relative difference between the CMB spectra in the Planck best-fit ΛCDM model and the M1 models best-fit to Planck+SH_0ES (blue dashed line) and Planck+BAO+SN+SH_0ES (orange solid line). In the case of TE, which is positively or negatively correlated depending on ℓ, we avoid divisions by zero by computing $(C_\ell - C_\ell^{\Lambda CDM})/C_\ell^{ref}$, where C_ℓ^{ref} is the absolute value of $C_\ell^{\Lambda CDM}$ convolved with a Gaussian of width $\sigma_\ell = 100$ centred at ℓ. The plot illustrates the importance of small-scale CMB anisotropies for constraining baryon clumping. Figure reproduced from [19]

Planck+SH_0ES and Planck+BAO+SN+SH_0ES best-fit M1 models. The enhancements in the temperature (TT) and E-mode polarization (EE) spectra at high ℓ is due to the smaller r_D. The same enhancement is also seen in the temperature-polarization cross-correlation (TE), which is negative due to the anti-correlation between T and E at high ℓ. This illustrates the fact that high resolution CMB measurements will be a key discriminant in constraining inhomogeneous recombination models.[3]

32.6 Relieving the Hubble Tension with PMFs

The extent to which PMFs can help relieve the Hubble tension is limited by the fact that reducing the sound horizon $r_\star$, by itself, can at best raise the value of H_0 to ~ 70 km/s/Mpc without causing new tensions between CMB, BAO and the large-scale structure clustering. In what follows, we first reproduce the general argument illustrating this point, originally made in [103], before describing the current observational status of the M1 baryon clumping model.

[3] Figure 32.2 also shows that at $\ell \lesssim 20$ the polarization is reduced, which is due to the lower best fit value of the optical depth τ.

32.6.1 Why Reducing $r_\star$ Helps, But Can Not Fully Resolve the Hubble Tension by Itself

The acoustic peaks in CMB anisotropy spectra provide a very accurate measurement of the angular size of the sound horizon at recombination,

$$\theta_\star = \frac{r_\star}{D(z_\star)}, \tag{32.22}$$

where $D(z_\star)$ is the comoving distance from a present day observer to the last scattering surface, where $z_\star$ is the redshift of the peak of the visibility function. In a given model, $r_\star$ and $D(z_\star)$ can be determined from $r_\star = \int_{z_\star}^{\infty} c_s(z)\mathrm{d}z/H(z)$ and $D(z_\star) = \int_0^{z_\star} c\,\mathrm{d}z/H(z)$, where $c_s(z)$ is the sound speed of the photon-baryon fluid, $H(z)$ is the redshift-dependent cosmological expansion rate and c is the speed of light. To complete the prescription, one also needs to determine $z_\star$ using a model of recombination.

The sound waves responsible for the acoustic peaks in the CMB spectra are also imprinted in the galaxy power spectra as Baryon Acoustic Oscillations (BAO) with a minor difference in this standard ruler. Rather than $r_\star$, which is the sound horizon at photon decoupling, the scale imprinted in the BAO is the sound horizon at the baryon decoupling r_d, also known as the "cosmic drag" epoch when the photon drag on baryons becomes unimportant. The latter takes place at a slightly lower redshift than recombination, so that $r_\mathrm{d} \approx 1.02 r_\star$ in ΛCDM. The proportionality factor between the two sound horizons does not change appreciably in alternative recombination scenarios.

While the difference between $r_\star$ and r_d is small, the difference in redshifts at which the acoustic features in the CMB and BAO are observed is significant, with the latter measured at $0 \lesssim z \lesssim 2.5$ accessible to galaxy redshift surveys. The angular scale of the BAO feature measured using galaxy correlations in the transverse direction to the line of sight is

$$\theta_\perp^{\mathrm{BAO}}(z_\mathrm{obs}) \equiv \frac{r_\mathrm{d}}{D(z_\mathrm{obs})}, \tag{32.23}$$

where z_obs is the redshift of the correlated galaxies.

Let us now consider Eqs. (32.22) and (32.23) while remaining agnostic about the particular model that determines the sound horizon. Namely, let us treat $r_\star$ as an independent parameter and assume $r_\mathrm{d} = 1.02 r_\star$. Let us also assume that after recombination the expansion of the universe is well-described by the ΛCDM model and, for simplicity, ignore the contribution of radiation to the distance integrals $D(z_\star)$ and $D(z_\mathrm{obs})$. Then, for the CMB acoustic feature, we can write

$$\theta_\star = \frac{r_\star}{2998\,\mathrm{Mpc}} \left(\int_0^{z_\star} \frac{\mathrm{d}z}{\omega_m^{1/2}\sqrt{(1+z)^3 + h^2/\omega_m - 1}} \right)^{-1}, \tag{32.24}$$

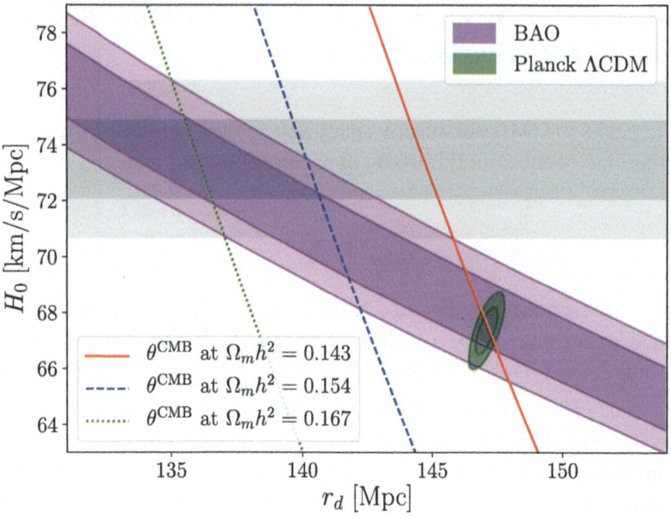

Fig. 32.3 A plot illustrating that achieving a full agreement between CMB, BAO and SH_0ES through a reduction of r_d requires a higher value of ω_m. Shown are the CMB lines of degeneracy between the sound horizon r_d and the Hubble constant H_0 at three different values of ω_m: 0.143, 0.155 and 0.167. Also shown are the 68% and 95% CL bands derived from the combination of all current BAO data marginalized over ω_m, the ΛCDM based bounds from Planck and the determination of the Hubble constant by SH_0ES. Figure reproduced from [103]

where $\omega_\mathrm{m} = \Omega_\mathrm{m} h^2$ is the matter density today, Ω_m is the fractional matter density, h is H_0 in units of 100 km/s/Mpc, and 2998 Mpc $= c/100\mathrm{km/s/Mpc}$. An analogous equation for the BAO is obtained by replacing $(r_\star, \theta_\star, z_\star)$ with $(r_\mathrm{d}, \theta_\perp^\mathrm{BAO}, z_\mathrm{obs})$.

For a given ω_m, and with the precisely measured $\theta_\star$, Eq. (32.24) defines a line in the r_d-H_0 plane[4]. Similarly, a BAO measurement at each different redshift also defines a respective line in the r_d-H_0 plane. However, the slopes of the CMB and BAO lines are very different due to the vast difference in redshifts at which the standard ruler is observed, $z_\star \approx 1100$ for CMB *vs* $z_\mathrm{obs} \sim 1$ for BAO. This is illustrated in Fig. 32.3 that shows the $r_\mathrm{d} - H_0$ degeneracy line for CMB at three different values of ω_m, as well as the 68% and 95% confidence level (CL) regions derived from the combination of all presently available BAO observations while treating r_d as as an independent parameter and marginalizing over ω_m (see [104] for details). Also shown are the ΛCDM constraint from Planck (in red) and the SH_0ES determination of H_0 (the grey band). On can see from Fig. 32.3 that BAO is consistent with either Planck or SH_0ES, depending on the value of r_d, but the latter two are in clear disagreement with each other for the Planck-ΛCDM-preferred $\omega_\mathrm{m} \approx 0.143$.

One can reconcile Planck with SH_0ES by reducing r_d and "moving up" the CMB $r_\mathrm{d} - H_0$ degeneracy line. However, doing so at a fixed ω_m would quickly move the values of r_d and H_0 out of the purple band in Fig. 32.3 creating a tension with BAO.

---

[4] From here on we will use r_d to represent the acoustic feature in both the CMB and BAO.

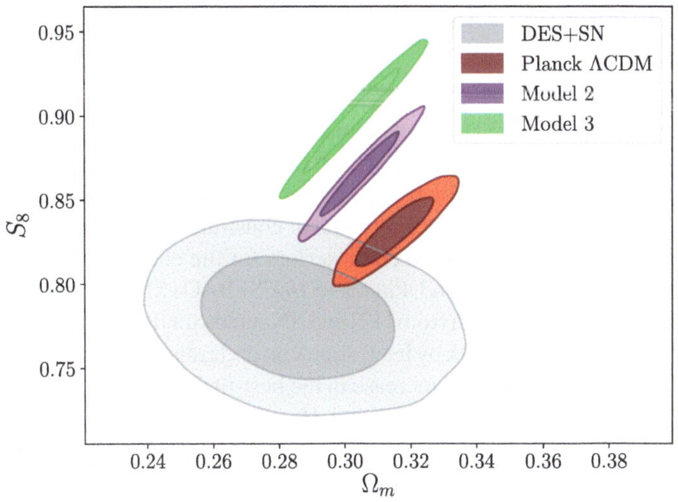

Fig. 32.4 The 68 and 95% confidence level bounds on S_8 and Ω_m. Shown are the constraints derived by fitting the ΛCDM model to a joint dataset of DES and SN and to Planck, along with the contours for Model 2 and Model 3 corresponding to the $\omega_m = 0.155$ and $\omega_m = 0.167$ $r_d - H_0$ lines in Fig. 32.3. Figure reproduced from [103]

In order to reconcile CMB, BAO and SH_0ES at the same time, one needs to reduce r_d and increase the matter density, as illustrated in the Figure. In particular, a full resolution of the tension would require $\omega_m \approx 0.167$, which is substantially higher than the best-fit ΛCDM value. A larger matter density results in more clustering quantified by the S_8 parameter. It is defined as $S_8 \equiv \sigma_8(\Omega_m/0.3)^{0.5}$, where σ_8 is the matter clustering amplitude on the scale of 8 h^{-1}Mpc. A larger S_8 would exacerbate the already existing mild tension between its Planck best-fit value and that measured from weak gravitational lensing of galaxies [105].

The reason why simply reducing $r_\star$ would not fully solve the Hubble tension is seen in Fig. 32.4. The figure shows the 68% and 95% CL constraints on S_8 and Ω_m by DES supplemented by the Pantheon SN sample [106]. Also shown is the Planck ΛCDM best fit, corresponding to $\omega_m = 0.143$, as well as the two cases corresponding to $\omega_m = 0.155$ (Model 2) and $\omega_m = 0.167$ (Model 3). The constraints for the latter two were obtained by simultaneously fitting BAO and CMB acoustic peaks at the corresponding fixed values of ω_m. One can see that when attempting to bring CMB, BAO and SH_0ES in agreement with each other by reducing r_d, one increases the S_8 tension.

Thus, any proposed solution to the Hubble tension that amounts to reducing r_d without other significant changes to ΛCDM would not be able to raise the CMB-extracted value of H_0 above 70 km/s/Mpc. Baryon clumping belongs to that category of models, hence we do not expect it to fully resolve the Hubble tension. But it could still help relieve it to a statistically acceptable level.

32.6.2 Observational Status of the M1 Baryon Clumping Model

Without knowing the detailed ionized fraction evolution obtained from MHD simulations with a PMF (see Sect. 32.7), the M1 three-zone model [13] was used in [17–20, 103] for exploring the ability of baryon clumping to resolve the Hubble tension. Below we summarize the current observational constraints of the M1 model.

Figure 32.5 shows the marginalized posteriors for the clumping factor b and H_0 from Planck, Planck+SH_0ES and Planck+SH_0ES+BAO+SN, where we used the eBOSS DR16 BAO compilation from [12] and SN stands for the Pantheon supernovae sample [106]. We see that Planck by itself shows no preference for clumping, with $b < 0.47$ at 95% CL and only a small increase in the best-fit H_0. Adding the SH_0ES prior (H_0) to Planck gives $b = 0.48 \pm 0.19$ and $H_0 = 70.32 \pm 0.85$ km/s/Mpc. Adding the BAO and SN data results in $b = 0.40^{+0.15}_{-0.19}$ and $H_0 = 69.68 \pm 0.67$ km/s/Mpc, a reduction due to the general difficulty in maintaining the agreement between CMB and BAO while decreasing the sound horizon, as discussed in the previous subsection.

Additional constraints on baryon clumping can be derived by adding the high-resolution CMB temperature and polarization data from the Atacama Cosmology Telescope fourth data release (ACT DR4) [108] and the South Pole Telescope Third Generation (SPT-3G) 2018 data [20, 109]. As discussed in Sect. 32.5 and demonstrated in Fig. 32.2, baryon clumping impacts the Silk damping tail of the CMB spectrum, making the spectra at multipoles $\ell > 2000$ a powerful probe of baryon

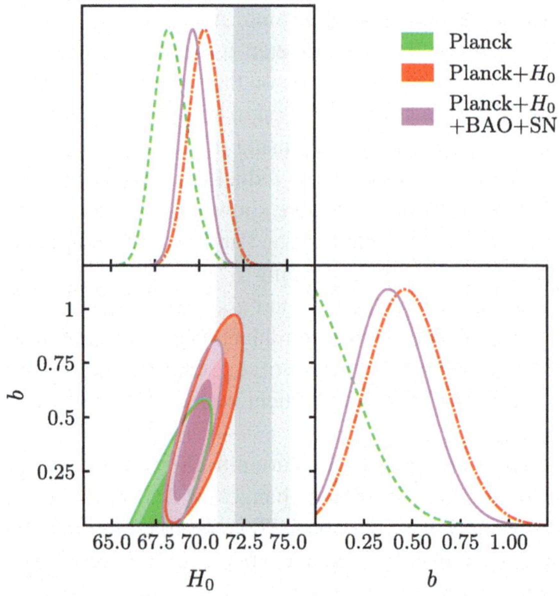

Fig. 32.5 Constraints on the clumping factor b and H_0 derived from Planck (lime), the combination of Planck with the SH_0ES prior on H_0 (red), and the combination of the Planck, SH_0ES, BAO and SN data (violet). With the H_0 prior, there is a preference for a nonzero clumping at $\sim 3\sigma$ level

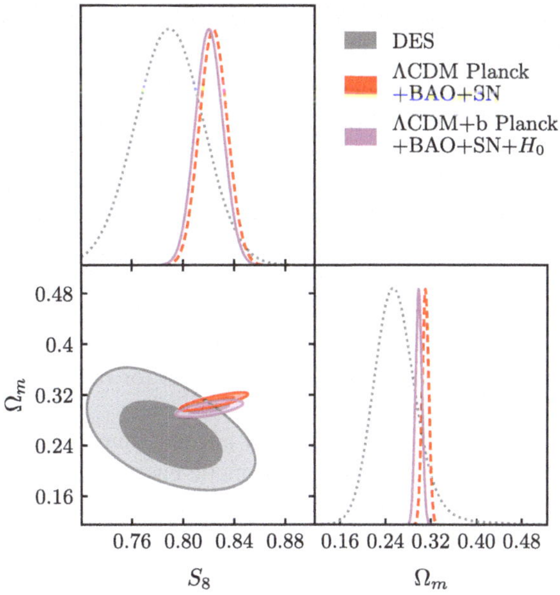

Fig. 32.6 Joint constraints in the S_8-Ω_m plane from Planck+BAO+SN in ΛCDM (red), and in the M1 model fit to Planck+BAO+SN with the SH_0ES-based H_0 prior (violate). The ΛCDM based constraints from the DES Y1 data [107] are shown in grey. There is a slight relief of the tension in M1 due to $\Omega_m h^2$ remaining largely the same as in the best-fit ΛCDM model, while h is increased

clumping. Combining Planck and ACT yields $b < 0.34$ at 95% CL [17, 19], while Planck+SPT gives $b < 0.38$ at 95% CL [20] in the M1 model.

Many models that resolve the Hubble tension tend to make the S_8 tension worse, mostly due to the reasons described in the previous subsection. In contrast, the baryon clumping generally reduces the value of S_8 deduced from CMB, thus also helping with another tension. The primary reason for the lower S_8 (and Ω_m) values is that the best fit $\Omega_m h^2$ value in the clumping model is largely the same as in ΛCDM, while h is increased. Figure 32.6 compares the S_8-Ω_m joint posteriors in the Planck+BAO+SN best-fit ΛCDM model to those in the M1 ΛCDM+b , together with the DES Year 1 contours.

Based on preliminary results from MHD simulations described in the next section, it is clear that one should not attribute too much weight to observational constraints derived from the three-zone model. However, for general reasons discussed in Sect. 32.6.1, it is clear that, even in the best-case scenario, baryon clumping could at best raise the CMB-based value of the Hubble constant to H_0 ~70 km/s/Mpc. This may or may not turn out to be enough but, even if the H_0 tension was not fully relieved, a detection of clumping would be highly interesting by itself, as it would provide (indirect) evidence of the PMF. If no clumping is detected, it would provide the tightest constraint on the PMF strength.

32.7 What to Expect from MHD Simulations

The three-zone model may capture much of the physics of baryon clumping due to a PMF before recombination. It is not, however, sufficiently accurate for comparison to current and future high precision CMB data as subtle details are neglected or simply assumed. The M1 three-zone model, employed in earlier sections, was originally introduced in [13] to derive the preliminary stringent limit on the PMF strength from the (older) CMB data at that time. It assumed a particular baryon PDF and neglected its evolution. Furthermore, at the time of publication of [13], not all of the effects that impact the ionization history in the presence of PMFs were known. Since then, it became apparent that a complete exhaustive analysis, employing numerical MHD simulations and Monte Carlo methods for radiation transfer is required, which is currently underway.

Figure 32.7 shows results of a recent numerical MHD simulation [98]. The left panel shows the projected magnetic energy density and the right panel shows the projected baryon overdensity $\rho/\langle\rho\rangle$ from a 256^3 simulation of a $(24\,\mathrm{kpc})^3$ volume permeated by a stochastic non-helical magnetic field at redshift $z = 1000$. The initial conditions were a homogeneous baryon density, vanishing peculiar velocities, and a field of root-mean-square amplitude $B_{\rm rms} = 0.53$ nG (corresponding to $c_{A,\rm rms} = 12c_s$) at $z = 4500$. One can see that substantial baryon density fluctuations $\delta\rho/\rho \sim 1$ on ~kpc scales are generated by redshift $z = 1000$. Most of the volume is occupied by underdense regions, whereas substantial overdensities exist in small pockets. A small volume fraction of $\sim 10^{-3}$ has overdensities above 10, that are not visible in the figure as it shows the projected density. In comparison, the magnetic field energy density is much more diffuse. The maximum clumping factor $b \approx 1.5$ is attained around $z \approx 1000$. The corresponding baryon PDF is very skew-positive, i.e. has a substantial positive third moment, and is significantly different than the PDF

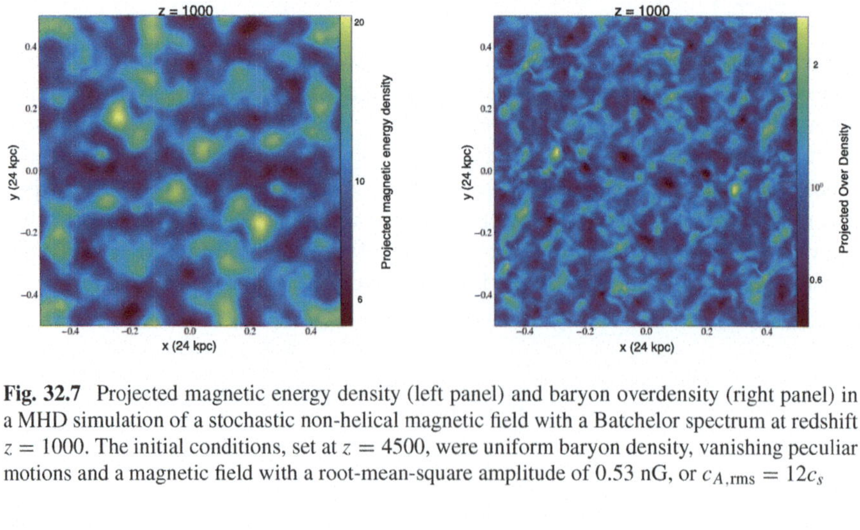

Fig. 32.7 Projected magnetic energy density (left panel) and baryon overdensity (right panel) in a MHD simulation of a stochastic non-helical magnetic field with a Batchelor spectrum at redshift $z = 1000$. The initial conditions, set at $z = 4500$, were uniform baryon density, vanishing peculiar motions and a magnetic field with a root-mean-square amplitude of 0.53 nG, or $c_{A,\rm rms} = 12c_s$

assumed in M1. At redshift $z = 1000$, the magnetic field has dissipated to $B_{\rm rms} = 0.2$ nG, whereas the "final" field is $B_{\rm rms} = 0.044$ nG at $z \approx 10$.

The study in [98] established several new insights:

- Even knowing the full evolution of the baryon PDF is not sufficient for an accurate computation of the ionization history. Rather, the density evolution of each fluid element needs to be known.
- The clumping factor receives large contributions from rare very high density regions. These, however, do not contribute much to the bulk of the recombination. Therefore, the clumping factor b is not a useful parameter for gauging the effect on the ionization history.
- A fraction of Lyman-α photons emitted during the recombination process locally may actually travel to other regions with different local conditions. The resultant cosmic average ionization is significantly changed due to this Lyman-α transport.
- The drop in the speed of sound during recombination by a factor of $1/\sqrt{2}$, due the diminishing electron pressure, leads to further baryon compression by the PMF during recombination.

The last two points are illustrated in Fig. 32.8, showing that all known effects have to be taken into account for a precision computation of the ionization history.

Ultimately, results of such numerical simulations should be used for precision tests against current and future CMB data. A meaningful comparison faces difficulties due to the required immense CPU time. In theory, given the PMF spectrum and helicity, for each magnetic field strength and cosmological parameters such as

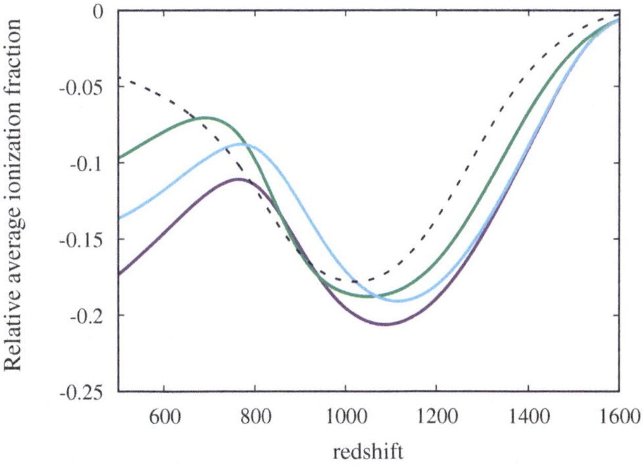

Fig. 32.8 Relative change in the average ionization fraction ΔX_e compared to ΛCDM for numerical MHD simulations with **a** all known effects taken into account, **b** neglecting the Lyman-α transport and, **c** neglecting the reduction of the speed of sound during recombination, as labeled. A relatively strong magnetic field with scale-invariant spectrum was assumed. The prediction of the M1 model with $b = 0.5$ is shown for comparison **d**

baryon density, dark matter density, etc., an independent large simulation should be performed. For the large number of cosmological parameters sampled in such tests, it would be impossible. Fortunately, the effects on X_e from changing the cosmological parameters is much smaller than that of the PMF parameters of appreciable amplitude, such that they may be added linearly to a good approximation. Given a particular set of cosmological parameters, numerical simulations still have to be performed for different initial PMF strengths. This can be done for a discrete set of strengths with subsequent interpolation. The numerical simulation have to be of a significant box size to cover a sufficient dynamical range to simulate all the scales which produce density fluctuations and dissipate afterwards between helium- and hydrogen- recombination, and for the particular realization of the stochastic field (i.e. the choice of the random number) to result in a small variance in the results. This points to fairly large simulations and requires extensive amounts of CPU time.

32.8 Summary

PMFs are often categorized as "exotic physics" because, on one hand, there is no firm theoretical prediction of their existence within standard models of particle physics and cosmology, and also because the observational evidence supporting the PMF hypothesis, while increasingly compelling, is still indirect and can, at least in principle, be explained via less exotic astrophysical processes. However, PMFs are a rather mature field of research sustained for over half-a-century by a small but capable community of researchers. The reason for this longevity is not (only) due to PMFs often playing the role of a "ghost fairy" (using the term much liked by a certain famous cosmologist) coming to the rescue whenever there is a new unexplained phenomenon. Rather, most of the researchers working on the PMFs are inspired by the fact that their existence in the early universe is inevitable. It is not a question of whether they were generated at some level, e.g. during the known phase transitions, but whether they would be of sufficient strength to make a detectable impact. If there is even a remote chance of detecting them, it is worth trying, as they would provide an invaluable insight and a new window into the physics of the early universe.

It is the above context that, in our view, distinguishes the PMF proposal for relieving the Hubble tension from others. It is not a "fairy" explanation invented to solve the puzzle of the day, but rather an observation that if the primordial universe was magnetized at a level sufficient to be relevant to the magnetic fields we see in galaxies and clusters, it could also help with alleviating the H_0 problem. This is a fully falsifiable proposal—there is no new physics to invent when it comes to working out the details of recombination in the presence of a PMF. The required MHD simulations are challenging but doable, and there will be a firm prediction in a reasonable time that can then be tested against the data.

Like other models that aim to relieve the Hubble tension by lowering the sound horizon at recombination, baryon clumping can not raise the value of H_0 beyond $\sim$70 km/s/Mpc, which is still over 2σ lower than the SH_0ES measurement. However, even if PMFs do not fully resolve the Hubble tension, finding evidence for them in the CMB would be a major discovery in its own right. Alternatively, if not detected, accounting for their impact on recombination when analyzing data from future CMB experiments, such as Simons Observatory and CMB-S4, would lead to the tightest constraints on a PMF.

Acknowledgements We thank Tom Abel, Lennart Balkenhol, Silvia Galli and Gong-Bo Zhao for collaborations and discussions. This research was enabled in part by support provided by the BC DRI Group and the Digital Research Alliance of Canada (alliancecan.ca), and the National Sciences and Engineering Research Council of Canada.

References

1. A.G. Riess et al., A comprehensive measurement of the local value of the Hubble constant with 1 km s$^{?1}$ Mpc$^{?1}$ uncertainty from the Hubble space telescope and the SH_0ES team. Astrophys. J. Lett. **934**, L7 (2022). https://doi.org/10.3847/2041-8213/ac5c5b arxiv.org/abs/2112.04510
2. Planck Collaboration, N. Aghanim et al., Planck 2018 results. VI. Cosmological parameters. A&A **641**, A6 (2020). https://doi.org/10.1051/0004-6361/201833910 arxiv.org/abs/1807.06209
3. E. Abdalla et al., Cosmology intertwined: a review of the particle physics, astrophysics, and cosmology associated with the cosmological tensions and anomalies. JHEAp Series **34**, 49 (2022)
4. D. Pesce et al., The Megamaser cosmology project. XIII. Combined Hubble constant constraints. Astrophys. J. **891**, L1 (2020). https://doi.org/10.3847/2041-8213/ab75f0 arxiv.org/abs/2001.09213
5. K.C. Wong et al., H0LiCOW XIII. A 2.4% measurement of H_0 from lensed quasars: 5.3σ tension between early and late-Universe probes. arxiv.org/abs/1907.04869
6. DES collaboration, STRIDES: a 3.9 percent measurement of the Hubble constant from the strong lens system DES J0408-5354. Mon. Not. Roy. Astron. Soc. **494**, 6072 (2020). https://doi.org/10.1093/mnras/staa828 arxiv.org/abs/1910.06306
7. D. Harvey, A 4% measurement of H_0 using the cumulative distribution of strong-lensing time delays in doubly-imaged quasars. arxiv.org/abs/2011.09488
8. M. Millon et al., TDCOSMO. I. An exploration of systematic uncertainties in the inference of H_0 from time-delay cosmography. Astron. Astrophys. **639**, A101 (2020). https://doi.org/10.1051/0004-6361/201937351 arxiv.org/abs/1912.08027
9. W.L. Freedman, B.F. Madore, T. Hoyt, I.S. Jang, R. Beaton, M.G. Lee et al., Calibration of the tip of the red giant branch (TRGB). arxiv.org/abs/2002.01550
10. M.M. Ivanov, M. Simonović, M. Zaldarriaga, Cosmological parameters from the BOSS galaxy power spectrum. JCAP Series **05**, 042 (2020)
11. shape ACT collaboration, The Atacama cosmology telescope: DR4 maps and cosmological parameters. JCAP Series **12**, 047 (2020)
12. eBOSS collaboration, The completed SDSS-IV extended Baryon oscillation spectroscopic survey: cosmological implications from two decades of spectroscopic surveys at the Apache point observatory. arxiv.org/abs/2007.08991
13. K. Jedamzik, T. Abel, Small-scale primordial magnetic fields and anisotropies in the cosmic microwave background radiation. JCAP Series **1310**, 050 (2013)
14. K. Jedamzik, A. Saveliev, Stringent limit on primordial magnetic fields from the cosmic microwave background radiation. Phys. Rev. Lett. Series **123**, 021301 (2019)
15. P.J.E. Peebles, Principles of physical cosmology (1994)

16. K. Jedamzik, L. Pogosian, Relieving the Hubble tension with primordial magnetic fields. Phys. Rev. Lett. Series **125**, 181302 (2020)
17. L. Thiele, Y. Guan, J.C. Hill, A. Kosowsky, D.N. Spergel, Can small-scale baryon inhomogeneities resolve the Hubble tension? An investigation with ACT DR4. arxiv.org/abs/2105.03003
18. M. Rashkovetskyi, J.B. Muñoz, D.J. Eisenstein, C. Dvorkin, Small-scale clumping at recombination and the Hubble tension. arXiv e-prints (2021) arXiv:2108.02747
19. S. Galli, L. Pogosian, K. Jedamzik, L. Balkenhol, Consistency of Planck, ACT, and SPT constraints on magnetically assisted recombination and forecasts for future experiments. Phys. Rev. D series **105**, 023513 (2022)
20. SPT-3G collaboration, A measurement of the CMB temperature power spectrum and constraints on cosmology from the SPT-3G 2018 TT/TE/EE data set. arxiv.org/abs/2212.05642
21. F. Hoyle, XI solvay congress. Brussells (1958)
22. Y. Zeldovich, Magnetic model of the universe. J. Exptl. Theoret. Phys. (U.S.S.R.) **48**, 986 (1965)
23. L.M. Widrow, Origin of galactic and extragalactic magnetic fields. Rev. Mod. Phys. Series **74**, 775 (2002)
24. L.M. Widrow, D. Ryu, D.R.G. Schleicher, K. Subramanian, C.G. Tsagas, R.A. Treumann, The first magnetic fields. Space Sci. Rev. Series **166**, 37 (2012)
25. T. Vachaspati, Progress on cosmological magnetic fields. arxiv.org/abs/2010.10525
26. A. Brandenburg, K. Subramanian, Astrophysical magnetic fields and nonlinear dynamo theory. Phys. Rept. Series **417**, 1 (2005)
27. R.M. Athreya, V.K. Kapahi, P.J. McCarthy, W. van Breugel, Large rotation measures in radio galaxies at Z > 2. Astron. Astrophys. Series **329**, 809 (1998)
28. A.M. Beck, K. Dolag, H. Lesch, P.P. Kronberg, Strong magnetic fields and large rotation measures in protogalaxies by supernova seeding. Mon. Not. Roy. Astron. Soc. Series **435**, 3575 (2013)
29. D. Seifried, R. Banerjee, D. Schleicher, Supernova explosions in magnetized, primordial dark matter haloes. Mon. Not. Roy. Astron. Soc. Series **440**, 24 (2014)
30. T. Vachaspati, Magnetic fields from cosmological phase transitions. Phys. Lett. Series **B265**, 258 (1991). https://doi.org/10.1016/0370-2693(91)90051-Q
31. M.S. Turner, L.M. Widrow, Inflation produced, large scale magnetic fields. Phys. Rev. Series **D37**, 2743 (1988). https://doi.org/10.1103/PhysRevD.37.2743
32. B. Ratra, Cosmological "seed" magnetic field from inflation. Astrophys. J. Series **391**, L1 (1992)
33. R. Durrer, A. Neronov, Cosmological magnetic fields: their generation, evolution and observation. Astron. Astrophys. Rev. Series **21**, 62 (2013)
34. K. Subramanian, The origin, evolution and signatures of primordial magnetic fields. Rept. Prog. Phys. Series **79**, 076901 (2016)
35. K. Jedamzik, V. Katalinic, A.V. Olinto, Damping of cosmic magnetic fields. Phys. Rev. D Series **57**, 3264 (1998)
36. K. Subramanian, J.D. Barrow, Magnetohydrodynamics in the early universe and the damping of noninear Alfven waves. Phys. Rev. Series **D58**, 083502 (1998)
37. A. Neronov, I. Vovk, Evidence for strong extragalactic magnetic fields from Fermi observations of TeV blazars. Sci. Series **328**, 73 (2010)
38. F. Tavecchio, G. Ghisellini, L. Foschini, G. Bonnoli, G. Ghirlanda, P. Coppi, The intergalactic magnetic field constrained by Fermi/LAT observations of the TeV blazar 1ES 0229+200. Mon. Not. Roy. Astron. Soc. Series **406**, L70 (2010)
39. A. Taylor, I. Vovk, A. Neronov, Extragalactic magnetic fields constraints from simultaneous GeV-TeV observations of blazars. Astron. Astrophys. **529**, A144 (2011) [arxiv.org/abs/1101.0932]
40. K. Dolag, M. Kachelriess, S. Ostapchenko, R. Tomas, Lower limit on the strength and filling factor of extragalactic magnetic fields. Astrophys. J. Lett. Series **727**, L4 (2011)

41. A.E. Broderick, P. Chang, C. Pfrommer, The cosmological impact of luminous TeV Blazars I: implications of plasma instabilities for the intergalactic magnetic field and extragalactic gamma-ray background. Astrophys. J. Series **752**, 22 (2012)
42. F. Govoni, E. Orrù, A. Bonafede, M. Iacobelli, R. Paladino, F. Vazza et al., A radio ridge connecting two galaxy clusters in a filament of the cosmic web. Sci. Series **364**, 981–984 (2019)
43. CTA Consortium collaboration, B.S. Acharya et al., Science with the Cherenkov telescope array. WSP **11** (2018). https://doi.org/10.1142/10986 [arxiv.org/abs/1709.07997]
44. A. Korochkin, O. Kalashev, A. Neronov, D. Semikoz, Sensitivity reach of gamma-ray measurements for strong cosmological magnetic fields. Astrophys. J. Series **906**, 116 (2021)
45. K. Subramanian, J.D. Barrow, Microwave background signals from tangled magnetic fields. Phys. Rev. Lett. Series **81**, 3575 (1998)
46. K. Subramanian, J.D. Barrow, Small-scale microwave background anisotropies due to tangled primordial magnetic fields. Mon. Not. Roy. Astron. Soc. Series **335**, L57 (2002)
47. A. Mack, T. Kahniashvili, A. Kosowsky, Microwave background signatures of a primordial stochastic magnetic field. Phys. Rev. Series **D65**, 123004 (2002)
48. A. Lewis, Observable primordial vector modes. Phys. Rev. Series **D70**, 043518 (2004)
49. T. Kahniashvili, B. Ratra, Effects of cosmological magnetic helicity on the cosmic microwave background. Phys. Rev. Series **D71**, 103006 (2005)
50. G. Chen, P. Mukherjee, T. Kahniashvili, B. Ratra, Y. Wang, Looking for cosmological Alfven waves in WMAP data. Astrophys. J. **611**, 655 (2004). https://doi.org/10.1086/422213 [arxiv.org/abs/astro-ph/0403695]
51. A. Lewis, CMB anisotropies from primordial inhomogeneous magnetic fields. Phys. Rev. **D70**, 043011 (2004). https://doi.org/10.1103/PhysRevD.70.043011 [arxiv.org/abs/astro-ph/0406096]
52. H. Tashiro, N. Sugiyama, R. Banerjee, Nonlinear evolution of cosmic magnetic fields and cosmic microwave background anisotropies. Phys. Rev. Series **D73**, 023002 (2006)
53. D. Yamazaki, K. Ichiki, T. Kajino, G.J. Mathews, Constraints on the evolution of the primordial magnetic field from the small scale CMB angular anisotropy. Astrophys. J. Series **646**, 719 (2006)
54. M. Giovannini, Entropy perturbations and large-scale magnetic fields. Class. Quant. Grav. Series **23**, 4991 (2006)
55. T. Kahniashvili, B. Ratra, CMB anisotropies due to cosmological magnetosonic waves. Phys. Rev. Series **D75**, 023002 (2007)
56. M. Giovannini, K.E. Kunze, Magnetized CMB observables: a dedicated numerical approach. Phys. Rev. Series **D77**, 063003 (2008)
57. D.G. Yamazaki, K. Ichiki, T. Kajino, G.J. Mathews, New constraints on the primordial magnetic field. Phys. Rev. Series **D81**, 023008 (2010)
58. D. Paoletti, F. Finelli, CMB constraints on a stochastic background of primordial magnetic fields. Phys. Rev. Series **D83**, 123533 (2011)
59. J.R. Shaw, A. Lewis, Constraining primordial magnetism. Phys. Rev. Series **D86**, 043510 (2012)
60. K.E. Kunze, CMB anisotropies in the presence of a stochastic magnetic field. Phys. Rev. Series **D83**, 023006 (2011)
61. D. Paoletti, F. Finelli, Constraints on a stochastic background of primordial magnetic fields with WMAP and south pole telescope data. Phys. Lett. Series **B726**, 45 (2013)
62. M. Ballardini, F. Finelli, D. Paoletti, CMB anisotropies generated by a stochastic background of primordial magnetic fields with non-zero helicity. JCAP Series **1510**, 031 (2015)
63. Planck collaboration, Planck 2015 results. XIX. Constraints on primordial magnetic fields. Astron. Astrophys. **594**, A19 (2016). https://doi.org/10.1051/0004-6361/201525821 [arxiv.org/abs/1502.01594]
64. D.R. Sutton, C. Feng, C.L. Reichardt, Current and future constraints on primordial magnetic fields. Astrophys. J. Series **846**, 164 (2017)

65. A. Zucca, Y. Li, L. Pogosian, Constraints on primordial magnetic fields from Planck combined with the south pole telescope CMB B-mode polarization measurements. Phys. Rev. Series **D95**, 063506 (2017)
66. K. Jedamzik, V. Katalinic, A.V. Olinto, A Limit on primordial small scale magnetic fields from CMB distortions. Phys. Rev. Lett. Series **85**, 700 (2000)
67. A. Zizzo, C. Burigana, On the effect of cyclotron emission on the spectral distortions of the cosmic microwave background. New Astron. Series **11**, 1 (2005)
68. K.E. Kunze, E. Komatsu, Constraining primordial magnetic fields with distortions of the black-body spectrum of the cosmic microwave background: pre- and post-decoupling contributions. JCAP Series **1401**, 009 (2014)
69. J. Ganc, M.S. Sloth, Probing correlations of early magnetic fields using mu-distortion. JCAP Series **1408**, 018 (2014)
70. D. Paoletti, J. Chluba, F. Finelli, J.A. Rubino-Martin, Improved CMB anisotropy constraints on primordial magnetic fields from the post-recombination ionization history. Mon. Not. Roy. Astron. Soc. Series **484**, 185 (2019)
71. S.K. Sethi, K. Subramanian, Primordial magnetic fields in the post-recombination era and early reionization. Mon. Not. Roy. Astron. Soc. Series **356**, 778 (2005)
72. K.E. Kunze, E. Komatsu, Constraints on primordial magnetic fields from the optical depth of the cosmic microwave background. JCAP **1506**, 027 (2015). https://doi.org/10.1088/1475-7516/2015/06/027 [arxiv.org/abs/1501.00142]
73. J. Chluba, D. Paoletti, F. Finelli, J.-A. Rubiño-Martín, Effect of primordial magnetic fields on the ionization history. Mon. Not. Roy. Astron. Soc. Series **451**, 2244 (2015)
74. R. Durrer, P.G. Ferreira, T. Kahniashvili, Tensor microwave anisotropies from a stochastic magnetic field. Phys. Rev. Series **D61**, 043001 (2000)
75. T.R. Seshadri, K. Subramanian, CMBR polarization signals from tangled magnetic fields. Phys. Rev. Lett. Series **87**, 101301 (2001)
76. K. Subramanian, T.R. Seshadri, J. Barrow, Small-scale CMB polarization anisotropies due to tangled primordial magnetic fields. Mon. Not. Roy. Astron. Soc. Series **344**, L31 (2003)
77. S. Mollerach, D. Harari, S. Matarrese, CMB polarization from secondary vector and tensor modes. Phys. Rev. Series **D69**, 063002 (2004)
78. C. Scoccola, D. Harari, S. Mollerach, B polarization of the CMB from Faraday rotation. Phys. Rev. Series **D70**, 063003 (2004)
79. A. Kosowsky, T. Kahniashvili, G. Lavrelashvili, B. Ratra, Faraday rotation of the cosmic microwave background polarization by a stochastic magnetic field. Phys. Rev. Series **D71**, 043006 (2005)
80. L. Pogosian, T. Vachaspati, A. Yadav, Primordial magnetism in CMB B-modes. Can. J. Phys. Series **91**, 451 (2013)
81. T. Kahniashvili, Y. Maravin, G. Lavrelashvili, A. Kosowsky, Primordial magnetic helicity constraints from WMAP nine-year data. Phys. Rev. Series **D90**, 083004 (2014)
82. L. Pogosian, A. Zucca, Searching for primordial magnetic fields with CMB B-modes. Class. Quant. Grav. Series **35**, 124004 (2018)
83. I. Brown, R. Crittenden, Non-Gaussianity from cosmic magnetic fields. Phys. Rev. Series **D72**, 063002 (2005)
84. T. Seshadri, K. Subramanian, CMB bispectrum from primordial magnetic fields on large angular scales. Phys. Rev. Lett. Series **103**, 081303 (2009)
85. C. Caprini, F. Finelli, D. Paoletti, A. Riotto, The cosmic microwave background temperature bispectrum from scalar perturbations induced by primordial magnetic fields. JCAP Series **0906**, 021 (2009)
86. R.-G. Cai, B. Hu, H.-B. Zhang, Acoustic signatures in the cosmic microwave background bispectrum from primordial magnetic fields. JCAP Series **1008**, 025 (2010)
87. P. Trivedi, K. Subramanian, T.R. Seshadri, Primordial magnetic field limits from cosmic microwave background bispectrum of magnetic passive scalar modes. Phys. Rev. Series **D82**, 123006 (2010)
88. I.A. Brown, Intrinsic bispectra of cosmic magnetic fields. Astrophys. J. Series **733**, 83 (2011)

89. M. Shiraishi, D. Nitta, S. Yokoyama, K. Ichiki, K. Takahashi, Cosmic microwave background bispectrum of vector modes induced from primordial magnetic fields. Phys. Rev. Series **D82**, 121302 (2010)
90. M. Shiraishi, D. Nitta, S. Yokoyama, K. Ichiki, K. Takahashi, Cosmic microwave background bispectrum of tensor passive modes induced from primordial magnetic fields. Phys. Rev. Series **D83**, 123003 (2011)
91. P. Trivedi, T.R. Seshadri, K. Subramanian, Cosmic microwave background trispectrum and primordial magnetic field limits. Phys. Rev. Lett. Series **108**, 231301 (2012)
92. M. Shiraishi, T. Sekiguchi, First observational constraints on tensor non-Gaussianity sourced by primordial magnetic fields from cosmic microwave background. Phys. Rev. Series **D90**, 103002 (2014)
93. P. Trivedi, K. Subramanian, T.R. Seshadri, Primordial magnetic field limits from the CMB trispectrum: scalar modes and Planck constraints. Phys. Rev. Series **D89**, 043523 (2014)
94. R. Durrer, C. Caprini, Primordial magnetic fields and causality. JCAP Series **11**, 010 (2003)
95. G.K. Batchelor, Small-scale variation of convected quantities like temperature in turbulent fluid. Part 1. General discussion and the case of small conductivity. J. Fluid Mech. **5**, 113 (1959). https://doi.org/10.1017/S002211205900009X
96. K. Jedamzik, G. Sigl, The evolution of the large-scale tail of primordial magnetic fields. Phys. Rev. Series **D83**, 103005 (2011)
97. R. Banerjee, K. Jedamzik, The evolution of cosmic magnetic fields: from the very early universe, to recombination, to the present. Phys. Rev. D Series **70**, 123003 (2004)
98. K. Jedamzik, T. Abel, Cosmological recombination in the presence of primordial magnetic fields (in preparation) (2023)
99. K. Dolag, M. Bartelmann, H. LeschAstron, Astrophys. Series **348**, 351 (1999)
100. K. Dolag, M. Bartelmann, H. LeschAstron, Astrophys. Series **387**, 383 (2002)
101. J.L.W.V. Jensen, Sur les fonctions convexes et les inégalités entre les valeurs moyennes. Acta Mathematica Series **30**, 175 (1906). https://doi.org/10.1007/BF02418571
102. M. Zaldarriaga, D.D. Harari, Analytic approach to the polarization of the cosmic microwave background in flat and open universes. Phys. Rev. D Series **52**, 3276 (1995)
103. K. Jedamzik, L. Pogosian, G.-B. Zhao, Why reducing the cosmic sound horizon can not fully resolve the Hubble tension. arxiv.org/abs/2010.04158
104. L. Pogosian, G.-B. Zhao, K. Jedamzik, Recombination-independent determination of the sound horizon and the Hubble constant from BAO. Astrophys. J. Lett. Series **904**, L17 (2020)
105. Kilo-Degree Survey, DES collaboration, DES Y3 + KiDS-1000: consistent cosmology combining cosmic shear surveys. arxiv.org/abs/2305.17173
106. D. Scolnic et al., The complete light-curve sample of spectroscopically confirmed SNe Ia from Pan-STARRS1 and cosmological constraints from the combined Pantheon sample. Astrophys. J. Series **859**, 101 (2018)
107. shape DES collaboration, Dark energy survey year 1 results: cosmological constraints from galaxy clustering and weak lensing. Phys. Rev. Series **D98**, 043526 (2018)
108. shape ACT collaboration, The Atacama cosmology telescope: a measurement of the cosmic microwave background power spectra at 98 and 150 GHz. JCAP Series **12**, 045 (2020)
109. SPT-3G collaboration, Measurements of the E-mode polarization and temperature-E-mode correlation of the CMB from SPT-3G 2018 data. arxiv.org/abs/2101.01684

Chapter 33
Varying Fundamental Constants Meet Hubble

Jens Chluba and Luke Hart

33.1 Why Varying Fundamental Constants?

The laws of nature depend on fundamental constants (FCs) such as Newton's constant, G, the speed of light, c, and elementary charge of an electron, e, to name just a few. The values of these constants have been determined experimentally, but generally should emerge directly from the underlying theory. As such, there is no reason to assume that the values of the FCs determined locally simply translate to other parts of the cosmos or to other eras in cosmic history (see, [59, 91, 92] for broad review). Studies of fundamental constants and their possible temporal and spatial variations are thus of utmost importance, and could provide a glimpse at physics beyond the standard model, possibly shedding light on the presence of additional scalar fields and their couplings to the standard sector.

Of the many fundamental constants, the fine-structure constant, $\alpha_{\rm EM}$, and electron rest mass, $m_{\rm e}$, are the most interesting to CMB studies (e.g., [6–8, 47, 60, 62, 72, 80]). This is because these constants play crucial roles in the way photons and baryons interact. Most importantly, they affect important atomic transition rates, which in turn control the cosmological recombination process and hence the Thomson visibility function (defining the last scattering surface) that is so crucial to the formation of the CMB temperature and polarization anisotropies [42, 43, 57, 65, 89].

Using *Planck* 2013 data, the values of $\alpha_{\rm EM}$ and $m_{\rm e}$ around recombination were proven to coincide with those obtained in the lab to within $\simeq 0.4\%$ for $\alpha_{\rm EM}$ and $\simeq 1\% - 6\%$ for $m_{\rm e}$ [67]. These limits are $\simeq 2-3$ orders of magnitude weaker than constraints obtained from other 'local' measurements [11, 12, 53, 73]; however, the CMB places limits during very different phases in the history of the Universe,

J. Chluba (✉) · L. Hart (✉)
Jodrell Bank Centre for Astrophysics, Alan Turing Building, University of Manchester, Manchester M13 9PL, UK
e-mail: jens.chluba@manchester.ac.uk

L. Hart
e-mail: luke.hart@tneigroup.com

L. Hart
TNEI Services Ltd., Bainbridge House, 88 London Road, Manchester M1 2PW, UK

E. Di Valentino and D. Brout (eds.), *The Hubble Constant Tension*, Springer Series in Astrophysics and Cosmology, https://doi.org/10.1007/978-981-99-0177-7_33

centered around the time of last scattering some 380, 000 years after the Big Bang, thereby complementing these low-redshift measurements. In addition, CMB measurements can be used to probe spatial variations of the FCs at cosmological distances [86], opening yet another avenue for exploration.

With the *Planck* 2013 results in mind, no significant surprises were expected from the analysis of improved CMB data of the *Planck* 2015 and 2018 releases. However, it turns out that when considering models with varying m_e, the geometric degeneracy becomes significant and can accommodate shifts in the value of the Hubble parameter when multiple probes are combined [36]. The same geometric freedom is not encountered when varying α_{EM} due to the modified dependence of the visibility function on this parameter. This finding has spurred an increased interest in studying VFCs in this context, with scenarios that allow for varying m_e [e.g., [41, 54, 83, 90] for additional extensions] ranking high in model comparisons [79]. Since VFCs can be caused by the presence of scalar fields (e.g., [9, 61]), a natural question is whether the same scalar fields could also be causing effects relating to EDE (see Chap. 22), indicating a 'two sides of the same coin' interplay. It will therefore be extremely important to ask how different measurements can be combined to shed light on the physical origin of the Hubble tension.

In this chapter we explain how α_{EM} and m_e enter the cosmological recombination problem and calculation of the CMB power spectra. We then briefly recap some of the constraints, highlighting important findings, before moving on to a discussion of their role in the Hubble tension. Modifications to the recombination process also affect the cosmological recombination radiation (CRR) [15, 88], implying that direct insight into the underlying physics could be gained by future measurements of CMB spectral distortions (e.g., [14]). We highlight how this avenue may even allow distinguishing models of EDE, PMFs and VFCs and possibly identify modifications to the recombination history as the main cause of the tension [38, 58].

33.2 Effects of VFCs on the Recombination Process

The ionization history of the Universe is one of the crucial theoretical ingredients in the computations of the CMB temperature and polarization anisotropies [42, 65, 89]. The first computations of this transition from the fully-ionized plasma to a neutral medium were carried out in the late 60 s, recognizing the important role of Lyman-α transport and the 2 s-1 s two-photon decay of Hydrogen [64, 94]. These early calculations reached a precision of $\simeq$20–30% in the free electron fraction, $X_e = N_e/N_H$, around the maximum of the Thomson scattering visibility function at redshift $z \simeq 1100$. Here N_e is the free electron number density and N_H denotes the total number density of hydrogen nuclei in the Universe. However, with the advent of precision CMB data from *WMAP* and *Planck* it became important to improve the modeling of the cosmological recombination process [18, 42, 57]. Initially, this led to the development of `Recfast` [81], which reached $\simeq$1–3% precision for Hydrogen and neutral Helium recombination. However, for the analysis of data from *Planck*, this precision was still insufficient, and many detailed atomic physics and radiative transfer effects had to be accounted for (see, [31, 33, 74, 88, 93] for overview), leading to changes at the level of several standard deviations in particular for the value of the spectral index of scalar perturbations, n_S [74, 84]. This necessitated the development of the highly-flexible and accurate recombination codes `CosmoRec` [22] and `HyRec` [5], which ensured that for the analysis of *Planck* none of the standard parameters were biased at a significant level [66, 68].

This short recap highlights the crucial role of the recombination history in the computations of the CMB anisotropies using standard Boltzmann solvers such as `CAMB` [56] and `CLASS` [55], and conversely, the impressive precision and sensitivity of the current measurements to subtle modifications in the ionization history. The statements assume standard physics during the recombination era at $z \simeq 10^2 - 10^4$. In particular, it is assumed that the atomic transition rates for Hydrogen and Helium are the same as those inferred in the lab.

From standard textbook atomic physics, it is well-understood how $\alpha_{\rm EM}$ and $m_{\rm e}$ affect the energy levels and transition probabilities of Hydrogen and Helium (e.g., [10, 25]). It is immediately clear that varying $\alpha_{\rm EM}$ and $m_{\rm e}$ inevitably create changes to the cosmological ionization history and hence CMB observables. Most importantly, the energy levels of hydrogen and helium depend on these constants as $E_i \propto \alpha_{\rm EM}^2 m_{\rm e}$, which directly affects the recombination redshift. In addition, the atomic bound-bound transition rates and photoionization/recombination rates are altered when varying $\alpha_{\rm EM}$ and $m_{\rm e}$. Lastly, the interactions of photons and electrons through Compton and resonance scattering modify the radiative transfer physics, which control the dynamics of recombination (e.g., [4, 16, 19, 21, 40, 50, 75]).

In an effective three-level atom approach [64, 82, 94], the individual dependencies can be summarized as (e.g., [34, 47, 67, 80])

$$\begin{array}{lll} \sigma_{\rm T} \propto \alpha_{\rm EM}^2 m_{\rm e}^{-2} & A_{2\gamma} \propto \alpha_{\rm EM}^8 m_{\rm e} & P_{\rm S} A_{1\gamma} \propto \alpha_{\rm EM}^6 m_{\rm e}^3 \\ \alpha_{\rm rec} \propto \alpha_{\rm EM}^2 m_{\rm e}^{-2} & \beta_{\rm phot} \propto \alpha_{\rm EM}^5 m_{\rm e} & T_{\rm eff} \propto \alpha_{\rm EM}^{-2} m_{\rm e}^{-1}. \end{array} \tag{33.1}$$

Here, $\sigma_{\rm T}$ denotes the Thomson scattering cross section; $A_{2\gamma}$ is the two-photon decay rate of the second shell; $\alpha_{\rm rec}$ and $\beta_{\rm phot}$ are the effective recombination and photoionization rates, respectively; $T_{\rm eff}$ is the effective temperature at which $\alpha_{\rm rec}$ and $\beta_{\rm phot}$ need to be evaluated (see explanation below); $P_{\rm S} A_{1\gamma}$ denotes the effective dipole transition rate for the main resonances (e.g., Lyman-α), which is reduced by the Sobolev escape probability, $P_{\rm S} \leq 1$ [82, 87] with respect to the vacuum rate, $A_{1\gamma}$. For a more detailed account of how the transition rates depend on the fundamental constants we refer to `CosmoSpec` [15] and the manual of `HyRec` [5].

The scalings of $\sigma_{\rm T}$, $A_{2\gamma}$ and $P_{\rm S} A_{1\gamma}$ directly follow from their explicit dependence on $\alpha_{\rm EM}$ and $m_{\rm e}$. For $\alpha_{\rm rec}$ and $\beta_{\rm phot}$, only the renormalisations of the transition rates is reflected, again stemming from their explicit dependencies on $\alpha_{\rm EM}$ and $m_{\rm e}$ (e.g., [48]). However, these rates also depend on the ratio of the electron and photon temperatures to the ionization threshold. This leads to an additional dependence on $\alpha_{\rm EM}$ and $m_{\rm e}$, which can be captured by evaluating these rates at a rescaled temperature, with a scaling indicated through $T_{\rm eff}$. Overall, this leads to the effective dependence $\alpha_{\rm rec} \propto$

$\alpha_{\rm EM}^{3.44}\, m_{\rm e}^{-1.28}$ around hydrogen recombination [15]. The required photoionization rate, $\beta_{\rm phot}$, is obtained using the detailed balance relation. Slightly different overall scalings for $\alpha_{\rm rec}$ and $\beta_{\rm phot}$ were used in [67], but the associated effect on the recombination history were found to be sub-dominant and limited to $z \lesssim 800$ [34].

For neutral helium, non-hydrogenic effects (e.g., fine-structure transitions, singlet-triplet couplings) become relevant [25, 26]. However, the corrections should be sub-dominant and are usually neglected. A detailed discussion of changes to the escape probabilities during helium recombination can be found in [38]. In [34], the changes to helium and hydrogen recombination were furthermore treated separately.

33.2.1 Ionization History Modifications Due to Variation of $\alpha_{\rm EM}$

Given the above ingredients, one can now answer the question about how various effects propagate to the ionization history. A detailed study that also directly demonstrated the validity of simpler three-level approximations against CosmoRec was carried out in [34]. The overall effect of varying $\alpha_{\rm EM}$ is illustrated in Fig. 33.1, assuming a constant variation parametrized as $\alpha_{\rm EM} = \alpha_{\rm EM,\,0}(1 + \Delta\alpha/\alpha)$, where $\alpha_{\rm EM,\,0} = 1/137$ denotes the standard value. Increasing the fine structure constant shifts the moment of recombination toward higher redshifts. This agrees with the results found earlier in [8, 47, 72] and can intuitively be understood in the following manner: $\Delta\alpha/\alpha > 0$ increases the transition energies between different atomic levels and the continuum. This increases the energy threshold at which recombination occurs, hence increasing the recombination redshift, an effect that is basically captured by an effective temperature rescaling in the evaluation of the photoionization and recombination rates (see below).

The relative changes to the ionization history, $\Delta X_{\rm e}/X_{\rm e}$, for the different quantities in Eq. (33.1) are illustrated in Fig. 33.2. We chose a value for $\Delta\alpha/\alpha = 10^{-3}$, which leads to a percent-level effect on $X_{\rm e}$. As expected, the biggest effect appears after rescaling the temperature for the evaluation of the photonionization and recombination rates. More explicitly, this can be understood when considering the net recombination rate to the second shell, which can be written as[1]

$$\Delta R_{\rm con} = N_{\rm e} N_{\rm p} \alpha_{\rm rec} - N_2\, \beta_{\rm phot} = \alpha_{\rm rec}[N_{\rm e} N_{\rm p} - g(T_\gamma) N_2],$$

where $g(T_\gamma) \propto {T_\gamma}^{3/2}\, {\rm e}^{-h\nu_{2c}/kT_\gamma}$ with continuum threshold energy, $E_{2c} = h\nu_{2c}$. Here, the exponential factor ($\leftrightarrow$ Boltzmann factor) is most important, leading to an exponential effect once we replace $T_\gamma{}'(z) = T_\gamma(z) \times (\alpha_{\rm EM}/\alpha_{\rm EM,\,0})^{-2}(m_{\rm e}/m_{\rm e,\,0})^{-1}$.

The second largest individual effect is due to the rescaling of the two-photon decay rate, $A_{2\gamma}$. This is expected since $\alpha_{\rm EM}$ appears in a high power, $A_{2\gamma} \propto \alpha_{\rm EM}^8$, and also because the 2 s-1 s two-photon channel plays such a crucial role for the recombination

[1] In full equilibrium, $\Delta R_{\rm con} = 0$.

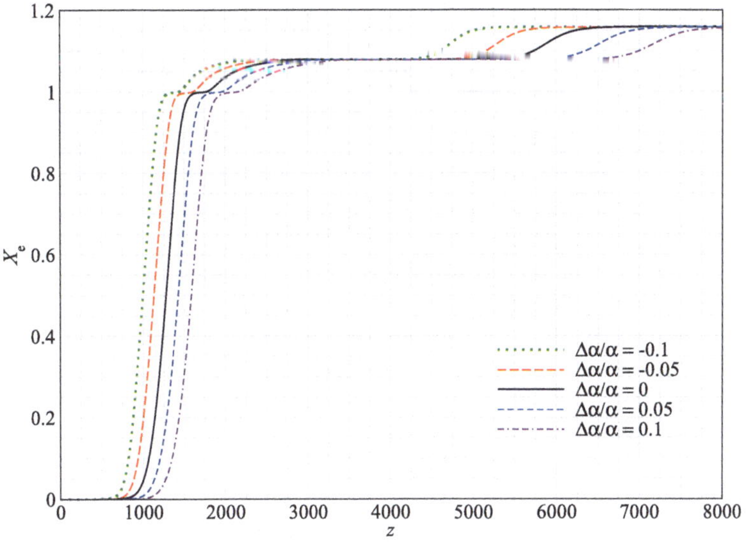

Fig. 33.1 Ionization histories for different values of $\alpha_{\rm EM}$. The dominant effect is caused by modifications of the ionization threshold, which implies that for increased $\alpha_{\rm EM}$ recombination finishes earlier. The curves were computed using the `Recfast++` module of `CosmoRec` [22]. The figure was taken from [34]

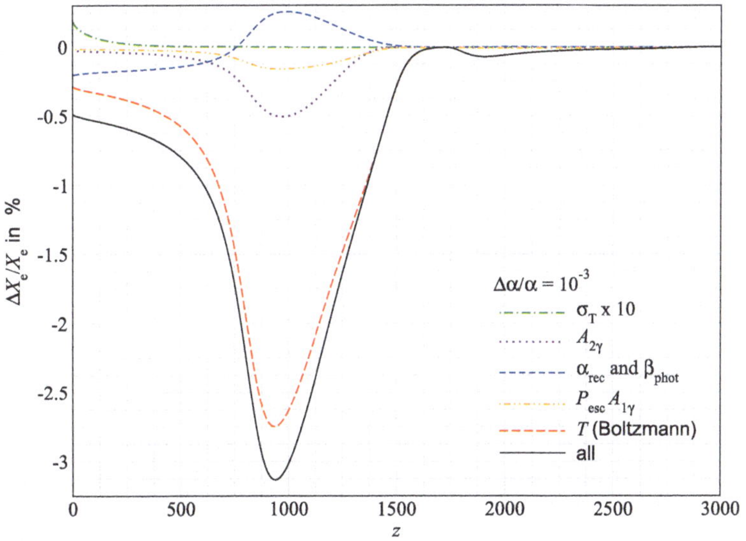

Fig. 33.2 The relative changes in the ionization history for $\Delta\alpha/\alpha = 10^{-3}$ with respect to the standard case caused by different effects. `Recfast++` was used for the computations. The rescaling of temperature ($\leftrightarrow$ mainly affecting the Boltzmann factors) yields $\Delta X_{\rm e}/X_{\rm e} \simeq -2.7\%$, dominating the total contributions, which peaks with $\simeq -3.1\%$ at $z \simeq 1000$. Note that the modification due to $\sigma_{\rm T}$ has been scaled by a factor of ten to make it visible. The figure was taken from [34]

dynamics [18, 64, 94], allowing $\simeq 58\%$ of all hydrogen atoms to become neutral through this route [17].

The normalizations of the recombination and photoionization rates (blue/dashed line in Fig. 33.2) give rise to a net delay of $\Delta X_e/X_e \simeq 0.3\%$ at $z \simeq 1000$, which partially cancels the acceleration due to $A_{2\gamma}$. This is due to the stronger scaling of $\beta_{\rm phot}$ with $\alpha_{\rm EM}$ than $\alpha_{\rm rec}$. At low redshifts ($z \lesssim 750$), recombination is again accelerated, indicating that a higher fraction of recombination events occurs, as the importance of photoionization ceases. The correction related to the Lyman-α channel is found to be $\simeq$3.3 times smaller than for the two-photon channel, yielding $\Delta X_e/X_e \simeq -0.15\%$ at $z \simeq 1000$. Finally, the effect of rescaling σ_T are very small and only becomes noticeable at low redshifts. At these redshifts, the matter and radiation temperature begins to depart, giving $T_e < T_\gamma$. For larger $\alpha_{\rm EM}$, this departure is delayed, such that T_e stays longer close to T_γ. Hotter electrons recombine less efficiently, so that a slight delay of recombination appears (cf., Fig. 33.2).

33.2.2 Ionization History Modifications Due to Variation of m_e

We now briefly illustrate the changes caused by the effective electron mass, with the variation parametrized as $m_e = m_{e,0}(1 + \Delta m_e/m_e)$ with respect to the standard value $m_{e,0}$. Inspecting the scalings of Eq. (33.1), one expects the overall effect to be smaller than for $\alpha_{\rm EM}$, but otherwise very comparable. For example, the effect of temperature rescaling should be roughly half as large. Similarly, the effect due to rescaling $A_{2\gamma}$ should be roughly 8 times smaller, and so on. This is in good agreement with our findings (cf. Fig. 33.3).

The net effect on X_e is about 2.5 times smaller than for $\alpha_{\rm EM}$ around $z \simeq 1000$. This suggests that the CMB constraint on m_e is weakened by a similar factor. However, adding the rescaling of the Thomson cross section for the computation of the visibility function strongly enhances the geometric degeneracy for m_e, such that the CMB only constraint on m_e is $\gtrsim$20 times weaker than for $\alpha_{\rm EM}$ (see Sect. 33.3). In addition, a small difference related to the renormalizations of the photoionization and recombination rates (blue/dashed line) appears. For $\Delta m_e/m_e > 0$, the photoionization rate is increased and the recombination rate is reduced for these contributions [cf. Eq. (33.1)]. Both effects delay recombination (see Fig. 33.3). Thus, around $z \simeq 10^3$ the net effect is slightly larger than for $\alpha_{\rm EM}$. In contrast to $\alpha_{\rm EM}$, at late times no net acceleration of recombination occurs. These effects slightly modify the overall redshift dependence of the total X_e change, in addition to lowering the effect in the freeze-out tail. At the level $\Delta m_e/m_e \simeq 1\%$, additional higher order terms become important, allowing one to break the near degeneracy between $\alpha_{\rm EM}$ and m_e in joint analyses (see also [67]). At the level of the current CMB only constraints, this aspect indeed is of relevance (Sect. 33.3.2).

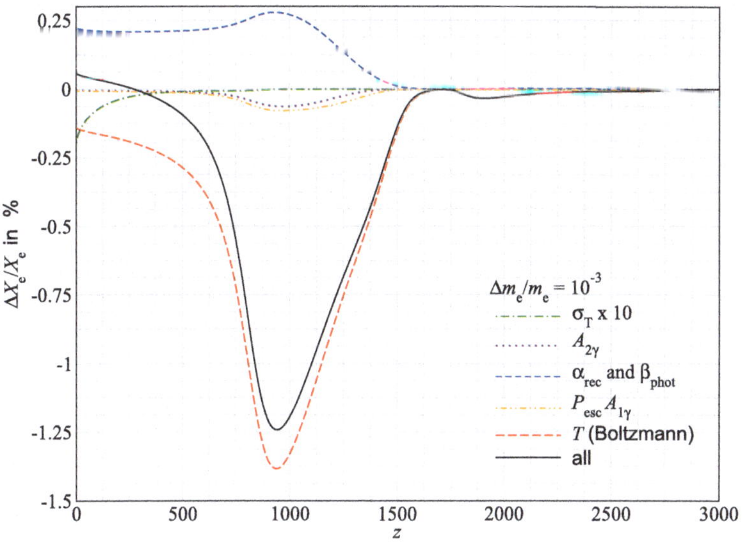

Fig. 33.3 Same as in Fig. 33.2 but for $\Delta m_e/m_e = 10^{-3}$. The effective temperature rescaling again dominates the total change. Around $z \simeq 1000$, the total effect is $\simeq 2.5$ times smaller than for $\Delta\alpha/\alpha = 10^{-3}$ mainly due to the weaker dependence of the level energy on m_e. The figure was taken from [34]

33.2.3 Comparison with `CosmoRec` *and Generalized VFC Models*

We highlight that [34] also explicitly demonstrated that a full treatment of the problem using the advanced recombination code `CosmoRec` yields results that are very similar to those from a simpler three-level treatment with correction function approach (e.g., [85] fordetails) to mimic the recombination physics corrections. This also allowed [34] to perform calculations with explicit time-dependence of α_{EM} and m_e, initially focusing on a simple power-law redshift dependence

$$\alpha_{EM}(z) = \alpha_{EM}(z_0)\left(\frac{1+z}{1100}\right)^p, \qquad (33.2)$$

for α_{EM} and similarly for m_e. Using a principal component analysis (PCA) [37], which extended the previous recombination perturbation framework developed in [29, 30, 35], general VFC variations around the recombination era were further studied. This showed that time-dependence can in fact be independently constrained already with existing data and also led to various generalized limits on VFCs during recombination (e.g., [54, 90]), which we will briefly highlight below.

33.3 CMB Anisotropy Constraints

33.3.1 Propagating the Effects to the CMB Anisotropies

The temperature and polarization power spectra of the CMB depend on the dynamics of recombination through the ionization history, which defines the Thomson visibility function and last scattering surface (e.g., [44, 65, 89]). Therefore, variations of α_{EM} and m_e can leave a direct imprint on the CMB power spectra. A detailed description of changes to the visibility function and various additional illustrations also for time-dependent VFCs can be found in [34, 37]. Here, we highlight the changes to the CMB temperature power spectra for variations of α_{EM} and m_e, noting that those in polarization show very similar features. The power spectra were computed using CAMB [56] for the standard cosmology [66].

To propagate the effect of VFCs to the CMB anisotropies, two changes are required. The standard recombination history has to be replaced as explained in the previous section. In addition, the Thomson scattering rate has be to updated in the Boltzmann code using the modified Thomson cross section. For the latter, two approaches are possible, one based on directly modifying the Boltzmann code, the other on mimicking the effect by rescaling the ionization history. Both seem to deliver consistent final results [34]. However, we stress how important it is to include the changes to the Thomson scattering rate, as without this effect α_{EM} and m_e variations essentially exhibit a very similar phenomenology [34, 36].

In Fig. 33.4, we illustrate the effect of α_{EM} on the CMB temperature power spectrum. Two main features are visible: Firstly, the peaks of the power spectrum are

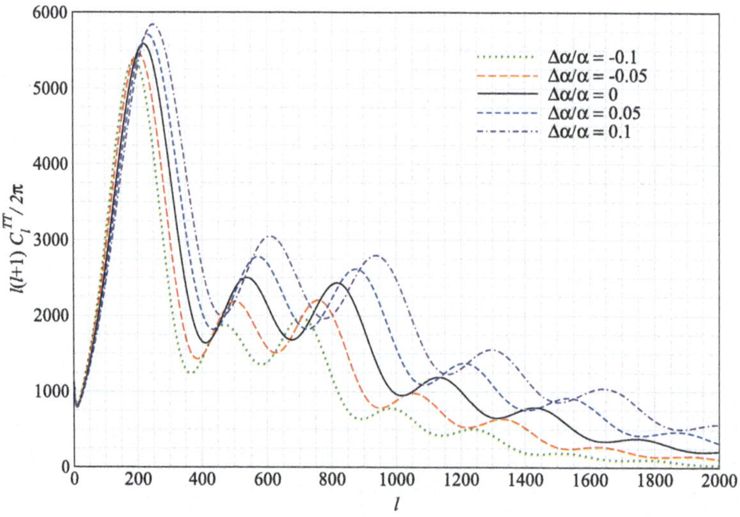

Fig. 33.4 The CMB temperature power spectra for different values of α_{EM}. This shows that as the fine structure constant increases, the anisotropies shift toward smaller scales and higher amplitudes. The figure was taken from [34]

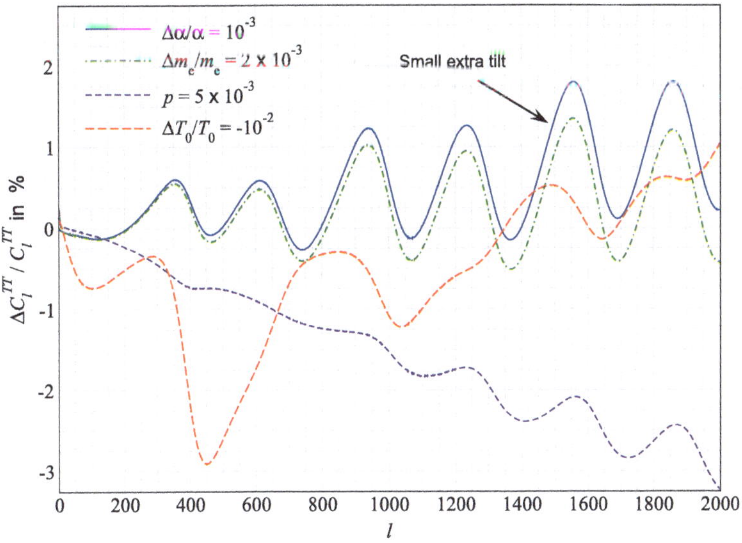

Fig. 33.5 Comparison of the CMB TT power spectrum deviations when varying $\alpha_{\rm EM}$, $m_{\rm e}$, T_0 and p. We chose $\Delta\alpha/\alpha = 10^{-3}$, $\Delta m_{\rm e}/m_{\rm e} = 2 \times 10^{-3}$, $\Delta T/T = -10^{-2}$ and $p = 5 \times 10^{-3}$ (simultaneously for $\alpha_{\rm EM}$ and $m_{\rm e}$) to obtain effects at a similar level. Notice the small extra tilt when comparing the case for $\alpha_{\rm EM}$ with $m_{\rm e}$, which helps when constraining $\alpha_{\rm EM}$. The figure was taken from [34]

shifted to smaller scales (larger ℓ) when $\Delta\alpha/\alpha > 0$. This happens because earlier recombination moves the last scattering surface towards higher redshifts, which decreases the sound horizon and increases the angular diameter distance to recombination [8, 47]. Secondly, for $\Delta\alpha/\alpha > 0$, the peak amplitudes are enhanced. This is mainly because earlier recombination suppresses the effect of photon diffusion damping on the anisotropies [8, 47]. For variations of $m_{\rm e}$, very similar responses are found, but with an amplitude that is reduced by a factor of $\simeq$2–3 [34].

For small $\Delta\alpha/\alpha$ and $\Delta m_{\rm e}/m_{\rm e}$, we illustrate the relative change of the temperature power spectrum in Fig. 33.5. The effect on the peak positions is more noticeable than the small overall tilt caused by changes related to diffusion damping. As expected, the changes to the CMB TT power spectra, $\Delta C_\ell/C_\ell(\Delta\alpha/\alpha)$ and $\Delta C_\ell/C_\ell(\Delta m_{\rm e}/m_{\rm e})$, become almost indistinguishable when using $\Delta m_{\rm e}/m_{\rm e} \approx (2-3)\,\Delta\alpha/\alpha$. This presents a quasi-degeneracy between the two parameters and also suggests that naively the analysis for $\Delta\alpha/\alpha$ could be sufficient to estimate the errors for a corresponding analysis of $m_{\rm e}$. However, when constraining $m_{\rm e}$, an enhanced geometric degeneracy, because of the differing effect of $\sigma_{\rm T}$, inflates the error to the percent level [34, 67]. In this case, higher order terms become important and the near degeneracy is broken. When also adding information from BAO, the error on $m_{\rm e}$ is strongly reduced, and a simple scaling of the errors, $\sigma(\Delta m_{\rm e}/m_{\rm e}) \simeq 3\sigma(\Delta\alpha/\alpha)$, is recovered [34].

In Fig. 33.5, we also illustrate the effect of varying the average CMB temperature, T_0, and a time-dependent model for both $\alpha_{\rm EM}$ and $m_{\rm e}$ with a phenomenological power-law as in Eq. (33.2) around a pivot redshift of $z = 1100$. Changes in the CMB monopole temperature show a different response pattern than VFCs, making these two effects principally distinguishable. This is because varying the CMB temperature affects the ionization history (at leading order like VFCs), but without changing the Thomson scattering rate. In addition, T_0 modifies the matter-radiation equality and therefore has a separate overall effect. Similarly, we conclude that the power-law index of the time-dependent model can be independently constrained [34].

33.3.2 Constraints from Planck

Now that we have developed a detailed understanding about how the CMB anisotropies are affected by changes of $\alpha_{\rm EM}$ and $m_{\rm e}$, we can directly consider some of the existing constraints from *Planck*. Early constraints were derived using *Planck* 2013 data in [67]. Aside from additional data (e.g., *Planck* polarization) and improvements in the understanding of systematics and calibration, the later analysis of the *Planck* 2015 data yielded similar constraints [34], although with slightly improved errors.[2] Here we highlight the latest *Planck* 2018 results, which were obtained in [36]. For details we refer the interested reader to that paper.

In Table 33.1, we summarize some of the *Planck* 2018 constraints on $\alpha_{\rm EM}$ and $m_{\rm e}$, including the 2018 baseline *Planck* data, with low-ℓ and high-ℓ data for temperature and E-mode polarisation power spectra, along with the lensing data from the same release [69, 70]. The addition of $\alpha_{\rm EM}$ marginally affects the values and errors of the six standard parameters, yielding $\alpha_{\rm EM}/\alpha_{\rm EM,0} = 1.0005 \pm 0.0024$. In contrast, varying $m_{\rm e}$ has a strong effect especially on H_0, shifting it to extremely low values and allowing for a low value of the electron rest mass, $m_{\rm e}/m_{\rm e,0} = 0.888 \pm 0.059$. This is caused by the small differences in the way $m_{\rm e}$ affects the CMB power spectra, with a crucial role played by adding changes to $\sigma_{\rm T}$ [34, 36]. As we discuss below, this large geometric degeneracy is one of the key ingredients for alleviating the Hubble tension when combined with supernova data.

In Fig. 33.6, we show the two-dimensional posteriors for the constraint on simultaneously varying values of $\alpha_{\rm EM}$ and $m_{\rm e}$. For comparison, the contours relating to the 2015 analysis and also combinations with baryon acoustic oscillation (BAO, [3]) data are illustrated. The constraints for *Planck* alone exhibit very wide posteriors, with slight differences in the centroids of the 2015 and 2018 constraints. We note that a narrower prior of $H_0 > 40\,{\rm km\,s^{-1}\,Mpc^{-1}}$ (conforming with the initial CMB analysis of *Planck* [67]) affects the posterior for the 2015 data, shrinking it in the $m_{\rm e}$-direction. This further highlights the significant differences in the role of $\alpha_{\rm EM}$ and $m_{\rm e}$ on the CMB anisotropies.

[2] For a detailed discussion of the effects of various analysis choices we refer to [34].

Table 33.1 Marginalised values of the fine structure constant and effective electron mass $\alpha_{\rm EM}$ and $m_{\rm e}$ using the *Planck* 2018 data alone. A very wide prior $\left(H_0 > 20\,\text{km s}^{-1}\,\text{Mpc}^{-1}\right)$ for H_0 was used to avoid biasing the marginalised $m_{\rm e}$ posterior, which affected some of the results of the 2013 analysis [34, 36]

Parameter	*Planck* 2018	*Planck* 2018 + varying $\alpha_{\rm EM}$	*Planck* 2018 + varying $m_{\rm e}$
$\Omega_b h^2$	0.02237 ± 0.00015	0.02236 ± 0.00015	$0.0199^{+0.0012}_{-0.0014}$
$\Omega_c h^2$	0.1199 ± 0.0012	0.1201 ± 0.0014	0.1058 ± 0.0076
$100\theta_{MC}$	1.04088 ± 0.00031	1.0416 ± 0.0034	0.958 ± 0.045
τ	0.0542 ± 0.0074	0.0540 ± 0.0075	0.0512 ± 0.0077
$\ln(10^{10} A_{\rm s})$	3.044 ± 0.014	3.043 ± 0.015	3.029 ± 0.017
$n_{\rm s}$	0.9649 ± 0.0041	0.9637 ± 0.0070	0.9640 ± 0.0040
$\alpha_{\rm EM}/\alpha_{\rm EM,0}$	–	1.0005 ± 0.0024	–
$m_{\rm e}/m_{\rm e,0}$	–	–	0.888 ± 0.059
H_0 [km s^{-1} Mpc^{-1}]	67.36 ± 0.54	67.56 ± 0.99	46^{+9}_{-10}

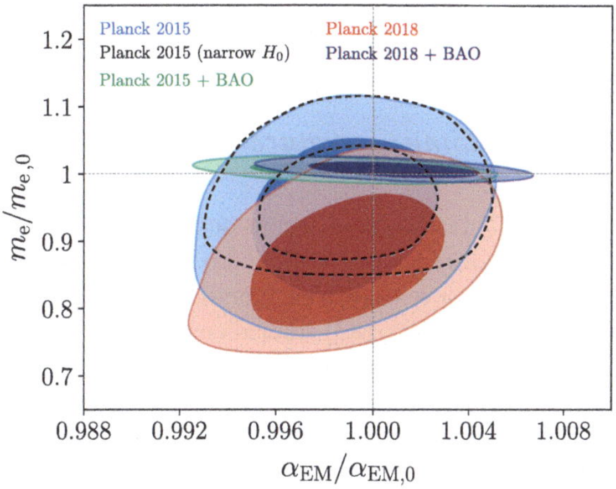

Fig. 33.6 Posterior contours between $\alpha_{\rm EM}$ and $m_{\rm e}$ for the *Planck* 2015 and 2018 data along with BAO contributions. Note that the dashed contour shows the 2015 contour but with a tighter prior on $H_0 \in \{40, 100\}$ km s^{-1} Mpc^{-1} to conform with the previous *Planck* 2013 analysis. The figure was taken from [36]

The introduction of BAO data significantly tightens the constraints on $m_{\rm e}$ and we can also observe a small drift in the centeral value of $\alpha_{\rm EM}$. The obtained value changes from $\alpha_{\rm EM}/\alpha_{\rm EM,\,0} = 0.9989 \pm 0.0026$ for the 2015 data to $\alpha_{\rm EM}/\alpha_{\rm EM,\,0} = 1.0010 \pm 0.0024$ with 2018 data, both with BAO included and when simultaneously varying $m_{\rm e}$. By contrast, there is no drift for $m_{\rm e}$, with $m_{\rm e}/m_{\rm e,\,0} = 1.0056 \pm 0.0080$ changing to $m_{\rm e}/m_{\rm e,\,0} = 1.0054 \pm 0.0080$ for *Planck* 2015+BAO and *Planck* 2018+BAO, respectively. These findings highlight a strong level of agreement of *Planck* with BAO data, which has been emphasized on many occasions (e.g., [68]).

33.4 Alleviating the Hubble Tension with VFCs

The attentive reader will already have noticed the route forward to alleviating the Hubble tension through varying the electron rest mass. This possibility was first noticed in [36], where the constraints on VFCs from different data combinations were studied, also adding supernova/Cepheid (R19, [71]) data.

A summary of the constraints is shown in Fig. 33.7 with particular focus on the interplay with H_0. For $\alpha_{\rm EM}$, we can notice broadly consistent constraints across all data combinations with only a small shift in the value of H_0 towards R19 in the combined constraints, indicating a resistance in terms of geometric freedom. For $m_{\rm e}$, two effects are found. First, as pointed out above, when adding BAO data, the error on $m_{\rm e}$ is significantly reduced in comparison to the CMB only constraints, bringing H_0 into agreement with the only CMB inference. Second, when also adding R19, a non-standard value of $m_{\rm e}$ (at $\simeq 3.5\sigma$ significance) is traded for a reduction of the Hubble tension. This begs the question whether $m_{\rm e}$ could indeed play a role in this problem. A model comparison study carried out in [79] indeed indicated that a simple variation of $m_{\rm e}$ provides a good contender in this respect, although not all issues could be resolved.

33.4.1 Adding Curvature

As Fig. 33.7 clearly shows, a constant variation of $m_{\rm e}$ does not fully solve the Hubble tension, albeit reducing it below 2 σ. As one possible extension, [36] also studied solutions with power-law VFC time-dependence, but found this to not further improve matters. A little later, [83] studied cosmologies with non-zero curvature in addition to variations of $m_{\rm e}$, finding these models to solve the Hubble tension. Indeed, this possibility was the winning finalist in the H_0-Olympics model comparison exercise [79]. However, allowing for non-zero curvature does open yet another non-standard direction in cosmology (in addition to accepting varying $m_{\rm e}$), with strong resistance in terms of standard inflationary predictions [24]. One could feel inclined to cling on to zero curvature cosmologies given the great successes of the inflationary paradigm, but ultimately observations and careful analysis will have to decide.

33.4.2 Time-Dependent VFCs Models

Most of the results presented above assumed constant (i.e., time-independent/single-valued) changes to the values of FCs in the early Universe. As highlighted in [34], explicitly time-dependent VFCs can in principle be constrained independently even with existing CMB data, given that the CMB responses are distinct (see Fig. 33.5). This idea was later generalized by applying a principal component analysis (PCA) to

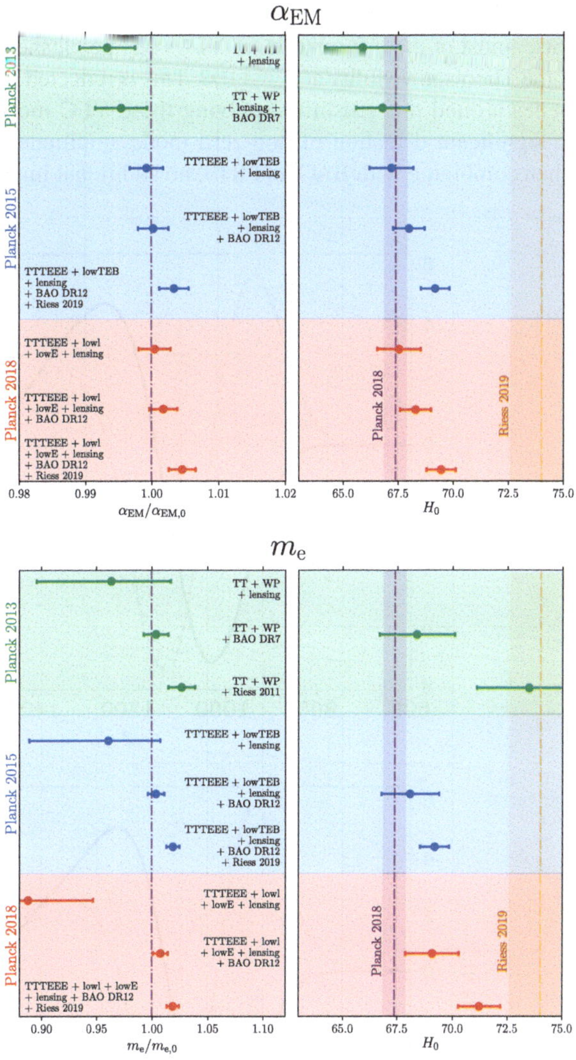

Fig. 33.7 Constraints on the fundamental constants (*left*) using various combinations of *Planck* data together with their H_0 values and errors (*right*). *Top:* results from the fine structure constant α_{EM}. *Bottom:* similar results but from the effective electron mass m_e. Here, we have redacted the constraint for H_0 from CMB data only because the error bars are so large. For the m_e MCMC analysis, we have widened the prior on the Hubble constant such that $H_0 > 20\,\mathrm{km\,s^{-1}\,Mpc^{-1}}$. Figure is from [36]

possible time-dependent VFC perturbations around the standard value in the hopes to further diminish the Hubble tension [37]. The idea is very simple: given the (CMB) observables, it is difficult to limit a specific change in the values of FCs at a given redshift. However, certain correlated variations/perturbations (i.e., VFC modes) across wider ranges of redshifts can indeed be analysized and constrained. To construct these VFC eigenmodes, one can use the Fisher information matrix to assess the observability. This yields a ranked system of possible VFC modes that can be best constrained by the data. Similar approaches were applied to studies of perturbations in the reionization [63] and recombination [29, 30, 35] histories.

The first few $\alpha_{\rm EM}$ and $m_{\rm e}$ eigenmodes are shown in Fig. 33.8. In a perturbative sense, most of the information on VFCs can be gleaned from around the maximum of the Thomson visiblity at $z \simeq 1100$. This is reflected in the localization of the VFC modes around this redshift. Applying these VFC modes to the *Planck* 2018 data no significant detection of non-zero mode amplitudes was found [37]. Similarly, when combining with BAO and R19, no additional improvement with respect to the

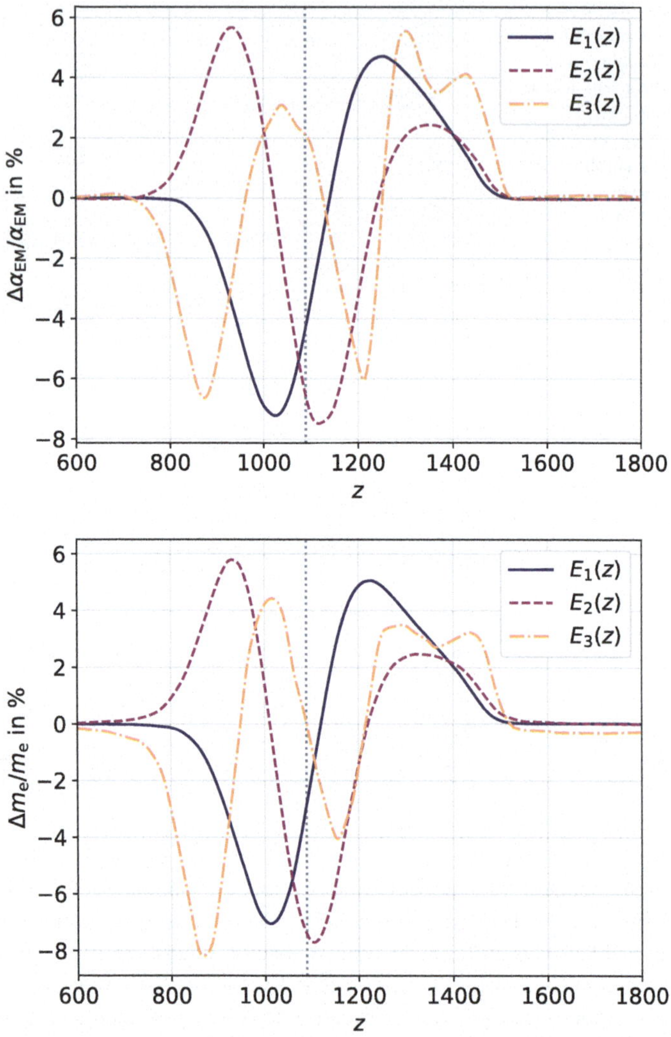

Fig. 33.8 The first three VFC principal components for $\alpha_{\rm EM}$ *(top)* and $m_{\rm e}$ *(bottom)* as constructed using *Planck* 2018 data. The eigenmodes are all normalised such that $\int |E_i^2(z)|{\rm d}z = 1$. The maximum of the ΛCDM Thomson visibility function has been marked as vertical dotted line. Figure is taken from [37]

Hubble tension over a constant variation was identified [37]. However, to exclude the possibility of a more general time-dependent VFC history, this would require further investigations of time-dependent models that interface between the standard candle, low-redshift era and the surface of last scattering.

The PCA approach inherently assumes a perturbative variation of the FCs. This assumption need not hold, and in addition higher order eigenmodes that individually fall below the detection threshold (due to their more rapid redshift-variablity) could together allow for more general modifications (see, [30]). In [54], an assessment of which time-dependent change to $\alpha_{\rm EM}$ and $m_{\rm e}$ would be required to reduce the Hubble tension was carried out. It was demonstrated that more general time-dependent variations of $\alpha_{\rm EM}$ and $m_{\rm e}$ around recombination can solve the Hubble tension (and even reduce the S_8 tension) when applied to CMB data alone. However, once BAO and supernova data is added, full solutions to the Hubble tension evade a perturbative treatment, although extension of the framework to the non-perturbative regime can be done [54]. This points towards the possibility that more general theoretical models could indeed help to restore consistency between various probes, although more work is certainly needed.

33.5 New Insights from the Cosmological Recombination Radiation

In the previous sections, we have primarily focused on illustrating the role of VFCs, and in particular of $m_{\rm e}$, in possible solutions to the Hubble tension. In this, one key ingredient is the associated modification to the cosmological recombination history, although it seems clear that a simple perturbative change does not provide a sufficient degree of freedom to fully resolve the tension. However, stepping back from a VFC driven solution, one important question is how we could tell if indeed a (strongly) modified recombination history is responsible for the Hubble tension? Furthermore, solutions based on PMFs (see Chap. 32) are also mainly successful because of their modifications to the average recombination history. It is thus important to ask how could we distinguish different options?

One important avenue forward is to use more accurate measurements with upcoming CMB experiments such as The Simons Observatory [2] and CMB-S4 [1] to improve the constraints on VFC (see, [36, 37] for some projections). However, ultimately access to new observables will be required.

The cosmological recombination processes is also associated with the emission of photons from the Hydrogen and Helium plasma (e.g., [13, 17, 27, 28, 49, 75–77]). The cosmological recombination radiation (CRR) can now be accurately computed using `CosmoSpec` [15], which was extended to also incorporate the effect of VFCs [38]. This CMB distortion signal may become observable [39] with future CMB spectrometers akin to *PIXIE* [14, 51, 52], opening the exciting possibility to directly study the dynamics of the recombination process [20, 88].

In Fig. 33.9, we illustrate the effect of varying $\alpha_{\rm EM}$ and $m_{\rm e}$ on the CRR. Firstly, one can clearly see that for the chosen parameters there is a noticeable effect on the amplitude and position of the CRR features. For $\alpha_{\rm EM}$, the modifications are more pronounced, which in part is due to a near degeneracy of $m_{\rm e}$ with modifications to the time of recombination [38]. Nevertheless, it is clear that precision measurements of the average CMB spectrum could principally identify these modifications in compar-

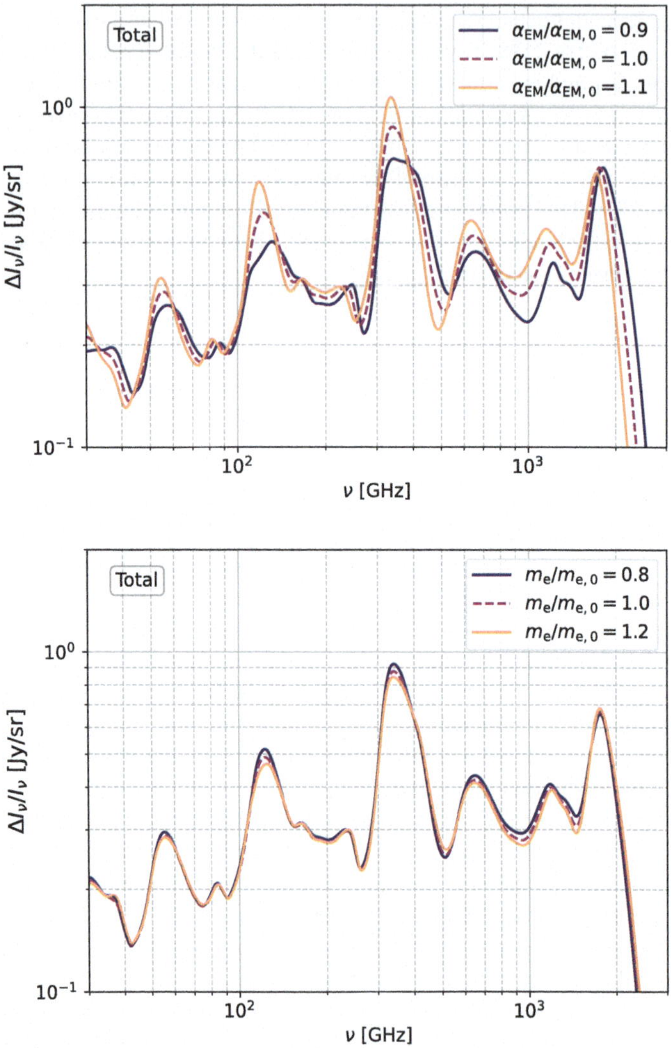

Fig. 33.9 The total impact on the recombination lines from the variations for $\Delta\alpha_{\rm EM}/\alpha_{\rm EM,0} = \pm 0.1$ *(top)*. The total variations due to changes in $\Delta m_{\rm e}/m_{\rm e,0} = \pm 0.2$ are shown for comparison *(bottom)*. Figure taken from [38]

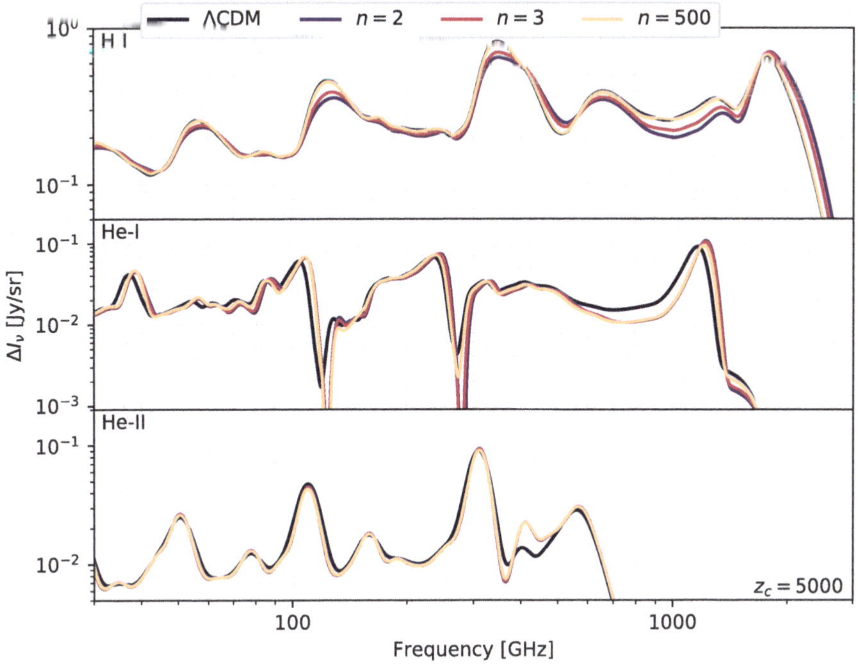

Fig. 33.10 Comparison for an EDE model that has different slopes for $z_c = 5000$ across the different atomic species. Hydrogen (*top*) is compared against the Helium I and II (*middle, bottom*) lines, respectively. For better clarity, this has been plotted with $f_{\rm ede} = 0.8$. Figure taken from [38]

ison to the standard ΛCDM prediction, thereby shedding new light on the physical cause of the problem.

However, we would like to highlight that these novel opportunities are not limited to searches for VFCs. Also EDE models (Chap. 22) indirectly affect the shape of the CRR [38]. In Fig. 33.10, we illustrate the effect of EDE on the CRR. Importantly, depending on the parameter choices (see, [38] for additional details), the redshifting between the Hydrogen and Helium recombination lines can be modified in addition to when the lines are created. This leads to principally observable features in the CRR, would a sufficient spectral sensitivity be reached. Finally, small-scale perturbations in the baryonic density, as possibly induced by PMFs (Chap. 32), can leave an imprint to the CRR that could be used to identify this process (see Fig. 33.11). Physically, the modifications are simply related to the fact that the average CRR is no longer given by the CRR of the average parameters, as non-linear (non-perturbative) corrections to the recombination process remain. Overall, all these example illustrate that future CMB spectral distortion measurements could in principle allow us to directly test the underlying recombination process. From the scientific point of view it would actually be crucial to investigate the recombination history directly, thereby eliminating one of the remaining theoretical ingredients of CMB cosmology. However, it is also

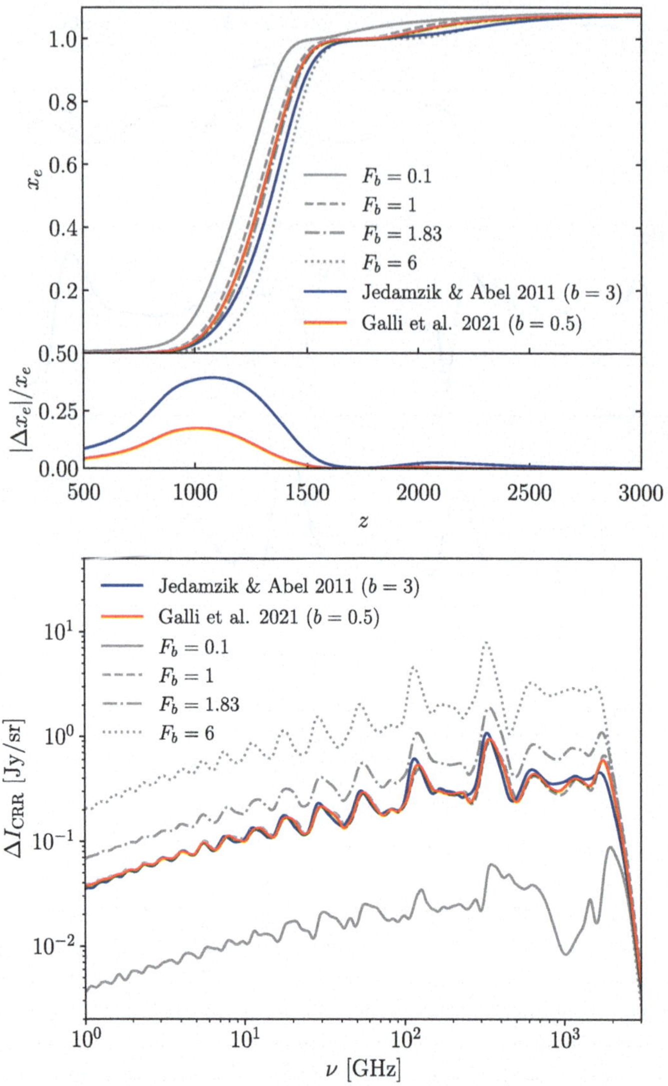

Fig. 33.11 Recombination histories (upper panel) and CRR spectra (lower panel) for the PMF models similar to those consider in [32, 45, 46]. The individual spectra for various values of the baryon density enhancement factor are shown as gray lines (the $F_b = 1$ case corresponds to ΛCDM). The averages are instead displayed in blue and red, and show significant second order contributions, manifesting in smearing of the lines and shifts in their position

clear that due to the presence of foregrounds the required spectral sensitivity and coverage are still quite futuristic [23, 39, 78]. In addition, a control of systematics and removal of foregrounds will have to be performed to unprecedented precision. These challenges will remain to be solved by generations of cosmologists to come.

33.6 Conclusions

In this chapter, we illustrated the role of VFCs in the Hubble tension. Previous studies have shown that varying m_e could indeed offer viable solutions, although no VFC scenario could fully resolve all aspects of the tension. It is however important to note that additional studies in the non-perturbative regime (i.e., significant changes with explicit time-dependence) could further improve the consistency. In addition, so far no attempt has been made to simultaneously treat VFCs and EDE, although both could principally originate from the same scalar field. With the advent of improved cosmological data, these lines of research might become very interesting.

One obvious question remains: how can we ultimately distinguish between various solutions to the Hubble tension. As highlighted in Sect. 33.5, EDE, PMFs and VFCs can in principle also be (directly) constrained through detailed measurements of the CRR. This guaranteed ΛCDM signal may become observable in the future with advanced CMB spectrometers and would open a way to confront our understanding of the cosmological recombination process with direct observational evidence. Should the Hubble tension not be resolved in the next decades, this future probe might be one of the important avenues towards a final primordial test.

Acknowledgements This work was supported by the ERC Consolidator Grant *CMBSPEC* (No. 725456). JC was furthermore supported by the Royal Society as a Royal Society University Research Fellow at the University of Manchester, UK (No. URF/R/191023). Part of this work was carried out in the stimulating and peaceful environment offered at the Aspen Center for Physics, which is supported by the National Science Foundation grant PHY-2210452.

References

1. K.N. Abazajian, P. Adshead, Z. Ahmed, S.W. Allen, D. Alonso, K.S. Arnold, C. Baccigalupi, J.G. Bartlett, N. Battaglia, B.A. Benson, C.A. Bischoff, J. Borrill, V. Buza, E. Calabrese, R. Caldwell, J.E. Carlstrom, C.L. Chang, T.M. Crawford, F.-Y. Cyr-Racine, F. De Bernardis, T. de Haan, S. di Serego Alighieri, J. Dunkley, C. Dvorkin, J. Errard, G. Fabbian, S. Feeney, S. Ferraro, J.P. Filippini, R. Flauger, G.M. Fuller, V. Gluscevic, D. Green, D. Grin, E. Grohs, J.W. Henning, J.C. Hill, R. Hlozek, G. Holder, W. Holzapfel, W. Hu, K.M. Huffenberger, R. Keskitalo, L. Knox, A. Kosowsky, J. Kovac, E.D. Kovetz, C.-L. Kuo, A. Kusaka, M. Le Jeune, A.T. Lee, M. Lilley, M. Loverde, M.S. Madhavacheril, A. Mantz, D.J.E. Marsh, J. McMahon, P.D. Meerburg, J. Meyers, A.D. Miller, J.B. Munoz, H.N. Nguyen, M.D. Niemack, M. Peloso, J. Peloton, L. Pogosian, C. Pryke, M. Raveri, C.L. Reichardt, G. Rocha, A. Rotti, E. Schaan, M.M. Schmittfull, D. Scott, N. Sehgal, S. Shandera, B.D. Sherwin, T.L. Smith, L. Sorbo, G.D.

Starkman, K.T. Story, A. van Engelen, J.D. Vieira, S. Watson, N. Whitehorn, W.L. Kimmy Wu. *CMB-S4 Science Book*, 1st edn. (2016) ArXiv:1610.0274

2. P. Ade, J. Aguirre, Z. Ahmed, S. Aiola, A. Ali, D. Alonso, M.A. Alvarez, K. Arnold, P. Ashton, J. Austermann, H. Awan, C. Baccigalupi, T. Baildon, D. Barron, N. Battaglia, R. Battye, E. Baxter, A. Bazarko, J.A. Beall, R. Bean, D. Beck, S. Beckman, B. Beringue, F. Bianchini, S. Boada, D. Boettger, J.R. Bond, J. Borrill, M.L. Brown, S.M. Bruno, S. Bryan, E. Calabrese, V. Calafut, P. Calisse, J. Carron, A. Challinor, G. Chesmore, Y. Chinone, J. Chluba, H.-M.S. Cho, S. Choi, G. Coppi, N.F. Cothard, K. Coughlin, D. Crichton, K.D. Crowley, K.T. Crowley, A. Cukierman, J.M. D'Ewart, R. Dünner, T. de Haan, M. Devlin, S. Dicker, J. Didier, M. Dobbs, B. Dober, C.J. Duell, S. Duff, A. Duivenvoorden, J. Dunkley, J. Dusatko, J. Errard, G. Fabbian, S. Feeney, S. Ferraro, P. Fluxà, K. Freese, J.C. Frisch, A. Frolov, G. Fuller, B. Fuzia, N. Galitzki, P.A. Gallardo, J. Tomas Galvez Ghersi, J. Gao, E. Gawiser, M. Gerbino, V. Gluscevic, N. Goeckner-Wald, J. Golec, S. Gordon, M. Gralla, D. Green, A. Grigorian, J. Groh, C. Groppi, Y. Guan, J.E. Gudmundsson, D. Han, P. Hargrave, M. Hasegawa, M. Hasselfield, M. Hattori, V. Haynes, M. Hazumi, Y. He, E. Healy, S.W. Henderson, C. Hervias-Caimapo, C.A. Hill, J.C. Hill, G. Hilton, M. Hilton, A.D. Hincks, G. Hinshaw, R. Hložek, S. Ho, S.-P.P. Ho, L. Howe, Z. Huang, J. Hubmayr, K. Huffenberger, J.P. Hughes, A. Ijjas, M. Ikape, K. Irwin, A.H. Jaffe, B. Jain, O. Jeong, D. Kaneko, E.D. Karpel, N. Katayama, B. Keating, S.S. Kernasovskiy, R. Keskitalo, T. Kisner, K. Kiuchi, J. Klein, K. Knowles, B. Koopman, A. Kosowsky, N. Krachmalnicoff, S.E. Kuenstner, C.-L. Kuo, A. Kusaka, J. Lashner, A. Lee, E. Lee, D. Leon, J.S.Y. Leung, A. Lewis, Y. Li, Z. Li, M. Limon, E. Linder, C. Lopez-Caraballo, T. Louis, L. Lowry, M. Lungu, M. Madhavacheril, D. Mak, F. Maldonado, H. Mani, B. Mates, F. Matsuda, L. Maurin, P. Mauskopf, A. May, N. McCallum, C. McKenney, J. McMahon, P.D. Meerburg, J. Meyers, A. Miller, M. Mirmelstein, K. Moodley, M. Munchmeyer, C. Munson, S. Naess, F. Nati, M. Navaroli, L. Newburgh, H.N. Nguyen, M. Niemack, H. Nishino, J. Orlowski-Scherer, L. Page, B. Partridge, J. Peloton, F. Perrotta, L. Piccirillo, G. Pisano, D. Poletti, R. Puddu, G. Puglisi, C. Raum, C.L. Reichardt, M. Remazeilles, Y. Rephaeli, D. Riechers, F. Rojas, A. Roy, S. Sadeh, Y. Sakurai, M. Salatino, M. Sathyanarayana Rao, E. Schaan, M. Schmittfull, N. Sehgal, J. Seibert, U. Seljak, B. Sherwin, M. Shimon, C. Sierra, J. Sievers, P. Sikhosana, M. Silva-Feaver, S.M. Simon, A. Sinclair, P. Siritanasak, K. Smith, S.R. Smith, D. Spergel, S.T. Staggs, G. Stein, J.R. Stevens, R. Stompor, A. Suzuki, O. Tajima, S. Takakura, G. Teply, D.B. Thomas, B. Thorne, R. Thornton, H. Trac, C. Tsai, C. Tucker, J. Ullom, S. Vagnozzi, A. van Engelen, J. Van Lanen, D.D. Van Winkle, E.M. Vavagiakis, C. Vergès, M. Vissers, K. Wagoner, S. Walker, J. Ward, B. Westbrook, N. Whitehorn, J. Williams, J. Williams, E.J. Wollack, Z. Xu, B. Yu, C. Yu, F. Zago, H. Zhang, N. Zhu, Simons observatory collaboration. The Simons observatory: science goals and forecasts. JCAP **2019**(2), 056 (2019)
3. S. Alam, F.D. Albareti, C.A. Prieto, F. Anders, S.F. Anderson, T. Anderton, B.H. Andrews, E. Armengaud, A. Aubourg, S. Bailey et al., The eleventh and twelfth data releases of the Sloan digital sky survey: Final data from SDSS-iii. The Astrophys. J. Suppl. Series **219**(1), 12 (2015)
4. Y. Ali-Haïmoud, D. Grin, C.M. Hirata. Radiative transfer effects in primordial hydrogen recombination. Phys. Rev. D **82**(12), 123502 (2010)
5. Y. Ali-Haïmoud, C.M. Hirata. HyRec: a fast and highly accurate primordial hydrogen and helium recombination code. Phys. Rev. D **83**(4), 043513 (2011)
6. P.P. Avelino, S. Esposito, G. Mangano, C.J.A.P. Martins, A. Melchiorri, G. Miele, O. Pisanti, G. Rocha, P.T.P. Viana, Early-universe constraints on a time-varying fine structure constant. Phys. Rev. D **64**(10), 103505 (2001)
7. P.P. Avelino, C.J.A.P. Martins, G. Rocha, P. Viana, Looking for a varying α in the cosmic microwave background. Phys. Rev. D **62**(12), 123508 (2000)
8. R.A. Battye, R. Crittenden, J. Weller, Cosmic concordance and the fine structure constant. Phys. Rev. D **63**(4), 043505 (2001)
9. J.D. Bekenstein, Fine-structure constant: is it really a constant? Phys. Rev. D **25**, 1527–1539 (1982)
10. H.A. Bethe, E.E. Salpeter. *Quantum Mechanics of One- and Two-Electron Atoms* (1957)

11. S. Bize, S.A. Diddams, U. Tanaka, C.E. Tanner, W.H. Oskay, R.E. Drullinger, T.E. Parker, T.P. Heavner, S.R. Jefferts, L. Hollberg, W.M. Itano, J.C. Bergquist, Testing the stability of fundamental constants with the $^{199}Hg^{+}$ single-ion optical clock. Phys. Rev. Lett. **90**, 150802 (2003)
12. P. Bonifacio, H. Rahmani, J.B. Whitmore, M. Wendt, M. Centurion, P. Molaro, R. Srianand, M.T. Murphy, P. Petitjean, I.I. Agafonova, S. D'Odorico, T.M. Evans, S.A. Levshakov, S. Lopez, C.J.A.P. Martins, D. Reimers, G. Vladilo, Fundamental constants and high-resolution spectroscopy. Astronomische Nachrichten **335**, 83 (2014)
13. M.S. Burgin, Hydrogen subordinate line emission at the epoch of cosmological recombination. Astron. Rep. **47**, 709–716 (2003)
14. J. Chluba, M.H. Abitbol, N. Aghanim, Y. Ali-Haïmoud, M. Alvarez, K. Basu, B. Bolliet, C. Burigana, P. de Bernardis, J. Delabrouille, E. Dimastrogiovanni, F. Finelli, D. Fixsen, L. Hart, C. Hernández-Monteagudo, J.C. Hill, A. Kogut, K. Kohri, J. Lesgourgues, B. Maffei, J. Mather, S. Mukherjee, S.P. Patil, A. Ravenni, M. Remazeilles, A. Rotti, J.A. Rubiño-Martin, J. Silk, R.A. Sunyaev, E.R. Switzer, New horizons in cosmology with spectral distortions of the cosmic microwave background. Exp. Astron. **51**(3), 1515–1554 (2021)
15. J. Chluba, Y. Ali-Haïmoud, COSMOSPEC: fast and detailed computation of the cosmological recombination radiation from hydrogen and helium. MNRAS **456**, 3494–3508 (2016)
16. J. Chluba, J. Fung, E.R. Switzer, Radiative transfer effects during primordial helium recombination. MNRAS **423**, 3227–3242 (2012)
17. J. Chluba, R.A. Sunyaev, Free-bound emission from cosmological hydrogen recombination. A&A **458**, L29–L32 (2006)
18. J. Chluba, R.A. Sunyaev, Induced two-photon decay of the 2s level and the rate of cosmological hydrogen recombination. A&A **446**, 39–42 (2006)
19. J. Chluba, R.A. Sunyaev, Cosmological hydrogen recombination: lyn line feedback and continuum escape. A&A **475**, 109–114 (2007)
20. J. Chluba, R.A. Sunyaev, Is there a need and another way to measure the cosmic microwave background temperature more accurately? A&A **478**, L27–L30 (2008)
21. J. Chluba, R.A. Sunyaev, Time-dependent corrections to the Ly α escape probability during cosmological recombination. A&A **496**, 619–635 (2009)
22. J. Chluba, R.M. Thomas, Towards a complete treatment of the cosmological recombination problem. MNRAS **412**, 748–764 (2011)
23. V. Desjacques, J. Chluba, J. Silk, F. de Bernardis, O. Doré, Detecting the cosmological recombination signal from space. MNRAS **451**, 4460–4470 (2015)
24. E. Di Dio, F. Montanari, A. Raccanelli, R. Durrer, M. Kamionkowski, J. Lesgourgues, Curvature constraints from large scale structure. JCAP **2016**(6), 013 (2016)
25. G.W.F. Drake. *Springer Handbook of Atomic, Molecular, and Optical Physics* (Springer, 2006)
26. G.W.F. Drake, D.C. Morton, A multiplet table for neutral Helium (^{4}He I) with transition rates. ApJS **170**, 251–260 (2007)
27. V.K. Dubrovich. Hydrogen recombination lines of cosmological origin. Soviet Astron. Lett. **1**, 196 (1975)
28. V.K. Dubrovich, V.A. Stolyarov. Fossil radio lines of hydrogen in the cosmic background radiation at decimeter and meter wavelengths. A&A **302**, 635 (1995)
29. M. Farhang, J.R. Bond, J. Chluba, Semi blind Eigen analyses of recombination histories using cosmic microwave background data. ApJ **752**, 88 (2012)
30. M. Farhang, J.R. Bond, J. Chluba, E.R. Switzer, Constraints on perturbations to the recombination history from measurements of the cosmic microwave background damping tail. ApJ **764**, 137 (2013)
31. W.A. Fendt, J. Chluba, J.A. Rubiño-Martín, B.D. Wandelt, RICO: a new approach for fast and accurate representation of the cosmological recombination history. ApJS **181**, 627–638 (2009)
32. S. Galli, L. Pogosian, K. Jedamzik, L. Balkenhol, Consistency of Planck, ACT, and SPT constraints on magnetically assisted recombination and forecasts for future experiments. Phys. Rev. D **105**(2), 023513 (2022)

33. S.C.O. Glover, J. Chluba, S.R. Furlanetto, J.R. Pritchard, D.W. Savin, Chapter three—atomic, molecular, and optical physics in the early universe: from recombination to reionization. Adv. Atomic Mol. Opt. Phys. **63**, 135–270 (2014)
34. L. Hart, J. Chluba, New constraints on time-dependent variations of fundamental constants using Planck data. MNRAS **474**(2), 1850–1861 (2018)
35. L. Hart, J. Chluba, Improved model-independent constraints on the recombination era and development of a direct projection method. MNRAS **495**(4), 4210–4226 (2020)
36. L. Hart, J. Chluba, Updated fundamental constant constraints from Planck 2018 data and possible relations to the Hubble tension. MNRAS **493**(3), 3255–3263 (2020)
37. L. Hart, J. Chluba, Varying fundamental constants principal component analysis: additional hints about the Hubble tension. MNRAS **510**(2), 2206–2227 (2022)
38. L. Hart, J. Chluba, Using the cosmological recombination radiation to probe early dark energy and fundamental constant variations. MNRAS **519**(3), 3664–3680 (2023)
39. L. Hart, A. Rotti, J. Chluba, Sensitivity forecasts for the cosmological recombination radiation in the presence of foregrounds. MNRAS **497**(4), 4535–4548 (2020)
40. C.M. Hirata. Two-photon transitions in primordial hydrogen recombination. Phys. Rev. D **78**(2), 023001 (2008)
41. K. Hoshiya, Y. Toda, Electron mass variation from dark sector interactions and compatibility with cosmological observations. Phys. Rev. D **107**(4), 043505 (2023)
42. W. Hu, D. Scott, N. Sugiyama, M. White, Effect of physical assumptions on the calculation of microwave background anisotropies. Phys. Rev. D **52**, 5498–5515 (1995)
43. W. Hu, N. Sugiyama, Anisotropies in the cosmic microwave background: an analytic approach. ApJ **444**, 489–506 (1995)
44. W. Hu, N. Sugiyama, Small-scale cosmological perturbations: an analytic approach. ApJ **471**, 542 (1996)
45. K. Jedamzik, T. Abel, Small-scale primordial magnetic fields and anisotropies in the cosmic microwave background radiation. JCAP **10**, 050 (2013)
46. K. Jedamzik, L. Pogosian, Relieving the Hubble tension with primordial magnetic fields. Phys. Rev. Lett. **125**(18), 181302 (2020)
47. M. Kaplinghat, R.J. Scherrer, M.S. Turner. Constraining variations in the fine-structure constant with the cosmic microwave background. Phys. Rev. D **60**(2), 023516 (1999)
48. W.J. Karza, R. Latter, Electron radiative transitions in a Coulomb field. ApJS **6**, 167 (1961)
49. E.E. Kholupenko, A.V. Ivanchik, D.A. Varshalovich, Gravitation Cosmol. **11**, 161–165 (2005)
50. E.E. Kholupenko, A.V. Ivanchik, D.A. Varshalovich, Rapid HeII-HeI recombination and radiation arising from this process. MNRAS **378**, L39–L43 (2007)
51. A. Kogut, D.J. Fixsen, Foreground Bias from parametric models of Far-IR dust emission. ApJ **826**, 101 (2016)
52. A. Kogut, D.J. Fixsen, D.T. Chuss, J. Dotson, E.Dwek, M. Halpern, G.F. Hinshaw, S.M. Meyer, S.H. Moseley, M.D. Seiffert, D.N. Spergel, E.J. Wollack. The Primordial Inflation Explorer (PIXIE): a nulling polarimeter for cosmic microwave background observations. JCAP **7**, 25 (2011)
53. S.M. Kotuš, M.T. Murphy, R.F. Carswell, High-precision limit on variation in the fine-structure constant from a single quasar absorption system. Monthly Notices of the Royal Astron. Soc. **464**(3), 3679 (2017)
54. N. Lee, Y. Ali-Haïmoud, N. Schöneberg, V. Poulin, What it takes to solve the Hubble tension through modifications of cosmological recombination. Phys. Rev. Lett. **130**(16), 161003 (2023)
55. J. Lesgourgues. The cosmic linear anisotropy solving system (CLASS) I: overview (2011). hyperimagehttp://arxiv.org/abs/1104.2932ArXiv:1104.2932
56. A. Lewis, A. Challinor, A. Lasenby, Efficient computation of cosmic microwave background anisotropies in closed Friedmann-Robertson-Walker models. ApJ **538**, 473–476 (2000)
57. A. Lewis, J. Weller, R. Battye, The cosmic microwave background and the ionization history of the Universe. MNRAS **373**, 561–570 (2006)
58. M. Lucca, J. Chluba, A. Rotti. CRRfast: an emulator for the cosmological recombination radiation with effects from inhomogeneous recombination. arXiv e-prints (2023). arXiv:2306.08085

59. C.J.A.P. Martins, The status of varying constants: a review of the physics, searches and implications. Rep. Progr. Phys. **80**(12), 126902 (2017)
60. C.J.A.P. Martins, A. Melchiorri, G. Rocha, R. Trotta, P.P. Avelino, P.T.P. Viana, WMAP constraints on varying α and the promise of reionization. Phys. Lett. B **585**, 29–34 (2004)
61. C.J.A.P. Martins, P.E. Vielzeuf, M. Martinelli, E. Calabrese, S. Pandolfi, Evolution of the fine-structure constant in runaway dilaton models. Phys. Lett. B **743**, 377–382 (2015)
62. E. Menegoni, M. Archidiacono, E. Calabrese, S. Galli, C.J.A.P. Martins, A. Melchiorri, Fine structure constant and the CMB damping scale. Phys. Rev. D **85**(10), 107301 (2012)
63. M.J. Mortonson, C. Dvorkin, H.V. Peiris, W. Hu, CMB polarization features from inflation versus reionization. Phys. Rev. D **79**, 103519 (2009)
64. P.J.E. Peebles, Recombination of the primeval plasma. ApJ **153**, 1 (1968)
65. P.J.E. Peebles, J.T. Yu, Primeval adiabatic perturbation in an expanding universe. ApJ **162**, 815 (1970)
66. Planck Collaboration, P.A.R. Ade, N. Aghanim, M. Arnaud, M. Ashdown, J. Aumont, C. Baccigalupi, A.J. Banday, R.B. Barreiro, J.G. Bartlett et al., Planck 2015 results. XIII. Cosmological parameters (2015). ArXiv:1502.01589
67. Planck Collaboration, P.A.R. Ade, N. Aghanim, M. Arnaud, M. Ashdown, J. Aumont, C. Baccigalupi, A.J. Banday, R.B. Barreiro, E. Battaner, K. Benabed, A. Benoit-Lévy, J.-P. Bernard, M. Bersanelli, P. Bielewicz, J.R. Bond, J. Borrill, F.R. Bouchet, C. Burigana, R.C. Butler, E. Calabrese, A. Chamballu, H.C. Chiang, P.R. Christensen, D.L. Clements, L.P.L. Colombo, F. Couchot, A. Curto, F. Cuttaia, L. Danese, R.D. Davies, R. J. Davis, P. de Bernardis, A. de Rosa, G. de Zotti, J. Delabrouille, J.M. Diego, H. Dole, O. Doré, X. Dupac, T. A. Enßlin, H.K. Eriksen, O. Fabre, F. Finelli, O. Forni, M. Frailis, E. Franceschi, S. Galeotta, S. Galli, K. Ganga, M. Giard, J. González-Nuevo, K.M. Górski, A. Gregorio, A. Gruppuso, F.K. Hansen, D. Hanson, D.L. Harrison, S. Henrot-Versillé, C. Hernández-Monteagudo, D. Herranz, S.R. Hildebrandt, E. Hivon, M. Hobson, W.A. Holmes, A. Hornstrup, W. Hovest, K.M. Huffenberger, A.H. Jaffe, W.C. Jones, E. Keihänen, R. Keskitalo, R. Kneissl, J. Knoche, M. Kunz, H. Kurki-Suonio, J.-M. Lamarre, A. Lasenby, C.R. Lawrence, R. Leonardi, J. Lesgourgues, M. Liguori, P.B. Lilje, M. Linden-Vørnle, M. López-Caniego, P.M. Lubin, J.F. Macías-Pérez, N. Mandolesi, M. Maris, P.G. Martin, E. Martínez-González, S. Masi, S. Matarrese, P. Mazzotta, P.R. Meinhold, A. Melchiorri, L. Mendes, E. Menegoni, A. Mennella, M. Migliaccio, M.-A. Miville-Deschênes, A. Moneti, L. Montier, G. Morgante, A. Moss, D. Munshi, J.A. Murphy, P. Naselsky, F. Nati, P. Natoli, H.U. Nørgaard-Nielsen, F. Noviello, D. Novikov, I. Novikov, C.A. Oxborrow, L. Pagano, F. Pajot, D. Paoletti, F. Pasian, G. Patanchon, O. Perdereau, L. Perotto, F. Perrotta, F. Piacentini, M. Piat, E. Pierpaoli, D. Pietrobon, S. Plaszczynski, E. Pointecouteau, G. Polenta, N. Ponthieu, L. Popa, G.W. Pratt, S. Prunet, J.P. Rachen, R. Rebolo, M. Reinecke, M. Remazeilles, C. Renault, S. Ricciardi, I. Ristorcelli, G. Rocha, G. Roudier, B. Rusholme, M. Sandri, G. Savini, D. Scott, L.D. Spencer, V. Stolyarov, R. Sudiwala, D. Sutton, A.-S. Suur-Uski, J.-F. Sygnet, J.A. Tauber, D. Tavagnacco, L. Terenzi, L. Toffolatti, M. Tomasi, M. Tristram, M. Tucci, J.-P. Uzan, L. Valenziano, J. Valiviita, B. Van Tent, P. Vielva, F. Villa, L.A. Wade, D. Yvon, A. Zacchei, A. Zonca, Planck intermediate results. XXIV. Constraints on variations in fundamental constants. A&A **580**, A22 (2015)
68. Planck Collaboration, N. Aghanim, Y. Akrami, M. Ashdown, J. Aumont, C. Baccigalupi, M. Ballardini, A.J. Banday, R.B. Barreiro, N. Bartolo, S. Basak, R. Battye, K. Benabed, J.P. Bernard, M. Bersanelli, P. Bielewicz, J.J. Bock, J.R. Bond, J. Borrill, F.R. Bouchet, F. Boulanger, M. Bucher, C. Burigana, R.C. Butler, E. Calabrese, J.F. Cardoso, J. Carron, A. Challinor, H.C. Chiang, J. Chluba, L.P.L. Colombo, C. Combet, D. Contreras, B.P. Crill, F. Cuttaia, P. de Bernardis, G. de Zotti, J. Delabrouille, J.M. Delouis, E. Di Valentino, J.M. Diego, O. Doré, M. Douspis, A. Ducout, X. Dupac, S. Dusini, G. Efstathiou, F. Elsner, T.A. Enßlin, H.K. Eriksen, Y. Fantaye, M. Farhang, J. Fergusson, R. Fernandez-Cobos, F. Finelli, F. Forastieri, M. Frailis, E. Franceschi, A. Frolov, S. Galeotta, S. Galli, K. Ganga, R.T. Génova-Santos, M. Gerbino, T. Ghosh, J. González-Nuevo, K.M. Górski, S. Gratton, A. Gruppuso, J.E. Gudmundsson, J. Hamann, W. Handley, D. Herranz, E. Hivon, Z. Huang, A.H. Jaffe, W.C. Jones,

A. Karakci, E. Keihänen, R. Keskitalo, K. Kiiveri, J. Kim, T.S. Kisner, L. Knox, N. Krachmalnicoff, M. Kunz, H. Kurki-Suonio, G. Lagache, J.M. Lamarre, A. Lasenby, M. Lattanzi, C.R. Lawrence, M. Le Jeune, P. Lemos, J. Lesgourgues, F. Levrier, A. Lewis, M. Liguori, P.B. Lilje, M. Lilley, V. Lindholm, M. López-Caniego, P.M. Lubin, Y.Z. Ma, J.F. Macías-Pérez, G. Maggio, D. Maino, N. Mandolesi, A. Mangilli, A. Marcos-Caballero, M. Maris, P.G. Martin, M. Martinelli, E. Martínez-González, S. Matarrese, N. Mauri, J.D. McEwen, P.R. Meinhold, A. Melchiorri, A. Mennella, M. Migliaccio, M. Millea, S. Mitra, M.A. Miville-Deschênes, D. Molinari, L. Montier, G. Morgante, A. Moss, P. Natoli, H.U. Nørgaard-Nielsen, L. Pagano, D. Paoletti, B. Partridge, G. Patanchon, H.V. Peiris, F. Perrotta, V. Pettorino, F. Piacentini, L. Polastri, G. Polenta, J.L. Puget, J.P. Rachen, M. Reinecke, M. Remazeilles, A. Renzi, G. Rocha, C. Rosset, G. Roudier, J.A. Rubiño-Martín, B. Ruiz-Granados, L. Salvati, M. Sandri, M. Savelainen, D. Scott, E.P.S. Shellard, C. Sirignano, G. Sirri, L.D. Spencer, R. Sunyaev, A.S. Suur-Uski, J.A. Tauber, D. Tavagnacco, M. Tenti, L. Toffolatti, M. Tomasi, T. Trombetti, L. Valenziano, J. Valiviita, B. Van Tent, L. Vibert, P. Vielva, F. Villa, N. Vittorio, B.D. Wandelt, I.K. Wehus, M. White, S.D.M. White, A. Zacchei, A. Zonca. Planck 2018 results. VI. Cosmological parameters (2018). ArXiv:1807.06209

69. Planck Collaboration, N. Aghanim, Y. Akrami, M. Ashdown, J. Aumont, C. Baccigalupi, M. Ballardini, A.J. Banday, R.B. Barreiro, N. Bartolo, S. Basak, K. Benabed, J.P. Bernard, M. Bersanelli, P. Bielewicz, J.J. Bock, J.R. Bond, J. Borrill, F.R. Bouchet, F. Boulanger, M. Bucher, C. Burigana, R.C. Butler, E. Calabrese, J.F. Cardoso, J. Carron, B. Casaponsa, A. Challinor, H.C. Chiang, L.P.L. Colombo, C. Combet, B.P. Crill, F. Cuttaia, P. de Bernardis, A. de Rosa, G. de Zotti, J. Delabrouille, J.M. Delouis, E. Di Valentino, J.M. Diego, O. Doré, M. Douspis, A. Ducout, X. Dupac, S. Dusini, G. Efstathiou, F. Elsner, T.A. Enßlin, H.K. Eriksen, Y. Fantaye, R. Fernand ez-Cobos, F. Finelli, M. Frailis, A.A. Fraisse, E. Franceschi, A. Frolov, S. Galeotta, S. Galli, K. Ganga, R.T. Génova-Santos, M. Gerbino, T. Ghosh, Y. Giraud-Héraud, J. González-Nuevo, K.M. Górski, S. Gratton, A. Gruppuso, J.E. Gudmundsson, J. Hamann, W. Handley, F.K. Hansen, D. Herranz, E. Hivon, Z. Huang, A.H. Jaffe, W.C. Jones, E. Keihänen, R. Keskitalo, K. Kiiveri, J. Kim, T.S. Kisner, N. Krachmalnicoff, M. Kunz, H. Kurki-Suonio, G. Lagache, J.M. Lamarre, A. Lasenby, M. Lattanzi, C.R. Lawrence, M. Le Jeune, F. Levrier, A. Lewis, M. Liguori, P.B. Lilje, M. Lilley, V. Lindholm, M. López-Caniego, P.M. Lubin, Y.Z. Ma, J.F. Macías-Pérez, G. Maggio, D. Maino, N. Mandolesi, A. Mangilli, A. Marcos-Caballero, M. Maris, P.G. Martin, E. Martínez-González, S. Matarrese, N. Mauri, J.D. McEwen, P.R. Meinhold, A. Melchiorri, A. Mennella, M. Migliaccio, M. Millea, M.A. Miville-Deschênes, D. Molinari, A. Moneti, L. Montier, G. Morgante, A. Moss, P. Natoli, H.U. Nørgaard-Nielsen, L. Pagano, D. Paoletti, B. Partridge, G. Patanchon, H.V. Peiris, F. Perrotta, V. Pettorino, F. Piacentini, G. Polenta, J.L. Puget, J.P. Rachen, M. Reinecke, M. Remazeilles, A. Renzi, G. Rocha, C. Rosset, G. Roudier, J.A. Rubiño-Martín, B. Ruiz-Granados, L. Salvati, M. Sandri, M. Savelainen, D. Scott, E.P.S. Shellard, C. Sirignano, G. Sirri, L.D. Spencer, R. Sunyaev, A.S. Suur-Uski, J.A. Tauber, D. Tavagnacco, M. Tenti, L. Toffolatti, M. Tomasi, T. Trombetti, J. Valiviita, B. Van Tent, P. Vielva, F. Villa, N. Vittorio, B.D. Wandelt, I.K. Wehus, A. Zacchei, A. Zonca. Planck 2018 results. V. CMB power spectra and likelihoods (2019). arXiv e-prints arXiv:1907.12875

70. Planck Collaboration, N. Aghanim, Y. Akrami, M. Ashdown, J. Aumont, C. Baccigalupi, M. Ballardini, A.J. Banday, R.B. Barreiro, N. Bartolo, S. Basak, K. Benabed, J.P. Bernard, M. Bersanelli, P. Bielewicz, J.J. Bock, J.R. Bond, J. Borrill, F.R. Bouchet, F. Boulanger, M. Bucher, C. Burigana, E. Calabrese, J.F. Cardoso, J. Carron, A. Challinor, H.C. Chiang, L.P.L. Colombo, C. Combet, B.P. Crill, F. Cuttaia, P. de Bernardis, G. de Zotti, J. Delabrouille, E. Di Valentino, J.M. Diego, O. Doré, M. Douspis, A. Ducout, X. Dupac, G. Efstathiou, F. Elsner, T.A. Enßlin, H.K. Eriksen, Y. Fantaye, R. Fernandez-Cobos, F. Forastieri, M. Frailis, A.A. Fraisse, E. Franceschi, A. Frolov, S. Galeotta, S. Galli, K. Ganga, R.T. Génova-Santos, M. Gerbino, T. Ghosh, J. González-Nuevo, K.M. Górski, S. Gratton, A. Gruppuso, J.E. Gudmundsson, J. Hamann, W. Hand ley, F.K. Hansen, D. Herranz, E. Hivon, Z. Huang, A.H. Jaffe, W.C. Jones, A. Karakci, E. Keihänen, R. Keskitalo, K. Kiiveri, J. Kim, L. Knox, N. Krachmalnicoff, M. Kunz, H. Kurki-Suonio, G. Lagache, J.M. Lamarre, A. Lasenby, M. Lattanzi, C.R. Lawrence,

M. Le Jeune, F. Levrier, A. Lewis, M. Liguori, P.B. Lilje, V. Lindholm, M. López-Caniego, P.M. Lubin, Y.Z. Ma, J.F. Macías-Pérez, G. Maggio, D. Maino, N. Mandolesi, A. Mangilli, A. Marcos-Caballero, M. Maris, P.G. Martin, E. Martínez-González, S. Matarrese, N. Mauri, J.D. McEwen, A. Melchiorri, A. Mennella, M. Migliaccio, M.A. Miville-Deschênes, D. Molinari, A. Moneti, L. Montier, G. Morgante, A. Moss, P. Natoli, L. Pagano, D. Paoletti, B. Partridge, G. Patanchon, F. Perrotta, V. Pettorino, F. Piacentini, L. Polastri, G. Polenta, J.L. Puget, J.P. Rachen, M. Reinecke, M. Remazeilles, A. Renzi, G. Rocha, C. Rosset, G. Roudier, J.A. Rubiño-Martín, B. Ruiz-Granados, L. Salvati, M. Sandri, M. Savelainen, D. Scott, C. Sirignano, R. Sunyaev, A.S. Suur-Uski, J.A. Tauber, D. Tavagnacco, M. Tenti, L. Toffolatti, M. Tomasi, T. Trombetti, J. Valiviita, B. Van Tent, P. Vielva, F. Villa, N. Vittorio, B.D. Wandelt, I.K. Wehus, M. White, S.D.M. White, A. Zacchei, A. Zonca. Planck 2018 results. VIII. Gravitational lensing (2018). arXiv e-prints arXiv:1807.06210
71. A.G. Riess, S. Casertano, W. Yuan, L.M. Macri, D. Scolnic, Large Magellanic cloud Cepheid standards provide a 1% foundation for the determination of the Hubble constant and stronger evidence for physics beyond ΛCDM. Astrophys. J. **876**(1), 85 (2019)
72. G. Rocha, R. Trotta, C.J.A.P. Martins, A. Melchiorri, P.P. Avelino, R. Bean, P.T.P. Viana, Measuring α in the early Universe: cosmic microwave background polarization, re-ionization and the Fisher matrix analysis. MNRAS **352**, 20–38 (2004)
73. T. Rosenband, D.B. Hume, P.O. Schmidt, C.W. Chou, A. Brusch, L. Lorini, W.H. Oskay, R.E. Drullinger, T.M. Fortier, J.E. Stalnaker, S.A. Diddams, W.C. Swann, N.R. Newbury, W.M. Itano, D.J. Wineland, J.C. Bergquist, Frequency ratio of al+ and hg+ single-ion optical clocks; metrology at the 17th decimal place. Science **319**(5871), 1808–1812 (2008)
74. J.A. Rubiño-Martín, J. Chluba, W.A. Fendt, B.D. Wandelt, Estimating the impact of recombination uncertainties on the cosmological parameter constraints from cosmic microwave background experiments. MNRAS **403**, 439–452 (2010)
75. J.A. Rubiño-Martín, J. Chluba, R.A. Sunyaev, Lines in the cosmic microwave background spectrum from the epoch of cosmological helium recombination. A&A **485**, 377–393 (2008)
76. J.A. Rubiño-Martín, C. Hernández-Monteagudo, R.A. Sunyaev, The imprint of cosmological hydrogen recombination lines on the power spectrum of the CMB. A&A **438**, 461–473 (2005)
77. G.B. Rybicki, I.P. dell'Antonio, The time development of a resonance line in the expanding universe. ApJ **427**, 603–617 (1994)
78. M. Sathyanarayana Rao, R. Subrahmanyan, N. Udaya Shankar, J. Chluba, On the detection of spectral ripples from the recombination epoch. ApJ **810**, 3 (2015)
79. N. Schöneberg, G.F. Abellán, A.P. Sánchez, S.J. Witte, V. Poulin, J. Lesgourgues, The H_0 Olympics: a fair ranking of proposed models. Phys. Rep. **984**, 1–55 (2022)
80. C.G. Scóccola, S.J. Landau, H. Vucetich. WMAP 5-year constraints on time variation of alpha and m_e. Memorie della Societ Astronomica Italiana **80**, 814 (2009)
81. S. Seager, D.D. Sasselov, D. Scott, A new calculation of the recombination epoch. APJL **523**, L1–L5 (1999)
82. S. Seager, D.D. Sasselov, D. Scott, How exactly did the universe become neutral? ApJS **128**, 407–430 (2000)
83. T. Sekiguchi, T. Takahashi, Early recombination as a solution to the H_0 tension. Phys. Rev. D **103**(8), 083507 (2021)
84. J.R. Shaw, J. Chluba, Precise cosmological parameter estimation using COSMOREC. MNRAS **415**, 1343–1354 (2011)
85. L.D. Shaw, D. Nagai, S. Bhattacharya, E.T. Lau, Impact of cluster physics on the Sunyaev-Zel'dovich power spectrum. ApJ **725**, 1452–1465 (2010)
86. T.L. Smith, D. Grin, D. Robinson, D. Qi, Probing spatial variation of the fine-structure constant using the CMB. Phys. Rev. D **99**(4), 043531 (2019)
87. V.V. Sobolev, *Moving Envelopes of Stars* (Harvard University Press, Cambridge, 1960), p.1960
88. R.A. Sunyaev, J. Chluba, Signals from the epoch of cosmological recombination (Karl Schwarzschild Award Lecture 2008). Astronomische Nachrichten **330**, 657 (2009)
89. R.A. Sunyaev, Y.B. Zeldovich, Small-scale fluctuations of relic radiation. ApSS **7**, 3 (1970)

90. H. Tohfa, J. Crump, E. Baker, L. Hart, D. Grin, M. Brosius, J. Chluba. A cosmic microwave background search for fine-structure constant evolution (2023). arXiv e-prints arXiv:2307.06768
91. J.-P. Uzan, The fundamental constants and their variation: observational and theoretical status. Rev. Modern Phys. **75**, 403–455 (2003)
92. J.-P. Uzan, Varying constants, gravitation and cosmology. Living Rev. Relativity **14**, 2 (2011)
93. W.Y. Wong, A. Moss, D. Scott, How well do we understand cosmological recombination? MNRAS **386**, 1023–1028 (2008)
94. Y.B. Zeldovich, V.G. Kurt, R.A. Syunyaev, Recombination of hydrogen in the hot model of the universe. Zhurnal Eksperimental noi i Teoreticheskoi Fiziki **55**, 278–286 (1968)

Chapter 34
Anomalies and Tensions in Cosmology and a Primordial Solution

Dhiraj Kumar Hazra and Arman Shafieloo

34.1 Introduction

Cosmological observations at scales probed by the Cosmic Microwave Background (CMB) observations (GigaParsecs to about a hundred kiloParsecs) are well explained by the 6-parameter ΛCDM model with power-law form of primordial spectrum. Different CMB observations from space-based missions and ground based surveys broadly agree with each other apart from moderate tensions appearing in different releases. With Planck [1] temperature and polarization data, the parameters of the model are constrained to an unprecedented precision. The ΛCDM also agrees with the large scale structure (LSS), lensing, supernovae and other cosmological observations.

However, over the last few years, within this concordance picture of cosmology, differences in the parameters estimated from different observations within the same standard model aggravated to become statistically significant. At present, the disagreement in the estimation of Hubble parameter between the CMB and the SH_0ES distance ladder measurements [2] is the most crucial tension in cosmology. Along similar times, mismatch in growth parameter between measurements from lensing surveys such as KiDS [3], DES [4] and Planck CMB have emerged as another impor-

D. K. Hazra (✉)
The Institute of Mathematical Sciences, HBNI, CIT Campus, Chennai 600113, India
e-mail: dhiraj@imsc.res.in

Homi Bhabha National Institute, Training School Complex, Anushakti Nagar, Mumbai 400085, India

INAF/OAS Bologna, Osservatorio di Astrofisica e Scienza dello Spazio, Area della ricerca CNR-INAF, via Gobetti 101, 40129 Bologna, Italy

A. Shafieloo
Korea Astronomy and Space Science Institute, Daejeon 34055, Korea
e-mail: shafieloo@kasi.re.kr

University of Science and Technology, Daejeon 34113, Korea

E. Di Valentino and D. Brout (eds.), *The Hubble Constant Tension*, Springer Series in Astrophysics and Cosmology, https://doi.org/10.1007/978-981-99-0177-7_34

tant tension. Within Planck CMB temperature data too, the preference for an ad-hoc lensing parameter or spatial curvature indicates certain anomalies within the data.

Significance of these tensions do not leave much room for discussions related to random fluctuations. At this stage, the scientific community is divided into two parts on the origin of these tensions, namely *systematic uncertainties* and *new physics*. Over the last couple of years, the extensive tests on the systematics have been performed by various groups which did not result any significant reduction of these tensions. Therefore, *new physics* is now considered as the most probable solution to these tensions. Most of the new physics developed in this aspect changes the distance scales by either changing the dark energy history [5, 6]. Majority of the solutions till now has attempted to address these tensions separately with different models.

In this chapter we will discuss the possibility of these anomalies and tensions being related. We will explore a particular solution to these in the spectrum of primordial quantum fluctuations and demonstrate the possible origin from a scalar field model.

34.2 Tensions and Anomalies: A Few Key Points

In this section, in order to distinguish the terms 'anomalies' and 'tensions', we provide two simple pictorial descriptions.

Based on a model (built with well established theories + certain supporting assumptions) we match our predictions to the observations and obtain bounds on the model parameters. These bounds refer to the best fit/mean predictions of the model parameters and statistical uncertainties (standard deviation) associated with them. For a model, consistent with the data from an observation, it is expected that the mean model prediction to the observation should be within the reach of a few standard deviations from the observed data. When that does not happen, we term this mismatch as an anomaly in the data as shown in the top plot of Fig. 34.1.

Now, thanks to the different satellite and ground based probes operating at different frequencies, we have access to several observing windows to the Universe. As light travels at a finite speed, through these windows we can witness different stages of the Universe. Now remember that the theoretical model we discussed as the ΛCDM model with power-law form of primordial spectrum (hereafter *baseline*), provides the origin and evolution of the Universe entirely. Therefore, it is possible to test the model against all observations. We expect that the bounds on the model parameters from different observations lie within the distance of a few standard deviations. Tension between the datasets arise when the distance becomes very large. The bottom plot of Fig. 34.1 describes tensions between observations in the determination of parameter A.

At this point, let's be more specific about the model. The standard model is based on Einstein's theory of general relativity (GR). It assumes a homogeneous and isotropic Universe at large scales (supported by observations). Ordinary matter (standard particles with mass), radiation (photon) and unknown type of matter (dark matter, interacts gravitationally) and unknown type of energy (dark energy with neg-

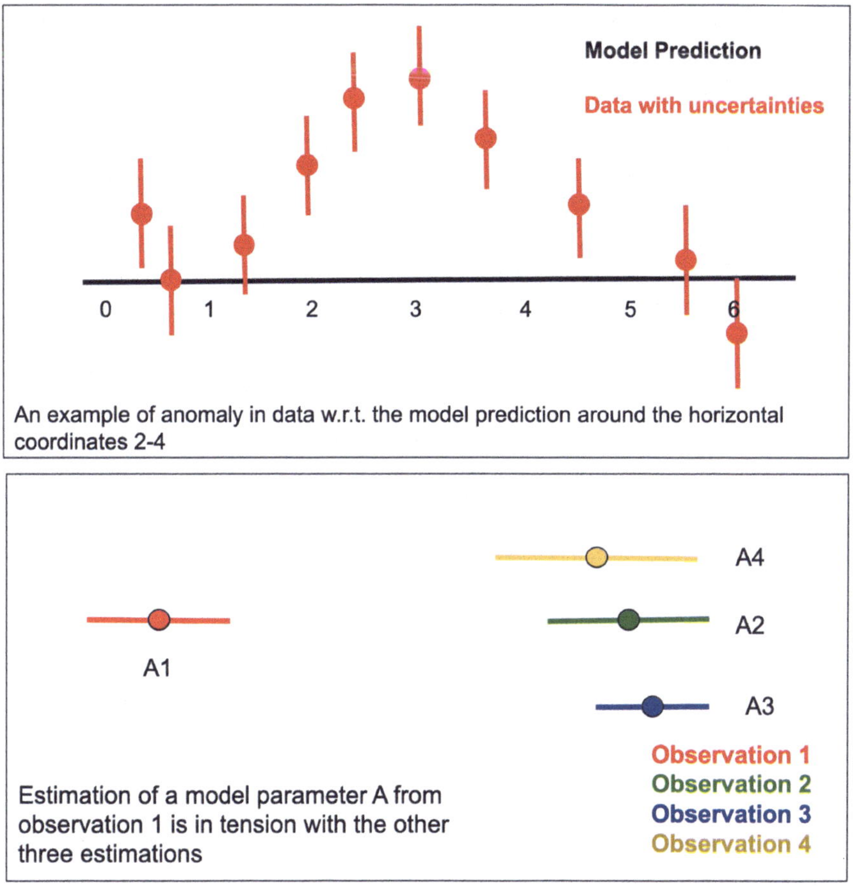

Fig. 34.1 [Top] Example of anomaly in a dataset. [Bottom] Example of tension between datasets

ative pressure) are the ingredients of this Universe. Dark matter and dark energy are two very crucial ingredients that we must take seriously as they claim about 95% of the entire energy budget. Moreover we assume the negative pressure of the dark energy follows an equation of state given by pressure = −energy density. This particular dark energy is known as the Cosmological Constant, Λ. With these constituents of the Universe, the background part of the standard model is known as ΛCold Dark Matter (ΛCDM) model. We referred to this as the background part, since we have not yet discussed the fluctuations (the structures that we observe in the Universe). These constituents dictate how a homogeneous and isotropic Universe should evolve from an early stage till today. Solving Einstein's GR, we understand that at very early stages the Universe expanded with deceleration governed by radiation and thereafter by matter. At present we are in a phase of dark energy driven acceleration. We also assume a flat Universe in the standard model (consistent with several observations).

Now, let's understand the fluctuations. The distribution of structures that we observe in the sky such as the galaxies, clusters etc. follows certain power spectrum. It tells us how the structures are correlated at different length scales. This power spectrum has evolved since its initial form through the evolution of the Universe and non-linear gravitational collapse. Mathematically we can think of this evolution as a differential equation. The differential equation can be solved with initial conditions. Therefore the natural question is: what was the initial power spectrum of the fluctuations? The seed of these structures has a quantum origin and we need to know the spectrum of these initial (primordial) quantum fluctuations. Without going into the details we can assume the spectrum to have a simple form that can be described with an amplitude and a tilt (power law). So across the length scales if we plot the spectrum it would look like a tilted line (as of now, we do not have any statistically compelling reason to go beyond a tilted line). If we combine these background and initial condition—a flat ΛCDM model with power law form of primordial spectrum is known as the standard model of the Universe which is used as the baseline of all our analyses. Therefore an anomaly or tension would mean inconsistencies with respect to the baseline predictions. This baseline model is parametrized with 6 paramcters—4 describing background and 2 describing initial conditions (amplitude and tilt). The 4 background parameters are baryon and cold dark matter densities, today's Hubble parameter (a measure of how two points in the Universe are receding from each other per unit time per unit distance) and the optical depth (free electron fraction integrated along the line of sight).

We will now be more specific about the observations in question. We will discuss three types of observations:

- The Cosmic Microwave Background (CMB) anisotropy observation from Planck satellite [1].
- Today's Hubble parameter measurement from SH_0ES [2].
- Large Scale Structure (LSS) surveys from Kilo-Degree Survey (KiDS [3]) and from the Dark Energy Survey (DES [4]).

Planck satellite mapped the CMB sky that captured the anisotropies in photon temperature and polarization to an unprecedented accuracy upto a very small scales corresponding to 0.07° (this is the resolution for temperature map, polarization maps has lower resolution). The power spectrum of the temperature anisotropy is plotted in Fig. 1 of [1]. With the data, the best fit baseline is plotted in green and at the bottom panel the data, residual to the best fit are also plotted to show the deviations. Here, we notice an oscillatory pattern between angles $\ell = 800 - 2000$ from the residual plot. At larger angular scales we also see other patterns but here we will not talk about those here as they are not directly relevant to the tensions in this study.

These residual patterns can be generated if we take the map of CMB fluctuations and smooth it in a particular way, which smears the original signals to some extent. This smearing effect is similar to the one produced by weak gravitational lensing of the CMB. One can show that if an unphysical lensing amplitude ($A_{\rm lens}$) parameter is introduced, and when it is more than 1 (in standard model, this must be 1), this excess can be explained. In Planck analysis [1] the significance of this parameter is reported to be nearly 3 standard deviations. This implies that inclusion of this parameter rejects the theoretical consistency of $A_L = 1$. Since this is an unphysical parameter, we turn to other physical sources to explain this excess smearing. If we relax the assumption that our Universe is flat, then also we can reject the standard model by better addressing the data and a closed Universe picture emerges (see, [1]). However, the closed Universe picture is also unacceptable since it invites unavoidable problems with initial conditions, and certain crucial cosmological parameters in this picture become inconsistent with all other cosmological measurements [7].

Now let is discuss the tensions between the cosmological observations *w.r.t. the standard model*. We turn to matter dominated epoch that follows the radiation dominated the Universe. During this epoch, the gravitational attraction brought the particles together and bound them to form structures. A structure (say, galaxy) that is closer to us, distorts the image of a background structure through gravitational lensing. Observations of these structures, lensing and their correlations also help us understand the Universe by constraining the model of the Universe. Here we will only talk about a particular parameter called S_8 which is a combination of matter density and the σ_8 which is related to the amplitude of the matter power spectrum (primordial power spectrum multiplied by a transfer function). It refers to the root mean square fluctuations of matter density within a spherical volume. The radius of that sphere is indicated by the number 8 in h^{-1} Mpc units. Surveys like the Kilo-Degree Survey and The Dark Energy Survey have observed galaxies over areas of 1500 and 5000 square degrees. Millions of galaxy observations help us in understanding gravitational lensing and galaxy clustering, which, in turn, put bounds on the model parameters. Hence, we get bounds on σ_8 and matter density or equivalently on S_8. The bounds obtained from KiDS and DES (Dark Energy Survey Year 3 Results are completely consistent. However, the bound derived from Planck CMB data shows an inconsistency at a level of 2–3 standard deviations, a tension of moderate level between CMB and LSS observations.

The most significant tension in cosmology, however, is noticed between the direct and indirect determinations of the Hubble parameter. In the indirect determination, CMB is the most precise observation which determines $H_0 = 67.4 \pm 0.5\,{\rm km/s/Mpc}$. This 0.7% determination of Hubble parameter comes from CMB temperature, polarization and lensing data in Planck analysis [1]. On the other hand, local Hubble measurements from the Cepheid and Supernovae observations, where SH_0ES collaboration [2] reports a the constraint $H_0 = 73.04 \pm 1.04$ km/s/Mpc which shows a 5σ tension with the CMB estimated constraint. In the prologue of this book "The Hubble Tension, Book Prologue: A Perspective Along Several Axes", plots the tension between the direct and indirect measurements of H_0.

34.3 The Idea of Primordial Solution

In this chapter, since we discuss the search for a common solution to the anomalies and tensions in cosmology, we keep this section for the historical development of primordial solution.

The attempt to capture CMB anomalies through features in primordial spectrum is not new and several models are available in the literature. Models have been designed to capture outliers at different scales in the angular power spectra of temperature and polarization [8–19]. While there are many interesting models, it will be fair to comment that we have not yet been able to find any statistically significant (more than 2σ) feature candidate from the CMB data.

Apart from forward modelling of features in the spectrum, direct model independent reconstruction techniques have been useful to identify the location and type of features [20–22]. In this context, using Planck data, Modified Richardson-Lucy (MRL) method was introduced to reconstruct the primordial spectrum assuming the baseline best-fit background cosmology [23, 24]. In Sect. 3.2 of [24], degeneracy between lensing effects and primordial features was explored in different Planck spectra of temperature anisotropy. It was found that certain oscillations can mimic the lensing signal to some extent. In Fig. 34.2 we quote the relevant plots from the publication that shows a damped oscillations in the primordial spectrum, peaking around 0.1/Mpc scales can generate limited lensing-like signals in the angular power spectrum.

Later, the role of primordial features in alleviating H_0, S_8 tensions was explored in [25, 26]. In [25] MRL was used to find features in the primordial spectrum that can address the CMB data as good as the baseline, but with the mean H_0 of SH_0ES measurements. In particular, here the reconstruction of power spectrum assumes a background cosmology that is not the best fit baseline model but a higher H_0 (keeping other parameters fixed). The reconstructed power spectrum contained oscillatory signals as demonstrated in Fig. 34.3 (top). It is needless to mention that this complicated form of power spectrum is difficult to parameterize and therefore, it is difficult to realize this in a model on inflation or bouncing cosmology. However, it is interesting to note that the power spectrum contains damped oscillations peaking around 0.1/Mpc, similar to what was obtained in [24]. When this spectrum is used as the primordial power spectrum instead of power law in fitting CMB data, we find substantial increase in the Hubble parameter and decrease in the matter density (bottom left plot). At the same time, we notice a significant decrease in the disagreement in the estimation of S_8 (see, $\Omega_m - \sigma_8$ plot in the bottom right).

Therefore, although the realization of the reconstructed spectrum seems implausible from theoretical modelling, it can be appreciated that certain features in the spectrum can indeed act as a common solution to the anomalies (by mimicking the lensing signal to certain extent) and tensions (similar primordial signals being able to resolve H_0 and S_8 tensions). In the next section we discuss the reconstruction of *One Spectrum* that combined the efforts of [24] and [25].

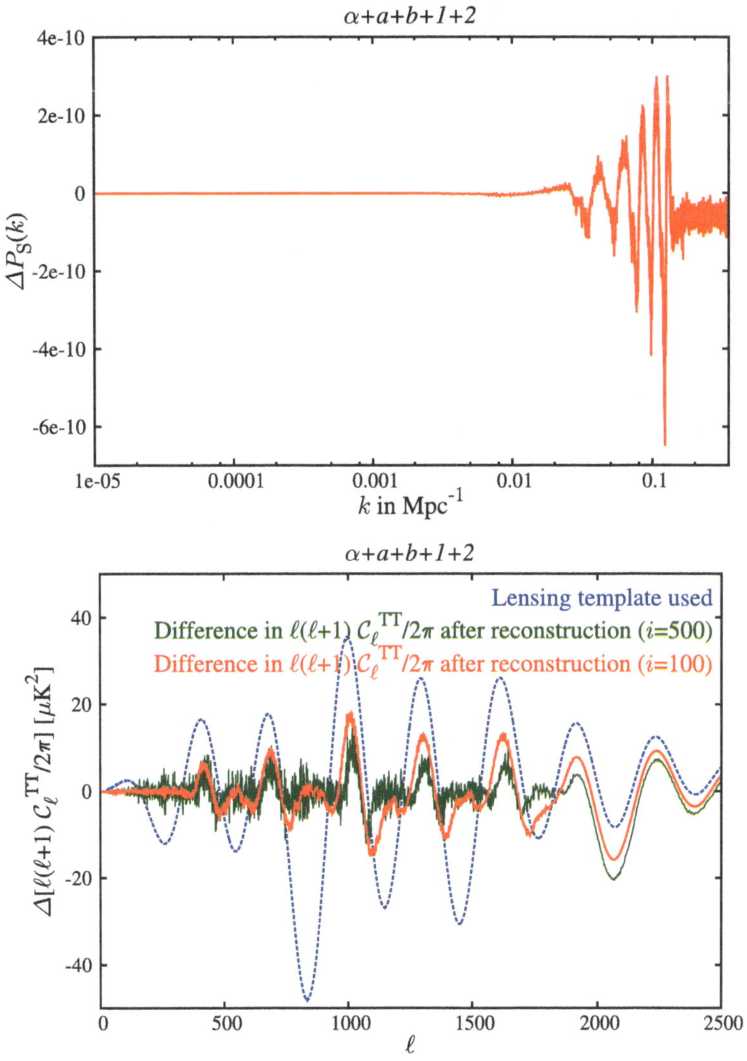

Fig. 34.2 Form of primordial power spectrum (top) that mimics the effect of lensing (bottom) to certain extent

34.4 *One Spectrum* reconstruction

In order to find a spectrum that simultaneously resolves the anomalies in the CMB data and the tensions between datasets, we use MRL reconstruction. The methodology has been elaborated in [27]. Here we outline certain key points.

The reconstruction is based on the temperature anisotropy power spectrum only, since the anomalies and tensions are not existent with the polarization data. The

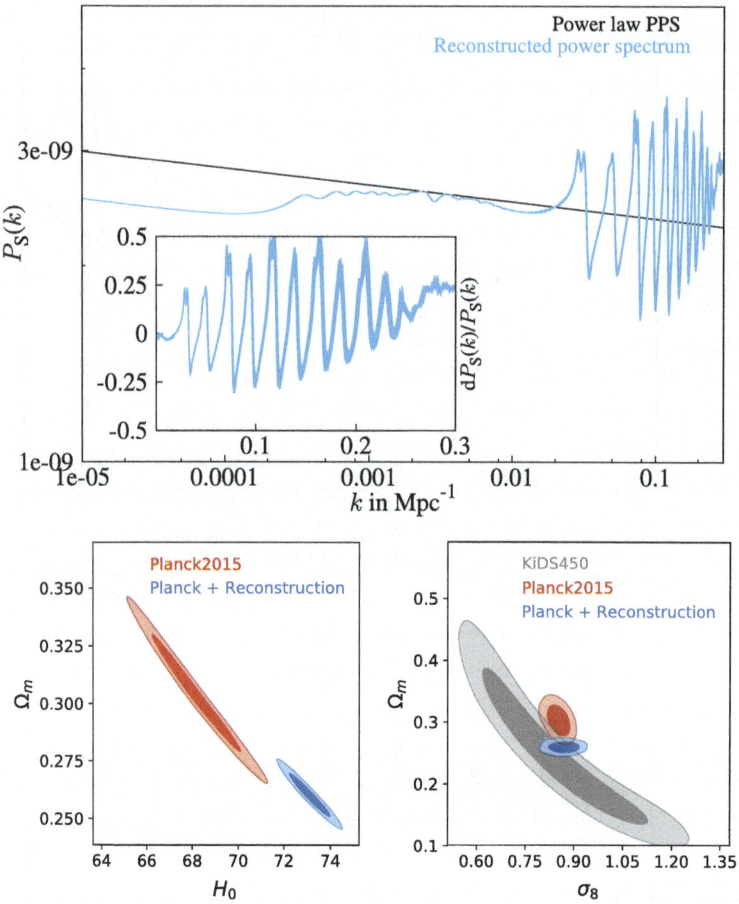

Fig. 34.3 [Top] Form of primordial power spectrum that prefers higher H_0 in CMB. [Bottom left]: Shifts in matter density and Hubble parameter when the reconstructed power spectrum is used. [Bottom right]: Shifts in the matter density and σ_8. The size of the contours in case of reconstructed power spectrum fit reduces as we use a fixed power spectrum and do not marginalize over the spectral tilt

angular power spectrum of temperature anisotropy ($\mathscr{C}_\ell^{\mathrm{TT}}$) is a convolution of transfer function ($G_{\ell k}^{\mathrm{TT}}$) dependent on the background parameters and the primordial power spectrum ($\mathscr{P}_k$) as:

$$\mathscr{C}_\ell^{\mathrm{TT}} = \sum_{k=k_{min}}^{k_{max}} \mathscr{G}_{\ell k} \mathscr{P}_k \tag{34.1}$$

The Richardson-Lucy deconvolution algorithm works iteratively and given an initial power spectrum $\mathscr{P}_k^i$ it calculates the $i+1$'th spectrum as:

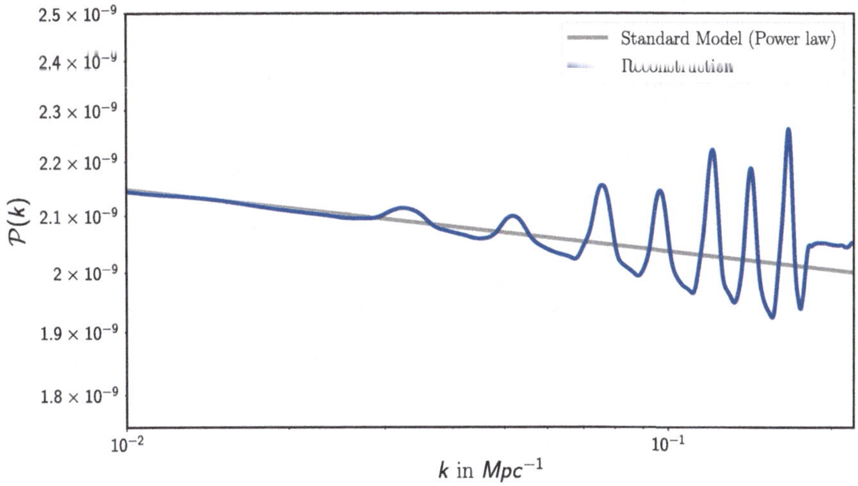

Fig. 34.4 *One Spectrum:* The reconstructed power spectrum that mimics excess smoothing

$$\mathscr{P}_k^{i+1} - \mathscr{P}_k^i = \mathscr{P}_k^i \times \sum_{\ell=\ell_{min}}^{\ell_{max}} \tilde{\mathscr{G}}_{\ell k} \left(\frac{\mathscr{C}_\ell^{\rm data} - \mathscr{C}_\ell^{\rm theory(i)}}{\mathscr{C}_\ell^{\rm theory(i)}} \right) \tag{34.2}$$

This algorithm converges the maximum likelihood solution and it was been demonstrated in earlier publication [22] that this algorithm, independent of the initial choice of spectrum deconvolves to the same resulting spectrum.

In obtaining *One Spectrum*, the choice of $\mathscr{C}_\ell^{\rm data}$ was of crucial importance. In all earlier works [22–24], $\mathscr{C}_\ell^{\rm data}$ was chosen to be the actual data from CMB. There, after reconstruction, post processing was necessary to identify the possible features in the power spectrum. In [27], however, the method aims for a smooth primordial spectrum apriori, and therefore, instead of using the data as in [25], we use the best fit angular spectrum from lensing amplitude extension of the baseline (baseline+$A_{\rm lens}$). The main idea behind this is to project the excess smoothing found in the extended model onto the primordial spectrum. Note that in Eq. 34.2, the $\mathscr{C}_\ell^{\rm data}$ refers to unlensed spectrum. Here, to obtain the best fit unlensed spectrum, the lensing contribution corresponding to $A_{\rm lens} = 1$ was subtracted so that the excess lensing contribution remains in the unlensed spectrum. Next the deconvolution reconstructs the *One Spectrum*, containing the primordial oscillations that mimics the excess smoothing in CMB temperature anisotropy spectrum, as plotted in Fig. 34.4.

Interestingly, compared to [25], *One Spectrum* appears to be much simpler and can be parametrized easily. The reconstructed spectrum, allowing an amplitude shift can be compared with the Planck temperature and polarization data. To begin with, we present the analysis with $A_{\rm lens}$ and $\Omega_{\rm K}$ separately against Planck temperature data. Since the spectrum is designed to mimic lensing effect, we expect that the spectrum can bring back cosmic concordance by solving lensing anomaly. Figure 34.5 plots

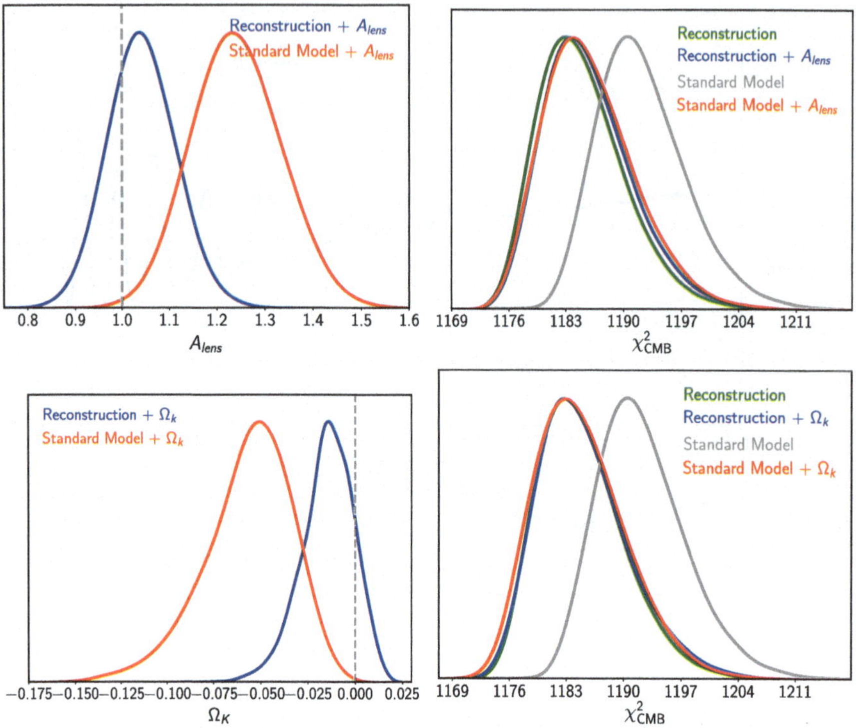

Fig. 34.5 [Top] Analysis with $A_{\rm lens}$ extension. Note that when *One Spectrum* is used, the lensing amplitude becomes consistent to 1, with significant improvement in fit compared to the Standard Model (baseline). [Bottom] Analysis including curvature parameter variation. Reconstructed spectrum brings back flat Universe $\Omega_{\rm K} = 0$, also with significant improvement in fit to the data

the results of such analyses. Note that both $A_{\rm lens} = 1$ and $\Omega_{\rm K} = 0$ comes within 1σ bounds from the mean, when reconstructed *One Spectrum* is used instead of power law primordial spectrum. Moreover, we notice the distribution of χ^2 moves significantly towards the left of standard model.

Analyses including Planck CMB polarization are detailed in [27]. Since the feature is dominantly located at small scales, we do not expect polarization data to constrain it better, given their low signal-to-noise ratio.

Apart from removing anomalies within the CMB data, we find that *One Spectrum* prefers a higher H_0 and lower matter density (Ω_m). Due to lower Ω_m, the spectrum also prefer lower $S_8 = \sigma_8\sqrt{\Omega_m/0.3}$. In Fig. 34.6 we plot the shifts in the parameters. The oscillations in *One Spectrum* provide the appropriate excess smoothing in the data compared to the baseline standard model. This in turn, requires certain shifts in the background parameter space. In fitting Planck temperature data (P18TT: denotes Plik high-ℓ TT + low-ℓ TT + simall-low-E likelihood), we find, $H_0 = 69.21 \pm 0.50$ and $S_8 = 0.786 \pm 0.015$, while for Planck temperature+polarization data (P18TP: denotes Plik high-ℓ TTTEEE + low-ℓ TT + simall-low-E likelihood), they shift to

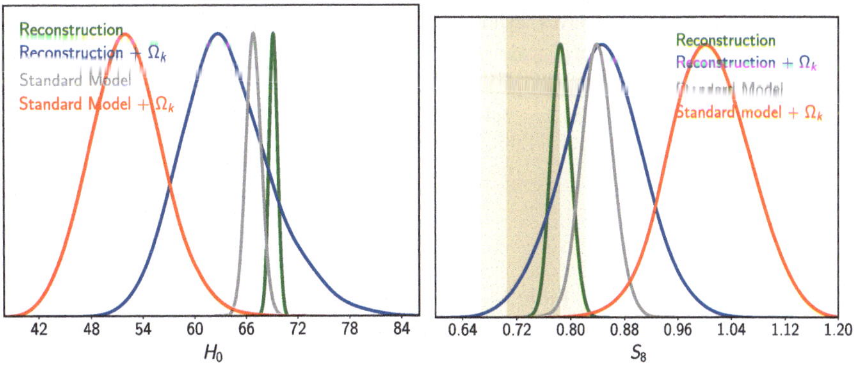

Fig. 34.6 Shifts in the parameters H_0 (left) and S_8 (right) when *One Spectrum* is used instead of power law primordial spectrum

$H_0 = 69.01 \pm 0.44$ and $S_8 = 0.798 \pm 0.013$. Similar to the analysis in [25], here too the constraints are tighter than the power-law baseline model, since the form of the spectrum used to fit here is fixed and in power law form the tilt is allowed to vary. Therefore, it should not be concluded from this analysis that the model is more constraining than the baseline model. Rather, we should note that *One Spectrum*, while providing a better fit to the data, prefers $A_{\text{lens}} = 1$ and $\Omega_K = 0$ and a high H_0 and lower S_8, which was our main goal of the reconstruction.

34.5 Parametrizing *One Spectrum*

In this section, we discuss the parametric form of the spectrum that captures the main features of *One Spectrum*. To assess the significance of the oscillations, parametrization is necessary. A four parameter model of primordial spectrum can be expressed as,

$$\mathcal{P}_{\text{Restricted}}(k) = \mathcal{P}_{\text{Power Law}}(k)\left[1 + \frac{\alpha_1 \sin\left(\omega(k - k_0)\right)}{1 + \beta(k - k_0)^4}\right], \tag{34.3}$$

with α_1 as amplitude, k_0 as position, ω as frequency and β as damping of the oscillations. This power spectrum can now used instead of numerical *One Spectrum* for the parameter estimation and the overall spectral amplitude and tilt can be varied through $\mathcal{P}_{\text{Power Law}}$.

In this parameter estimation, we will focus on two numbers, the significance of α_1. The significance refers to the percentage with which the marginalized posterior distribution of α_1 rejects $\alpha = 0$. Simply, it denotes the significance of preference for the oscillations compared to a featureless spectrum. Secondly, we also note the Bayes' factor, or the ratio of the evidences between the model with and without oscillatory features. In our convention, a positive logarithm of Bayes' factor indicates preference

Table 34.1 Logarithm of the Bayes' factor *w.r.t.* to the baseline model (first row) and the significance of oscillations (second row)

Model/Data	P18TT	P18TT+HST	P18TP	P18TP+HST
Parametrized model	-0.58 ± 0.52	3.4 ± 0.53	-0.01 ± 0.54	1.46 ± 0.55
	85%	99.9%	95%	99.5%

for the oscillatory model. In Table 34.1 we list the logarithm of Bayes' factor (in first row) and the evidences (in second row) for different dataset combinations. Only with CMB data we notice 1-2σ preference for oscillations. The significance increase when Hubble prior from SH_0ES is used.

When HST prior is used, while H_0 posterior is expected to shift towards higher values, we also find S_8 moves towards lower values, making the model more consistent to the lensing and clustering surveys.

The marginalized posterior distributions of the parameters describing the oscillations are plotted in Fig. 34.7. The dashed lines represent the CMB only results while the solid lines are obtained upon using the HST prior. In CMB only analyses, when polarization is added we get marginal improvement in significance for the features, indicating a better fit to both temperature and polarization data.

This parametrized power spectrum has been tested with several other combinations of datasets. Small scale CMB temperature and polarization data from the Atacama Cosmology Telescope are in agreement with the oscillations. Clustering and lensing data from the DES also support the lower of S_8 preference of this parametrization. Summary statistics from the Baryon Acoustic Oscillations data and distance

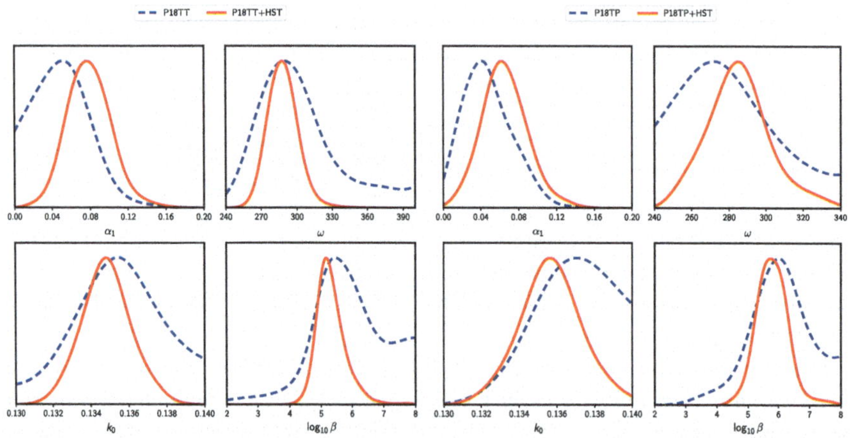

Fig. 34.7 Significance of the feature parameters. 2×2 plots on the left side shows results with CMB temperature while the plots on the right side demonstrates the effect of adding polarization data

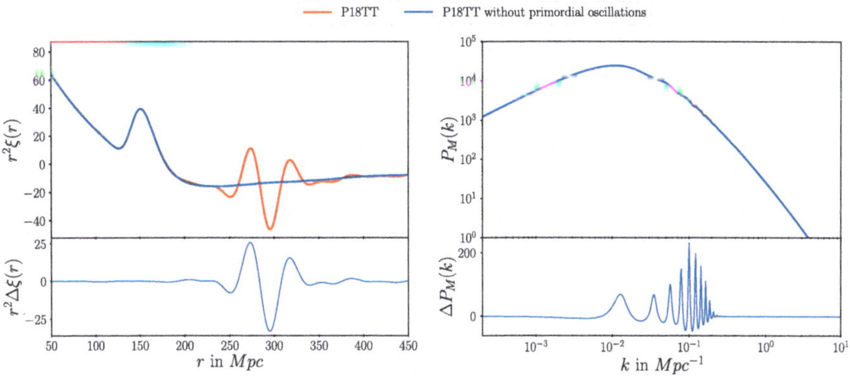

Fig. 34.8 Correlation function (left) and the matter power spectrum (right) obtained from the best fit primordial power spectrum parametrized in Eq. 34.3

modulus from Supernovae data jointly decrease the significance of these features as they prefer higher matter density.

Since the spectrum exhibits oscillations at a particular frequency (300 Mpc as can be understood from Fig. 34.7), the correlation function has a bump and a dip around that length scales as shown in Fig. 34.8. In this figure we also plot the matter power spectrum. With ongoing and future large scale structure observations such as Euclid [28], DESI [29] and LSST [30], significance of these oscillations can be tested jointly with CMB.

We close this section by summarizing that the features in the primordial spectrum while fitting the CMB data, prefers changes in the cosmological parameters to certain extent compared to the baseline mean values. Therefore, it brings CMB closer to the other observations. Given the degeneracy between the background and the power spectrum parameters, we also observe an increase in the spectral tilt. Since, we have access to the small scale CMB data from ACT, joint estimations are useful for consistency check with this increased spectral tilt. The ACT data prefers a higher spectral tilt even in the baseline parametrization. Therefore, the analysis with P18TP+ACT+DES data, the spectral tilt is found to be 0.9742 ± 0.0035. This shift in spectral tilt is crucial to model inflationary potentials that we discuss in the next section.

34.6 A Model of Inflation

The simple parametrization of *One Spectrum* opens up the window to explore an inflationary solution. We model the modification in inflationary dynamics over the standard slow roll inflationary models. This work has been discussed in the letter [31]. Note that for a potential, the spectral tilt (n_s) and tensor-to-scalar ratio (r) in models of inflation are usually fixed (for example, in quadratic model, we find $n_s \sim 0.96$ and $r \sim 0.16$). Variations in these two observables are possible only within limited windows for a fixed potential form. Theoretically, they are also correlated within an inflationary model. In the last section we mentioned the importance of change in n_s. Therefore the restrictions imposed by the choice potential form make the potential parametrization unsuitable for the parameter estimation with combined datasets.

In order to allow free variations in the $n_s - r$ alongside other inflationary parameters, we use a baseline slow-roll parametrization given by,

$$\epsilon_H^{baseline}(N) = \epsilon_1 \exp\left[\epsilon_2(N - N_*)\right] \tag{34.4}$$

Here, ϵ_1, ϵ_2 are free parameters and N_* is chosen to be 20 *e-folds*. Note that we impose the pivot scale $k = 0.05$/Mpc leaves the Hubble radius 50 *e-folds* before the end of inflation which directly relates N_* with the pivot scale. This parametrization is equivalent to the power law parametrization in the power spectrum as the scalar power spectrum amplitude, tilt and the tensor-to-scalar ratio can be varied through the variation of H_i^2/ϵ_1 (H_i represents the Hubble parameter at the onset of inflation), ϵ_1 and ϵ_2. Therefore, it would be fair to say that this slow roll parametrization allows us to marginalize over all minimal slow roll potential parametrizations that generate power law primordial spectrum without running and higher derivatives of spectral index. The power spectrum parameters can be approximated as,

$$\begin{aligned} n_s &\simeq 1 - 2\epsilon_1 - \epsilon_2, \\ r &\simeq 16\epsilon_1 \end{aligned} \tag{34.5}$$

Note, however, we do not use slow roll approximations to solve for the primordial spectrum and BINGO [32] is used to solve the Mukhanov-Sasaki equations with Bunch-Davies initial condition.

Using Eq. 34.4 as the baseline model, we add modifications as,

$$\epsilon_H(N) = \epsilon_H^{baseline}(N)\left(1 + \frac{\alpha \cos\left[\omega(N - N_0)\right]}{1 + \beta(N - N_0)^2}\right), \tag{34.6}$$

which introduces oscillatory phase (both local and non-local) in the slow roll dynamics with amplitude α and frequency ω. The scalar field rolling slowly down the potential temporarily fast rolls around $N = N_0$. A non-zero β parameter helps to damp the oscillation around that *e-folds*. The power spectrum resulting from such inflationary dynamics produces an envelope of oscillations in the order of sinusoidal ($sin(\tilde{\omega}k)$), sinusoidal logarithmic ($\sin(\tilde{\omega}\log k)$) and sinusoidal forms.

We compare this inflationary dynamics with combinations of Planck, BICEP-KECK18 (BK18) and SH_0ES data. BK18 data is directly relevant here as in this case we are constraining inflationary dynamics where BK18 data, jointly with Planck provides an upper bound on ϵ_1 through the upper bounds on the B-mode anisotropy spectrum. The Planck temperature and polarization data has been used in the com-

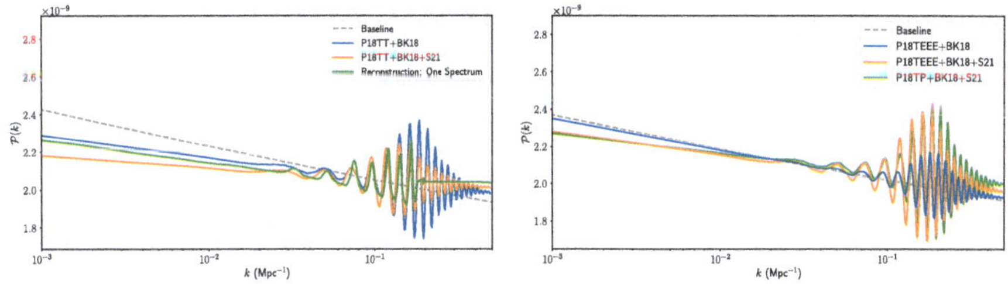

Fig. 34.9 Best fit primordial power spectra to different data combinations

binations of TT, TEEE, TTTEEE (TP). Lensing likelihood has also been used in a separate analysis with TP.

Markov Chain Monte Carlo through CosmoMC [33] has been used for parameter estimation and BOBYQA minimization [34] starting from MCMC minimum has been used to find the best fit for each data combination. Selected best fits are plotted in Fig. 34.9. All the best fit to different power spectra are consistent to each other in frequency and phases and they significantly overlap with the *One Spectrum*. The best fit feature model spectrum also prefers a higher spectral tilt than baseline as can be noticed in all plots where temperature auto-correlation data has been used.

In Fig. 34.10 the best fit residuals of feature models compared to the baseline best fit are plotted in blue. In red, the best fit baseline+A_L residuals are plotted. The oscillations in the primordial spectrum mimic the excess smoothing effect obtained from *ad-hoc* lensing amplitude extension (red). Interestingly, in the polarization spectrum, these two models exhibit out of phase property. Therefore, with future small scale CMB polarization observations (projected CMB-S4 error-bands are plotted in EE and TE panels) it will be possible to distinguish them.

The posteriors of the inflationary dynamics are plotted in Fig. 34.11. Here the 1-3σ bounds on the evolution of the slow roll parameters (top panel), Hubble parameter (bottom left) and the reconstructed potential (bottom right) are plotted. We highlight only the fractional change by factoring out the baseline model. The potential is obtained using $V(\phi[N]) = 3M_{\rm Pl}^2 H[N]^2(1-\epsilon_H[N]/3)$ as a function of *e-folds*. The relevant oscillations can be noticed for 0.5 *e-folds*.

The numerically reconstructed potential can be well approximated by the following functional form,

$$\frac{\Delta V(\phi)}{V_{\rm baseline}(\phi)} = \frac{\alpha\cos[\omega_\phi(\phi-\phi_0)]}{1+\beta(\phi-\phi_0)^2}\,, \tag{34.7}$$

where $V_{\rm baseline}(\phi)$ represents a baseline potential satisfying strict slow roll inflation. A damped sinusoidal oscillation in the potential introduces a temporary fast roll phase which, in turn, creates wiggles in the primordial spectrum with characteristic oscillations similar to *One Spectrum*.

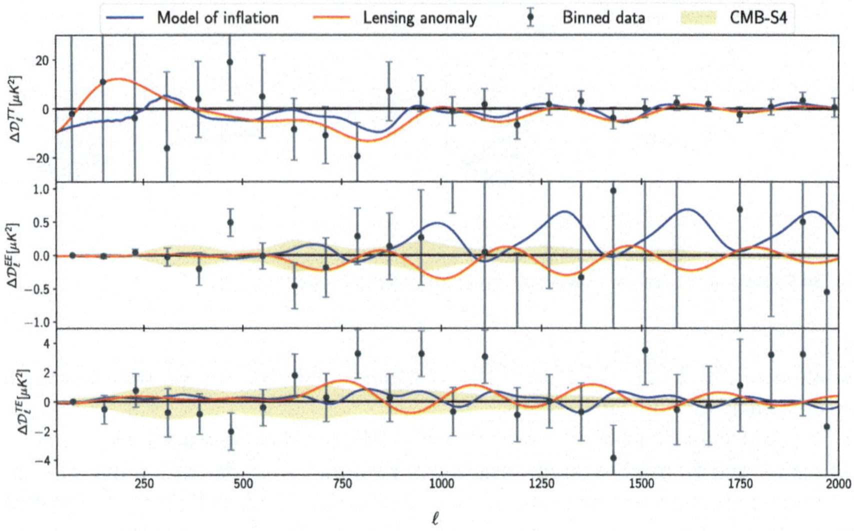

Fig. 34.10 Power spectrum best fit, residual from the baseline best fit

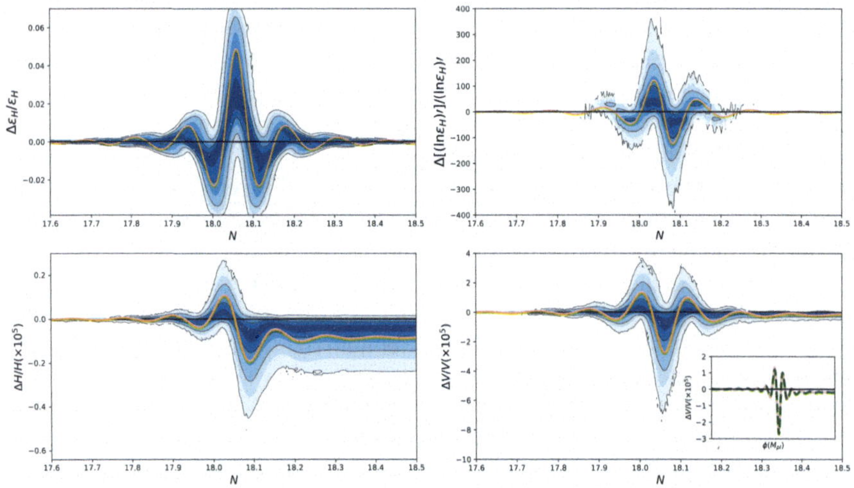

Fig. 34.11 Constraints on inflationary dynamics obtained using Planck (temperature + polarization) + BK18 + SH_0ES (S21) data

34.7 Summary and Outlook

Anomalies and tensions within and between cosmological datasets are not new findings. In several occasions, apart from statistical fluctuations, systematic uncertainties were found to be responsible for these. However, in certain cases, anomalies and tensions were found to be indicative of new physics beyond the standard model. The articles presented in this chapter build the possibility of a non-standard inflationary dynamics as a common answer to the anomalies and tensions. At the time of writing this chapter, the envelope of oscillations in the primordial spectrum is the only such common solution. While compared to other models this primordial solution does not completely resolve the Hubble tension, importantly, it shifts the cosmological parameters in directions that reduces the tensions in major parameters. Moreover, the shifts are obtained by fitting only to the CMB data, with noticeable improvement in fit compared to the baseline, resulting in 2σ support for the oscillatory features. Addition of other datasets such as DES or SH_0ES data increases the extents of the shifts further. The oscillations being generated in the inflationary epoch do not interfere with the expansion histories in the radiation, matter and dark energy dominated epochs and therefore do not degrade the constraints on parameters compared to the baseline. The uncertainties in the estimation of parameters (such as H_0, n_s, S_8) increase by 10% compared to their baseline estimations. This is an important characteristic to note in the model building exercise where the discordance is reduced due to the shift in the posterior volume and not by the enlargement of posterior volume by significantly reducing the predictability.

Given the ongoing and future observations, observables calculated from this primordial solution can soon be tested. With the present CMB three-point statistics [35], primordial bispectrum can be compared. With the small scale CMB data, the small scale oscillations can be probed in the polarization. The matter power spectrum and the correlation function can be tested with upcoming large scale structure data. Finally with future ground based CMB polarization observations, the primordial signal can be distinguished from other possible effects.

Acknowledgements DKH acknowledge support through the India-Italy mobility program INT/Italy/P-39/2022 (ER). D.K.H. is supported by CEFIPRA Grant No. 6704-4.

References

1. N. Aghanim et al., Planck 2018 results. VI. Cosmological parameters. Astron. Astrophys. **641**, A6 (2020) [Erratum: Astron.Astrophys. 652, C4 (2021)]
2. Adam G. Riess et al., A comprehensive measurement of the local value of the Hubble constant with 1 $km\ s^{-1}\ Mpc^{-1}$ uncertainty from the Hubble space telescope and the SH_0ES team. Astrophys. J. Lett. **934**(1), L7 (2022)
3. Marika Asgari et al., KiDS-1000 cosmology: cosmic shear constraints and comparison between two point statistics. Astron. Astrophys. **645**, A104 (2021)

4. L.F. Secco et al., Dark energy survey year 3 results: cosmology from cosmic shear and robustness to modeling uncertainty. Phys. Rev. D **105**(2), 023515 (2022)
5. Xiaolei Li, Arman Shafieloo, A simple phenomenological emergent dark energy model can resolve the Hubble tension. Astrophys. J. Lett. **883**(1), L3 (2019)
6. Vivian Poulin, Tristan L. Smith, Tanvi Karwal, Marc Kamionkowski, Early dark energy can resolve the Hubble tension. Phys. Rev. Lett. **122**(22), 221301 (2019)
7. Eleonora Di Valentino, Alessandro Melchiorri, Joseph Silk, Planck evidence for a closed Universe and a possible crisis for cosmology. Nature Astron. **4**(2), 196–203 (2019)
8. Xingang Chen, Richard Easther, Eugene A. Lim, Generation and characterization of large non-Gaussianities in single field inflation. JCAP **04**, 010 (2008)
9. Raphael Flauger, Liam McAllister, Enrico Pajer, Alexander Westphal, Xu. Gang, Oscillations in the CMB from axion monodromy inflation. JCAP **06**, 009 (2010)
10. Xingang Chen, Folded resonant non-Gaussianity in general single field inflation. JCAP **12**, 003 (2010)
11. M. Aich, D.K. Hazra, L. Sriramkumar, T. Souradeep, Oscillations in the inflaton potential: complete numerical treatment and comparison with the recent and forthcoming CMB datasets. Phys. Rev. D **87**, 083526 (2013)
12. Jennifer A. Adams, Bevan Cresswell, Richard Easther, Inflationary perturbations from a potential with a step. Phys. Rev. D **64**, 123514 (2001)
13. D.K. Hazra, M. Aich, R.K. Jain, L. Sriramkumar, T. Souradeep, Primordial features due to a step in the inflaton potential. JCAP **10**, 008 (2010)
14. D.K. Hazra, A. Shafieloo, G.F. Smoot, A.A. Starobinsky, Wiggly whipped inflation. JCAP **1408**, 048 (2014)
15. Jens Chluba, Jan Hamann, Subodh P. Patil, Features and new physical scales in primordial observables: theory and observation. Int. J. Mod. Phys. D **24**(10), 1530023 (2015)
16. M. Ballardini, F. Finelli, C. Fedeli, L. Moscardini, Probing primordial features with future galaxy surveys. JCAP **10**, 041 (2016). [Erratum: JCAP 04, E01 (2018)]
17. D.K. Hazra, D. Paoletti, M. Ballardini, F. Finelli, A. Shafieloo, G.F. Smoot, A.A. Starobinsky, Probing features in inflaton potential and reionization history with future CMB space observations. JCAP **02**, 017 (2018)
18. D.K. Hazra, D. Paoletti, I. Debono, A. Shafieloo, G.F. Smoot, A.A. Starobinsky, Inflation story: slow-roll and beyond. JCAP 12(12), 038 (2021)
19. Akhil Antony, Shweta Jain, Effect of damped oscillations in the inflationary potential. Eur. Phys. J. C **82**(8), 687 (2022)
20. William Hadley Richardson, Bayesian-based iterative method of image restoration∗. J. Opt. Soc. Am. **62**(1), 55–59 (1972)
21. L.B. Lucy, An iterative technique for the rectification of observed distributions. Astron. J. **79**, 745 (1974)
22. Arman Shafieloo, Tarun Souradeep, Primordial power spectrum from WMAP. Phys. Rev. D **70**, 043523 (2004)
23. D.K. Hazra, A. Shafieloo, T. Souradeep, Primordial power spectrum: a complete analysis with the WMAP nine-year data. JCAP **07**, 031 (2013)
24. D.K. Hazra, A. Shafieloo, T. Souradeep, Primordial power spectrum from Planck. JCAP **11**, 011 (2014)
25. D.K. Hazra, A. Shafieloo, T. Souradeep, Parameter discordance in Planck CMB and low-redshift measurements: projection in the primordial power spectrum. JCAP **04**, 036 (2019)
26. MiaoXin Liu, Zhiqi Huang, Band-limited features in the primordial power spectrum do not resolve the Hubble tension. Astrophys. J. **897**, 166 (2020)
27. D.K. Hazra, A. Antony, A. Shafieloo, One spectrum to cure them all: signature from early universe solves major anomalies and tensions in cosmology. JCAP **08**(08), 063 (2022)
28. G.D. Racca, R. Laureijs, L. Stagnaro, J.-C. Salvignol, J.L. Alvarez, G.S. Criado, L.G. Venancio, A. Short, P. Strada, T. Bönke, C. Colombo, A. Calvi, E. Maiorano, O. Piersanti, S. Prezelus, P. Rosato, J. Pinel, H. Rozemeijer, V. Lesna, P. Musi, M. Sias, A. Anselmi, V. Cazaubiel, L. Vaillon, Y. Mellier, J. Amiaux, M. Berthé, M. Sauvage, R. Azzollini, M. Cropper, S. Pottinger,

K. Jahnke, A. Ealet, T. Maciaszek, F. Pasian, A. Zacchei, R. Scaramella, J. Hoar, R. Kohley, R. Vavrek, A. Rudolph, M. Schmidt, The Euclid mission design. **9904**, 99040O (2016)
29. A. Aghamousa et al., The DESI experiment part I: science, targeting, and survey design, 10 (2016)
30. Željko Ivezić et al., LSST: from science drivers to reference design and anticipated data products. Astrophys. J. **873**(2), 111 (2019)
31. A. Antony, F. Finelli, D.K. Hazra, A. Shafieloo, Discordances in cosmology and the violation of slow-roll inflationary dynamics. Phys. Rev. Lett. **130**(11), 111001 (2023)
32. D.K. Hazra, L. Sriramkumar, J. Martin, BINGO: a code for the efficient computation of the scalar bi-spectrum. JCAP **05**, 026 (2013)
33. Antony Lewis, Sarah Bridle, Cosmological parameters from CMB and other data: a Monte Carlo approach. Phys. Rev. D **66**, 103511 (2002)
34. M. Powell, The bobyqa algorithm for bound constrained optimization without derivatives. *Technical Report, Department of Applied Mathematics and Theoretical Physics*, 01 (2009)
35. Y. Akrami et al., Planck 2018 results. IX. Constraints on primordial non-Gaussianity. Astron. Astrophys. **641**, A9 (2020)

Part V
Complications with Theoretical Solutions

Chapter 35
The Tension in the Absolute Magnitude of Type Ia Supernovae

David Camarena and Valerio Marra

35.1 Introduction

In the study of cosmology, one of the key quantities of interest is the Hubble constant, denoted as H_0. It represents the current rate of expansion of the universe and plays a crucial role in determining its age and evolution. Over the years, various methods have been employed to measure H_0, including observations of the cosmic microwave background (CMB), baryon acoustic oscillations (BAO), and the use of astronomical distance indicators such as Type Ia supernovae and Cepheids variables.

In recent years, the tension between different measurements of H_0 has garnered significant attention. While the Planck satellite mission [1], which analyzed the CMB, reported a value of H_0 around 67 km/s/Mpc, independent measurements using the local distance ladder technique, such as the SH_0ES collaboration [2], have consistently obtained higher values, closer to 73 km/s/Mpc. This discrepancy, significant at the 5σ level and often referred to as the "Hubble tension," has led to considerable interest in exploring cosmological models that can reconcile these conflicting results.

Here, we aim to understand the implications of the local calibration of the luminosity of Type Ia supernovae. After reviewing the basic background in Sect. 35.2, in Sect. 35.3 we will start with the 46-dimensional analysis by SH_0ES and extend it to an arbitrary cosmographic model. This will allow us to obtain a determination of local H_0 that only assumes the validity of the standard flat ΛCDM model, and another one that only assumes the FLRW metric, that is, large-scale homogeneity and isotropy.

D. Camarena (✉)
Department of Physics and Astronomy, University of New Mexico, Albuquerque, NM 87106, USA
e-mail: dacato115@gmail.com

V. Marra
Núcleo de Astrofísica e Cosmologia & Departamento de Física, Universidade Federal do Espírito Santo, 29075-910 Vitória, ES, Brazil
e-mail: valerio.marra@me.com

E. Di Valentino and D. Brout (eds.), *The Hubble Constant Tension*, Springer Series in Astrophysics and Cosmology, https://doi.org/10.1007/978-981-99-0177-7_35

We proceed in Sect. 35.4 to highlight the importance of considering the calibration of the supernova magnitude M_B during cosmological inference, emphasizing the necessity to test the consistency of a given cosmological model with respect to M_B. Finally, in Sect. 35.5 we discuss the "M_B tension," the fact that the supernova luminosity either gets calibrated by CMB and BAO observations or by the first two rungs of the cosmic distance ladder, providing inconsistent constraints. Our conclusions are presented in Sect. 35.6.

35.2 Curved *w*CDM Cosmography

Given the luminosity distance d_L, the apparent magnitude is given by:

$$m_B(z) = 5\log_{10}\left[\frac{d_L(z)}{1\text{Mpc}}\right] + 25 + M_B\,, \tag{35.1}$$

and the distance modulus by:

$$\mu(z) = m_B(z) - M_B\,, \tag{35.2}$$

where M_B is the absolute magnitude of Type Ia supernovae. The luminosity distance is a prediction of a cosmological model and, assuming the FLRW metric, is:

$$d_L(z) = (1+z)\frac{c}{H_0}\Omega_{k0}^{-1/2}\sinh\left[\Omega_{k0}^{1/2}\int_0^z \frac{d\bar{z}}{E(\bar{z})}\right], \tag{35.3}$$

where $H = \dot{a}(t)/a(t)$ describes the expansion history of the universe and $a(t)$ is the scale factor. For a curved *w*CDM model it is:

$$\frac{H^2(z)}{H_0^2} = E^2(z) = \Omega_{\text{m}0}(1+z)^3 + \Omega_{\text{de}0}(1+z)^{3(1+w)} + \Omega_{k0}(1+z)^2\,, \tag{35.4}$$

where we neglected radiation as it is inconsequential at redshifts lower than 100. In the previous equation the subscript 0 denotes the present-day value of the corresponding quantity, $\Omega_{\text{m}0}$ is the density parameter for cold dark matter (zero pressure), $\Omega_{\text{de}0}$ the one for a dark energy fluid with equation of state parameter w, and $\Omega_{k0} = -kc^2/H_0^2$ represents the spatial-curvature contribution to the Friedmann equation ($a(t_0) = 1$). It is $\Omega_{\text{m}0} + \Omega_{\text{de}0} + \Omega_{k0} = 1$.

Alternatively, using a cosmographic approach one has [3]:

$$d_L(z) = \frac{cz}{H_0}\left[1 + \frac{1-q_0}{2}z - \frac{1 - q_0 - 3q_0^2 + j_0 - \Omega_{k0}}{6}z^2 + O(z^3)\right], \tag{35.5}$$

where the Hubble constant H_0, the deceleration parameter q_0 and the jerk parameter j_0 are defined, respectively, according to:

$$H_0 = \left.\frac{\dot{a}(t)}{a(t)}\right|_{t_0}, \qquad q_0 = \left.\frac{-\ddot{a}(t)}{H^2(t)a(t)}\right|_{t_0}, \qquad j_0 = \left.\frac{\dddot{a}(t)}{H^3(t)a(t)}\right|_{t_0}. \tag{35.6}$$

Cosmography is a model-independent approach in the sense that it does not assume a specific model as it is based on the Taylor expansion of the scale factor. However, this does not mean that its parameters do not contain cosmological information. For example, in the case of curved wCDM cosmologies, q_0 and j_0 are given by:

$$q_0 = \frac{\Omega_{\rm m0}}{2} + \frac{1+3w}{2}\Omega_{\rm de0} \overset{\rm flat}{=} \frac{1}{2} + \frac{3}{2}w(1-\Omega_{\rm m0}) \overset{\Lambda}{=} \frac{3}{2}\Omega_{\rm m0} - 1 \overset{\rm fid}{=} -0.55\,, \tag{35.7}$$

$$j_0 = \Omega_{\rm de0} + \Omega_{\rm m0} + \frac{9}{2}w(1+w)\Omega_{\rm de0} \overset{\rm flat}{=} 1 + \frac{9}{2}w(1+w)(1-\Omega_{\rm m0}) \overset{\Lambda}{=} 1\,, \tag{35.8}$$

where in the last equality of (35.7) the concordance value of $\Omega_{\rm m0} = 0.3$ was adopted.

Finally, when determining the luminosity distance, one should adopt both the peculiar velocity-corrected CMB-frame redshift z (also called Hubble diagram redshift) and the heliocentric redshift $z_{\rm hel}$ so that the luminosity distance is given by [4]:

$$d_L(z, z_{\rm hel}) = \frac{1+z_{\rm hel}}{1+z} d_L(z)\,. \tag{35.9}$$

35.3 Local Determination of H_0

The SH_0ES collaboration measures the Hubble constant in the redshift range $0.023 < z < 0.15$, where the minimum redshift is large enough in order to reduce the impact of cosmic variance by the local large-scale structure [5, 6] and the maximum redshift is small enough to provide a late-time local measurement of the Hubble constant and reduce the impact of cosmology in its determination. Ideally, one would want a cosmology-independent local constraint. SH_0ES traditionally fits the value of H_0 in (35.5), but fixes the values of the higher-order parameters to their fiducial values of $q_0^{\rm fid} = -0.55$ and $j_0^{\rm fid} = 1$, using information from observations beyond the local universe, like higher-z supernovae.

In [7, 8] we obtained a local determination of H_0 without fixing the deceleration parameter q_0. Here, we will improve upon our previous results by extending the analysis of [2] using their latest data.[1] Specifically, we adopt the χ^2 statistic:

$$\chi^2(\theta, q_0, j_0) = \big(y(q_0, j_0) - L\theta\big)^{\rm T} C_y^{-1} \big(y(q_0, j_0) - L\theta\big)\,, \tag{35.10}$$

[1] https://github.com/PantheonPlusSH0ES/DataRelease.

where y is the 3492-dimensional data vector, θ is the 46-d parameter vector considered by SH_0ES, L is the 46×3492 equation matrix, so that the model is $L\theta$, and C_y is the 3492×3492 covariance matrix. The last elements of y are about the 277 Hubble flow supernovae (nhf) that are used to determine H_0:

$$\begin{aligned} y_i^{\text{nhf}} &= m_{B,i} - 5\log_{10} H_0 d_L - 25 \\ &= m_{B,i} - 5\log_{10} c z_i \left[1 + \frac{1-q_0}{2} z_i - \frac{1 - q_0 - 3q_0^2 + j_0 - \Omega_{k0}}{6} z_i^2\right] - 25 \,, \end{aligned} \tag{35.11}$$

and in Eq. (35.10) we let the components y_i^{nhf} be functions of q_0 and j_0.[2]

The baseline analysis by SH_0ES adopts $\chi^2(\theta, q_0^{\text{fid}}, j_0^{\text{fid}})$, so that y is constant and the model is linear. In this case the best-fit model and its covariance matrix are:

$$\theta_{\text{bf}}(q_0, j_0) = \left(L^{\text{T}} C_y^{-1} L\right)^{-1} L^{\text{T}} C_y^{-1}\, y(q_0, j_0)\,, \tag{35.12}$$

$$C_\theta = \left(L^{\text{T}} C_y^{-1} L\right)^{-1}, \tag{35.13}$$

from which we can note that the best fits depend on q_0 and j_0, but their covariance matrix does not. Using the properties of multi-variate Gaussian distributions (the model is linear), it is then straightforward to plot the marginalized constraints on any combination of parameters by considering the eigenvectors and eigenvalues of the covariance matrix. This is equivalent to performing a Bayesian analysis with wide flat priors, as done in [2]. In Fig. 35.1 we show the marginalized constraints on M_B and H_0.

If we now do not fix q_0 and j_0, the model used in Eq. (35.10) becomes nonlinear and a full Bayesian analysis is necessary. One possible approach is to approximate the posterior as a Gaussian distribution via its Fisher approximation, $F_{\alpha,ij} = \frac{1}{2}\frac{\partial^2 \chi^2}{\partial\alpha_i \partial\alpha_j}$, where $\alpha = \{\theta, q_0, j_0\}$ is the 48-d parameter vector, and $C_\alpha = F_\alpha^{-1}$ is the covariance matrix on the parameters (the inverse of the Fisher matrix).[3] However, the posterior will likely show a non-Gaussian character in the j_0 direction as the latter is expected to be poorly constrained by supernovae at $z < 0.15$. Consequently one should perform a computationally demanding 48-d Bayesian analysis.

Fortunately, we can exploit the fact that the model is linear in all parameters but q_0 and j_0 to substantially speed up the estimation of the posterior. We can indeed marginalize analytically on all the parameters we are not interested in:

$$\chi^2_{\text{marg}}(M_B, H_0, q_0, j_0) = \chi^2\big(\theta_{\text{bf}}(q_0, j_0), q_0, j_0\big) + V C_{M_B,H_0}^{-1} V\,, \tag{35.14}$$

$$V = \{M_B, H_0\} - \{M_B^{\text{bf}}(q_0, j_0), H_0^{\text{bf}}(q_0, j_0)\}\,, \tag{35.15}$$

where M_B^{bf} and H_0^{bf} comes from the corresponding entries of $\theta_{\text{bf}}(q_0, j_0)$ [2], and C_{M_B,H_0} is obtained by removing all the rows and columns of C_θ expect the ones

[2] We set $\Omega_{k0} = 0$, but the latter is degenerated with j_0, see discussion below.

[3] Note that $\frac{1}{2}\frac{\partial^2 \chi^2}{\partial\theta_i \partial\theta_j}$ exactly gives C_θ^{-1} as, in this case, the model is linear.

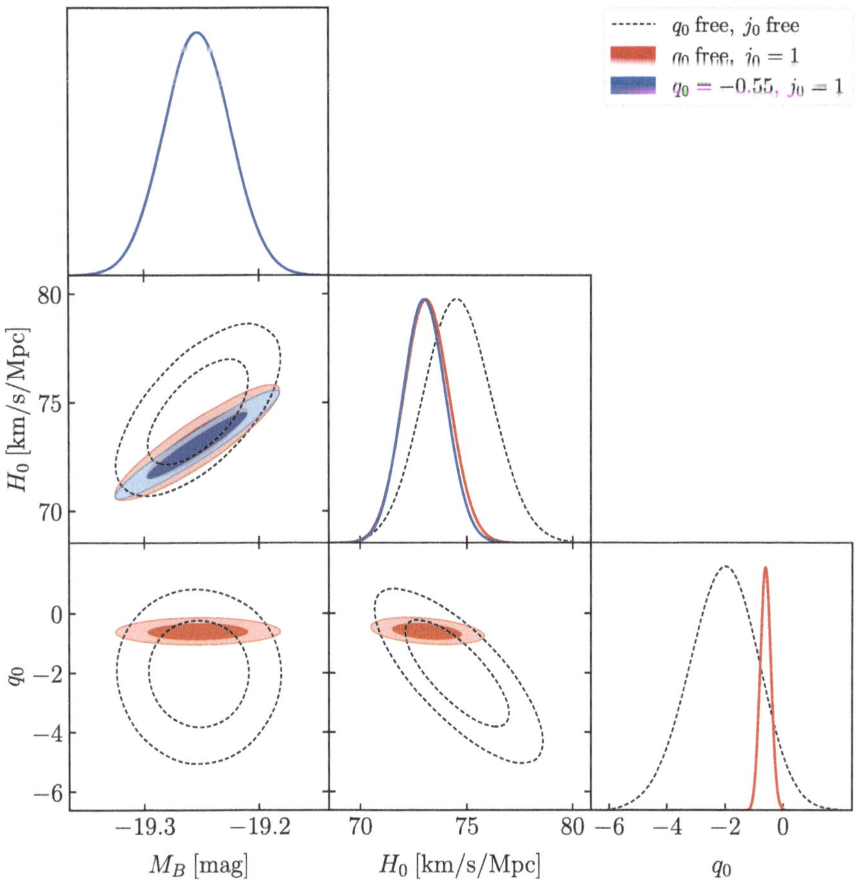

Fig. 35.1 Marginalized 68.3 and 95.4% credible level constraints on the supernova Ia absolute magnitude M_B, the Hubble constant H_0 and the deceleration parameter q_0 from the 48-dimensional analysis of Eq. (35.14) using the latest data from the SH_0ES collaboration, that is, the first two rungs of the distance ladder together with the 277 Hubble flow supernovae in $0.023 < z < 0.15$. Blue shows the SH_0ES baseline analysis, red the analysis that only assumes the flat ΛCDM model and promote q_0 (or, equivalently, Ω_{m0}) to a free parameter, and the empty contours show the analysis that only assumes the FLRW metric, that is, large-scale homogeneity and isotropy (both q_0 and j_0 are free, the latter is marginalized over)

relative to M_B and H_0, as this operation is equivalent to marginalizing the posterior over all the other linear parameters.

In Fig. 35.1 and Table 35.1 we compare the baseline results by SH_0ES with our results. It is interesting to note that the analysis relative to q_0 free and $j_0 = 1$ provides a very competitive constraint that is independent of the particular ΛCDM model as it assumes just flat FLRW metric and cosmological constant. It features a 1.5% uncertainty, as compared with the 1.4% uncertainty of the baseline analysis. In particular, marginalizing over q_0 is equivalent to marginalizing over Ω_{m0}. Indeed, their relation

Table 35.1 Marginalized constraints (mode and 68% uncertainty) for the three cosmographic analyses of Fig. 35.1. The first line corresponds to the baseline analysis by SH_0ES [2], providing the same constraints. The second analysis gives a constraint that is marginalized over any flat ΛCDM model. The last analysis only assumes the FLRW metric. Following [2], an additional systematic error of 0.3 km/s/Mpc was added to the uncertainty of H_0

Case	M_B (mag)	H_0 (km/s/Mpc)	q_0
$q_0 = -0.55$, $j_0 = 1$	-19.253 ± 0.029	73.04 ± 1.04 (1.4%)	–
q_0 free, $j_0 = 1$	-19.253 ± 0.029	73.14 ± 1.10 (1.5%)	-0.61 ± 0.18
q_0 and j_0 free	-19.252 ± 0.029	74.56 ± 1.61 (2.2%)	-2.0 ± 1.2

is linear, so that the same wide flat prior is adopted, and the relative difference of ΛCDM with respect to the corresponding cosmographic model, see Eq. (35.7), is $\approx 10^{-6}$ at $z = 0.15$.

Next, by adopting an improper flat prior for j_0, one obtains a measurement that only assumes the FLRW metric, that is, large-scale homogeneity and isotropy. It is interesting to note that in Eq. (35.5) the degenerate combination $j_0 - \Omega_{k0}$ appears so that by marginalizing over j_0, we are effectively marginalizing also over spatial curvature. The data poorly constrain this combination, giving $j_0 - \Omega_{k0} = 28^{+39}_{-28}$.

Finally, we found that the deceleration parameter is in good agreement with the value relative to the standard model, contrary to what we found in [7, 8] with the Pantheon and Supercal datasets. The difference is likely due to the improved modeling of uncertainty and redshift corrections in the Pantheon+ dataset [9].

35.4 Cosmological Inference with the Local Prior on M_B

The latest combined data release by the SH_0ES and Pantheon+ collaborations significantly improved the robustness and consistency of cosmological analyses that include a local determination of the Hubble constant. Now, the distance calibration from 37 Cepheids is provided for the 42 supernovae in the Cepheid host galaxies,[4] together with the combined supernova and Cepheid covariance matrix [9]. This provides an absolute calibration for the supernova Ia absolute magnitude M_B as the likelihood is now given by:

$$V_i = \begin{cases} m_{B,i} - M_B - \mu(z_i) & \text{if Hubble diagram supernova,} \\ m_{B,i} - M_B - \mu_{\text{ceph},i} & \text{if supernova in Cepheid host,} \end{cases}$$
$$\chi^2_{\text{sne}} = V^{\text{T}} \left(\Sigma_{\text{sne}} + \Sigma_{\text{ceph}}\right)^{-1} V \,, \tag{35.16}$$

[4] Some galaxies hosted more than one supernova. Note also that some supernovae have been observed by more than one telescope so that in Pantheon+ there are actually 47 supernova entries that are calibrated by Cepheid distances.

where Σ_{ceph} has nonzero entries only for the supernovae in Cepheid hosts. The "Hubble diagram" (HD) supernovae are the 1466 supernovae—corresponding to 1580 measurements—that are used to constrain cosmological models but not to calibrate M_B. These have a redshift $z > 0.01$, again to remove the cosmic variance from the local structure. The 42 supernovae (47 light curves) in Cepheid host galaxies are instead used to calibrate M_B, whose calibration is then propagated to the HD supernovae. This breaks the degeneracy between M_B and the $5 \log_{10} H_0$ term inside the distance modulus $\mu(z)$, see Eq. (35.2).

The above methodology is ideal as it includes the correlations between Cepheid distances and calibrating and HD supernovae, ensuring an accurate representativeness of the cosmic distance ladder in cosmological inference [7, 8, 10–12]. Indeed, if one just uses HD supernovae, together with a local prior on H_0, one will (i) double count the 277 low-redshift supernova measurements that were used to obtain the prior on H_0, (ii) assume the validity of cosmography, in particular fixing the deceleration parameter to the standard model value of $q_0 = -0.55$, (iii) not include in the analysis the fact that M_B is constrained by the local calibration, an information which would otherwise be neglected in the analysis, biasing both model selection and parameter constraints [8].

The approach of Eq. (35.16) allows to carry out a variety of analyses. For example, one can only consider the 238 HD supernovae (but 277 supernova measurements) that are used to determine the Hubble constant and determine H_0 according to various fiducial models, that is, varying q_0 and j_0 in the case of cosmography or varying Ω_{m0} and w in the case of wCDM models. This approach is equivalent to the one of the previous Section, which only uses the quantities L, y and C from SH_0ES. Here, we preferred to adopt the methodology of the previous Section because we noticed that the Pantheon+ dataset produces a value of H_0 which is 0.5 km/s/Mpc higher than the SH_0ES baseline value when considering only the 277 low-redshift supernova measurements that are used to determine the Hubble constant.[5]

The most interesting use of Eq. (35.16) is the possibility of constraining the Hubble constant conditionally to a particular cosmological model, bypassing completely the use of cosmography. In this case the tension in the high- and low-redshift constraint on H_0 is traduced into a tension between the local calibration of M_B and the one induced by the CMB via BAO observations. This highlights the important of using the new Pantheon+ and SH_0ES likelihood.

In order to offer an easy way to implement the absolute calibration of M_B, here we apply the demarginalization method introduced in [7] to obtain an effective prior on M_B. As target for the demarginalization procedure, we consider both the constraint $H_0 = 73.6 \pm 1.1$ km/s/Mpc for the flat ΛCDM model from [9] and a value 0.5 km/s/Mpc lower, because of the discrepancy mentioned above when using the Pantheon+ & SH_0ES likelihood.[6] We use the 1580 HD supernova entries and obtain:

[5] We contacted the authors of [2] and [9]; according to Adam Riess there may be an inconsistency between the production and application of the covariance matrix of Cepheid/SN Ia host distances in the Pantheon+ fitting code. This issue is under investigation.

[6] This approach was suggested by Adam Riess.

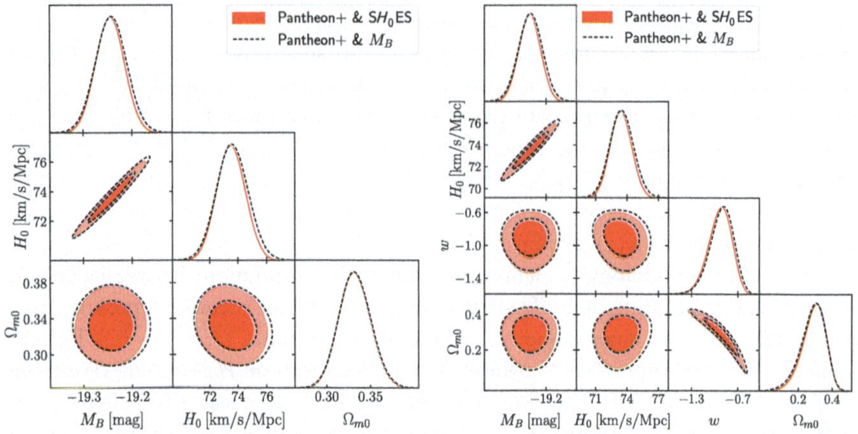

Fig. 35.2 Comparison between the analysis that uses the Pantheon+ & SH_0ES likelihood and the one that uses the Pantheon+ likelihood with the Gaussian prior on M_B from Eq. (35.17), for the flat ΛCDM model (left) and the flat wCDM model (right)

$$M_B = -19.243 \pm 0.032 \text{ mag} \quad \text{for} \quad H_0 = 73.6 \pm 1.1 \text{km/s/Mpc}\,, \tag{35.17}$$
$$M_B = -19.258 \pm 0.032 \text{ mag} \quad \text{for} \quad H_0 = 73.1 \pm 1.1 \text{km/s/Mpc}\,. \tag{35.18}$$

As a cross-check, using Eq. (35.17) together with the Pantheon+ likelihood, we obtain a good agreement with the results from [9]:[7]

$$H_0 = 73.6 \pm 1.1 \;\; \Omega_{m0} = 0.333 \pm 0.018 \qquad\qquad (\Lambda\text{CDM})\,,$$
$$H_0 = 73.5 \pm 1.1 \quad \Omega_{m0} = 0.306^{+0.063}_{-0.076} \quad w = -0.88 \pm 0.15 \;\; (w\text{CDM})\,.$$

Figure 35.2 shows a full comparison between the analysis that uses the Pantheon+ & SH_0ES likelihood and the one that uses the Pantheon+ likelihood with the Gaussian prior on M_B. Constraints and correlations are reproduced accurately. Furthermore, the use of the Gaussian prior has the advantage that one can marginalize analytically over M_B, see [8] for details.

35.5 M_B Tension Between SH_0ES and Planck

Several theoretical models have been proposed to resolve the discrepancy between the CMB determination of the Hubble constant by Planck [1] and the local measurement by SH_0ES [2]. These models modify the ΛCDM cosmology either in the late or early universe. Early-time modifications of the standard cosmological model typically accommodate a larger H_0 value by reducing the sound horizon scale r_s

[7] Ref. [9] adopted a prior $\Omega_{m0} > 0.1$, affecting their results in that parameter space region.

Table 35.2 The M_B tension. Top: marginalized constraints adopting the ΛCDM model. The first line corresponds to the results by [9], providing the same constraints. The second line shows the constraints that are obtained when including the full CMB likelihood together with the latest BAO measurements instead of the local prior on Cepheid host galaxies by SH_0ES. Bottom: as above but for the Pantheon dataset [17] and adopting the model-independent (inverse) distance ladder of [16]

ΛCDM analysis	M_B (mag)	H_0 (km/s/Mpc)	Ω_{m0}
Pantheon+ & SH_0ES 22	-19.244 ± 0.029	73.55 ± 1.02 (± 0.3)	0.332 ± 0.018
Pantheon+ & CMB + BAO	-19.438 ± 0.007	67.39 ± 0.19	0.315 ± 0.002
Tension	6.5σ	5.7σ	–
Model-independent analysis	M_B (mag)	H_0 (km/s/Mpc)	q_0
Pantheon & SH_0ES 21 [8]	-19.244 ± 0.037	74.30 ± 1.45	-0.91 ± 0.22
Pantheon & CMB + BAO [16]	-19.401 ± 0.027	69.71 ± 1.28	-1.09 ± 0.29
Tension	3.4σ	2.4σ	–

during recombination, whereas late-time modifications typically increase the Hubble rate at low redshifts [13–15]. While no compelling evidence for physics beyond the ΛCDM model has been found, late-time modifications of the standard paradigm face an additional problem: the M_B tension.

The local determination of the Hubble constant provided by the SH_0ES collaboration relies on a set of astrophysical distances that are used to calibrate the absolute magnitude of supernovae. In [16] we built a model-independent version of the inverse distance ladder, which uses BAO distances and a prior on r_s, coming from the Planck 2018 data release [1], to effectively calibrate the supernovae. Our analysis demonstrated that the absolute magnitude obtained from the cosmic distance ladder [7] is in tension with the underlying calibration on M_B provided by the combination of CMB + BAO, see Table 35.2. Since the latter constitutes a standard ruler that relies on the value adopted for r_s, the discrepancy in M_B suggests that, independent of the physics, models that solely change the Hubble flow, while keeping a sound horizon distance consistent with the one inferred from CMB under the assumption of a ΛCDM cosmology, will fail at explaining the discrepancy between the early and late times.

Indeed, the inability of late-time models to adequately explain the Hubble tension has been clearly demonstrated by several analyses [10, 11], including the analysis presented in [8]. This particular analysis employs a toy model, referred to as the 'hockey-stick' model, wherein the dark energy component rapidly undergoes a phantom transitions at low redshift. This illustrates that the M_B tension must be addressed in order to resolve the Hubble discrepancy effectively. As shown in [8], when analyzed with a prior on H_0 together with CMB and BAO data, the hockey-stick model, hsCDM, provides a value of the Hubble constant consistent with the local observation. Nevertheless, the underlying calibration of M_B that is obtained in the analysis still is in tension with the calibration inferred from the cosmic distance ladder, showing that hsCDM does not really restore the agreement between early and late times distances. This issue, and spurious conclusions, can be avoided if instead a prior on M_B is assumed.[8] Indeed, the analysis of the hsCDM model under the assumption of this prior explicitly shows the failure of the model to alleviate the H_0 (M_B) tension. Similar conclusions hold for void models, which try to explain the local higher expansion rate by placing the observer inside a large void. As far as the observed luminosity distance-redshift relation is concerned, these models are very similar to the hsCDM model and are ruled out for the very same reasons [18].

A category of models that potentially resolve the M_B tension encompasses those that feature a transition in supernova brightness at the extremely low redshifts of the first two rungs—specifically, before entering the Hubble flow with supernovae in the range $0.023 < z < 0.15$. This transition could be realized by a sudden shift in the effective gravitational constant by approximately 10% around 50–150 million years ago, consequently altering the Chandrasekhar mass and thereby the supernova luminosity [19, 20]. While this proposal conveniently resolves the Hubble tension, it also yields a series of testable predictions that could either support or disprove it [21]. Physical mechanisms that might trigger such an ultra-late gravitational transition include a first-order scalar-tensor theory phase transition from an early false vacuum, corresponding to the measured value of the cosmological constant, to a new vacuum with lower or zero vacuum energy [22].

We update the current significance of the M_B tension using the latest data and adopting the fiducial ΛCDM model. Using BAO data from BOSS DR12 [23], eBOSS ELG [24], eBOSS LRG [25, 26], SDSS MGS [27], and WiggleZ [28], high-ℓ TT+TE+EE, low-ℓ TT, and low-ℓ EE CMB data from Planck 2018 [1],[9] and the latest Pantheon+ & SH_0ES release, we constrain the ΛCDM model considering the cosmological equivalents to the cosmic distance ladder (Pantheon+ & SH_0ES data) and the inverse distance ladder (Pantheon+ in combination with BAO and CMB). The results of this analysis are presented in Fig. 35.3 and Table 35.2. We used CosmoSIS and CAMB [29, 30] to carry out the analyses. We find that the tension on M_B exceeds the 6σ level.

35.6 Conclusions

In this study, we have examined three distinct yet interconnected topics. Firstly, we extended the analysis conducted by the SH_0ES collaboration to determine the local value of the Hubble constant. We broadened their 46-dimensional analysis to cosmography with arbitrary values of the deceleration and jerk parameters.

[8] Alternatively, one should build a likelihood containing all the rungs of the cosmic distance ladder [12]. For a more detailed discussion, see Sect. 35.4.

[9] For simplicity, we used the compressed version of high-ℓ CMB data.

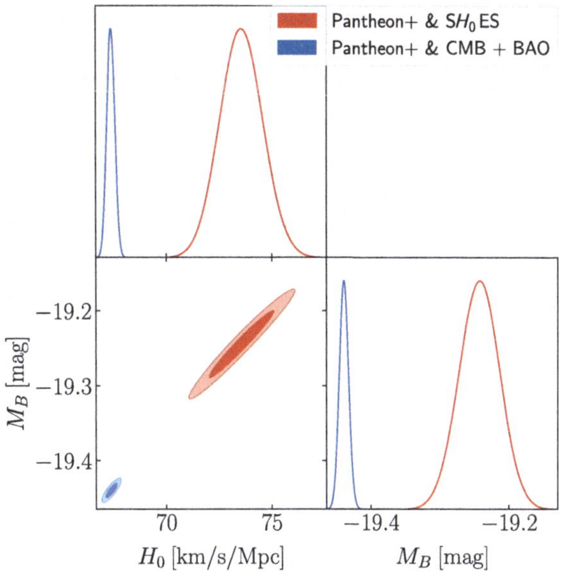

Fig. 35.3 Marginalized constraints from the cosmological equivalents to the cosmic distance ladder (Pantheon+ & SH_0ES data) and the inverse distance ladder (Pantheon+ in combination with BAO and CMB) when assuming the ΛCDM model as fiducial model of the universe. This figures shows how the tension on the Hubble constant can be interpreted as a tension in the calibration of the supernova brightness

We found that the latest Pantheon+ & SH_0ES data provides a competitive constraint that solely assumes the validity of the standard flat ΛCDM model:

$$73.14 \pm 1.10 \ (1.5\%) \ \text{km/s/Mpc},$$

and a model-independent constraint that only assumes the FLRW metric:

$$74.56 \pm 1.61 \ (2.2\%) \ \text{km/s/Mpc}.$$

Next, we underscored the importance of considering the calibration of the supernova magnitude M_B by the first two rungs of the distance ladder when conducting cosmological inference. Indeed, M_B is a crucial astrophysical parameter and should not be integrated out, but examined to understand the consistency of a given cosmological model.

This discussion led us to the issue of the "M_B tension." The supernova luminosity either gets calibrated by CMB and BAO observations or by the first two rungs of the cosmic distance ladder.

We found that, assuming the standard flat ΛCDM model, the two constraints on M_B are in tension at the 6.5σ level.

The discrepancy in M_B suggests that, independent of the physics involved, models that solely change the Hubble flow, while maintaining a sound horizon distance consistent with the one inferred from CMB, fail to explain the discrepancy between the early- and late-time measurements of H_0.

Acknowledgements We are grateful to Adam Riess for his valuable feedback regarding the SH_0ES data release. DC thanks the Robert E. Young Origins of the Universe Chair fund for its generous support. VM thanks CNPq (Brazil, 307969/2022-3) and FAPES (Brazil, TO 365/2022, 712/2022, 976/2022, 1020/2022, 1081/2022) for partial financial support. The authors acknowledge the use of the computational resources of the Sci-Com Lab of the Department of Physics–UFES.

References

1. Planck Collaboration, N. Aghanim et al., Planck 2018 results. VI. Cosmological parameters. Astron. Astrophys. **641**, A6 (2020). https://doi.org/10.1051/0004-6361/201833910 arXiv:1807.06209 [astro-ph.CO]
2. A.G. Riess et al., A comprehensive measurement of the local value of the Hubble constant with 1 km s^{-1} Mpc^{-1} uncertainty from the Hubble space telescope and the SH_0ES team. Astrophys. J. Lett. **934**, (1), L7 (2022). https://doi.org/10.3847/2041-8213/ac5c5b arXiv:2112.04510 [astro-ph.CO]
3. J.-Q. Xia, V. Vitagliano, S. Liberati, M. Viel, Cosmography beyond standard candles and rulers. Phys. Rev. D **85**, 043520 (2012). https://doi.org/10.1103/PhysRevD.85.043520 arXiv:1103.0378 [astro-ph.CO]
4. J. Calcino, T. Davis, The need for accurate redshifts in supernova cosmology. JCAP **01**, 038 (2017). https://doi.org/10.1088/1475-7516/2017/01/038 arXiv:1610.07695 [astro-ph.CO]
5. V. Marra, L. Amendola, I. Sawicki, W. Valkenburg, Cosmic variance and the measurement of the local Hubble parameter. Phys. Rev. Lett. **110**, 241305 (2013). https://doi.org/10.1103/PhysRevLett.110.241305 arXiv:1303.3121 [astro-ph.CO]
6. D. Camarena, V. Marra, The impact of the cosmic variance on H_0 on cosmological analyses. Phys. Rev. **D98**, 023537 (2018). https://doi.org/10.1103/PhysRevD.98.023537 arXiv:1805.09900 [astro-ph.CO]
7. D. Camarena, V. Marra, Local determination of the Hubble constant and the deceleration parameter. Phys. Rev. Res. **2**(1), 013028 (2020). https://doi.org/10.1103/PhysRevResearch.2.013028 arXiv:1906.11814 [astro-ph.CO]
8. D. Camarena, V. Marra, On the use of the local prior on the absolute magnitude of Type Ia supernovae in cosmological inference. Mon. Not. Roy. Astron. Soc. **504**, 5164–5171 (2021). https://doi.org/10.1093/mnras/stab1200 arXiv:2101.08641 [astro-ph.CO]
9. D. Brout et al., The Pantheon+ analysis: cosmological constraints. Astrophys. J. **938**(2), 110 (2022). https://doi.org/10.3847/1538-4357/ac8e04 arXiv:2202.04077 [astro-ph.CO]
10. G. Benevento, W. Hu, M. Raveri, Can late dark energy transitions raise the Hubble constant? Phys. Rev. D **101**(10), 103517 (2020). https://doi.org/10.1103/PhysRevD.101.103517 arXiv:2002.11707 [astro-ph.CO]
11. G. Efstathiou, To H_0 or not to H_0?. Mon. Not. Roy. Astron. Soc. **505**(3), 3866–3872 (2021). https://doi.org/10.1093/mnras/stab1588 arXiv:2103.08723 [astro-ph.CO]
12. K.L. Greene, F.-Y. Cyr-Racine, Hubble distancing: focusing on distance measurements in cosmology. JCAP **06**(06), 002 (2022). https://doi.org/10.1088/1475-7516/2022/06/002 arXiv:2112.11567 [astro-ph.CO]
13. L. Knox, M. Millea, Hubble constant hunter's guide. Phys. Rev. D **101**(4), 043533 (2020). https://doi.org/10.1103/PhysRevD.101.043533 arXiv:1908.03663 [astro-ph.CO]

14. N. Schöneberg, G. Franco Abellán, A. Pérez Sánchez, S. J. Witte, V. Poulin, J. Lesgourgues, The H_0 Olympics: a fair ranking of proposed models. Phys. Rept. **984**, 1–55 (2022). https://doi.org/10.1016/j.physrep.2022.07.001 arXiv:2107.10291 [astro-ph.CO]
15. P. Shah, P. Lemos, O. Lahav, A buyer's guide to the Hubble constant. Astron. Astrophys. Rev. **29**(1), 9 (2021). https://doi.org/10.1007/s00159-021-00137-4 arXiv:2109.01161 [astro-ph.CO]
16. D. Camarena, V. Marra, A new method to build the (inverse) distance ladder. Mon. Not. Roy. Astron. Soc. **495**, 2630–2644 (2020). https://doi.org/10.1093/mnras/staa770 arXiv:1910.14125 [astro-ph.CO]
17. D. Scolnic et al., The complete light-curve sample of spectroscopically confirmed SNe Ia from Pan-STARRS1 and cosmological constraints from the combined Pantheon sample. Astrophys. J. **859**(2), 101 (2018). https://doi.org/10.3847/1538-4357/aab9bb arXiv:1710.00845 [astro-ph.CO]
18. D. Camarena, V. Marra, Z. Sakr, C. Clarkson, A void in the Hubble tension? The end of the line for the Hubble bubble. Class. Quant. Grav. **39**(18), 184001 (2022). https://doi.org/10.1088/1361-6382/ac8635 arXiv:2205.05422 [astro-ph.CO]
19. V. Marra, L. Perivolaropoulos, Rapid transition of Geff at zt≃0.01 as a possible solution of the Hubble and growth tensions. Phys. Rev. D **104**(2), L021303 (2021). https://doi.org/10.1103/PhysRevD.104.L021303 arXiv:2102.06012 [astro-ph.CO]
20. G. Alestas, D. Camarena, E. Di Valentino, L. Kazantzidis, V. Marra, S. Nesseris, L. Perivolaropoulos, Late-transition versus smooth $H(z)$-deformation models for the resolution of the Hubble crisis. Phys. Rev. D **105**(6), 063538 (2022). https://doi.org/10.1103/PhysRevD.105.063538 arXiv:2110.04336 [astro-ph.CO]
21. L. Perivolaropoulos, Is the Hubble crisis connected with the extinction of dinosaurs? Universe **8**(5), 263 (2022). https://doi.org/10.3390/universe8050263 arXiv:2201.08997 [astro-ph.EP]
22. E. Abdalla et al., Cosmology intertwined: a review of the particle physics, astrophysics, and cosmology associated with the cosmological tensions and anomalies. JHEAp **34**, 49–211 (2022). https://doi.org/10.1016/j.jheap.2022.04.002 arXiv:2203.06142 [astro-ph.CO]
23. BOSS Collaboration, S. Alam et al., The clustering of galaxies in the completed SDSS-III Baryon oscillation spectroscopic survey: cosmological analysis of the DR12 galaxy sample. Mon. Not. Roy. Astron. Soc. **470**(3), 2617–2652 (2017). https://doi.org/10.1093/mnras/stx721 arXiv:1607.03155 [astro-ph.CO]
24. A. de Mattia et al., The completed SDSS-IV extended Baryon oscillation spectroscopic survey: measurement of the BAO and growth rate of structure of the emission line galaxy sample from the anisotropic power spectrum between redshift 0.6 and 1.1. Mon. Not. Roy. Astron. Soc. **501**(4), 5616–5645 (2021). https://doi.org/10.1093/mnras/staa3891 arXiv:2007.09008 [astro-ph.CO]
25. J.E. Bautista et al., The completed SDSS-IV extended Baryon oscillation spectroscopic survey: measurement of the BAO and growth rate of structure of the luminous red galaxy sample from the anisotropic correlation function between redshifts 0.6 and 1. Mon. Not. Roy. Astron. Soc. **500**(1), 736–762 (2020). https://doi.org/10.1093/mnras/staa2800 arXiv:2007.08993 [astro-ph.CO]
26. H. Gil-Marin et al., The completed SDSS-IV extended Baryon oscillation spectroscopic survey: measurement of the BAO and growth rate of structure of the luminous red galaxy sample from the anisotropic power spectrum between redshifts 0.6 and 1.0. Mon. Not. Roy. Astron. Soc. **498**(2), 2492–2531 (2020). https://doi.org/10.1093/mnras/staa2455 arXiv:2007.08994 [astro-ph.CO]
27. A.J. Ross, L. Samushia, C. Howlett, W.J. Percival, A. Burden, M. Manera, The clustering of the SDSS DR7 main Galaxy sample—I. A 4 per cent distance measure at $z = 0.15$. Mon. Not. Roy. Astron. Soc. **449**(1), 835–847 (2015). https://doi.org/10.1093/mnras/stv154 arXiv:1409.3242 [astro-ph.CO]
28. E.A. Kazin et al., The WiggleZ dark energy survey: improved distance measurements to z = 1 with reconstruction of the baryonic acoustic feature. Mon. Not. Roy. Astron. Soc. **441**(4), 3524–3542 (2014). https://doi.org/10.1093/mnras/stu778 arXiv:1401.0358 [astro-ph.CO]

29. A. Lewis, A. Challinor, A. Lasenby, Efficient computation of CMB anisotropies in closed FRW models. Astrophys. J. **538**, 473–476 (2000). https://doi.org/10.1086/309179 arXiv:astro-ph/9911177 [astro-ph]
30. C. Howlett, A. Lewis, A. Hall, A. Challinor, CMB power spectrum parameter degeneracies in the era of precision cosmology. JCAP **04**, 027 (2012). https://doi.org/10.1088/1475-7516/2012/04/027 arXiv:1201.3654 [astro-ph.CO]

Chapter 36
CMB Anomalies and the Hubble Tension

William Giarè

36.1 CMB Anomalies: A Brief Multi-experiment Overview

The most recent measurements of the Cosmic Microwave Background temperature and polarization anisotropies released by the Planck Collaboration [4] have achieved sub-percent accuracy in extracting most of the cosmological parameters [5]. While this is undoubtedly a blessing, it also represents a challenge: as precision improves, any deviations or anomalies may become more statistically significant and potentially point to tensions in our current understanding of the Universe [3, 20, 45].

One notable example is the higher lensing amplitude at about 2.8σ observed in the Planck data [24, 25, 43]. Since more lensing is expected with more Cold Dark Matter (CDM), the lensing anomaly immediately recast a preference for a closed Universe, which in turn helps to explain other large-scale anomalies in the data, such as the deficit of amplitude in the quadrupole and octupole modes. Consequently, the final Planck indication for a closed Universe becomes very significant, reaching the level of $3.4\,\sigma$ [23, 30].

Excitingly, the lensing and curvature anomalies, as well as any other observed discrepancies, may indicate the presence of new physics beyond the standard ΛCDM model of cosmology. However, it is also entirely plausible that anomalies simply reflect observational systematic errors and/or statistical fluctuations in the data. One effective way to discriminate between the two possibilities is to compare the results of multiple CMB measurements: any persistent deviations across independent probes may strongly hint at the presence of new physics. Fortunately, this is now possible thanks to the observations of the CMB angular power spectra recently released by the Atacama Cosmology Telescope (ACT) [6, 17] and the South Pole Telescope (SPT) [9, 12, 27]. While these experiments may not be as constraining as the Planck

W. Giarè (✉)
School of Mathematics and Statistics, University of Sheffield, Hounsfield Road, Sheffield S3 7RH, UK
e-mail: w.giare@sheffield.ac.uk

E. Di Valentino and D. Brout (eds.), *The Hubble Constant Tension*, Springer Series in Astrophysics and Cosmology, https://doi.org/10.1007/978-981-99-0177-7_36

satellite, they are still able to produce cosmological bounds that are accurate enough for independent tests of the Planck results [6–9].

Interestingly, both the ACT and SPT data have provided full support for the inflationary prediction of a spatially flat Universe and for a lensing amplitude consistent with ΛCDM [6, 8, 9]. This suggests that both the curvature and lensing anomalies may be explained in terms of statistical fluctuations or systematic errors. However, the same datasets have revealed other relevant deviations from the standard cosmological model. For instance, the the Data Release 4 of the Atacama Cosmology Telescope shows preference for a unitary spectral index of primordial perturbations ($n_s = 1.009 \pm 0.015$) [6, 29], in agreement with the phenomenological scale-invariant spectrum proposed by Harrison, Zel'dovich, and Peebles [33, 44, 51], and in tension with Planck with a significance of 99.3% Confidence Level (CL). Other mild yet relevant anomalies in the ACT data concern the value of the effective number of relativistic degrees of freedom in the early Universe ($N_{\rm eff} = 2.46 \pm 0.26$) [6, 21], in disagreement at more than 2 σ with the expected value of $N_{\rm eff} = 3.04$ predicted by the Standard Model of particle physics for three active neutrinos [11]; a $\sim 2\sigma$ preference for a positive running of the scalar spectral index of inflationary perturbations ($\alpha_s \equiv dn_s/d\log k = 0.058 \pm 0.028$) [6, 28, 29]; and a $\sim 3\sigma$ indication in favor of early Dark Energy [34, 46].

In turn, these deviations are not supported by Planck, which instead shows $\sim 8\sigma$ evidence for $n_s \neq 1$, a good agreement with the predictions of the standard model for $N_{\rm eff}$, and no evidence for Early Dark Energy [5, 34, 46]. Therefore, also in this case, the lack of cross-validation between independent CMB measurements may suggest statistical fluctuations and/or as of yet unknown systematic errors.

Regardless of whether these anomalies arise from observational systematic effects, statistical fluctuations, or new physics, if we aim to achieve precision cosmology and test fundamental physics, it is important to acquire a better comprehension of the collective agreement between independent experiments and to have a clear understanding of the limitations and uncertainties underlying both current data and the cosmological model. This becomes a crucial need in relation to the H_0-tension [13, 49] as many proposed solutions[1] call for a new paradigm shift in cosmology, while relying almost entirely on the resilience of such observations.

In this chapter, we address this problem and discuss the global consistency of independent CMB experiments. In Sect. 36.2, we investigate potential reasons for discrepancies by exploring both the limitations inherent in current data (Sect. 36.2.1) and the uncertainties of the standard cosmological model (Sect. 36.2.2). We eventually argue that anomalies seem to indicate a potential mismatch between the early and late Universe (Subsect. 36.2.3). In Sect. 36.3, we discuss the implications for the H_0-tension and its most common solutions.

[1] See, e.g., Refs. [3, 26, 36, 39, 40, 46] for recent reviews.

36.2 Global Consistency of CMB Experiments

Based on the overview provided in the previous section, it is evident that what makes CMB anomalies difficult to interpret *individually* is that different experiments often point in discordant directions, and none of the most relevant deviations can be cross-validated through independent probes. For this reason, accurate statistical methods have been developed to quantify the *global* consistency between independent experiments under a given model of cosmology, for instance by starting from the constraints on the parameter space of the model itself [15, 16, 22, 31, 41, 42].

Assuming a standard ΛCDM cosmology, the data release 4 of the Atacama Cosmology Telescope can be evaluated in mild-to-moderate tension with Planck and the South Pole Telescope at a gaussian equivalent level of 2.6 σ and 2.4 σ, respectively [22, 32]. Depending on one's level of tolerance, these controversial tensions ranging between 2 and 3 σ may be taken more or less seriously. If our objective is to accomplish "precision cosmology" and effectively assess fundamental physics, it is reasonable to conclude that they warrant further investigations. However, it remains crucial to understand the causes of these differences and shed light on their potential implications for the numerous proposed resolutions of the H_0-tension that are emerging from these observations.

36.2.1 The Limitations of Current Data

Assuming a baseline ΛCDM cosmology, the main source of tension between ACT and Planck arises from the measurements of the scalar spectral index, n_s, with Planck reporting a value of $n_s = 0.9649 \pm 0.0044$ [5] and ACT producing a value of $n_s = 1.008 \pm 0.015$ [6]; in tension at approximately 2.7 σ. Moreover, the discrepancy between the two experiments extends to the measurement of the baryon energy density with Planck and ACT giving respectively $\Omega_b h^2 = 0.02236 \pm 0.00015$ [5] and $\Omega_b h^2 = 0.02153 \pm 0.00030$ [6]; in tension at about 2.5 σ.

If we believe these differences to emerge from limitations in the data, then a logical step is to identify which (missing) part of the dataset is responsible for the discrepancy and eventually discard (include) it in the cosmological parameter inference analyses. Several possible explanations have been attempted to explain the disagreement.

- **Temperate data**: One suggested explanation could be attributed to the limited amount of ACT data at low multipoles in the TT Spectrum where ACT has a minimum sensitivity in multipole of $\ell = 600$ [6]. In absence of data around the first two acoustic peaks, a strong degeneracy exists between $\Omega_b h^2$ and n_s as a lower value of the former can be mimicked by a larger value of the latter, tilting the spectrum in the opposite direction [6]. Typically, the lack of information is filled by using measurements from Planck or a third independent experiment such as WMAP [10, 35] in the multipole range $\ell \lesssim 600$. While this helps in diluting the tension, it is worth noting that there is no a-priori reason to combine ACT with low-ℓ data from another experiment: despite the degeneracy, accurate data at the scales probed by ACT are enough to achieve precise constraints on both n_s and $\Omega_b h^2$. For instance, considering the Planck data only in the multipole range in common with ACT, the final results still indicate $n_s \neq 1$ at more than 4 σ [29]. This cast doubts on the explanation of the mismatch in terms of lack of information on the first acoustic peaks as one would expect a reasonable agreement between two experiments measuring an overlapping range of multipoles.

- **Polarization data**: Another possibility is that the limited amount of ACT low multipoles polarization data (which cover only multipoles $\ell \gtrsim 350$ in TE and EE) is responsible for the discrepancy. However, incorporating low-multipole polarization data from the Planck measurements of E-modes at $2 \leq \ell \leq 30$ fails to explain these anomalies [29]. Nonetheless, it was also suggested that an overall TE calibration could resolve the mismatch [6]. Neglecting any information from ACT polarization measurements (TE EE) and combining ACT temperature anisotropies (TT) with SPT polarization data (TE EE) the tension is in fact reduced below 2σ, restoring the agreement between the experiments for $\Omega_b h^2$ [29]. However, the same combination of ACT and SPT data still favors a value of n_s around unity [29], indicating that ACT polarization may only partially contribute to the tension, without fully reconciling it.
- **Non-CMB data**: Yet another possible avenue to increase the data constraining power is to break the geometrical degeneracy among cosmological parameters by exploiting CMB-independent data, such as astrophysical measurements of Baryon Acoustic Oscillation (BAO) and Redshift Space Distortion (RSD) [18, 19], as well as galaxy clustering and cosmic shear observations [1, 2]. While astrophysical data have been found to alleviate several tensions and restore the parameter values to those predicted by ΛCDM, this is not the case. Including them does not shift the results significantly but leads to tighter errors, ultimately increasing the tension [6, 29].

36.2.2 The Unknowns of the Cosmological Model

Although observational systematic errors and/or statistical fluctuations in the data may provide a simple explanation for the discrepancies between different CMB experiments, we cannot discount the possibility that part of these anomalies are rooted in the limitations of the standard cosmological model to accurately explain precise observations at very different angular scales.

From a theoretical standpoint, many ingredients of ΛCDM lack solid understanding in fundamental physics. The nature of Dark Matter and Dark Energy remains unknown in the context of the Standard Model of fundamental interactions as well as the physical origin of the inflationary epoch is also unclear. To account for all these unknowns, the ΛCDM model adopts a principle of simplicity and incorporates these components into a streamlined theoretical framework consisting of just six

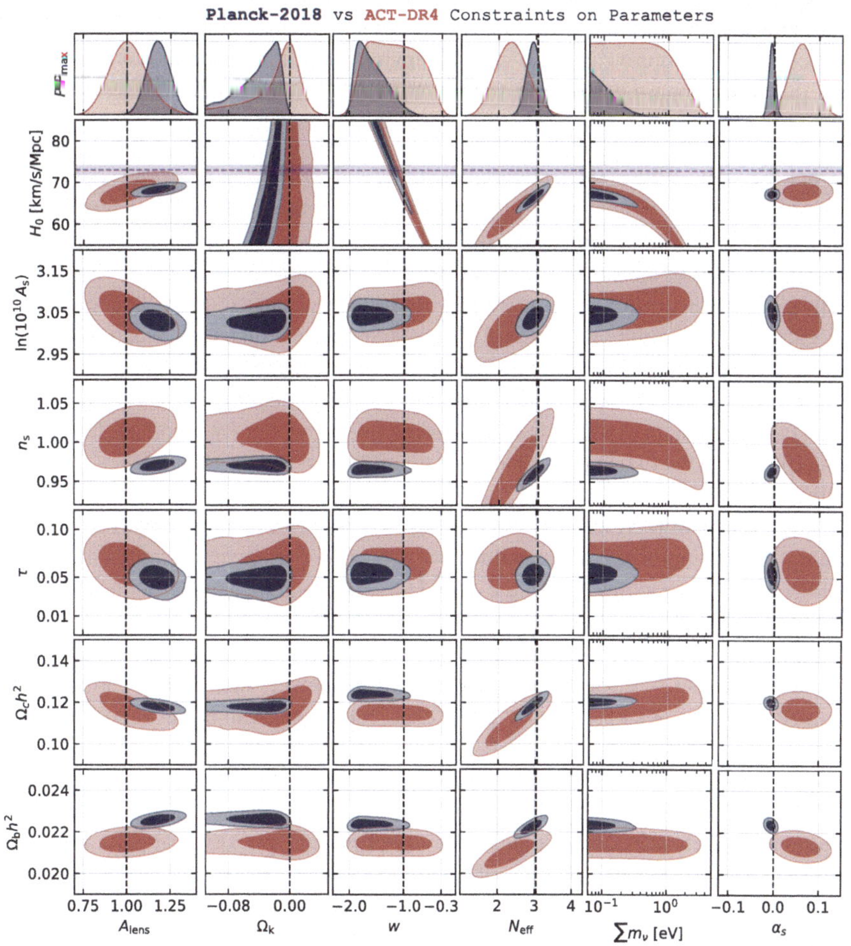

Fig. 36.1 Constraints on cosmological parameters in various minimal extensions to ΛCDM derived from the analysis of the Planck 2018 [5] (grey) and ACT-DR4 [6] (red) temperature and polarization data. The vertical dashed black lines indicate the reference values of parameters in ΛCDM. The horizontal magenta band in the 2D-plots involving the Hubble constant shows the direct measurements obtained by the SH_0ES Collaboration ($H_0 = 73 \pm 1$ km/s/Mpc) [47]

free parameters. However, the value of cosmological parameters inferred from CMB measurements clearly depends on the global parameterization adopted for the background cosmology and—ultimately—on the model's assumptions. For this reason, a possibility usually explored when finding anomalies in the cosmological parameter values, is to extend the baseline model and study how the overall tension changes.

Figure 36.1 provides a comprehensive summary of the constraints on cosmological parameters derived from Planck and ACT in various minimal extensions to ΛCDM, including those involving the anomalous parameters discussed so far. Directing our attention to the enigmatic foundations of the standard cosmological model, a careful inspection on the figure can yield valuable insights into the uncertainties underlying their theoretical parameterization and potentially reveal any missing ingredients in the theory able to account for the observed discrepancies.

- **Dark Energy**: Within the standard cosmological model, the late-time accelerated expansion of the Universe is attributed to a cosmological constant term in the Einstein equations, resulting in a constant energy-density component with a negative equation of state, $w = -1$. Relaxing this assumption we can use the Planck and ACT data to directly constrain w and test its consistency with the cosmological constant value. Interestingly, Planck suggests a phantom equation of state ($w < -1$) at about 2σ, while ACT is consistent with $w = -1$, albeit with large uncertainties that cannot exclude phantom or quintessential ($w > -1$) behaviors, see Fig. 36.1. In any case, late-time modifications can hardly reconcile the difference between the two experiments, as it mainly arises from a mismatch of the early Universe.
- **Dark Matter and Neutrinos**: As the tension between Planck and ACT arises primary from a mismatch in the values of early-time parameters, interesting extensions are those associated with the early Universe content, such as neutrinos, Dark Matter and Dark radiation. For instance, relaxing the value of total neutrino mass (fixed at $\sum m_\nu \sim 0.06$ eV within ΛCDM) produces a shift in the value of n_s measured by ACT towards the Planck result. However, such a shift can be associated with the ACT's peculiar preference for higher values of neutrino mass, which contradicts the tight constraints placed by the Planck data. A second an interesting avenue involves the radiation energy-density in the early Universe. In the case of three active neutrinos, the radiation energy density can be calculated quite accurately within the standard model of particle physics and parameterized in terms of the effective number of relativistic degrees of freedom, $N_{\rm eff} = 3.044$. By relaxing this theoretical constraint and leaving $N_{\rm eff}$ as a free parameter to be inferred by data, we can use the ACT and Planck observations to directly verify the consistency with the Standard Model's predictions as well as to test any theoretical extensions involving additional (thermal) degrees of freedom in the early Universe. From Fig. 36.1 we see that Planck is in good agreement with the Standard Model predictions, while the ACT's result ($N_{\rm eff} = 2.46 \pm 0.26$) is anomalous and in disagreement with the expected value for three active neutrinos by about $2.5\,\sigma$. Interestingly, when the energy density in the early Universe is significantly lower than the expected value, the disagreement between the two experiments gets diluted by the simultaneous effect of widening the uncertainties and shifting the values of n_s towards that measured by Planck. The interpretation of this result highly depends on our attitude regarding the reasons for the disagreement. If we attribute the observed discrepancy between ACT and Planck to new physics, this may suggest a potential role for the neutrino sector (and generally for new physics at early times) in accounting for at least part of this difference. On the other hand, if we rely primarily on the strength of the theoretical predictions, this may indicate that observational systematic effects are introducing biases into the determination

of N_{eff}, raising concern on the results obtained when considering new physics at early times.

- **Inflation.** One of the most common results from single-field models of inflation is the production of nearly scale-invariant perturbations that can be well described by a power-law form of the adiabatic scalar components. In the simplest scenarios, any residual scale dependence is parameterized solely in terms of the scalar spectral index n_{s}, where $n_s = 1$ corresponds to exact scale-invariance. However, additional parameterizations that incorporate higher-order terms in the power-law expansion should also be considered. In this regard, another important clue that can be inferred from Fig. 36.1 is the tendency for the value of n_s predicted by ACT to decrease significantly when extending the inflationary sector of the theory by accounting also for the possibility of a running in the scalar spectral index $\alpha_{\mathrm{s}} = dn_{\mathrm{s}}/d\log k$. In this case, the difference on the spectral index n_{s} is solved, but the disagreement maps into a strong controversy in the values of α_{s}: while Planck prefers a slightly negative running ($\alpha_{\mathrm{s}} = -0.0119 \pm 0.0079$), ACT favors a positive running ($\alpha_{\mathrm{s}} = 0.058 \pm 0.028$).

36.2.3 *How Many Early-Late Universe Mismatches Are There?*

To draw some general conclusions from the precedent analysis and gain a fresh perspective on the CMB anomalies, we want to argue that the global tension emerging between ACT and Planck always maps into mismatches between the early and late Universe.

Considering all the minimal extensions in Fig. 36.1, we can notice that the Planck anomalies *always* involve parameters associated with the local Universe such as the lensing amplitude, the spacetime geometry, and the dark energy equation of state. Instead, the ACT anomalies *always* involve parameters related to the early Universe such as the baryon energy density, the spectral index, its running, and a free-to-vary neutrino sector. Therefore, the picture emerging is the following one: Planck shows mild-to-moderate deviations from ΛCDM when extending the late time sector of the theory, remaining consistent for extensions involving modifications at early-times. In contrast, the Atacama Cosmology Telescope agrees pretty well with ΛCDM at late-times, but shows anomalies in the early Universe parameters. However, considering the large experimental uncertainties obtained when extending the late-time sector of the theory, the difference between the two probes remains mostly driven by a mismatch in the early Universe.

36.3 Implications for the H_0-Tension

Any determination of the Hubble constant inferred from observations of the cosmic microwave background necessarily depends on the cosmological model. For this reason, several alternative models have been proposed to address the widely known tension between the *direct* value of H_0 measured by the SH_0ES collaboration using luminosity distances of Type Ia supernovae [47] ($H_0 = 73 \pm 1$ Km/s/Mpc) and the *indirect* value inferred by the Planck satellite [5] ($H_0 = 67.4 \pm 0.5$ Km/s/Mpc).

If we intend to explain the mismatch in terms of new physics (and clarify the potential role of the CMB anomalies in this matter), it is important to understand what aspects of the theory are involved in the game when inferring H_0 from observations of the CMB. It is well known that measuring the positions of the acoustic peaks in the CMB angular power spectrum, we can determine the angular size of the sound horizon (θ_s) which is the cosmological parameter most accurately measured by Planck [5]. On the other hand, we can fix the baryon density $\Omega_b h^2$ from the relative heights of even and odd peaks, while we can obtain the cold dark matter density $\Omega_c h^2$ from the absolute heights of acoustic peaks. Once these parameters are known, we can pick a model of the early Universe (i.e., pre-recombination) and calculate the physical dimension of sound horizon (r_s). Given r_s and θ_s, we can easily infer the angular diameter distance from the CMB, $D_A(z_{ls}) = r_s/\theta_s$ (with $z_{ls} \simeq 1100$). Finally, by assuming a cosmological model for the subsequent late-time expansion (i.e., post-recombination), we eventually infer the value of H_0 from $D_A(z_{ls})$.

Therefore, any theoretical attempts aimed at solving the H_0-tension should fall into one or both these categories:

- **Late time solutions**: The first possibility is to modify the late time expansion history and change the value of the Hubble constant inferred from the angular diameter distance $D_A(z_{ls})$. This is typically done by introducing new late-time physics and/or extending the dark sector of the theory. Based on Fig. 36.1, one might expect these solutions to be preferred by data, given the significant room left by the CMB observations for new physics at late-times. However, they should be approached with some caution as late-time modifications often leads to a strong geometric degeneracy among parameters which needs to be broken in order to genuinely test new physics. In practice, this can be done by incorporating local Universe observations, such as baryon acoustic oscillations [18, 19] and/or supernovae measurements [14, 48]. They often clean away any late-time anomalies and restore the parameter values to ΛCDM. Therefore, contrary to what one might expect from CMB data, there is very little room to accommodate new physics at late-times. However, it is worth considering that the inclusion of certain local probes becomes a subject of debate when testing theoretical models that significantly deviate from ΛCDM. For instance, this objection is sometimes raised against BAO measurements, which assume a ΛCDM-like cosmology as an ansatz from the outset, but are in tension with CMB in many extended theoretical frameworks [23]. Therefore, taking a conservative approach, we cannot discount the possibility that (part of) the H_0-tension may be alleviated by late-time solutions.

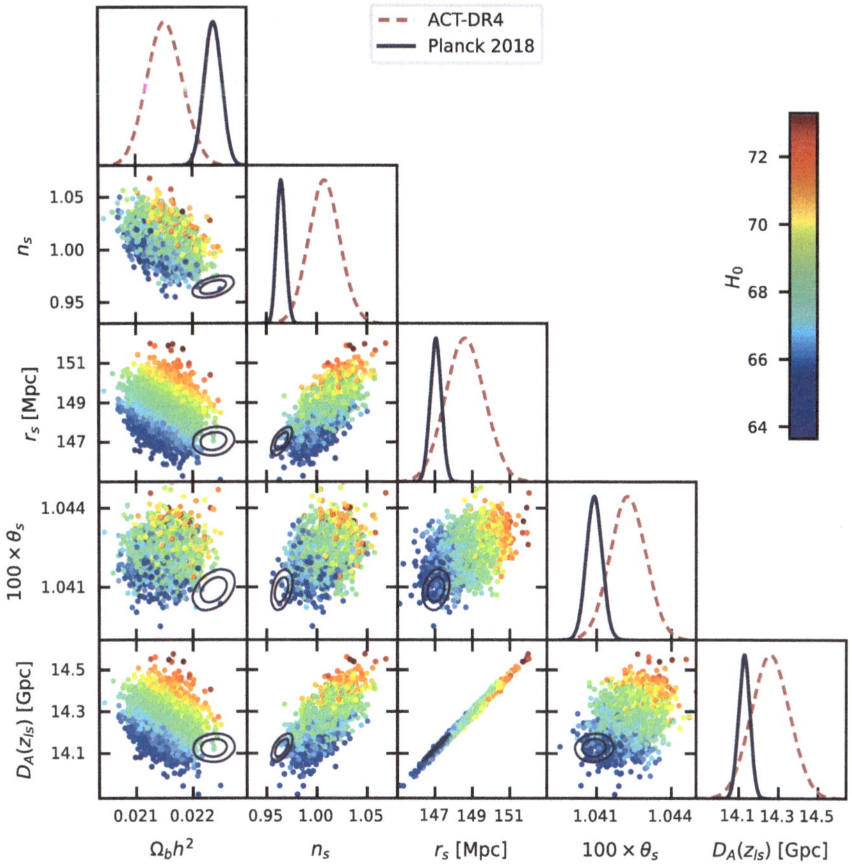

Fig. 36.2 Constraints on the early Universe parameters and their relation with H_0 (in the color-bar) as obtained by the ACT-DR4 data release [6]. The black empty contours define the region of the parameter space preferred by the Planck 2018 data [5]

In any case, it is unlikely that the tension between ACT and Planck will have a significant impact on these solutions since these experiments primarily disagree at early times, as seen in Sect. 36.2.

- **Early time solutions**: The second possibility is to consider new physics in early Universe to increase the physical size of the sound horizon r_s. This is typically done by incorporating new light degrees of freedom before recombination. However, these solutions come with their own caveats. For instance, many indications of this kind of new early-time physics arise when combining multiple CMB measurements at different angular scales (such as Planck and ACT), without finding clear cross-validation when these experiments are considered separately. Therefore, this may cast doubt on whether these indications genuinely reflect new physics or if they emerge from the controversial tension between these probes, as also seen in

Sect. 36.2.2 for many simpler minimal early-Universe extensions. To explore this issue further, we remain agnostic and refer to Fig. 36.2. Assuming a background ΛCDM cosmology, the figure depicts the constraints on the early Universe parameters of interest for early-time solutions, showing a consistent shift between ACT and Planck. Notably, ACT allows for greater flexibility in accommodating higher values of the sound horizon, while Planck peaks where ACT prefers very low values of H_0. This means that increasing H_0 requires moving towards the region of the parameter space where the disagreement between the two probes becomes more significant. In addition, the sound horizon, the spectral index, and the Hubble constant are all positively correlated: increasing r_s to increase H_0 will naturally push n_s towards higher values. This is why many potential solutions to the H_0 tension suggest a prominent role of n_s, often pointing towards values of $n_s \sim 1$ as measured by ACT, bringing the tension around this parameter back on the table, see, *e.g.*, Refs. [37, 38, 50]. However, one might argue that relaxing the early Universe sector of the theory will change the correlation among parameters. As seen in Sect. 36.2.2, considering a free-to-vary $N_{\rm eff}$ can partially accommodate the disagreement between ACT and Planck. While this is true, the results in Fig. 36.1 show a significant shift of $N_{\rm eff}$ towards smaller values, which is in the opposite direction of what we need to solve the H_0 tension. This also raises concerns about some proposed solutions whose underlying philosophy consists of incorporating *large* additional degrees of freedom as they appear to be disfavored in the first place.

References

1. T.M.C. Abbott et al., Dark energy survey year 1 results: cosmological constraints from galaxy clustering and weak lensing. Phys. Rev. D **98**(4), 043526 (2018)
2. T.M.C. Abbott et al., Dark energy survey year 3 results: cosmological constraints from galaxy clustering and weak lensing. Phys. Rev. D **105**(2), 023520 (2022)
3. E. Abdalla et al., Cosmology intertwined: a review of the particle physics, astrophysics, and cosmology associated with the cosmological tensions and anomalies. JHEAp **34**, 49–211 (2022)
4. N. Aghanim et al., Planck 2018 results. I. Overview and the cosmological legacy of Planck. Astron. Astrophys. **641**, A1 (2020)
5. N. Aghanim et al., Planck 2018 results. VI. Cosmological parameters. Astron. Astrophys. **641**, A6 (2020). [Erratum: Astron.Astrophys. 652, C4 (2021)]
6. S. Aiola et al., The Atacama cosmology telescope: DR4 maps and cosmological parameters. JCAP **12**, 047 (2020)
7. K. Aylor et al., A comparison of cosmological parameters determined from CMB temperature power spectra from the south pole telescope and the Planck satellite. Astrophys. J. **850**(1), 101 (2017)
8. L. Balkenhol et al., Constraints on ΛCDM extensions from the SPT-3G 2018 EE and TE power spectra. Phys. Rev. D **104**(8), 083509 (2021)
9. L. Balkenhol et al., A measurement of the CMB temperature power spectrum and constraints on cosmology from the SPT-3G 2018 TT/TE/EE Data Set
10. C.L. Bennett et al., Nine-year Wilkinson microwave anisotropy probe (WMAP) observations: final maps and results. Astrophys. J. Suppl. **208**, 20 (2013)

11. J.J. Bennett, G. Buldgen, P.F. De Salas, M. Drewes, S. Gariazzo, S. Pastor, Y.Y.Y. Wong, Towards a precision calculation of $N_{\rm eff}$ in the standard model II: neutrino decoupling in the presence of flavour oscillations and finite-temperature QED. JCAP **04**, 073 (2021)
12. B.A. Benson et al., SPT-3G: a next-generation cosmic microwave background polarization experiment on the south pole telescope. Proc. SPIE Int. Soc. Opt. Eng. **9153**, 91531P (2014)
13. J.L. Bernal, L. Verde, A.G. Riess, The trouble with H_0. JCAP **10**, 019 (2016)
14. D. Brout et al., The Pantheon+ analysis: cosmological constraints. Astrophys. J. **938**(2), 110 (2022)
15. R. Calderón, A. Shafieloo, D.K. Hazra, W. Sohn, On the consistency of ΛCDM with CMB measurements in light of the latest Planck. ACT, and SPT data
16. T. Charnock, R.A. Battye, A. Moss, Planck data versus large scale structure: methods to quantify discordance. Phys. Rev. D **95**(12), 123535 (2017)
17. S.K. Choi et al., The Atacama cosmology telescope: a measurement of the cosmic microwave background power spectra at 98 and 150 GHz. JCAP **12**, 045 (2020)
18. K.S. Dawson et al., The Baryon oscillation spectroscopic survey of SDSS-III. Astron. J. **145**, 10 (2013)
19. K.S. Dawson et al., The SDSS-IV extended Baryon oscillation spectroscopic survey: overview and early data. Astron. J. **151**, 44 (2016)
20. E. Di Valentino, Challenges of the standard cosmological model. Universe **8**(8), 399 (2022)
21. E. Di Valentino, W. Giarè, A. Melchiorri, J. Silk, Health checkup test of the standard cosmological model in view of recent cosmic microwave background anisotropies experiments. Phys. Rev. D **106**(10), 103506 (2022)
22. E. Di Valentino, W. Giarè, A. Melchiorri, J. Silk, Quantifying the global 'CMB tension' between the Atacama cosmology telescope and the Planck satellite in extended models of cosmology. Mon. Not. Roy. Astron. Soc. **520**(1), 210–215 (2023)
23. E. Di Valentino, A. Melchiorri, J. Silk, Planck evidence for a closed Universe and a possible crisis for cosmology. Nature Astron. **4**(2), 196–203 (2019)
24. E. Di Valentino, A. Melchiorri, J. Silk, Cosmological constraints in extended parameter space from the Planck 2018 legacy release. JCAP **01**, 013 (2020)
25. E. Di Valentino, A. Melchiorri, J. Silk, Investigating cosmic discordance. Astrophys. J. Lett. **908**(1), L9 (2021)
26. E. Di Valentino, O. Mena, S. Pan, L. Visinelli, W. Yang, A. Melchiorri, D.F. Mota, A.G. Riess, J. Silk, In the realm of the Hubble tension—a review of solutions. Class. Quant. Grav. **38**(15), 153001 (2021)
27. D. Dutcher et al., Measurements of the E-mode polarization and temperature-E-mode correlation of the CMB from SPT-3G 2018 data. Phys. Rev. D **104**(2), 022003 (2021)
28. M. Forconi, W. Giarè, E. Di Valentino, A. Melchiorri, Cosmological constraints on slow roll inflation: an update. Phys. Rev. D **104**(10), 103528 (2021)
29. W. Giarè, F. Renzi, O. Mena, E. Di Valentino, A. Melchiorri, Is the Harrison-Zel'dovich spectrum coming back? ACT preference for $n_s \sim 1$ and its discordance with Planck
30. W. Handley, Curvature tension: evidence for a closed universe. Phys. Rev. D **103**(4), L041301 (2021)
31. W. Handley, P. Lemos, Quantifying tensions in cosmological parameters: interpreting the DES evidence ratio. Phys. Rev. D **100**(4), 043504 (2019)
32. W. Handley, P. Lemos, Quantifying the global parameter tensions between ACT. SPT and Planck. Phys. Rev. D **103**(6), 063529 (2021)
33. E.R. Harrison, Fluctuations at the threshold of classical cosmology. Phys. Rev. D **1**, 2726–2730 (1970)
34. J.C. Hill et al., Atacama cosmology telescope: constraints on prerecombination early dark energy. Phys. Rev. D **105**(12), 123536 (2022)
35. G. Hinshaw et al., Nine-Year Wilkinson microwave anisotropy probe (WMAP) observations: cosmological parameter results. Astrophys. J. Suppl. **208**, 19 (2013)
36. K. Jedamzik, L. Pogosian, G.-B. Zhao, Why reducing the cosmic sound horizon alone can not fully resolve the Hubble tension. Commun. in Phys. **4**, 123 (2021)

37. J.-Q. Jiang, Y.-S. Piao, Toward early dark energy and ns=1 with Planck, ACT, and SPT observations. Phys. Rev. D **105**(10), 103514 (2022)
38. J.-Q. Jiang, G. Ye, Y.-S. Piao, Return of Harrison-Zeldovich spectrum in light of recent cosmological tensions
39. M. Kamionkowski, A.G. Riess, The Hubble tension and early dark energy
40. L. Knox, M. Millea, Hubble constant hunter's guide. Phys. Rev. D **101**(4), 043533 (2020)
41. A. La Posta, U. Natale, E. Calabrese, X. Garrido, T. Louis, Assessing the consistency between CMB temperature and polarization measurements with application to Planck, ACT, and SPT data. Phys. Rev. D **107**(2), 023510 (2023)
42. W. Lin, M. Ishak, A Bayesian interpretation of inconsistency measures in cosmology. JCAP **05**, 009 (2021)
43. P. Motloch, W. Hu, Tensions between direct measurements of the lens power spectrum from Planck data. Phys. Rev. D **97**(10), 103536 (2018)
44. P.J.E. Peebles, J.T. Yu, Primeval adiabatic perturbation in an expanding universe. Astrophys. J. **162**, 815–836 (1970)
45. L. Perivolaropoulos, F. Skara, Challenges for ΛCDM: an update, 5 (2021)
46. V. Poulin, T.L. Smith, T. Karwal, The ups and downs of early dark energy solutions to the Hubble tension: a review of models, hints and constraints circa (2023)
47. A.G. Riess et al., A comprehensive measurement of the local value of the Hubble constant with 1 km s^{-1} Mpc^{-1} uncertainty from the Hubble space telescope and the SH_0ES team. Astrophys. J. Lett. **934**(1), L7 (2022)
48. D. Scolnic et al., The Pantheon+ analysis: the full data set and light-curve release. Astrophys. J. **938**(2), 113 (2022)
49. L. Verde, T. Treu, A.G. Riess, Tensions between the early and the late universe. Nature Astron. **3**, 891 (2019)
50. G. Ye, J.-Q. Jiang, Y.-S. Piao, Toward inflation with ns=1 in light of the Hubble tension and implications for primordial gravitational waves. Phys. Rev. D **106**(10), 103528 (2022)
51. Y.B. Zeldovich, A hypothesis, unifying the structure and the entropy of the universe. Mon. Not. Roy. Astron. Soc. **160**, 1P-3P (1972)

GPSR Compliance

The European Union's (EU) General Product Safety Regulation (GPSR) is a set of rules that requires consumer products to be safe and our obligations to ensure this.

If you have any concerns about our products, you can contact us on ProductSafety@springernature.com

In case Publisher is established outside the EU, the EU authorized representative is:

Springer Nature Customer Service Center GmbH
Europaplatz 3
69115 Heidelberg, Germany

The manufacturer's authorised representative in the EU is Springer Nature Customer Service Centre GmbH, Europaplatz 3, 69115 Heidelberg, Germany. If you have any concerns regarding our products, please contact ProductSafety@springernature.com

Printed and bound by CPI Group (UK) Ltd, Croydon, CR0 4YY
01/10/2026
02225862-0016